First Semester Calculus

For Students of Mathematics and Related Disciplines

First Edition

Michael M. Dougherty and John D. Gieringer
Southwestern Oklahoma State University and Alvernia University

Bassim Hamadeh, CEO and Publisher
Angela Schultz, Acquisitions Editor
Michelle Piehl, Senior Project Editor
Berenice Quirino, Associate Production Editor
Jess Estrella, Senior Graphic Designer
Stephanie Kohl, Licensing Associate
Natalie Piccotti, Director of Marketing
Kassie Graves, Vice President of Editorial
Jamie Giganti, Director of Academic Publishing

Copyright © 2020 by Cognella, Inc. All rights reserved. No part of this publication may be reprinted, reproduced, transmitted, or utilized in any form or by any electronic, mechanical, or other means, now known or hereafter invented, including photocopying, microfilming, and recording, or in any information retrieval system without the written permission of Cognella, Inc. For inquiries regarding permissions, translations, foreign rights, audio rights, and any other forms of reproduction, please contact the Cognella Licensing Department at rights@cognella.com.

Trademark Notice: Product or corporate names may be trademarks or registered trademarks, and are used only for identification and explanation without intent to infringe.

Cover image copyright© 2013 Depositphotos/tolokonov.

Printed in the United States of America

Contents

Preface ix

Acknowledgments xiii

Introduction xvii

Table of Greek Letters xix

1 Mathematical Logic and Sets 1
   1.1 Logic Symbols and Truth Tables 2
      1.1.1 Lexicographical Listings of Possible Truth Values 3
      1.1.2 The Logic Operations 4
      1.1.3 Constructing Further Truth Tables 10
      1.1.4 Tautologies and Contradictions, A First Look 12
   1.2 Valid Logical Equivalences as Tautologies 17
      1.2.1 The Idea and Definition of Logical Equivalence 17
      1.2.2 Equivalences for Negations 19
      1.2.3 Equivalent Forms of the Implication 20
      1.2.4 Other Valid Equivalences 22
      1.2.5 Circuits and Logic 25
      1.2.6 The Statements $\mathscr{T}$ and $\mathscr{F}$ 29
   1.3 Valid Implications and Arguments 33
      1.3.1 Valid Implications Defined 33
      1.3.2 Partial List of Valid Implications 36
      1.3.3 Fallacies and Valid Arguments 37
      1.3.4 Analyzing Arguments Without Truth Tables 45
   1.4 Sets and Quantifiers 50
      1.4.1 Sets 50
      1.4.2 Quantifiers 52
      1.4.3 Statements with Multiple Quantifiers 53
      1.4.4 Detour: Uniqueness as an Independent Concept 54
      1.4.5 Negating Universally and Existentially Quantified Statements 56
      1.4.6 Negating Statements Containing Mixed Quantifiers 57
   1.5 Sets Proper (Optional) 61
      1.5.1 Subsets and Set Equality 61

		1.5.2	Intervals and Inequalities in $\mathbb{R}$	63

      1.5.2 Intervals and Inequalities in $\mathbb{R}$ 63
      1.5.3 Most General Venn Diagrams 66
      1.5.4 Set Operations 67
      1.5.5 More on Subsets 73

2 Continuity and Limits of Functions   76
  2.1 Definition of Continuity at a Point 77
  2.2 Continuity Theorems 89
      2.2.1 Basic Theorems 89
      2.2.2 One-Sided Continuity 96
      2.2.3 Essential versus Removable Discontinuities 101
  2.3 Continuity on Intervals 105
      2.3.1 Continuity on Open Intervals: Main Definition and Theorem 105
      2.3.2 Continuity on Closed Intervals: Definition, EVT, and IVT 106
      2.3.3 Simple Applications of the Intermediate Value Theorem (IVT) 110
      2.3.4 Polynomial Inequalities 111
      2.3.5 Rational Inequalities 116
  2.4 Finite Limits at Points 121
      2.4.1 Definition, Theorems, and Examples 122
      2.4.2 Further Limit Notation 133
      2.4.3 Proofs of Limit Theorems 134
  2.5 One-Sided Finite Limits at Points 138
  2.6 Infinite Limits at Points 146
  2.7 Sandwich, Composition, and Trigonometric Theorems 158
      2.7.1 Sandwich Theorem 158
      2.7.2 Visualizing "Approaches" for Independent
             versus Dependent Variables 161
      2.7.3 "Sandwiching" from One Side, or "Pushing" 162
      2.7.4 Limits with Compositions of Functions 164
      2.7.5 Continuity Considerations for Trigonometric Functions 166
      2.7.6 Proofs 169
  2.8 Limits "At Infinity" 174
  2.9 Further Limit Theorems and Trigonometric Limits 189
      2.9.1 Simple Limit Theorems 189
      2.9.2 A Trigonometric Limit 194
      2.9.3 Limits by Substitution 198
      2.9.4 Epilogue 205

3 The Derivative   207
  3.1 Introduction to Derivatives 207
      3.1.1 An Example: Position Gives Rise to Velocity 207
      3.1.2 Definition of Derivative 212
      3.1.3 Slope of a Curve 214
      3.1.4 Standard Types of Difference-Quotient Derivative Computations 218
      3.1.5 Marginal Cost 220

CONTENTS

- 3.1.6 Some Further Applications 223
- 3.1.7 Graphical Cases of Nonexistent Derivatives 226
- 3.2 First Differentiation Rules; Leibniz Notation 228
  - 3.2.1 Differentiability Implies Continuity 228
  - 3.2.2 Positive Integer Power Rule 230
  - 3.2.3 Leibniz Notation 231
  - 3.2.4 Sum and Constant Derivative Rules 232
  - 3.2.5 Increasing and Decreasing Functions; Graphing Polynomials 237
  - 3.2.6 Derivatives of Sine and Cosine 242
  - 3.2.7 Limits That Are Derivative Forms 246
  - 3.2.8 Absolute Value and Other Piecewise-Defined Functions . . . 248
  - 3.2.9 Velocities and Slopes 252
- 3.3 First Applications of Derivatives 257
  - 3.3.1 One- and Two-Dimensional Velocity: Vertical and Horizontal 258
  - 3.3.2 Volume and Fluid Flow in Liquids 260
  - 3.3.3 Zero-Order Chemical Reactions 261
  - 3.3.4 Energy, Work, and Power 261
  - 3.3.5 Electrical Charge and Current 266
  - 3.3.6 Ohm's Law 269
- 3.4 Chain Rule I 273
  - 3.4.1 Leibniz Notation to Explain the Chain Rule 273
  - 3.4.2 Chain Rule Versions of the Power, Sine, and Cosine Rules 275
  - 3.4.3 More Complicated Functions: Rules Calling Other Rules . . . 276
  - 3.4.4 More on "Matching" of Variables 279
  - 3.4.5 Power Rule for Rational Powers 280
  - 3.4.6 Chain Rule Proof of the Power Rule for Rational Powers 283
  - 3.4.7 Applied Examples 284
  - 3.4.8 Chain Rule in Prime Notation 288
  - 3.4.9 Further Examples 290
  - 3.4.10 Partial Proof of Chain Rule 291
  - 3.4.11 Simple Graphical Examples of Chain Rule 292
- 3.5 Product, Quotient, and Remaining Trigonometric Rules 297
  - 3.5.1 Product Rule Stated and First Applied 297
  - 3.5.2 Product Rule Proof 302
  - 3.5.3 Quotient Rule 303
  - 3.5.4 Tangent, Cotangent, Secant, and Cosecant Rules 305
  - 3.5.5 More on Putting Rules Together—Carefully 308
  - 3.5.6 Extended Product Rule 311
- 3.6 Chain Rule II: Implicit Differentiation 315
  - 3.6.1 Review of Chain Rule and Other Differentiation Rules 315
  - 3.6.2 Implicit Functions and Their Derivatives 318
  - 3.6.3 A Mistake to Avoid 329
  - 3.6.4 An Application 330
- 3.7 Arctrigonometric Functions and Their Derivatives 334
  - 3.7.1 One-to-One Functions and Inverses, Reviewed 334

        3.7.2   The Arctrigonometric Functions and Their Derivatives 335
        3.7.3   The Arcsine Function and Its Derivative 340
        3.7.4   The Arccosine and Its Derivative 343
        3.7.5   The Arctangent Function and Its Derivative 345
        3.7.6   The Arcsecant Function and Its Derivative 347
        3.7.7   Geometric and Algebraic Manipulations Aiding Computations 349
        3.7.8   Alternative Definitions of Arcsecant, Arccosecant, Arccotangent 353
    3.8 Exponential Functions 356
        3.8.1   Exponential Functions 356
        3.8.2   Derivative of a Special Exponential Function 359
        3.8.3   A Note on Differences between Polynomials, Exponentials 363
    3.9 The Natural Logarithm I 366
        3.9.1   Algebra of Logarithms 366
        3.9.2   Graph of the Natural Logarithm Function 369
        3.9.3   Derivative of the Natural Logarithm 370
    3.10 The Natural Logarithm II: Further Results 378
        3.10.1 Logarithmic Differentiation 378
        3.10.2 Bases Other Than e: Logarithms 383
        3.10.3 Bases Other Than e: Exponential Functions 384
        3.10.4 Derivative of $(f(x))^{g(x)}$ 385
    3.11 Further Interpretations of the Derivative 391
        3.11.1 Velocity and Acceleration 391
        3.11.2 First-Order and Second-Order Chemical Reactions 393
        3.11.3 Power 394
        3.11.4 Population Growth and Decay 395
    3.12 Hyperbolic Functions (Optional) 399
        3.12.1 Definitions of Hyperbolic Functions 399
        3.12.2 Derivatives of Hyperbolic Functions 400
        3.12.3 Derivatives of Inverse Hyperbolic Functions 401
        3.12.4 Formulas for Inverse Hyperbolic Functions 402
        3.12.5 Graphs of Hyperbolic Functions and Inverses 403

4 **Using Derivatives to Analyze Functions; Applications**   **407**
    4.1 Higher-Order Derivatives and Graphing 408
        4.1.1   Higher-Order Derivatives (Derivatives of Derivatives) 408
        4.1.2   Graphical Significance of the Second Derivative 410
        4.1.3   Detailed Graphs Showing All Discussed Behaviors 413
    4.2 Extrema on Closed Intervals 431
        4.2.1   Main Theorem and Examples 431
        4.2.2   Applications (Max/Min Problems) 435
        4.2.3   Argument for Main Theorem (for Completeness) 441
    4.3 The Mean Value Theorem 448
        4.3.1   Linear Interpolation and Average Rate of Change 448
        4.3.2   Mean Value Theorem 449
        4.3.3   Applications of MVT to Graphing Theorems 451

CONTENTS

- 4.3.4 Numerical Applications 452
- 4.3.5 Verifying the MVT for Particular Cases 453
- 4.3.6 An Application of Linear Interpolation 454
- 4.4 Differentials and the Linear Approximation Method 459
  - 4.4.1 Linear (Tangent Line) Approximation 460
  - 4.4.2 Linear Approximations and Implicit Functions 463
  - 4.4.3 Differentials 464
  - 4.4.4 Applications of Differentials 467
  - 4.4.5 Some Further Approximation Problems 470
- 4.5 Newton's Approximation Method 475
  - 4.5.1 Approximation Methods for Solving Equations 475
  - 4.5.2 More on Newton's Method 479
- 4.6 Chain Rule III: Related Rates 485
  - 4.6.1 Simple Examples of the Principle 485
  - 4.6.2 Choosing the Form of the Relationship 488
  - 4.6.3 Finding a Relationship among Variables 491
  - 4.6.4 Further Examples 494
  - 4.6.5 An Approximate Modeling Example 497

5 Basic Integration and Applications 507

- 5.1 Indefinite Integrals (Antiderivatives): Introduction 508
  - 5.1.1 Indefinite Integrals and Constants of Integration 508
  - 5.1.2 Power Rule for Integrals 511
  - 5.1.3 Finding C (Where Possible) 515
  - 5.1.4 First Trigonometric Rules 518
  - 5.1.5 Integrals Yielding Inverse Trigonometric Functions 520
  - 5.1.6 Integrals Yielding Exponential Functions 521
  - 5.1.7 Integrals Yielding Inverse Hyperbolic Functions 522
- 5.2 Summation (Sigma) Notation 525
- 5.3 Riemann Sums and the Second Fundamental Theorem (FTC) 532
  - 5.3.1 Error, Absolute Error, Relative Error, and Percent Error 533
  - 5.3.2 A Physics Example to Motivate Riemann Sums and the FTC 534
  - 5.3.3 General Riemann Sums and the Fundamental Theorem of Calculus . . 537
  - 5.3.4 Computing Basic Definite Integrals 539
  - 5.3.5 Geometric Interpretation of Definite Integrals 540
  - 5.3.6 Computing Areas Using FTC or Riemann Sums 542
  - 5.3.7 Total Area 547
  - 5.3.8 Physics Application: Net Displacement versus Distance Traveled . . . 549
  - 5.3.9 A Word on Units 550
- 5.4 First Fundamental Theorem of Calculus; Further Results 554
  - 5.4.1 First FTC: Explanations 554
  - 5.4.2 Derivative Computations Using the First (and Second) FTC 556
  - 5.4.3 Other General Theorems of Definite Integrals 559
  - 5.4.4 Linear Combinations of Integrands 561
  - 5.4.5 Integrating Odd and Even Functions over Symmetric Intervals 562

		5.4.6 Average Function Value and MVT for Integrals	564
		5.4.7 Integral Bounds	567
		5.4.8 Last Remarks on These Topics	569
	5.5	Infinitesimals and Definite Integrals in Geometry	573
		5.5.1 Infinitesimals and Areas under Curves	574
		5.5.2 Areas Bounded by Curves	575
		5.5.3 Area of a Circle	579
		5.5.4 Volume and Cavalieri's Principle	579
		5.5.5 Volumes of Revolution: Disc/Washer Method	580
		5.5.6 Volumes of Revolution: Cylindrical Shell Method	584
		5.5.7 Arc Length (and a Cautionary Note Regarding Infinitesimals)	588
		5.5.8 Surface Areas of Revolution	592
	5.6	Physical Applications of Infinitesimals, Definite Integrals	597
		5.6.1 Velocity and Position, Revisited	597
		5.6.2 Work and Energy	598
		5.6.3 Hydrostatic Pressure and Total Force	605
		5.6.4 Torque	609
		5.6.5 Torque and Work	611
		5.6.6 Power and Work	612
		5.6.7 Current and Charge	613
	5.7	Numerical Integration: Approximating Definite Integrals	616
		5.7.1 Previous Uses of Geometry for Integrals	616
		5.7.2 Riemann Sums; Computations by Spreadsheets	618
		5.7.3 Trapezoidal Rule	621
		5.7.4 Simpson's Rule	624
		5.7.5 Numerical Integration for Tabular Data	626
		5.7.6 Error Bounds	627
	5.8	Substitution with the Power Rule	629
		5.8.1 The Technique	630
		5.8.2 A Useful Twist on the Method	636
		5.8.3 Other Miscellaneous Power Rule Substitutions	638
		5.8.4 Substitution in Definite Integrals	639
	5.9	Second Trigonometric Rules	645
		5.9.1 Antiderivatives of the Six Trigonometric Functions	645
		5.9.2 Substitution and the Basic Trigonometric Functions, Part I	647
	5.10	Substitution with All Basic Forms	655
		5.10.1 List of Basic Forms	656
	5.11	Further Arctrigonometric Forms	663
		5.11.1 Factoring to Achieve "1"	663
		5.11.2 Completing the Square	666

Answers to Odd-Numbered Problems   **671**

Index   **703**

Preface

This text is devoted to **first-semester calculus as a standalone topic,** in a way in which it fits nicely into either a classical liberal arts curriculum or an applied sciences curriculum. By focusing our efforts this way, we believe we in fact better prepare students for subsequent calculus courses, as well as courses in other fields where calculus is applied.

This text is also devoted to that student who is looking for **a textbook that is meant to be read,** where all of the important insights are included, spelled out in painstaking detail, allowing more effort to be devoted to reading through these explanations than in trying to extract understanding from extremely concise or incomplete explanations. In other words, rather than having readers discover the details on their own with minimal guidance (which is a defensible pedagogical technique in some settings), we give very rich guidance. This also allows the instructor to give a more summary outline presentation of the topics in class, and to spend less time flushing out every relevant fact, since those facts can be found in the text.

We do include numerous examples of the application of first-semester calculus to other topics to give context and motivation, with the understanding that no text can be all things to all people, and that too much effort to prove otherwise can be distracting and counterproductive. Lately there has been a school of thought that students should spend more time on real-world applications in mathematics. But with only finite time and energy available to the student, in the opinion of the authors this has often gone too far, at an expense in students' time to practice, ponder and understand the underlying mathematics. When that understanding is more limited, mistakes in applications are also more likely, and sometimes more spectacular.

So here we strive to emphasize and deepen the mathematics coverage so that the reader can then enter courses in applied topics much better-versed in the underlying mathematics, and thus better able to concentrate on the subtleties of those other subjects. We tried to choose mostly those applications that give strong and transparent illustrations of the calculus and its importance to applied problems, although a few, more involved applications are also included to hint at the depths of some of these topics. For instance, much physics is introduced here, but we do not wish to "steal the thunder" from the physics courses, nor distract too much from the topic at hand: the calculus as a subject in its own right.

The main defining qualities of this text are a more rigorous treatment of early calculus topics, and more readable, detailed explanations in general.

We do not aspire to include many "tricky" problems, but do present problems that require and (hopefully) inspire some added depth of understanding. To better explain our approach, some context is helpful.

If we assume that undergraduate calculus is broken into the usual three-semester sequence of Calculus I–III, arguably Calculus I, the subject of this text, is the most important of the three. In that course, students are introduced to the most important calculus topics: limits, derivatives, and integrals. Calculus II and III are then further developments of these three topics, especially derivatives and integrals, in much wider and more varied contexts. Of these three topics—limits, derivatives, and integrals—limits form the foundation for the others.

Calculus I is also traditionally the "mushiest" of the three courses to teach, because students have to deal with so many varied examples that it is difficult for them to achieve a comprehensive understanding of the topics even at the Calculus I level, particularly for limits, but that weakness visits itself in the other topics. This mushiness is what we attempt to rectify here, so that **Calculus I can become as theoretically straightforward** as the other two (though few would say the other two are "easy").

But derivatives and integrals also require time to comprehend, and there are many examples where it is tempting to accept limited success, which can occur despite a lack of understanding of limits, just as people might believe it is acceptable if "they can drive the car without first learning what goes on under the hood." That is somewhat defensible, but as with the car analogy, when things go wrong because of limited understanding of the foundational theory, they can go wrong spectacularly. Of course, in the study of calculus the consequences are usually less alarming, and are perhaps more akin to a young student learning to play an instrument who is unaware when he or she plays incorrect notes or rhythms (which is painful for the informed listener to hear), and one is often content with the successes that can occur despite some common misunderstandings.

So, in practice, limits as a standalone topic usually gets a rather hurried treatment, justified by instructors in a number of different ways beyond the car-driving analogy, including: "students just don't really 'get' limits" (even if they can compute most of them correctly, again with spectacular exceptions); the truly interested students will fix misconceptions as they see limits in action later (echoed in the notion that "Calculus II is the best Calculus I course they'll ever take"); and the fact that, to many students and instructors, the calculus really comes alive in the derivatives and integrals, so there is some motivation to get to those as soon as possible.

One of this text's authors (Dougherty) was inspired to fight this kind of thinking after noticing in his graduate teaching assistant days that when a particularly difficult limit topic, namely **epsilon-delta (ε-δ) proofs,** was deleted from the Calculus I list of topics covered at his institution, subsequent students had more difficulty with derivative topics such as the Chain Rule, even though the theoretical path from ε-δ concepts to the Chain Rule is long enough that one is unlikely to be thinking of the former when performing calculations with the latter. Certainly more can be done with later applications of inequalities (e.g., with Mean Value Theorem applications) if ε-δ–type proofs are mastered. Perhaps most predictably, the ease of explaining the various limit computation examples, after ε-δ proofs were attempted, was quite evident, whereas without an ε-δ (or similar context), explanations of limit phenomena often have to be made on an *ad hoc* basis, making it difficult for students to see coherence in the topic.

The observation that students seemed to receive both short-term and long-term benefit from being immersed in ε-δ–type thinking for a time, despite (or perhaps because of) its frustrations, motivated the structure of the first two chapters of this textbook. Indeed, one could simply argue that the challenge of "wrapping their heads around" ε-δ proofs made the Chain Rule seem straightforward by comparison.

But there are reasons that ε-δ proofs are often omitted or given a cursory treatment, the main reason likely being because they are quite difficult as normally presented. In part to lessen such frustrations, the first chapter steps back even further to introduce **symbolic logic**, the inclusion of which alone would set this text apart. Symbolic logic clarifies the ε-δ definitions and others, which can be quite confusing when written in prose but are much more clear in symbolic logic. Symbolic logic also allows for other arguments to be clarified throughout the text. It is a topic students "get," as they can relate it to everyday life, though this topic does take some time (perhaps one week) to cover in a semester-long course.

Including symbolic logic also nicely ties the study of calculus to philosophy and other subjects of a **classical, liberal arts education,** letting the student experience what it would be like to think like a modern-day student of Newton, Descartes, and other great minds in that sense, so the study of calculus can transcend being simply technical. As a bonus, it is great help for students who will study computer science or advanced mathematics in later courses.

As the economists tell us—unlike in mathematics as a subject—in practical matters there are no (pure) solutions, only trade-offs. While adding symbolic logic and delving deeper into limits do delay the study of derivatives and integrals somewhat, we would argue that these later topics can in turn be sped up in the classroom setting, because of the stronger foundation, once limits are finished albeit after the extra time is spent "getting them right." One innovation in accomplishing this in particular is in our adoption of a **"forms" approach to limits,** which students find appealing (e.g., $-4/0^+$ yields a limit of $-\infty$). The approach also much better prepares students for topics in Calculus II such as L'Hôpital's Rule, improper integrals, and many series arguments.

It should be mentioned we also use the **continuity-first** approach to limits, since continuity is arguably more intuitive than limits (especially in its ε-δ definition) and limits can be in part described as a tool especially useful when continuity is "broken." The connection between finite limits at a point and continuity at that point is rather clear when the two ε-δ definitions are juxtaposed.

An equally important, and defining, quality of this text is its **high level of detail** not found in other texts, reflected in its higher page count for the same topics. Examples are more numerous, and are not only explained as proto-exercises, but are then further analyzed to offer the student greater insight and context. **Footnotes** also abound, adding yet further context without necessarily interrupting the flow.

So it is a text that is meant to be read, giving enough detail that the instructor can present an outline in lectures and refer to the text for more details. In contrast, most current textbooks have gotten so concise (some might say "terse") that the lecture is often where students see the details. This text therefore is a better resource for those wishing to fully or somewhat **"flip"** their classes (i.e., have the students do more of the discovery in their reading before coming to class, rather than experiencing more of the discovery in the lectures).

The detail of exposition and reasonable diversity of exercise sets make the text a good candidate for a **spiral approach** to homework assignments. Students have much appreciated having the exercise sets for each section broken into one to four smaller assignments (some quite short). Though this may mean homework from multiple sections is due simultaneously, they benefit significantly from a structurally-imposed revisiting of major topics several days in a row, in less intimidating quantities per topic.

Much positive feedback received regarding previously online drafts was from students using other texts and needing a **supplement.** We are happy to serve such students as well.

It is hoped that this book will also **help students to become better readers of technical texts,** by providing them one with a **gentler but richer approach,** in which most of their theoretical questions will be answered in the readings. It should be rare that the student needs to have the dreaded stack of paper and rows of pencils next to the text to figure out how the author got from one computation or statement to the next.

That said, homework is necessarily included, for which a lot of paper may be necessary, and it is likely that both ends of numerous pencils will get plenty of use, but that is the nature of how one comes to "own" a knowledge of calculus. We simply believe the explanatory portion of the text should not require that same level of frustration.

The number of homework problems on a particular topic is, for the most part, lower than found in the mainstream texts (most of which have had a double-digit number of editions published). We believe the number here to be sufficient, and that a much larger number can be unnecessarily intimidating. However, if a student wishes more practice, good sources for further problems include the *Schaum's Outline of Calculus* (McGraw Hill), other mainstream calculus texts (of which there are numerous older editions available at very low prices), and many of the large number of books devoted to calculus problems. However, we believe that our text is second-to-none in the overall detail and clarity of the explanations.

Incidentally, the title of this textbook is not meant to exclude students who are in fields of study other than mathematics. Indeed, it is hoped that **anyone who pursues calculus for whatever reason can do so as a "student of mathematics."** It is not uncommon for a gathering of individuals to contain some who may be considered "students of Shakespeare" but never formally completed a degree in literature or a related field. The same can be said of "students of history." The phrase here simply means that such individuals care enough about the subject to take personal time to examine it thoughtfully and continuingly, and to become respectably articulate in the subject, at least when among one's peers.

Of course, this textbook is intended to be thorough for those whose major field of study is mathematics. However, it is hoped that *First Semester Calculus for Students of Mathematics and Related Disciplines* will inspire each reader—whose study may be any field—to become, for a while if not for a lifetime, a true "student of mathematics."

A sequel covering Calculus II topics is also in preparation, using the same level of detail but otherwise resembling a more standard (though more verbose) treatment of those topics.

The entire textbook is typeset in LaTeX by the first author (who is therefore responsible for any errors), using the LaTeX book style, with graphics handled by the LaTeX pstricks package (and a few by the graphics package and the m4 scripting language). The main fonts are from the LaTeX-included Fourier New Century School Book family. Several other LaTeX packages were also used, mainly for modifying the format. No graphics were imported but are all generated using LaTeX code from these packages. Some style cues were taken from the very polished, open-source trigonometry textbook of Michael Corral, which the reader is encouraged to investigate.

Acknowledgments

Dougherty

First I would like to thank anyone who reads any part of this book. It is written for you! Even if you do not read it cover to cover, I very much appreciate your interest. And I would also like very much to hear back, regardless of your opinion.

A large share of the thanks naturally goes to my coauthor, John Gieringer, whom I met early in my career as an assistant professor while working at my first institution out of graduate school and then lost touch with as "life happened." Others in my closer proximity had been invited to join the effort, and I had mostly given up finding help with the project, when a chance encounter with new faculty at my current institution who knew John reminded me that he could be a good fit and inspired me to ask him if he was up for it. I am grateful in the extreme that he agreed to join the effort. His wealth of diverse teaching experience brought many new examples—which spawned several pages of background explanation—to the text, as well as many of the exercises and a tremendous editing effort. Most of the physics examples, and all of the economic examples, are due to him.

John's energy and work ethic were also very welcome additions to the project. Without his focus and determination, I have no doubt this project would take many more years to complete, and its breadth would have suffered greatly. Our organizational skills and general instincts seemed to complement each other well.

I am forever grateful to my best friend and wife of twenty-four years, Hung-Chieh Chang (aka Joy Dougherty), who never showed me any doubt in her mind that this work would eventually be finished, and who put up with the seemingly countless hours I spent bonding with several computers to finish this book. Her opinions on things mathematical, pedagogical, and artistic were invaluable. My children Rebekah and Anne (Annie) also deserve credit for sacrifices such a long-term project inevitably requires. Encouragement from my parents and other family members was also appreciated.

I am also very grateful to Southwestern Oklahoma State University, particularly all in the Department of Mathematics, for their encouragement and support for this project. Few institutions would allow junior faculty—as I was when the project first gained traction—so much freedom to undertake a calculus textbook, which inevitably saps time and energy from other scholarly efforts usually expected of new faculty. In particular I was given much release time and was allowed to use photocopied excerpts of the work in progress in my calculus courses. This was both risky and expensive for the department, and I sincerely hope my colleagues find that it was worth it. My students who gamefully followed my lead, learning from this material in their classes and giving me some feedback (including some welcome corrections), also deserve my gratitude.

This text is strongly influenced by James Phelan, my own high school calculus teacher in the mid 1980s, at Boylan Central Catholic High School in Rockford, Illinois. Though at the time (before Choate got hold of him) he also taught at a local college, he did not teach specifically to prepare us for the Advanced Placement (AP) test—as so many high school instructors are directed to do—but instead taught what *he* thought was a solid course. (With much credit due to him, I passed the AP test with a score of 5 out of 5 anyway, which gave his methods much validation!) His mission was to make us literate in mathematics in as many ways as possible while teaching us as much calculus as we could absorb as high school seniors. My desire to be a "student of mathematics," that is, to acquire some mathematical sophistication but not necessarily to earn a degree in the field, came first from him.[1]

The professor who, by example, convinced me to become a mathematics major was Shih-Chuan Cheng at Creighton University. The coherence and sophistication of his lecture notes first convinced me of the beauty of mathematics and probably constitute the single greatest influence on the style of this text. Other coursework under John Mordeson and James Carlson at Creighton convinced me that there is a style of teaching and learning mathematics that stresses depth and coherence first, and breadth second. This is far away from the common "sink or swim" approach, and still allows conscientious students to hold their own among their peers from top-tiered schools.

I must also thank all who made desktop publishing of mathematics possible. In particular, those in the TeX and LaTeX developers community who brought us not only LaTeX but some amazing supplemental packages, particularly pstricks but also multicols, enumerate, and caption, and for Adobe for inventing Postscript and PDF standards and, as importantly, keeping them "open" so the TeX community could exploit them for producing publication-quality mathematics. With these things I was able to produce textbook-quality copy to give to my students in class, as well as online when the text was still in preliminary form, and to present camera-ready copy to a publisher in the form of a PDF file. This ability has been enormously helpful to the development of this text.

Over the years I had early drafts of the manuscript available online, and would be remiss in not thanking those who found it and gave me encouraging feedback, and even pointed out a few early errors. There are too many to mention, though it was particularly encouraging when Professor Sarah Koskie of IUPUI found the first chapter material on symbolic logic to be of high enough quality that she included it in one of her graduate engineering courses.

My good friend and colleague Gerard East kindly read over each chapter and offered many editorial and mathematical suggestions. I was grateful for every suggestion and error correction he made, as if they were nuggets of gold, greatly enriching the final quality of the manuscript. I have no doubt any errors still found in the text were introduced long after Gerry's corrections and are thus my own responsibility.[2]

Of course the good folks at Cognella, who saw potential in this project, were also crucial

---

[1] When I was interviewing for my first position as an assistant professor, and was asked about my teaching philosophy, I explained that the students had been reading too much Stephen King mathematics. I wanted them to think they were reading Tolstoy! (A friend later added, "Yes, and in Russian!") I got the job.

[2] I am reminded of a line in the acknowledgments in an impressive and difficult book by Joseph Doob, an important American mathematician, in which he thanks his typist as "usually faithful, sometimes accurate." Purdue Professor Rodrigo Bañuelos told me he asked Professor Doob, "How can you write such a terrible thing?" Professor Doob answered, "I typed it!" I also have to admit I typed this textbook myself, so all errors are indeed mine.

and cannot be thanked enough. It is one thing to produce content, and another thing to turn it then into a respectable publication, meaning: to polish the text, to package it, and to lay the legal and logistic framework for publication. I am grateful for their tremendous help in all these things. These professionals include (in reverse order of their appearance in my inbox): Berenice Quirino, Abbey Hastings, Jess Estrella, Danielle Menard, Michelle Piehl, and Angela Schultz. Much copy editing by Michele Mitchell rescued us from numerous embarrassing errors, for which I am eternally grateful, and Allie Griffin's final proofreading, which caught a couple dozen remaining (or newly introduced) errors, was the icing on the cake. Any errors that remain after this final effort were despite these folks' (and my coauthor's) Herculean efforts, and can only be due to my unreasonable stubbornness, tired eyes, or inability to resist making "one last change."

Gieringer

My first thank you goes to anyone reading this book. While there are several calculus books on the market, we find this book to be unique and we hope you agree. We thank you for choosing the book and we hope that you take the time to read the book in depth. We hope you enjoy the difference.

My next thank you goes to Michael Dougherty, the principle author of this book. This calculus book is his brainchild and he has been involved with it from the start. I thank Mike for allowing me to join midway through the project. It was good to be reunited with him for a mathematics endeavor after our initial work in calculus together several years ago in Pennsylvania.

I thank my family for the support that they have given me these past few years while this book has been taking shape. I thank my wife, Majida, for allowing me to take time away from the family to travel to Oklahoma to work on the book. She has been nothing but supportive during the entire process.

I also wish to thank my children, Kathleen, Christine, and Colleen, for being understanding when I needed to take time to focus on the book. Many times they have asked me when the book would be done. The time is here! Enjoy reading it.

Thank you to my colleagues at Alvernia University who have been supportive of this project. A final thank you goes to Cognella for bringing this project to a successful conclusion.

Introduction

The discovery of calculus was one of the most important and exciting achievements in the history of intellectual progress. Virtually every field that deals with quantities has benefited from calculus. It allowed Sir Isaac Newton to derive the laws of planetary motion, Professor Albert Einstein to derive relativity, engineers and scientists to design space vehicles, many an economist to model and analyze market variables, disease control specialists to model and therefore best fight epidemics, and countless other specialists to further their achievements. In particular, it is reasonable to estimate that physics would be centuries behind its present maturity were it not for the availability of calculus.

Despite the technicalities involved in the lofty fields already mentioned, the fundamental principles of calculus are quite accessible, especially now that the subject has been distilled into more a coherent and accessible form with the passing of centuries since its initial discoveries. Generations of researchers, educators, and authors have refined the presentations to be understandable to motivated college students with a variety of interests. Some larger universities have separate calculus courses specifically for majors in business, biology, agriculture, forestry, and liberal arts, as well as the generic courses for engineering, mathematics, and science.

Calculus is the marquee mathematical subject for many of these programs of study, particularly science and engineering, and its importance cannot be overstated. Even the algebra-trigonometry courses at our institutions have been fashioned largely to groom students for eventual study of calculus.

So, what is calculus? The short answer is that it is the field of mathematics which deals with change, both instantaneous and cumulative. Respectively, this means calculus is mainly—but certainly not exclusively—interested in solving the following two problems:

(1) Given algebraic relationships among variables, compute their rates of change with respect to each other.

(2) Given the rates of change of variables with respect to each other, find the algebraic relationships among the variables.

Indeed, the second is simply the first in reverse. For a simple, though abstract example, consider the following questions:

(i) If we know the position s of an object at every time t, can we know the velocity v of the object at every time t? (The answer is yes.)

(ii) If we know the velocity v of an object at every time t, can we know the position s of the object at every time t? (The answer is again yes, given a little more information, such as where the object started.)

There are other problems that mimic these, but from very different fields. For instance, if we know the amount of fuel inside of a rocket engine at any time, do we know the rate of fuel consumption? Or, if we know the pressure at every point on the wall of a dam, and know the geometry of the dam, do we know the total force on the wall? Again, the answers are yes and yes. These may seem trivial until one realizes that the rate of fuel consumption can vary with time, and the pressure on the wall of a dam varies with depth.

Problems of type (1) are part of *differential calculus*, also known as calculus of *derivatives*. Problems of type (2) are part of *integral calculus* where, perhaps predictably, we will compute many *antiderivatives*. Problems of this second type tend to be more (sometimes much more) difficult than problems of the first type, and so we rely on more clever approaches.

Calculus is also important in finding sophisticated approximations for complex problems. Using the same principles that answer the questions already posed, with a bit of creativity we can write a simple program to teach a computer to compute the sine of an angle to any degree of accuracy we like, even if the programming language only understands addition, subtraction, multiplication, and division. In practical terms, sometimes a fast algorithm that is "accurate enough" is preferable to a slow but very precise algorithm, for instance if we wish to program a small missile to chase and destroy a larger missile that was programmed to evade interceptors. A fast algorithm, which can be run repeatedly to help the interceptor adjust to the larger missile's trajectory changes, may indeed be preferable to a more ponderous algorithm whose output may be irrelevant by the time it is achieved.

Before we can work on these problems, we will need some preliminaries. We will begin with Chapter 1 on symbolic logic so that we can employ that language throughout the text to greatly clarify definitions and mathematical arguments.

In Chapter 2 we consider the very foundational preliminaries specific to calculus in the concepts of continuity and limits. Because their technicalities appear repeatedly in differential and integral calculus, we will spend considerable effort on these. The payoff is that the calculus that follows will benefit from every effort spent with the concepts of limits and continuity.

Chapter 3 begins the discussion of derivatives, including their nature and how to compute them, and some of their applications.

Chapter 4 is dedicated to applications—mathematical and real-world—of derivatives.

Chapter 5 discusses antiderivatives, their meaning, their computation in many cases, and their applications. Techniques of computing larger classes of antiderivatives is a main topic of second-semester calculus, but a strong foundation in first-semester techniques is necessary for success with the later techniques.

Throughout the text we will explore many applications of limits, derivatives, and antiderivatives to the extent they are reasonable for a text of this scope. The development of the analytical tools is our main goal. The student well-versed in the mechanics of those tools will surely (and, it is hoped, easily) find numerous other uses for the methods developed here.

Table of Greek Letters

We will often use Greek letters in this text, as is standard for technical writing. In calculus, Δ (upper-case delta), δ (lower-case delta), ε (lower-case epsilon), and Σ (upper-case sigma) have particularly special roles, as do θ (lower-case theta) and ϕ (lower-case phi), among others, in trigonometry. Furthermore, we will also use other Greek letters when they are either appropriate stand-ins for their English/Latin counterparts, or when we want them to be conspicuous in mathematical expressions. Finally, we would like to reinvigorate the tradition that a reasonably informed student in any technical discipline is expected to eventually know all the letters of the Greek alphabet. For these reasons we include a table of Greek letters.

A	α	alpha		N	ν	nu
B	β	beta		Ξ	ξ	xi
Γ	γ	gamma		O	o	omicron
Δ	δ	delta		Π	π	pi
E	ε	epsilon		P	ρ	rho
Z	ζ	zeta		Σ	σ	sigma
H	η	eta		T	τ	tau
Θ	θ	theta		Y	υ	upsilon
I	ι	iota		Φ	ϕ	phi
K	κ	kappa		X	χ	chi
Λ	λ	lambda		Ψ	ψ	psi
M	μ	mu		Ω	ω	omega

Chapter 1

Mathematical Logic and Sets

In this chapter we introduce symbolic logic and set theory. These are not specific to calculus, but are shared among all branches of mathematics. There are various symbolic logic systems, and indeed mathematical logic is its own branch of mathematics, but here we look at that portion of mathematical logic that should be understood by any professional mathematician or advanced student. The set theory is a natural extension of logic and provides further useful notation as well as some interesting insights of its own.

The importance of logic to mathematics cannot be overstated.[1] No conjecture in mathematics is considered fact until it has been logically proven, and truly valid mathematical analysis is done only within the rigors of logic. Because of this dependence, mathematicians have carefully developed and formalized logic beyond some of the murkier "common sense" notions we learn from childhood and given it the precision required to explore, manipulate, and communicate mathematical ideas unambiguously. Part of that development is the codification of mathematical logic into symbols. With logic symbols and their rules for use, we can analyze and rewrite complicated logic statements much like we do with algebraic statements.

Symbolic logic is a powerful tool for analysis and communication, but we will not abandon written English altogether. In fact, most of our ideas will be expressed in sentences that mix English, mathematical equations or inequalities, and symbolic logic. We will strive for a pleasant style of mixed prose, but we will always keep in mind the formal logic on which we base our arguments and resort to the symbolic logic when the logic in prose is complicated or can be illuminated by a symbolic representation.

Because we will use English phrases as well as symbolic logic, it is important that we clarify exactly what we mean by the English versions of our logic statements. Part of our effort in this chapter is devoted toward that end.

The symbolic language developed here is used throughout the text. It is descriptive and

---

[1]In fact, Bertrand Russell (1872–1970)—one of the greatest mathematicians of the twentieth century—argued successfully that mathematics and logic are exactly the same discipline. Indeed, they seem to be supersets of each other, implying they are the same set. It just happens that to many a lay person, mathematics may be associated only with numbers and computations while logic deals with argument. The field of geometry belies this categorization, but there are many other vast mathematical disciplines that are not so interested in our everyday number systems. These include graph theory (useful for network design and analysis), topology (used to study surfaces, relativity), and abstract algebra (used, for instance, in coding theory) to name a few. Indeed, both mathematics and logic can be defined as interested in abstract, coherent structural systems. Thus, to a modern mathematician, logic versus mathematics may be considered a "distinction without a difference."

Symbol	Read	Example	Also Read
$\sim$	not	$\sim P$	P is not true P is false
$\wedge$	and	$P \wedge Q$	P and Q both P and Q are true
$\vee$	or	$P \vee Q$	P or Q P is true or Q is true (or both)
$\longrightarrow$	implies	$P \longrightarrow Q$	if P then Q P only if Q
$\longleftrightarrow$	if and only if	$P \longleftrightarrow Q$	P bi-implies Q P if and only if (iff) Q

Table 1.1: Some basic logic notation.

precise, and learning its correct use forces clarity in thinking and presentation. It is not common for a calculus textbook to include a study of logic, since authors have more than enough to accomplish in trying to offer a respectably complete treatment of the calculus itself. However, it is quite common for teachers and professors to insert some of the logic notation into the class lectures because of its usefulness for presenting and explaining calculus to students. Unfortunately, a casual or "on-the-fly" introduction to these devices can cause as many problems as it solves. In this text we will instead commit early to developing and using the symbolic logic notation so we can take advantage of its correct use.

We begin with the first section (Section 1.1) devoted to the construction of truth tables, which ultimately define our first group of logic symbols. Subsequent sections in this chapter will explore valid logical equivalences (Section 1.2), valid implications and some general argument types (Section 1.3), quantifiers (Section 1.4), and sets (Section 1.5).

1.1 Logic Symbols and Truth Tables

The first logic symbols we develop in the text are listed in Table 1.1. In what follows we will explain their meanings and give their English versions, while also pointing out where casual English interpretations often differ, from each other as well as from their formal meanings. It is useful to learn to read the symbols as they would usually be said out loud. For instance, $P \wedge Q$ can be read, "P and Q," while $P \longrightarrow Q$ is usually read "P implies Q." One reads $\sim P$ as "not P," while more elaborate means for verbalizing, say, $\sim (P \vee Q)$ would include "It is not the case that P or Q." In fact, if P is any statement, such as "it is raining," then we can graft the words "it is the case that" and have a new statement with exactly the same meaning: "It is the case that it is raining." This allows us more flexibility to read negations in a more natural order: $\sim P$ becomes "it is not the case that it is raining."[2]

Now we look again at the symbols in Table 1.1. The symbol $\sim$ is called a *unary logic*

---

[2]Occasionally, some of these are verbalized using what amounts to their typographical descriptions, so for instance $P \wedge Q$ becomes "P wedge Q," while $P \vee Q$ becomes "P vee Q."

1.1. LOGIC SYMBOLS AND TRUTH TABLES

operation because it operates on one (albeit possibly compound) statement, e.g., P. The symbols $\wedge, \vee, \longrightarrow, and \longleftrightarrow$ are called *connectives* or *binary logic operations*, connecting two statements, such as P and Q. Both types will be developed in detail in this chapter.

1.1.1 Lexicographical Listings of Possible Truth Values

In the next subsection we develop the logic operators listed previously in Table 1.1, page 2. These operators connect *statements*, that are meaningful declarative sentences that are either true or false (such as "all men are mortal," "I eat pizza," and "$2x+1=9$"). For simplicity and to explore the abstract nature of logical operators, we sometimes give these statements variable names such as P, Q, R and so on. The logical operators combine or otherwise operate on these to form new, *compound* statements $P \longrightarrow Q$, $\sim P$, $P \wedge Q$, etc. In doing so, we analyze the truth or falsity of the compound statements based on the truth or falsity of the underlying, *component* statements P, Q, etc.[3]

We always assume a particular statement can be either true or false, but not simultaneously both.[4] We signify these possibilities by the *truth values*, T or F, respectively. Note that for n independent statements $P_1, P_2, \ldots, P_n$ there are 2^n different combinations of T and F.[5] Thus for a single statement P, we have $2^1 = 2$ truth value possibilities, T or F. For two independent statements P and Q, we have $2^2 = 4$ possible combinations of truth values: TT, TF, FT or FF (i.e., P and Q both true, P true and Q false, P false and Q true, or P and Q both false). For three statements P, Q, R, the possibilities are $2^3 = 8$-fold. To list exhaustively all possible combinations, we will employ a *lexicographical order*, as shown in Figure 1.1, page 4. If there are $n \geq 2$ independent statements, then for the first we write half (2^{n-1}) T's and the same number of F's. For the next statement we write half (2^{n-2}) T's, and the same number of F's, and then repeat. If there is a third, we simply alternate T's with F's twice as fast (i.e., 2^{n-3} T's, and as many F's), and then repeat following that pattern until we fill out 2^n entries. The last statement's entries are TFTF$\cdots$TF, until 2^n entries are made. Figure 1.1 illustrates this pattern, for $n = 1, 2, 3$.

---

[3] In our analyses, the *component* statements will consist of single letters P, Q, and so on, and be allowed truth value T or F. *Compound* statements are not necessarily allowed either truth value, but their truth values are determined by those of the underlying component statements. For instance, we will see $P \vee (\sim P)$ can only have truth value T, and $P \wedge (\sim P)$ can have only F, while $P \longrightarrow Q$ is sometimes T, sometimes F.

[4] This is sometimes referred to as the "law of the excluded middle." It is useful in future discussions since it is often easier to prove P is not false (i.e., $(\sim P)$ is false) than to prove P is true.

[5] This is a simple counting principle. For another example, suppose we have four shirts and three pairs of pants, and we want to know how many different combinations of these we can wear, assuming we will wear exactly one shirt and one pair of pants. Since we can wear any of the shirts with any of the pants, the choices—for counting purposes—are independent. We have four choices of shirts, and for each of those we have three choices of pants. It is not difficult to see that we have $4 \cdot 3 = 12$ possible combinations to choose from.

If we also include two choices of belts, and assume we will wear exactly one belt, then we have $4 \cdot 3 \cdot 2 = 24$ possible combinations of shirt, pants, and belt.

Here, there are two choices for the truth values (T or F) of each of the $P_1, \ldots, P_n$, so there are 2^n possible truth-value combinations.

Whole textbooks are written regarding this and other counting principles, but in this text we will only encounter a few. For undergraduates, some of these principles are often found embedded in probability courses, or courses relying on probability, such as genetics, or in combinatorics which appears especially in computer science and electrical engineering.

4   CHAPTER 1. MATHEMATICAL LOGIC AND SETS

P
T
F

P	Q
T	T
T	F
F	T
F	F

P	Q	R
T	T	T
T	T	F
T	F	T
T	F	F
F	T	T
F	T	F
F	F	T
F	F	F

Figure 1.1: Lexicographical ordering of the possibilities for one, two, or three independent statements' truth values. The extra horizontal line in the third table is for ease of reading only.

P	~P
T	F
F	T

P	Q	$P \wedge Q$	$P \vee Q$	$P \to Q$	$P \leftrightarrow Q$
T	T	T	T	T	T
T	F	F	T	F	F
F	T	F	T	T	F
F	F	F	F	T	T

Table 1.2: The basic logic operations defined for all possible truth values of their arguments

1.1.2 The Logic Operations

The basic five logic operations we will use in this text are given in Table 1.2 for every possible truth value of underlying component statements. We say that the operation $\sim$ takes one *argument* (not to be confused with the colloquial meaning of the term), that argument being P in Table 1.2 above. The other operators $\wedge$, $\vee$, $\to$, and $\leftrightarrow$ each take two arguments, dubbed P and Q in the table above.

We begin with the logical negation $\sim$, which is a unary operation, i.e., acting on one (possibly compound) statement. For example consider the statement $\sim P$, usually read "not P." This is the negation of the statement P. Of course $\sim P$ is not independent of P, but its truth value is based on that of P; stating that $\sim P$ is true is the same as stating that P is false, and stating that $\sim P$ is false is the same as stating that P is true. We can completely describe the relationship between P and $\sim P$ in the following truth table diagram:[6]

P	~P
T	F
F	T

This also completely describes the action of the operation $\sim$: it takes a statement with truth value T and returns a statement with truth value F, and vice-versa. For an English example,

---

[6]We will always use double lines to separate the independent component statements P, Q, etc., from compound statements based on them.

1.1. LOGIC SYMBOLS AND TRUTH TABLES

if we define a statement P by

$$P : \text{I will go to the store,}$$

then the resultant statement for $\sim P$ is simply

$$\sim P : \text{I will not go to the store.}$$

We can also read $\sim$ as "it is not the case that," so our example above could read

$$\sim P : \text{It is not the case that I will go to the store.}$$

Even in an English example, it is not difficult to see that P is true exactly when $\sim P$ is false, and P is false exactly when $\sim P$ is true. For truth value computations, we can summarize the action of $\sim$ as follows:

$\sim$ switches the truth value of the statement on which it operates, from T to F or from F to T.

It will be interesting to see how $\sim$ operates on compound statements as we proceed. How it interacts "with itself" is rather straightforward. Indeed, it is not difficult to see that the statement $\sim(\sim P)$ is the same as the statement P. We might read

$$\sim(\sim P) : \text{It is not the case that it is not the case that I will go to the store.}$$

Perhaps a better English translation of $\sim(\sim P)$ here would be, "It is not the case that I will not go to the store," which clearly states that I will go to the store, i.e., P. In the next section we will look at ways to *calculate* when two logic statements in fact mean the same thing, such as P and $\sim(\sim P)$.[7]

We next turn our attention to the binary operation $\wedge$. This is called the logical *conjunction*, or just simply *and*; the statement $P \wedge Q$ is usually read "P and Q." This compound statement $P \wedge Q$ is true exactly when both P and Q are true, and false if a component statement is false. Thus its truth table is given by the following:

P	Q	$P \wedge Q$
T	T	T
T	F	F
F	T	F
F	F	F

As an operation, $\wedge$ returns T if both statements it connects have truth value T, and returns F otherwise, i.e., if either of the statements connected by $\wedge$ is false.

---

[7] It will be taken for granted throughout the text that the reader has some familiarity with the use of parentheses ( ), brackets [ ], and similar devices for grouping quantities—logical, numerical, or otherwise—to be treated as single quantities. For instance, $\sim(\sim P)$ means that the "outer" (or first) $\sim$ will operate on the statement $\sim P$, treated as a single, albeit "compound" statement. Thus we first find $\sim P$, and then its logical negation is $\sim(\sim P)$. This type of device is used throughout the chapter and the rest of the text.

Example 1.1.1

Suppose we set P and Q to be the statements

$$P : \text{I will eat pizza,}$$
$$Q : \text{I will drink soda.}$$

Connecting these with $\wedge$ gives

$$P \wedge Q : \text{I will eat pizza and I will drink soda.}$$

This is true exactly when I do both, eat pizza and drink soda, and is false if I fail to do one, the other, or both.

Next we look at the binary operation $\vee$, called the logical *disjunction*, or simply *or*. The statement $P \vee Q$ is usually read "P or Q." For $P \vee Q$ to be true, we only need one of the underlying component statements to be true; for $P \vee Q$ to be false, we need both P and Q to be false. The truth table for $P \vee Q$ is thus as follows:

P	Q	$P \vee Q$
T	T	T
T	F	T
F	T	T
F	F	F

It is important to note that $P \vee Q$ is not an *exclusive or*,[8] so we still take $P \vee Q$ to be true for the case that both P and Q are true. At times it is not interpreted this way in spoken English, but our standard for a statement being false (i.e., having truth value F) is that it is in fact contradicted. If we state "P or Q," to a logician we are only taken to be lying if both P and Q are false.[9]

Example 1.1.2

For the P and Q from the previous example, we have

$$P \vee Q : \text{I will eat pizza or I will drink soda.}$$

Again, this is still true if I do both, eat pizza and drink soda, or just do one of these; it is sufficient that one be true, but it is not contradicted if both are true. Note that this is false exactly when both P and Q are false, i.e., for the case that I do not eat pizza and do not drink soda.

---

[8] The case where we have P or Q but not both is called an *exclusive or*. Computer scientists and electrical engineers know this as **XOR**. For our purposes the *inclusive* or $\vee$ will suffice, and anyhow is much simpler to deal with computationally in symbolic logic manipulations, though **XOR** will appear in the exercises.

[9] To see how English understanding is context-driven, consider the following situations. First, suppose a parent tells the child to "clean the bedroom or the garage" before dinner. If the child does both, the parent will likely take the request to be fulfilled. Next, suppose instead that parent tells the child to "take a cookie or a brownie" after dinner, and the child takes one of each. In this second context the parent may have a very different understanding of the child's compliance to the parental instructions (see note on **XOR** above). To a logician (and perhaps to any self-respecting smart aleck) $\vee$ must be context-independent.

1.1. LOGIC SYMBOLS AND TRUTH TABLES

Next we consider $\longrightarrow$. Arguably the most common and therefore important logic statements in mathematics are of the form $P \longrightarrow Q$, read "P implies Q" or "if P then Q." These are also the most misunderstood by novice mathematics students, and so we will discuss them at length. As before, a truth table summarizes the action of this (binary) operation:

P	Q	$P \longrightarrow Q$
T	T	T
T	F	F
F	T	T
F	F	T

Note that the only circumstance in which we take $P \longrightarrow Q$ to be false is when P is true, but Q is false. As before, our standard for falsity is that the statement must be actually contradicted, and that occurs exactly when we have the truth of the *antecedent* P, but not of the *consequent* Q. In particular, if P is false, then $P \longrightarrow Q$ cannot be contradicted, so we take those two cases to be true, dubbing $P \longrightarrow Q$ *vacuously true* for those two cases where P is false.

In summary, the connection $\longrightarrow$ returns T for all cases *except* when the first statement is true, but the second is false.

The importance of the implication extends beyond mathematics and into philosophy and other studies. Because of its ubiquity, logical implication has several syntaxes which all mean the same to a logician. It is interesting to compare the various phrases, but first we will look at an example in the same spirit as we had for $\sim$, $\wedge$, and $\vee$.

Example 1.1.3 _____

For the P, Q in the previous examples, we have

$P \longrightarrow Q$: If I will eat pizza, then I will drink soda.

It is useful to see when this is clearly false: when P is true but Q is false, which for these P, Q would be the case that I eat pizza but do not drink soda. In fact, it is important that that is the **only** case in which we consider $P \longrightarrow Q$ to be false. In particular, if P is false, then $P \longrightarrow Q$ is vacuously true. The idea is that if I do not eat pizza, then whether or not I drink soda I do not contradict the statement, "If I will eat pizza, then I will drink soda."

_____

There are several English phrases that mean $P \longrightarrow Q$. Below are five equivalent ways to write the corresponding English version of $P \longrightarrow Q$ for the P, Q in the examples. (That the fourth and fifth versions are equivalent will be proved in the next section.)

1. My eating pizza implies my drinking soda (P implies Q).

2. If I will eat pizza, then I will drink soda (if P then Q).

3. I will eat pizza only if I will drink soda (P only if Q).

4. I will drink soda or I will not eat pizza (Q or not P).

5. If I will not drink soda, then I will not eat pizza (if not Q then not P).

These five ways of stating $P \longrightarrow Q$ might not all be immediately obvious, and so are worth reflection and eventual commitment to memory. Two other common—and rather elegant—ways of stating the same thing are given below by example and in the abstract:

6. My drinking soda is necessary for my eating pizza (Q is *necessary* for P).

7. My eating pizza is sufficient for my drinking soda (P is *sufficient* for Q).

The kind of diction in 6 and 7 is very common in philosophical as well as mathematical discussions. We will return to implication after next discussing bi-implication, since a very common mistake for novice mathematics students is to confuse the two.

The bi-implication is denoted $P \longleftrightarrow Q$, and often read "P if and only if Q." This is sometimes also abbreviated "P iff Q". It states that P implies Q and Q implies P simultaneously. Thus truth of P gives truth of Q, while truth of Q would give truth of P. Furthermore, if P is false, then so must be Q, because Q being true would have forced P to be true as well. Similarly Q false would imply P false (since if P were instead true, so would be Q). The truth table for the bi-implication is the following:

P	Q	$P \longleftrightarrow Q$
T	T	T
T	F	F
F	T	F
F	F	T

An important, alternative way to describe the operation $\longleftrightarrow$ is to note that $P \longleftrightarrow Q$ is true exactly when P and Q have the same truth values (TT or FF). Thus the connective $\longleftrightarrow$ can be used to detect when the connected statements' truth values match, and when they do not. This will be crucial in the next section.

Example 1.1.4

Consider the statement $P \longleftrightarrow Q$ for our earlier P and Q, for which we have

$P \longleftrightarrow Q$: I will eat pizza if and only if I will drink soda.

This is the idea that I cannot have one without the other: If I have the pizza, I must also have the soda ("only if"), **and** I will have the pizza if I have the soda ("if"). This is false for the cases that I have one but not the other. It is important to note that it is not false if I have neither pizza nor soda.

In fact a bi-implication $P \longleftrightarrow Q$ is well-named since it is the same as $(P \longrightarrow Q) \wedge (Q \longrightarrow P)$. (The proof of this fact is given in the next section.) Note that we can switch the order of statements connected by $\wedge$ (and), so we can instead write $(Q \longrightarrow P) \wedge (P \longrightarrow Q)$, i.e., $Q \longleftrightarrow P$. In prose we can write "P is necessary and sufficient for Q," for $P \longleftrightarrow Q$, which is then the same as "Q is necessary and sufficient for P," i.e., $Q \longleftrightarrow P$.

At this point we will make a few more observations concerning the differences between the English and formal logic uses of terms they have in common. The cases that follow illustrate how casual English users are often unclear about which of "if," "only if," or "if

1.1. LOGIC SYMBOLS AND TRUTH TABLES

and only if" is meant in both speaking and listening. Again, mathematics requires absolute precision in what we mean by these things.

The first difference involves the phrase "only if." This is often misunderstood to mean "if and only if" in everyday speech. When we combine the two words "only if," the standard logic meaning is not the same as "if" modified by the adverb "only." Taken together, the words "only if" have a different, but precise meaning in logic. Consider the following statements:

- You can drive that car only if there is gasoline in the tank.

- You can drive that car only if there is air in the tires.

- You can drive that car only if the ignition system is working.

Clearly it is not the case that you can drive that car if and only if there is gasoline in the tank, since the gasoline is *necessary but not sufficient* for running the car; you also need the air, ignition, etc., or the car still will not drive regardless of the state of the gasoline tank. Similarly, a father who tells his teenaged child, "You can go out with your friends only if your homework is finished" might justifiably find another reason to keep the child from joining the friends even after the homework is done. (Sudden severe weather, inappropriate activities planned, mechanical problems, and several other reasons quickly come to mind.) Note that these are all mathematical implications $\longrightarrow$: that you can drive the car would *imply* there is gas, air and ignition; that the child can go out *implies* that the homework is done. One has to be careful not to read bi-implications (if and only if) into any of these statements, which are only implications.

The other difference deals with another way to state implications: if/then. This is also often misunderstood to mean if and only if. Consider the following colloquial English statements:[10]

(a) If it stops raining, I'll go to the store.

(b) If I win the lottery, I'll buy a new car.

Unfortunately, the "if" in statement (a) might be intended to mean "if and only if." Thus, by stating (a) the speaker leads the listener to believe the speaker will definitely go to the store if it stops raining, but also that the speaker will go to the store *only if* it stops raining (and thus *will not* go if it does not stop raining). To the strict logician (a) is not violated in the case it does not stop raining, but the speaker still goes to the store. Recall that in such a case (a) is vacuously true.

On the other hand, it seems somewhat more likely (b) is understood the same by the logician and the casual user of English; though we are tempted to understand the speaker to mean if and only if, on reflection we would not consider him or her a liar for buying the car without having first won the lottery.[11]

---

[10]This is related by Steven Zucker, from Johns Hopkins University, writing in the appendix of Steven Krantz's *How to Teach Mathematics*, second edition, 1999 (Providence: American Mathematical Society).

[11]In everyday English, context may be important to our interpretations. For instance, if the first person was asked, "Will you go to the store," we might interpret (a) as an if and only if. For the second person, if asked, "What will you do if you win the lottery," then (b) might be interpreted as an "if," while if they were instead asked, "Will you buy a new car," this answer might be interpreted as an "only if."

In both (a) and (b) the personalities and shared experiences of the speaker and listener will likely play roles in what was meant by the speaker and what was understood by the listener. In mathematics we cannot have this kind of subjectivity.

1.1.3 Constructing Further Truth Tables

Here we look at truth tables of more complicated compound statements. To do so, we first list the underlying component statements P, Q, and so on in lexicographical order. We then proceed—working "inside-out" and step by step—to construct the resulting truth values of the desired compound statement for each possible truth value combination of the component statements.

It will be necessary to recall the actions of each of the operations introduced earlier. These are completely summarized by their truth tables in the previous subsection, but we can summarize the actions in words:

$\sim$ changes T to F, and F to T.

$\wedge$ returns F unless both statements it connects are true, in which case it returns T.

$\vee$ returns T if either statement it connects is true, and F exactly when both statements are false.

$\longrightarrow$ returns T except when the first statement is true and the second false. In particular, if the first statement is false, then this returns T (vacuously).

$\longleftrightarrow$ returns T if truth values of both statements match, and F if they differ.

Example 1.1.5

Construct a truth table for $\sim (P \longrightarrow Q)$.

Solution: The underlying component statements are P and Q, so we first list these, and then their possible truth value combinations in lexicographical order. In order to construct the resulting truth table values for $\sim (P \longrightarrow Q)$, we build this statement one step at a time with the operations, in an "inside-out" fashion. By this we mean that we write the truth table column for $P \longrightarrow Q$, and then apply the negation to get the truth table column for $\sim (P \longrightarrow Q)$:

P	Q	$P \longrightarrow Q$	$\sim (P \longrightarrow Q)$
T	T	T	F
T	F	F	T
F	T	T	F
F	F	T	F

This reflects a fact we had before: that we have $\sim (P \longrightarrow Q)$ true—we have $P \longrightarrow Q$ false—exactly when we have P true but Q t false.

Note that in this example the third column, which represents $P \longrightarrow Q$, essentially connects the statements represented by the first and second columns with the connective $\longrightarrow$, while the last column applied the operation $\sim$ to the statement represented by that third column. Thus the example reads easily from left to right without interruption. It is not always

1.1. LOGIC SYMBOLS AND TRUTH TABLES

possible (or easiest) to do so; often we will add a column connecting statements from previous columns, which are some distance from where we want to place our new column, though our style here will always have our final column representing the desired compound statement.

Example 1.1.6

Compute the truth table for $(P \vee Q) \longrightarrow (P \wedge Q)$.

<u>Solution</u>: Our "inside-out" strategy is still the same. Here we list P and Q, construct $P \vee Q$ and $P \wedge Q$ respectively, and then connect these with $\longrightarrow$:

P	Q	$P \vee Q$	$P \wedge Q$	$(P \vee Q) \longrightarrow (P \wedge Q)$
T	T	T	T	T
T	F	T	F	F
F	T	T	F	F
F	F	F	F	T

Some texts refer to the $\longrightarrow$ above as the "major connective," since ultimately the statement $(P \vee Q) \longrightarrow (P \wedge Q)$ is an implication, albeit connecting two already-compound statements $(P \vee Q)$ and $(P \wedge Q)$. Thus "major connective" can be seen as referring to the last operator whose action was computed in making the truth table for the statement as a whole. (In the previous example, $\sim$ would be the major connective, though we do not refer to unary operators as "connectives.")

After constructing such a truth table step by step, it is also instructive to step back and examine the result. In particular, it is always useful to see which circumstances render the whole statement false, which here are the second and third combinations. In those, we have (the antecedent) $P \vee Q$ true since one of the P, Q is true, but (the consequent) $P \wedge Q$ is not true, since P and Q are not both true.

Example 1.1.7

Construct the truth table for $P \longrightarrow [\sim (Q \vee R)]$.

<u>Solution</u> Here we need $2^3 = 8$ different combinations of truth values for the underlying component statements P, Q and R. Once we list these combinations in lexicographical order, we then compute $Q \vee R$, $\sim (Q \vee R)$, and then compute $P \longrightarrow [\sim (Q \vee R)]$, in essence computing the major connective $\longrightarrow$.

P	Q	R	$Q \vee R$	$\sim (Q \vee R)$	$P \longrightarrow [\sim (Q \vee R)]$
T	T	T	T	F	F
T	T	F	T	F	F
T	F	T	T	F	F
T	F	F	F	T	T
F	T	T	T	F	T
F	T	F	T	F	T
F	F	T	T	F	T
F	F	F	F	T	T

Note that the last four cases are true vacuously. The cases where this is false are when we have P true, but $\sim (Q \vee R)$ is false, i.e., when we have both P and $Q \vee R$ being true, i.e., when P is true and either Q or R is true.

Example 1.1.8

Construct the truth table for $P \wedge (Q \vee R)$.

Solution: Again we need $2^3 = 8$ rows. The column for P is repeated, though this is not necessary or always desirable (see the previous example), but here was done so that the last column represents the major connective operating on the two columns immediately preceding it.

P	Q	R	P	$Q \vee R$	$P \wedge (Q \vee R)$
T	T	T	T	T	T
T	T	F	T	T	T
T	F	T	T	T	T
T	F	F	T	F	F
F	T	T	F	T	F
F	T	F	F	T	F
F	F	T	F	T	F
F	F	F	F	F	F

In fact it is not difficult to spot which entries in the final column should have value T, since what was required was that P be true, and at least one of the Q or R also be true. Later we will see that this has exactly the same truth value, in all circumstances, as $(P \wedge Q) \vee (P \wedge R)$, which is true if we have both P and Q true, or we have both P and R true. That these two compound statements, $P \wedge (Q \vee R)$ and $(P \wedge Q) \vee (P \wedge R)$ basically state the same thing will be explored in Section 1.2, as will other "logical equivalences."

1.1.4 Tautologies and Contradictions, A First Look

Two very important classes of compound statements are those that form tautologies, and those that form contradictions. As we will see throughout the text, the tautologies loom especially large in our study and use of logic. We will study both tautologies and contradictions further in the next section. Here we introduce the concepts and begin to develop an intuition for these types of statements. We begin with the definitions and most obvious examples.

Definition 1.1.1 *A compound statement formed by the component statements $P_1, P_2, \ldots, P_n$ is called a* **tautology** *iff its truth table column consists entirely of entries with truth value T for each of the 2^n possible truth value combinations (T and F) of the component statements.*

Definition 1.1.2 *A compound statement formed by the component statements $P_1, P_2, \ldots, P_n$ is called a* **contradiction** *iff its truth table column consists entirely of entries with truth value F for each of the 2^n possible truth value combinations (T and F) of the component statements.*

Example 1.1.9

Consider the statement $P \vee (\sim P)$, which is a tautology:

P	$\sim P$	$P \vee (\sim P)$
T	F	T
F	T	T

1.1. LOGIC SYMBOLS AND TRUTH TABLES

Example 1.1.10

Next consider the statement $P \wedge (\sim P)$, which is a contradiction:

P	$\sim P$	$P \wedge (\sim P)$
T	F	F
F	T	F

We see that the statement $P \vee (\sim P)$ is always true, whereas $P \wedge (\sim P)$ is always false. There are other interesting tautologies, as well as other interesting contradictions. For the moment let us concentrate on the tautologies.[12]

That the statement $P \vee (\sim P)$ is a tautology—especially when a particular example is examined—should be obvious when we consider what the statement says: P is true or $(\sim P)$ is true. If P is the statement that I will eat pizza, then we get the always true statement

$$P \vee (\sim P): \text{I will eat pizza or I will not eat pizza.}$$

In some contexts, tautologies seem to provide no useful information. Indeed, there are times in formal speech that declaring a statement to be a tautology is meant to be demeaning. However, we will see that there are many nontrivial tautologies, and it can be quite useful to recognize complex statements that are always true. For the moment we will look at the most basic of tautologies. For instance, the next tautology is obvious if we can read and understand its symbolic representation.

Example 1.1.11

$(P \wedge Q) \longrightarrow P$ is a tautology:

P	Q	$P \wedge Q$	P	$(P \wedge Q) \longrightarrow P$
T	T	T	T	T
T	F	F	T	T
F	T	F	F	T
F	F	F	F	T

Note that the last three cases are true vacuously.

A simple English example shows how this is clearly a tautology. If we take P as before, and Q as the statement, "I will drink soda," then $(P \wedge Q) \longrightarrow P$ becomes, "If I will eat pizza and drink soda, then I will eat pizza." Looking at it abstractly, if we have both P and Q true, then we have P true. Note that we cannot replace the implication $\longrightarrow$ with a bi-implication $\longleftrightarrow$ and still have a tautology.

In the next section we will be very much interested in tautologies in which the major connective is the bi-implication $\longleftrightarrow$. In fact we will develop a variation of the notation for just those cases. The following tautology is one such example.

---

[12]We will essentially devote the next two sections to classes of tautologies, though we use different terms there.

Example 1.1.12

Show that $[\sim(P \vee Q)] \longleftrightarrow [(\sim P) \wedge (\sim Q)]$ is a tautology.[13]

Solution:

P	Q	$P \vee Q$	$\sim(P \vee Q)$	$\sim P$	$\sim Q$	$(\sim P) \wedge (\sim Q)$	$[\sim(P \vee Q)] \longleftrightarrow [(\sim P) \wedge (\sim Q)]$
T	T	T	F	F	F	F	T
T	F	T	F	F	T	F	T
F	T	T	F	T	F	F	T
F	F	F	T	T	T	T	T

Later we will get into the habit of simply pointing out how the two columns representing, say, $\sim(P \vee Q)$ and $(\sim P) \wedge (\sim Q)$ have the same truth values, so when connected by $\longleftrightarrow$ we get a tautology. That will be more convenient, as the entire statement does not fit easily into a relatively narrow truth table column heading.

With reflection, the various tautologies and contradictions become intuitive and easy to identify. (For the previous example, consider the discussion of when $P \vee Q$ is false, as in Example 1.1.2, page 6.) However, not all things that may at first appear to be contradictions are in fact contradictions. At the heart of the problem in such examples is usually the nature of the implication operation $\longrightarrow$. Consider the following:

Example 1.1.13

Write a truth table for $P \longrightarrow (\sim P)$ to demonstrate that it is **not** a contradiction.

Solution: There is only one independent statement P, so we need only $2^1 = 2$ rows.

P	$\sim P$	$P \longrightarrow (\sim P)$
T	F	F
F	T	T

Note that the second case is true vacuously. It is also interesting to note that the statement $P \longrightarrow (\sim P)$ has the same truth values as $\sim P$, which can be interpreted as saying $\sim P$ is the same as $P \longrightarrow (\sim P)$. That is worth pondering, but for the moment we will not elaborate.

It is perhaps easier to spot contradictions that do not involve the implication. For instance, $P \longleftrightarrow (\sim P)$ is a contradiction, but the demonstration of that (by truth tables) is left to the reader. (Simply replace $\longrightarrow$ with $\longleftrightarrow$ in the truth table above.)

---

[13]Some texts employ a strict hierarchy of "precedence" or "order of operations" on logic operations. It is akin to arithmetic, where $4 \cdot 5^2 + 3/5$ has us computing 5^2, multiplying that by 4, separately computing $3/5$, and then adding our two results. With grouping symbols one might write $[4(5^2)] + [3/5]$, but through conventions designed for convenience the grouping is understood in the original expression. For our example above, some texts will simply write
$$\sim(P \vee Q) \longleftrightarrow \sim P \wedge \sim Q,$$
understanding the precedence to be $\sim$, then $\wedge$ and $\vee$, and then $\longrightarrow$ and $\longleftrightarrow$ in the order of appearance. Thus we need the first parentheses to override the precedence of the $\sim$, else we would interpret $\sim P \vee Q$ to mean $(\sim P) \vee Q$. As this text is not for a course in logic *per se*, we will continue to use grouping symbols rather than spend the effort to develop and practice a procedure for precedence. (Besides, the authors find the clearly grouped statements easier and more pleasing to read and write.)

1.1. LOGIC SYMBOLS AND TRUTH TABLES

Exercises

1. A very useful way to learn the nuances of the logic operations is to consider when their compound statements are false. For each of the following compound statements, discuss all possible circumstances in which the given statement is false.

    For example, $P \longleftrightarrow Q$ is false exactly when P is true and Q false, or Q true and P false.

    (a) $\sim P$
    (b) $P \wedge Q$
    (c) $P \vee Q$
    (d) $P \longrightarrow Q$
    (e) $P \longleftrightarrow Q$
    (f) $P \longrightarrow (\sim Q)$.

2. Repeat 1(a)–(f), except use truth tables for each to answer the question of when each statement is false. Compare and reconcile your answers to Exercise 1 above. (You can use one truth table that includes each of these.)

3. Consider the statement
    $$(\sim Q) \longrightarrow (\sim P).$$
    (a) When is it false?
    (b) Now consider $P \longrightarrow Q$. When is it false?
    (c) Do you believe these two compound statements mean the same thing?
    (d) Construct the truth table for the statement $(\sim Q) \longrightarrow (\sim P)$. Then revisit your answer to (c).

4. Construct the truth table for $P \textbf{ XOR } Q$. (See Footnote 8, page 6.)

5. Construct the truth table for the statement $\sim (P \longleftrightarrow Q)$. Compare your answer to the previous exercise.

6. Find six other English statements that are equivalent to the statement,

    "You can go out with your friends only if your homework is finished."

    (See page 7. Some of your answers may seem very formal.)

Construct truth tables for Exercises 7–14.

7. $(\sim P) \longleftrightarrow (\sim Q)$ (Compare to $P \longleftrightarrow Q$.)
8. $[P \vee (\sim Q)] \longrightarrow P$
9. $\sim [P \wedge (Q \vee R)]$
10. $\sim (P \wedge Q)$
11. $P \vee (Q \wedge R)$
12. $P \vee (Q \vee R)$
13. $(P \vee Q) \vee R$ (Compare to the previous statement.)
14. $(P \longrightarrow Q) \wedge (Q \longrightarrow P)$

For Exercises 15–26, decide which are tautologies, which are contradictions, and which are neither. Try to decide using intuition, and then check with truth tables.

15. $P \longrightarrow P$
16. $P \longleftrightarrow P$
17. $P \vee (\sim P)$
18. $P \wedge (\sim P)$
19. $P \longleftrightarrow (\sim P)$
20. $P \longrightarrow (\sim P)$
21. $((P) \wedge (\sim P)) \longrightarrow Q$
22. $(P \longrightarrow (\sim P)) \longrightarrow (\sim P)$
23. $(P \wedge Q) \longrightarrow P$
24. $(P \vee Q) \longrightarrow P$
25. $P \longrightarrow (P \wedge Q)$

26. $P \longrightarrow (P \vee Q)$

27. Some confuse implication $\longrightarrow$ with causation, interpreting $P \longrightarrow Q$ as "P causes Q." However, the implication is in fact weaker than the layman's concept of causation. Answer the following:

    (a) Show that $(P \longrightarrow Q) \vee (Q \longrightarrow P)$ is a tautology.

    (b) Explain why, replacing $\longrightarrow$ with the phrase "causes" clearly does not give us a tautology.

    (c) On the other hand, if P being true **causes** Q to be true, can we say $P \longrightarrow Q$ is true?

28. Write the lexicographical ordering of the possible truth value combinations for four statements P, Q, R, S.

1.2 Valid Logical Equivalences as Tautologies

1.2.1 The Idea and Definition of Logical Equivalence

In lay terms, two statements are logically equivalent when they say the same thing, albeit perhaps in different ways. To a mathematician, two statements are called logically equivalent when they will always be simultaneously true or simultaneously false. To see that these notions are compatible, consider an example of a man named John N. Smith who lives alone at 12345 North Fictional Avenue in Miami, Florida, and has a United States Social Security number 987-65-4325.[14] Of course there should be exactly one person with a given Social Security number. Hence, when we ask any person the questions, "Are you John N. Smith of 12345 North Fictional Avenue in Miami, Florida?" and "Is your US Social Security number 987-65-4325?" in essence we would be asking the same question in both cases. Indeed, the answers to these two questions would always be both yes, or both no, so the statements "You are John N. Smith of 12345 North Fictional Avenue in Miami, Florida," and "Your US Social Security number is 987-65-4325," are logically equivalent. The notation we would use is the following:

$$\text{You are John N. Smith of 12345 North Fictional Avenue in Miami, Florida}$$
$$\Longleftrightarrow \text{Your US Social Security number is 987-65-4325.}$$

The motivation for the notation "$\Longleftrightarrow$" will be explained shortly.

On a more abstract note, consider the statements $\sim(P \vee Q)$ and $(\sim P) \wedge (\sim Q)$. We compute both of these compound statements' truth values in one table:

P	Q	$P \vee Q$	$\sim(P \vee Q)$	$\sim P$	$\sim Q$	$(\sim P) \wedge (\sim Q)$
T	T	T	F	F	F	F
T	F	T	F	F	T	F
F	T	T	F	T	F	F
F	F	F	T	T	T	T

the same

We see that these two statements are both true or both false, under any of the $2^2 = 4$ possible circumstances, those being the possible truth value combinations of the underlying, independent component statements P and Q. Thus the statements $\sim(P \vee Q)$ and $(\sim P) \wedge (\sim Q)$ are indeed logically equivalent in the sense of always having the same truth value. Having established this, we would write

$$\sim(P \vee Q) \Longleftrightarrow (\sim P) \wedge (\sim Q).$$

Note that in logic, this symbol "$\Longleftrightarrow$" is similar to the symbol "=" in algebra and elsewhere.[15] There are a couple of ways it is read out loud, which we will consider momentarily. For now we take the occasion to give the formal definition of logical equivalence:

---

[14] This is all, of course, fictitious.

[15] Some texts use the symbol "≡" for logical equivalence. However, there is another standard use for this symbol, used for "is identical to" in function notation, as in $f(x) \equiv 1$ meaning $f(x) = 1$ for all x in the domain of the function. For instance in trigonometry one could write $\cos^2 x + \sin^2 x \equiv 1$.

Definition 1.2.1 *Given n independent statements $P_1,\ldots,P_n$, and two statements R,S which are compound statements of the $P_1,\ldots,P_n$, we say that R and S are **logically equivalent**, which we then denote $R \iff S$, if and only if their truth table columns have the same entries for each of the 2^n distinct combinations of truth values for the $P_1,\ldots,P_n$. When R and S are logically equivalent, we will also call $R \iff S$ a **valid equivalence**.*

Again, this is consistent with the idea that to say statements R and S are logically equivalent is to say that, under any circumstances, they are both true or both false, so that asking if R is true is—functionally—exactly the same as asking if S is true. (Recall our example of John N. Smith's Social Security number.)

Note that if two statements' truth values always match, then connecting them with $\longleftrightarrow$ yields a tautology. Indeed, the bi-implication yields T if the connected statements have the same truth value, and F otherwise. Since two logically equivalent statements will have matching truth values in all cases, connecting with $\longleftrightarrow$ will always yield T, and we will have a tautology. On the other hand, if connecting two statements with $\longleftrightarrow$ forms a tautology, then the connected statements must have always-matching truth values, and thus be equivalent. This argument yields our first theorem:[16]

Theorem 1.2.1 *Suppose R and S are compound statements of $P_1\ldots,P_n$. Then R and S are logically equivalent if and only if $R \longleftrightarrow S$ is a tautology.*

This theorem gives us the motivation behind the notation $\iff$. Assuming R and S are compound statements built upon component statements $P_1\ldots,P_n$, then

$$R \iff S \quad \text{means that} \quad R \longleftrightarrow S \text{ is a tautology.} \tag{1.1}$$

To be clear, when we write $R \longleftrightarrow S$ we understand that this might have truth value T or F (i.e., it might be true or false). However, when we write $R \iff S$, we mean that $R \longleftrightarrow S$ is always true (i.e., a tautology), which partially explains why we call $R \iff S$ a *valid equivalence*.[17]

To prove $R \iff S$, we could (but usually will not) construct $R \longleftrightarrow S$ and show that it is a tautology. We do so next to prove

$$\underbrace{\sim(P \vee Q)}_{\text{``}R\text{''}} \iff \underbrace{(\sim P) \wedge (\sim Q)}_{\text{``}S\text{''}}.$$

---

[16] A **theorem** is a fact that has been proven to be true, particularly dealing with mathematics. We will state numerous theorems in this text. Most we will prove, though occasionally we will include a theorem that is too relevant to omit, but whose proof is too technical to include in an undergraduate calculus book. Such proofs are left to courses with titles such as mathematical (or real) analysis, topology, or advanced calculus.

Some theorems are also called lemmas (or, more archaically, lemmata) when they are mostly useful as steps in larger proofs of the more interesting results. Still others are called corollaries if they are themselves interesting, but follow with very few extra steps after the underlying theorem is proved.

[17] Depending upon the author, both $R \longleftrightarrow S$ and $R \iff S$ are sometimes verbalized "R is equivalent to S," or "R if and only if S." We distinguish the cases by using the term "equivalent" for the double-lined arrow, and "if and only if" for the single-lined arrow. To help avoid confusion, we emphasize this more restrictive use of "equivalences" (denoted with $\iff$) by calling them "valid equivalences."

1.2. VALID LOGICAL EQUIVALENCES AS TAUTOLOGIES

P	Q	$P \vee Q$	$\overbrace{\sim(P \vee Q)}^{R}$	$\sim P$	$\sim Q$	$\overbrace{(\sim P) \wedge (\sim Q)}^{S}$	$\overbrace{[\sim(P \vee Q)] \longleftrightarrow [(\sim P) \wedge (\sim Q)]}^{R \longleftrightarrow S}$
T	T	T	F	F	F	F	T
T	F	T	F	F	T	F	T
F	T	T	F	T	F	F	T
F	F	F	T	T	T	T	T

However, our preferred method will be as in the previous truth table, where we simply show that the truth table columns for R and S have the same entries at each horizontal level (i.e., for each truth value combination of the component statements). That approach saves space and reinforces our original notion of equivalence (matching truth values). However, it is still important to understand the connection between $\longleftrightarrow$ and $\Longleftrightarrow$, as given in (1.1).

1.2.2 Equivalences for Negations

Much of the intuition achieved from studying symbolic logic comes from examining various logical equivalences. Indeed, we will make much use of these, for the theorems we use throughout the text are often stated in one form, and then used in a different, but logically equivalent form. When we prove a theorem, we may prove even a third, logically equivalent form.

The first logical equivalences we will look at here are the negations of the basic operations. We already looked at the negations of $\sim P$ and $P \vee Q$. Next we also look at negations of $P \wedge Q$, $P \longrightarrow Q$, and $P \longleftrightarrow Q$. Historically, (1.3) and (1.4) below are called *De Morgan's Laws*, but each basic negation is important. We now list these negations.

$$\sim (\sim P) \Longleftrightarrow P \tag{1.2}$$

$$\sim (P \vee Q) \Longleftrightarrow (\sim P) \wedge (\sim Q) \tag{1.3}$$

$$\sim (P \wedge Q) \Longleftrightarrow (\sim P) \vee (\sim Q) \tag{1.4}$$

$$\sim (P \longrightarrow Q) \Longleftrightarrow P \wedge (\sim Q) \tag{1.5}$$

$$\sim (P \longleftrightarrow Q) \Longleftrightarrow [P \wedge (\sim Q)] \vee [Q \wedge (\sim P)]. \tag{1.6}$$

Fortunately, with a well-chosen perspective these are intuitive. Recall that any statement R can also be read "R is true," while the negation asserts the original statement is false. For example $\sim R$ can be read as the statement "R is false," or a similar wording (such as "it is not the case that R"). Similarly the statement $\sim (P \vee Q)$ is the same as "'P or Q' is false." With that it is not difficult to see that for $\sim (P \vee Q)$ to be true requires both that P be false *and* Q be false. For a specific example, consider our earlier P and Q:

P : I will eat pizza.
Q : I will drink soda.
$P \vee Q$: I will eat pizza or I will drink soda.
$\sim (P \vee Q)$: It is not the case that (either) I will eat pizza or I will drink soda.
$(\sim P) \wedge (\sim Q)$: It is not the case that I will eat pizza, and it is not the case that I will drink soda.

That these last two statements essentially have the same content, as stated in (1.3), should be intuitive. An actual *proof* of (1.3) is best given by truth tables, and can be found on page 17.

Next we consider (1.5). This states that $\sim(P \longrightarrow Q) \iff P \wedge (\sim Q)$. Now we can read $\sim(P \longrightarrow Q)$ as "it is not the case that $P \longrightarrow Q$," or "$P \longrightarrow Q$ is false." Recall that there was only one case for which we considered $P \longrightarrow Q$ to be false, which was the case that P was true but Q was false, which itself can be translated to $P \wedge (\sim Q)$. For our earlier example, the negation of the statement "If I eat pizza, then I will drink soda" is the statement "I will eat pizza but (and) I will not drink soda." While this discussion is correct and may be intuitive, the actual proof of (1.5) is by truth table:

P	Q	$P \to Q$	$\sim(P \to Q)$	P	$\sim Q$	$P \wedge (\sim Q)$
T	T	T	F	T	F	F
T	F	F	T	T	T	T
F	T	T	F	T	F	F
F	F	T	F	F	T	F

the same

We leave the proof of (1.6) by truth tables to the exercises. Recall that $P \longleftrightarrow Q$ states that we have P true if and only if we also have Q true, which we further translated as the idea that we cannot have P true without Q true, and cannot have Q true without P true. Now $\sim(P \longleftrightarrow Q)$ is the statement that $P \longleftrightarrow Q$ is false, which means that P is true and Q false, or Q is true and P false, which taken together form the statement $[P \wedge (\sim Q)] \vee [Q \wedge (\sim P)]$, as reflected in (1.6) above. For our example P and Q from before, $P \longleftrightarrow Q$ is the statement "I will eat pizza if and only if I will drink soda," the negation of which is "I will eat pizza and not drink soda, or I will drink soda and not eat pizza."

Another intuitive way to look at these negations is to consider the question of exactly when someone uttering the original statement is lying (assuming they know it is false). For instance, if someone states $P \wedge Q$ (or some English equivalent), when is this a lie? Since they stated "P *and* Q," it is not difficult to see they are lying exactly when at least one of the statements P, Q is false (i.e., when P is false or Q is false), i.e., when we can truthfully state $(\sim P) \vee (\sim Q)$. That is the kind of thinking one should employ when examining (1.4), that is $\sim(P \wedge Q) \iff (\sim P) \vee (\sim Q)$, intuitively.[18]

1.2.3 Equivalent Forms of the Implication

In this subsection we examine two statements that are equivalent to $P \longrightarrow Q$. The first is more important conceptually, and the second is more important computationally. We list them both now before contemplating them further:

$$P \longrightarrow Q \iff (\sim Q) \longrightarrow (\sim P) \qquad (1.7)$$

$$P \longrightarrow Q \iff (\sim P) \vee Q. \qquad (1.8)$$

---

[18]Note that here as always we use the inclusive "or," so when we write "P is false or Q is false," we include also the case in which both P and Q are false. (See Footnote 8, page 6 for remarks on the other, exclusive "or.")

1.2. VALID LOGICAL EQUIVALENCES AS TAUTOLOGIES

We will combine the proofs into one truth table, where we compute $P \longrightarrow Q$, followed in turn by $({\sim} Q) \longrightarrow ({\sim} P)$ and $({\sim} P) \vee Q$.

P	Q	$P \to Q$	${\sim} Q$	${\sim} P$	$({\sim} Q) \to ({\sim} P)$	${\sim} P$	Q	$({\sim} P) \vee Q$
T	T	T	F	F	T	F	T	T
T	F	F	T	F	F	F	F	F
F	T	T	F	T	T	T	T	T
F	F	T	T	T	T	T	F	T

the same

The form (1.7) is important enough that it warrants a name:

Definition 1.2.2 *Given any implication $P \longrightarrow Q$, we call the (logically equivalent) statement $({\sim} Q) \longrightarrow ({\sim} P)$ its* **contrapositive** *(and vice-versa, see the following).*

In fact, note that the contrapositive of $({\sim} Q) \longrightarrow ({\sim} P)$ would be $[{\sim} ({\sim} P)] \longrightarrow [{\sim} ({\sim} Q)]$, i.e., $P \longrightarrow Q$, and so $P \longrightarrow Q$ and $({\sim} Q) \longrightarrow ({\sim} P)$ are contrapositives *of each other*.

We have proved that $P \longrightarrow Q$, its contrapositive form $({\sim} Q) \longrightarrow ({\sim} P)$, and the other form $({\sim} P) \vee Q$ are equivalent using a truth table, but developing the intuition that these *should* be equivalent can require some effort. Some examples can help to clarify this.

P : I will eat pizza.

Q : I will drink soda.

$P \longrightarrow Q$: If I eat pizza, then I will drink soda.

$({\sim} Q) \longrightarrow ({\sim} P)$: If I do not drink soda, then I will not eat pizza.

$({\sim} P) \vee Q$: I will not eat pizza, or I will drink soda.

Perhaps more intuition can be found when Q is a more natural consequence of P. Consider the following P, Q combination, which might be used by parents communicating to their children.

P : You leave your room messy.

Q : You get spanked.

$P \longrightarrow Q$: If you leave your room messy, then you get spanked.

$({\sim} Q) \longrightarrow ({\sim} P)$: If you do not get spanked, then you do (did) not leave your room messy.

$({\sim} P) \vee Q$: You do not leave your room messy, or you get spanked.

A mathematical example could look like the following (assuming x is a "real number," as discussed later in this text):

$P : x = 10$.

$Q : x^2 = 100$.

$P \longrightarrow Q$: If $x = 10$, then $x^2 = 100$.

$({\sim} Q) \longrightarrow ({\sim} P)$: If $x^2 \neq 100$, then $x \neq 10$.

$({\sim} P) \vee Q : x \neq 10$ or $x^2 = 100$.

The contrapositive is very important because many theorems are given as implications, but are often used in their logically equivalent, contrapositive forms. However, it is equally important to avoid confusing $P \longrightarrow Q$ with either of the statements $P \longleftrightarrow Q$ or $Q \longrightarrow P$. For instance, in the second example the child may get spanked without leaving the room messy, as there are quite possibly other infractions that would result in a spanking. Thus leaving the room messy does not follow from being spanked, and leaving the room messy is not necessarily connected with the spanking by an "if and only if." In the last algebraic example, all the forms of the statement are true, but $x^2 = 100$ does not imply $x = 10$. Indeed, it is possible that $x = -10$, and so the correct equivalence one usually notes for this case is $x^2 = 100 \Longleftrightarrow [(x = 10) \vee (x = -10)]$.

1.2.4 Other Valid Equivalences

While negations and equivalent alternatives to the implication are arguably the most important of our valid logical equivalences, there are several others. Some are rather trivial, such as

$$P \wedge P \Longleftrightarrow P \Longleftrightarrow P \vee P. \tag{1.9}$$

Also rather easy to see are the "commutativities" of $\wedge$, $\vee$ and $\longleftrightarrow$:

$$P \wedge Q \Longleftrightarrow Q \wedge P, \qquad P \vee Q \Longleftrightarrow Q \vee P, \qquad P \longleftrightarrow Q \Longleftrightarrow Q \longleftrightarrow P. \tag{1.10}$$

There are also associative rules. The latter was in fact a topic in the previous exercises:

$$P \wedge (Q \wedge R) \Longleftrightarrow (P \wedge Q) \wedge R \tag{1.11}$$

$$P \vee (Q \vee R) \Longleftrightarrow (P \vee Q) \vee R. \tag{1.12}$$

However, it is not so clear when we mix together $\vee$ and $\wedge$. In fact, these "distribute over each other" in the following ways:

$$P \wedge (Q \vee R) \Longleftrightarrow (P \wedge Q) \vee (P \wedge R), \tag{1.13}$$

$$P \vee (Q \wedge R) \Longleftrightarrow (P \vee Q) \wedge (P \vee R). \tag{1.14}$$

We prove the first of these distributive rules next, and leave the other for the exercises.

P	Q	R	$Q \vee R$	$P \wedge (Q \vee R)$	$P \wedge Q$	$P \wedge R$	$(P \wedge Q) \vee (P \wedge R)$
T	T	T	T	T	T	T	T
T	T	F	T	T	T	F	T
T	F	T	T	T	F	T	T
T	F	F	F	F	F	F	F
F	T	T	T	F	F	F	F
F	T	F	T	F	F	F	F
F	F	T	T	F	F	F	F
F	F	F	F	F	F	F	F

the same

1.2. VALID LOGICAL EQUIVALENCES AS TAUTOLOGIES

To show that this is reasonable, consider the following:

P : I will eat pizza.

Q : I will drink cola.

R : I will drink lemon-lime soda.

Then our logically equivalent statements become

$P \wedge (Q \vee R)$: I will eat pizza, and drink cola or lemon-lime soda.

$(P \wedge Q) \vee (P \wedge R)$: I will eat pizza and drink cola, or

I will eat pizza and drink lemon-lime soda.

Table 1.3, page 25, gives these and some further valid equivalences. It is important to be able to read these and, through reflection and the exercises, to be able to see their reasonableness. Each can be proved using truth tables.

For instance, we can prove that $P \longleftrightarrow Q \iff (P \longrightarrow Q) \wedge (Q \longrightarrow P)$, justifying the choice of the double-arrow symbol $\longleftrightarrow$:

P	Q	$P \longleftrightarrow Q$	$P \longrightarrow Q$	$Q \longrightarrow P$	$(P \longrightarrow Q) \wedge (Q \longrightarrow P)$
T	T	T	T	T	T
T	F	F	F	T	F
F	T	F	T	F	F
F	F	T	T	T	T

the same

This was discussed in Example 1.1.4 on page 8.

For another example of such a proof, we next demonstrate the following interesting equivalence:

$$P \longrightarrow (Q \wedge R) \iff (P \longrightarrow Q) \wedge (P \longrightarrow R)$$

P	Q	R	$Q \wedge R$	$P \longrightarrow (Q \wedge R)$	$P \longrightarrow Q$	$P \longrightarrow R$	$(P \longrightarrow Q) \wedge (P \longrightarrow R)$
T	T	T	T	T	T	T	T
T	T	F	F	F	T	F	F
T	F	T	F	F	F	T	F
T	F	F	F	F	F	F	F
F	T	T	T	T	T	T	T
F	T	F	F	T	T	T	T
F	F	T	F	T	T	T	T
F	F	F	F	T	T	T	T

the same

This should be somewhat intuitive: if P is to imply $Q \wedge R$, that should be the same as P implying Q and P implying R. This equivalence is later given as (1.33), page 25. There is also (1.34), in which we replace $\wedge$ with $\vee$ and get another valid equivalence.

Still, one must be careful about declaring two statements to be equivalent. These are all ultimately intuitive, but intuition must be informed by experience. The next two are left to the exercises, and left off Table 1.3 of valid equivalences because they are somewhat obscure. We include them here to illustrate that not all equivalences are immediately transparent.[19]

$$(P \vee Q) \longrightarrow R \iff (P \longrightarrow R) \wedge (Q \longrightarrow R), \tag{1.15}$$

$$(P \wedge Q) \longrightarrow R \iff (P \longrightarrow R) \vee (Q \longrightarrow R). \tag{1.16}$$

On reflection one can see how these are reasonable. For instance, we can look more closely at (1.15) with the following P, Q and R:

$$P : \text{I eat pizza.}$$
$$Q : \text{I eat chicken.}$$
$$R : \text{I drink cola.}$$

Then the left and right sides of (1.15) become

$(P \vee Q) \longrightarrow R$: If I eat pizza or chicken, then I drink cola.

$(P \longrightarrow R) \wedge (Q \longrightarrow R)$: If I eat pizza then I drink cola, and if I eat chicken then I drink cola.

In fact (1.16) is perhaps more difficult to see.

There are techniques where one proves more sophisticated equivalences in steps which refer to a small number of simpler, established equivalences, rather than always resorting to truth tables. With those techniques one can quickly prove (1.15) and (1.16), again without truth tables. It is analogous to proving complicated trigonometric identities using simpler ones, or the leap from memorizing single-digit multiplication tables to applying them to several-digit problems. For a glimpse of the process, we can look at such a proof of the equivalence of the contrapositive: $P \longrightarrow Q \iff (\sim Q) \longrightarrow (\sim P)$. To do so, we begin with (1.29), that $P \longrightarrow Q \iff (\sim P) \vee Q$. The proof runs as follows:

$$P \longrightarrow Q \iff (\sim P) \vee Q$$
$$\iff Q \vee (\sim P)$$
$$\iff [\sim (\sim Q)] \vee (\sim P)$$
$$\iff (\sim Q) \longrightarrow (\sim P).$$

The first line used (1.29), the second used commutativity (1.23), the third used (1.18) (that $Q \iff \sim (\sim Q)$), and the fourth used (1.29) again but with the part of "P" played by $(\sim Q)$ and the part of "Q" played by $(\sim P)$. This proof is not much more efficient than a truth table proof, but for (1.15) and (1.16), this technique of proofs without truth tables is much faster. However, that technique assumes that the more primitive equivalences used in the proof are valid, and those are ultimately proved using truth tables.

---

[19] It is a truism in mathematics and other fields that, while one part of learning is discovering what is true, another part is discovering what is not true, especially when the latter seems reasonable at first glance.

1.2. VALID LOGICAL EQUIVALENCES AS TAUTOLOGIES

$$P \wedge P \iff P \iff P \vee P \tag{1.17}$$
$$\sim(\sim P) \iff P \tag{1.18}$$
$$\sim(P \vee Q) \iff (\sim P) \wedge (\sim Q) \tag{1.19}$$
$$\sim(P \wedge Q) \iff (\sim P) \vee (\sim Q) \tag{1.20}$$
$$\sim(P \longrightarrow Q) \iff P \wedge (\sim Q) \tag{1.21}$$
$$\sim(P \longleftrightarrow Q) \iff [P \wedge (\sim Q)] \vee [Q \wedge (\sim P)] \tag{1.22}$$
$$P \vee Q \iff Q \vee P \tag{1.23}$$
$$P \wedge Q \iff Q \wedge P \tag{1.24}$$
$$P \vee (Q \vee R) \iff (P \vee Q) \vee R \tag{1.25}$$
$$P \wedge (Q \wedge R) \iff (P \wedge Q) \wedge R \tag{1.26}$$
$$P \wedge (Q \vee R) \iff (P \wedge Q) \vee (P \wedge R) \tag{1.27}$$
$$P \vee (Q \wedge R) \iff (P \vee Q) \wedge (P \vee R) \tag{1.28}$$
$$P \longrightarrow Q \iff (\sim P) \vee Q \tag{1.29}$$
$$P \longrightarrow Q \iff (\sim Q) \longrightarrow (\sim P) \tag{1.30}$$
$$P \longrightarrow Q \iff \sim[P \wedge (\sim Q)] \tag{1.31}$$
$$P \longleftrightarrow Q \iff (\sim P) \longleftrightarrow (\sim Q) \tag{1.32}$$
$$P \longrightarrow (Q \wedge R) \iff (P \longrightarrow Q) \wedge (P \longrightarrow R) \tag{1.33}$$
$$P \longrightarrow (Q \vee R) \iff (P \longrightarrow Q) \vee (P \longrightarrow R) \tag{1.34}$$
$$(P \longrightarrow Q) \wedge (Q \longrightarrow P) \iff P \longleftrightarrow Q \tag{1.35}$$
$$(P \longrightarrow Q) \wedge (Q \longrightarrow R) \wedge (R \longrightarrow P) \iff (P \longleftrightarrow Q) \wedge (Q \longleftrightarrow R)$$
$$\wedge (P \longleftrightarrow R) \tag{1.36}$$

Table 1.3: Table of common valid logical equivalences.

1.2.5 Circuits and Logic

While we will not develop this next theory deeply, it is worthwhile to consider a short introduction. The idea is that we can model compound logic statements with electrical switching circuits.[20] When current is allowed to flow across a switch, the switch is considered "on" and the statement it represents has truth value T; it is considered "off" when not allowing current to flow through, and the truth value is F. We can decide if the compound circuit is "on" or "off" based upon whether or not current could flow from one end to the other, based on whether the compound statement has truth value T or F. The analysis can be complicated if the switches are not necessarily independent (P is "on" when $\sim P$ is "off," for instance), but this approach is interesting nonetheless.

For example, the statement $P \vee Q$ is represented by a parallel circuit:

---

[20]Technically a *circuit* would allow current to flow from a source, through components, and back to the source. Here we only show part of the possible path. We will encounter some complete circuits later in the text.

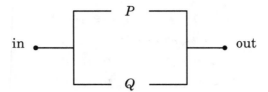

If either P or Q is on (T), then the current can flow from the "in" side to the "out" side of the circuit. On the other hand, we can represent $P \wedge Q$ by a series circuit:

Of course $P \wedge Q$ is only true when both P and Q are true, and the circuit reflects this: current can flow exactly when both "switches" P and Q are "on."

It is interesting to see diagrams of some equivalent compound statements, illustrated as circuits. For instance, (1.27), the distributive-type equivalence

$$P \wedge (Q \vee R) \iff (P \wedge Q) \vee (P \wedge R)$$

can be seen as the equivalence of the two circuits that follow:

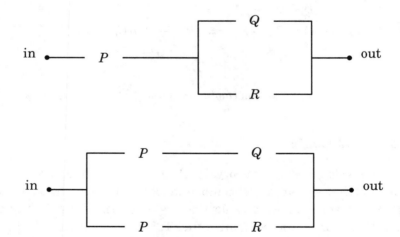

In both circuits, we must have P "on," and also either Q or R for current to flow. Note that in the second circuit, P is represented in two places, so it is either "on" in both places, or "off" in both places. Situations such as these can complicate analyses of switching circuits, but this one is relatively simple.

We can also represent negations of simple statements. To represent $\sim P$ we simply put "$\sim P$" into the circuit, where it is "on" if $\sim P$ is true, i.e., if P is false. This allows us to construct circuits for the implication by using (1.29), which states $P \longrightarrow Q \iff (\sim P) \vee Q$:

1.2. VALID LOGICAL EQUIVALENCES AS TAUTOLOGIES

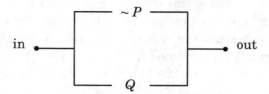

We see that the only time the circuit does *not* flow is when P is true ($\sim P$ is false) and Q is false, so this matches what we know of when $P \longrightarrow Q$ is false. From another perspective, if P is true, then the top part of the circuit won't flow, so Q must be true to allow flow for the circuit as a whole to be "on," or "true."

When negating a whole circuit it gets even more complicated. In fact, it is arguably easier to look at the original circuit and simply note when current will *not* flow. For instance, we know $\sim(P \wedge Q) \iff (\sim P) \vee (\sim Q)$, so we can construct $P \wedge Q$:

in •————— P ————— Q ————— • out

and note that it is *off* exactly when either P is off or Q is off. We then note that that is exactly when the circuit for $(\sim P) \vee (\sim Q)$ is *on*.

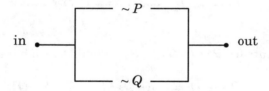

There are, in fact, electrical/mechanical means by which one can take a circuit and "negate" its truth value, for instance with relays or reverse-position switch levers, but that subject is more complicated than we wish to pursue here.

It is interesting to consider $P \longleftrightarrow Q$ as a circuit. It will be "on" if P and Q are both "on" or both "off," and the circuit will be "off" if P and Q do not match. Such a circuit is actually used commonly, such as for a room with two light switches for the same light. To construct such a circuit we note that

$$P \longleftrightarrow Q \iff (P \longrightarrow Q) \wedge (Q \longrightarrow P)$$
$$\iff [(\sim P) \vee Q] \wedge [(\sim Q) \vee P]$$

We will use the last form to draw our diagram:

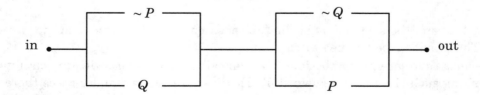

The reader is invited to study this diagram (also reproduced below) to be convinced it represents $P \longleftrightarrow Q$, perhaps most easily in the sense that "you cannot have one (P or Q) being true without the other, but you can have neither." While the diagram does represent $P \longleftrightarrow Q$ by the more easily diagrammed $[(\sim P) \vee Q] \wedge [(\sim Q) \vee P]$, it also suggests another equivalence, since the next two circuits below seem to be functionally equivalent. In the first, we can add two more wires to replace the "center" wire, and also switch the $\sim Q$ and P, since $(\sim Q) \vee P$ is the same as $P \vee (\sim Q)$:

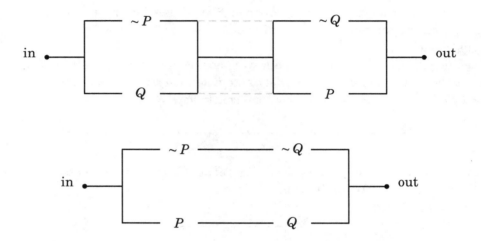

This circuit represents $[(\sim P) \wedge (\sim Q)] \vee [P \wedge Q]$, and so we have (as the reader can check)

$$P \longleftrightarrow Q \iff [(\sim P) \wedge (\sim Q)] \vee [P \wedge Q], \qquad (1.37)$$

which could be added to our previous Table 1.3, page 25, of valid equivalences. It is also consistent with a more colloquial way of expressing $P \longleftrightarrow Q$, such as "neither or both."

Incidentally, this circuit is used in applications where we wish to have two switches within a room that can both change a light (or other device) from on to off or vice versa. When switch P is "on," switch Q can turn the circuit on or off by matching P or being its negation. It is also the case when P is "off." Mechanically this is accomplished with two "single pole, double throw (SPDT)" switches.

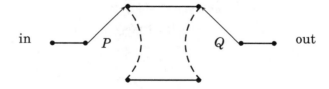

In this diagram, the switch P is in the "up" position when P is true, and "down" when P is false. The same is true for the switch Q.

Because there are many possible "mechanical" diagrams for switching circuits, reading and writing such circuits is its own skill. However, for many simpler cases there is a relatively easy connection to our symbolic logic.

1.2.6 The Statements $\mathscr{T}$ and $\mathscr{F}$

Just as there is a need for zero in addition, we have use for a symbol representing a statement that is always true, and for another symbol representing a statement that is always false. For convenience, we will make the following definitions:

Definition 1.2.3 *Let $\mathscr{T}$ represent any compound statement that is a **tautology**, i.e., whose truth value is always* T. *Similarly, let $\mathscr{F}$ represent any compound statement that is a **contradiction**, i.e., whose truth value is always* F.

We will assume there is a universal $\mathscr{T}$ and a universal $\mathscr{F}$, i.e., statements that are respectively true regardless of any other statements' truth values, and false regardless of any other statements' truth values. In doing so, we consider any tautology to be logically equivalent to $\mathscr{T}$, and any contradiction similarly equivalent to $\mathscr{F}$.[21]

So, for any given $P_1\ldots,P_n$, we have that $\mathscr{T}$ is exactly that statement whose column in the truth table consists entirely of T's, and $\mathscr{F}$ is exactly that statement whose column in the truth table consists entirely of F's. For example, we can write

$$P \vee (\sim P) \iff \mathscr{T}; \tag{1.38}$$
$$P \wedge (\sim P) \iff \mathscr{F}. \tag{1.39}$$

These are easily seen by observing the truth tables.

P	$\sim P$	$P \vee (\sim P)$	$P \wedge (\sim P)$
T	F	T	F
F	T	T	F

We see that $P \vee (\sim P)$ is always true, and $P \wedge (\sim P)$ is always false. Anything that is always true we can dub $\mathscr{T}$, and anything that is always false we can dub $\mathscr{F}$. In this truth table for verifying (1.38) and (1.39), the third column can represent $\mathscr{T}$, and the last column can represent $\mathscr{F}$.

---

[21]In fact it is not difficult to see that all tautologies are logically equivalent. Consider the tautologies $P \vee (\sim P)$, $(P \longrightarrow Q) \longleftrightarrow [(\sim Q) \longrightarrow (\sim P)]$, and $R \longrightarrow R$. A truth table for all three must contain independent component statements P,Q,R, and the abridged version of the table would look like the following:

P	Q	R	$P \vee (\sim P)$	$(P \longrightarrow Q) \longleftrightarrow [(\sim Q) \longrightarrow (\sim P)]$	$R \longrightarrow R$
T	T	T	T	T	T
T	T	F	T	T	T
T	F	T	T	T	T
T	F	F	T	T	T
F	T	T	T	T	T
F	T	F	T	T	T
F	F	T	T	T	T
F	F	F	T	T	T

So here, when all possible underlying independent component statements are included, we see the truth table columns of these tautologies are indeed the same (all T's!). Similarly, all contradictions are mutually equivalent.

From the definitions we can also eventually get the following:

$$P \vee \mathcal{T} \iff \mathcal{T} \tag{1.40}$$
$$P \wedge \mathcal{T} \iff P \tag{1.41}$$
$$P \vee \mathcal{F} \iff P \tag{1.42}$$
$$P \wedge \mathcal{F} \iff \mathcal{F}. \tag{1.43}$$

To demonstrate how one would prove these, we prove here the first two, (1.40) and (1.41), using a truth table. Notice that all entries for $\mathcal{T}$ are simply T, which means that there are only $2 \cdot 1 = 2$ possible truth value combinations for P and $\mathcal{T}$ (two possible values for P, and one possible value for $\mathcal{T}$):

P	$\mathcal{T}$	$P \vee \mathcal{T}$	$P \wedge \mathcal{T}$
T	T	T	T
F	T	T	F

Equivalence (1.40) is demonstrated by the equivalence of the second and third columns, while (1.41) is shown by the equivalence of the first and fourth columns. The others are left as exercises.

These are also worth some reflection. Consider the equivalence $P \wedge \mathcal{T} \iff P$. When we use $\wedge$ to connect P to a statement that is always true, then the truth of the compound statement only depends upon the truth of P. There are similar explanations for the rest of (1.40)–(1.43).

Some other interesting equivalences involving these are the following:

$$\mathcal{T} \longrightarrow P \iff P \tag{1.44}$$
$$P \longrightarrow \mathcal{F} \iff \sim P. \tag{1.45}$$

We leave the proofs of these for the exercises. These are, in fact, interesting to interpret. The first says that having a true statement imply P is true is the same as having P be true. The second says that having P imply a false statement is the same as having $\sim P$ true, i.e., as having P false. Both types of reasoning are useful in mathematics and other disciplines.

If a statement contains *only* $\mathcal{T}$ or $\mathcal{F}$, then that statement itself must be a tautology ($\mathcal{T}$) or a contradiction ($\mathcal{F}$). This is because there is only one possible combination of truth values. For instance, consider the statement $\mathcal{T} \longrightarrow \mathcal{F}$, which is a contradiction. One proof is in the table:

$\mathcal{T}$	$\mathcal{F}$	$\mathcal{T} \longrightarrow \mathcal{F}$
T	F	F

Since the component statement $\mathcal{T} \longrightarrow \mathcal{F}$ always has truth value F, it is a contradiction. Thus $\mathcal{T} \longrightarrow \mathcal{F} \iff \mathcal{F}$.

1.2. VALID LOGICAL EQUIVALENCES AS TAUTOLOGIES

Exercises

Some of these were solved within the section. It is useful to attempt them here again, in the context of the other problems. Unless otherwise specified, all proofs should be via truth tables.

1. Prove (1.18): $\sim(\sim P) \iff P$.

2. Prove (1.32):
   $P \longleftrightarrow Q \iff (\sim P) \longleftrightarrow (\sim Q)$.

3. Prove the logical equivalence of the contrapositive (1.30):
   $P \longrightarrow Q \iff (\sim Q) \longrightarrow (\sim P)$.

4. Prove (1.29): $P \longrightarrow Q \iff (\sim P) \vee Q$.

5. Show that $P \longrightarrow Q$ and $Q \longrightarrow P$ are not equivalent.

6. Prove De Morgan's Laws (1.19) and (1.20), which are listed again:
   (a) $\sim(P \vee Q) \iff (\sim P) \wedge (\sim Q)$.
   (b) $\sim(P \wedge Q) \iff (\sim P) \vee (\sim Q)$.

7. Use truth tables to prove the *distributive-type* laws (1.27) and (1.28):
   (a) $P \wedge (Q \vee R) \iff (P \wedge Q) \vee (P \wedge R)$.
   (b) $P \vee (Q \wedge R) \iff (P \vee Q) \wedge (P \vee R)$.

8. Repeat the previous problem but using circuit diagrams.

9. Prove (1.33): $P \longrightarrow (Q \wedge R)$
   $\iff (P \longrightarrow Q) \wedge (P \longrightarrow R)$.

10. Prove (1.34): $P \longrightarrow (Q \vee R)$
    $\iff (P \longrightarrow Q) \vee (P \longrightarrow R)$.

11. Prove (1.21):
    $\sim(P \longrightarrow Q) \iff P \wedge (\sim Q)$.

12. Prove (1.35), which we write below as
    $P \longleftrightarrow Q \iff (P \longrightarrow Q) \wedge (Q \longrightarrow P)$.
    Note that this justifies the choice of the double-arrow notation $\longleftrightarrow$.

13. Prove (1.22): $\sim(P \longleftrightarrow Q)$
    $\iff [P \wedge (\sim Q)] \vee [Q \wedge (\sim P)]$.

14. Recall the description of **XOR** in Footnote 8, page 6.
    (a) Construct a truth table for P **XOR** Q.
    (b) Compare to the previous problem. Can you make a conclusion?
    (c) Find an expression for P **XOR** Q using P, Q, $\sim$, $\wedge$, and $\vee$.

15. Prove that $(P \vee Q) \longrightarrow (P \wedge Q)$ is equivalent to $P \longleftrightarrow Q$. How would you explain in words why this is reasonable? (Perhaps you can think of a colloquial way to verbalize the statement so it will sound equivalent to $P \longleftrightarrow Q$.)

16. Prove (1.31). How would you explain in words why this is reasonable?

17. Prove the following:
    (a) (1.40): $P \vee \mathscr{T} \iff \mathscr{T}$.
    (b) (1.41): $P \wedge \mathscr{T} \iff P$.
    (c) (1.42): $P \vee \mathscr{F} \iff P$.
    (d) (1.43): $P \wedge \mathscr{F} \iff \mathscr{F}$.

18. For each of the following, find a simple, equivalent statement, using truth tables if necessary.
    (a) $\mathscr{T} \vee \mathscr{T}$ (f) $\mathscr{T} \wedge \mathscr{F}$
    (b) $\mathscr{F} \vee \mathscr{F}$ (g) $\mathscr{T} \longrightarrow \mathscr{T}$
    (c) $\mathscr{T} \vee \mathscr{F}$ (h) $\mathscr{F} \longrightarrow \mathscr{F}$
    (d) $\mathscr{T} \wedge \mathscr{T}$ (i) $\mathscr{T} \longrightarrow \mathscr{F}$
    (e) $\mathscr{F} \wedge \mathscr{F}$ (j) $\mathscr{F} \longrightarrow \mathscr{T}$

19. Repeat the previous exercise for the following:

    (a) $\mathscr{T} \longrightarrow P$
    (b) $\mathscr{F} \longrightarrow P$
    (c) $P \longrightarrow \mathscr{T}$
    (d) $P \longrightarrow \mathscr{F}$
    (e) $P \longleftrightarrow \mathscr{T}$
    (f) $P \longleftrightarrow \mathscr{F}$
    (g) $P \longleftrightarrow (\sim P)$
    (h) $P \longrightarrow (\sim P)$

20. Prove the associative rules (1.25) and (1.26), page 25.

21. Prove (1.36), page 25.

22. Show $(P \vee Q) \to R \iff [(P \to R) \wedge (Q \to R)]$. Try to explain why this makes sense.

23. Show $(P \wedge Q) \to R \iff [(P \to R) \vee (Q \to R)]$. (This is not so easily explained as is the previous exercise.)

24. There is a notion in logic theory regarding "strong" versus "weak" statements, the stronger ones claiming in a sense more information regarding the underlying statements such as P, Q. For instance, $P \wedge Q$ is considered "stronger" than $P \vee Q$, because $P \wedge Q$ tells us more about P, Q (both are declared true) than $P \vee Q$ (at least one is true but both may be). Similarly $P \leftrightarrow Q$ is stronger than $P \to Q$.

    Each of the following statements may appear "strong" but in fact give little interesting content regarding P, Q. Construct a truth table for each and use the truth tables to then explain why they are not terribly "interesting" statements to make about P, Q. (Hint: What are these equivalent to?)

    (a) $(P \to Q) \vee (Q \to P)$
    (b) $(P \to Q) \to (P \leftrightarrow Q)$

1.3. VALID IMPLICATIONS AND ARGUMENTS

$$(\text{Assumptions}) \iff [\,(\text{Assumptions}) \land \mathscr{T}\,] \implies (\text{Conclusions})$$
$$\Updownarrow$$
$$\text{Theorems}$$

Figure 1.2: A model for mathematical reasoning. To show the conclusions follow from the assumptions, one usually attaches some theorems that are relevant to the assumptions. A *theorem* is a mathematical truth that we know to be always true because we have some logical proof for it. This model is valid since every statement P is equivalent to $P \land \mathscr{T}$, and every theorem is equivalent to $\mathscr{T}$.

1.3 Valid Implications and Arguments

Most theorems in this text are in the form of implications, rather than the more rigid equivalences of the last section. Indeed, our theorems are usually of the form "hypotheses imply conclusion." So we have need of an analog to our valid equivalences, namely a notion of valid implications.

1.3.1 Valid Implications Defined

Our definition of valid implications is similar to our previous definition of valid equivalences:

Definition 1.3.1 *Suppose that R and S are compound statements of some independent component statements $P_1, \ldots, P_n$. If $R \longrightarrow S$ is a tautology (always true), then we write*

$$R \implies S, \tag{1.46}$$

which we then call a **valid logical implication**.[22]

Example 1.3.1 ──

Perhaps the simplest example is the following: $P \implies P$. This seems obvious enough on its face. It can be proved using a truth table (note the vacuous case):[23]

P	P	$P \longrightarrow P$
T	T	T
F	F	T

──

---

[22] In this text we will differentiate between implications as statements, such as $P \longrightarrow Q$, which may be true or false, and **valid** implications, which are declarations that a particular implication is always true. For example, $R \implies S$ means $R \longrightarrow S$ is a tautology. (We similarly differentiated $\longleftrightarrow$ from $\iff$.)

[23] We could also show that $P \longrightarrow P$ is a tautology by way of previously proved results. For instance, with $P \longrightarrow Q \iff (\sim P) \lor Q$ ((1.29), page 25), with the part of Q played by P, we have

$$P \longrightarrow P \iff (\sim P) \lor P \iff \mathscr{T},$$

the second statement being in effect our original example of a tautology (Example 1.1.9, page 12).

Thus we see that there are logical implications that are tautologies. A slightly more complicated—and very instructive—example is the following:

Example 1.3.2

The following is a valid implication:

$$(P \wedge Q) \implies P. \tag{1.47}$$

To prove this, we will use a truth table to show that the following is a tautology:

$$(P \wedge Q) \longrightarrow P.$$

P	Q	$P \wedge Q$	P	$(P \wedge Q) \longrightarrow P$
T	T	T	T	T
T	F	F	T	T
F	T	F	F	T
F	F	F	F	T

Notice that three of the four cases have the implication true vacuously.

This example is fairly easy to interpret: If P and Q are true, then (of course) P is true. Next we show that bi-implication implies implication. Another intuitive example follows, basically stating that if we have a bi-implication, then we have an implication.

Example 1.3.3

The following is a valid implication:

$$P \longleftrightarrow Q \implies P \longrightarrow Q. \tag{1.48}$$

As before, we prove that replacing $\implies$ with $\longrightarrow$ gives us a tautology.

P	Q	$P \longleftrightarrow Q$	$P \longrightarrow Q$	$(P \longleftrightarrow Q) \longrightarrow (P \longrightarrow Q)$
T	T	T	T	T
T	F	F	F	T
F	T	F	T	T
F	F	T	T	T

Also as before, we see the importance of the vacuous cases in the final implication. In fact, this is just an application of the previous implication (1.47), if we remember that $P \longleftrightarrow Q$ is equivalent to $(P \longrightarrow Q) \wedge (Q \longrightarrow P)$:

$$\underbrace{(P \longrightarrow Q)}_{\text{``}P\text{''}} \wedge \underbrace{(Q \longrightarrow P)}_{\text{``}Q\text{''}} \implies \underbrace{(P \longrightarrow Q)}_{\text{``}P\text{''}}.$$

The quotes indicate what roles in (1.47) are played by the parts of (1.48).

Another interesting valid implication—worth reflecting on—is given next.

Example 1.3.4

$$(P \longrightarrow Q) \wedge (Q \longrightarrow R) \implies (P \longrightarrow R).$$

Note that to prove this, we must show that the following statement is a tautology:

$$[(P \longrightarrow Q) \wedge (Q \longrightarrow R)] \longrightarrow (P \longrightarrow R).$$

1.3. VALID IMPLICATIONS AND ARGUMENTS

P	Q	R	$P \to Q$	$Q \to R$	$(P \to Q) \wedge (Q \to R)$	$P \to R$	$[(P \to Q) \wedge (Q \to R)]$ $\longrightarrow (P \to R)$
T	T	T	T	T	T	T	T
T	T	F	T	F	F	F	T
T	F	T	F	T	F	T	T
T	F	F	F	T	F	F	T
F	T	T	T	T	T	T	T
F	T	F	T	F	F	T	T
F	F	T	T	T	T	T	T
F	F	F	T	T	T	T	T

In other words, if P implies Q, and Q implies R, then P implies R.

At this stage in the development, it is perhaps best to check such things using truth tables, however unwieldy they can be. Of course with practice comes intuition, informed memory, and many shortcuts, but for now we will use this brute force method of truth tables to determine if an implication is valid.

For a simple algebraic perspective of the difference between a valid implication and a valid equivalence, consider the following list of algebraic facts:

$$x = 5 \implies x^2 = 25,$$
$$x = -5 \implies x^2 = 25,$$
$$x^2 = 25 \iff (x = 5) \vee (x = -5).$$

Knowing $x = 5$ is not *equivalent* to knowing $x^2 = 25$. That is because there is an alternative explanation for $x^2 = 25$, namely that perhaps $x = -5$. But it is true that knowing $x = 5$ *implies* knowing—at least in principle—that $x^2 = 25$ (just as knowing $x = -5$ implies knowing $x^2 = 25$). If an equivalence is desired, a valid one is that knowing $x^2 = 25$ *is* equivalent to knowing that x must be either 5 or -5.

Later in the text we will briefly focus on algebra in earnest, bringing our symbolic logic to bear on that topic. In algebra (and in calculus), it is often important to know when we have an equivalence and when we have only an implication. For some algebraic problems, the implication often means we need to check our initial "answer," while the equivalence means there is no need (except to detect errors). For an example of this phenomenon, consider

$$\sqrt{x+2} = x \implies x + 2 = x^2 \iff 0 = x^2 - x - 2 \iff 0 = (x-2)(x+1)$$
$$\iff (x - 2 = 0) \vee (x - 1 = 0) \iff (x = 2) \vee (x = -1).$$

We lost the equivalence at the first step, and so we can only conclude from the logic that

$$\sqrt{x+2} = x \implies (x = 2) \vee (x = -1).$$

All this tells us is that *if* there is a number x so that $\sqrt{x+2} = x$, then the number must be either $x = 2$ or $x = -1$ (or perhaps both work; recall that we always interpret *or* inclusively). When we check $x = 2$ in the original equation, we get $\sqrt{4} = 2$, which is true. However, $x = -1$ gives $\sqrt{1} = -1$, which is not true. Since we have now solved the original equation, we can say that

$$\sqrt{x+2} = x \iff x = 2.$$

$$P \wedge Q \Longrightarrow P \qquad (1.49)$$
$$P \Longrightarrow P \vee Q \qquad (1.50)$$
$$P \longleftrightarrow Q \Longrightarrow P \longrightarrow Q \qquad (1.51)$$
$$(P \longrightarrow Q) \wedge (Q \longrightarrow R) \Longrightarrow P \longrightarrow R \qquad (1.52)$$
$$(P \longleftrightarrow Q) \wedge (Q \longrightarrow R) \Longrightarrow (P \longrightarrow R) \qquad (1.53)$$
$$(P \longrightarrow Q) \wedge (Q \longleftrightarrow R) \Longrightarrow (P \longrightarrow R) \qquad (1.54)$$
$$(P \longleftrightarrow Q) \wedge (Q \longleftrightarrow R) \Longrightarrow (P \longleftrightarrow R) \qquad (1.55)$$

$$(P \longrightarrow Q) \wedge P \Longrightarrow Q \qquad (1.56)$$
$$(P \longrightarrow Q) \wedge (\sim Q) \Longrightarrow \sim P \qquad (1.57)$$
$$(P \vee Q) \wedge (\sim Q) \Longrightarrow P \qquad (1.58)$$
$$P \longrightarrow (\sim P) \Longrightarrow \sim P \qquad (1.59)$$
$$P \longrightarrow \mathscr{F} \Longrightarrow \sim P \qquad (1.60)$$
$$\mathscr{T} \longrightarrow P \Longrightarrow P \qquad (1.61)$$

Table 1.4: Table of valid logical implications. If we replace $\Longrightarrow$ with $\longrightarrow$ in each of these implications (perhaps enclosing each side in brackets $[\cdots]$), we would have tautologies.

For another example, consider how one can solve a linear equation:

$$2x + 1 = 3 \iff 2x = 2 \iff x = 1.$$

Here we subtracted 1 from both sides, and then divided by 2, neither of which break the logical equivalence. We do not have to check the answer (unless we believe our arithmetic or reasoning may be faulty).

1.3.2 Partial List of Valid Implications

Table 1.4 lists some basic valid equivalences and implications. All can be proved using truth tables. However, it is important to learn to recognize validity without always resorting to truth tables. Each can be viewed in light of English examples. Still, it is the rigorous *mathematical* framework that gives us the precise rules for rewriting and analyzing statements.[24]

It is useful to see why (1.49)–(1.58) are *not* equivalences.[25] For instance, a little reflection should make clear that $P \not\Longrightarrow P \wedge Q$[26] (unless there is some underlying relationship between P and Q which is not stated), and so we cannot replace $\Longrightarrow$ with $\iff$ in (1.49). Similarly, in (1.52) having $P \longrightarrow R$ in itself says nothing about Q, so there is no reason to believe $(P \longrightarrow Q) \wedge (Q \longrightarrow R)$ is implied by $P \longrightarrow R$.

Implicit in this discussion is the fact that having $\iff$ is the same as simultaneously having both $\Longrightarrow$ and $\Longleftarrow$. Put another way, $R \iff S$ is the same as collectively having both $R \Longrightarrow S$ and $S \Longrightarrow R$.[27] In fact it is important to note that all valid logical equivalences (for instance those in Table 1.3, page 25) can be also considered to be combinations of two valid implications, one with $\Longrightarrow$, and the other with $\Longleftarrow$, replacing $\iff$. We do not list them all here, but rather list the most commonly used implications that are not equivalences, except

---

[24]Similarly, we use the rules of algebra to rewrite and analyze equations in hopes of solving for the variables.

[25]The exceptions in the table are (1.60) and (1.61), which are valid if we replace $\Longrightarrow$ with $\iff$, as was discussed in the previous section. See (1.45), page 30, and the exercises of that section.

[26]Note that it is common to negate a statement by including a "slash" through the main symbol, as in for example $\sim (x = 3) \iff x \ne 3$. What we mean by $R \not\Longrightarrow S$ is that it is not true that $R \longrightarrow S$ is a tautology.

[27]Note that $R \Longleftarrow S$ would be interpreted as $S \Longrightarrow R$. We will not make extensive use of "$\Longleftarrow$."

1.3. VALID IMPLICATIONS AND ARGUMENTS

for the last three in the table (which are valid equivalences if we replace $\Longrightarrow$ with $\Longleftrightarrow$, but are more useful as implications).

1.3.3 Fallacies and Valid Arguments[28]

*The name **fallacy** is usually reserved for typical faults in arguments that we nevertheless find persuasive. Studying them is therefore a good defense against deception.*
—Peter Suber, Department of Philosophy, Earlham College, Richmond, Indiana

Here we look at some classical argument styles, some of which are valid, and some of which are invalid and therefore called *fallacies* (whether or not they may seem persuasive at first glance). The valid styles will mostly mirror the valid logical implications of Table 1.4.

A common method for diagramming simple arguments is to have a horizontal line separating the *premises*[29] from the *conclusions*. Usually we will have multiple premises and a single conclusion. For style considerations, the conclusion is often announced with the symbol $\therefore$, which is read "therefore."[30]

In this subsection we look at several of these arguments, both valid and fallacious. Many are classical, with classical names. We will see how to analyze arguments for validity. In all cases here, it will amount to determining if a related implication is valid.

Example 1.3.5

Our first example we consider is the argument form classically known as **modus ponens**, or **law of detachment**. It is outlined as follows:[31]

$$\begin{array}{c} P \longrightarrow Q \\ P \\ \hline \therefore Q \end{array}$$

The idea is that if we **assume** $P \longrightarrow Q$ and P are true, then we must **conclude** that Q is also true. This is ultimately an implication. The key is that checking to see if this is valid is the same as checking to see if

$$(P \longrightarrow Q) \wedge P \Longrightarrow Q,$$

i.e., that $[(P \longrightarrow Q) \wedge P] \longrightarrow Q$ is a tautology. We know this to be the case already, as this is just (1.56), though we should **prove** this by producing the relevant truth table to show that $[(P \longrightarrow Q) \wedge P] \longrightarrow Q$ is indeed a tautology, meaning that it has truth value T for all cases of truth values of P and Q:

P	Q	$P \to Q$	$(P \to Q) \wedge P$	Q	$[(P \to Q) \wedge P] \to Q$
T	T	T	T	T	T
T	F	F	F	F	T
F	T	T	F	T	T
F	F	T	F	F	T

---

[28] Valid argument forms are also called *rules of inference*.

[29] Premises are also called *hypotheses*. The singular forms are premise and hypothesis.

[30] In fact most texts use either the horizontal line or the symbol $\therefore$, but not both. We use both to emphasize where the hypotheses end and the conclusion begins.

[31] *Modus ponens* is short for *modus ponendo ponens*, which is Latin for "the way that affirms by affirming." It is important enough that it has been extensively studied through the ages, and thus has many names, another being "affirmation argument." As for "law of detachment," it is pointed out in J.E. Rubin's *Mathematical Logic: Applications and Theory*, 1990 (Philadelphia: Saunders), that the idea is that we can validly "detach" the consequent Q of the conditional $P \longrightarrow Q$ when we also assume the antecedent P.

Thus to test the validity of an argument is to test whether or not the argument, written as an implication, is a tautology. This gives us a powerful, *computational* tool to analyze the classical argument styles. It also connects some of our symbolic logic to this style of diagramming arguments, so the intuition of these two flavors of logic can illuminate each other.

To repeat and emphasize the criterion for validity we give the following definition:

Definition 1.3.2 *A* **valid argument** *is one which, when diagrammed as an implication, represents a tautology. In other words, if the premises are $\mathscr{P}_1, \mathscr{P}_2 \ldots, \mathscr{P}_m$ and the conclusion is $\mathscr{Q}$ (where $\mathscr{P}_1, \mathscr{P}_2, \ldots, \mathscr{P}_m$ and $\mathscr{Q}$ are compound statements based upon some underlying independent statements $P_1, \ldots, P_n$), then the argument is valid if and only if*

$$\mathscr{P}_1 \wedge \mathscr{P}_2 \wedge \cdots \wedge \mathscr{P}_m \implies \mathscr{Q},$$

meaning if and only if $[\mathscr{P}_1 \wedge \mathscr{P}_2 \wedge \cdots \wedge \mathscr{P}_m] \longrightarrow \mathscr{Q}$ is a tautology. If not, then the argument is a called a **fallacy**.

Note that the validity of any argument does not depend upon the truth or falsity of the conclusion. Indeed the *modus ponens* argument in Example 1.3.5 is perfectly valid, regardless of whether Q is true. That is because we do not know—or even ask for purposes of discovering if the logic is valid—whether the premises are true. What we do know is that, *if the premises are true, then so is the conclusion*. In other words, the statement $[(P \rightarrow Q) \wedge P] \longrightarrow Q$ is always true. (If one or more of the premises are false, the implication is true vacuously.)

One example often used to shed light on the law of detachment, and other argument styles as well, uses the following choices for P and Q.

P : It rained

Q : The ground is wet

We diagram this argument using the English statements represented by P, Q:

If it rained, then the ground is wet.
It rained.
∴ The ground is wet.

This is a perfectly valid argument, meaning that if we accept the premises we must accept the conclusion. In other words, the logic is flawless. That said, one need not necessarily accept the conclusion just because the argument is valid, since one can always debate the truthfulness of the premises. Again the key is that the logic here is valid, even if the premises may be faulty.[32]

---

[32] In mature philosophical discussions, the logic is rarely in question because the valid models of argument are well known. When a conclusion seems unacceptable or just questionable, it is usually the premises that then come under scrutiny.

1.3. VALID IMPLICATIONS AND ARGUMENTS

Next we look at an example of an invalid argument, i.e., a fallacy. The following is called the **fallacy of the converse**:[33]

Example 1.3.6

Show that the following argument is a fallacy:

$$\begin{array}{c} P \longrightarrow Q \\ Q \\ \hline \therefore P \text{ (Invalid)} \end{array}$$

As before, we analyze the corresponding implication, in this case $(P \to Q) \wedge Q \implies P$, by constructing the truth table for $[(P \to Q) \wedge Q] \longrightarrow P$, this time noting we do not get a tautology:

P	Q	$P \to Q$	$(P \to Q) \wedge Q$	P	$[(P \to Q) \wedge Q] \longrightarrow P$
T	T	T	T	T	T
T	F	F	F	T	T
F	T	T	T	F	F
F	F	T	F	F	T

It is always useful to review an invalid argument to see which conditions were problematic. In the third row of our truth table, $P \to Q$ is vacuously true, and Q is true so the premises hold true, but the conclusion P is false (which was why $P \to Q$ was vacuously true!). From a more common-sense standpoint, while $P \to Q$ is assumed, P may not be the only condition that forces Q to be true. (If it were, we would instead have $P \longleftrightarrow Q$.) Consider again our previous choices for P and Q:

$$\begin{array}{c} \text{If it rained, then the ground is wet.} \\ \text{The ground is wet.} \\ \hline \therefore \text{It rained. (Invalid)} \end{array}$$

Even if the premises are correct in Example 1.3.6, the ground being wet does not guarantee that it rained. Perhaps it is wet from dew, or a sprinkler, or flooding from some other source. Here one can accept the premises, but the conclusion given is not valid.

There is a subtle—perhaps difficult—general point in this subsection that bears repeating: The truth table associated with an argument reflects the validity or invalidity of the logic of the **argument** (i.e., the validity of the corresponding implication), regardless of the truthfulness of the premises. Indeed, note how the truth table for the valid form *modus ponens* of Example 1.3.5, page 37, contains cases where the premises, $P \longrightarrow Q$ and P, can have truth value F as well as T.[34]

---

[33]The *converse* of an implication $R \longrightarrow S$ is the statement $S \longrightarrow R$ (or $R \longleftarrow S$ if we want to preserve the order). An implication and its converse are not logically equivalent, as a quick check of their truth tables would reveal. However, it is a common mistake to forget which direction an implication follows, or to be careless and mistake an implication for a bi-implication. The "fallacy of the converse" refers to a state of mind in which one mistakenly believes the converse true, based upon the assumption that the original implication is true. (Note that if we mistakenly replace $P \longrightarrow Q$ with $Q \longrightarrow P$ in Example 1.3.6, the new argument would be valid. In fact, it is *modus ponens*.)

[34] Another, perhaps more subtle point here is the extensive role of the vacuous cases in the underlying implication of an argument. If the premises are not true, then the conclusion cannot contradict them. This gives

For completeness, we mention that some use the adjective **sound** to describe an argument that is not only valid, but whose premises (and therefore conclusions) are in fact true. Of course in reality those are usually the arguments that we seek, but (arguably) one must first understand validity before probing the soundness of arguments, and so for this text, we are mostly interested in abstract, valid arguments, and worry about soundness only in context.

The next example is also very common. It is a valid form of argument often known by its Latin name **modus tollens**.[35]

Example 1.3.7

Analyze the following (**modus tollens**) argument.

$$\begin{array}{c} P \longrightarrow Q \\ \sim Q \\ \hline \therefore \sim P \end{array}$$

Solution: As before, we analyze the following associated implication (which we leave as a single-arrow implication until we establish it is a tautology):

$$[(P \to Q) \wedge (\sim Q)] \longrightarrow (\sim P).$$

P	Q	$P \to Q$	$\sim Q$	$(P \to Q) \wedge (\sim Q)$	$\sim P$	$[(P \to Q) \wedge (\sim Q)] \longrightarrow (\sim P)$
T	T	T	F	F	F	T
T	F	F	T	F	F	T
F	T	T	F	F	T	T
F	F	T	T	T	T	T

That the final column is all T's thus establishes the argument's validity. In fact, we see that the argument above is just a re-diagrammed version of (1.57), page 36.

A short chapter could be written just on the insights that can be found studying the *modus tollens* argument. For instance, for a couple of reasons one could make the case that *modus tollens* and *modus ponens* are the same type of argument. We will see one of these reasons momentarily, but first we will look at *modus tollens* by itself. Inserting our previous P and Q into this form, we would have

$$\begin{array}{c} \text{If it rained, then the ground is wet.} \\ \text{The ground is not wet.} \\ \hline \therefore \text{It did not rain.} \end{array}$$

---

rise to some strange forms of argument indeed, for the premises can be self-contradictory and therefore, taken as a group of statements joined by $\wedge$, can be equivalent to $\mathscr{F}$. (Recall $\mathscr{F} \to P$ is a tautology.) However, the only practical uses of arguments come when we know the premises to be true, or we think they are false and demonstrate it by showing the valid conclusions they imply are demonstrably false. The latter use is often called *proof by contradiction* or *indirect proof*, but there are many structures that use the same idea. *Modus tollens* (Example 1.3.7) is one permutation of the idea behind indirect proof.

[35]Short for *modus tollendo tollens,* Latin for "the way that denies by denying." It is also called "denying the consequent," which contrasts it to "affirming the consequent," another name for the fallacy of the converse, Example 1.3.6, page 39. As the reader can deduce, most of these common arguments—valid or not—have many names, inspired by different contexts and considerations. Computationally they are simple enough, once seen as implications to be analyzed and found to be tautologies (in cases of valid implications) or non-tautologies (in cases of fallacies).

1.3. VALID IMPLICATIONS AND ARGUMENTS

The validity of this argument should be intuitive. A common way of explaining it is that it must not have rained, because (first premise) if it had rained the ground would be wet, and (second premise) it is not wet. Of course that explanation is probably no simpler than just reading the argument as it stands.

Another way to look at it is to recall the equivalence of the implication to the contrapositive ((1.30), page 25 and elsewhere):

$$P \longrightarrow Q \iff (\sim Q) \longrightarrow (\sim P).$$

Thus we can replace in the *modus tollens* argument the first premise, $P \to Q$, with its (equivalent) contrapositive:

$$\begin{array}{c} (\sim Q) \longrightarrow (\sim P) \\ \sim Q \\ \hline \therefore \sim P \end{array}$$

This is valid by *modus ponens* (Example 1.3.5, page 37), with the part of P there played by the statement $\sim Q$ here, and the part of Q by $\sim P$. In fact the next valid argument form further unifies *modus ponens* and *modus tollens*, as will be explained next, though this next form is interesting in its own right.

Example 1.3.8

Consider the following form of argument, called *disjunctive syllogism*, which is valid.[36]

$$\begin{array}{c} P \vee Q \\ \sim P \\ \hline \therefore Q \end{array}$$

The proof of this is left as an exercise. To prove this one needs to show

$$(P \vee Q) \wedge (\sim P) \Longrightarrow Q,$$

that is, to show $[(P \vee Q) \wedge (\sim P)] \longrightarrow Q$ is a tautology. Of course the idea of this argument style is that when we assume "P or Q" to be true, and then further assume P is false (by assuming "$\sim P$" is true), we are forced to conclude Q must be true. Note that this is just a re-diagrammed version of (1.58), page 36, with P and Q exchanging roles. For an example, we will use a different pair of statements P and Q:

P : I will eat pizza.
Q : I will eat spaghetti.

This argument becomes (after minor colloquial adjustment):

$$\begin{array}{c} \text{I will eat pizza or spaghetti.} \\ \text{I will not eat pizza.} \\ \hline \therefore \text{I will eat spaghetti.} \end{array}$$

---

[36] A lay person might call this a form of "process of elimination."

To see how this unifies both *modus ponens* and *modus tollens* as two manifestations of the same principle, recall the following (easily proved by a truth table):

$$P \longrightarrow Q \iff (\sim P) \vee Q.$$

This appeared as (1.29), page 25, for instance. Thus the *modus ponens* and *modus tollens* become, respectively,

$$\begin{array}{c} (\sim P) \vee Q \\ \sim(\sim P) \\ \hline \therefore Q, \end{array} \qquad \text{and} \qquad \begin{array}{c} (\sim P) \vee Q \\ \sim Q \\ \hline \therefore \sim P. \end{array}$$

Compare these to the original forms, respectively, to see they are the same:

$$\begin{array}{c} P \longrightarrow Q \\ P \\ \hline \therefore Q, \end{array} \qquad \text{and} \qquad \begin{array}{c} P \longrightarrow Q \\ \sim Q \\ \hline \therefore \sim P. \end{array}$$

Let us now consider the **fallacy of the inverse**:[37]

Example 1.3.9 _____

Show that the following statement is a fallacy:

$$\begin{array}{c} P \longrightarrow Q \\ \sim P \\ \hline \therefore \sim Q \text{ (Invalid)} \end{array}$$

Solution: We check to see if the statement

$$[(P \to Q) \wedge (\sim P)] \longrightarrow (\sim Q)$$

is a tautology (in which case we could replace the major operation $\longrightarrow$ with $\Longrightarrow$).

P	Q	$P \to Q$	$\sim P$	$(P \to Q) \wedge (\sim P)$	$\sim Q$	$[(P \to Q) \wedge (\sim P)] \longrightarrow (\sim Q)$
T	T	T	F	F	F	T
T	F	F	F	F	T	T
F	T	T	T	T	F	F
F	F	T	T	T	T	T

We note that the argument—as an implication—is false in the case P is false and Q is true. In that case we have both premises of the argument ($P \to Q$ vacuously, and $\sim P$ obviously), but not the conclusion. Let us return to our rain and wet ground statements from before:

$$\begin{array}{c} \text{If it rained, then the ground is wet.} \\ \text{It did not rain.} \\ \hline \therefore \text{The ground is not wet. (Invalid)} \end{array}$$

---

[37]The **inverse** of an implication $R \longrightarrow S$ is the statement $(\sim R) \longrightarrow (\sim S)$. It is **not** equivalent to the original implication. In fact, it is equivalent to the converse (see Footnote 33, page 39), the proof of which is left to the exercises. A course on logic would emphasize these two statements, which are related to the implication. However, they are a source of some confusion, so we do not elaborate extensively here. It is much more important to realize that $R \longrightarrow S$ is equivalent to its contrapositive $(\sim S) \longrightarrow (\sim R)$, and **not** equivalent to these other two related implications, namely the converse $S \longrightarrow R$ or the inverse $(\sim R) \longrightarrow (\sim S)$. These facts should become more self-evident as the material is studied and utilized.

1.3. VALID IMPLICATIONS AND ARGUMENTS 43

For another example, consider the following obviously fallacious argument:

> If you drink the hemlock, then you will die.
> You do not drink the hemlock.
> ∴ You will not die. (Invalid)

There are, of course, other reasons why you might die (or why the ground might be wet). The case (seen also in the truth table) which ruins the bid for the corresponding implication to be a tautology is that case in which you do not drink the hemlock and still die, contradicting the conclusion but not the hypotheses.

There are many other forms of argument, both valid and invalid. A rudimentary strategy for detecting whether an argument is valid is to look at the corresponding implication and see if it is a tautology. If so then the argument is valid, and if not then the argument is a fallacy. In the next subsection, we introduce a slightly more sophisticated method, in which we use previously established valid implications and equivalences to make shorter work of some complicated arguments, but for now we will continue to use the truth table test of the validity of the underlying implication.

Our next example was proved earlier in the form of a valid implication. As such it was the subject of Example 1.3.4, page 34, and was listed as (1.52), page 36. We will not rework the truth table here.

Example 1.3.10

The following is a valid form of argument:

$$\begin{array}{c} P \longrightarrow Q \\ Q \longrightarrow R \\ \hline \therefore P \longrightarrow R \end{array}$$

For an example of an application, consider the following assignments for P, Q and R:

> P : I am paid.
> Q : I will buy you a present.
> R : You will be happy.

The argument then becomes the following, the validity of which is reasonably clear:

> If I am paid, then I will buy you a present.
> If I will buy you a present, then you will be happy.
> ∴ If I am paid, then you will be happy.

Next we look at an example that contains three underlying component statements, and three premises. Before doing so, we point out that we can compute the truth tables for $P \wedge Q \wedge R$ and $P \vee Q \vee R$ relatively quickly; the former is true whenever all three are true (and false if at least one is false), and the latter is true if any of the three are true (and false only if all three are false). The reason we can do this is that there is no ambiguity in computing, for instance, $P \wedge (Q \wedge R)$ or $(P \wedge Q) \wedge R$, as these are known to be equivalent (see page 25). It is similar for $\vee$. To be clearer, we note the truth tables for these.[38]

P	Q	R	$P \wedge Q \wedge R$	$P \vee Q \vee R$
T	T	T	T	T
T	T	F	F	T
T	F	T	F	T
T	F	F	F	T
F	T	T	F	T
F	T	F	F	T
F	F	T	F	T
F	F	F	F	F

With this observation, we can more easily analyze arguments with more than two premises.

Example 1.3.11

To determine the validity of the argument

$$P \longrightarrow (Q \vee R)$$
$$P$$
$$\sim R$$
$$\therefore Q$$

we need to see if the following conditional—which we dub ARG (for "argument") for space considerations—is a tautology:

$$ARG: \quad \{[P \longrightarrow (Q \vee R)] \wedge P \wedge (\sim R)\} \longrightarrow Q.$$

P	Q	R	$Q \vee R$	$P \to (Q \vee R)$	$\sim R$	$(P \to (Q \vee R)) \wedge P \wedge (\sim R)$	Q	ARG
T	T	T	T	T	F	F	T	T
T	T	F	T	T	T	T	T	T
T	F	T	T	T	F	F	F	T
T	F	F	F	F	T	F	F	T
F	T	T	T	T	F	F	T	T
F	T	F	T	T	T	F	T	T
F	F	T	T	T	F	F	F	T
F	F	F	F	T	T	F	F	T

Note that "ARG" connects the two immediately preceding columns with the logical operation $\longrightarrow$. Also notice that most of the cases are true vacuously (as often happens when we have more premises to be met, connected by $\wedge$), and eventually we see that the argument is valid.

---

[38]This is similar to the arithmetic rules that $A + B + C = A + (B + C) = (A + B) + C$. The first expression is not at first defined *per se*, but because the second and third are the same, we allow for the first. Similarly $A \cdot B \cdot C = A \cdot (B \cdot C) = (A \cdot B) \cdot C$. However, this does not extend to all operations, such as subtraction: Usually $A - (B - C) \neq (A - B) - C$, so when we write $A - B - C$ we have to choose one, and we choose the latter. Similarly, we have to be careful with other logical operations besides $\wedge$ and $\vee$. For instance, $P \longrightarrow Q \longrightarrow R$ would probably be interpreted $(P \longrightarrow Q) \longrightarrow R$, but it would depend upon the author. When we write $P \Longrightarrow Q \Longrightarrow R$ we mean $P \Longrightarrow Q$ and $Q \Longrightarrow R$ are both valid implications. If so, then so too is $P \Longrightarrow R$, as the argument in Example 1.3.10 indicates.

1.3. VALID IMPLICATIONS AND ARGUMENTS

Reading more examples, and perhaps some trial and error, one's intuition for what is valid and what is not should develop. With English examples one might be able to see that this argument style is reasonable. For instance, consider the following.

$$P : \text{I will eat pizza.}$$
$$Q : \text{I will drink soda.}$$
$$R : \text{I will drink beer.}$$

Then the premises become

$$P \longrightarrow (Q \vee R) : \text{If I eat pizza, then I will drink soda or beer.}$$
$$P : \text{I will eat pizza.}$$
$$\sim R : \text{I will not drink beer.}$$

It is reasonable to believe that, after declaring that if I eat pizza then I will drink soda or beer, and that I indeed will eat pizza, but not drink the beer, then I must drink the soda.[39]

1.3.4 Analyzing Arguments Without Truth Tables

In fact there is another approach for analyzing complicated arguments such as the previous one. The strategy is manipulative, and there are two tools we can make use of. The first is the somewhat obvious fact that we can always replace one of the hypotheses with a logically equivalent statement. That is because the truth table column entries are what matter in our computations there. But there is another strategy that is not quite so obvious. It relies on the following logical equivalence, which is left as an exercise:

$$\mathscr{P} \longrightarrow \mathscr{Q} \iff \mathscr{P} \longleftrightarrow (\mathscr{P} \wedge \mathscr{Q}). \tag{1.62}$$

For our purposes, it means that if our premises $\mathscr{P}$ imply $\mathscr{Q}$, then we can replace them with $\mathscr{P} \wedge \mathscr{Q}$, in effect *attaching* Q to the list of hypotheses. Note also this is true if $\mathscr{P}$ is instead just a single hypothesis in the list of hypotheses, which are joined by "ands" under $\wedge$, which is commutative and associative, so $\mathscr{Q}$ can be attached anywhere in the list as another hypothesis.

If one or more of the hypotheses taken collectively validly imply a statement $\mathscr{Q}$, then $\mathscr{Q}$ can be attached to the list of hypotheses, again due to the commutative and associative nature of $\wedge$.

Example 1.3.12

Let us re-examine the previous example, but this time we append some intermediate conclusions.

$$\left. \begin{array}{c} P \longrightarrow (Q \vee R) \\ P \\ \sim R \\ \hline \therefore Q \end{array} \right\} \therefore Q \vee R \quad \Big\} \therefore Q$$

---

[39] Or else I was lying when I recited my premises. The argument, anyway, is valid.

The first conclusion $Q \vee R$ is implied by the first two premises by a simple *modus ponens*, and the second came from the third original hypothesis $\sim R$ and our newly attached hypothesis $(Q \vee R)$ by way of the disjunctive syllogism (Example 1.3.8, page 41, and (1.57), page 36). It requires more creativity than a truth table verification, but is clearly less tedious.

Example 1.3.13 _____

Consider the following argument style and determine if it is valid:
$$\begin{array}{c}(P \vee Q) \longrightarrow R \\ \sim R \\ \hline \therefore \sim P.\end{array}$$

Solution: Here again we can work backwards.
$$\left.\begin{array}{c}(P \vee Q) \longrightarrow R \\ \sim R \\ \hline \therefore \sim P.\end{array}\right\} \therefore \sim (P \vee Q) \Big\} \therefore (\sim P) \wedge (\sim Q) \Big\} \therefore \sim P$$

A reasonable alternative style may skip some steps.

Note that in this latest example we could have appended our original hypotheses to include $(\sim P) \wedge (\sim Q)$, which would be the same as appending $\sim P$ and $\sim Q$ separately.

This method is often quite useful for verifying validity, but it is not, by itself, a way to detect a fallacy. However, since common fallacies rest upon the fallacy of the inverse (page 42) or the fallacy of the converse (page 39), we can sometimes detect when an argument tempts us to agree with the conclusion because of such invalid, but common, reasoning.

Example 1.3.14 _____

Consider the following argument:
$$\begin{array}{c}(P \vee Q) \longrightarrow R \\ \sim Q \\ R \\ \hline \therefore P. \text{ (Invalid!)}\end{array}$$

What is tempting (but invalid) to do with this argument, is to reason that the first and third hypotheses imply $P \vee Q$, and with the second reading $\sim Q$, we would then (seemingly validly) conclude P. However, we cannot conclude P from our premises. Indeed, R does not imply anything about P and Q, and if nothing else, a truth table will prove this style invalid. An errant diagram might look like the following:

$$\left.\begin{array}{c}\boxed{(P \vee Q) \longrightarrow R} \\ \sim Q \\ \boxed{R} \\ \hline \therefore P\end{array}\right\} \therefore (P \vee Q) \quad \text{(INVALID!)} \quad \left.\begin{array}{c}\sim Q\end{array}\right\} \therefore P$$

An example in English might help to see that indeed this is invalid. If we define P, Q and R as

1.3. VALID IMPLICATIONS AND ARGUMENTS

follows, it seems pretty clear this style is a fallacy.

$$P : \text{I shot you.}$$
$$Q : \text{I stabbed you.}$$
$$R : \text{You died.}$$

This particular argument then reads:

$$\begin{array}{l} \text{If I shot you or stabbed you, then you died.} \\ \text{I did not stab you.} \\ \underline{\text{You died.}} \\ \quad \therefore \text{I shot you. (Invalid!)} \end{array}$$

Clearly this is fallacious reasoning. The "alternative explanation" for the consequent, i.e., $\mathscr{Q}$ where $\mathscr{P} \longrightarrow \mathscr{Q}$, should always be looming when we look at implications and try to read them backwards. (In this case, there is nothing in the premises that do not allow for other causes of your dying.)

From this discussion, one might conclude that we can often intuitively detect the likelihood of an argument being invalid, but unless we can rewrite it as a known fallacy—or clearly see a case where the premises hold true and the conclusion does not[40]—we would need to go back to our primitive but absolutely reliable method of constructing the truth table for the underlying implication to see if we have a tautology. If we do not have a tautology, then the argument is a fallacy; if we do, then the argument is valid.[41]

Still, this new method of proving validity for arguments can be very useful, especially in lieu of long truth tables, but of course it necessarily rests on the styles we previously proved to be valid, or the valid equivalences or implications from before. (It also requires a bit of creativity in using them.) The more known valid styles, equivalences and implications one has available, the larger the number of argument styles that can be proven without resorting to truth tables. Of course it was the truth tables that allowed us to prove the preliminary equivalences, implications, and argument styles to be valid, so ultimately all these things rest upon the truth tables. They are the one device at our disposal that allowed logic to be rendered computational (and at a very fundamental level), rather than just intuitive.

---

[40] Some texts have the reader check the validity of an argument by checking only the cases in which **all** premises are true to see if the conclusions are also true for those cases. Indeed it is only the cases in which the premises are true and the conclusion is false that invalidate an argument style. However, that kind of analysis de-emphasizes the connection between valid argument styles and valid implications, and the role of tautologies, so we prefer to include these ideas, at the cost of looking at every case of truth values for the underlying statements P, Q, and so on when analyzing an argument for validity.

[41] By now the reader, having encountered abstract and applied implications on many abstract levels, should be aware of the reason to emphasize both sides of the semicolon in the sentence above, namely, "If we do not have a tautology, then the argument is a fallacy; if we do, then the argument is valid." To spell this out better, consider

$$P : \text{We have a tautology (in the form of the argument as an implication).}$$
$$Q : \text{The argument is valid.}$$

Then the sentence in quotes reads $[(\sim P) \to (\sim Q)] \wedge [P \to Q]$, which is equivalent to $[Q \to P] \wedge [P \to Q]$, i.e., $P \leftrightarrow Q$. If we wrote only the first part, $(\sim P) \to (\sim Q)$, that alone would not declare the second part $P \to Q$, though many casual readers would erroneously assume that it would (in words if not symbols).

Exercises

Some of these were proved previously in the text. The reader should attempt to prove these first for himself or herself without referring to the text for the proofs.

1. Prove $P \longrightarrow Q \iff P \longrightarrow (P \wedge Q)$. This is essentially (1.62), page 45.

2. Consider the statement $P \longrightarrow (\sim P)$.

    (a) Use a truth table to prove the validity of $P \longrightarrow (\sim P) \implies (\sim P)$. Is this reasonable?

    (b) If possible without truth tables, and using a short string of known equivalences, show that in fact
    $$P \longrightarrow (\sim P) \iff (\sim P).$$
    (See (1.29), page 25, and (1.17), page 25.)

3. Use truth tables to show (1.49) and (1.50):
    $$P \wedge Q \implies P,$$
    $$P \implies P \vee Q.\text{[42]}$$

4. Prove the following without using truth tables.
    $$P \implies (P \vee Q)$$
    by proving that $P \longrightarrow (P \vee Q)$ is a tautology. (Again, see (1.29), page 25.)

5. Prove the following two valid implications:

    (a) (1.56): $(P \longrightarrow Q) \wedge P \implies Q$.

    (b) (1.57): $(P \longrightarrow Q) \wedge (\sim Q) \implies \sim P$.

6. Prove the following using truth tables. Note that there are $2^4 = 16$ different combinations of truth values for P, Q, R and S.

    $$(P \longrightarrow R) \wedge (Q \longrightarrow S)$$
    $$\implies (P \wedge Q) \longrightarrow (R \wedge S), \quad (1.63)$$

    $$(P \longrightarrow R) \wedge (Q \longrightarrow S)$$
    $$\implies (P \vee Q) \longrightarrow (R \vee S) \quad (1.64)$$

    These are useful because we often have a string of statements $A_1,\ldots,A_n$, connected entirely by $\wedge$ or entirely by $\vee$, and wish to replace them with $B_1,\ldots,B_n$ (or just some of them), where $A_1 \implies B_1$, $A_2 \implies B_2$, ..., $A_n \implies B_n$.[43] With (1.63) and (1.64) we can generalize and write

    $$A_1 \wedge A_2 \wedge \cdots \wedge A_n$$
    $$\implies B_1 \wedge B_2 \wedge \cdots \wedge B_n, \quad (1.65)$$
    $$A_1 \vee A_2 \vee \cdots \vee A_n$$
    $$\implies B_1 \vee B_2 \vee \cdots \vee B_n. \quad (1.66)$$

    Valid implications (1.63) and (1.64) would be quite laborious to prove without truth tables.

---

[42] Some would characterize $P \wedge Q \implies P \implies P \vee Q$ to be a progression from the strongest statement, $P \wedge Q$, to the weakest, $P \vee Q$, of the three. A similar form would be $P \wedge Q \implies P \implies P \vee R$.

[43] Here the parts of $P \longrightarrow R$ and $Q \longrightarrow S$ are played by the $A_i \implies B_i$, but the idea is the same. We could instead attach $\mathscr{T} \iff (A_1 \longrightarrow B_1) \wedge \cdots \wedge (A_n \longrightarrow B_n)$ to the left-hand sides of (1.65) and (1.66) with the wedge operation, since $\mathscr{T} \wedge U \iff U$.

1.3. VALID IMPLICATIONS AND ARGUMENTS

For Exercises 7–13, decide if you believe it is a valid argument or a fallacy. Then check by constructing the corresponding truth table.

7. $\begin{array}{c} P \longrightarrow Q \\ Q \longrightarrow P \\ \hline \therefore P \longleftrightarrow Q \end{array}$

8. $\begin{array}{c} P \longleftrightarrow Q \\ P \\ \hline \therefore Q \end{array}$

9. $\begin{array}{c} P \longrightarrow Q \\ Q \\ \hline \therefore P \end{array}$

10. $\begin{array}{c} P \longleftrightarrow Q \\ Q \\ \hline \therefore P \end{array}$

11. $\begin{array}{c} \sim (P \longrightarrow Q) \\ P \\ \hline \therefore \sim Q \end{array}$

12. $\begin{array}{c} P \longrightarrow Q \\ P \longrightarrow (\sim Q) \\ \hline \therefore \sim P \end{array}$

13. $\begin{array}{c} P \\ \sim P \\ \hline \therefore Q \end{array}$

(For this last case, see Footnote 34, page 39.)

For Exercises 14–34, decide if it is valid or a fallacy. For those marked "(Prove)," offer a truth table proof or a manipulation of the hypotheses to justify your answer.

14. $\begin{array}{c} (P \wedge Q) \longrightarrow R \\ P \\ Q \\ \hline \therefore R \end{array}$

15. $\begin{array}{c} (P \wedge Q) \longrightarrow R \\ P \\ \sim R \\ \hline \therefore \sim Q \end{array}$

16. $\begin{array}{c} (P \wedge Q) \longrightarrow R \\ Q \\ \sim P \\ \hline \therefore \sim R \text{ (Prove)} \end{array}$

17. $\begin{array}{c} P \longrightarrow Q \\ Q \longrightarrow R \\ \sim R \\ \hline \therefore \sim P \text{ (Prove)} \end{array}$

18. $\begin{array}{c} P \longrightarrow Q \\ Q \longrightarrow R \\ \sim P \\ \hline \therefore \sim R \end{array}$

19. $\begin{array}{c} P \longrightarrow Q \\ Q \longrightarrow R \\ R \longrightarrow P \\ \hline \therefore (P \longleftrightarrow R) \end{array}$

20. $\begin{array}{c} P \vee Q \vee R \\ \sim P \\ \sim R \\ \hline \therefore Q \text{ (Prove)} \end{array}$

21. $\begin{array}{c} P \longrightarrow Q \\ Q \longrightarrow (\sim R) \\ R \\ \hline \therefore \sim P \end{array}$

22. $\begin{array}{c} P \vee Q \vee R \\ P \longrightarrow Q \\ \sim Q \\ \hline \therefore R \end{array}$

23. $\begin{array}{c} P \longrightarrow (Q \wedge R) \\ \sim R \\ \hline \therefore \sim P \text{ (Prove)} \end{array}$

24. $\begin{array}{c} P \longrightarrow (Q \vee R) \\ P \\ \sim R \\ \hline \therefore Q \end{array}$

25. $\begin{array}{c} (\sim P) \longrightarrow (\sim Q) \\ \sim Q \\ \hline \therefore \sim P \end{array}$

26. $\begin{array}{c} (P \wedge Q) \longrightarrow R \\ S \longrightarrow P \\ U \longrightarrow Q \\ S \wedge U \\ \hline \therefore R \end{array}$

27. $\begin{array}{c} (P \wedge Q) \longrightarrow R \\ S \longrightarrow P \\ U \longrightarrow Q \\ (\sim S) \wedge (\sim U) \\ \hline \therefore \sim R \end{array}$

28. $\begin{array}{c} (P \wedge Q) \longrightarrow R \\ S \longrightarrow P \\ U \longrightarrow Q \\ \sim R \\ \hline \therefore (\sim S) \vee (\sim U) \end{array}$

29. $\begin{array}{c} \mathscr{F} \\ \hline \therefore P \end{array}$

30. $\begin{array}{c} P \\ \hline \therefore \mathscr{T} \end{array}$

31. $\begin{array}{c} \mathscr{T} \longrightarrow P \\ \hline \therefore P \end{array}$

32. $\begin{array}{c} P \longrightarrow \mathscr{F} \\ \hline \therefore \sim P \end{array}$

33. $\begin{array}{c} P \longrightarrow R \\ R \longrightarrow P \\ \hline \therefore P \longleftrightarrow R \text{ (Prove)} \end{array}$

34. $\begin{array}{c} P \longleftrightarrow R \\ Q \longleftrightarrow R \\ \hline \therefore P \longleftrightarrow Q \end{array}$

1.4 Sets and Quantifiers

In this section we introduce quantifiers which form the last class of logic symbols we will consider in this text. To use quantifiers, we also need some notions and notations from set theory. This section introduces sets and quantifiers to the extent required for our study of calculus. For the interested reader, Section 1.5 will extend this introduction, though even with that section we would be only just beginning to delve into these topics if studying them for their own sakes. Fortunately, what we need of these topics for our study of calculus is contained in this section.

1.4.1 Sets

Put simply, a set is a collection of objects, which are then called *elements* or *members* of the set. We give sets names just as we do variables and statements. For an example of the notation, consider a set A defined by

$$A = \{2, 3, 5, 7, 11, 13, 17\}.$$

We usually define a particular set by describing or listing the elements between "curly braces" { } (so the reader understands it is indeed a *set* we are discussing). The defining of A was accomplished by a complete listing, but some sets are too large for that to be possible, let alone practical. As an alternative, the set A can also be written

$$A = \{x \mid x \text{ is a prime number less than } 18\}.$$

This equation is usually read, "A is the set of all x such that x is a prime number less than 18." Here x is a "dummy variable," used only briefly to describe the set.[44] Sometimes it is convenient to simply write

$$A = \{\text{prime numbers between 2 and 17, inclusive}\}.$$

(Usually, "inclusive" is meant by default, so here we would include 2 and 17 as possible elements, if they also fit the rest of the description.) Of course there are often several ways of describing a list of items. For instance, we can replace "between 2 and 17, inclusive" with "less than 18," as before.

Often an ellipsis "..." is used when a pattern should be understood from a partial listing. This is particularly useful if a complete listing is either impractical or impossible. For instance, the set B of integers from 1 to 100 could be written

$$B = \{1, 2, 3, \ldots, 100\}.$$

To note that an object is in a set, we use the symbol $\in$. For instance we may write $5 \in B$, read "5 is an element of B." To indicate concisely that 5, 6, 7, and 8 are in B, we can write $5, 6, 7, 8 \in B$.

---

[44]"Dummy variables" are also used to describe the actions of functions, as in $f(x) = x^2 + 1$. In this context, the function is considered to be the *action* of taking an input number, squaring it, and adding 1. The x is only there so that we can easily trace the action on an arbitrary input. We will revisit functions later.

1.4. SETS AND QUANTIFIERS

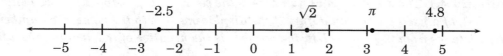

Figure 1.3: The number line representing the set $\mathbb{R}$ of real numbers, with a few points plotted. On this graph, the hash marks fall at the integers.

Just as we have use for zero in addition, we also define the *empty set* or *null set* as the set that has no elements. We denote that set $\varnothing$. Note that $x \in \varnothing$ is always false, that is,

$$x \in \varnothing \iff \mathscr{F},$$

because it is impossible to find any element of any kind inside $\varnothing$. We will revisit this set repeatedly in the optional, more advanced Section 1.5.

For calculus we are mostly interested in sets of numbers. While not the most important, the following three sets will occur from time to time in this text:

$$\underline{\text{Natural numbers}}^{45}\colon \quad \mathbb{N} = \{1, 2, 3, 4, \ldots\}, \tag{1.67}$$

$$\underline{\text{Integers}}\colon \quad \mathbb{Z} = \{\ldots, -3, -2, -1, 0, 1, 2, 3, \ldots\}, \tag{1.68}$$

$$\underline{\text{Rational numbers}}\colon \quad \mathbb{Q} = \left\{ \frac{p}{q} \;\middle|\; (p, q \in \mathbb{Z}) \wedge (q \neq 0) \right\}. \tag{1.69}$$

Here we again use the ellipsis to show that the established pattern continues forever in each of the cases $\mathbb{N}$ and $\mathbb{Z}$. The sets $\mathbb{N}$, $\mathbb{Z}$, and $\mathbb{Q}$ are examples of *infinite* sets, i.e., sets that do not have a finite number of elements. The rational numbers are those that are *ratios* of integers, except that division by zero is not allowed, for reasons we will consider later.[46]

For calculus, the most important set is the set $\mathbb{R}$ of *real numbers*, which cannot be defined by a simple listing or by a simple reference to $\mathbb{N}$, $\mathbb{Z}$, or $\mathbb{Q}$. One intuitive way to describe the real numbers is to consider the horizontal *number line*, where geometric points on the line are represented by their *displacements* (meaning distances, but counted as positive if to the right and negative if to the left) from a fixed point, called the *origin* in this context. That fixed point is represented by the number 0, since the fixed point is a displacement of zero units from itself. In Figure 1.3 the number line representation of $\mathbb{R}$ is shown. Hash marks at convenient intervals are often included. In this case, they are at the integers. The arrowheads indicate the number line is an actual line and thus infinite in both directions. The points -2.5 and 4.8 on the graph are not integers, but are rational numbers, since they can be written $-25/10 = -5/2$, and $48/10 = 24/5$, respectively. The points $\sqrt{2}$ and π are real, but not rational, and so are called *irrational*. To summarize,

---

[45]The natural numbers are also called *counting* numbers in some texts.

[46]For a hint, think about what should be $x = 1/0$. If we multiply both sides by zero, we might think we get $0x = (1/0) \cdot 0$, giving $0 = 1$, which is absurd. In fact there is no such x, so $x = 1/0 \longrightarrow 0 = 1$, which is of the form $P \longrightarrow \mathscr{F}$ which we recall to be equivalent to $\sim P$.

Definition 1.4.1 *The set of all **real** numbers is the set $\mathbb{R}$ of all possible displacements, to the right or left, of a fixed point 0 on a line. If the displacement is to the right, the number is the positive distance from 0. If to the left, the number is the negative of the distance from 0.*[47]

Thus,
$$\mathbb{R} = \{\text{displacements from 0 on the number line}\}. \tag{1.70}$$

This is not a rigorous definition, not least because "right" and "left" require a fixed perspective. Even worse, the definition is really a kind of "circular reasoning," since we are effectively defining the number line in terms of $\mathbb{R}$, and then defining $\mathbb{R}$ in terms of (displacements on) the number line.

1.4.2 Quantifiers

The three quantifiers used by nearly every professional mathematician are as follows:

Universal quantifier: $\forall$, read, "for all," or "for every;"
Existential quantifier: $\exists$, read, "there exists;"
Uniqueness quantifier: $!$, read, "unique."

The first two are of equal importance, and are far more important than the third, which is usually only found after the second. Quantified statements are usually found in forms such as:

$(\forall x \in S)P(x)$,   (i.e., for all $x \in S$, $P(x)$ is true);
$(\exists x \in S)P(x)$,   (i.e., there exists an $x \in S$ such that $P(x)$ is true);
$(\exists! x \in S)P(x)$,   (i.e., there exists a unique (exactly one) $x \in S$ such that $P(x)$ is true).

Here S is a set and $P(x)$ is some statement about x. The meanings of these quickly become straightforward. For instance, consider

$(\forall x \in \mathbb{R})(x + x = 2x)$: for all $x \in \mathbb{R}$, $x + x = 2x$;
$(\exists x \in \mathbb{R})(x + 2 = 2)$: there exists (an) $x \in \mathbb{R}$ such that $x + 2 = 2$;
$(\exists! x \in \mathbb{R})(x + 2 = 2)$: there exists a unique $x \in \mathbb{R}$ such that $x + 2 = 2$.

All three quantified statements above are true. In fact they are true under any circumstances, and can thus be considered tautologies. Unlike unquantified statements P, Q, R, etc., from our first three sections, a quantified statement is either true always or false always, and is thus, for our purposes, equivalent to either $\mathcal{T}$ or $\mathcal{F}$. Each has to be analyzed

---

[47] It should be noted that we have to choose a direction to call "right," the other then being "left." It will depend upon our perspective. When we look at the Cartesian Plane, the horizontal axis measures displacements as right (positive horizontal) or left (negative horizontal), and the vertical axis measures displacements as upward (positive vertical) or downward (negative vertical). In that context, the *origin* is where the axes intersect.

1.4. SETS AND QUANTIFIERS

on its face, based on known mathematical principles; we do not have a brute-force mechanism analogous to truth tables to analyze these systematically.[48] For a couple more short examples, consider the following cases from algebra, which should be clear enough:

$$(\forall x \in \mathbb{R})(0 \cdot x = 0) \iff \mathscr{T};$$
$$(\exists x \in \mathbb{R})(x^2 = -1) \iff \mathscr{F}.$$

The optional advanced section shows how we can still find equivalent or implied statements from quantified statements in many circumstances.

1.4.3 Statements with Multiple Quantifiers

Many of the interesting statements in mathematics contain more than one quantifier. To illustrate the mechanics of multiply quantified statements, we will first turn to a more worldly setting. Consider the following sets:

$$M = \{\text{men}\},$$
$$W = \{\text{women}\}.$$

In other words, M is the set of all men, and W the set of all women. Consider the statement[49]

$$(\forall m \in M)(\exists w \in W)[w \text{ loves } m]. \tag{1.71}$$

Set to English, (1.71) could be written, "For every man there exists a woman who loves him."[50] So if (1.71) is true, we can in principle arbitrarily choose a man m, and then know that there is a woman w who loves him. It is important that the man m was quantified *first*. A common syntax that would be used by a logician or mathematician would be to say here that, once our choice of a man is *fixed*, we can in principle find a woman who loves him. Note that (1.71) allows that different men may need different women to love them, and also that a given man may be loved by more than (but not less than) one woman.

Alternatively, consider the statement

$$(\exists w \in W)(\forall m \in M)[w \text{ loves } m]. \tag{1.72}$$

A reasonable English interpretation would be, "There exists a woman who loves every man." Granted that is a summary, for the word-for-word English would read more like "There exists a woman such that, for every man, she loves him." This says something very different from (1.71), because that earlier statement does not assert that we can find a woman who, herself, loves every man, but that for each man there is a woman who loves him.[51]

---

[48]This is part of what makes quantified statements interesting!

[49]Note that this is of the form $(\forall m \in M)(\exists w \in W)P(m,w)$, that is, the statement P says something about both m and w. We will avoid a protracted discussion of the difference between statements regarding one variable object—as in $P(x)$ from our previous discussion—and statements that involve more than one, as in $P(m,w)$ here. Statements of multiple (variable) quantities will recur in subsequent examples.

[50]At times it seems appropriate to translate "$\forall$" as "for all," and at other times it seems better to translate it as "for every." Both mean the same.

[51]We do not pretend to know the truth values of either (1.71) or (1.72).

We can also consider the statement

$$(\forall m \in M)(\forall w \in W)[w \text{ loves } m]. \tag{1.73}$$

This can be read, "For every man and every woman, the woman loves the man." In other words, every man is loved by every woman. In this case, we can reverse the order of quantification:

$$(\forall w \in W)(\forall m \in M)[w \text{ loves } m]. \tag{1.74}$$

In fact, if the two quantifiers are the same type—both universal or both existential—then the order does not matter. Thus,

$$(\forall m \in M)(\forall w \in W)[w \text{ loves } m] \iff (\forall w \in W)(\forall m \in M)[w \text{ loves } m],$$
$$(\exists m \in M)(\exists w \in W)[w \text{ loves } m] \iff (\exists w \in W)(\exists m \in M)[w \text{ loves } m].$$

In both representations in the existential statements, we are stating that there is at least one man and one woman such that she loves him. In fact that equivalence is also valid if we replace $\exists$ with $\exists!$, though it would mean then that there is exactly one man and exactly one woman such that the woman loves the man, but we will not delve too deeply into uniqueness here.

Note that in cases where the sets are the same, we can combine two similar quantifications into one, as in

$$(\forall x \in \mathbb{R})(\forall y \in \mathbb{R})[x + y = y + x] \iff (\forall x, y \in \mathbb{R})[x + y = y + x]. \tag{1.75}$$

It is similar with existence.

However, we repeat the point at the beginning of the subsection, which is that the order does matter if the types of quantification are different.

For another short example, which is algebraic in nature, consider

$$(\forall x \in \mathbb{R})(\exists K \in \mathbb{R})(x = 2K). \qquad \text{(True.)} \tag{1.76}$$

This is read, "For every $x \in \mathbb{R}$, there exists $K \in \mathbb{R}$ such that $x = 2K$." That $K = x/2$ exists (and is actually unique) makes this true, while it would be false if we were to reverse the order of quantification:

$$(\exists K \in \mathbb{R})(\forall x \in \mathbb{R})(x = 2K). \qquad \text{(False.)} \tag{1.77}$$

Statement (1.77) claims (erroneously) that there exists $K \in \mathbb{R}$ so that, for every $x \in \mathbb{R}$, $x = 2K$. That is impossible, because no value of K is half of *every* real number x. For example, the value of K that works for $x = 4$ is not the same as the value of K that works for $x = 100$.

1.4.4 Detour: Uniqueness as an Independent Concept

We will have occasional statements in the text that include uniqueness. However, most of those will not require us to rewrite the statements in ways that require actual manipulation of the uniqueness quantifier. Still, it is worth noting a couple of interesting points about this quantifier.

First we note that uniqueness can be formulated as a separate concept from existence, interestingly instead requiring the universal quantifier.

1.4. SETS AND QUANTIFIERS

Definition 1.4.2 *Uniqueness is the notion that if $x_1, x_2 \in S$ satisfy the same particular statement $P(\ )$, then they must in fact be the same object. That is, if $x_1, x_2 \in S$ and $P(x_1)$ and $P(x_2)$ are true, then $x_1 = x_2$. This may or may not be true, depending on the set S and the statement $P(\ )$.*

Note that there is the vacuous case where nothing satisfies the statement $P(\ )$, in which case the uniqueness of any such hypothetical object is proved but there is actually no existence. Consider the following, symbolic representation of the uniqueness of an object x which satisfies $P(x)$:[52]

$$(\forall x, y \in S)[(P(x) \wedge P(y)) \longrightarrow x = y]. \tag{1.78}$$

Finally we note that a proof of a statement such as $(\exists! x \in S)P(x)$ is thus usually divided into two separate proofs:

(1) *Existence:* $(\exists x \in S)P(x)$

(2) *Uniqueness:* $(\forall x, y \in S)[(P(x) \wedge P(y)) \longrightarrow x = y]$

For example, if we were to give the usual rigorous, axiomatic definition of the set of real numbers $\mathbb{R}$, one of the axioms[53] defining the real numbers is the existence of an additive identity:

$$(\exists z \in \mathbb{R})(\forall x \in \mathbb{R})(z + x = x). \tag{1.79}$$

In fact it follows quickly that such a "z" must be unique, so we have

$$(\exists! z \in \mathbb{R})(\forall x \in \mathbb{R})(z + x = x). \tag{1.80}$$

To prove (1.80), we need to prove (1) existence and (2) uniqueness. In this setting, the existence is an axiom, so there is nothing to prove. We turn then to the uniqueness. A proof is best written in prose, but it is based upon proving that the following is true:

$$(\forall z_1, z_2 \in \mathbb{R})[(z_1 \text{ an additive identity}) \wedge (z_2 \text{ an additive identity}) \longrightarrow z_1 = z_2].$$

---

[52] The statement indeed says that any two elements $x, y \in S$, which both satisfy P, must be the same. Note that we use a single arrow here, because the statement between the brackets [ ] is not likely to be a tautology, but may be true for enough cases for the entire quantified statement to be true. Indeed, the symbols $\Longrightarrow$ and $\Longleftrightarrow$ belong *between* quantified statements, not *inside* them.

[53] Recall that an *axiom* is an assumption, usually self-evident, from which we can logically argue towards theorems. Axioms are also known as *postulates*. If we attempt to argue only using "pure logic" (as a mathematician does when developing theorems, for instance), it eventually becomes clear that we still need to make some assumptions because one cannot argue "from nothing." Indeed, some "starting points" from which to argue towards the conclusions are required. These are then called axioms.

The word "axiomatic" is often used colloquially to mean clearly evident and therefore not requiring proof. That is not always the case with mathematical axioms. Indeed, the postulates required for defining the real numbers seem rather strange at first. They were developed to be a minimal number of assumptions required to give the real numbers their apparent properties which could be observed. In that case, it seemed we worked towards a foundation, after seeing the outer structure. A similar phenomenon can be seem in Einstein's Special Relativity, where his two simple—yet at the time quite counterintuitive—axioms were able to completely replace a much larger set of postulates required to explain many of the electromagnetic phenomena discovered before and around his time, and predict many new phenomena that were later observed.

Now we prove this. Suppose z_1 and z_2 are additive identities, i.e., they can stand in for z in (1.79), which could also read $(\exists z \in \mathbb{R})(\forall x \in \mathbb{R})(x = z + x)$. Note the order there, where the identity z (think "zero") is placed on the left of x in the equation $x = z + x$. So, assuming z_1, z_2 are additive identities, we have

$$\begin{aligned} z_1 &= z_2 + z_1 && \text{(since } z_2 \text{ is an additive identity)} \\ &= z_1 + z_2 && \text{(since addition is commutative—order is irrelevant)} \\ &= z_2 && \text{(since } z_1 \text{ is an additive identity).} \end{aligned}$$

This argument showed that if z_1 and z_2 are any real numbers that act as additive identities, then $z_1 = z_2$. In other words, if there are any additive identities, there must be only one. Of course, assuming its existence we call that unique real number *zero*. (It should be noted that the commutativity used above is another axiom of the real numbers. Some texts list fourteen in all.)

The distinction between existence and uniqueness of an object with some property P is often summarized as follows:

(1) Existence asserts that there is at least one such object.

(2) Uniqueness asserts that there is at most one such object.

If both hold, then there is exactly one such object.

1.4.5 Negating Universally and Existentially Quantified Statements

For statements with a single universal or existential quantifier, we have the following negations.

$$\sim [(\forall x \in S)P(x)] \iff (\exists x \in S)[\sim P(x)], \tag{1.81}$$
$$\sim [(\exists x \in S)P(x)] \iff (\forall x \in S)[\sim P(x)], \tag{1.82}$$

The left side of (1.81) states that it is not the case that $P(x)$ is true for all $x \in S$; the right side states that there is an $x \in S$ for which $P(x)$ is false. We could ask when it is a lie that for all x, $P(x)$ is true. The answer is when there is an x for which $P(x)$ is false, i.e., $\sim P(x)$ is true.

The left side of (1.82) states that it is not the case that there exists an $x \in S$ so that $P(x)$ is true; the right side says that $P(x)$ is false for all $x \in S$. When is it a lie that there is an x making $P(x)$ true? When $P(x)$ is false for all x.

Thus when we negate such a statement as $(\forall x)P(x)$ or $(\exists x)P(x)$, we change $\forall$ to $\exists$ or vice-versa, and negate the statement after the quantifiers.

Example 1.4.1

Negate $(\forall x \in S)[P(x) \longrightarrow Q(x)]$.

Solution: We will need (1.21), page 25, namely $\sim (P \longrightarrow Q) \iff P \wedge (\sim Q)$.

$$\sim [(\forall x \in S)(P(x) \longrightarrow Q(x))] \iff (\exists x \in S)[\sim (P(x) \longrightarrow Q(x))]$$
$$\iff (\exists x \in S)[P(x) \wedge (\sim (Q(x)))].$$

Summarizing, $\sim [(\forall x \in S)[P(x) \longrightarrow Q(x)]] \iff (\exists x \in S)[P(x) \wedge (\sim (Q(x)))]$.

1.4. SETS AND QUANTIFIERS

This example should also be intuitive, despite all of the nested grouping symbols. To say that it is not the case that for all $x \in S$ we have $P(x) \longrightarrow Q(x)$ is to say there exists an x so that we do have $P(x)$ but not the consequent $Q(x)$.

Example 1.4.2

Negate $(\exists x \in S)[P(x) \wedge Q(x)]$.

<u>Solution</u>: Here we use $\sim(P \wedge Q) \iff (\sim P) \vee (\sim Q)$, so we can write

$$\sim[(\exists x \in S)(P(x) \wedge Q(x))] \iff (\forall x)[(\sim P(x)) \vee (\sim(Q(x)))].$$

This last example shows that if it is not the case that there exists an $x \in S$ so that $P(x)$ and $Q(x)$ are both true, that is the same as saying that for all x, either $P(x)$ is false or $Q(x)$ is false.

1.4.6 Negating Statements Containing Mixed Quantifiers

Here we simply apply (1.81) and (1.82) two or more times, as appropriate. For a typical case of a statement first quantified by $\forall$, and then by $\exists$, we note that we can group these as follows:[54]

$$(\forall x \in R)(\exists y \in S)P(x,y) \iff (\forall x \in R)[(\exists y \in S)P(x,y)].$$

(Here "R" is another set, not to be confused with the set of real numbers $\mathbb{R}$.) Thus,

$$\sim[(\forall x \in R)(\exists y \in S)P(x,y)] \iff \sim\{(\forall x \in R)[(\exists y \in S)P(x,y)]\}$$
$$\iff (\exists x \in R)\{\sim[(\exists y \in S)P(x,y)]\}$$
$$\iff (\exists x \in R)(\forall y \in S)[\sim P(x,y)].$$

Ultimately we have, in turn, the $\forall$'s become $\exists$'s, the $\exists$'s become $\forall$'s, the variables are quantified in the same order as before, and finally the statement P is replaced by its negation $\sim P$. The pattern would continue no matter how many universal and existential quantifiers arise. (The uniqueness quantifier is left for the exercises.) To summarize for the case of two quantifiers,

$$\sim[(\forall x \in R)(\exists y \in S)P(x,y)] \iff (\exists x \in R)(\forall y \in S)[\sim P(x,y)] \qquad (1.83)$$
$$\sim[(\exists x \in R)(\forall y \in S)P(x,y)] \iff (\forall x \in R)(\exists y \in S)[\sim P(x,y)]. \qquad (1.84)$$

Example 1.4.3

Consider the following statement, which is false:

$$(\forall x \in \mathbb{R})(\exists y \in \mathbb{R})[xy = 1].$$

---

[54]This may not be the most transparent fact, and indeed there are somewhat deep subtleties involved, but eventually this should be clear. The subtleties lie in the idea that once a variable is quantified, it is *fixed* for that part of the statement that follows it. For instance, that part $(\exists y \in S)P(x,y)$ treats x as if it were "constant."

One could say that the statement says every real number x has a real number reciprocal y. This is false, but before that is explained we compute the negation, which must be true:

$$\sim [(\forall x \in \mathbb{R})(\exists y \in \mathbb{R})(xy = 1)] \iff (\exists x \in \mathbb{R})(\forall y \in \mathbb{R})(xy \neq 1).$$

Indeed, there exists such an x, namely $x = 0$, such that $xy \neq 1$ for all y.[55]

In this we borrowed one of the many convenient mathematical notations for the negations of various symbols. Some common negations follow:

$$\sim (x = y) \iff x \neq y$$
$$\sim (x < y) \iff x \geq y$$
$$\sim (x \leq y) \iff x > y$$
$$\sim (x \in S) \iff x \notin S$$

Of course, we can negate both sides of any one of these and get, for example, $x \in S \iff \sim (x \notin S)$. Reading one of these backwards, we can have $\sim (x \geq y) \iff x < y$.

---

[55] Note that in using English, the quantification often follows after the variable quantified, as in Example 1.4.3. That can become quite confusing when statements get complicated. Indeed, much of the motivation of this section is so that we can use the notation to, in essence, diagram the logic of such statements, and analyze them to see if they may be false (by seeing if their negations ring true).

Exercises

For Exercises 1–6, consider the sets

$$P = \{\text{prisons}\},$$
$$M = \{\text{methods of escape}\}.$$

For Exercises 1–6, write a short English version of the given statement.

1. $(\forall p \in P)(\exists m \in M)[m \text{ will get you out of } p]$

2. $(\exists m \in M)(\forall p \in P)[m \text{ will get you out of } p]$

3. $(\exists p \in P)(\forall m \in M)[m \text{ will get you out of } p]$

4. $(\forall m \in M)(\exists p \in P)[m \text{ will get you out of } p]$

5. $(\exists m \in M)(\exists p \in P)[m \text{ will get you out of } p]$

6. $(\forall m \in M)(\forall p \in P)[m \text{ will get you out of } p]$

7. Write a negation for the statement in Exercise 1 in both symbolic logic and English.

8. Write a negation for the statement in Exercise 2 in both symbolic logic and English.

9. Write a negation for the statement in Exercise 3 in both symbolic logic and English.

10. Write a negation for the statement in Exercise 4 in both symbolic logic and English.

11. Write a negation for the statement in Exercise 5 in both symbolic logic and English.

12. Write a negation for the statement in Exercise 6 in both symbolic logic and English.

For Exercises 13–16, write negations of the given quantified statement. If in the process a logical statement within the quantified statement is negated, write that negation in the clearest possible form. For instance, instead of writing $\sim (P \to Q)$, write $P \wedge (\sim Q)$. Similarly, instead of writing $\sim (x > y)$, write $x \leq y$.

13. $(\forall x \in R)[x \in S]^*$

14. $(\forall x \in R)(\exists y \in S)[y \leq x]$

15. $(\forall x, y \in R)(\exists r, t \in S)[rx + ty = 1]$

    Hint: This can also be written $(\forall x \in R)(\forall y \in R)(\exists r \in S)(\exists t \in S)[rx + ty = 1]$.

16. $(\forall \varepsilon \in \mathbb{R}^+)(\exists \delta \in \mathbb{R}^+)(\forall x \in \mathbb{R})\left[|x - a| < \delta \longrightarrow |f(x) - f(a)| < \varepsilon\right]$

    Hint: Consider $P : |x - a| < \delta$ and $Q : |f(x) - f(a)| < \varepsilon$. Then consider the usual negation of $P \to Q$, with these statements inserted literally, and then rewrite it in a more "understandable" way.

17. For each of the following, write the negation of the statement and decide which is true, the original statement or its negation.

    (a) $(\exists x \in \mathbb{R})(x^2 < 0)$

    (b) $(\forall x \in \mathbb{R})(|-x| \neq x)$

    (c) $(\forall x \in \mathbb{R})(\exists y \in \mathbb{R})(y = 2x + 1)$

18. Consider $(\forall c \in C)(\exists b \in B)(b \text{ would buy } c)$. Here $C = \{\text{cars}\}$ and $B = \{\text{buyers}\}$.

    (a) Write in English what this statement says.

    (b) Write in English what the negation of the statement should be.

    (c) Write in symbolic logic what the negation of the statement should be.

19. Consider the statement $(\forall x, y \in \mathbb{R})(x < y \longrightarrow x^2 < y^2)$.

    (a) Write the negation of this statement.

    (b) In fact it is the negation that is true. Can you explain why?

20. Using the fact that

    $$(\exists! x \in S)P(x) \iff \underbrace{\left[(\exists x \in S)P(x)\right]}_{\text{existence}} \wedge \underbrace{\left[(\forall x, y \in S)[(P(x) \wedge P(y)) \longrightarrow x = y]\right]}_{\text{uniqueness}},$$

    compute the form of the negation of the unique existence:

    $$\sim [(\exists! x \in S)P(x)].$$

1.5 Sets Proper (Optional)

In this section, we introduce set theory in its own right. We also apply the earlier symbolic logic to the theory of sets (rather than vice-versa). We also approach set theory visually and intuitively, while simultaneously introducing all the set-theoretic notation we will use throughout the text. To begin, we make the following definition:

Definition 1.5.1 *A* **set** *is a well-defined collection of objects.*

By well-defined, we mean that once we define the set, the objects contained in the set are totally determined, and so any given object is either in the set or not in the set. We might also note that, in a sense, a set is defined (or *determined*) by its elements; sets that are different collections of elements are different sets, while sets with exactly the same elements are the same set. We can also define equality by means of quantifiers:

Definition 1.5.2 *Given two sets A and B, we defined the statement $A = B$ as being equivalent to the statement $(\forall x)[(x \in A) \longleftrightarrow (x \in B)]$:*

$$A = B \iff (\forall x)[(x \in A) \longleftrightarrow (x \in B)]. \tag{1.85}$$

If we allow ourselves to understand that x is quantified universally (that is, we assume "$(\forall x)$" is understood) unless otherwise stated, we can write, instead of $A = B$, that $x \in A \iff x \in B$.

When we say a set is well–defined we also mean that once defined the set is *fixed*, and does not change. If elements can be listed in a table (finite or otherwise),[56] then the order we list the elements is not relevant; sets are defined by exactly which objects are elements, and which are not. Moreover, it is also irrelevant if objects are listed more than once in the set, such as when we list $\mathbb{Q} = \{x \mid x = p/q, \ p, q \in \mathbb{Z}, \ q \neq 0\}$. In that definition $2 = 2/1 = 4/2 = 6/3$ is "listed" infinitely many times, but it is simply one element of the set of rational numbers $\mathbb{Q}$. While it actually is possible to "list" the elements of $\mathbb{Q}$ if we allow for the ellipsis ($\cdots$), it is more practical to describe the set, as we did, using some *defining property* of its elements (here they were ratios of integers, without dividing by zero), as long as it is exactly those elements in the set—no more and no fewer—that share that property. One usually uses a "dummy variable" such as x and then describes what properties all such x in the set should have. We could have just as easily used z or any other variable.[57]

1.5.1 Subsets and Set Equality

When all the elements of a set A are also elements of another set B, we say A is a *subset* of B. To express this in set notation, we write $A \subseteq B$. In this case, we can also take another

---

[56]Note that not all sets can be listed in a table, even if it is infinitely long. We can list $\mathbb{N} = \{1, 2, 3, \ldots\}$ and $\mathbb{Z} = \{0, 1, -1, 2, -2, 3, -3, \ldots\}$, and even with a little ingenuity list the elements of $\mathbb{Q}$, but we cannot do so with $\mathbb{R}$ or with the set of irrational numbers $\mathbb{R} - \mathbb{Q}$ (see the notation in (1.92), page 68). Those sets whose elements can be listed in a table are called *countable*, and the others *uncountable*. All sets with a finite number of elements are also countable. Of the others, some are countably infinite, and the others uncountably infinite (or simply uncountable, as the "infinite" in "uncountably infinite" is redundant).

[57]Of course we would not use fixed elements of the set as "variables," which they are not since each has a unique identity.

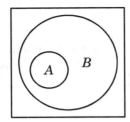 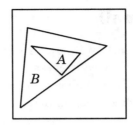 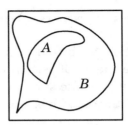

Figure 1.4: Three possible abstract Venn Diagrams illustrating $A \subseteq B$. (Note that in the first figure, for example, B is the set of all elements within the interior of the larger circle.) What is important is that all elements of A are necessarily contained in B as well.

perspective, and say B is a *superset* of A, written $B \supseteq A$. Both symbols represent types of *set inclusions*, i.e., they show one set is contained in another.

A useful graphical device that can illustrate the notion that $A \subseteq B$ and other set relations is the *Venn Diagram*, as in Figure 1.4. There we see a visual representation of what it means for $A \subseteq B$. The sets are represented by enclosed areas in which we imagine the elements reside. In each representation given in Figure 1.4, all the elements inside A are also inside B.

Using symbolic logic, we can *define* subsets, and the notation, as follows:

$$A \subseteq B \iff (\forall x)(x \in A \longrightarrow x \in B). \tag{1.86}$$

The role of the implication, which is the main feature of (1.86), should seem intuitive. Perhaps less intuitive are some of the statements that are therefore logically equivalent to (1.86):

$$\begin{aligned} A \subseteq B &\iff (\forall x)(x \in A \longrightarrow x \in B) \\ &\iff (\forall x)[(\sim (x \in A)) \vee (x \in B)] \\ &\iff (\forall x)[(x \notin A) \vee (x \in B)], \end{aligned}$$

which uses the fact that $P \longrightarrow Q \iff (\sim P) \vee Q$, and

$$\begin{aligned} A \subseteq B &\iff (\forall x)[(\sim (x \in B)) \longrightarrow (\sim (x \in A))] \\ &\iff (\forall x)[(x \notin B) \longrightarrow (x \notin A)], \end{aligned}$$

which uses the contrapositive $P \longrightarrow Q \iff (\sim Q) \longrightarrow (\sim P)$. Note that we used the shorthand notation $\sim (x \in A) \iff x \notin A$. With the definition (1.86), we can quickly see two more, technically interesting facts about subsets:

Theorem 1.5.1 *For any sets A and B, the following hold true:*

$$A \subseteq A, \quad \text{and} \tag{1.87}$$
$$A = B \iff (A \subseteq B) \wedge (B \subseteq A). \tag{1.88}$$

Now we take a moment to remind ourselves of what is meant by *theorem*:

Definition 1.5.3 A **theorem** *is a statement that we know to be true because we have a proof of it. We can therefore accept it as a tautology.*

1.5. SETS PROPER (OPTIONAL)

A theorem's scope may be very limited (the previous theorem only applies to sets and subsets as we have defined them.) Furthermore, a theorem's scope and "truth" depend on the axiomatic system on which it rests, such as the definitions we gave our symbolic logic symbols (which might not have always been completely obvious to the novice, as in our definitions of "$\vee$" and "$\longrightarrow$"). For another example, there is Euclidean Geometry, the theorems of which rest upon Euclid's postulates (or axioms, or original assumptions), while other geometric systems begin with different postulates.

Nonetheless, once we have the definitions and postulates we can say that a theorem is a statement that is always true (i.e., valid, demonstrated by some form of proof), and in fact therefore equivalent to $\mathcal{T}$ (introduced on page 29). We will use that fact in the proof of (1.87), but for (1.88) we will instead demonstrate the validity of the equivalence ($\Longleftrightarrow$). For the first statement's proof, we have

$$A \subseteq A \iff (\forall x)[(x \in A) \longrightarrow (x \in A)] \iff \mathcal{T}.$$

Note that this proof is based on the fact that $P \longrightarrow P$ is a tautology (i.e., equivalent to $\mathcal{T}$). A glance at a Venn Diagram with a set A can also convince one of this fact, that any set is a subset of itself. For the proof of (1.88) we offer the following:

$$\begin{aligned} A = B &\iff (\forall x)[(x \in A) \longleftrightarrow (x \in B)] \\ &\iff (\forall x)[((x \in A) \longrightarrow (x \in B)) \wedge ((x \in B) \longrightarrow (x \in A))] \\ &\iff [(\forall x)[(x \in A) \longrightarrow (x \in B)] \wedge [(\forall x)[(x \in B) \longrightarrow (x \in A)] \\ &\iff (A \subseteq B) \wedge (B \subseteq A), \text{ q.e.d.}^{58} \end{aligned}$$

A consideration of Venn diagrams also leads one to believe that for all the area in A to be contained in B and vice versa, it must be the case that $A = B$. That $A = B$ implies they are mutual subsets and is perhaps easier to see.

Note that these arguments can also be made with supersets instead of subsets, with $\supseteq$ replacing $\subseteq$ and $\longleftarrow$ replacing $\longrightarrow$.

One needs to be careful with quantifiers and symbolic logic, as we saw in Section 1.4, but in what we did here the $(\forall x)$ effectively "went along for the ride."

Of course, Venn Diagrams can accommodate more than two sets. For example, we can illustrate the chain of set inclusions

$$\mathbb{N} \subseteq \mathbb{Z} \subseteq \mathbb{Q} \subseteq \mathbb{R} \tag{1.89}$$

using a Venn Diagram, as in Figure 1.5. Note that this is a compact way of writing six different set inclusions: $\mathbb{N} \subseteq \mathbb{Z}$, $\mathbb{N} \subseteq \mathbb{Q}$, $\mathbb{N} \subseteq \mathbb{R}$, $\mathbb{Z} \subseteq \mathbb{Q}$, $\mathbb{Z} \subseteq \mathbb{R}$, and $\mathbb{Q} \subseteq \mathbb{R}$.

1.5.2 Intervals and Inequalities in $\mathbb{R}$

The number line, which we will henceforth dub the *real line*, has an inherent order in which the numbers are arranged. Suppose we have two numbers $a, b \in \mathbb{R}$. Then the order relation between a and b has three possibilities, each with its own notation:

---

[58]Latin, *quod erat demonstrandum*), the traditional ending of a proof meaning *that which was to be proved*.

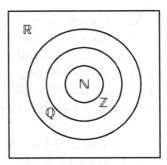

Figure 1.5: Venn Diagram illustrating $\mathbb{N} \subseteq \mathbb{Z} \subseteq \mathbb{Q} \subseteq \mathbb{R}$.

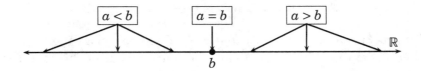

Figure 1.6: For any two real numbers a and b, we have the three cases concerning their relative positions on the real line: $a < b$, $a = b$, $a > b$. Arrows indicate the possible positions of a for the three cases.

1. a is to the left of b, written $a < b$ and spoken "a is *less than* b."

2. a is to the right of b, written $a > b$ and spoken "a is *greater than* b."

3. a is at the same location as b, written $a = b$ and spoken "a equals b."

Figure 1.6 shows these three possibilities. Note that "less than" and "greater than" refer to relative positions on the real line, not how "large" or "small" the numbers are. For instance, $4 < 5$ but $-5 < -4$, though it is natural to consider -5 to be a "larger" number than -4. Similarly $-1000 < 1$. Of course $a < b \iff b > a$. We have further notation that describes when a is left of or at b, and when a is right of or at b:[59]

4. a is at or left of b, written $a \leq b$ and spoken "a is less than or equal to b."

5. a is at or right of b, written $a \geq b$ and spoken "a is greater than or equal to b."

Using inequalities, we can describe *intervals* in $\mathbb{R}$, which are exactly the *connected* subsets of $\mathbb{R}$, meaning those sets that can be represented by darkening the real line at only those points that are in the subset, and where doing so can be theoretically accomplished without lifting our pencils as we darken. In other words, these are "unbroken" subsets of $\mathbb{R}$. Later we will see that intervals are subsets of particular interest in calculus.

---

[59]Later we will refer to a number's *absolute size*, or *absolute value*, in which context we will describe -1000 as "larger" than 1.

1.5. SETS PROPER (OPTIONAL)

Intervals can be classified as finite or infinite (referring to their lengths), and open, closed or half-open (referring to their "endpoints"). The finite intervals are of three types: closed, open and half-open. Intervals of these types, with real *endpoints* a and b, where $a < b$ (though the idea extends to work with $a \leq b$) are shown as follows, respectively by graphical illustration in *interval notation*, and using earlier set-theoretic notation:

open:	⊕———⊕ at a, b	(a,b)	$\{x \in \mathbb{R} \mid a < x < b\}$
closed:	●———● at a, b	$[a,b]$	$\{x \in \mathbb{R} \mid a \leq x \leq b\}$
half-open:	●———⊕ at a, b	$[a,b)$	$\{x \in \mathbb{R} \mid a \leq x < b\}$
half-open:	⊕———● at a, b	$(a,b]$	$\{x \in \mathbb{R} \mid a < x \leq b\}$

Note that $a < x < b$ is short for $(a < x) \wedge (x < b)$, i.e., $(x > a) \wedge (x < b)$. The others are similar.

We will concentrate on the open and closed intervals in calculus. For the finite open interval, we see that we do not include the endpoints a and b in the set, denoting this fact with parentheses in the interval notation and an "open" circle at each endpoint on the graph. What is crucial to calculus is that immediately surrounding any point $x \in (a,b)$ are only other points still inside the interval; if we pick a point x *anywhere* in the interval (a,b), we see that just left and just right of x are only points in the interval. Indeed, we have to travel some distance—albeit possibly short—to leave the interval from a point $x \in (a,b)$. Thus no point inside of (a,b) is on the boundary, and so each point in (a,b) is "safely" in the interior of the interval. This will be crucial to the concepts of continuity, limits and (especially) derivatives later in the text.

For a closed interval $[a,b]$, we *do* include the endpoints a and b, which are not surrounded by other points in the interval. For instance, immediately left of a is outside the interval $[a,b]$, though immediately right of a is in the interior.[60] We denote this fact with brackets in the interval notation, and a "closed" circle at each endpoint when we sketch the graph. Half-open (or half-closed) intervals are simple extensions of these ideas, as illustrated above.

For infinite intervals, we have either one or no endpoints. If there is an endpoint it is either not included in the interval or it is, the former giving an open interval and the latter a closed interval. An open interval that is infinite in one direction will be written (a, ∞) or $(-\infty, a)$, depending upon the direction in which it is infinite. Here ∞ (infinity) means that we can move along the interval to the right "forever," and $-\infty$ means we can move left without end. For infinite closed intervals, the notation is similar: $[a, \infty)$ and $(-\infty, a]$. The whole real line is also considered an interval, which we denote $\mathbb{R} = (-\infty, \infty)$.[61] When an interval

---

[60] For a closed interval $[a,b]$, later we will sometimes refer to the *interior* of the interval, meaning all points whose immediate neighbors left and right are also in the interval. This means that the interior of $[a,b]$ is simply (a,b).

[61] For technical reasons, $\mathbb{R}$ is considered to be both an open and a closed interval. Roughly, it is open because every point is interior, but closed because every point that can be approached as close as we want from the interior is contained in the interval. Those are the topological factors that characterize open and closed intervals as such. Topology as a subject is rarely taught before the junior level of college, or even graduate school, though advanced calculus usually includes some topology of $\mathbb{R}$.

continues *without bound* in a direction, we also darken the arrow in that direction. Thus we have the following:

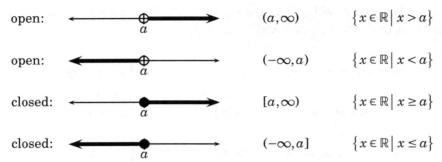

Note that we never use brackets to enclose an infinite "endpoint," since $-\infty, \infty$ are not actual boundaries but rather are concepts of unending continuance. Indeed, $-\infty, \infty \notin \mathbb{R}$, i.e., they are not points on the real line, so they cannot be boundaries of subsets of $\mathbb{R}$; there are no elements "beyond" them.

1.5.3 Most General Venn Diagrams

Before we get to the title of this subsection, we introduce a notion for which we will have occasional use, namely the concept of *proper subset*.

Definition 1.5.4 *If $(A \subseteq B) \wedge (A \neq B)$, we call A a **proper** subset of B, and write $A \subset B$*[62].

Thus $A \subset B$ means A is contained in B, but A is not all of B. Note that $A \subset B \implies A \subseteq B$ (just as $P \wedge Q \implies P$). When we have that A is a subset of B and are not interested in emphasizing whether $A \neq B$ (or are not sure if this is true), we will use the "inclusive" notation $\subseteq$. In fact, the inclusive case is less complicated logically (just as $P \vee Q$ is easier than P XOR Q) and so we will usually opt for it even when we do know that $A \neq B$. We mention the exclusive case here mainly because it is useful in explaining the most general Venn Diagram for two sets A and B.

Of course it is possible to have two sets, A and B, where neither is a subset of the other. Then A and B may share some elements, or no elements. In fact, for any given sets A and B, exactly one of the following will be true:

Case 1: $A = B$

Case 2: $A \subset B$, i.e., A is a proper subset of B

Case 3: $B \subset A$, i.e., B is a proper subset of A

Case 4: A and B share common elements, but neither is a subset of the other

Case 5: A and B have no common elements. In such a case the two sets are said to be *disjoint*

---

[62]The notation has changed over the years. Many current texts use "⊂" the way we use "⊆" here. This is unfortunate, because the notations "⊆, ⊂" here are strongly analogous to the notations ≤, < from arithmetic. One has to take care to know how notation is being used in a given context. (A few authors even use ⊆, ⊊ !)

1.5. SETS PROPER (OPTIONAL)

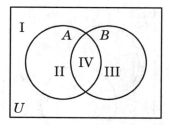

Figure 1.7: Most general Venn diagram for two arbitrary sets A and B. Here U is some superset of both A and B.

Even if we do not know which of the five cases is correct, we can use a single illustration that covers all of these. That illustration is given in Figure 1.7, with the various regions labeled. (We will explain the meaning of U in the next subsection.) To see that this covers all cases, we take them in turn:

case 1: $A = B$: all elements of A and B are in Region IV; there are no elements in Regions II and III.

case 2: $A \subset B$: there are elements in Regions III and IV, and no elements in Region II.

case 3: $B \subset A$: there are elements in Regions II and IV, and no elements in Region III.

case 4: A and B share common elements, but neither is a subset of the other: there are elements in Regions II, III and IV.

case 5: A and B have no common elements: there are no elements in Region IV.

Note that whether Region I has elements is irrelevant in the discussion above, though it will become important shortly.

The most general Venn diagram for three sets is given in Figure 1.8, though we will not exhaustively show this to be the most general. It is not important that the sets are represented by circles, but only that there are sufficiently many separate regions and that every case of an element being, or not being, in A, B and C is represented. Note that there are three sets for an element to be or not to be a member of, and so there are $2^3 = 8$ subregions needed.

1.5.4 Set Operations

When we are given two sets, A and B, it is natural to combine or compare their memberships with each other and the universe of all elements of interest. In particular, we form new sets called the union and intersection of A and B, the difference of A and B (and of B and A), and the complement of A (and of B). The first three are straightforward, but the fourth requires some clarification. Usually A and B contain only objects of a certain class like numbers, colors, etc. Thus we take elements of A and B from a specific *universal set U* of objects rather than an all-encompassing universe of all objects. It is unlikely in mathematics that we would need, for instance, to mix numbers with persons and planets and verbs, so we find

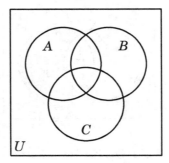

Figure 1.8: The most general Venn Diagram for three sets A, B and C.

it convenient to limit our universe U of considered objects. With that in mind (but without presently defining U), the notations for these new sets are as follows:

Definition 1.5.5

$$A \cup B = \{x \mid (x \in A) \vee (x \in B)\} \tag{1.90}$$

$$A \cap B = \{x \mid (x \in A) \wedge (x \in B)\} \tag{1.91}$$

$$A - B = \{x \mid (x \in A) \wedge (x \notin B)\} \tag{1.92}$$

$$A' = \{x \in U \mid (x \notin A)\}. \tag{1.93}$$

These are read "A union B," "A intersect B," "A minus B," and "A complement," respectively. Note that in the first three, we could have also written $\{x \in U \mid \cdots\}$, but since $A, B \subseteq U$, there it is unnecessary. Also note that one could define the complement in the following way, though (1.93) is more convenient for symbolic logic computations:

$$A' = \{x \mid (x \in U) \wedge (x \notin A)\} = U - A. \tag{1.94}$$

These operations are illustrated by the Venn diagrams of Figure 1.9, where we also construct B' and $B - A$. Note the connection between the logical $\vee$ and $\wedge$, and the set-theoretical $\cup$ and $\cap$.[63]

---

[63]While not as intuitive as the connections between $\cup$ and $\vee$, and $\cap$ and $\wedge$, the set-theoretical "$-$" could be interpreted as "$\wedge \sim \cdots \in$," and if we always assume we know what the universal set is, we can interpret the complement symbol "$'$" as "$\sim \cdots \in$."

1.5. SETS PROPER (OPTIONAL)

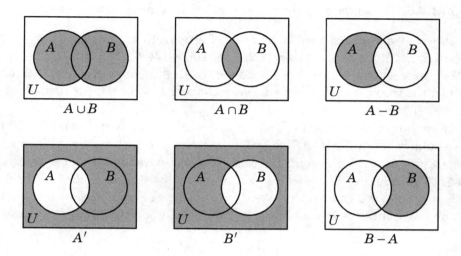

Figure 1.9: Some Venn Diagrams involving two sets A and B inside a universal set U, which is represented by the whole "box."

Example 1.5.1

Find $A \cup B$, $A \cap B$, $A - B$ and $B - A$ if

$$A = \{1, 2, 3, 4, 5, 6, 7\}$$
$$B = \{5, 6, 7, 8, 9, 10\}.$$

Solution: Though not necessary (and often impossible), we will list these set elements in a table from which we can easily compare the membership.

$$\begin{array}{rlllllllllll} A & = \{ & 1, & 2, & 3, & 4, & 5, & 6, & 7, & & & & \}, \\ B & = \{ & & & & & 5, & 6, & 7, & 8, & 9, & 10 & \}. \end{array}$$

Now we can compare the memberships using the operations defined earlier.

$$A \cup B = \{1, 2, 3, 4, 5, 6, 7, 8, 9, 10\},$$
$$A \cap B = \{5, 6, 7\},$$
$$A - B = \{1, 2, 3, 4\},$$
$$B - A = \{8, 9, 10\}.$$

The complements depend upon the identity of the assumed universal set. If in the example we had $U = \mathbb{N}$, then $A' = \{8, 9, 10, 11, \ldots\}$ and $B' = \{1, 2, 3, 4, 11, 12, 13, 14, 15, \ldots\}$. If instead we took $U = \mathbb{Z}$, we have $A' = \{\ldots, -3, -2, -1, 0, 8, 9, 10, 11, \ldots\}$, for instance. (We leave B' to the interested reader.)

Just as it is important to have a zero element in $\mathbb{R}$ for arithmetic and other purposes, it is also useful in set theory to define a set that contains no elements:

Definition 1.5.6 *The set with no elements is called the* **empty set**,[64] *denoted* $\emptyset$.

One reason we need such a device is for cases of intersections of disjoint sets. If $A = \{1, 2, 3\}$ and $B = \{4, 5, 6, 7, 8, 9, 10\}$, then $A \cup B = \{1, 2, 3, \cdots, 10\}$, while $A \cap B = \emptyset$. Notice that regardless of the set A, we will always have $A - A = \emptyset$, $A - \emptyset = A$, $A \cup \emptyset = A$, $A \cap \emptyset = \emptyset$, and $\emptyset \subseteq A$. The last statement is true because, after all, every element of $\emptyset$ is also an element of A.[65] Note also that $\emptyset' = U$ and $U' = \emptyset$.

The set operations for two sets A and B can only give us finitely many combinations of the areas enumerated in Figure 1.7, page 67. In fact, since each such area is either included or not, there are $2^4 = 16$ different diagram shadings possible for the general case, as in Figure 1.7. The situation is more interesting if we have three sets A, B and C. Using Figure 1.8, page 68, we can prove several interesting set equalities. First we have some fairly obvious commutative laws (1.95) and (1.96), and associative laws (1.97) and (1.98):

$$A \cup B = B \cup A \tag{1.95}$$
$$A \cap B = B \cap A \tag{1.96}$$
$$A \cup (B \cup C) = (A \cup B) \cup C \tag{1.97}$$
$$A \cap (B \cap C) = (A \cap B) \cap C \tag{1.98}$$

Next are the following two *distributive laws*, which are the set-theory analogs to the logical equivalences (1.27) and (1.28), found on page 25.

$$A \cap (B \cup C) = (A \cap B) \cup (A \cap C), \tag{1.99}$$
$$A \cup (B \cap C) = (A \cup B) \cap (A \cup C). \tag{1.100}$$

Example 1.5.2 _____

We will show how to prove (1.99) using our previous symbolic logic, and then give a visual proof using Venn diagrams. Similar techniques can be used to prove (1.100). For the proof that $A \cap (B \cup C) = (A \cap B) \cup (A \cap C)$, we use definitions, and (1.27), page 25, to get the following:

$$x \in A \cap (B \cup C) \iff (x \in A) \land (x \in B \cup C)$$
$$\iff (x \in A) \land [(x \in B) \lor (x \in C)]$$
$$\iff [(x \in A) \land (x \in B)] \lor [(x \in A) \land (x \in C)]$$
$$\iff [x \in (A \cap B)] \lor [x \in (A \cap C)]$$
$$\iff x \in [(A \cap B) \cup (A \cap C)], \text{ q.e.d.}$$

We proved that $(\forall x)[(x \in A \cap (B \cup C)) \longleftrightarrow (x \in (A \cap B) \cup (A \cap C))]$, which is the definition for the sets in question to be equal. The visual demonstration of $A \cap (B \cup C) = (A \cap B) \cup (A \cap C)$ is given in Figure 1.10, page 71 where we construct both sets of the equality in stages.

To construct the left-hand side of the equation, in the first box we color A, then $B \cup C$ in the second, and finally look at the intersection of these colored areas from the first two boxes to color the appropriate area in the third. This gives us our Venn Diagram for $A \cap (B \cup C)$. To construct the right-hand side of the equation, we color $A \cap B$ and $A \cap C$ in separate boxes, and finally color in the

---

[64] It is also called the *null set*. Some older texts use empty braces $\emptyset = \{\}$.

[65] This is precisely because there are no elements of $\emptyset$; the statement $x \in \emptyset \longrightarrow x \in A$ is vacuously true because $x \in \emptyset$ is false, regardless of x.

1.5. SETS PROPER (OPTIONAL)

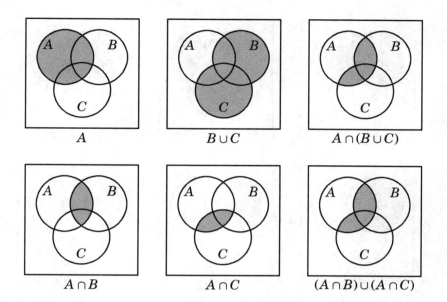

Figure 1.10: Venn Diagrams for Example 1.5.2 verifying one of the distributive laws, specifically $A \cap (B \cup C) = (A \cap B) \cup (A \cap C)$. It is especially important to note how one constructs the third box in each line from the first two.

third box the union of these, which includes all areas colored in either of the previous two boxes. This gives us our Venn Diagram for $(A \cap B) \cup (A \cap C)$. We see that the left- and right-hand sides are the same, and conclude the equality is valid.

The next two are distributive in nature also:

$$A - (B \cup C) = (A - B) \cap (A - C) \qquad (1.101)$$
$$A - (B \cap C) = (A - B) \cup (A - C). \qquad (1.102)$$

Finally, if we replace A with U, we get the set-theoretic version of *De Morgan's Laws*:

$$(B \cup C)' = B' \cap C' \qquad (1.103)$$
$$(B \cap C)' = B' \cup C'. \qquad (1.104)$$

Note that these are very much like our earlier de Morgan's laws, and indeed use the previous versions (1.3) and (1.4), page 19 (also see page 25) in their proofs. For instance, assuming $x \in U$ where U is fixed, we have

$$\begin{aligned} x \in (B \cup C)' &\iff \sim (x \in B \cup C) \\ &\iff \sim ((x \in B) \vee (x \in C)) \\ &\iff [\sim (x \in B)] \wedge [\sim (x \in C)] \\ &\iff [x \in B'] \wedge [x \in C'] \\ &\iff x \in B' \cap C', \text{ q.e.d.} \end{aligned}$$

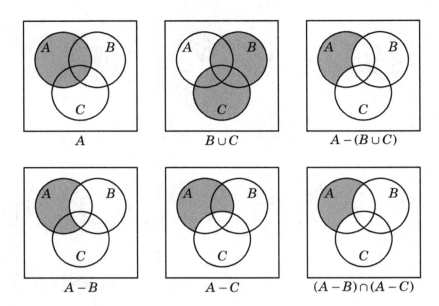

Figure 1.11: Venn Diagrams for Example 1.5.3 verifying that $A-(B\cup C)=(A-B)\cap(A-C)$.

That proves (1.103), and (1.104) has a similar proof. It is interesting to prove these using Venn Diagrams as well (see exercises).

Example 1.5.3

Another example of how to prove these using logic and Venn diagrams is in order. We will prove (1.101) using both methods. First, with symbolic logic:

$$\begin{aligned}
x\in A-(B\cup C) &\iff (x\in A)\wedge[\sim(x\in B\cup C)]\\
&\iff (x\in A)\wedge[\sim((x\in B)\vee(x\in C))]\\
&\iff (x\in A)\wedge[(\sim(x\in B))\wedge(\sim(x\in C))]\\
&\iff (x\in A)\wedge(\sim(x\in B))\wedge(\sim(x\in C))\\
&\iff (x\in A)\wedge(\sim(x\in B))\wedge(x\in A)\wedge(\sim(x\in C))\\
&\iff [(x\in A)\wedge(\sim(x\in B)]\wedge[(x\in A)\wedge(\sim(x\in C))]\\
&\iff (x\in A-B)\wedge(x\in A-C)\\
&\iff x\in(A-B)\cap(A-C),\ \text{q.e.d.}
\end{aligned}$$

If we took the steps in turn, we used the definition of set subtraction (1.92), the definition of union (1.90), one of de Morgan's logic rules (1.19), the associative property of $\wedge$; added a redundant $(x\in A)$; regrouped; used the definition of set subtraction; and finally the definition of intersection.

Now we will see how we can use Venn diagrams to prove (1.101). As before, we will do this by constructing Venn Diagrams for the sets $A-(B\cup C)$ and $(A-B)\cap(A-C)$ separately and verify that we get the same sets. We do this in Figure 1.11. (If it is not visually clear how we proceed from one diagram to the next "all at once," a careful look at each of the $2^3 = 8$ distinct regions can verify the constructions.)

1.5. SETS PROPER (OPTIONAL)

1.5.5 More on Subsets

Before closing this section, a few more remarks should be included on the subject of subsets. Consider for instance the following:

Example 1.5.4 _____

Let $A = \{1, 2\}$. List all subsets of A.

Solution: As $A = \{1, 2\}$ has two elements, it can have subsets that contain zero elements, one element, or two elements. The subsets are thus $\varnothing$, $\{1\}$, $\{2\}$, and $\{1, 2\} = A$.

_____

It is common for novices studying sets to forget that $\varnothing \subseteq A$, and $A \subseteq A$, though by definition,

$$x \in \varnothing \implies x \in A \quad \text{(vacuously)},$$
$$x \in A \implies x \in A \quad \text{(trivially)}.$$

If one wanted only *proper* subsets of A, those would be $\varnothing, \{1\}, \{2\}$ (we omit the set A).

Note that with our set $A = \{1, 2\}$, we can rephrase the question of which subset we might refer to into a question of exactly which elements are in it, from the choices 1 and 2. In other words, given a subset $B \subseteq A$, which (if any) of the following are true: $1 \in B$, $2 \in B$? From these statements, we can construct a truth table-like structure to describe every possible subset of A:

$A = \{1, 2\}$

$1 \in B$	$2 \in B$	subset B
T	T	$\{1,2\} = A$
T	F	$\{1\}$
F	T	$\{2\}$
F	F	$\varnothing$

Similarly, a question about subsets B of $A = \{a, b, c\}$ can be placed in context of a truth table-like construct:

$A = \{a, b, c\}$

$a \in B$	$b \in B$	$c \in B$	subset B
T	T	T	$\{a,b,c\} = A$
T	T	F	$\{a,b\}$
T	F	T	$\{a,c\}$
T	F	F	$\{a\}$
F	T	T	$\{b,c\}$
F	T	F	$\{b\}$
F	F	T	$\{c\}$
F	F	F	$\varnothing$

It would not be too difficult to list the subsets of $A = \{1,2,3\}$ by listing subsets with zero, one, two and three elements separately (i.e., $\varnothing$, $\{a\}$, $\{b\}$, $\{c\}$, $\{a,b\}$, $\{a,c\}$, $\{b,c\}$, $\{a,b,c\}$), but if we were to need to list subsets of a set with significantly more elements, it might be easier to use the lexicographical order embedded in the truth table format to exhaust all the possibilities. The only disadvantage is that the order in which subsets are listed might not be quite as natural as the order we would likely find if we listed subsets with zero, one, two elements, and so on.

Exercises

For Exercises 1–6, prove the set equation using both symbolic logic **and** Venn Diagrams, as in Examples 1.5.2 and 1.5.3.

1. (1.97): $A \cup (B \cup C) = (A \cup B) \cup C$
2. (1.98): $A \cap (B \cap C) = (A \cap B) \cap C$
3. (1.102): $A - (B \cap C) = (A - B) \cup (A - C)$
4. (1.99): $A \cap (B \cup C) = (A \cap B) \cup (A \cap C)$
5. (1.103): $(B \cup C)' = B' \cap C'$
6. (1.104): $(B \cap C)' = B' \cup C'$

For Exercises 7–9, use Venn Diagrams to draw and determine a simpler way of writing the following sets:

7. $A - (B - A) =$
8. $A - (A - B) =$
9. $(A - B) \cap (B - A) =$

Answer each of the following.

10. If $A \subseteq B$, what is $A - B$?
11. If $A \subset B$, what can you say about $B - A$?
12. Referring to Figure 1.9, what are U' and $\varnothing'$?
13. Suppose $A \subseteq B$. How are A' and B' related?
14. Suppose $A - B = A$. What is $A \cap B$?

15. Suppose $A - B = B$. What is B? What is A?
16. Is it possible that $A \subseteq B$ and $B \subseteq A$?
17. The set of *irrational numbers* is the set

$$\mathbb{I} = \{x \in \mathbb{R} \mid x \text{ is not rational}\}.$$

Using previously defined sets and set notation, find a concise definition of $\mathbb{I}$.

Many books define the *symmetric difference* between two sets A and B by

$$A \triangle B = (A - B) \cup (B - A). \qquad (1.105)$$

Exercises 18–23 involve this set operation.

18. Use a Venn Diagram to show that $A \triangle B = (A \cup B) - (A \cap B)$.
19. Is it true that $A \triangle B = B \triangle A$?
20. Use a Venn Diagram to show that $A \triangle (B \triangle C) = (A \triangle B) \triangle C$.
21. Calculate $A \triangle A$, $A \triangle U$, and $A \triangle \varnothing$.
22. If $A \subseteq B$, what is $A \triangle B$? What is $B \triangle A$?
23. Define $A \triangle B$ in a manner similar to the definitions (1.90)–(1.93). That is, replace the dots with a description of x in the following:

$$A \triangle B = \{x \mid \cdots\}.$$

24. Redraw the Venn Diagram of Figure 1.8 and label each of the eight disjoint areas I–VIII. Then use the sets U, A, B, and C, together with the operations $\cup$, $\cap$, $-$ and $'$, to find a definition of each of these sets I–VIII.

25. How many different shading combinations are there for the general Venn Diagram for 3 sets A, B, and C? (See Figure 1.8, page 68.) Speculate about how many combinations there are for 4 sets, 5 sets, and n sets. Test your hypothesis for $n = 0$ and $n = 1$.

26. Show that $A \subseteq B \iff B' \subseteq A'$. (See the discussion immediately following (1.86), page 62.)

A useful concept in set theory is *cardinality* of a set S, which we denote $n(S)$, defined to be the number of elements in a set if the set is finite. Thus $n(\{1,2,3,8,9,10\}) = 6$.

27. Use a Venn Diagram to show the reasonableness of the counting principle that follows:

$$n(S \cup T) = n(S) + n(T) - n(S \cap T). \quad (1.106)$$

28. Using (1.106), show that if $n(S \cup T) = n(S) + n(T)$, then $S \cap T = \varnothing$.

For Exercises 29–33, assume the following:

$$U = \{1, 2, 3, \ldots, 12\},$$
$$A = \{1, 2, 3, 4, 5, 6\},$$
$$B = \{2, 3, 4, 6, 7, 8\},$$
$$C = \{3, 5, 7\},$$
$$D = \{3, 5, 7, 11\}.$$

Find each of the following:

29. $(A - B) \cup (B - A)$

30. $(A - D)'$

31. The number of subsets C

32. The number of subsets of D

33. Use the results from the previous exercises to determine the number of subsets that a set of 5 elements should have.

34. Draw all of the $2^4 = 16$ possible shadings for Figure 1.7, page 67. Then use the sets A, B, and U, together with unions, intersections, complements, and set differences ($\cup, \cap, ', -$) to write a corresponding expression for each of the shaded areas. Note that Figure 1.9 illustrates six of them. Also note that there may be more than one way of representing a set. For example, $A' = U - A$.

Chapter 2

Continuity and Limits of Functions

The concept of continuity is an important first step in the analysis leading to differential and integral calculus. It is also an important analytical tool in its own right, with significant practical applications. Fortunately, the main theorems are intuitive, though their proofs can be technically challenging. Nonetheless, we prove most of the continuity theorems we state, while the remaining theorems we discuss and apply without proof, since they are intuitive and useful but their proofs require background material from at least a junior-level real analysis or topology course.

Limits are sophisticated tools for describing trends in outputs of functions. Initially related to continuity, limits are crucial for calculus, putting it on the same rigorous footing as other mathematical disciplines such as algebra and geometry. In our more modern times it has further conceptual appeal, as it is often only possible to approximate the solution to some problem, even though our method of approximation may be arbitrarily close to the actual solution if given enough computing resources. However, the real value of limits lies in confronting an interesting phenomenon. This is the fact that in mathematics we at times find our analysis (algebraic, geometric, or otherwise) breaking down at exactly the value of some variable where we would like to compute something; that is, we are allowed to let that variable "approach" the desired value as closely as we would like, but it cannot equal that value according to our classical, pre-calculus mathematics. The relatively modern mathematical tool we call "limits" can often break through the analytic barrier at that value, in turn opening us to the extensive and spectacularly useful field we call calculus.

Many examples of continuity and limits in action seem straightforward enough, but without a sufficiently deep understanding it is all too easy for novices to fall victim to common errors. For this reason, we introduce the rather technical definition of continuity here, and develop a method to prove continuity in cases that may seem obvious. Some powerful continuity theorems follow, as do applications. We then employ limits for cases in which continuity is "broken," and in numerous other contexts to make calculus possible, but we also develop limits as analytical tools in their own right.

Understanding limits and continuity sufficiently to avoid common mistakes requires a care and depth of thought which we attempt to foster in this chapter. After continuity, we use a "forms" approach to limits, those forms themselves being ultimately intuitive but nonetheless requiring students to study them carefully and extensively to achieve satisfactory proficiency.[1]

2.1 Definition of Continuity at a Point

The function $f(x)$ is continuous at $x = a$ if and only if we can guarantee $f(x)$ to be close to the value $f(a)$ by restricting x to be close to a. To rephrase, we say $f(x)$ is continuous at $x = a$ if, given any positive tolerance $\varepsilon > 0$ we choose for $f(x)$ as an approximation for $f(a)$, we can then find a positive tolerance $\delta > 0$ for x as an approximation for a so that δ-tolerance in x allows at most ε-tolerance in $f(x)$. The definition is very technical, but through reflection and exposure to examples, one eventually sees that this is exactly what is required.

Definition 2.1.1 *The function $f(x)$ is continuous at the point $x = a$ if and only if*[2]

$$(\forall \varepsilon > 0)(\exists \delta > 0)(\forall x)[\,|x-a| < \delta \longrightarrow |f(x)-f(a)| < \varepsilon\,]. \tag{2.1}$$

This is sometimes called the **epsilon-delta** (ε-δ) definition of continuity. Now let us examine the various parts of the definition.

$$|f(x) - f(a)| \;<\; \varepsilon \tag{2.2}$$
$$|x - a| \;<\; \delta \tag{2.3}$$
$$(\forall x)[\,|x-a| < \delta \;\longrightarrow\; |f(x)-f(a)| < \varepsilon\,] \tag{2.4}$$
$$(\forall \varepsilon > 0) \qquad (\exists \delta > 0) \tag{2.5}$$

(2.2): $f(x)$ will be *within ε of $f(a)$*. In other words, the function value at x will be near in value to the value of the function at a. How near? Less than ε distance away.

(2.3): x is within δ of a. (Otherwise the implication holds true vacuously, but that case is useless. What is important is what occurs when $|x-a| < \delta$.)

(2.4): Having x within δ of a would force $f(x)$ to be within ε of $f(a)$. In other words, allowing x to stray by less than δ from a keeps $f(x)$ within ε of $f(a)$. By controlling x by allowing it a tolerance of less than δ, we control $f(x)$ to have a tolerance of less than ε.

---

[1] Unfortunately, limit computations can be deceiving in that perhaps the majority of problems one first encounters do not require a deep understanding in order to "guess" their correct answers. However, the interesting (and more advanced) cases tend to lie outside of those that are easily guessed, and so computing the correct answers in such interesting cases requires a much deeper understanding. We take the approach here that it is better to heavily analyze the simpler cases, so that the later cases are more easily learned.

[2] Here we assume all numbers and variables ($\varepsilon, \delta, x, a, f(x), f(a)$) are real, and thus abbreviate, for instance, $(\forall x \in \mathbb{R})$ by $(\forall x)$, and $(\forall \varepsilon \in (0, \infty))$ by $(\forall \varepsilon > 0)$. Many texts further abbreviate statements like (2.1) as follows:

$$(\forall \varepsilon > 0)(\exists \delta > 0)[\,|x-a| < \delta \longrightarrow |f(x)-f(a)| < \varepsilon\,],$$

the idea being that the $(\forall x)$ is understood in the unquantified (in x) statement $|x-a| < \delta \longrightarrow |f(x)-f(a)| < \varepsilon$. For a similar example in English, consider the following two statements, usually deemed equivalent:

All Americans have trouble speaking English.

If X is an American, then X has trouble speaking English.

Most see both as false exactly when we can find one American (counterexample) who has no such trouble. In other words, the second statement is as much a "blanket" statement as the first, and is in fact equivalent to the first. We leave the $\forall x$ in (2.1) for precision and to aid in negating the statement, using rules from Section 1.4.

(2.5): Whatever positive value of ε we choose, we can find a δ that satisfies (2.4). In particular, *no matter how small (but positive) we choose $\varepsilon > 0$, we can find a positive δ so that (2.4) is satisfied. We can force $f(x)$ to be as close to $f(a)$ as we like—though we must allow there to be* **some** *positive tolerance ($\varepsilon > 0$) in the output of $f(x)$—and accomplish this by holding x to be within some positive tolerance ($\delta > 0$) of a.*

For a final rephrasing, we have the statement that **we can control the tolerance ε in the output $f(x)$ as much as we would like, so long as $\varepsilon > 0$, by controlling the tolerance δ (which must also be positive[3]) in the input variable x.** This is illustrated graphically in Figure 2.1, page 79, along with an illustration of a case where $f(x)$ is **not** continuous at $x = a$. Note that the negation of our definition would read $(\exists \varepsilon > 0)(\forall \delta > 0)(\exists x)[(|x-a|<\delta) \wedge \sim (|f(x)-f(a)|<\varepsilon)]$, an example of which is given in that figure.

One can also illustrate the implication $|x-a| < \delta \longrightarrow |f(x) - f(a)| < \varepsilon$ using the function diagram on the right. Note that we would expect δ to shrink as we shrink ε, and so when the interval at the bottom shrinks, so might the interval on the top, though neither is allowed to shrink down to a point, or the definition is violated.

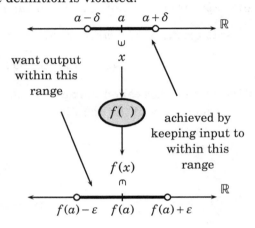

Example 2.1.1

Prove that the function $f(x) = 5x - 9$ is continuous at the point $x = 2$, according to the definition ((2.1)).

Solution: First we notice that $a = 2$ in our definition, and so $f(a) = f(2) = 1$, so we are trying to prove the truth of the statement that

$$(\forall \varepsilon > 0)(\exists \delta > 0)(\forall x)(|x-2| < \delta \longrightarrow |f(x) - 1| < \varepsilon).$$

Before completing this example, we insert the following general strategy for all such proofs.

Strategy for Writing ε-δ Proofs

1. Use the statement $|f(x) - f(a)| < \varepsilon$ to see how $|f(x) - f(a)|$ can be controlled by $|x - a|$ (which is controlled by δ), in particular if $|x - a|$ is a factor of $|f(x) - f(a)|$.

2. If necessary, assume *a priori*[4] that δ is smaller than some fixed positive number to control the

---

[3]Notice that $\delta = 0$ would be worthless for several reasons. First, the implication would be vacuously true and all functions would be continuous everywhere, since $|x - a| < 0$ would never be satisfied. Second, when—in reality—do we ever have a tolerance of zero in a measurement? Finally, as we explore the implications of continuity, we will see that having positive δ is central to the spirit of what follows, particularly with regards to limits; $\delta > 0$ allows for some "wiggle room" for x near $x = a$, and this wiggle room is crucial to the concept of continuity.

[4]Presumptive; before observations. When Step 2 is necessary, we make such suppositions not necessarily based upon observation, but to help focus our search for δ. If continuity is true, (as we will see) we will find that a legitimate δ is still available even with the restriction. In fact, if the limit definition holds for a value of $\delta > 0$, it holds for any smaller positive value δ, so this is not a fatal restriction at all. Note that if $0 < \delta_1 < \delta_2$, then $|x - a| < \delta_1 \implies |x - a| < \delta_2$, so we can always take a smaller value for δ in our proof. The upshot is that *(a priori)* restricting the size of $\delta > 0$ from the start never jeopardizes our ability to prove continuity.

2.1. DEFINITION OF CONTINUITY AT A POINT

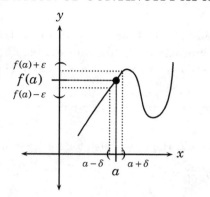
Figure 2.1a: Continuous case.

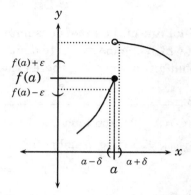
Figure 2.1b: A discontinuity.

Figure 2.1: Figure 2.1a shows a function continuous at $x = a$, illustrating that we can force $f(x)$ to be within any fixed $\varepsilon > 0$ of $f(a)$ by keeping x to within some $\delta > 0$ of a (depending upon the ε). By contrast, in Figure 2.1b we see an $\varepsilon > 0$ for which no positive δ-tolerance in x can force $f(x)$ to be within ε-tolerance of $f(a)$, and so $f(x)$ is not continuous at $x = a$.

other factors contained in $|f(x) - f(a)|$.

3. Find δ as a function of ε, which we could write

$$\delta = \delta(\varepsilon) \tag{2.6}$$

with $\varepsilon \in (0, \infty)$—technically, the *domain* of this function $\delta(\varepsilon)$—and $\delta \in (0, \infty)$, and such that the preliminary (or exploratory) analysis indicates that choice of δ satisfies the definition, i.e., (2.1).

4. Verify that (2.1) holds with this choice of δ, and in doing so write the actual proof.

Steps 1–3 form the preliminary analysis, or "scratch-work," to determine the form of δ. The final Step 4 is the actual proof, though elements of it are often contained in the analytical scratch-work. Let us apply this strategy to the problem at hand.

<u>Scratch-work</u>: We want $|f(x) - f(a)| < \varepsilon$ to follow from our choice of δ. We work backward from that statement, with $f(x) = 5x - 9$, $a = 2$, and $f(a) = f(2) = 1$, to see what choice of δ might achieve this, when $|x - a| < \delta$:

$$\begin{array}{rrcl}
& |f(x) - f(a)| & < & \varepsilon \quad \text{(what we need)} \\
\Longleftrightarrow & |f(x) - 1| & < & \varepsilon \\
\Longleftrightarrow & |5x - 9 - 1| & < & \varepsilon \\
\Longleftrightarrow & |5x - 10| & < & \varepsilon \\
\Longleftrightarrow & 5|x - 2| & < & \varepsilon \\
\Longleftrightarrow & |x - 2| & < & \tfrac{1}{5}\varepsilon \quad \text{(how to get it).}
\end{array}$$

We can see from this that, if we take $\delta = \frac{1}{5}\varepsilon$, then $\delta > 0$ (since $\varepsilon > 0$ is assumed), and the last inequality could be written $|x - 2| < \delta$. Then the bi-implications could be read as implications in the direction $\Longleftarrow$ from that last statement upwards to get $|f(x) - 1| < \varepsilon$. We summarize this in the actual proof.

Proof: For any $\varepsilon > 0$ choose $\delta = \frac{1}{5}\varepsilon$. Then clearly this $\delta > 0$ exists, and moreover satisfies

$$|x - 2| < \delta \Longrightarrow |f(x) - f(2)| = |(5x - 9) - 1|$$
$$= |5x - 10| = \underbrace{5|x - 2| < 5\delta}_{\text{since } |x-2|<\delta} = 5 \cdot \frac{1}{5}\varepsilon = \varepsilon, \text{ q.e.d.}$$

Note that the final line of the proof does imply that $|f(x)-f(2)| < \varepsilon$, with intermediate calculations, some of which one may wish to omit with practice. Furthermore, the general outline of the proof is that given any $\varepsilon > 0$, there exists $\delta > 0$ such that $|x-2| < \delta \implies |f(x)-f(2)| < \varepsilon$.

Example 2.1.2

Show that $f(x) = 2x+3$ is continuous at $x = -5$.

<u>Scratch-work</u>: Here $a = -5$ and $f(a) = f(-5) = -7$. Hence we wish to find $\delta > 0$ so that

$$|x-(-5)| < \delta \implies |f(x)-(-7)| < \varepsilon,$$

which we can rewrite

$$|x+5| < \delta \implies |f(x)+7| < \varepsilon.$$

Again we work backward from the conclusion we wish to justify.

$$\begin{aligned}
& & |f(x)+7| & < \varepsilon \\
\iff & & |2x+3+7| & < \varepsilon \\
\iff & & |2x+10| & < \varepsilon \\
\iff & & 2|x+5| & < \varepsilon \\
\iff & & |x+5| & < \tfrac{1}{2}\varepsilon.
\end{aligned}$$

This time we take $\delta = \tfrac{1}{2}\varepsilon$, and write the proof. (We readily write $|x+5|$ instead of $|x-(-5)|$.)

Proof: For $\varepsilon > 0$, set $\delta = \tfrac{1}{2}\varepsilon$. Then $\delta > 0$ exists and satisfies

$$|x+5| < \delta \implies |f(x)+7| = |2x+3+7| = |2x+10| = 2\underbrace{|x+5|}_{<\delta} < 2\delta = 2 \cdot \frac{\varepsilon}{2} = \varepsilon, \text{ q.e.d.}$$

Proving continuity for first-degree polynomials becomes routine at any $x = a$. The strategy is always the same, with possible complications coming from the signs of the values in question.

Example 2.1.3

Show that $f(x) = 9-4x$ is continuous at $x = 2$.

<u>Scratch-work</u>: Here $a = 2$, $f(a) = f(2) = 1$. We will see that we must be a little more careful here in factoring a negative number within the absolute values, but can use the fact that $|a \cdot b| = |a| \cdot |b|$.

$$\begin{aligned}
& & |f(x)-1| & < \varepsilon \\
\iff & & |9-4x-1| & < \varepsilon \\
\iff & & |-4x+8| & < \varepsilon \\
\iff & & |(-4)(x-2)| & < \varepsilon \\
\iff & & 4|x-2| & < \varepsilon \\
\iff & & |x-2| & < \tfrac{1}{4}\varepsilon.
\end{aligned}$$

Proof: For $\varepsilon > 0$, choose $\delta = \tfrac{1}{4}\varepsilon$. Then $\delta > 0$ exists and

$$|x-2| < \delta \implies |f(x)-1| = |9-4x-1| = |-4x+8|$$

$$= |(-4)(x-2)| = 4|x-2| < 4\delta = 4 \cdot \frac{\varepsilon}{4} = \varepsilon, \text{ q.e.d.}$$

2.1. DEFINITION OF CONTINUITY AT A POINT

Functions that represent lines are the easiest for which to confirm continuity at every point. If $f(x) = mx + b$, where $m \neq 0$, it is clear from the geometric meaning of slope m that a variation (absolute value of "rise") of less than ε in height $f(x)$ can be achieved by allowing a variation (absolute value of "run") of less than $\frac{1}{|m|}\varepsilon$ in x. Thus for a given $\varepsilon > 0$, clearly $\delta = \frac{1}{|m|}\varepsilon$ is the largest δ which satisfies the definition of continuity for such a function $f(x)$. (See our three previous, "linear" examples, and compare their slopes with our choices of δ.)

Example 2.1.4

Show that $f(x) = x^2$ is continuous at $x = 0$.

<u>Scratch-work</u>: Here $a = 0$ and $f(a) = f(0) = 0$. We therefore want to choose $\delta > 0$ such that
$$|x - 0| < \delta \implies |f(x) - 0| < \varepsilon, \quad \text{i.e.,}$$
$$|x| < \delta \implies |x|^2 < \varepsilon.$$

Again we begin with the inequality we would like to result, and see how we might achieve it. Here
$$|x|^2 < \varepsilon \iff |x| < \sqrt{\varepsilon}.$$

We have $\iff$ here because we are dealing with only positive quantities (recall that $\sqrt{\;\;}$ is an increasing function on $[0,\infty)$, i.e., $(\forall a, b \in [0,\infty))(a < b \iff \sqrt{a} < \sqrt{b})$). We can thus take $\delta = \sqrt{\varepsilon}$.

Proof: For $\varepsilon > 0$, set $\delta = \sqrt{\varepsilon}$. Then $\delta > 0$ exists and satisfies
$$|x - 0| < \delta \implies |f(x) - f(0)| = |x^2| = |x|^2 < \delta^2 = \left(\sqrt{\varepsilon}\right)^2 = \varepsilon, \quad \text{q.e.d.}$$

It should be clear that we could easily modify this example to show that $f(x) = x^n$ is continuous at $x = 0$ for $n \in \mathbb{N}$. From there it is not hard to show $f(x) = x^{m/n}$ is also continuous at $x = 0$, as long as n is odd and $m, n \in \mathbb{N}$.[5] Once we stray from $x = 0$, we begin to have more difficulties, as illustrated in the next example.

Example 2.1.5

Show that $f(x) = x^2$ is continuous at $x = 4$.

<u>Scratch-work</u>: This has a complication that the previous problem did not. To see this, we first attempt to proceed as in Example 2.1.4. Here $a = 4$ and $f(a) = 16$, so we wish to find $\delta > 0$ such that
$$|x - 4| < \delta \implies |f(x) - 16| < \varepsilon.$$

Working backward as before we get
$$|f(x) - 16| < \varepsilon \iff |x^2 - 16| < \varepsilon$$
$$\iff |x + 4| \cdot |x - 4| < \varepsilon.$$

We would like to be able to divide both sides by $|x + 4|$, except that **it is not constant**. Here Step 2 in our strategy comes into play. We will control the $|x + 4|$ term by assuming *a priori* that (in all cases)

---
[5]It is false if n is even, assuming m/n is a reduced fraction. The trouble there is that $x^{m/n} = \left(\sqrt[n]{x}\right)^m$ is undefined for $x < 0$, so the second part of $(|x - 0| < \delta) \longrightarrow (|f(x) - f(0)| < \varepsilon)$ is false for such values of x.

$\delta \leq 1$.[6]
$$(|x-4|<\delta) \wedge (\delta \leq 1) \implies |x-4|<1 \implies -1 < x-4 < 1 \implies 3 < x < 5.$$

Now we add 4 to both sides of this last inequality to get

$$(|x-4|<\delta) \wedge (\delta \leq 1) \implies 7 < x+4 < 9.$$

With $x+4$ between 7 and 9, its absolute size is strictly bounded by the number with the largest absolute value, 9, i.e.,

$$(|x-4|<\delta) \wedge (\delta \leq 1) \implies 7 < x+4 < 9 \implies |x+4| < 9.$$

Continuing the scratch-work, we would get

$$(|x-4|<\delta) \wedge (\delta \leq 1) \implies |f(x)-16| = |x+4||x-4| \leq 9|x-4| < 9\delta,$$

(where we only had $\leq$ in our inequality $|x+4||x-4| \leq 9|x-4|$ because of the case $x=4$, where we have the equality $0=0$) and now this looks similar to the earlier examples. To achieve $|f(x)-16|<\varepsilon$ it would seem at first glance to be sufficient to have $9\delta \leq \varepsilon$, which is to say $\delta \leq \frac{1}{9}\varepsilon$.

However, picking $\delta = \frac{1}{9}\varepsilon$ **is not quite enough**, since we assumed $\delta \leq 1$ (to get our estimate $|x+4|<9$ that led to our requirement that $\delta \leq \varepsilon/9$), and this would be false if $\varepsilon > 9$. To ensure both requirements are met for δ for every given ε (as the definition requires), we choose $\delta = \min\{1, \frac{1}{9}\varepsilon\}$, i.e., the *minimum* of the two positive numbers $1, \varepsilon/9$, which will still be positive. Now we write the proof.

Proof: For $\varepsilon > 0$, choose $\delta = \min\{1, \frac{1}{9}\varepsilon\}$. Then $\delta > 0$ exists and satisfies

$$|x-4|<\delta \implies |f(x)-16| = \underbrace{|x+4|}_{<9} \cdot |x-4| \leq 9|x-4| < 9\delta \leq 9 \cdot \frac{1}{9}\varepsilon = \varepsilon, \text{ q.e.d.}$$

To be completely correct, we wrote $|x+4|\cdot|x-4| \leq 9|x-4|$ even though we know $|x+4|<9$, because in the case $x=4$ this becomes $0=0$, while for all other cases we can replace $\leq$ with $<$. Even for the case $x=4$, the next inequality, $9|x-4|<9\delta$, is still correct since $\delta > 0$.

An experienced reader of mathematical proofs, perhaps with minimal extra work, would be able to make sense of the proof without the note about the $|x+4|$ term, but for our purposes we leave the note, and mention that much of the explanation of the proof can be also found in the scratch-work. A key observation is that since $\delta = \min\{1, \frac{1}{9}\varepsilon\}$, we have both $\delta \leq 1$ and $\delta \leq \frac{1}{9}\varepsilon$. Some "steps" in our proof rely on related implications of these in turn, namely $(|x-4|<\delta) \wedge (\delta \leq 1) \implies |x+4| < 9$, which is believable on its face (as it is not hard to see that $|x-4|<1 \implies x \in (3,5) \implies x+4 \in (7,9) \implies |x+4|<9$), and $\delta \leq \frac{1}{9}\varepsilon \implies 9\delta \leq 9 \cdot \frac{1}{9}\varepsilon = \varepsilon$. As in previous examples, it is useful that in the statement of our proof we can see the outline of the definition of continuity (2.1), page 77. Also important to note is that while we could have made a different choice for δ, the obvious possibilities would still require δ to be defined as a similar type of minimum (depending upon the *a priori* restriction of the form $\delta \leq M$).

The next example requires not only some *a priori* bound on δ, but also the triangle inequality $|a+b| \leq |a|+|b|$, or in that case, $|a+b+c| \leq |a|+|b|+|c|$.

---

[6]Here we chose $\delta \leq 1$, but we could have chosen any positive number for the maximum we allow δ to be. We only needed to restrict δ (while keeping it positive) to control the other factors of $|f(x)-16|$.

2.1. DEFINITION OF CONTINUITY AT A POINT

Example 2.1.6

Show that $f(x) = -x^3$ is continuous at $x = -2$.

<u>Scratch-work</u>: Here $a = -2$, and $f(a) = f(-2) = 8$, so we hope for every $\varepsilon > 0$ to find $\delta > 0$ such that $|x - (-2)| < \delta \implies |f(x) - 8| < \varepsilon$, i.e., $|x + 2| < \delta \implies |f(x) - 8| < \varepsilon$. Working backward as before, we see (using $a^3 + b^3 = (a + b)(a^2 - ab + b^2)$) that

$$\begin{aligned} |f(x) - f(2)| < \varepsilon &\iff \left|-x^3 - 8\right| < \varepsilon \\ &\iff \left|(-1)(x^3 + 8)\right| < \varepsilon \\ &\iff 1 \cdot \left|x^3 + 8\right| < \varepsilon \\ &\iff \underbrace{|x + 2|}_{<\delta} \cdot \left|x^2 - 2x + 4\right| < \varepsilon. \end{aligned}$$

Here we will make a rather crude estimate on the size of $|x^2 - 2x + 4|$, eventually using an extended form of the triangle inequality, namely $|a + b + c| \leq |a| + |b| + |c|$, but to do so will require as before an *a priori* bound on δ. So for simplicity we will again assume in all cases that $\delta \leq 1$ for what follows (and incorporate that restriction into our proof as well).[7]

$$(|x + 2| < \delta) \wedge (\delta \leq 1) \implies |x + 2| \leq 1 \iff -1 < x + 2 < 1 \iff -3 < x < -1$$
$$\implies |x| < 3,$$

from which we can summarize (in the first line below) and extend using the triangle inequality (in the second) to get

$$\begin{aligned} (|x + 2| < \delta) \wedge (\delta \leq 1) &\implies |x| < 3 \\ &\implies |x^2 - 2x + 4| \leq |x^2| + |-2x| + |4| = |x|^2 + 2|x| + 4 \\ & < 3^2 + 2(3) + 4 = 19, \end{aligned}$$

and so $\quad (|x + 2| < \delta) \wedge (\delta \leq 1) \implies |x^2 - 2x + 4| \leq 19.$

So far, under the assumptions that $|x + 2| < \delta$ and $\delta \leq 1$ we have (slightly rearranged)

$$|f(x) - 8| = |x^2 - 2x + 4| \cdot |x + 2| \leq 19|x + 2| < \underbrace{19\delta}_{\text{want } \leq \varepsilon}.$$

If we *also* have $\delta \leq \varepsilon/19$, then we will have $|f(x) - 8| < 19\frac{\varepsilon}{19} \leq \varepsilon$, and our proof would be complete. However we must remember that this was possible because we already assumed $\delta \leq 1$. We accomplish both of these equally important restrictions by setting $\delta = \min\left\{1, \frac{\varepsilon}{19}\right\}$. (Again, note that this implies δ exists, $\delta > 0$, $\delta \leq 1$, and $\delta \leq \frac{\varepsilon}{19}$.) The inequalities involved in the proof (after the " $\implies$ ") are implied by, respectively, the triangle inequality, that $\delta \leq 1$, that $|x + 2| < \delta$, and finally that $\delta \leq \frac{\varepsilon}{19}$.

Proof: Let $\varepsilon > 0$, and set $\delta = \min\left\{1, \frac{\varepsilon}{19}\right\}$. Then $\delta > 0$ exists and

$$\begin{aligned} |x + 2| < \delta \implies |f(x) - f(-2)| &= |x^2 - 2x + 4| \cdot |x + 2| \leq \left(|x|^2 + 2|x| + 4\right)|x + 2| \\ &\leq (3^2 + 2(3) + 4)|x + 2| \\ &= 19|x + 2| < 19\delta \leq 19\frac{\varepsilon}{19} = \varepsilon, \quad \text{q.e.d.} \end{aligned}$$

---

[7]In mathematical analysis, the term *estimate* often refers to a bound on the size of a quantity. For instance, $|x| < 100$ means $-100 < x < 100$, giving a lower and upper bound for x. In common usage, the word *estimate* often refers instead to what mathematicians and other scientists would call *approximation*.

The next example requires a somewhat different bit of cleverness, but also an *a priori* restriction on δ.

Example 2.1.7

Show that $f(x) = \dfrac{1}{x}$ is continuous at $x = 5$.

<u>Scratch-work</u>: Here $a = 5$ and $f(a) = f(5) = \frac{1}{5}$. Again we work backward:

$$\left|f(x) - \frac{1}{5}\right| < \varepsilon \iff \left|\frac{1}{x} - \frac{1}{5}\right| < \varepsilon$$
$$\iff \left|\frac{5-x}{5x}\right| < \varepsilon$$
$$\iff \frac{1}{5} \cdot \frac{1}{|x|} \cdot |x - 5| < \varepsilon.$$

As before, the $|x-5|$ will be controlled by δ, but we need to also use δ to control the factor $\frac{1}{|x|}$. Again we will assume *a priori* that $\delta \leq 1$.

$$|x - 5| < \delta \implies |x - 5| < 1 \iff -1 < x - 5 < 1 \iff 4 < x < 6 \iff 4 < |x| < 6.$$

With these bounds on $|x|$, we also get bounds on $\frac{1}{|x|}$ (noting that if $z_1, z_2 > 0$ and $z_1 < z_2$ then $1/z_1 > 1/z_2$, which we note by intuition or by dividing each side of $z_1 < z_2$ by $z_1 z_2$ and rearranging terms):

$$|x - 5| < 1 \implies 4 < |x| < 6 \iff \frac{1}{6} < \frac{1}{|x|} < \frac{1}{4}.$$

With these assumptions $\delta \leq 1$, $|x - 5| < \delta$, we have (continuing from before, taking care for the $x = 5$ case), we have

$$\left|f(x) - \frac{1}{5}\right| = \cdots = \frac{1}{5} \cdot \frac{1}{|x|} \cdot |x - 5| \leq \frac{1}{5} \cdot \frac{1}{4} |x - 5| < \frac{1}{20} \delta.$$

So $\left|f(x) - \frac{1}{5}\right|$ is less than ε if we also have $\frac{1}{20}\delta \leq \varepsilon$, i.e., if $\delta \leq 20\varepsilon$. But recall that this analysis also assumed $\delta \leq 1$, so we take

$$\delta = \min\{1, 20\varepsilon\}.$$

Now we state the proof.

Proof: For $\varepsilon > 0$, choose $\delta = \min\{1, 20\varepsilon\}$. Then $\delta > 0$ exists and satisfies

$$|x - 5| < \delta \implies |f(x) - f(5)| = \left|\frac{1}{x} - \frac{1}{5}\right| = \left|\frac{5-x}{5x}\right|$$
$$= \frac{1}{5} \cdot \frac{1}{|x|} \cdot |x - 5| < \frac{1}{5} \cdot \frac{1}{4} \cdot \delta = \frac{1}{20}\delta \leq \frac{1}{20} \cdot 20\varepsilon = \varepsilon, \quad \text{q.e.d.}$$

We should note here that if we had chosen $a = 0.5$, then we could not use 1 as the upper bound for δ, since the function is undefined at a point within 1 of 0.5 (namely at $x = 0$). For such an a we should instead assume *a priori* that, say, $\delta < 0.25$, or a similar number to be sure to avoid any problems with the definition of the function for any values of x in which $|x - a| < \delta$. Indeed, we would wish to be sure that $f(x)$ is defined within the "wiggle room" allowed by δ in the continuity definition (2.1), page 77, at the start of this section.

Example 2.1.8

Show that $f(x) = x^4$ is continuous at $x = -2$.

Scratch-work: Here $a = -2$ and $f(a) = 16$. Again we will attempt to work backward.

$$|f(x) - f(-2)| < \varepsilon$$
$$\iff |x^4 - 16| < \varepsilon$$
$$\iff |x^2 + 4| \cdot |x - 2| \cdot |x + 2| < \varepsilon.$$

Now our "$|x - a|$", namely $|x + 2|$, is controlled by δ, so we need to control the other two factors. Again let us assume that $\delta \le 1$. Then

$$|x + 2| < \delta \implies |x + 2| < 1 \iff -1 < x + 2 < 1 \iff -3 < x < -1 \implies |x| < 3.$$

We note for later reference that $|x + 2| < 1 \implies |x| < 3$.

For the $|x - 2|$ factor we can subtract 2 from $-3 < x < -1$ to get $|x + 2| < 1 \implies -5 < x - 2 < -3 \implies |x - 2| < 5$.[8] For the $|x^2 + 4|$ term, we have $x^2 + 4 > 0$, and so absolute values are redundant, and thus

$$\delta \le 1 \implies |x^2 + 4| = x^2 + 4 = |x|^2 + 4 < (3)^2 + 4 = 13.$$

(Note that this was because $|x| < 3$.) So far we have

$$\delta \le 1 \implies |f(x) - f(-2)| = (x^2 + 4)|x - 2| \cdot |x + 2| < \underbrace{13 \cdot 5 \cdot \delta}_{\text{want} \le \varepsilon}.$$

Taking $\delta = \min\left\{\frac{\varepsilon}{65}, 1\right\}$ should give us a proof.

Proof: Let $\varepsilon > 0$ and choose $\delta = \min\left\{\frac{\varepsilon}{65}, 1\right\}$. Then $\delta > 0$ exists and

$$|x - (-2)| < \delta \implies |f(x) - f(-2)| = |x^4 - 16| = (x^2 + 4)|x - 2| \cdot |x + 2|$$
$$= (x^2 + 4)|x - 2| \cdot |x - (-2)|$$
$$< 13 \cdot 5 \cdot \delta \le 65 \cdot \frac{\varepsilon}{65} = \varepsilon, \qquad \text{q.e.d.}$$

For our final example, we look at a case where we introduce a new factor in our computation to extract the $|x - a|$ factor. In this case we use

$$\left(\sqrt{a} - \sqrt{b}\right)\left(\sqrt{a} + \sqrt{b}\right) = a - b,$$

useful particularly when $a - b$ is a desired factor.

---

[8] We could also use the triangle inequality to get $|x - 2| \le |x| + |-2| < 3 + 2 = 5$. This happens to give the same bound, but in more complicated cases that might not happen. Either bound would then work. For an engineering application, one would likely prefer whatever gives the larger δ, which indicates less sensitivity to tolerance in x to achieve ε-tolerance in $f(x)$.

Example 2.1.9

Show that $f(x) = \sqrt{x}$ is continuous at $x = 4$.

Scratch-work: As always, we begin with $|f(x) - f(a)| < \varepsilon$ and work backward to find a case where this is implied by $|x - a| < \delta$ for a strategically chosen $\delta > 0$.

$$|f(x) - f(4)| < \varepsilon \iff |\sqrt{x} - 2| < \varepsilon$$
$$\iff \left|\frac{\sqrt{x} - 2}{\sqrt{x} + 2} \cdot (\sqrt{x} + 2)\right| < \varepsilon$$
$$\iff \frac{1}{\sqrt{x} + 2} \underbrace{|x - 4|}_{<\delta} < \varepsilon.$$

Note that the fraction $1/(\sqrt{x} + 2)$ is (1) positive wherever it is defined, which is where $x \geq 0$, and therefore (2) maximized when the denominator is minimized, which will happen when the square root term is minimized, i.e., when x itself is minimized, but nonnegative, lest $\sqrt{x}$ be undefined. For this case, we will assume *a priori* that $\delta \leq 4$, so that x is in the domain of f—namely $[0, \infty)$—and so

$$|x - 4| < \delta \implies |x - 4| < 4 \implies x \in (4 - 4, 4 + 4) = (0, 8)$$
$$\implies \frac{1}{\sqrt{x} + 2} \in \left(\frac{1}{\sqrt{8} + 2}, \frac{1}{2}\right)$$
$$\implies \frac{1}{\sqrt{x} + 2} < \frac{1}{2}.$$

Using the previous computation we can say that

$$(|x - 4| < \delta) \wedge (\delta \leq 4) \implies |f(x) - f(4)| = \frac{1}{\sqrt{x} + 2}|x - 4| \leq \frac{1}{2}|x - 4| < \frac{1}{2}\delta$$

(the $\leq$ to accommodate the $x = 4$ case), and so we can accomplish having $|f(x) - f(4)|$ be less than ε if we also have $\delta \leq 2\varepsilon$. We will therefore take $\delta = \min\{4, 2\varepsilon\}$ in our proof.

Proof: For $\varepsilon > 0$, let $\delta = \min\{4, 2\varepsilon\}$, so $\delta > 0$ exists and

$$|x - 4| < \delta \implies |f(x) - f(4)| = |\sqrt{x} - 2| = \frac{1}{\sqrt{x} + 2}|x - 4|$$
$$\leq \frac{1}{2}|x - 4| < \frac{1}{2}\delta \leq \frac{1}{2} \cdot 2\varepsilon = \varepsilon, \text{ q.e.d.}$$

While this section discusses the technical definition of continuity at a point $x = a$, the reader is invited to consider where else this idea of continuity may apply. Anytime we wish to control the output of some process, the presence of continuity with respect to the input would mean that we can in principle guarantee that we could keep change (or tolerance) in the output small (of size less than ε, whatever we choose that to be as long as it is positive) by controlling the changes in the input (keeping them within δ, also positive). For instance, if an audio amplifier's gain changes continuously with the position of its volume control, it is much easier to make minor changes in gain by carefully moving the volume control. If the volume control's effect is not always continuous (as in the case with old or dirty internal contacts, or "stepper" controls with only finitely many choices), it becomes much more difficult or

2.1. DEFINITION OF CONTINUITY AT A POINT

impossible to control the gain "continuously." There are numerous other practical examples where continuity is desirable, but for now we leave it to the reader's imagination.

However, the proofs are fairly technical, and the exercises somewhat difficult. Nonetheless, the reader should be encouraged that the study of subsequent sections will benefit to the extent time is spent studying this section, even if it is not mastered immediately. Furthermore, this section can and should be revisited after study of future sections, so it can be seen in the larger context and with the benefit of further practice. Of course this is true of all sections in this or any other textbook whose subject is the least bit challenging.

Exercises[9]

For Exercises 1–15, prove that $f(x)$ is continuous at the given value of x by the ε-δ definition.

1. $f(x) = 9x - 11$ is continuous at $x = 2$.
2. $f(x) = 9x - 11$ is continuous at $x = -2$.
3. $f(x) = 3x + 1$ is continuous at $x = 5$.
4. $f(x) = 6 - 2x$ is continuous at $x = -8$.
5. $f(x) = mx + b$ is continuous at every $x = a$. Assume $m \neq 0$. (See Exercise 11 for the case $m = 0$.)
6. $f(x) = x^3$ is continuous at $x = 0$.
7. $f(x) = x^2$ is continuous at $x = 9$.
8. $f(x) = x^2$ is continuous at $x = -3$.
9. $f(x) = 5x^2 - 3$ is continuous at $x = 2$.
10. $f(x) = \frac{1}{x^2}$ is continuous at $x = 5$.
11. $f(x) = b$ (i.e., a line with slope zero) is continuous at every point $x = a$. (Hint: Choosing any $\delta > 0$ will work for the continuity definition.)
12. $f(x) = x^3$ is continuous at $x = 1$.
13. $f(x) = x^3$ is continuous at $x = -3$.
14. $f(x) = \sqrt{x}$ is continuous at $x = 9$.
15. $f(x) = \sqrt{x}$ is continuous at any $a > 0$.
16. Show that $f(x) = \frac{1}{x}$ is continuous at any $a \neq 0$. (Letting $\delta \leq 1$ as in Example 2.1.7 works fine until $0 < |a| \leq 1$. For this more general case desired here, one approach is to assume a pri-

---

[9]In all fairness, it should be pointed out to the reader that some of the exercises here are likely to be quite difficult, particularly for beginning calculus students. This is because the proofs are very involved and use a large variety of methods. Furthermore, reading the proofs of the examples is very different from producing one's own proofs from scratch.

With these exercises, students are thus advised to adopt the following general approach:

(a) Attempt as many problems as possible, looking back on earlier examples for ideas;

(b) Move on to the rest of the text even if few problems are completed on the first attempt; but also

(c) Revisit this section and its problems from time to time to attempt complete proofs for any results that were not finished previously.

With further calculus experience, the ideas and techniques should become clearer, just as the inner workings of an automobile likely make more sense—and seem more important—as one gains experience from actually driving.

ori that $\delta \leq \frac{1}{2}|a|$, so that

$$|x-a| < \delta \implies |x-a| < \frac{1}{2}|a|$$
$$\implies |x| > \frac{1}{2}|a| \implies \frac{1}{|x|} < \frac{2}{|a|}.$$

There is some detail to showing the second implication. From this a δ can be chosen, using some minimum of two values, for each $\varepsilon > 0$.)

17. Consider $f(x) = \sqrt[3]{x}$.

    (a) Show that $f(x)$ is continuous at $x = 0$.

    (b) Show that $f(x)$ is continuous at $x = 8$.

18. Recall that $x^n - a^n = (x-a)(x^{n-1} + ax^{n-2} + a^2 x^{n-3} + \cdots + a^{n-1})$. Use this to show that $f(x) = \sqrt[n]{x}$ is continuous for all $x > 0$ if $n \in \mathbb{N}$ is even, and continuous for all $x \in \mathbb{R}$ if $n \in \mathbb{N}$ is odd. (For the latter you should do the case $x = 0$ separately.) Notice why the even case does not allow $x = 0$ or $x < 0$. <u>Hint</u>: The triangle inequality is needed, as is the fact that $\sqrt[n]{|x_1|} < \sqrt[n]{|x_2|}$ if $|x_1| < |x_2|$.

19. If $f(a)$ is undefined, can $f(x)$ be continuous at $x = a$? Refer to the definition of continuity to explain your answer.

20. If the domain of f is $[a,b]$ only, can $f(x)$ be continuous at $x = a$? Refer to the definition of continuity to explain your answer.

2.2 Continuity Theorems

Though fundamental and technically important, using the ε-δ definition to locate each point—or even a single point—where a function is continuous can be an unwieldy approach. Fortunately, there are many very general theorems that allow us to see where a function is continuous by simple inspection. The theorems are of great use, collectively and individually, for doing just that. Here we list and prove these theorems, and later in the section we will show how to make use of them. The proofs are interesting and introduce some new techniques, but are not crucial for most of what we do. We include them here for completeness. *The reader is advised to first concentrate on the theorems; the proofs are interesting and arguably important for full understanding, but are secondary to the results of the theorems.*

2.2.1 Basic Theorems

For each theorem, it is useful to reflect upon some intuitions regarding what makes a function $f(x)$ continuous at a point $x = a$. While ultimately all continuity refers to the technical definition found in (2.1), page 77, namely

$$f(x) \text{ is continuous at } x = a \iff (\forall \varepsilon > 0)(\exists \delta > 0)(\forall x)(|x - a| < \delta \longrightarrow |f(x) - f(a)| < \varepsilon),$$

a seemingly less precise interpretation can nonetheless be of some use, namely that if $f(x)$ is continuous at $x = a$, we expect that:

> *if the input x of the function $f(\ )$ is changed very little from the value at $x = a$, then the output $f(x)$ will change very little from the value $f(a)$; sufficiently small (but nonzero) changes from $x = a$ in the value of the input of $f(\ )$ will result in only small changes (possibly zero) from $f(a)$ in the output of $f(x)$. (This is not true if $f(\ )$ is not continuous at $x = a$. See Figure 2.1, page 79.)*

It is the ε-δ definition that makes this notion precise, that we can control the size of the change (less than ε, where $\varepsilon > 0$) in the output by controlling the size of the change (less than δ, for some $\delta > 0$ relying on ε) in the input.

The theorems extend this to many combinations of functions. For instance, if $f(x)$ and $g(x)$ are both continuous at $x = a$, then so is $f(x) + g(x)$; if $f(x)$ and $g(x)$ change very little if x changes a small amount from the value $x = a$, it is natural to expect the sum $f(x) + g(x)$ to change little as well. That is the content of this first theorem.

Theorem 2.2.1 *Suppose that $f(x)$ and $g(x)$ are continuous at $x = a$. Then so is $f(x) + g(x)$.*

> **Proof:** We are beginning under the assumption that $f(x)$ and $g(x)$ are each (separately) continuous at $x = a$. This means that
>
> $$(\forall \varepsilon_1 > 0)(\exists \delta_1 > 0)(\forall x)(|x - a| < \delta_1 \longrightarrow |f(x) - f(a)| < \varepsilon_1),$$
> $$(\forall \varepsilon_2 > 0)(\exists \delta_2 > 0)(\forall x)(|x - a| < \delta_2 \longrightarrow |g(x) - g(a)| < \varepsilon_2).$$
>
> Define $h(x) = f(x) + g(x)$. We want to show that $h(x)$ is also continuous at $x = a$. For a given $\varepsilon > 0$, set both $\varepsilon_1, \varepsilon_2 = \varepsilon/2 > 0$. Next find corresponding $\delta_1, \delta_2 > 0$

satisfying the definitions of continuity for $f(x)$ and $g(x)$ at $x = a$, respectively. Finally, pick $\delta = \min\{\delta_1, \delta_2\}$. Recalling the triangle inequality, $|A+B| \le |A|+|B|$, with $A = f(x) - f(a)$ and $B = g(x) - g(a)$, we have

$$|x-a| < \delta \implies |h(x) - h(a)| = |f(x) + g(x) - f(a) - g(a)|$$
$$= |f(x) - f(a) + g(x) - g(a)| \le |f(x) - f(a)| + |g(x) - g(a)|$$
$$< \varepsilon_1 + \varepsilon_2 = \frac{\varepsilon}{2} + \frac{\varepsilon}{2} = \varepsilon, \quad \text{q.e.d.}$$

It is somewhat interesting to diagram $f(x) + g(x)$ and note the flow of the tolerances that appear in the proof. We do this next, with tolerances in gray:

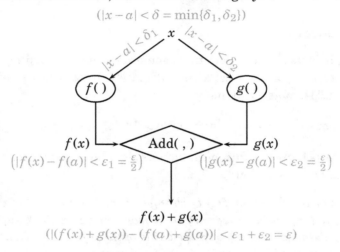

It is not difficult to see that this can be extended to include sums of more functions. Indeed, if $i(x) = f(x) + g(x) + h(x)$ where $f(x)$, $g(x)$, and $h(x)$ are all continuous at $x = a$, then so is the sum $[f(x) + g(x)]$ by the previous theorem, and then again so will be the sum of this sum and $h(x)$, i.e., so is $i(x) = [f(x) + g(x)] + h(x)$ continuous. A proof "from scratch" can be done as well, following the same pattern but using $\varepsilon_1, \varepsilon_2, \varepsilon_3 = \varepsilon/3$ and $\delta = \min\{\delta_1, \delta_2, \delta_3\}$, but it is faster to "bootstrap" later results from theorems that were already proved.

The next theorem's proof is less difficult, and is left as an exercise (Exercise 23, page 104):

Theorem 2.2.2 *If $f(x)$ is continuous at $x = a$, then so is $Cf(x)$ for any constant C.*

The next theorem is the most difficult of these to prove, but will also be one of the most useful in practice.

Theorem 2.2.3 *Suppose that $f(x)$ and $g(x)$ are continuous at $x = a$. Then so is $f(x)g(x)$.*

Proof: Let $\varepsilon > 0$, for the definition of continuity for the product $h(x) = f(x)g(x)$. Now we carefully construct a $\delta > 0$ so that we can prove $h(x)$ continuous at $x = a$, meaning that

$$|x - a| < \delta \implies |h(x) - h(a)| = |f(x)g(x) - f(a)g(a)| < \varepsilon \quad \text{(to prove!).}$$

2.2. CONTINUITY THEOREMS

As before we are assuming that $f(x), g(x)$ are continuous at $x = a$, so that

$$(\forall \varepsilon_1 > 0)(\exists \delta_1 > 0)(\forall x)(|x - a| < \delta_1 \longrightarrow |f(x) - f(a)| < \varepsilon_1), \qquad (2.7)$$
$$(\forall \varepsilon_2 > 0)(\exists \delta_2 > 0)(\forall x)(|x - a| < \delta_2 \longrightarrow |g(x) - g(a)| < \varepsilon_2). \qquad (2.8)$$

First we note that what we need to control is $|f(x)g(x) - f(a)g(a)|$, which can be cleverly expanded and then bounded using the triangle inequality as follows:

$$\begin{aligned}
|f(x)g(x) - f(a)g(a)| &= \frac{1}{2}|(f(x) - f(a))(g(x) + g(a)) + (f(x) + f(a))(g(x) - g(a))| \\
&\leq \frac{1}{2}|(f(x) - f(a))(g(x) + g(a))| + \frac{1}{2}|(f(x) + f(a))(g(x) - g(a))| \\
&= \frac{1}{2}|f(x) - f(a)| \cdot |g(x) + g(a)| + \frac{1}{2}|f(x) + f(a)| \cdot |g(x) - g(a)|.
\end{aligned}$$

It is enough that the expression in the final line be less than ε.

We will choose $\varepsilon_1, \varepsilon_2$ based on the choice of ε. Let us first assume *a priori* that any $\varepsilon_1, \varepsilon_2 \leq 1$ in (2.7) and (2.8), and so with $|x - a| < \min\{\delta_1, \delta_2\}$ we get by the triangle inequality

$$|f(x) - f(a)| < \varepsilon_1 \leq 1 \implies |f(x)| = |f(a) + (f(x) - f(a))| < |f(a)| + 1, \qquad (2.9)$$
$$|g(x) - g(a)| < \varepsilon_2 \leq 1 \implies |g(x)| = |g(a) + (g(x) - g(a))| < |g(a)| + 1. \qquad (2.10)$$

Now define L and M as follows, and note the following bounds since $\varepsilon_1, \varepsilon_2 \leq 1$:

$$L = |f(a)| + 1 \geq 1 > 0, \qquad (2.11)$$
$$M = |g(a)| + 1 \geq 1 > 0. \qquad (2.12)$$

From (2.9) and (2.10), and then (2.11) and (2.12), we get $|x - a| < \min\{\delta_1, \delta_1\} \implies$

$$|f(x)| \leq |f(a)| + 1 = L, \qquad |f(a)| = L - 1 < L, \qquad (2.13)$$
$$|g(x)| \leq |g(a)| + 1 = M, \qquad |g(a)| = M - 1 < M. \qquad (2.14)$$

Now we prove the statement of the continuity of $f(x)g(x)$ at $x = a$, within the context of the quantities already defined:

For any $\varepsilon > 0$ define (recalling $L, M \geq 1$):

$$\varepsilon_1 = \min\left\{1, \varepsilon, \frac{\varepsilon}{2M}\right\} = \min\left\{1, \frac{\varepsilon}{2M}\right\}, \qquad (2.15)$$
$$\varepsilon_2 = \min\left\{1, \varepsilon, \frac{\varepsilon}{2L}\right\} = \min\left\{1, \frac{\varepsilon}{2L}\right\}, \qquad (2.16)$$

and choose $\delta_1, \delta_2 > 0$ from the continuity conditions on $f(x)$ and $g(x)$ at $x = a$, namely (2.7) and (2.8). Finally, choose $\delta = \min\{\delta_1, \delta_2\}$. Then $\delta > 0$ exists and

$|x - a| < \delta \implies$

$$|f(x)g(x) - f(a)g(a)| \leq \frac{1}{2}|f(x) - f(a)| \cdot |g(x) + g(a)| + \frac{1}{2}|f(x) + f(a)| \cdot |g(x) - g(a)|$$
$$< \frac{1}{2}\varepsilon_1 \cdot (|g(x)| + |g(a)|) + \frac{1}{2}(|f(x)| + |f(a)|)\varepsilon_2$$
$$\leq \frac{1}{2}\varepsilon_1(2M) + \frac{1}{2}\varepsilon_2(2L)$$
$$\leq \frac{1}{2} \cdot \frac{\varepsilon}{2M} \cdot (2M) + \frac{1}{2} \cdot \frac{\varepsilon}{2L} \cdot (2L)$$
$$= \frac{\varepsilon}{2} + \frac{\varepsilon}{2} = \varepsilon, \qquad \text{q.e.d.}$$

We could diagram $f(x)g(x)$ as we did with $f(x) + g(x)$, though the tolerances are more complicated. In fact this proof is more difficult than most in this text, and requires careful examination to understand completely, or to be able to reproduce it. Contained within it lies a common method for controlling $|f(x)g(x) - f(a)g(a)|$ by controlling $|x - a|$, which is worth tracing. In particular, we used the following identity, which is worth verifying:

$$f(x)g(x) - f(a)g(a) = \frac{1}{2}(f(x) - f(a))(g(x) + g(a)) + \frac{1}{2}(f(x) + f(a))(g(x) - g(a)).$$

All three continuity theorems so far can be described intuitively as saying, respectively (for functions continuous at $x = a$):

- **A sum of functions that are continuous at a will also be continuous at a;**
- **A constant multiple of a function that is continuous at a will also be continuous at a;**
- **A product of functions that are continuous at a will also be continuous at a.**

Rephrased, if we can control how far functions stray from their values at $x = a$—by controlling how far x strays from a—then we can control how far their sums, products, and constant multiples of the functions stray from their respective values at $x = a$ as well. For the next theorem, we first need the following:

Lemma 2.2.1 $(\forall a \in \mathbb{R})[f(x) = x \text{ is continuous at } x = a]$.

This is fairly trivial to prove, because we would just set $\delta = \varepsilon$ in the definition of continuity. Details are left to the exercises, though clearly in this case output and input tolerances are equal.

Theorem 2.2.4 *Polynomial functions* $f(x) = a_n x^n + a_{n-1} x^{n-1} + \cdots + a_1 x + a_0$ *are continuous at every* $a \in \mathbb{R}$.

Hence the functions $f(x) = x^2 + 1$, $g(x) = 55x^{39} + 101x - 10{,}000{,}000$, and $h(x) = (9 - 23x)^{15}$ are all continuous. Certainly this theorem gives a welcome relief from trying to prove these functions are continuous using the ε-δ definition, although it can be done. Instead we will use previous results for a proof, which is simple enough it may even appear flippant.

2.2. CONTINUITY THEOREMS

Proof: Any polynomial can be written

$$f(x) = a_n x^n + a_{n-1} x^{n-1} + \cdots + a_1 x + a_0.$$

Now each x^k can be written as a product of k factors of the continuous function $g(x) = x$, so the x^k are all continuous for every $x = a$, as are constant multiples of these, i.e., the $a_k x^k$ terms. Of course $h(x) = a_0$ is a constant function and is therefore continuous (see Exercise 11, page 87), so a polynomial is just the sum of continuous functions, and is therefore continuous.

The next theorem is also very useful, and deals with compositions of functions. Compared with, say, Theorem 2.2.3, this is perhaps surprisingly simple to prove.

Theorem 2.2.5 *Suppose that $f(x)$ is continuous at $x = L = g(a)$, and that $g(x)$ is continuous at $x = a$. Then $f(g(x))$ is continuous at $x = a$.*

Put concisely, the composition of continuous functions is continuous.

Proof: Let $\varepsilon > 0$. We need to construct a $\delta > 0$ such that

$$|x - a| < \delta \implies |f(g(x)) - f(g(a))| < \varepsilon.$$

By our continuity assumptions, we know that

$$(\forall \varepsilon_1 > 0)(\exists \delta_1 > 0)(\forall x)(|x - L| < \delta_1 \longrightarrow |f(x) - f(L)| < \varepsilon_1),$$
$$(\forall \varepsilon_2 > 0)(\exists \delta_2 > 0)(\forall x)(|x - a| < \delta_2 \longrightarrow |g(x) - g(a)| < \varepsilon_2).$$

So for this ε, choose $\varepsilon_1 = \varepsilon$, which gives a $\delta_1 > 0$ so that

$$|x - L| < \delta_1 \implies |f(x) - f(L)| < \varepsilon.$$

Next set $\varepsilon_2 = \delta_1 > 0$. Then continuity of g gives a $\delta_2 > 0$ so that

$$|x - a| < \delta_2 \implies |g(x) - g(a)| < \varepsilon_2 = \delta_1.$$

Finally, let $\delta = \delta_2$, corresponding to ε_2 in the continuity requirement for g. This gives $\delta > 0$ and the following (note the only necessary directions are $\implies$, but we note the steps for which we actually have $\iff$):

$|x - a| < \delta \iff |x - a| < \delta_2$
$\qquad \implies |g(x) - g(a)| < \varepsilon_2$ (since the input x of g is δ_2-close to a)
$\qquad \iff |g(x) - L| < \delta_1$ (since $\varepsilon_2 = \delta_1$)
$\qquad \implies |f(g(x)) - f(L)| < \varepsilon_1$ (since the input $g(x)$ of f is δ_1-close to L)
$\qquad \iff |f(g(x)) - f(g(a))| < \varepsilon,$ (since $\varepsilon_1 = \varepsilon$ and $g(a) = L$)

q.e.d.

Intuitively, if $g(x)$ is changing continuously at $x = a$, and $f(\ )$ changes continuously at its input value $g(a)$, it seems reasonable that $f(g(x))$ should change continuously at $x = a$. Even less precisely, with the assumed continuity conditions of $f(\ )$ and $g(\ )$ we might expect that if $x \approx a$ then $g(x) \approx g(a)$, and then $f(g(x)) \approx f(g(a))$.

Now we have a proof, constructed as usual backward from the function's flow, illustrated on the right. First we choose $\varepsilon > 0$ tolerance for the final output. With $\varepsilon_1 = \varepsilon > 0$ we have $\delta_1 > 0$ tolerance for the input of $f(\ )$ (near the value $g(a)$) guaranteeing ε tolerance in the output of $f(\ )$. Letting $\varepsilon_2 = \delta_1 > 0$, we have $\delta_2 > 0$ tolerance in x near $x = a$, allowing for at most ε_2 tolerance in the output of $g(x)$ to be near $g(a)$, guaranteeing ε-tolerance when this is fed to $f(\ )$.

We next work towards a simple theorem regarding quotients, which we argue deductively towards rather than stating and proving the steps in turn.[10]

First, recall from Exercise 16, page 87, that we have the following (stated here without proof):

Theorem 2.2.6
$$f(x) = \frac{1}{x} \quad \text{is continuous for all } x \neq 0. \tag{2.17}$$

Continuing an argument which leads to a theorem on quotients, next we suppose that $g(x)$ is continuous at $x = a$, and that $g(a) \neq 0$. Then (2.17) and the previous theorem (Theorem 2.2.5 above) with $f(x) = \frac{1}{x}$ conspire to give us that $f(g(x)) = 1/g(x)$ is thereby continuous at $x = a$:

Theorem 2.2.7 $(g(x)$ continuous at $x = a) \wedge (g(a) \neq 0) \implies \dfrac{1}{g(x)}$ continuous at $x = a$.

For arbitrary functions $f(x)$ and $g(x)$ which are continuous at $x = a$, if $g(a) \neq 0$ we can always write $\frac{f(x)}{g(x)} = f(x) \cdot \frac{1}{g(x)}$, a product of two functions now known to be continuous at $x = a$, and so by our Theorem 2.2.3, page 90, that product is also continuous and we get another theorem:

Theorem 2.2.8 *If $f(x)$ and $g(x)$ are continuous at $x = a$, and $g(a) \neq 0$, then $f(x)/g(x)$ is also continuous at $x = a$.*

This gives us a quick result on rational functions:

---

[10] When a thread contains theorems that are each difficult to prove, it is common (as we did previously) for the discussion to proceed with a structure of the form theorem-proof, theorem-proof, etc. When theorems flow from each other with minimal argument needed, the style often changes to short argument-result, short argument-result, etc., and so a proof can be contained within the arguments which lead to the results, rather than setting aside separate proofs as we did for most of the earlier theorems.

2.2. CONTINUITY THEOREMS

Theorem 2.2.9 *If $f(x) = \dfrac{p(x)}{q(x)}$, where p and q are polynomials, then $f(x)$ is continuous at every $a \in \mathbb{R}$ except where $q(a) = 0$, at which points $f(x)$ is undefined and therefore discontinuous.*

In other words, *rational functions are continuous where defined.* (The proofs of both of these are contained in the previous discussion.) Next we mention the following reasonable definition:

Definition 2.2.1 *If $f(x)$ is not continuous at some point c, then $f(x)$ is called **discontinuous** at $x = c$, and c is called a **point of discontinuity** of $f(x)$.*[11]

It is sometimes easier to list those points at which a function is discontinuous than the set of points at which it is continuous. Of course continuity at a point, and discontinuity at a point, are negations of each other.

Example 2.2.1

Find where the function $f(x) = \dfrac{3x-5}{x^2-1}$ is continuous.

Solution: As a rational function, $f(x)$ is continuous **except** where $x^2 - 1 = 0$, i.e., except where $x^2 = 1$, i.e., except where $x = -1, 1$. We can effectively describe where $f(x)$ is continuous in the following ways (all of which are equivalent):

a. $f(x)$ is continuous except at $x = -1, 1$;

b. $f(x)$ is continuous at each $x \in (-\infty, -1) \cup (-1, 1) \cup (1, \infty)$;

c. $f(x)$ is continuous for all $x \neq \pm 1$.

It is important that we not simplify $f(x)$ in any way before describing where it is continuous, lest problem points in the definition of $f(x)$ be (deceptively) masked by a simplification. For instance, consider the following:

Example 2.2.2

Find where $f(x) = \dfrac{x^2 - 25}{x - 5}$ is continuous.

Solution: Clearly $f(x)$ is not defined at $x = 5$, and therefore cannot possibly be continuous there. (The way in which $f(a)$ appears in the definition of continuity requires that it exists.) Simplifying is desirable, but we must note that $x = 5$ is not in the domain. To be clear on this we can write

$$f(x) = \frac{x^2 - 25}{x - 5} = \frac{(x+5)(x-5)}{x-5} = x + 5, \qquad x \neq 5.$$

Note that the cancellation step indeed required that $x - 5 \neq 0$, i.e., $x \neq 5$ because there we would essentially be dividing numerator and denominator by zero—not a valid arithmetic operation—when canceling the $x - 5$ factors. This function is defined and continuous for $x \in (-\infty, 5) \cup (5, \infty)$, i.e., for $x \neq 5$. A more precise simplification would read $f(x) = x + 5$, $x \neq 5$.

---

[11] The word *point* sometimes refers to (among other things) a value (point) on the real line (or an element of $\mathbb{R}$), and other times refers to a point in the xy-plane (or other geometric set). For instance, one may describe $f(x) = x^2$ "at the point $x = 3$" or "at the point $(3,9)$." In most cases it is an imprecision in the language which is cleared up by an understanding of the context, though occasionally different authors will have different but strong opinions on how to make the language precise.

The proof of the next theorem was the subject of Exercise 18, page 88.

Theorem 2.2.10 *Suppose that $f(x) = \sqrt[n]{x} = x^{1/n}$, with $n \in \mathbb{N}$.*

(i) If n is odd, then $f(x)$ is continuous at each $x \in \mathbb{R}$.

(ii) If n is even, then $f(x)$ is continuous at each $x > 0$, and discontinuous otherwise.

Thus $f(x) = \sqrt[3]{x}$ is continuous at each $x \in \mathbb{R}$, while $g(x) = \sqrt{x} = \sqrt[2]{x}$ is only continuous at each $x > 0$. Though $g(x) = \sqrt{x}$ is defined at $x = 0$, there is no room for tolerance anywhere left of zero, so any interval $(0-\delta, 0+\delta)$, with $\delta > 0$, will have points outside of the domain of $g(x)$, namely for $x \in (0-\delta, 0)$. Thus with even roots (or any function) it is not enough to look where the function is defined to determine continuity.

On the other hand for $g(x) = \sqrt{x}$ if we consider only $x \geq 0$, then for any $\varepsilon > 0$ we can find a $\delta > 0$ so that we can at least say $(|x - 0| < \delta) \wedge (x \geq 0) \implies |g(x) - g(0)| < \varepsilon$, and this is occasionally interesting, and so we have use for so-called one-sided continuity, the types of which are defined below.

2.2.2 One-Sided Continuity

At this point it is useful to introduce the following concepts.

Definition 2.2.2 *We call $f(x)$ **left-continuous** at $x = a$ if and only if*

$$(\forall \varepsilon > 0)(\exists \delta > 0)(\forall x)(x \in (a-\delta, a] \longrightarrow |f(x) - f(a)| < \varepsilon). \tag{2.18}$$

*Similarly, we call $f(x)$ **right-continuous** at $x = a$ if and only if*

$$(\forall \varepsilon > 0)(\exists \delta > 0)(\forall x)(x \in [a, a+\delta) \longrightarrow |f(x) - f(a)| < \varepsilon). \tag{2.19}$$

Left-continuity at $x = a$ only considers points at or to the left of a ($x \leq a$), while right-continuity considers those at or to the right of a ($x \geq a$). In the definition of continuity (2.1), page 77, both sides of $x = a$ are considered. In fact that original definition of continuity can be rewritten

$$(\forall \varepsilon > 0)(\exists \delta > 0)(\forall x)(x \in (a-\delta, a] \cup [a, a+\delta) \longrightarrow |f(x) - f(a)| < \varepsilon), \tag{2.20}$$

from which we can see how left-continuity (2.18) and right-continuity (2.19) are related to continuity. Thus these one-sided continuity conditions are effectively two halves of the continuity requirement as written in (2.20).

One aspect of this theory which has been mentioned in passing is that each type of continuity (the original two-sided continuity, left-continuity and right-continuity) at $x = a$ requires that $f(a)$ must be defined, in order that $|f(x) - f(a)|$ be true in the implications within the definitions. A function $f(x)$ can have none of these continuity properties at $x = a$ if $f(a)$ is undefined. This will be an issue in identifying where some functions are continuous.

Example 2.2.3 _____

Consider the function $f(x)$ defined only on $x \in [0, \infty)$, but on that interval given by $f(x) = x^2$. In other words, $f(x) = \begin{cases} x^2 & \text{if } x \geq 0, \\ \text{undefined} & \text{if } x < 0. \end{cases}$

2.2. CONTINUITY THEOREMS

- Then $f(x)$ is continuous at $x = a$ for any $a > 0$.

Intuitively, we can see from the graph that we have "wiggle room" in x as long as $a > 0$; we can guarantee positive tolerances as small as we like in $f(x)$ being close to $f(a)$ by forcing x to be close to, but not necessarily equal to, a if $a > 0$. This is not the case if $a = 0$, as we would have no "wiggle room" to the left of $x = 0$.

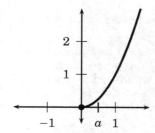

More technically, if $a > 0$, then in the definition of continuity we can assume *a priori* that $\delta \leq a$, and then $|x - a| < \delta \implies f(x)$ is defined, and so we can borrow continuity results from our earlier theorem on the continuity of polynomials. To do so properly, we should consider the polynomial function $g(x) = x^2$ defined on all of $\mathbb{R}$. Now $g(x)$ is continuous for any $x = a$, with $a \in \mathbb{R}$. Thus for any $a \in \mathbb{R}$, if we let $\varepsilon > 0$, there exists $\delta_g > 0$ ("the delta for g") such that $|x - a| < \delta_g \implies |g(x) - g(a)| < \varepsilon$. Now for our function $f(x)$, and any $a > 0$, we can take $\delta = \min\{\delta_g, a\}$, so $\delta > 0$ exists and

$$|x-a| < \delta \implies \left[\begin{array}{c} (|x-a| < \delta \leq \delta_g) \\ \wedge \\ (x > 0) \end{array} \right] \implies \underbrace{|f(x) - f(a)|}_{\text{exists}} = \underbrace{|g(x) - g(a)|}_{\text{since } x, a > 0} \underbrace{< \varepsilon}_{\text{since } \delta \leq \delta_g}.$$

- $f(x)$ is both left-continuous and right-continuous (as well as continuous in our original two-sided sense) at each $a > 0$. That should be clear because continuity is a stronger condition than the one-sided continuities. (See Theorem 2.2.11 below.)

- $f(x)$ is right-continuous (but neither continuous nor left-continuous) at $x = 0$. We only need to allow tolerance ("wiggle room") allowing for x to vary to the right of $x = 0$ (see graph). In fact, $\delta = \sqrt{\varepsilon}$ proves right-continuity: Let $\varepsilon > 0$, and choose $\delta = \sqrt{\varepsilon}$, so $\delta > 0$ exists and

$$x \in [0, \delta) \implies |f(x) - f(0)| = x^2 < \delta^2 = \varepsilon, \text{ q.e.d.}$$

- $f(x)$ has no continuity properties for $x < 0$, because the function is not defined there.

Continuity is, in fact, a "local" phenomenon; what matters for $x = a$ is the behavior of the function "near $x = a$." This is because we can always restrict $\delta > 0$ to be as small as we like *a priori*, in essence ignoring what occurs outside of $(a - \delta, a + \delta)$ no matter how small we make δ, while keeping it positive. Even so restricted, we are still allowing for (and in fact requiring) a *continuum* of values including a, extending from a in one or both directions, depending upon if we are considering one-sided continuity or our original, two-sided continuity.

Of course the different types of continuities are related. We leave for an exercise the following theorem:

Theorem 2.2.11 *$f(x)$ is continuous at $x = a$ if and only if $f(x)$ is both left-continuous and right-continuous at $x = a$.*

There are settings in which we check continuity by separately checking each one-sided continuity condition, and other occasions in which we require only one-sided continuity but have that easily from the theorem because we have two-sided continuity.

Example 2.2.4

Consider the function $f(x) = \begin{cases} (x-2)^2 - 1 & \text{if } x \geq 2, \\ 1-x & \text{if } x < 2. \end{cases}$

- At $x = 0$ this function is continuous, because it is essentially $f(x) = 1 - x$ for $x \in (-\infty, 2)$, and $x = 0$ is safely inside this interval (i.e., in and surrounded by a continuum within the interval). If we wished to prove continuity at $x = 0$ from the definition, we can always choose $\delta \leq 2$ so that when we write $|x - 0| < \delta$ we know we are within an interval in which $f(x)$ is a fixed polynomial and therefore continuous. (In fact $\delta = \min\{2, \varepsilon\}$ would work in the continuity definition.)

- At $x = 3$, this function is again continuous, because we are safely inside of $[2, \infty)$, and we could if necessary take $\delta \leq 1$ so that $|x - 3| < \delta \implies x \in [2, \infty) \implies f(x) = (x-2)^2 - 1$, i.e., $f(x) = x^2 - 4x + 4 - 1 = x^2 - 4x + 3$, ultimately a polynomial, albeit one which requires a more careful choice of $\delta > 0$ for a given $\varepsilon > 0$ in the definition of continuity.

- At $x = 2$, we have to be more careful. Note that $f(2) = (2-2)^2 - 1 = -1$, which coincides with $1 - x$ where $x = 2$ as well. This means we could have used $x \leq 2$ as the condition for which $f(x) = 1 - x$, because it did not matter which formula we applied at $x = 2$. Thus,

    (i) $x \in [2, \infty) \implies f(x) = (x-2)^2 - 1$,

    (ii) $x \in (-\infty, 2] \implies f(x) = 1 - x$.

    From (i) we know that $f(x)$, being a polynomial on $[2, \infty)$, is right-continuous at $x = 2$; and from (ii) we know that $f(x)$, being a polynomial on $(-\infty, 2]$, is left-continuous at $x = 2$. Since $f(x)$ is both left-continuous and right-continuous at $x = 2$, we conclude that $f(x)$ is continuous (in the two-sided sense) at $x = 2$.

In essence, if a function is defined piecewise and the defining formulas agree with $f(a)$ in their outputs at a point a on the boundary of the formulas' respective intervals, and if the defining formulas would themselves define appropriately continuous functions (left-continuous at a in the left-piece formula, and right-continuous at a in the right-piece formula), then $f(x)$ will be continuous at $x = a$. In the previous example, we could simply say that the two formulas to the left and right of $x = 2$ would both yield $f(2)$. Visually, the two pieces should "meet" at $(2, f(2))$, as in the graph.

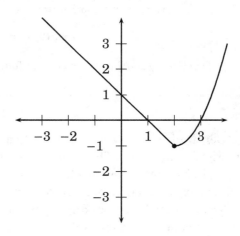

While it is useful to consider the graphs when determining continuity, so far we have stressed the *analysis* without resorting to graphs. We will utilize graphs more in the next sections, but for now we will attempt to develop a "number sense" and "function sense" in a manner somewhat independent from the graphs. For instance, if we return again to the even roots $f(x) = \sqrt[n]{x}$, (where n is even) we see that, though these functions are not continuous at $x = 0$, they are right-continuous at $x = 0$. When dealing with functions that contain radicals, for continuity we often need only see where the functions are defined, and where we have

2.2. CONTINUITY THEOREMS

some "wiggle room" on both sides of a particular $x = a$. Odd roots are defined and (two-sided) continuous for any real input, so they do not in themselves limit where a function is continuous. Consider the following:

Example 2.2.5

For each function, find where it is continuous.

$$f(x) = \frac{1}{\sqrt[3]{x-1}}$$
$$g(x) = \sqrt{2x+1}$$
$$h(x) = \sqrt{20-4x}$$

$$i(x) = \sqrt{x^2+1}$$
$$j(x) = \frac{1}{\sqrt{x^2-1}},$$
$$k(x) = \sqrt[3]{x^9 - 27x^4 + x - 11}.$$

<u>Solution</u>: We take each of these in turn, making use of our previous continuity theorems on sums, products, compositions and quotients of functions, as well as results on polynomials and roots.

- The function $f(x)$ is a ratio of functions that are continuous everywhere (the denominator being a composition of functions that are continuous everywhere, see Theorem 2.2.5, page 93), so we need only check where the denominator is zero. Thus $f(x)$ is discontinuous only at $x = 1$. In other words, $f(x)$ is continuous on $(-\infty, 1) \cup (1, \infty)$, but we will be content in such a case to simply state where it is discontinuous.

- The function $g(x)$ is definitely continuous where $2x + 1 > 0$, i.e., $2x > -1$, i.e., $x > -1/2$. It is not continuous at $x = -1/2$ because $g(x)$ is undefined to the left of $x = -1/2$, as a quick check will show. ($g(x)$ is right-continuous at $x = -1/2$, but not continuous there.) Conclude $g(x)$ is continuous for $x > -1/2$.

- The function $h(x)$ is definitely continuous for $20 - 4x > 0$, i.e., $-4x > -20$, i.e., $x < 5$. It is not continuous at $x = 5$ since it is undefined to the right of that point. (It is left-continuous at $x = 5$, but again, that was not the question.) Conclude $h(x)$ is continuous where $x < 5$.

- The function $i(x)$ is continuous everywhere, since $x^2 + 1$ is continuous everywhere, and for all $x \in \mathbb{R}$ we also have $x^2 + 1 > 0$ is a point of continuity of the square root function. See again Theorem 2.2.5, page 93.

- The function $j(x)$ is definitely discontinuous at $x = \pm 1$ because the denominator is zero there. With that consideration, and that of the square root disallowing its input being negative, for continuity we need $x^2 - 1 > 0$, i.e., $x^2 > 1$, which occurs for $x \in (-\infty, -1) \cup (1, \infty)$.

- The function $k(x)$ is continuous everywhere, being an odd root of a (continuous) polynomial.

Example 2.2.6

Find where the function $f(x) = \dfrac{\sqrt{9-x^2}}{x^2-1}$ is continuous.

<u>Solution</u>: For denominator considerations, we need only that $x \ne \pm 1$. The numerator is definitely continuous for $9 - x^2 > 0$, i.e., $9 > x^2$, i.e., $-3 < x < 3$. It is not continuous at $x = \pm 3$ since both are on the edge of the domain (no "wiggle room"). Putting this together, we see $f(x)$ is continuous for $x \in (-3, 3) - \{-1, 1\} = (-3, -1) \cup (-1, 1) \cup (1, 3)$.

To be clear, we next look at a function that is the reciprocal of the previous function to illustrate the differences.

Example 2.2.7

Find where the function $f(x) = \dfrac{x^2-1}{\sqrt{9-x^2}}$ is continuous.

Solution: For this function, the numerator is continuous for all $x \in \mathbb{R}$, so there is no restriction on x implied by the form of the numerator. (Numerators can be equal to zero, while denominators cannot, for a quotient to be defined.) Clearly $x \neq -3, 3$ due to the denominator, lest we attempt to divide by zero. Requiring the input of the square root to be nonnegative forces $-3 \leq x \leq 3$, but we already disallowed $x = \pm 3$ due to the restriction against dividing by zero, though an alternative explanation here would be the lack of "wiggle room" at either value, -3 to the left and 3 to the right, avoiding again a negative input for the square root function $\sqrt{(\ )}$. Our conclusion is that this $f(x)$ is continuous for $x \in (-3, 3)$.

A useful corollary to Theorem 2.2.10, page 96, is the fact that

$$f(x) = |x| = \sqrt{x^2}$$

is continuous for all $x \in \mathbb{R}$. To see this, note that clearly $|x|$ is continuous for $x \neq 0$, for which $x^2 > 0$. But there is "wiggle room" at $x = 0$ as well. In fact it is easier to see that $|x|$ is both left- and right-continuous at $x = 0$ if we write

$$|x| = \begin{cases} x & \text{if } x \geq 0, \\ -x & \text{if } x \leq 0. \end{cases} \tag{2.21}$$

As a matter of form, we usually write definitions such as (2.21) with the "if" conditions being mutually exclusive (non-overlapping, as if we were instructing a computer), but we can see that at $x = 0$, both formulas give the same result. We can then see that for $x \in [0, \infty)$, we can write $|x| = x$, which is obviously right-continuous at $x = 0$. Similarly, for $x \in (-\infty, 0]$, we can write $|x| = -x$, which is obviously left-continuous at $x = 0$. Since $|x|$ as a function is both left-continuous and right-continuous at $x = 0$, it is continuous there.

Furthermore, at any $x \neq 0$ we are "safely" within either branch: where $x > 0 \implies |x| = x$ which is continuous, or where $x < 0 \implies |x| = -x$, which is also continuous. From these three cases ($x = 0$, $x > 0$, and $x < 0$), we can see that $|x|$ is continuous at each $x \in \mathbb{R}$, as we state next:

Theorem 2.2.12 *The function $f(x) = |x|$ is continuous for all $x \in \mathbb{R}$.*

By our theorem on function compositions (Theorem 2.2.5 on page 93), we have that the absolute value of a continuous function is also continuous:

Theorem 2.2.13 *If $g(x)$ is continuous at $x = a$, then $f(x) = |g(x)|$ is also continuous at $x = a$.*

Example 2.2.8

$f(x) = \sqrt{x^2 - 2x + 1}$ is continuous on $\mathbb{R}$ since $f(x) = \sqrt{(x-1)^2} = |x-1|$ is the absolute value of a continuous function $(x-1)$ and is therefore continuous.

2.2. CONTINUITY THEOREMS

2.2.3 Essential versus Removable Discontinuities

Many functions that we encounter will have a discontinuity at a particular point, but the discontinuity may be *removable*. This means that if we could redefine the function at that point to have a well chosen output value, it would cease to be discontinuous there.

Example 2.2.9

Consider the function $f(x) = \frac{x^2}{|x|}$. Clearly this function is undefined at $x = 0$, because we cannot divide by zero. Note that this function can instead be defined piecewise:

$$f(x) = \begin{cases} x^2/x & \text{if} \quad x > 0 \\ x^2/(-x) & \text{if} \quad x < 0 \end{cases} = \begin{cases} x & \text{if} \quad x > 0 \\ -x & \text{if} \quad x < 0. \end{cases}$$

In other words, $f(x) = |x|$ for $x \neq 0$. We say that $x = 0$ is a **removable discontinuity** because if we were to simply redefine f so that $f(0) = 0$, we would have a continuous function at $x = 0$.

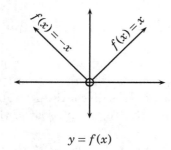

$y = f(x)$

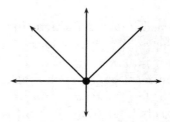

$y = f(x)$ (redefined at $x = 0$)

Clearly the only value we could use to redefine the output of $f(x)$ at $x = 0$ is 0, i.e., so $f(0) = 0$, to render $f(x)$ continuous at $x = 0$.

Example 2.2.10

Consider the function $f(x) = \dfrac{x}{|x|}$. This function is also undefined at $x = 0$. Defined piecewise, we have the easily graphed function

$$f(x) = \begin{cases} x/x & \text{if} \quad x > 0 \\ x/(-x) & \text{if} \quad x < 0 \end{cases} = \begin{cases} 1 & \text{if} \quad x > 0 \\ -1 & \text{if} \quad x < 0. \end{cases}$$

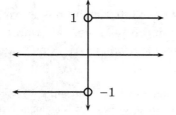

This is graphed on the left. In order for the function to be right-continuous at $x = 0$ we would require $f(0) = 1$, but for left-continuity we would require $f(0) = -1$. Clearly we cannot have both (or we would not have a function!) and so the discontinuity at $x = 0$ is called **essential**.[12]

---

[12] Another common term for essential discontinuities is **nonremovable**. Both terms have their respective advantages.

It is usually not difficult to spot removable and essential discontinuities from the graph of the function: If we can "fill in a hole" or "move one point" on the graph to produce a function that is continuous at the point, then the discontinuity is removable; otherwise, it is essential. However, one goal of the analysis is to be able to make such determinations based on formulas for the functions, rather than based on a requirement that we always produce a graph first. At this point, we are ready for our definitions:[13]

Definition 2.2.3 *Given a function $f(x)$, and a discontinuity $x = a$ of $f(x)$, then*

- *we call $x = a$ a* **removable discontinuity** *if there exists $y_0 \in \mathbb{R}$ such that, if $f(x)$ were redefined at $x = a$ so that $f(a) = y_0$, then $f(x)$ would be continuous at $x = a$;*

- *we call $x = a$ an* **essential (or nonremovable) discontinuity** *if no such y_0 exists. That is, no matter how we (re)define $f(x)$ at $x = a$, i.e., regardless of any redefinition $f(a) = y_0$, the function $f(x)$ is still discontinuous at $x = a$.*

It is assumed in the definition that $f(x)$ is only redefined at $x = a$, and so the output of $f(x)$ will be the same as before for all $x \ne a$.

Note that if $f(x)$ has a removable discontinuity at $x = a$, then the redefinition is unique; there can be at most one y_0 such that the redefinition $f(a) = y_0$ removes the discontinuity of $f(x)$ at $x = a$. That may seem intuitive from various graphs of removable discontinuities, but we will find tools later (namely limit theorems in later sections) that make a proof of this fact easier.

Examples of essential discontinuities include *jump discontinuities* and *vertical asymptotes*:

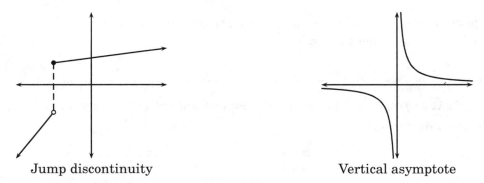

Jump discontinuity   Vertical asymptote

We see that with a jump discontinuity, the function can be either left-continuous or right-continuous, but not both. For this case, if we filled in the remaining "hole," we would no longer have a function, as we would have two outputs (y-values) for that particular input a. Vertical asymptotes of any kind are always essential discontinuities.

---

[13]There are approaches for learning calculus based on exploring most problems with graphical calculators or software first—in essence, approaching calculus first as a visual exercise—though it does not take long for one to encounter functions complicated enough that a reliance on electronically produced graphs is cumbersome or misleading. The calculator-based calculus instruction is especially popular in high school calculus courses, where there is an assumption that algebraic skills may be lacking, and that students will be more interested when they have a visual reference. One drawback is that the user can be fooled by limitations of a calculator display's resolution, or by the analytical shortcomings of today's common graphing calculators.

2.2. CONTINUITY THEOREMS

Some nonremovable discontinuities we encounter are not terribly interesting. For instance, $x = -1$ is not a removable discontinuity of $f(x) = \sqrt{x}$ because it is well outside of the domain of $\sqrt{x}$. For that matter, $x = 0$ is an essential discontinuity of $f(x) = \sqrt{x}$ by our definition because it is on the boundary of the domain of $\sqrt{x}$ (and no redefinition at $x = 0$ will change that it is a discontinuity). However, $\sqrt{x}$ is right-continuous there. We could define left-removable or right-removable discontinuities, but we will not bother to do so here. Where applicable, we will later have other language for such phenomena.

Example 2.2.11

Consider the function $f(x) = \begin{cases} \frac{x^2+x-6}{x-2}, & \text{if } x \neq 2, \\ 1, & \text{if } x = 2. \end{cases}$

Note that for $x \neq 2$ we can simplify

$$x \neq 2 \implies f(x) = \frac{(x+3)(x-2)}{x-2} = x+3,$$

and so in fact

$$f(x) = \begin{cases} x+3, & \text{if } x \neq 2 \\ 1, & \text{if } x = 2. \end{cases}$$

From the graph on the right we can see that $f(x)$ is discontinuous at $x = 2$, yet that discontinuity is "removable."

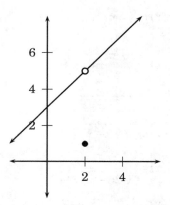

Indeed a simple redefining of $f(2)$ to be 5 would "fill the hole" and make $f(x)$ continuous at $x = 2$. Of course if we did redefine $f(2) = 5$, then this would be a *new function*, albeit one that is continuous at $x = 2$ (as well as the rest of $\mathbb{R}$). So to be more formal, we define

$$g(x) = \begin{cases} f(x), & \text{if } x \neq 2, \\ 5, & \text{if } x = 2, \end{cases}$$

and we will have $f(x)$ and $g(x)$ disagreeing only at $x = 2$, but the latter will be continuous at $x = 2$ and the former will not.

Removable discontinuities are one motivation behind the upcoming topic of limits, which in such a context would answer the question of what value a function is tending toward as its input tends towards some number, without necessarily reaching it. However, it is important to spend one more section (after this current one) discussing more aspects of continuous functions before being properly equipped to make efficient progress through a development of limit concepts.

Exercises

For 1–16, find all x for which the function is continuous.

1. $f(x) = \dfrac{x^2 - 4}{x^2 + 4}$

2. $f(x) = \dfrac{1 + x}{1 - x^2}$

3. $f(x) = \sqrt{3x - 7}$

4. $f(x) = \sqrt{9 - 3x}$

5. $f(x) = \sqrt{16 - x^2}$

6. $f(x) = \sqrt{16 + x^2}$

7. $f(x) = \sqrt{x^2 + 6x + 9}$

8. $f(x) = \dfrac{1}{\sqrt{x^2 + 6x + 9}}$

9. $f(x) = \sqrt[3]{25 - x^2}$

10. $f(x) = \dfrac{x^2 - 1}{\sqrt[3]{25 - x^2}}$

11. $f(x) = |x^2 - 7x + 12|$

12. $f(x) = \sqrt{|x|}$

13. $f(x) = \sqrt{\dfrac{1 - x}{1 + x^2}}$

14. $f(x) = \sqrt[4]{\dfrac{1 - x}{1 + x^2}}$

15. $f(x) = \sqrt[5]{\dfrac{1 - x}{1 + x^2}}$

16. $f(x) = \dfrac{1}{\sqrt[3]{x^2 - 1}}$

For Exercises 17–20, determine if the function given is

(a) left-continuous at the the indicated point,

(b) right-continuous there,

(c) continuous there, and if not,

(d) whether the discontinuity is essential or removable.

17. $f(x) = \begin{cases} -6 & \text{if } x < 0 \\ 0 & \text{if } x = 0 \\ 6 & \text{if } x > 0 \end{cases}$

at $x = 0$

18. $f(x) = \begin{cases} \dfrac{x^2 - 1}{x + 1} & \text{if } x \neq -1 \\ -2 & \text{if } x = -1, \end{cases}$

at $x = -1$

19. $f(x) = \begin{cases} x^3 + 2 & \text{if } x < 2 \\ 8 & \text{if } x = 2 \\ 5x & \text{if } x > 2 \end{cases}$

at $x = 2$

20. $f(x) = \begin{cases} -x^2 & \text{if } x < 0 \\ 1 - \sqrt{x} & \text{if } x \geq 0 \end{cases}$

at $x = 0$

21. For what value of A is the following function continuous on all of $\mathbb{R}$?

$$f(x) = \begin{cases} 2x^3 & \text{if } x < 1 \\ Ax - 3 & \text{if } x \geq 1 \end{cases}$$

22. Using ε-δ and the definition of right-continuity, show that $f(x) = \sqrt{x}$ is right-continuous at $x = 0$.

23. Prove Theorem 2.2.2, page 90, i.e., given $C \in \mathbb{R}$, we have $f(x)$ continuous at $x = a \implies Cf(x)$ is continuous at $x = a$. Here is the general strategy:

    (a) For $C = 0$, one can refer to Exercise 11, page 87.

    (b) If $C \neq 0$, let $\varepsilon_1 = \dfrac{\varepsilon}{|C|}$, and find the δ_1 corresponding to this ε_1 satisfying the definition of continuity for $f(x)$. Then take $\delta = \delta_1$.

24. Prove Lemma 2.2.1, page 92, i.e., that $f(x) = x$ is continuous at each $a \in \mathbb{R}$.

25. Prove Theorem 2.2.11, page 97. Hint: Let $\delta = \min\{\delta_1, \delta_2\}$ where δ_1 and δ_2 come from the one-sided continuity conditions when proving the "if," and let $\delta_1, \delta_2 = \delta$ for the "only if."

2.3 Continuity on Intervals

"A function $\phi(x)$ is said to be continuous between any limiting values of x, such as a and b, when to each value of x between those limits there corresponds a finite value of the function, and when an indefinitely small change in the value of x produces only an indefinitely small change in the function. In such cases the function in its passage from any one value to any other between the limits receives every intermediate value, and does not become infinite. This continuity can be readily illustrated by taking $\phi(x)$ as the ordinate of a curve, whose equation may then be written $y = \phi(x)$."

—"Infinitesimal Calculus," in *Encyclopædia Britannica: A Dictionary of Arts, Sciences, and General Literature*, Volume XIII, Ninth Edition (1881)

2.3.1 Continuity on Open Intervals: Main Definition and Theorem

The significance of continuity is perhaps best understood when applied to whole intervals (a,b), or $[a,b]$, etc., rather than single points. We will define what it means for $f(x)$ to be continuous on (a,b), and continuous on $[a,b]$. (Other cases like $[a,b)$ would be defined in ways that would be obvious extensions once we have the (a,b) and $[a,b]$ cases.) Continuity on open intervals is rather trivial to define, but nonetheless has interesting consequences. In practice, we will focus more on continuity on finite closed intervals $[a,b]$, properly defined, with $a < b$. But first we look at the open intervals.

Definition 2.3.1 *A function $f(x)$ is said to be* **continuous on the open interval** (a,b) *if and only if $f(x)$ is continuous at each value $x_0 \in (a,b)$.*[14]

What is interesting about continuity on (a,b) is it implies $f((a,b))$ is an interval of some kind—(c,d), $[c,d]$, $(c,d]$, $[c,d)$, possibly infinite (a,∞), $[a,\infty)$, $(-\infty,b)$, $(-\infty,b]$, $(-\infty,\infty) = \mathbb{R}$, or even a single point $[d,d]$—and that the curve $y = f(x)$ is a *connected* graph for $a < x < b$.[15]

Theorem 2.3.1 *If $f(x)$ is continuous on (a,b), then $f((a,b))$ is an interval (possibly containing only a single point), and the graph of $y = f(x)$, $a < x < b$ is a connected curve.*

Unfortunately, the proof is a few steps beyond the scope of this textbook and is left, for the interested reader, to a course in advanced calculus, real analysis, or topology.[16] Still, with the previous continuity sections behind us we should at least hear the ring of truth in the notion that the graph of $f(x)$, with domain restricted to $x \in (a,b)$, would be connected if f

---

[14] Of course, each x_0 may require a different δ for a given ε in the original definition of continuity given in Section 2.1, but the definition only requires that each individual $x_0 \in (a,b)$ is a point at which $f(x)$ satisfies the ε-δ definition. To summarize, $(\forall x_0 \in (a,b))(\forall \varepsilon > 0)(\exists \delta > 0)(\forall x)(|x - x_0| < \delta \longrightarrow |f(x) - f(x_0)| < \varepsilon)$.

[15] If S is a subset of the domain of $f(x)$, then we define $f(S) = \{y \mid (\exists x \in S)(y = f(x))\}$. In other words, $f(S)$ is the set of all possible outputs of $f(\ )$ if the inputs are taken from S. For instance, if $f(x) = 2x+3$, then $f((0,1)) = (3,5)$, since $x \in (0,1) \implies y \in (3,5)$, as can easily be seen by graphing $y = 2x+3$ on the interval $x \in (0,1)$. Similarly, if $f(x) = \sin x$ then $f(\mathbb{R}) = [-1,1]$.

[16] The preliminaries needed to prove these are not terribly difficult but would require a distracting amount of effort for a first-semester calculus course.

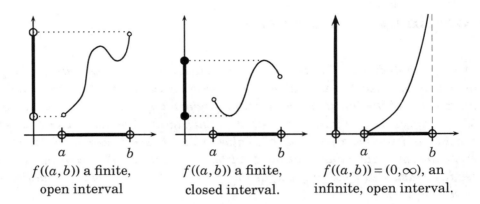

| $f((a,b))$ a finite, open interval | $f((a,b))$ a finite, closed interval. | $f((a,b)) = (0,\infty)$, an infinite, open interval. |

Figure 2.2: Some examples of functions continuous on open intervals (a,b), with images $f((a,b))$ drawn on the vertical axis. The image will always be an interval of some kind—finite, infinite, open, closed, half-open, or a "single-point" interval $[c,c]$ if f is a constant function.

is continuous on (a,b). Put another way, we could in principle draw the graph without lifting our pen from the paper. This is consistent with the elegant quote given at the beginning of this section, particularly the words, "...when an indefinitely small change in the value of x produces only an indefinitely small change in the function"; our pens would not need to "jump" off of the page to restart drawing some other piece of the curve than where we left off, as we move slowly along the curve by increasing x through the interval (a,b). As a consequence, it is then reasonable that the range of outputs $f((a,b)) = \{y \mid (\exists x \in (a,b)(y = f(x))\}$ will be a (connected) interval, as the theorem also states.

Figure 2.2 shows sample cases for continuity on open intervals, illustrating that the image is always an interval.

2.3.2 Continuity on Closed Intervals: Definition, EVT, and IVT

Now we turn our attention to continuity on closed intervals $[a,b]$. The definition is slightly different because there is not a two-sided "wiggle room" at either of the boundary points a,b (though we do have that freedom in the interior points of (a,b)). So for instance, if we have a function "continuous on $[a,b]$," then $|x-a| < \delta$ will include points outside of the interval for any $\delta > 0$ we choose. The same difficulty occurs for $|x-b| < \delta$.

Though the definition of $f : [a,b] \longrightarrow \mathbb{R}$ being continuous on $[a,b]$ is more complicated, the theorems are much stronger than those where we replace $[a,b]$ with (a,b), as in the previous subsection.

Recall that in Subsection 2.2.2 (starting on page 96), we also had a definitions for right-continuity and left-continuity—that is, for the two types of one-sided continuity—where we used, for instance, that $|x-a| < \delta \iff x \in (a-\delta, a+\delta) = (a-\delta, a] \cup [a, a+\delta)$. We replaced $|x-a| < \delta$ with $x \in [a, a+\delta)$ to define right-continuity (or "continuity from the right"), and replaced $|x-b| < \delta$ with $x \in (b-\delta, b]$ for left-continuity (or "continuity from the left"). These notions allow us to extend the idea of continuity to include closed intervals as well:

2.3. CONTINUITY ON INTERVALS

Definition 2.3.2 *A function $f:[a,b] \longrightarrow \mathbb{R}$, with $a < b$ is called* **continuous on the (finite) closed interval** $[a,b]$ *if and only if each of the following are true:*

1. *$f(x)$ is continuous on the open interval (a,b), i.e., continuous at each $x \in (a,b)$;*
2. *$f(x)$ is right-continuous at $x = a$;*
3. *$f(x)$ is left-continuous at $x = b$.*

In other words, in the *interior* of $[a,b]$, i.e., in (a,b), we require our original two-sided continuity at every point, but we require only right-continuity at a, and left-continuity at b. In this way, we strictly ignore the behavior of $f(\ )$ on inputs outside of $[a,b]$.

Note that we can also define continuity on $(a,b]$ as requiring our original (two-sided) continuity on (a,b) and left-sided continuity at b. It is similar for continuity on $[a,b)$.

Two very important results that follow from continuity on closed intervals $[a,b]$ are the Intermediate Value Theorem (IVT) and the Extreme Value Theorem (EVT). Both are wrapped up nicely in the following analog to Theorem 2.3.1, the difference being that the previous theorem required continuity on (a,b), while this theorem requires continuity on $[a,b]$. That minor difference gives us a much stronger theorem (though again we omit the proof):

Theorem 2.3.2 *If $f(x)$ is continuous on $[a,b]$, where $a < b$, then $f([a,b])$ is a closed interval of the form $[c,d]$ (with the possibility that $c = d$ in the case $f(x)$ is constant on $[a,b]$).*

In other words, the continuous image of a closed and bounded interval $[a,b]$ is also a closed and bounded interval $[c,d]$. If we again think about graphing such a function, we can see that this is also believable. After all, we would have to "pin down" the first point $(a,f(a))$, and draw the graph continuously until we end by "pinning down" the last point $(b,f(b))$. In doing so we would somewhere draw the highest and lowest points of our curve, with heights d and c, respectively, from our theorem. Albeit we may hit those high and low vertical levels repeatedly, and maybe not at $x = a$ or $x = b$, but they will be achieved nonetheless.

The proof of Theorem 2.3.2 is also beyond the scope of this text, but the EVT and IVT follow very quickly from the theorem. We list these very important facts as corollaries.

Corollary 2.3.1 (Extreme Value Theorem (EVT)) *If $f:[a,b] \longrightarrow \mathbb{R}$ is continuous on $[a,b]$, then the output of $f(x)$ achieves its maximum and minimum value of $f([a,b])$, say y_{max}, y_{min}, respectively, at some input values $x_{max}, x_{min} \in [a,b]$. In other words,*

$$\left(\exists x_{min}, x_{max} \in [a,b] \right) \left(f([a,b]) = [f(x_{min}), f(x_{max})] = [y_{min}, y_{max}] \right).$$

This says that if $f(x)$ is continuous on $[a,b]$, then there will be a "highest point" y_{max} and a "lowest point" y_{min} in the range of output values, and that these will occur at some inputs $x_{max}, x_{min} \in [a,b]$. Some verbalize this by saying that $f(x)$ will achieve its maximum and minimum values *for* $[a,b]$ at some points *in* $[a,b]$. That does not necessarily happen when the set of inputs is an open interval (a,b), as Figure 2.2, page 106, demonstrates.

The other, next corollary is used often in algebra and other subjects where we might be interested in producing "sign charts," showing where a function is positive or negative. It is reasonably intuitive on reflection.

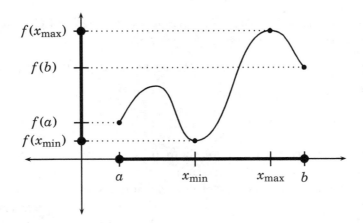

Figure 2.3: A typical graph of a function that is continuous on a finite, closed interval $[a,b]$. The image $f([a,b])$ is also a finite, closed interval (Theorem 2.3.2), containing a maximum value $f(x_{\max})$ and a minimum value $f(x_{\min})$, which are achieved at some $x_{\max}, x_{\min} \in [a,b]$ (Extreme Value Theorem, Corollary 2.3.1). Also note that all values between $f(a)$ and $f(b)$ (and more, for this particular graph) are in the image $f([a,b])$, and thus achieved for some x-values between a and b (Intermediate Value Theorem, Corollary 2.3.2).

Corollary 2.3.2 (Intermediate Value Theorem (IVT)) *If $f(x)$ is continuous on $[a,b]$, then $f(x)$ achieves every value between $f(a)$ and $f(b)$. In other words, if y_0 is between $f(a), f(b)$, then there exists $x_0 \in [a,b]$ such that $f(x_0) = y_0$.*

(Note that the statement of IVT is contained in the second sentence of the quote from the *Encyclopædia Britannica* on page 105, and the EVT is also hinted at in the quote.) A sample function that demonstrates the theorem and two corollaries is given in Figure 2.3.

It is crucial that the function in question be continuous, and that the domain in question be closed and bounded, i.e., of the form $[a,b]$, to guarantee that the image is another closed, bounded interval. The closed criterion is clearly necessary, as shown in the the first and third graphs in Figure 2.2, page 106. In each graph in that figure, a function $f(x)$ defined on an *open* interval (a,b) is graphed: In the first graph, the image is an open interval and so no maximum or minimum is actually achieved; in the third graph, the image is unbounded from above, so no maximum is achieved, and no minimum is achieved either. It may happen that $f((a,b))$ is a closed and bounded interval, as in the second graph in that figure, but it clearly is not guaranteed (as the other two graphs in the figure illustrate). Continuity is also required in these theorems, as we see in Figure 2.4, page 109.

Note that the first function in the figure is continuous on $[-1,1)$ because it is continuous on $(-1,1)$ and right-continuous at -1. Similarly, it is continuous on $(1,3]$. The second function is continuous on $[-1,1)$ and $[1,3.5]$. That is *not* to say it is continuous at each $x \in [1,3.5]$, but rather that the "piece" drawn on that interval is a continuous "piece," in the sense of Definition 2.3.2, page 107, and the discussion following that.

2.3. CONTINUITY ON INTERVALS

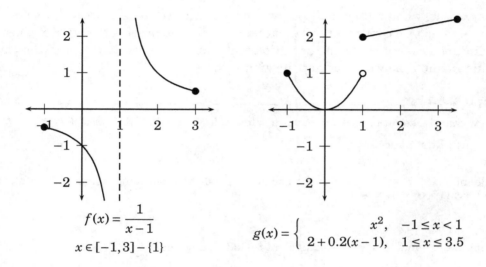

Figure 2.4: Theorems 2.3.1 and 2.3.2—and their corollaries EVT and IVT—require continuity of f, without which the image can be disconnected or unbounded. In the first graph, there is a vertical asymptote at $x = 1$, which is a point of discontinuity. The image $f([-1,1) \cup (1,3]) = (-\infty, -1/2] \cup [1/2, \infty)$ is disconnected (so no IVT conclusion) and unbounded (so no EVT conclusion). In the second case, there is a "jump" discontinuity at $x = 1$, and the image $g([0,1] \cup [2,2.5]) = [0,1] \cup [2,2.5]$ is also disconnected (no IVT conclusion), though bounded. It happens that the second graph does have maximum and minimum values $g(0) = 0$ and $g(3.5) = 2.5$, though this was not guaranteed because $g(x)$ is not continuous on all of $[-1, 3.5]$. These examples do not violate the corollaries IVT and EVT since both corollaries claim the truth of tautologies of the form $P \to Q$, which is true here vacuously. (Alternatively, $P \to Q \iff (\sim P) \vee Q$, which is trivially true if $\sim P$ is true, as it is here.)

2.3.3 Simple Applications of the Intermediate Value Theorem (IVT)

We will return to the Extreme Value Theorem (EVT) and its applications later in the text. Here we will instead look at the Intermediate Value Theorem (IVT) and its usefulness in algebra. The following simple theorem can often be useful when we look at continuity considerations:

Theorem 2.3.3 *If I and J are intervals of any kind except for single points, with $I \subseteq J$, and $f: J \longrightarrow \mathbb{R}$ is continuous on J, then $f: I \longrightarrow \mathbb{R}$ is continuous on I.*

In other words, when a function is known to be continuous on an interval, its restriction to a subinterval is also continuous on that subinterval. The proof is a matter of chasing down the definitions of continuity on the various intervals and checking each of the cases. The following example shows a quick application.

Example 2.3.1 _____

Show that the equation $x^5 + 7 = x^2$ has a solution in $\mathbb{R}$, i.e., that $(\exists x \in \mathbb{R})(x^5 + 7 = x^2)$.

Solution: First notice that
$$x^5 + 7 = x^2 \iff x^5 - x^2 + 7 = 0.$$

Next, define $f(x) = x^5 - x^2 + 7$, which is a polynomial and thus continuous on all of $\mathbb{R}$ (think of $\mathbb{R}$ as the open interval $(-\infty, \infty)$). Now,
$$x^5 + 7 = x^2 \iff x^5 - x^2 + 7 = 0 \iff f(x) = 0,$$
so solving the original equation is equivalent to finding x so that $f(x) = 0$. Next, notice that
$$f(-2) = (-2)^5 - (-2)^2 + 7 = -29, \quad \text{and}$$
$$f(1) = 1 - 1 + 7 = 7.$$

Because f is continuous in $\mathbb{R}$, it is also continuous on $[-2, 1] \subseteq \mathbb{R}$. Since $f(-2) = -29$ while $f(1) = 7$, there exists some x_0 between -2 and 1 such that $f(x_0) = 0$ (by IVT), and this x_0 therefore solves the original equation, q.e.d. (We did not *find* x_0 exactly, but did detect its existence via the IVT.)
_____

Example 2.3.1 used a standard technique of rewriting the problem, wherein we analyze an equation of the form $g(x) = h(x)$ by instead analyzing the logically equivalent equation $g(x) - h(x) = 0$. It is often convenient to define $f(x) = g(x) - h(x)$ and solve (or, as here, simply detect the presence of a solution of) the logically equivalent equation $f(x) = 0$. As we learn from algebra, solving $f(x) = 0$ is often simpler than finding where $g(x) = h(x)$.

We can apply the IVT more than once to a single function, as in the following example:

Example 2.3.2 _____

Show that the equation $x^3 - 8x^2 + 15x = -1$ has at least three solutions.

Solution: This is equivalent to the equation $x^3 - 8x^2 + 15x + 1 = 0$ having at least three solutions. Defining $f(x) = x^3 - 8x^2 + 15x + 1$, we thus want to prove that $f(x) = 0$ occurs at least three times.

First we notice that $f(x)$ is a polynomial, and therefore continuous on $\mathbb{R}$. Next we see that

2.3. CONTINUITY ON INTERVALS

$$f(-1) = -23,$$
$$f(0) = 1,$$
$$f(4) = -3,$$
$$f(5) = 1.$$

We see that $f(x)$ must be zero for some value $x \in (-1, 0)$, another $x \in (0, 4)$, and yet another value $x \in (4, 5)$, therefore proving that there must be at least three solutions of $f(x) = 0$, or equivalently, of $x^3 - 8x^2 + 15x = -1$, q.e.d.

2.3.4 Polynomial Inequalities

A very important consequence of the Intermediate Value Theorem is that, for a continuous function's output to change sign (positive to negative or vice versa) on an interval, its values must pass through zero. We can exploit this to solve polynomial and rational inequalities. The application of the IVT to polynomial inequalities is particularly straightforward, as the next two examples illustrate.

Example 2.3.3

Solve the inequality $x^3 - 9x^2 + 20x < 0$.

<u>Solution</u>: Define $f(x) = x^3 - 9x^2 + 20x$, which is a polynomial and therefore continuous on all of $\mathbb{R}$. We wish to see where $f(x) < 0$. First we will see where $f(x) = 0$, to detect the possible points at which $f(x)$ changes signs (from positive to negative or vice-versa).

$$\begin{aligned} & f(x) = 0 \\ \iff & x^3 - 9x^2 + 20x = 0 \\ \iff & x(x^2 - 9x + 20) = 0 \\ \iff & x(x-4)(x-5) = 0 \\ \iff & x = 0, 4, 5. \end{aligned}$$

These points, $x = 0, 4, 5$, are the **only** points in $\mathbb{R}$ where $f(x)$ can change signs as x passes through values along the x-axis continuum, so $f(x)$ will **not** change signs anywhere **within** the intervals $(-\infty, 0)$, $(0, 4)$, $(4, 5)$, or $(5, \infty)$, as we will explain for one interval of the four intervals (and the argument works for the others as well).

For example, we can use the IVT to help us detect the sign of $f(x)$ for all of $(-\infty, 0)$, as follows. If we know the sign of $f(x)$ at, say, $x_1 = -10$, then $f(x)$ will have the same sign for all $x \in (-\infty, 0)$. This is because if $f(x)$ had the opposite sign as $f(-10)$ at some $x_2 \in (-\infty, 0)$, then (from continuity and the Intermediate Value Theorem) there would have to be *another* value x_0 in between x_1 (i.e., -10) and x_2 for which $f(x_0) = 0$, but that is impossible because x_0 must itself be in $(-\infty, 0)$, though all such values where $f(x) = 0$ were already accounted for in the list $x = 0, 4, 5$, none of which are within that interval. This is a contradiction, and so we must conclude that whatever is the sign of $f(-10)$ is the sign of $f(x)$ for all $x \in (-\infty, 0)$.

Similarly, if we know the sign of $f(x)$ for one value of $x \in (0, 4)$, then we know it for all values of $x \in (0, 4)$. The same argument holds separately for the remaining intervals $(4, 5)$ and $(5, \infty)$.

A useful visual device for determining where $f(x)$ is positive, and where it is negative, is a **sign chart**, as we will illustrate. For this particular function, the signs are shown on their respective intervals, with the points $x = 0, 4, 5$ as boundary points. In the style of sign chart given, we use $\oplus$ to represent a positive quantity and $\ominus$ to represent a negative quantity. We also look at the signs of the factors to determine the sign of the function. For instance, $f(4.5) = 4.5(4.5 - 4)(4.5 - 5) = 4.5(0.5)(-0.5)$, with respective factor signs $\oplus \oplus \ominus$, which yield a negative value for $f(4.5)$. With some practice, the signs of the factors can be detected easily by quick inspection.

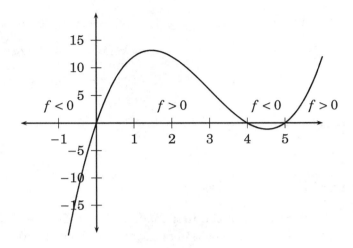

Figure 2.5: Actual graph of $f(x) = x(x-4)(x-5)$, showing where $f(x) > 0$, where $f(x) < 0$, and how the transitions between these occur: in this case by the function's height passing through the value zero. Compare to the sign chart in Example 2.3.3.

Function:		$f(x) = x(x-4)(x-5)$		
Test $x =$	-10	2	4.5	10
Sign Factors:	$\ominus\ominus\ominus$	$\oplus\ominus\ominus$	$\oplus\oplus\ominus$	$\oplus\oplus\oplus$

$$\text{Sign } f(x): \quad \ominus \quad 0 \quad \oplus \quad 4 \quad \ominus \quad 5 \quad \oplus$$

From the chart we see that $f(x) < 0$ on the first and third intervals, i.e., for $x \in (-\infty, 0) \cup (4, 5)$. Since $f(x) < 0$ is equivalent to our original inequality, this is also the solution of that inequality.

The information at hand will not give us a complete picture of the graph of $f(x)$, but it is instructive to see what the graph looks like, and how an accurate enough (for our purposes here) picture can be easily imagined from the sign chart. For this reason the graph is given in Figure 2.5.

The logic that was used in constructing the sign chart bears repeating. Since $f(x)$ is continuous on $\mathbb{R}$, the only way its output can change sign (as its input increases along the continuum of $\mathbb{R}$) is for the output to pass through the value zero (by IVT, see also Figure 2.5), so we chart all the x-values for which $f(x) = 0$. These mark boundaries of subintervals of $\mathbb{R}$ on which f does not change sign (within the intervals). For each such interval, knowing the sign of $f(x)$ at any value in the interval gives us the sign of $f(x)$ for the whole interval (since, again, $f(x)$ being continuous cannot change sign in the interval without there being another zero-valued output in the interval, and all such points are accounted for). If $f(x)$ happens to be factored, we only need to check the signs of each factor to see if the negative factors "cancel" completely to leave a positive function, or if we have an odd number of negative factors to make f negative on the interval in question.

Example 2.3.3 was relatively straightforward. There can be complications, and we have

2.3. CONTINUITY ON INTERVALS

to be careful to answer the given question. For instance, we do not always have *strict inequalities* <, >, but may have *inclusive inequalities* ≤, ≥.

Example 2.3.4

Solve $x^2 \geq x+1$.

Solution: First we subtract, and then define $f(x) = x^2 - x - 1$, so that

$$x^2 \geq x+1 \iff x^2 - x - 1 \geq 0 \iff f(x) \geq 0.$$

Solving $f(x) = 0$ requires the quadratic formula or completing the square. We will opt for the former. Recall first that $f(x) = 0 \iff x^2 - x - 1 = 0$, so

$$f(x) = 0 \iff x = \frac{-(-1) \pm \sqrt{(-1)^2 - 4(1)(-1)}}{2(1)} = \frac{1}{2} \pm \frac{1}{2}\sqrt{5} \approx -0.61803, 1.61803.$$

Whenever possible, we will use exact values, but the approximate ones are also useful since we need to know where to find our test points.[17]

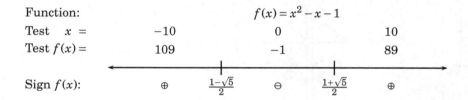

Recall that we are searching for all points for which $f(x) \geq 0$. These include the cases where $f(x) = 0$ as well as where $f(x) > 0$. Therefore we include the endpoints when we report the solution: $x \in \left(-\infty, \frac{1-\sqrt{5}}{2}\right] \cup \left[\frac{1+\sqrt{5}}{2}, \infty\right)$.

The function $f(x) = x^2 - x - 1$ is graphed in Figure 2.6, page 114. Compare the graph with the sign chart.

In the first two examples, we had the function $f(x)$ switch signs at every point where $f(x) = 0$. This is not always the case. It is possible for the graph to touch the axis, and retreat back to the same side. Consider the following example.

Example 2.3.5

Solve the inequality $x^3 - 6x^2 + 9x > 0$.

Solution: This is already a question about sign, so we will simply take $f(x) = x^3 - 6x^2 + 9x$ and solve $f(x) > 0$. Now,

$$f(x) = x^3 - 6x^2 + 9x = x(x^2 - 6x + 9) = x(x-3)^2.$$

We see that $f(x) = 0$ at $x = 0, 3$. This gives us the following sign chart:

---

[17]Alternatively but somewhat less efficiently, in Example 2.3.4 we could factor $f(x)$ based on the solutions to $f(x) = 0$, namely $\frac{1}{2} \pm \frac{1}{2}\sqrt{5}$:

$$f(x) = \left(x - \frac{1+\sqrt{5}}{2}\right)\left(x - \frac{1-\sqrt{5}}{2}\right).$$

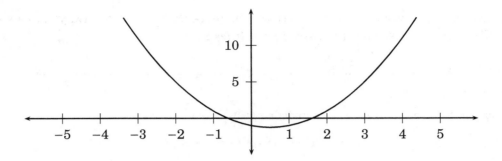

Figure 2.6: Actual graph of $f(x) = x^2 - x - 1$. The function is zero at $\frac{1}{2} \pm \frac{1}{2}\sqrt{5} \approx -0.61803, 1.61803$. Compare to the sign chart in Example 2.3.4, page 113.

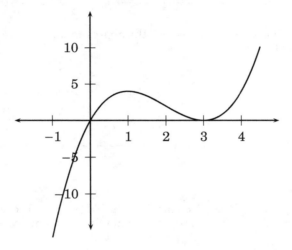

Figure 2.7: Graph of $f(x) = x(x-3)^2$ from Example 2.3.5, starting on page 113, illustrating that a function's height can be zero without the function changing signs there.

Function:		$f(x) = x(x-3)^2$	
Test $x =$	-1	2	4
Sign factors:	$\ominus\ominus^2$	$\oplus\ominus^2$	$\oplus\oplus^2$
Sign $f(x)$:	$\ominus$ 0	$\oplus$ 3	$\oplus$

We see from the sign chart that $f(x) > 0$ for $x \in (0,3) \cup (3,\infty)$.

We see from these examples that we can get an idea of how a function's graph might look from its sign chart. This is particularly true with polynomial functions, but also true with other functions if we are given further information, as we will see as the text unfolds.

An interesting phenomenon regarding *zeros* of polynomials becomes apparent in constructing these sign charts: If $(x - a)$ appears as a factor in a polynomial $f(x)$ to an odd

2.3. CONTINUITY ON INTERVALS

power, the polynomial changes signs at $x = a$, while if $(x-a)$ appears to an even power, the polynomial does not change signs at $x = a$. Some terminology helps to express this.

Definition 2.3.3 *If there is a polynomial $g(x)$ and a positive integer $k \in \mathbb{N}$ so that $f(x)$ can be written*

$$f(x) = (x-a)^k g(x),$$

*but $x - a$ is not a factor of $g(x)$—in other words $(x-a)^k$ is a factor of $f(x)$ but $(x-a)^{k+1}$ is not—then $x = a$ is called a **zero**, or **root**, **of multiplicity** or **degree** k of $f(x)$.*

Then we can observe the following for a polynomial $f(x)$:

1. If $x = a$ is a zero of odd multiplicity, then $f(x)$ changes sign at $x = a$;

2. If $x = a$ is a zero of even multiplicity, then $f(x)$ does not change sign at $x = a$.

In either case, $f(a) = 0$ and so $x = a$ is an x-intercept of the function.

In Example 2.3.5, where $f(x) = x(x-3)^2$, we had $x = 3$ was a zero of multiplicity 2, which is even and so $f(x)$ did not change sign (in the sense of changing from positive to negative or vice versa) as x passed through the value 3, while $x = 0$ was a zero of multiplicity 1, which is odd and so the function did change sign there. If one knew the degrees of each zero, and the sign on one interval, one could construct the sign chart from that information alone, depending on whether the function changes sign (or not) while x passes from one interval to the next.

It is sometimes the case that, even when $f(x)$ is factored, some of the factors (which are themselves polynomials) might never be zero. Such factors therefore never change signs. One common such type of factor is of the form $x^{2k} + a$, where $k \in \mathbb{N}$ and $a > 0$. Such a factor is always positive regardless of $x \in \mathbb{R}$. Another type is a factor of the form $ax^2 + bx + c$ where $b^2 - 4ac < 0$, in which case $ax^2 + bx + c = 0$ has no real solutions. By the Intermediate Value Theorem, if we determine a factor has no real zeros, we know it will not change signs. The following example illustrates one case.

Example 2.3.6 _____

Solve the inequality $x^5 \leq 25x$.

Solution: As before, we construct $f(x) = x^5 - 25x$, so that

$$x^5 \leq 25x \iff x^5 - 25x \leq 0 \iff f(x) \leq 0.$$

Next we factor[18] $f(x) = x(x^4 - 25) = x(x^2 + 5)(x^2 - 5)$. Now the first factor (x) is zero for $x = 0$, the second factor $(x^2 + 5)$ is never zero, and the third factor $(x^2 - 5)$ is zero for $x = \pm\sqrt{5} \approx \pm 2.23607$. Hence we get the following sign chart:

---

[18]For this example, we could continue factoring $f(x) = x(x^2 + 5)(x^2 - 5) = x(x^2 + 5)(x - \sqrt{5})(x + \sqrt{5})$. However, here we will work with the partially factored form, as this will be sufficient. It is really a matter of personal taste.

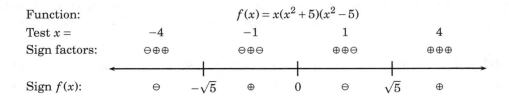

We see that $f(x) \leq 0$ for $x \in (\infty, -\sqrt{5}] \cup [0, \sqrt{5}]$.

2.3.5 Rational Inequalities

When applying the Intermediate Value Theorem, we do have to be careful that we actually have continuity. For instance, if we define $f(x) = 1/(x-1)$, we see that $f(-1) = -1/2$, while $f(3) = 1/2$. However, there are no values between -1 and 3 at which $f(x) = 0$. The reason the IVT did not apply is that $f(x) = 1/(x-1)$ is not continuous on $[-1,3]$, since it is discontinuous at $x = 1$. If we allow for discontinuities, then a function's output can "jump" past a particular value, and the image be disconnected (i.e., not an interval). See the first graph in Figure 2.4, page 109, for an illustration of this phenomenon with this particular function. (The second graph in the figure also shows how discontinuity allows "jumps" in some functions.)

However, we can still use the IVT to solve rational inequalities. We simply need to analyze them and the IVT further. For instance, we can use the following corollary to that theorem:

Corollary 2.3.3 *Suppose that $f(a) = A$ and $f(b) = B$, where $a < b$, and suppose further that C is between A and B. Then at least one of the following must hold:*

(i) *There exists c between a and b so that $f(c) = C$;*

(ii) *$f(x)$ has at least one discontinuity on $[a,b]$.*[19]

In other words, for a function to pass from the height A to the height B, it must either pass through every height in between, or must be discontinuous, as proved next (see "q.e.d.").

> **Proof:** We can use some symbolic logic notation to demonstrate this. We can start with the statement of the Intermediate Value Theorem,
>
> $$f \text{ continuous on } [a,b] \implies \text{(i)}.$$
>
> Thus, f continuous on $[a,b] \implies$ (i), meaning that f continuous on $[a,b] \longrightarrow$ (i) is a tautology. Since $P \longrightarrow Q \iff (\sim P) \vee Q$, our valid implication is equivalent to another statement of the corollary (see footnote):
>
> $$\sim (f \text{ continuous on } [a,b]) \vee \text{(i)}, \qquad \text{q.e.d.}$$

---

[19] By "$f(x)$ has at least one discontinuity on $[a,b]$," we mean simply that $f(x)$ does not satisfy the definition of a function continuous on $[a,b]$. This can happen if (1) there is a point in (a,b) at which f is not continuous in the two-sided sense, or (2) f is not right-continuous at a, or (3) f is not left-continuous at b.

2.3. CONTINUITY ON INTERVALS

The corollary should be intuitive: As we move the value of x along the interval $[a,b]$, for the point $(x, f(x))$ to pass continuously from one height to another, it must pass through all intermediate heights; if it is not going to pass through all intermediate heights, it has to somehow "jump" over any missed heights, requiring a discontinuity.

One implication for rational function or any other function is immediate. *For a function's output to change signs as we vary its input, that output must pass through zero or be discontinuous.* This has implications for sign charts for more general functions:

> **When we construct a sign chart for a function f, we must include as boundary points all those points where f is zero or discontinuous.**

For polynomial or other everywhere-continuous functions, there are no discontinuities and it is enough to consider those values for which the function is zero. For rational functions (ratios of polynomials), we must look at where the numerator is zero (at which points the function has zero value, unless the denominator is also zero there) and where the denominator is zero (at which points the function is discontinuous because it is undefined).

Example 2.3.7

Solve $\dfrac{x}{x^2-1} \le 0$.

Solution: Defining $f(x) = \dfrac{x}{x^2-1} = \dfrac{x}{(x+1)(x-1)}$, we see that f is zero at $x=0$ and discontinuous at $x = \pm 1$. We now use all three of these points to construct the sign chart.

Function:		$f(x) = \dfrac{x}{(x+1)(x-1)}$		
Test $x=$	-2	-0.5	0.5	2
Sign factors:	$\dfrac{\ominus}{\ominus\ominus}$	$\dfrac{\ominus}{\oplus\ominus}$	$\dfrac{\oplus}{\oplus\ominus}$	$\dfrac{\oplus}{\oplus\oplus}$
Sign $f(x)$:	$\ominus \quad -1 \quad \oplus \quad 0 \quad \ominus \quad 1 \quad \oplus$			

Since we are interested in the points where $f(x) \le 0$, we need the open intervals on which $f(x) < 0$, i.e., the first and third intervals, and all points where $f(x) = 0$, i.e., $x = 0$. Collecting all these, we get $x \in (-\infty, -1) \cup [0, 1)$.

We include a graph of $f(x) = \dfrac{x}{(x+1)(x-1)}$ in Figure 2.8 to illustrate how f changes signs by passing through or leaping over zero. There are other types of discontinuities besides vertical asymptotes, but for rational functions f in which the numerator and denominator have no common factors, vertical asymptotes are the only type of discontinuity that can occur. Notice from the graph why we include $x = 0$, but not $x = \pm 1$, when we solve $f(x) \le 0$.

For another perspective justifying the technique for sign charts for rational functions, consider that a ratio of functions can only change signs if the numerator or denominator changes signs. Since both numerator and denominator are polynomials and hence continuous, they can only change signs by passing through zero. Summarizing, we conclude that

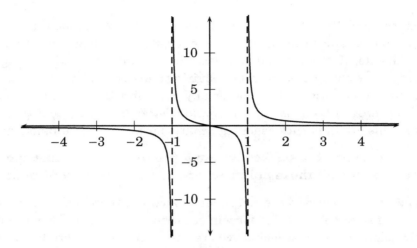

Figure 2.8: Graph of $f(x) = \dfrac{x}{(x+1)(x-1)}$ from Example 2.3.7. Dashed lines are vertical asymptotes at $x = \pm 1$, which are those values of x outside the domain of $f(x)$. Vertical asymptotes will be properly developed later in the text.

a ratio of polynomials can only change signs if the numerator passes through zero or the denominator passes through zero. However, for more general functions we have to consider all possible types of discontinuities.

This method can generalize for solving any rational inequality the same way we generalized for the polynomial case, as we simply rewrite any inequality into an equivalent statement about signs $(+/-)$.

Example 2.3.8

Solve the inequality $\dfrac{x}{x^2-7} \leq \dfrac{2x}{x^2-9}$.

Solution: First we do as before—make this into a question about signs—by subtracting the right-hand side from the inequality, and define the difference to be $f(x)$. Hence we have

$$f(x) = \frac{x}{x^2-7} - \frac{2x}{x^2-9} = \frac{x(x^2-9) - 2x(x^2-7)}{(x^2-7)(x^2-9)} = \frac{-x^3+5x}{(x^2-7)(x^2-9)} = \frac{-x(x^2-5)}{(x^2-7)(x^2-9)},$$

and we are trying to find where $f(x) \leq 0$. We see that $f(x) = 0$ for $x = 0$ and $x = \pm\sqrt{5} \approx \pm 2.23607$, and is discontinuous at $x = \pm 3$ and $x = \pm\sqrt{7} \approx \pm 2.64575$. The sign chart follows:

Function:				$f(x) = \dfrac{(-x)(x^2-5)}{(x^2-7)(x^2-9)}$				
Test $x =$	-10	-2.9	-2.5	-1	1	2.5	2.9	10
Sign factors:	$\dfrac{\oplus\oplus}{\oplus\oplus}$	$\dfrac{\oplus\oplus}{\oplus\ominus}$	$\dfrac{\oplus\oplus}{\ominus\ominus}$	$\dfrac{\oplus\ominus}{\ominus\ominus}$	$\dfrac{\ominus\ominus}{\ominus\ominus}$	$\dfrac{\ominus\oplus}{\ominus\ominus}$	$\dfrac{\ominus\oplus}{\ominus\ominus}$	$\dfrac{\ominus\oplus}{\oplus\oplus}$

Number line: $-3,\ -\sqrt{7},\ -\sqrt{5},\ 0,\ \sqrt{5},\ \sqrt{7},\ 3$

Sign $f(x)$: $\oplus\ \ \ominus\ \ \oplus\ \ \ominus\ \ \oplus\ \ \ominus\ \ \oplus\ \ \ominus$

2.3. CONTINUITY ON INTERVALS

It is important to note that the sign chart gives us the signs of f on the various **open** intervals. At each endpoint of these intervals the function is either zero or discontinuous (in this case because it is undefined). Since for this case we want our solution to be those points where $f(x) \leq 0$, we have to include those endpoints where $f(x) = 0$, namely $x = 0, \sqrt{5}, -\sqrt{5}$. Putting all this together, we see that

$$x \in \left(-3, -\sqrt{7}\right) \cup \left[-\sqrt{5}, 0\right] \cup \left[\sqrt{5}, \sqrt{7}\right) \cup (3, \infty).$$

Note where f changes signs continuously (i.e., passing through zero height) at $0, \pm\sqrt{5}$ and discontinuously (in fact, via vertical asymptotes) at $x = \pm\sqrt{7}, \pm 3$. We do not include a graph of $f(x)$ here, but much of its behavior is evident by the sign chart and the way f changes signs by passing through height zero at $x = 0, \pm\sqrt{5}$ and by discontinuity (in fact, via vertical asymptotes) at $x = \pm 3, \pm\sqrt{7}$.

Fortunately, most standard calculus problems require only that we know where certain functions (or more precisely, derivatives of functions) are positive (> 0) and where they are negative ($<$). For completeness, here we also discussed the inclusive inequalities ($\geq, \leq$). We will conclude this section with the following example.

Example 2.3.9

Solve the inequality $\dfrac{x^2 + x + 1}{x^2 - 7x + 12} > 0$.

Solution: This is already a question of the sign of a rational function. We begin by defining $f(x)$ to be that function on the left-hand side of the inequality, namely $f(x) = \frac{x^2+x+1}{x^2-7x+12}$, and answer the question of where $f(x) > 0$. As is usual for rational functions, we first check to see where the numerator is zero (and so $f(x) = 0$ or is undefined if the denominator is also zero there), and where the denominator is zero (where $f(x)$ is undefined and therefore discontinuous). For the numerator (NUM), we need the quadratic formula:

$$\text{NUM} = 0 \iff x = \frac{-1 \pm \sqrt{1^2 - 4(1)(1)}}{2(1)} = \frac{-1 \pm \sqrt{-3}}{2} \notin \mathbb{R}.$$

We see that the numerator is never zero for $x \in \mathbb{R}$, and so the numerator does not contribute any points to include in making the sign chart. For the denominator (DEN) we have

$$\text{DEN} = 0 \iff x^2 - 7x + 12 = 0 \iff (x-3)(x-4) = 0 \iff (x=3) \lor (x=4).$$

We use these two points to bound the regions of the sign chart:

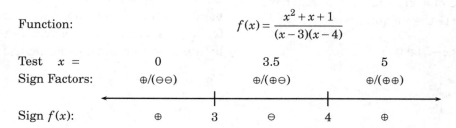

From the sign chart, we see the solution is $x \in (-\infty, 3) \cup (4, \infty)$.

Exercises

1. For each of the following, draw a continuous function $f(x)$ whose domain is $x \in (2,5)$, and whose image of that set (i.e., whose range) is given.

    (a) $(1,4)$
    (b) $\{3\}$
    (c) $\mathbb{R}$
    (d) $[1,4]$
    (e) $(1,4]$
    (f) $(-\infty, 1)$
    (g) $(-\infty, 1]$

2. Repeat the previous problem except assume the domain for $f(x)$ is $x \in [2,5]$. Some of the cases are impossible (but are interesting to attempt anyway). See Theorem 2.3.2, page 107.

3. Draw the graph of a function $y = f(x)$, defined for $x \in [2,5]$, where $f(x)$ is continuous on $(2,5)$ but not on $[2,5]$.

4. Show that $x^3 - 6x^2 + 4x + 10 = 0$ has at least three solutions, by checking values of $f(x) = x^3 - 6x^2 + 4x + 10$ at various x-values.

5. Show that $x^5 - 8x^2 + 15x = 97$ has at least one solution in $\mathbb{R}$.

For each of the following, solve the inequality by means of a sign chart. You may have to first rewrite the inequality.

6. $(x+1)(x-3) \geq 0$

7. $x^2 - 9 < 0$

8. $x^2 + 8 \leq 6x$

9. $x^2 + 3x \geq 2$

10. $x^2 - 15 > 0$

11. $x^2 + 15 > 0$

12. $x^2 + 18 > 11x$

13. $\dfrac{x}{x+5} < 0$

14. $\dfrac{x-7}{x+6} \geq 0$

15. $\dfrac{x^2 - 16}{x^2 + 16} > 0$

16. $\dfrac{2x+1}{x^2 - 16} \leq 0$

17. $\dfrac{27x - x^2}{x^2 + 11x + 30} < 0$

18. $\dfrac{(x+2)^2}{x^2 + 4} < 1$

19. $\dfrac{x}{x+5} > \dfrac{1}{x-7}$

20. $x^3 + 2x^2 \leq 15x$

21. $\dfrac{x^2 - 1}{x^4 - 3x^2 - 10} \leq 0$

22. $\dfrac{2x}{x+5} < \dfrac{3x}{x+6}$

Use a sign chart to graph the following functions, to the extent that continuity and sign of f are illustrated. You can assume a vertical asymptote will pass through any point in which the denominator is zero (after any cancellation of factors common to both numerator and denominator).

23. $f(x) = x(x-1)^2(x^2 - 4)$

24. $f(x) = x^3 - x^2 - 2x$

25. $f(x) = \dfrac{x}{x^2 - 9}$

26. $f(x) = \dfrac{x-1}{x^2 - 9x + 8}$

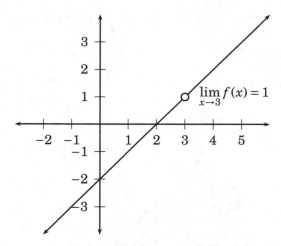

Figure 2.9: Graph of the function $f(x) = (x^2 - 5x + 6)/(x-3) = (x-3)(x-2)/(x-3)$. Except at $x = 3$, this simplifies to $f(x) = x - 2$. Thus the graph of $y = f(x)$ is the same as the line $y = x - 2$, except for a "hole" at $x = 3$. Though $f(x)$ is undefined at $x = 3$, we can still say $\lim_{x \to 3} f(x) = 1$, meaning that the function's output "approaches" 1 as the input "approaches" 3. Verbally we say, "the limit as x approaches 3, of $f(x)$ equals 1."

2.4 Finite Limits at Points

Limits describe many trends in functions, but in very specific ways. Figure 2.9 is one such example, with a limit describing that the output of $f(x)$ "approaches 1" as the input x "approaches 3." The value $f(3)$, whether there is one, is not within the scope of the limit; only the trend is. We will eventually describe precisely what we mean by this.

There are many times when available algebraic analysis regarding equality and such breaks down precisely where our conceptual development of calculus pushes us. The language of limits can "break through" those barriers and give us the insights of calculus. Hence, limits truly form the basis for calculus.

Before the development of limits, mathematicians were able to observe many of the apparent truths of calculus but were unable to actually *prove* them. The development of a theory of limits, and the parallel development of a rigorous definition of $\mathbb{R}$, helped to bridge many important theoretical gaps in calculus between what could be observed and what could be proved.[20]

For our simplest cases, limits are just a small step away from continuity; we can compute many limits quickly based on our knowledge of continuous functions. However, the concept of limit has been greatly expanded to have meaning in many other contexts. This section is devoted to the simplest case—closest to continuity—which is the case of a finite limit at a point.

As with continuity, we will have many theorems that should be remembered and understood, and that are intuitive in their statements, even though their proofs are somewhat

---

[20] In fact even the ancient Greek mathematician and physicist Archimedes of Syracuse (287–212 BC) used many arguments that are considered calculus today for his mathematical discoveries. Without the foundations of calculus, his arguments were very convincing but fell short of proofs.

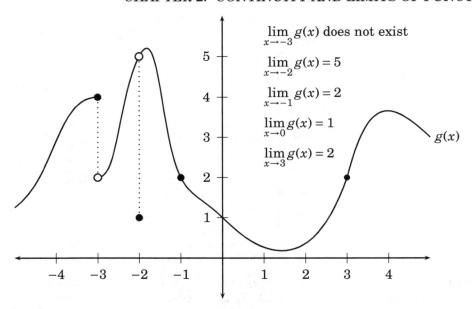

Figure 2.10: A function $g(x)$ to illustrate the concept of limit. (Circles and dotted lines are drawn as visual aids.) See text.

technical. Because of this, we leave the proofs until the end of the section.[21]

2.4.1 Definition, Theorems, and Examples

Before we give the technical definition of a finite limit at a point, we will sharpen the idea further by considering limits arising from the function $g(x)$ graphed in Figure 2.10.

Example 2.4.1

Consider the graph in Figure 2.10. Some of the following conclusions may seem strange at first, but will make more sense as we continue to develop the concept.

- $\lim_{x \to 0} g(x) = 1$. That should be clear from the graph; as x nears zero, the output (height) indeed approaches the value 1. We will note later how continuity, as we have at $x = 0$, has implications for the limit. Similarly, $\lim_{x \to -1} g(x) = 2$.

- $\lim_{x \to -2} g(x) = 5$. This limit expresses the value to which the function $g(x)$ *approaches* as x approaches -2—from both the left and the right of -2—but not *at* -2. This limit is *by design* oblivious to what actually occurs *at $x = -2$*.

- $\lim_{x \to -3} g(x)$ *does not exist*. As the input x approaches the value -3 from the left (i.e., using only values less than -3) the function's value nears 4, but as x approaches -3 from the right (i.e., using only values greater than -3) the function's value approaches 2. For such an ambiguous—indeed contradictory—case, we simply decline to assign a value to the limit, instead declaring that the limit *does not exist* (often written "DNE"). Though not actually relevant, we note that $g(-3) = 4$.

---

[21] To be sure, the proofs are always worth reading and understanding, and the techniques involved are accessible and relevant to our reading here, but for now it is more important to be able to understand and apply the principles enunciated by the theorems and to understand the limit definition and examples.

2.4. FINITE LIMITS AT POINTS

- $\lim_{x \to -1} g(x) = 2$ and (coincidentally) $\lim_{x \to 3} g(x) = 2$. These should be clear from the graph.

Next we will give the technical definition of a finite limit at (or "about") a point.

Definition 2.4.1 *Given a function $f(x)$ defined on the set $0 < |x - a| < d$, for some $d > 0$ (and possibly defined elsewhere), and some $L \in \mathbb{R}$, we define $\lim_{x \to a} f(x) = L$ as follows:*

$$\lim_{x \to a} f(x) = L \iff (\forall \varepsilon > 0)(\exists \delta > 0)(\forall x)(0 < |x-a| < \delta \longrightarrow |f(x) - L| < \varepsilon). \tag{2.22}$$

*If there is no such L satisfying (2.22), we say that $\lim_{x \to a} f(x)$ **does not exist (DNE)**.*[22]

This is sometimes called the *epsilon-delta* (ε-δ) definition of limit. It differs crucially from the definition of continuity in the implication part of the definition:

$$\textbf{continuity:} \quad |x - a| < \delta \longrightarrow |f(x) - f(a)| < \varepsilon; \tag{2.23}$$
$$\textbf{limit:} \quad 0 < |x - a| < \delta \longrightarrow |f(x) - L| < \varepsilon. \tag{2.24}$$

A simplistic interpretation would see that the part of $f(a)$ is played by L in the limit, so L is, in some sense, where $f(a)$ seems it should be, at least if the function is to be continuous (which it may or may not be) at $x = a$.

Equally crucial is that in the limit definition, the implication in (2.24) is silent on the behavior at $x = a$, i.e., when $0 = |x - a|$. Indeed, neither $x = a$ nor $f(a)$ play a role in the limit definition's implication (2.24), while both are crucial in continuity's implication (2.23). (See again Figure 2.10, page 122.)

This definition (2.22) of limit is nontrivial and justifies periodic revisiting. From the cases we will examine in this section, it will become more and more clear that (2.22) is exactly what we need. However, it would be unwieldy indeed to use this definition to prove every limit computation, particularly since such computations are ubiquitous in the rest of this text. Fortunately, we know something about identifying continuous functions, and indeed will be able to bootstrap much of our knowledge of continuity to complete nearly all our limit calculations without working with the ε-δ definition directly. The limit-specific theorems of this section will also assist in developing intuition and computational methods.

Our first theorem is on the uniqueness of the limit, if it exists.

Theorem 2.4.1 *If $\lim_{x \to a} f(x)$ exists, it is unique. In other words,*

$$\left(\lim_{x \to a} f(x) = L\right) \wedge \left(\lim_{x \to a} f(x) = M\right) \implies L = M.$$

---

[22] Note that we will use arrows such as "$\to$" and "$\longrightarrow$" for both implication ($P \longrightarrow Q$) and "approaches" ($x \to a$). The meanings should be clear from the contexts. The valid implication "$\implies$" will keep its earlier meaning throughout.

Theorem 2.4.1 really says the obvious (though the proof is not so transparent): that a function cannot simultaneously be within arbitrarily small ε tolerance of two different values and still conform to the limit definition. Eventually, the function has to "choose" between approaching L or approaching M (or other values, or no values, but never more than one value), or the limit definition is violated. (Recall the discussion of Figure 2.10, particularly as $x \to -3$.) The proof is given at the end of the section.

The following theorem is very important for computing limits.[23]

Theorem 2.4.2 *$f(x)$ is continuous at $x = a$ if and only if $\lim\limits_{x \to a} f(x) = f(a)$.*

Figure 2.9, page 121, makes a good case for the reasonableness of this theorem, as do the limits in Figure 2.10, page 122. The proof is given at the end of the section. In imprecise lay terms: If the value that the function approaches is also where the function "ends up," then we have continuity; if these are different, then we do not. With this theorem, we can compute many limits by simply evaluating the functions at the limit points, *if the functions are continuous there*. (To help decide continuity, we had several theorems in Section 2.2.) For now, we will denote where we use Theorem 2.4.2 by writing "CONT" above an elongated equality symbol, meaning that we can compute the limit by simple evaluation of the function within the limit at the "limit point" due to the continuity of the function there (when this applies).

Example 2.4.2

Consider the following limits, where we can evaluate the functions at their limit points because the functions are continuous at those respective limit points.

- $\lim\limits_{x \to 5} (x^2 + 6x) \stackrel{\text{CONT}}{=\!=\!=} 5^2 + 6(5) = 25 + 30 = 55$,

- $\lim\limits_{x \to 4} \sqrt{26 - x^2} \stackrel{\text{CONT}}{=\!=\!=} \sqrt{26 - 4^2} = \sqrt{10}$,

- $\lim\limits_{x \to 9} \dfrac{1}{x-5} \stackrel{\text{CONT}}{=\!=\!=} \dfrac{1}{9-5} = \dfrac{1}{4}$,

- $\lim\limits_{x \to -3} \dfrac{x}{x^2+1} \stackrel{\text{CONT}}{=\!=\!=} \dfrac{-3}{(-3)^2+1} = \dfrac{-3}{10}$.

Again, we used "CONT" when a function was continuous at the limit point, and therefore we could simply evaluate the function at that point to find the limit. If the function is not continuous at that point, we cannot do so.[24]

We do need to be very careful not to draw a wrong conclusion from Example 2.4.2. By contrast, consider the limits in the following example:

---

[23] Many calculus texts define limits before continuity, using an ε-δ definition for limits only, and then use the *statement* of Theorem 2.4.2 as the *definition* of continuity of $f(x)$ at $x = a$. This is valid since the ε-δ definition of limit stands alone without reference to continuity, and (as the proof of the theorem shows) their limit definition of continuity is equivalent to our earlier ε-δ definition. We instead first developed the arguably more intuitive concept of continuity, defined using ε-δ and independent of limits. This avoids the need to first properly develop limits—with their many extra subtleties (stemming from limits using L instead of $f(a)$ and $0 < |x - a| < \delta$ instead of $|x - a| < \delta$)—in order to develop the more intuitive concept of continuity. Our numerous continuity theorems have analogs in the other texts' limit theorems, from which they derive our continuity theorems. We will eventually state their stand-alone limit theorems (Section 2.9) for when they are immediately useful.

[24] In textbooks which introduce limits before continuity and use the statement of Theorem 2.4.2 as the definition of continuity, the usual way to decide continuity is to check $\lim_{x \to a} f(x) = f(a)$ by verifying that (1) $\lim_{x \to a} f(x)$ exists, (2) $f(a)$ is defined, and finally (3) that these are equal. If any of these are false, the function is not continuous at $x = a$. It is a common technique whose spirit we will use in later sections.

2.4. FINITE LIMITS AT POINTS

Example 2.4.3

Consider the following limit computations:

- $\lim_{x \to 3} \sqrt{9-x^2}$ does not exist because we cannot approach from the right side of 3; if $x > 3$, then $9 - x^2 < 0$ and so $\sqrt{9-x^2}$ is undefined. This does not contradict Theorem 2.4.2, since the function is not continuous at $x = 3$, but only left-continuous there.

- $\lim_{x \to 2} \sqrt{9-x^2} \overset{\text{CONT}}{=\!=\!=\!=} \sqrt{9-2^2} = \sqrt{5}$, since $\sqrt{9-x^2}$ is continuous at each $x \in (-3,3)$, and $2 \in (-3,3)$.

- $\lim_{x \to -3} \sqrt{9-x^2}$ does not exist because we cannot approach from the left side of -3. The function is not continuous at $x = -3$, but only right-continuous there.

- $\lim_{x \to 0} \sqrt{9-x^2} \overset{\text{CONT}}{=\!=\!=\!=} \sqrt{9-0^2} = 3$.

Continuity at $x = a$ is a stronger condition than the limit existing at $x = a$: With continuity the limit must exist *and* equal the function value, which itself must also exist (be defined) at the limit point $x = a$.

Where limits are truly useful then is with functions that might not be continuous at a given point, but might nonetheless have a limit there. In computing limits, we will make repeated use of the following theorem, which lets us examine limits by considering functions that are equal to the given function in question near, but not necessarily at, the limit point.

Theorem 2.4.3 *If $f(x) = g(x)$ on a set $0 < |x - a| < d$, where $d > 0$, then*

$$\lim_{x \to a} f(x) = \lim_{x \to a} g(x),$$

or both limits do not exist.

Sometimes, albeit somewhat casually, this is phrased, "If $f(x)$ and $g(x)$ agree near, but not necessarily at, $x = a$, then their limits as x approaches a will be the same or both not exist." The previous theorem clarifies what we mean by "near," meaning

$$(\exists d > 0)(\forall x)[0 < |x - a| < d \longrightarrow (f(x) = g(x))].$$

Theorem 2.4.3 follows quickly from the definition of limit, in which we can replace f with g, assuming $\delta \leq d$, which is the sort of thing we did in proving several continuity theorems. This proof is short and so we will insert it here instead of at the end of the section.

Proof: Assume $\lim_{x \to a} g(x) = L$ and there exists $d > 0$ so that $f(x) = g(x)$ whenever $0 < |x - a| < d$.

Let $\varepsilon > 0$. Then there exists $\delta_g > 0$ so that $0 < |x - a| < \delta_g \implies |g(x) - L| < \varepsilon$. Now take $\delta = \min\{d, \delta_g\}$. Then

$$0 < |x - a| < \delta \implies \underbrace{|f(x) - L| = |g(x) - L|}_{\text{since } \delta \leq d} \underbrace{< \varepsilon.}_{\text{since } \delta \leq \delta_g}$$

We showed that under the theorem's hypotheses, if $\lim_{x\to a} g(x)$ exists, it must be equal to $\lim_{x\to a} f(x)$. A similar argument with f and g reversed shows the converse. Their contrapositives show that if one of the limits does not exist, then the other limit cannot exist, q.e.d.

Besides having several applications, this theorem again illustrates two important features of the limit: its local nature and its built-in blind spot at the limit point. As for its local nature, note that f and g could be wildly different farther away from $x = a$ (i.e., for $|x-a| \geq d$) and it would be of no importance to the limit. As for its blind spot, again the case $|x - a| = 0$ (i.e., $x = a$) has no bearing on the limit computation (though the function at $x = a$ and the limit for $x \to a$ coincide when, and only when, the function is continuous there).

The most common occasion to use Theorem 2.4.3 is when we calculate limits by simplifying the given function to one that is the same near $x = a$ and has a more obvious limit there, in particular if the replacement function is continuous at the limit point. We will consider several examples next. For our first example, we revisit the limit that began this section. (See again Figure 2.9, page 121.)

Example 2.4.4

Compute the limit $\lim\limits_{x\to 3} \dfrac{x^2 - 5x + 6}{x - 3}$.

Solution: Assuming $f(x) = \dfrac{x^2 - 5x + 6}{x - 3}$, we see that

$$f(x) = \frac{(x-3)(x-2)}{x-3} = x - 2, \qquad x \neq 3. \tag{2.25}$$

In other words, for $0 < |x - 3| < \infty$ (thus $0 < |x - 3| < d$ for any positive number d in the theorem), we have $f(x) = g(x)$, where $g(x) = x - 2$, a function continuous at $x = 3$ (and everywhere else). Now

$$\lim_{x\to 3} g(x) = \lim_{x\to 3}(x - 2) \stackrel{\text{CONT}}{=\!=\!=} 3 - 2 = 1,$$

and since f and g agree except at $x = 3$, we conclude that $\lim\limits_{x\to 3} f(x) = 1$ as well. (See the summary explanation that follows for how we would normally—and more efficiently—compute this limit.)

The explanation in this example is complete and correct but rather long-winded. Since it is one of the simpler limit problems we will come across, a less verbose explanation can suffice while still being faithful in spirit to the theorems used. For example, a summary version of the limit computation for this case can be written in the following style:

$$\lim_{x\to 3} \frac{x^2 - 5x + 6}{x - 3} \stackrel{0/0}{\underset{\text{ALG}}{=\!=\!=}} \lim_{x\to 3} \frac{(x-3)(x-2)}{x-3} \stackrel{0/0}{\underset{\text{ALG}}{=\!=\!=}} \lim_{x\to 3}(x - 2) \stackrel{\text{CONT}}{=\!=\!=} 3 - 2 = 1.$$

Writing "0/0" (usually read "zero divided by zero," or "zero over zero") over the equality symbols "=" signifies that the limit is of a particular *form*, namely 0/0 form, which we will introduce in Definition 2.4.2. The "ALG" underneath the second equality symbol signifies that an algebraic rewriting was carried out, which was legitimate near, but not necessarily at, the limit point (again, (2.25)). That particular step is where we used Theorem 2.4.3, with

2.4. FINITE LIMITS AT POINTS

the original function playing the part of $f(x)$ and the function $(x-2)$ playing the part of $g(x)$. (The "ALG" under the first equality symbol just signifies that we algebraically rewrote the function, so we will use "ALG" in both contexts.) We will have other comments to write in the convenient spaces above and below the equality symbols for bookkeeping purposes, so we can concisely organize, justify, and check our work.[25]

The previous example is a limit of a certain *form*, specifically the "0/0 form," as described next.

Definition 2.4.2 *A limit* $\lim_{x \to a} f(x)$ *is of form* **0/0** *if and only if $f(x)$ is written in the form* $f(x) = \frac{g(x)}{h(x)}$, *where* $\lim_{x \to a} g(x) = 0$ *and* $\lim_{x \to a} h(x) = 0$. *It is* **indeterminate,** *in that knowing this limit behavior of $g(x)$ and $h(x)$ is not enough to determine the limit of $f(x)$, or even if it exists.*

The form 0/0 is indeterminate—meaning the form does not imply the limit's value—because

- a numerator shrinking in absolute size by itself tends to shrink a fraction in absolute size, while

- a denominator shrinking in absolute size by itself tends to make a fraction grow in absolute size.

Exceptions to each of these occur when the other part of the fraction is zero already or undefined.

Knowing these trends are happening simultaneously (i.e., knowing we have 0/0 form) does not tell us the relative rates at which the two influences are operating. In other words, which competing influence (numerator shrinking or denominator shrinking) dominates? Or is there an equilibrium approached, as in the previous example?

We will encounter several other indeterminate forms in this text, and even more *determinate* forms in which the limit's value is indeed implied by the form. Most of the interesting limits we will find here begin as indeterminate forms. Fortunately, we can often simplify an indeterminate form—particularly if we ignore the limit point itself, as we are allowed to—and thus produce a new, equal limit with a determinate form. We did so in the previous example, finding an equivalent polynomial limit, though we invoked continuity rather than another "form" for the final computation.

For now, it is important to note that when we have 0/0 form, we cannot evaluate the function at the limit point as we can with a continuous function, but instead have more work to do, often algebraic, ideally finding a continuous function that is equal to the original function within the limit near (or even everywhere), except possibly at, the limit point itself. This is our strategy in the following examples. In particular, note how we work towards canceling terms in the fraction which cause the denominator to approach zero as the input (in our cases, x) approaches the limit point.

---

[25]The comment system we use here was developed by the authors (who doubt that it is original or unique). It has been very useful to calculus students wishing to follow the instructor's thinking and to clarify their own. One would usually omit such comments in professional publications, where readers are expected to have sufficient knowledge and experience to fully understand—or be able to verify for themselves—each step without further explanation. Their knowledge and experience, of course, come from having practice in solving problems themselves as they learned and later applied these principles.

Example 2.4.5

Compute $\lim\limits_{x\to 9}\dfrac{\sqrt{x}-3}{x-9}$.

<u>Solution</u>: Though $x = 9$ is outside of the domain of f, namely $[0,9)\cup(9,\infty)$, we can certainly approach $x = 9$ from both directions. More casually, we can say that we can let x venture small distances to the left or right of $x = 9$ and the function will be defined. The usual technique for a problem such as this is to algebraically rewrite it by multiplying by $(\sqrt{x}+3)/(\sqrt{x}+3)$:

$$\lim_{x\to 9}\frac{\sqrt{x}-3}{x-9}\stackrel{0/0}{\underset{\text{ALG}}{=\!=\!=}}\lim_{x\to 9}\frac{\sqrt{x}-3}{x-9}\cdot\frac{\sqrt{x}+3}{\sqrt{x}+3}\stackrel{0/0}{\underset{\text{ALG}}{=\!=\!=}}\lim_{x\to 9}\frac{x-9}{(x-9)(\sqrt{x}+3)}$$
$$\stackrel{0/0}{\underset{\text{ALG}}{=\!=\!=}}\lim_{x\to 9}\frac{1}{\sqrt{x}+3}\stackrel{\text{CONT}}{=\!=\!=}\frac{1}{\sqrt{9}+3}=\frac{1}{6}.$$

As before, we took a 0/0 form and algebraically manipulated it until we found a function equal to the original near, but not at, $x = 9$ and could then evaluate the new function at that point since it was continuous there. A quick alternative method for this limit uses some slightly clever factoring of the denominator:

$$\lim_{x\to 9}\frac{\sqrt{x}-3}{x-9}\stackrel{0/0}{\underset{\text{ALG}}{=\!=\!=}}\lim_{x\to 9}\frac{\sqrt{x}-3}{(\sqrt{x}-3)(\sqrt{x}+3)}\stackrel{0/0}{\underset{\text{ALG}}{=\!=\!=}}\lim_{x\to 9}\frac{1}{\sqrt{x}+3}\stackrel{\text{CONT}}{=\!=\!=}\frac{1}{6}.$$

Example 2.4.6

Compute $\lim\limits_{x\to 4}\dfrac{\frac{1}{x}-\frac{1}{4}}{x-4}$.

<u>Solution</u>: This is of form 0/0 as well. We need to simplify the fraction and see how things might cancel. To be rid of "fractions in the numerator," we will multiply by $\frac{4x}{4x}$.

$$\lim_{x\to 4}\frac{\frac{1}{x}-\frac{1}{4}}{x-4}\stackrel{0/0}{\underset{\text{ALG}}{=\!=\!=}}\lim_{x\to 4}\frac{\frac{1}{x}-\frac{1}{4}}{x-4}\cdot\frac{4x}{4x}\stackrel{0/0}{\underset{\text{ALG}}{=\!=\!=}}\lim_{x\to 4}\frac{\frac{1}{x}\cdot 4x-\frac{1}{4}\cdot 4x}{(x-4)(4x)}\stackrel{0/0}{\underset{\text{ALG}}{=\!=\!=}}\lim_{x\to 4}\frac{4-x}{(x-4)(4x)}$$
$$\stackrel{0/0}{\underset{\text{ALG}}{=\!=\!=}}\lim_{x\to 4}\frac{(-1)(x-4)}{(x-4)(4x)}\stackrel{0/0}{\underset{\text{ALG}}{=\!=\!=}}\lim_{x\to 4}\frac{-1}{4x}\stackrel{\text{CONT}}{=\!=\!=}\frac{-1}{16}.$$

In this example, the domain of the original function was $x\ne 0,4$, so we could approach $x = 4$ locally inside the domain. Such technicalities are important, but one usually does not make mention of them while working a problem unless a quick inspection shows they need to be considered. An alternative algebraic method of simplifying the expression in this limit is to combine the fractions in the numerator: $1/x - 1/4 = (4-x)/(4x)$, so the function simplifies

$$\frac{\frac{1}{x}-\frac{1}{4}}{x-4}=\frac{\frac{4-x}{4x}}{x-4}=\frac{4-x}{4x}\cdot\frac{1}{x-4}=\frac{4-x}{4x(x-4)}.$$

The first method gives the same simplification a step (or two) sooner. Next consider the following rational limit.

2.4. FINITE LIMITS AT POINTS

Example 2.4.7

Compute $\lim_{x\to 3} \dfrac{x^2-9}{x^6-243x}$.

<u>Solution</u>: This time we will factor both numerator and denominator, using $243 = 3^5$ along the way. Recall $(a^n - b^n) = (a-b)(a^{n-1} + a^{n-2}b + \cdots + ab^{n-2} + b^{n-1})$.

$$\lim_{x\to 3} \frac{x^2-9}{x^6-243x} \underset{\text{ALG}}{\overset{0/0}{=\!=}} \lim_{x\to 3} \frac{(x+3)(x-3)}{x(x^5-3^5)} \underset{\text{ALG}}{\overset{0/0}{=\!=}} \lim_{x\to 3} \frac{(x+3)(x-3)}{x(x-3)(x^4+3x^3+9x^2+27x+81)}$$

$$\underset{\text{ALG}}{\overset{0/0}{=\!=}} \lim_{x\to 3} \frac{x+3}{x(x^4+3x^3+9x^2+27x+81)}$$

$$\overset{\text{CONT}}{=\!=} \frac{6}{3(81+81+81+81+81)} = \frac{3\cdot 2}{3\cdot(5\cdot 81)} = \frac{2}{405}.$$

In this example we used the factoring $a^5 - b^5 = (a-b)(a^4 + a^3b + a^2b^2 + ab^3 + b^4)$. Recall earlier we used a version of $a^2 - b^2 = (a-b)(a+b)$ in the form of $x - 9 = (\sqrt{x}-3)(\sqrt{x}+3)$. Similarly, we can use $a^3 - b^3 = (a-b)(a^2 + ab + b^2)$, or the related form $a^3 + b^3 = (a+b)(a^2 - ab + b^2)$. These and related techniques often work with limits containing radicals.

Example 2.4.8

Compute $\lim_{x\to -8} \dfrac{\sqrt[3]{x}+2}{x+8}$.

<u>Solution</u>: This can be accomplished by completing the factorization $a^3 + b^3 = (a+b)(a^2 - ab + b^2)$ in the numerator, with $a = \sqrt[3]{x}$ and $b = 2$:

$$\lim_{x\to -8} \frac{\sqrt[3]{x}+2}{x+8} \underset{\text{ALG}}{\overset{0/0}{=\!=}} \lim_{x\to -8} \frac{x^{1/3}+2}{x+8} \cdot \frac{x^{2/3}-2x^{1/3}+4}{x^{2/3}-2x^{1/3}+4}$$

$$\underset{\text{ALG}}{\overset{0/0}{=\!=}} \lim_{x\to -8} \frac{x+8}{(x+8)(x^{2/3}-2x^{1/3}+4)} \underset{\text{ALG}}{\overset{0/0}{=\!=}} \lim_{x\to -8} \frac{1}{x^{2/3}-2x^{1/3}+4}$$

$$\overset{\text{CONT}}{=\!=} \frac{1}{(-8)^{2/3}-2(-8)^{1/3}+4} = \frac{1}{4-2(-2)+4} = \frac{1}{12}.$$

Note that $1/(x^{2/3} - 2x^{1/3} + 4)$ is continuous at $x = -8$, since this is just $1/\left(\left(\sqrt[3]{x}\right)^2 - 2\sqrt[3]{x} + 4\right)$, the denominator's terms being continuous and the denominator not approaching zero. An algebraic alternative is to factor the denominator using $a^3 + b^3 = (a+b)(a^2 - ab + b^2)$, with $a = \sqrt[3]{x}$ and $b = 2$. The ultimate replacement function is the same, as of course is the limit's value:

$$\lim_{x\to -8} \frac{\sqrt[3]{x}+2}{x+8} \underset{\text{ALG}}{\overset{0/0}{=\!=}} \lim_{x\to -8} \frac{\sqrt[3]{x}+2}{(\sqrt[3]{x}+2)(x^{2/3}-2x^{1/3}+4)} \underset{\text{ALG}}{\overset{0/0}{=\!=}} \lim_{x\to -8} \frac{1}{x^{2/3}-2x^{1/3}+4} \overset{\text{CONT}}{=\!=} \frac{1}{12}.$$

Even when the function is defined at the limit point, we do have to be careful that the limit actually exists, and there are many instances in which it will not. We will discuss a few next, and add more instances in the next section. For the moment, these will tend to involve even roots. The first is similar to Example 2.4.3, page 125.

Example 2.4.9

Consider the limit $\lim_{x\to 0}\dfrac{\sqrt{x}-3}{x-9}$. This is the same function as in Example 2.4.5, page 128, but with a different limit point. However, this limit does not exist, because the function is not defined for $x < 0$, and for the limit to exist we need to be able to approach $x = 0$ from the immediate left as well as the right. Indeed, the domain is $x \in [0,9) \cup (9,\infty)$, and so we cannot have x approaching 0 from the left and expect the function (which is undefined there) to approach a real number in any meaningful way.

The previous example shows again that we cannot always compute a limit by inputting the limit point even though the function is defined there. Recall that we can do so if and only if the function is continuous at that point. The function in Example 2.4.9 is certainly defined at $x = 0$, but has no "wiggle room" to the left of that input value. This fact renders the function discontinuous at $x = 0$, and renders the limit nonexistent there, because for any $\varepsilon > 0$ (though one example is enough) we have no $\delta > 0$ where we can say $|x - 0| < \delta \implies |f(x) - f(0)| < \varepsilon$, or an $L \in \mathbb{R}$ where we can say $0 < |x - 0| < \delta \implies |f(x) - L| < \varepsilon$, because $f(x)$ is undefined for some values of x in either case.

Example 2.4.10

Consider the limit $\lim_{x\to 0}\dfrac{x}{\sqrt{x^2}}$. This limit is of 0/0 form, but ultimately does not exist. For the sake of argument, if it did we would have

$$\lim_{x\to 0}\frac{x}{\sqrt{x^2}} = \lim_{x\to 0}\frac{x}{|x|}.$$

Although $f(x)$ is undefined at $x = 0$, we can use the piecewise definition of the absolute value function to rewrite the function for $x \ne 0$:

$$f(x) = \frac{x}{|x|} = \begin{cases} x/(x) & \text{for } x > 0 \\ x/(-x) & \text{for } x < 0 \end{cases} = \begin{cases} 1 & \text{for } x > 0 \\ -1 & \text{for } x < 0. \end{cases}$$

This function has height -1 for $x < 0$ and height 1 for $x > 0$, so we get different heights when we approach from different sides. Thus the limit does not exist. This function is graphed in Figure 2.11.

If we stay away from the "problem" points of a function, we may have continuity, making our work simpler, as we see next.

Example 2.4.11

Next consider $\lim_{x\to -2}\dfrac{\sqrt{x^2}}{x} = \lim_{x\to -2}\dfrac{|x|}{x}$.

Here we look at two methods of computing this. First we note that the function is continuous at $x = -2$ (and any other point other than at $x = 0$), so we can simply evaluate the function there:

$$\lim_{x\to -2}\frac{\sqrt{x^2}}{x} = \lim_{x\to -2}\frac{|x|}{x} \overset{\text{CONT}}{=\!=\!=} \frac{|-2|}{-2} = \frac{2}{-2} = -1.$$

2.4. FINITE LIMITS AT POINTS

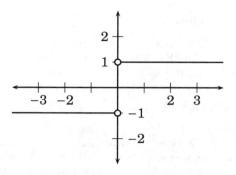

Figure 2.11: Partial graph of $f(x) = x/|x|$. See Example 2.4.10.

Another useful method is to replace the function by another, which is equal to the original near $x = -2$:

$$\lim_{x \to -2} \frac{\sqrt{x^2}}{x} = \underbrace{\lim_{x \to -2} \frac{|x|}{x} = \lim_{x \to -2} \frac{-x}{x}}_{\substack{\text{since } x \to -2 < 0 \\ (\therefore |x| = -x)}} = \lim_{x \to -1} (-1) \overset{\text{CONT}}{=\!=\!=} -1.$$

We include the comment below the brace for explanatory purposes and would normally omit it within the actual computation. The point made there is that in the limit computation, we are not claiming that $|x|/x = -x/x$ for all x, but merely that we can make that substitution here because it **is** true "near $x = -2$"; that is, we can replace the original function by $g(x) = (-x)/x$ because $|x| = -x$ for $x \in (-\infty, 0)$ and -2 is "safely" inside of $(-\infty, 0)$. (Here we used Theorem 2.4.3, page 125, with $d \le 2$.) Furthermore, that function could be then be replaced by the **constant** function $h(x) = -1$—which is trivially continuous—near $x = -2$.

A key fact used in this example, that *constant* functions are continuous, is often overlooked at first by students as they compute limits. Many calculus textbooks emphasize this fact in the limit context by enshrining it in a theorem, but we simply note it as a somewhat trivial fact that is nonetheless useful to remember (here, $a, K \in \mathbb{R}$ are any constants):

$$\lim_{x \to a} K = K. \tag{2.26}$$

This is obvious when its meaning is understood: that if we define $f(x) = K$, where $K \in \mathbb{R}$ is a constant, then $\lim_{x \to a} f(x) = K$ as well. A quick glance at such a function—whose graph is a horizontal line at height K—shows that such a function is obviously continuous, so we can evaluate the limit by evaluating the (constant) function. To be more technical, *any* $\delta > 0$ can satisfy the definition of continuity: Let $\varepsilon > 0$. Choose *any* $\delta > 0$. Then $|x - a| < \delta \implies |f(x) - f(a)| = |K - K| = 0 < \varepsilon$. So, for instance, $f(x) = 8$ is continuous at all $a \in \mathbb{R}$, and we could write for example, $\lim_{x \to 27} 8 = 8$.

Example 2.4.12

Consider the function $f(x) = \begin{cases} 2x + 3 & \text{if } x > 4, \\ x + 4 & \text{if } -3 < x \le 4, \\ x^2 - 8 & \text{if } x < -3. \end{cases}$

- $\lim\limits_{x\to 5} f(x) = \lim\limits_{x\to 5}(2x+3) = 2(5)+3 = 13$.

  This is because $5 \in (4,\infty)$, with room to the left and right in which $f(x) = 2x+3$. More precisely (with $d \le 1$ for instance), $(\exists d > 0)(\forall x)[(0 < |x-5| < d) \longrightarrow (f(x) = 2x+3)]$. The important point is that 5 is well within the interior of an interval on which $f(x)$ is defined to be the continuous function $2x+3$.

- $\lim\limits_{x\to -2.9} f(x) = \lim\limits_{x\to -2.9}(x+4) = -2.9 + 4 = 1.1$.

  Here $-2.9 \in (-3,4)$, and -2.9 is safely within the interval on which $f(x) = x+4$, a continuous function there. Though -2.9 is "only" 0.1 units from where the function's formula changes, that is a positive distance and we can always assume $\delta \le 0.1$ in our limit and continuity proofs to show $f(x)$ is continuous at $x = -2.9$, and thus we can evaluate $x+4$ at $x = -2.9$ in our limit computation.

- $\lim\limits_{x\to 4} f(x)$ does not exist.

  This limit is problematic in that $f(x)$ approaches one number as x approaches 4 from the left, and another as x approaches 4 from the right.

  Later, we will have use for a notation which should be somewhat self-explanatory here, though there are technical matters to clarify: From the left-hand side of $x = 4$, we see $f(x) = x+4 \to 8$, but from the right-hand side of $x = 4$ we have $f(x) = 2x+3 \to 11$. (Here "$\to$" is read "approaches," and not as an implication arrow. This is explained later.)

- $\lim\limits_{x\to -3} f(x) = 1$, but this must be checked.

  To be sure, this requires us to again look at both left-hand side and right-hand side of $x = -3$. From the left we have $f(x) = x^2 - 8 \to 9 - 8 = 1$, and from the right we have $f(x) = x+4 \to -3+4 = 1$. Thus $\lim\limits_{x\to -3} f(x) = 1$ exists.

The arguments made and notation used in the limits as $x \to 4$ and $x \to -3$ will be made more systematic in the next section, and we are only showing a glimpse of them here. In the next section we will deal more with piecewise-defined functions, particularly at points on the boundaries of the "pieces." It is noteworthy that we did not require a graph to analyze any of the limits in the previous example, though we may have had aspects of the graph in mind.

We include another example here, which involves a function that could be piecewise-defined, showing another use for what was the second method used in Example 2.4.11, page 130. As happened in that example, we have use for rewriting an absolute value based upon the location of the limit point.

Example 2.4.13

Consider the function $f(x) = \dfrac{x^2 - 9}{|x+3|(x-3)}$. Then

$$\lim_{x\to 3} f(x) = \lim_{x\to 3} \frac{(x+3)(x-3)}{(x+3)(x-3)} \xrightarrow[\text{ALG}]{0/0} \lim_{x\to 3} 1 = 1;$$

$$\lim_{x\to -5} f(x) = \lim_{x\to -5} \frac{(x+3)(x-3)}{-(x+3)(x-3)} \xrightarrow[\text{ALG}]{0/0} \lim_{x\to -5} (-1) = -1.$$

We used the fact that $|x+3| = x+3$ for x near 3, while $|x+3| = -(x+3)$ for x near -5. A quick check shows this limit does not exist for $x \to -3$ (in a way similar to Example 2.4.10, page 130).

2.4. FINITE LIMITS AT POINTS

Example 2.4.14

Consider the limit $\lim_{x \to 0} \frac{1}{x}$.

This limit does not exist as a finite number. The graph is given at the start of Section 2.6, page 146. It is easy to see that small inputs into this function quickly return large outputs. For instance, $f(10^{-2}) = 10^2$, $f(10^{-3}) = 10^3$, and so on, while $f(-10^{-2}) = -10^2$, $f(-10^{-3}) = -10^3$, and so on. This is sometimes verbally described as $f(x)$ "blowing up" as x gets nearer to zero: $1/x$ is unbounded and negative as x approaches zero from the left and unbounded and positive as x approaches zero from the right. The geometric result is a vertical asymptote at $x = 0$. When we have vertical asymptotes at our limit points, we do not have finite limits.[26]

2.4.2 Further Limit Notation

We next take the opportunity to introduce some convenient notation, already alluded to previously. Unfortunately, it resembles some notation from logic, but it is usually obvious which meaning is intended from the context. The notation is defined as follows:

$$\Big[(x \to a) \longrightarrow (f(x) \to L)\Big] \iff \lim_{x \to a} f(x) = L. \qquad (2.27)$$

For reasons of style, it is often less awkward to write, "$x^2 \longrightarrow 9$ as $x \longrightarrow 3$," than to write "$\lim_{x \to 3} x^2 = 9$." We might also write "as $x \longrightarrow 3$, $x^2 \longrightarrow 9$," or "$x \longrightarrow 3 \implies x^2 \longrightarrow 9$." (Arrow lengths are often chosen for convenience or aesthetics, both here and in logic notation.) It is important, but usually obvious, where the expressions would be placed in our original limit notation. The usefulness of this notation will become more apparent in the next section. One nice use we can put it to here is with the following, perhaps more elegant, restatement of Theorem 2.4.2:

$$f(x) \text{ continuous at } x = a \iff (x \to a) \longrightarrow (f(x) \to f(a)). \qquad (2.28)$$

We will have much more use for this kind of notation as we look at other limit forms and other contexts where we use limits. It is only mentioned here so that the reader will be aware of it well before we make more widespread use of it.

However, there are technicalities that must be considered. For instance, there are differences between what is meant by $x \to a$ and $f(x) \to f(a)$ in our somewhat imprecisely self-explanatory notation for continuity in (2.28). For instance, $x \to a$ requires x be allowed—and indeed required—to assume a continuum of values surrounding a, approaching a as closely as could ever be desired from both sides of a, but never actually equal to a. By contrast, $f(x)$ can approach (in this setting) $f(a)$ from either or both sides, and $f(x)$ can even be equal to $f(a)$ repeatedly.

In fact, such ideas are useful but not sufficient to describe the exact nature of limits, and so when we write "$f(x)$ continuous at $x = a \iff (x \to a) \longrightarrow (f(x) \to f(a))$," we actually mean in the sense of ε-δ definitions in both cases. But this less precise visualization described in the preceding paragraph is somewhat ubiquitous in mathematical analysis, and is useful

---

[26] In subsequent sections we will define infinite limits, and left- and right-side limits. For this section we only concern ourselves with finite limits at (or "about," i.e., from both the left and the right) a point. Nothing we do here contradicts subsequent sections.

enough that we will use it in this text as well, but we will always fall back on the precise definitions if we require clarification.

2.4.3 Proofs of Limit Theorems

First we prove Theorem 2.4.2, which states that $f(x)$ is continuous at $x = a$ if and only if $\lim_{x \to a} f(x) = f(a)$.

Proof: We prove this in two parts. First we show the "only if" part ($\Longrightarrow$):

Suppose that $f(x)$ is continuous at $x = a$. Then by definition,

$$(\forall \varepsilon > 0)(\exists \delta > 0)(\forall x)(|x - a| < \delta \longrightarrow |f(x) - f(a)| < \varepsilon).$$

Now, for a pair ε, δ from continuity, we get

$$0 < |x - a| < \delta \Longrightarrow |x - a| < \delta \longrightarrow |f(x) - f(a)| < \varepsilon,$$

and so the definition of limit works here with the ε, δ from (assumed) continuity, and $f(a)$ for L.

For the "if" part ($\Longleftarrow$), suppose $\lim_{x \to a} f(x) = f(a)$. Thus

$$(\forall \varepsilon > 0)(\exists \delta > 0)(\forall x)(0 < |x - a| < \delta \longrightarrow |f(x) - f(a)| < \varepsilon).$$

We only need to show that

$$0 = |x - a| \Longrightarrow |f(x) - f(a)| < \varepsilon.$$

But that is obvious from the definition of a function, since

$$|x - a| = 0 \Longleftrightarrow (x = a) \Longrightarrow f(x) = f(a) \Longrightarrow |f(x) - f(a)| = 0 < \varepsilon.$$

Thus the case $0 < |x - a| < \delta$ is taken care of by the limit assumption, and the case $0 = |x - a|$ by the definition of function, giving us the full case $|x - a| < \delta$, as required by the continuity definition, q.e.d.

Now we prove Theorem 2.4.1, that $\left(\lim_{x \to a} f(x) = L\right) \wedge \left(\lim_{x \to a} f(x) = M\right) \Longrightarrow L = M$.

Proof: We will prove this by contradiction. Suppose that the theorem is false, i.e., that there are two numbers, $L, M \in \mathbb{R}$, different, which satisfy the definition of the limit at $x = a$ (if necessary see (1.21), page 25). In other words,

$$(\forall \varepsilon > 0)(\exists \delta_1 > 0)(\forall x)(0 < |x - a| < \delta_1 \longrightarrow |f(x) - L| < \varepsilon), \tag{2.29}$$

$$(\forall \varepsilon > 0)(\exists \delta_2 > 0)(\forall x)(0 < |x - a| < \delta_2 \longrightarrow |f(x) - M| < \varepsilon). \tag{2.30}$$

Since we are assuming $L \neq M$, we must have $|L - M| > 0$. Choose

$$\varepsilon = \frac{1}{3}|L - M| > 0.$$

2.4. FINITE LIMITS AT POINTS

Next pick $\delta_1 > 0$ which satisfies the implication in (2.29) for this particular ε, and $\delta_2 > 0$ which satisfies the implication in (2.30) for this particular ε. Now define

$$\delta = \min\{\delta_1, \delta_2\}. \tag{2.31}$$

Now choose any x satisfying $0 < |x-a| < \delta$. This also gives $0 < |x-a| < \delta_1$ and $0 < |x-a| < \delta_2$ by (2.31). This ultimately implies

$$|L-M| = |L - f(x) + f(x) - M| \leq |L - f(x)| + |f(x) - M| = |f(x) - L| + |f(x) - M|$$
$$< \varepsilon + \varepsilon = 2\varepsilon = 2 \cdot \frac{1}{3}|L-M| = \frac{2}{3}|L-M| < |L-M| \implies \mathscr{F}.$$

The reason that this is a contradiction is that we ultimately showed that if the conclusion of the theorem can be false (under the hypotheses of the theorem), i.e., if $L \neq M$, then this implied $|L-M| < |L-M|$, which is impossible. Hence the assumption that the theorem is false is itself false, proving that the theorem must be true. This completes the proof of Theorem 2.4.1, q.e.d.[27]

From the symbolic logic point of view, it is interesting to analyze this proof, which an advanced student would recognize immediately as a classic "proof by contradiction," which is a minor variation of *modus tollens*. We could also outline it as

$$(\sim P) \to \mathscr{F} \implies \sim(\sim P) \iff P,$$

where P is the statement of Theorem 2.4.1.

Of course, we have to refer to our discussion of quantifiers to analyze what exactly would be the statement of $\sim P$, and that sets up our proof that $\sim P \implies \mathscr{F}$.

Note also that $(\sim P) \to \mathscr{F} \iff [\sim(\sim P)] \vee \mathscr{F} \iff P \vee \mathscr{F} \iff P$, so we actually have logical equivalence $\iff$ as well as $\implies$ in the previous symbolic logic computation. We showed $(\sim P) \to \mathscr{F}$ is a tautology (by proving the truth of it), hence proving P.

---

[27] We need to be careful about the crucial last inequality in our proof. It was because (under our assumption $L \neq M$) $|L - M| > 0$ that we could write $\frac{2}{3}|L-M| < |L-M|$, which ultimately led to the contradiction $|L-M| < |L-M|$.

As a general rule, if we do not know that $L \neq M$ (and thus $|L-M| > 0$), then we can only say $\frac{2}{3}|L-M| \leq |L-M|$ (because we have equality when $L = M$).

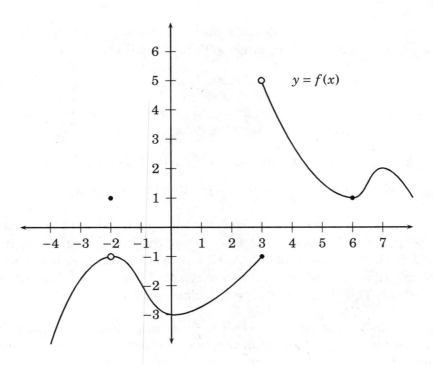

Figure 2.12: Figure for Exercises 1–3. Dotted lines like those found in Figure 2.10, page 122, are omitted here.

Exercises

Consider the graph of $y = f(x)$ given in Figure 2.12.

1. For each limit, see if it exists. If not, state so. If so, evaluate it.

    (a) $\lim_{x \to -2} f(x)$

    (b) $\lim_{x \to 0} f(x)$

    (c) $\lim_{x \to 3} f(x)$

    (d) $\lim_{x \to 6} f(x)$

2. Find $f(-2)$, $f(0)$, $f(3)$ and $f(6)$.

3. Decide if $f(x)$ is continuous at the given point. Explain why or why not, in light of Theorem 2.4.2, page 124.

    (a) $x = -2$

    (b) $x = 0$

    (c) $x = 3$

    (d) $x = 6$

Compute the following limits, if they exist, using Theorem 2.4.2, page 124. If the limit does not exist, explain why.

4. $\lim\limits_{x \to 0} \dfrac{x+2}{x-5}$

5. $\lim\limits_{x \to 5} \sqrt{x^2 - 16}$

6. $\lim\limits_{x \to 4} \sqrt{x^2 - 16}$

7. $\lim\limits_{x \to 3} \sqrt{x^2 - 16}$

8. $\lim\limits_{x \to 81} \sqrt[4]{x}$

2.4. FINITE LIMITS AT POINTS

9. $\lim_{x \to -1} (x^{100} - x^2 + 3)$

10. $\lim_{x \to 0} \sqrt{x^2 + 1}$

11. $\lim_{x \to 0} \sqrt[3]{x}$

12. $\lim_{x \to 7} ((x^2 - 9)(x + 4))$

13. Here we consider why the 0/0-form is indeterminate, i.e., why knowing the numerator and denominator both approach zero does not tell us immediately what the limit is. Compute each of the following limits, where possible.

    (a) $\lim_{x \to 0} \dfrac{x}{2x}$
    (b) $\lim_{x \to 0} \dfrac{x}{3x}$
    (c) $\lim_{x \to 0} \dfrac{x^2}{x}$
    (d) $\lim_{x \to 0} \dfrac{x}{x^2}$
    (e) $\lim_{x \to 0} \dfrac{x}{|x|}$
    (f) $\lim_{x \to 0} \dfrac{x^2}{|x|}$

    (g) Explain why these computations show that the 0/0-form is indeed indeterminate.

Compute the given limits where possible. Be sure to indicate any 0/0 forms and any algebraic steps (preferably using the spaces above and below the "=", as shown in previous examples). If a particular limit does not exist, state so and explain why.

14. $\lim_{x \to 3} \dfrac{x^2 + x - 12}{2x^2 - 5x - 3}$

15. $\lim_{x \to 25} \dfrac{x - 25}{\sqrt{x} - 5}$

16. $\lim_{x \to \sqrt{3}} \dfrac{x^4 - 9}{x^2 - 3}$

17. $\lim_{x \to 0} \dfrac{\frac{1}{x+3} - \frac{1}{3}}{x}$

18. $\lim_{x \to 0} \dfrac{\sqrt{x+1} - 1}{x}$

19. $\lim_{x \to 1} \dfrac{x^8 - 1}{x^2 - 1}$

20. $\lim_{x \to 4} \dfrac{\frac{1}{\sqrt{x}} - \frac{1}{2}}{x - 4}$

21. $\lim_{x \to 16} \dfrac{\sqrt[4]{x} - 2}{x - 16}$

22. $\lim_{x \to -3} \dfrac{x^2 - 9}{x^3 + 27}$

23. Let $f(x) = \dfrac{(x-2)|2x-5|}{(x-1)(2x-5)}$. Compute the following where possible.

    (a) $\lim_{x \to 2} f(x)$
    (b) $\lim_{x \to 0} f(x)$
    (c) $\lim_{x \to 3} f(x)$
    (d) $\lim_{x \to 5/2} f(x)$

24. For each of the following, compute the given limit if it exists. Otherwise, state and explain why it does not.

    (a) $\lim_{x \to 3} \sqrt{x^2 - 4}$
    (b) $\lim_{x \to -2} \sqrt{x^2 - 4}$

25. Compute the limit
$$\lim_{x \to 0} \dfrac{\sqrt{1+x} + \sqrt{1-x} - 2}{x^2}.$$
It can help to rewrite the limit
$$\lim_{x \to 0} \dfrac{1}{x} \left[\dfrac{\sqrt{1+x} - 1}{x} + \dfrac{\sqrt{1-x} - 1}{x} \right].$$

26. Retrace the proof of Theorem 2.4.1, page 123 (that proof actually starting on page 134), and show that the proof still works if we instead take $\varepsilon = \dfrac{1}{2}|L - M|$ to arrive at a contradiction. (You should be able to then observe why the factor 1/2 is the largest factor that works in the proof of the theorem.)

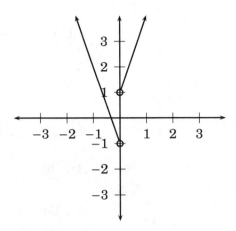

Figure 2.13: Partial graph of $f(x) = (3x^2 + x)/|x|$.

2.5 One-Sided Finite Limits at Points

Just as we found a use for one-sided continuity, so we also have use for so-called left-side and right-side limits. For instance, consider the function $f(x) = (3x^2+x)/|x|$. Now $|x|$ can be defined piecewise to be $-x$ for $x \leq 0$, and x for $x > 0$, for instance. Although $f(x)$ is undefined at $x = 0$, for $x \neq 0$ we can write[28]

$$f(x) = \frac{3x^2 + x}{|x|} = \begin{cases} (3x^2+x)/(x) & \text{for } x > 0 \\ (3x^2+x)/(-x) & \text{for } x < 0 \end{cases} = \begin{cases} 3x+1 & \text{for } x > 0 \\ -3x-1 & \text{for } x < 0. \end{cases}$$

The function is graphed in Figure 2.13. Of course this function is undefined at $x = 0$, but we might be interested in what occurs when we approach zero from one side or the other. We see that approaching zero from the left side (thus moving right towards zero) we travel along a linear piece whose height is approaching -1, while from the right of zero (moving left) we travel a different line whose height approaches 1. The notation we use to reflect this is the following:

$$\lim_{x \to 0^-} f(x) = -1, \qquad \lim_{x \to 0^+} f(x) = 1.$$

We read the notation $x \to 0^-$ as "x approaches zero from the left," and $x \to 0^+$ as "x approaches zero from the right." To compute such limits, in particular without resorting to a graph, we could write the following, as the two cases reflect where $|x| = -x$ and where $|x| = x$:

$$\lim_{x\to 0^-} f(x) = \lim_{x\to 0^-}\frac{3x^2+x}{|x|} \overset{0/0}{\underset{\text{ALG}}{=\!=\!=}} \lim_{x\to 0^-}\frac{3x^2+x}{-x} \overset{0/0}{\underset{\text{ALG}}{=\!=\!=}} \lim_{x\to 0^-}(-3x-1) \overset{\text{CONT}}{=\!=\!=} -1,$$

$$\lim_{x\to 0^+} f(x) = \lim_{x\to 0^+}\frac{3x^2+x}{|x|} \overset{0/0}{\underset{\text{ALG}}{=\!=\!=}} \lim_{x\to 0^+}\frac{3x^2+x}{x} \overset{0/0}{\underset{\text{ALG}}{=\!=\!=}} \lim_{x\to 0^+}(3x+1) \overset{\text{CONT}}{=\!=\!=} 1.$$

Notice that $\lim_{x\to 0} f(x)$ does not exist, since approaching the limit point (zero) from one side gives us one value (-1), while approaching from the other side gives another value (1). We next make several clarifications regarding one-sided limits.

---

[28]Of course, we can let $|x|$ equal x or $-x$ for the case $x = 0$, but that is irrelevant to either limit computation.

2.5. ONE-SIDED FINITE LIMITS AT POINTS

1. When we analyze a left-side limit, we disregard behavior at and to the right of the limit point; when we analyze a right-side limit, we disregard behavior at and to the left of the limit point.

2. We have analogs of previous theorems (Theorem 2.2.11, page 97, Theorem 2.4.2, page 124, and Theorem 2.4.3, page 125) in the following theorem, except that here it is enough to have one-sided continuity and equality of the replacement function in (d) and (e). Consequently, "CONT" written over an equality symbol in the context of one-sided limit computations (using (b)–(e)) refers only to the relevant one-sided continuity. We also have the new, very important theorem in (a) that follows.

Theorem 2.5.1 *The following hold (where we assume that $d > 0$ where it appears):*

(a) $\left(\lim\limits_{x \to a^-} f(x) = L \right) \wedge \left(\lim\limits_{x \to a^+} f(x) = L \right) \iff \lim\limits_{x \to a} f(x) = L.$

(b) $\lim\limits_{x \to a^-} f(x) = f(a) \iff f(x)$ *is left-continuous at* $x = a.$

(c) $\lim\limits_{x \to a^+} f(x) = f(a) \iff f(x)$ *is right-continuous at* $x = a.$

(d) $(\exists d > 0)(\forall x \in (a-d,a))[f(x) = g(x)] \implies \lim\limits_{x \to a^-} f(x) = \lim\limits_{x \to a^-} g(x)$ *(or both do not exist).*

(e) $(\exists d > 0)(\forall x \in (a,a+d))[f(x) = g(x)] \implies \lim\limits_{x \to a^+} f(x) = \lim\limits_{x \to a^+} g(x)$ *(or both do not exist).*

These should become clear as we proceed but are rather technical to prove. In fact (b) and (c) are sometimes given as the *definitions* of left- and right-continuity, but here we defined those first. The interested reader can somewhat easily modify the previous proofs of similar results to arrive at each of these. For now we will pursue these ideas in several examples.

Example 2.5.1

If possible, find $\lim\limits_{x \to 5^-} \sqrt{x^2 - 25}$ and $\lim\limits_{x \to 5^+} \sqrt{x^2 - 25}.$

Solution: Note that the domain of $\sqrt{x^2 - 25}$ is $(-\infty, -5] \cup [5, \infty)$. Thus we cannot approach $x = 5$ from the (immediate) left, so $\lim\limits_{x \to 5^-} \sqrt{x^2 - 25}$ does not exist.

On the other hand, the function is right-continuous at $x = 5$, so we can follow the height at x-values on the right of $x = 5$ and move left in x, and the height will change continuously to the height at $x = 5$ as the input drops to that x-value.

$$\lim\limits_{x \to 5^+} \sqrt{x^2 - 25} \stackrel{\text{CONT}}{=\!=\!=\!=} \sqrt{5^2 - 25} = 0.$$

Of course, it is often helpful to consider the graph as well:

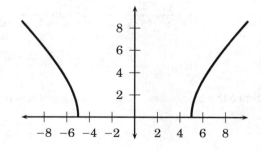

The graph also illustrates that $\lim\limits_{x \to 5^-} \sqrt{x^2 - 25}$ DNE, $\lim\limits_{x \to 5^+} \sqrt{x^2 - 25} = 0$, and (since these do not match) $\lim\limits_{x \to 5} \sqrt{x^2 - 25}$ DNE.

What gave away the fact that we could just evaluate $f(x)$ at $x = 5$ for the right-side limit was that the values inside the square root were nonnegative for $x \in [5,\infty)$ (though the crucial fact was that this held for $x \in [5,d)$ for some $d > 0$). That was not the case as $x \to 5^-$; we have "wiggle room" within the domain of $\sqrt{x^2 - 25}$ to the right, but not to the left, of $x = 5$ (see also its graph). For a more concise presentation, we could write

$$\lim_{x \to 5^-} \sqrt{\underbrace{x^2 - 25}_{<0}} \quad \text{does not exist,}$$

$$\lim_{x \to 5^+} \sqrt{\underbrace{x^2 - 25}_{>0}} \quad \overset{\text{CONT}}{=\!=\!=} \sqrt{5^2 - 25} = 0.$$

Note that the example $f(x) = (3x^2 + x)/|x|$ at the beginning of this section did not allow us to "plug in" $x = 0$ for either limit because the function is neither left-continuous nor right-continuous (nor even defined) at $x = 0$. But in both cases we replaced the given function with functions that were $g(x) = -3x - 1$ for the left-side limit, and $g(x) = 3x + 1$ for the right-side limit.

Example 2.5.2

Compute if possible $\lim\limits_{x \to 3} \dfrac{(x+2)\sqrt{x^2 - 6x + 9}}{x^2 - 7x + 12}$.

Solution: First we will do some algebraic simplification:

$$\lim_{x \to 3} \frac{(x+2)\sqrt{x^2 - 6x + 9}}{x^2 - 7x + 12} \overset{0/0}{\underset{\text{ALG}}{=\!=\!=}} \lim_{x \to 3} \frac{(x+2)\sqrt{(x-3)^2}}{(x-4)(x-3)} \overset{0/0}{\underset{\text{ALG}}{=\!=\!=}} \lim_{x \to 3} \frac{(x+2)|x-3|}{(x-4)(x-3)}.$$

Because we have an absolute value which is zero at our limit point, the (continuous) function inside the absolute value may change sign there and we should check left and right limits.

$$\lim_{x \to 3^-} \frac{(x+2)|\overbrace{x-3}^{<0}|}{(x-4)(x-3)} \overset{0/0}{\underset{\text{ALG}}{=\!=\!=}} \lim_{x \to 3^-} \frac{(x+2)[-(x-3)]}{(x-4)(x-3)} \overset{0/0}{\underset{\text{ALG}}{=\!=\!=}} \lim_{x \to 3^-} \frac{-(x+2)}{(x-4)} \overset{\text{CONT}}{=\!=\!=} \frac{-5}{-1} = 5,$$

$$\lim_{x \to 3^+} \frac{(x+2)|\overbrace{x-3}^{>0}|}{(x-4)(x-3)} \overset{0/0}{\underset{\text{ALG}}{=\!=\!=}} \lim_{x \to 3^+} \frac{(x+2)[(x-3)]}{(x-4)(x-3)} \overset{0/0}{\underset{\text{ALG}}{=\!=\!=}} \lim_{x \to 3^-} \frac{(x+2)}{(x-4)} \overset{\text{CONT}}{=\!=\!=} \frac{5}{-1} = -5.$$

Since the left and right limits do not agree, the original (two-sided) limit does not exist.

With absolute value problems like this example, it is often possible to see quickly if it should be replaced by the quantity inside or its opposite (additive inverse). One only needs to check if it is positive or negative just left or just right of the limit point (depending on the side from which we are approaching). Since $x - 3$ is negative for x-values just left of 3, we can use $|x - 3| = -(x - 3)$ as $x \to 3^-$. (Recall the piecewise definition of $|x|$.)

It is not the case that the two-sided limit will not exist whenever an absolute value is involved. Consider the next two examples.

2.5. ONE-SIDED FINITE LIMITS AT POINTS

Example 2.5.3

Compute, if possible, $\lim\limits_{x\to 0}\dfrac{x^2}{|x|}$.

Solution: As before, we will see what limits we get from both sides.

$$\left.\begin{array}{l}\lim\limits_{x\to 0^-}\dfrac{x^2}{|x|}\overset{0/0}{\underset{\text{ALG}}{=\!=\!=}}\lim\limits_{x\to 0^-}\dfrac{x^2}{-x}\overset{0/0}{\underset{\text{ALG}}{=\!=\!=}}\lim\limits_{x\to 0^-}(-x)=0,\\[2mm]\lim\limits_{x\to 0^+}\dfrac{x^2}{|x|}\overset{0/0}{\underset{\text{ALG}}{=\!=\!=}}\lim\limits_{x\to 0^+}\dfrac{x^2}{x}\overset{0/0}{\underset{\text{ALG}}{=\!=\!=}}\lim\limits_{x\to 0^+}(x)\;\;=0.\end{array}\right\}\therefore\lim\limits_{x\to 0}\dfrac{x^2}{|x|}=0.$$

As indicated, because the left and right limits were the same (zero), the usual two-sided limit must also equal that value, from Theorem 2.5.1(a), page 139.[29]

Example 2.5.4

Find $\lim\limits_{x\to 9}\dfrac{(2x-18)|2x-23|}{x^2-11x+18}$.

Solution: For this function, the expression inside the absolute value does not change signs at $x=9$ (but remains negative at and near $x=9$), and so we can use older techniques:

$$\lim\limits_{x\to 9}\dfrac{(2x-18)|2x-23|}{x^2-11x+18}\overset{0/0}{\underset{\text{ALG}}{=\!=\!=}}\lim\limits_{x\to 9}\dfrac{2(x-9)[-(2x-23)]}{(x-9)(x-2)}\overset{0/0}{\underset{\text{ALG}}{=\!=\!=}}\lim\limits_{x\to 9}\dfrac{2(23-2x)}{x-2}\overset{\text{CONT}}{=\!=\!=}\dfrac{2(23-18)}{9-2}=\dfrac{10}{7}.$$

In fact we could have left $|2x-23|$ as written instead of "simplifying" it to $23-2x$ in the limit, because it is continuous and not used to cancel any factors that contribute to this limit being in 0/0 form.

Example 2.5.5

Suppose $f(x)=\begin{cases}x+1 & \text{for } x>3 \\ 5x-11 & \text{for } x\in[-2,3] \\ x+3 & \text{for } x<-2.\end{cases}$

Find all left-, right-, and two-sided limits, where possible, at $x=3$, $x=1$ and $x=-2$.

Solution: First we look at $x=3$. In each case, the key is to see which "piece" x is on in its approach to the limit point. (See Theorem 2.5.1(d),(e), page 139.)

$$\left.\begin{array}{l}\lim\limits_{x\to 3^-}f(x)=\lim\limits_{x\to 3^-}(5x-11)\overset{\text{CONT}}{=\!=\!=}5(3)-11=4,\\[2mm]\lim\limits_{x\to 3^+}f(x)=\lim\limits_{x\to 3^+}(x+1)\overset{\text{CONT}}{=\!=\!=}4.\end{array}\right\}\therefore\lim\limits_{x\to 3}f(x)=4.$$

Next are the limits at $x=1$. In all cases the computation is the same, but we will go ahead and write them out here. (With practice, for such a case only the third would likely be computed, the other two following.)

$$\left.\begin{array}{l}\lim\limits_{x\to 1^-}f(x)=\lim\limits_{x\to 1^-}(5x-11)\overset{\text{CONT}}{=\!=\!=}5(1)-11=-6,\\[2mm]\lim\limits_{x\to 1^+}f(x)=\lim\limits_{x\to 1^+}(5x-11)\overset{\text{CONT}}{=\!=\!=}5(1)-11=-6,\end{array}\right\}\therefore\lim\limits_{x\to 1}f(x)=-6.$$

---

[29] Alternatively, one could note that $x^2=|x|^2$, and so $\lim\limits_{x\to 0}\dfrac{x^2}{|x|}\overset{0/0}{\underset{\text{ALG}}{=\!=\!=}}\lim\limits_{x\to 0}\dfrac{|x|^2}{|x|}\overset{0/0}{\underset{\text{ALG}}{=\!=\!=}}\lim\limits_{x\to 0}|x|\overset{\text{CONT}}{=\!=\!=}0.$

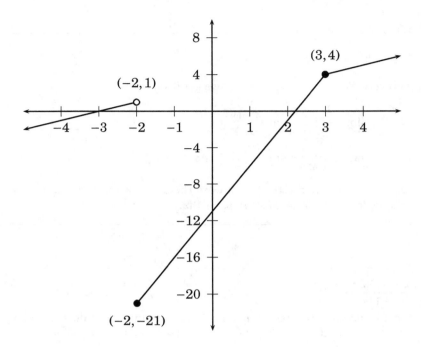

Figure 2.14: Partial graph of function from Example 2.5.5, page 141.

Those were easier because $x = 1$ is safely inside the interval $[-2, 3]$, on which f is defined by the (continuous) function $5x - 12$. In fact, we could have computed the two-sided limit first, and then made our conclusions about the one-sided limits:

$$\lim_{x \to 1} f(x) = \lim_{x \to 1}(5x - 11) \overset{\text{CONT}}{=\!=\!=} 5(1) - 11 = -6 \implies \begin{cases} \lim_{x \to 1^-} f(x) = -6, \\ \lim_{x \to 1^+} f(x) = -6. \end{cases}$$

Finally, we turn our attention to $x = -2$:

$$\left. \begin{array}{l} \lim_{x \to -2^-} f(x) = \lim_{x \to -2^-}(x + 3) = -2 + 3 = 1, \\ \lim_{x \to -2^+} f(x) = \lim_{x \to -2^+}(5x - 11) = 5(-2) - 11 = -21. \end{array} \right\} \therefore \lim_{x \to -2} f(x) \text{ DNE.}$$

If we collect the information in the previous example, along with the values of $f(x)$ at the limit points, we can use Theorem 2.5.1(b)(c), page 139, and the two-sided analogs to make some conclusions regarding various types of continuity at these points:

- At $x = 3$, we have $\lim_{x \to 3} f(x) = 4 = f(3)$, so $f(x)$ is continuous at $x = 3$.

- At $x = 1$, we have $\lim_{x \to 1} f(x) = -6 = f(1)$, so $f(x)$ is continuous at $x = 1$.

- $\lim_{x \to -2^-} f(x) = 1 \neq f(-2) = -21$, so $f(x)$ is not left-continuous at $x = -2$.

2.5. ONE-SIDED FINITE LIMITS AT POINTS

- $\lim_{x \to -2^+} f(x) = -21 = f(-2)$, so $f(x)$ **is** right-continuous at $x = -2$.

- $f(x)$ is not continuous at $x = -2$, since it is **not both** left- and right-continuous at $x = -2$. Furthermore, since $\lim_{x \to -2} f(x)$ does not exist, it cannot equal $f(-2)$, so Theorem 2.4.2, page 124, also gives us a discontinuity at $x = -2$ (since $\lim_{x \to -2} f(x) \neq f(-2)$, the limit being nonexistent).

- Since the function is "piecewise linear," it is continuous at all other x-values (which are not boundaries of the "pieces"). Thus it is continuous for $x \neq -2$, i.e., for all $x \in \mathbb{R} - \{-2\} = (-\infty, -2) \cup (-2, \infty)$.

Just as with continuity and limits, the graph of a function can often indicate one-sided continuity and the values of one-sided limits. If we can "ride along" the graph towards the limit point from the prescribed direction, we can visually observe if some height is approached. While the function in Example 2.5.5 is graphed in Figure 2.14, page 142, we should be able to read many of the limit and continuity properties from the graph itself.

- $\lim_{x \to -2^-} f(x) = 1$
- $\lim_{x \to -2^+} f(x) = -21$
- $\lim_{x \to -2} f(x)$ DNE
- $f(-2) = -21$
- $f(x)$ is (only) right-continuous at $x = -2$
- $\lim_{x \to -3} f(x) = 0$

- $f(-3) = 0$.
- $f(x)$ is continuous at $x = -3$
- $\lim_{x \to 3} f(x) = 4$
- $f(3) = 4$
- $f(x)$ is continuous at $x = 3$
- $f(x)$ is continuous on $(-\infty, -2)$
- $f(x)$ is continuous on $[-2, \infty)$

Recall that the continuity on $[-2, \infty)$, by definition, means $f(x)$ is continuous at each $x \in (-2, \infty)$ (the open interval) and right-continuous at $x = -2$. This is Definition 2.3.2, page 107. Continuity on the open interval $(-\infty, -2)$ simply means (two-sided) continuity at each $x \in (-\infty, -2)$, as in Definition 2.3.1, page 105.

Now that we have discussed one-sided limits, the previous section can be somewhat clarified on those points where the function, say $f(x)$, approached one value as x approached a value a from one side, but $f(x)$ may have approached another value as x approached a from the other side. The discussion in both sections assumed limits that existed were finite, and the inputs x approached finite numbers themselves (i.e., "points" in $\mathbb{R}$). In future sections, we will allow x to "approach infinity" and make sense of that, as well as allow for infinite limits. In this way, the language of limits will be of great help in describing more and more behaviors of functions. We will also see many more "forms" besides 0/0.

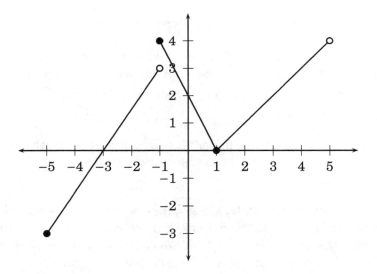

Figure 2.15: Graph of the function $f(x)$ given in Exercises 1 and 2.

Exercises

For Exercises 1–11 consider the function graphed in Figure 2.15. *From looking at the graph,* compute each limit. (It is possible that a requested limit does not exist.) For some problems, you may wish to work the limits in a different order (e.g., two-sided limit first).

1. $\lim_{x \to -5^+} f(x)$, $\lim_{x \to -5^-} f(x)$, $\lim_{x \to -5} f(x)$

2. $\lim_{x \to 5^+} f(x)$, $\lim_{x \to 5^-} f(x)$, $\lim_{x \to 5} f(x)$

3. $\lim_{x \to -1^-} f(x)$, $\lim_{x \to -1^+} f(x)$, $\lim_{x \to -1} f(x)$

4. $\lim_{x \to 0^-} f(x)$, $\lim_{x \to 0^+} f(x)$, $\lim_{x \to 0} f(x)$

5. $\lim_{x \to 1} f(x)$, $\lim_{x \to 1^-} f(x)$, $\lim_{x \to 1^+} f(x)$

6. $\lim_{x \to -3} f(x) =$

7. Is $f(x)$ continuous at $x = -5$? Is $f(x)$ left-continuous at $x = -5$? Is $f(x)$ right-continuous at $x = -5$?

8. Repeat for $x = -1$.

9. Repeat for $x = 1$.

10. Repeat for $x = 5$.

11. Repeat for $x = 0$.

The function in Figure 2.15 can be defined piecewise as follows:

$$f(x) = \begin{cases} x - 1 & \text{for } x \in (1, 5) \\ -2(x-1) & \text{for } x \in [-1, 1] \\ \frac{3}{2}(x+3) & \text{for } x \in [-5, -1). \end{cases}$$

Using this definition of $f(x)$, and not referring to the graph, compute the limits in Exercises 12–16. (For example, $\lim_{x \to 2} f(x) = \lim_{x \to 2}(x-1) = 2 - 1 = 1$, though left and right limits might require more care. For continuity and other questions here, see also Theorem 2.5.1, page 139, and the previous examples.)

12. $\lim_{x \to -5^+} f(x)$, $\lim_{x \to -5^-} f(x)$, $\lim_{x \to -5} f(x)$. Is $f(x)$ continuous at $x = -5$? Left-continuous? Right-continuous?

2.5. ONE-SIDED FINITE LIMITS AT POINTS

13. $\lim_{x\to -1^+} f(x)$, $\lim_{x\to -1^-} f(x)$, $\lim_{x\to -1} f(x)$. Is $f(x)$ continuous at $x = -1$? Left-continuous? Right-continuous?

14. $\lim_{x\to 0^+} f(x)$, $\lim_{x\to 0^-} f(x)$, $\lim_{x\to 0} f(x)$. Is $f(x)$ continuous at $x = 0$? Left-continuous? Right-continuous?

15. $\lim_{x\to 1^+} f(x)$, $\lim_{x\to 1^-} f(x)$, $\lim_{x\to 1} f(x)$. Is $f(x)$ continuous at $x = 1$? Left-continuous? Right-continuous?

16. $\lim_{x\to 3^+} f(x)$, $\lim_{x\to 3^-} f(x)$, $\lim_{x\to 3} f(x)$. Is $f(x)$ continuous at $x = 3$? Left-continuous? Right-continuous?

17. Consider the function $f(x) = \sqrt{x^2 - 16}$. Compute (if possible) and determine

    (a) $\lim_{x\to 4^-} f(x)$, $\lim_{x\to 4^+} f(x)$, $\lim_{x\to 4} f(x)$.

    (b) Do we have continuity, left-continuity, right-continuity or neither at $x = 4$?

    (c) $\lim_{x\to -4^-} f(x)$, $\lim_{x\to -4^+} f(x)$, $\lim_{x\to -4} f(x)$.

    (d) Do we have continuity, left-continuity, right-continuity or neither at $x = -4$?

18. Using the same function and directions as in the previous example,

    (a) $\lim_{x\to 5^-} f(x)$, $\lim_{x\to 5^+} f(x)$, $\lim_{x\to 5} f(x)$. (You may wish to compute these in a different order.)

    (b) Do we have continuity, left-continuity, right-continuity or neither at $x = 5$?

For each of the following limits 19–23, compute its value or (if appropriate) state why it does not exist. Show details. (For some, you may need to check both left-side and right-side limits.)

19. $\lim_{x\to 3} \dfrac{|x-5|}{x+4}$

20. $\lim_{x\to 1^-} \dfrac{|x-1|}{x^2 + 3x - 4}$

21. $\lim_{x\to -4} \dfrac{|x+4|}{x^2 - 16}$

22. $\lim_{x\to 4} \dfrac{|x^2 - 16|}{x - 4}$

23. $\lim_{x\to 4} \dfrac{x^2 - 8x + 16}{|x - 4|}$

24. Consider the following function (where m and b will be determined later).

$$f(x) = \begin{cases} \sqrt{x} & \text{for } x > 4 \\ mx + b & \text{for } -4 \leq x \leq 4 \\ -3 - (x+4)^2 & \text{for } x < -4. \end{cases}$$

   (a) Find $\lim_{x\to 4^+} f(x)$.

   (b) Find $\lim_{x\to 4^-} f(x)$.

   (c) Find $\lim_{x\to -4^+} f(x)$.

   (d) Find $\lim_{x\to -4^-} f(x)$.

   (e) Use these to find m and b so that $f(x)$ is continuous on $\mathbb{R}$.

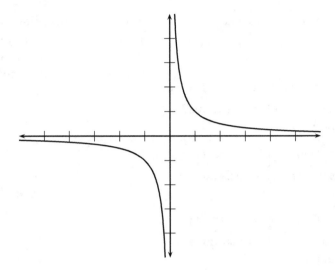

Figure 2.16: Partial graph of $f(x) = 1/x$. We see $f(x) \longrightarrow \infty$ as $x \longrightarrow 0^+$, while $f(x) \longrightarrow -\infty$ as $x \longrightarrow 0^-$.

2.6 Infinite Limits at Points

In this section, we extend the notion of limit so it can describe quantities that are growing without bound in many circumstances. The prototype function to help define what we mean here will be the familiar $f(x) = 1/x$, which we graph in Figure 2.16.

As x approaches zero from the right, we have $1/x$ returning larger and larger positive numbers, and unbounded growth in $1/x$. Similarly, as x approaches zero from the left we have $1/x$ returning larger and larger negative numbers, growing in absolute size without bound. For this function we would write

$$\lim_{x \to 0^+} \frac{1}{x} \overset{1/0^+}{=\!=\!=} +\infty \text{ (or just } \infty\text{),} \tag{2.32}$$

$$\lim_{x \to 0^-} \frac{1}{x} \overset{1/0^-}{=\!=\!=} -\infty, \tag{2.33}$$

$$\lim_{x \to 0} \frac{1}{x} \overset{1/0^\pm}{=\!=\!=} \text{DNE.} \tag{2.34}$$

These limits are the simplest prototypical examples of infinite limits at points. The first two limits are particular *determinate* forms, $1/0^+$ and $1/0^-$, respectively, which we will discuss later in this section. The third limit does not exist because the left-side and right-side limits do not agree, as can also happen with finite limits. We will still label its form $1/0^\pm$. For completeness and future reference, we now give definitions of what it means for a limit at a point, either one-sided or two-sided, to be ∞:

2.6. INFINITE LIMITS AT POINTS

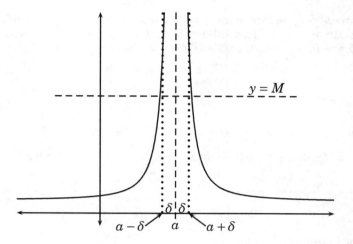

Figure 2.17: A function $f(x)$ with $\lim_{x\to a} f(x) = \infty$. By (2.35), for any height M we can find $\delta > 0$ so that $0 < |x-a| < \delta \implies f(x) > M$. In other words, we can choose any height, and then force $f(x)$ to be still higher than that height by forcing x to be within some δ of a but (as always with limits) never actually equal to a. Notice that since $f(x)$ is continuous near $x = a$ (but not at $x = a$), the limit as $x \to a$ being infinite gives a vertical asymptote there (given by the dashed vertical line at $x = a$).

Definition 2.6.1 *For $a \in \mathbb{R}$, we say*

$$\lim_{x\to a} f(x) = \infty \iff (\forall M \in \mathbb{R})(\exists \delta > 0)(\forall x)(0 < |x-a| < \delta \longrightarrow f(x) > M), \tag{2.35}$$

$$\lim_{x\to a^+} f(x) = \infty \iff (\forall M \in \mathbb{R})(\exists \delta > 0)(\forall x)(x \in (a, a+\delta) \longrightarrow f(x) > M), \tag{2.36}$$

$$\lim_{x\to a^-} f(x) = \infty \iff (\forall M \in \mathbb{R})(\exists \delta > 0)(\forall x)(x \in (a-\delta, a) \longrightarrow f(x) > M). \tag{2.37}$$

In other words, to say that a limit is ∞ is to say that we can force $f(x)$ to be greater than any previously chosen number M by forcing x to be within δ of a (but not equal to a), from one or both directions depending upon if it is a right-, left-, or two-sided limit. (Of course the choice of δ depends upon the choice of M.) A graphical example of (2.35) is given in Figure 2.17.

Proofs using these definitions are interesting, and we will include one here, but we will have numerous shortcuts based on general observations.

Example 2.6.1

Prove that $\lim_{x \to 0^+} \dfrac{1}{x} = \infty$.

Solution: Here $a = 0$ in our definition. For scratch-work, we note that we wish to find $\delta > 0$ so that

$$x \in (0, \delta) \implies f(x) = \frac{1}{x} > M.$$

One complication here is that solving this last inequality depends on whether or not M is positive. Fortunately, if $M_1 < M_2$ and a value for δ gives us the desired implication for $M = M_2$, then that δ also gives us the implication for $M = M_1$, since $(f(x) > M_2) \wedge (M_2 > M_1) \implies f(x) > M_1$.

So let us assume $M>0$, and then any $\delta>0$ we disover for $M>0$ will suffice for lesser values of M, including nonpositive values. (This will be clarified further within the proof.) Working backward as we did with ε-δ proofs, we note that for $M,x>0$ we have $\frac{1}{x}>M \iff x<\frac{1}{M}$, and so we will take $\delta=\frac{1}{M}$. Note that $x\in(0,\delta) \iff 0<x<\delta \implies \frac{1}{x}>\frac{1}{\delta}$.

Proof: Let $M\in\mathbb{R}$.

1. Suppose $M>0$. Let $\delta=\frac{1}{M}$, so $\delta>0$ exists, and $x\in(0,\delta) \implies f(x)=\frac{1}{x}>\frac{1}{\delta}=M$.
2. Suppose $M\leq 0$. Take $\delta>0$. Then $x\in(0,\delta) \implies f(x)=\frac{1}{x}>\frac{1}{\delta}>0\geq M$,   q.e.d.

Another proof might read as follows:

Without loss of generality, assume $M>0$. Take $\delta=\frac{1}{M}$. Then $\delta>0$ exists and

$$x\in(0,\delta) \implies f(x)=\frac{1}{x}>\frac{1}{\delta}=M, \qquad q.e.d.$$

The experienced reader of such proofs would be expected to notice that due to the nature of the inequality we are trying to imply by our choice of δ, if we can prove the inequality for any value of M, then it also holds with that δ for any lesser values of M. Thus, if we can prove the existence of δ for any positive value of M, that value of δ is valid for zero and negative values of M as well. That is what is meant by "without loss of generality," namely that we can prove one case (e.g., $M>0$) and from the nature of the result we are trying to prove, the other cases will follow from that case or in a very similar manner.

In this section we will not further emphasize proofs in which we would often need to show that a particular limit conforms to a particular limit definition. While interesting in their own rights, from the previous example we can see that such proofs will often contain illuminating but perhaps distracting technicalities.

Instead we will emphasize the limit forms $1/0^+$, $1/0^-$, $1/0^\pm$, and their variants, and argue intuitively what such limits' values will be: real (finite), infinite, or nonexistent.

However, there is value in being aware of the various limit definitions as they are given precisely, and therefore what exactly we are claiming when we write, for instance, $\lim_{x\to 0^-}\frac{1}{x}=-\infty$. Therefore, for completeness we next list the similar definitions for the limits at a point being $-\infty$ and invite the reader to make some sense of these as well as the previous definitions.

Definition 2.6.2 *For $a\in\mathbb{R}$, we say*

$$\lim_{x\to a}f(x)=-\infty \iff (\forall N\in\mathbb{R})(\exists \delta>0)(\forall x)(0<|x-a|<\delta \longrightarrow f(x)<N), \qquad (2.38)$$

$$\lim_{x\to a^+}f(x)=-\infty \iff (\forall N\in\mathbb{R})(\exists \delta>0)(\forall x)(x\in(a,a+\delta) \longrightarrow f(x)<N), \qquad (2.39)$$

$$\lim_{x\to a^-}f(x)=-\infty \iff (\forall N\in\mathbb{R})(\exists \delta>0)(\forall x)(x\in(a-\delta,a) \longrightarrow f(x)<N). \qquad (2.40)$$

2.6. INFINITE LIMITS AT POINTS

Thus, to say a limit is $-\infty$ is to say that the function's height can be forced to be below any given level by forcing x to be within some distance of a, ignoring the case $x = a$ as we always do with limits since limits are about *trends* and not the actual values of functions at the limit points.

Earlier, we had the *indeterminate* limit form 0/0 (meaning that knowing the numerator and denominator both approach zero does not tell us the value of the limit). In many of these next cases, we will have variations of forms $1/0^+$, $1/0^-$, and $1/0^\pm$, which are *not indeterminate*, i.e., are *determinate*, and as such should become intuitive on reflection. In the definitions that follow, we will write the functions NUM for the numerator and DEN for the denominator. (Note that here "$\longrightarrow$" means "approaches.")

Definition 2.6.3 *For a function* $f(x) = \dfrac{\text{NUM}(x)}{\text{DEN}(x)}$, *define the following forms, as x approaches the limit point in the prescribed way:*

1. $1/0^+$: *any limit in which* $\text{NUM}(x) \longrightarrow 1$ *and* $\text{DEN}(x) \longrightarrow 0^+$, *the latter meaning the denominator is positive but approaching zero.*

2. $1/0^-$: *any limit in which* $\text{NUM}(x) \longrightarrow 1$ *and* $\text{DEN}(x) \longrightarrow 0^-$, *the latter meaning the denominator is negative but approaching zero.*

3. $1/0^\pm$: *any limit in which* $\text{NUM}(x) \longrightarrow 1$ *and* $\text{DEN}(x) \longrightarrow 0$, *with* $\text{DEN}(x) \neq 0$ *but sometimes* $\text{DEN}(x)$ *is positive, other times negative, no matter how close the input is to its limit point.*

Theorem 2.6.1 *Forms* $1/0^+$, $1/0^-$ *and* $1/0^\pm$ *are determinate forms: Any limit of the form* $1/0^+$ *will return* $+\infty$; *any limit of the form* $1/0^-$ *will return* $-\infty$; *and any limit of the form* $1/0^\pm$ *will not exist.*

This is well illustrated in the behavior of $f(x) = 1/x$ as zero is approached, as in Figure 2.16 at the opening of this section (page 146), and the corresponding limits (2.32), (2.33), and (2.34).

Example 2.6.2

Consider the following limits and their methods of computation:

1. $\lim\limits_{x \to 5^+} \dfrac{1}{\sqrt{x^2 - 25}} \stackrel{1/0^+}{=\!=\!=} \infty.$

2. $\lim\limits_{x \to 9^+} \dfrac{10 - x}{9 - x} \stackrel{1/0^-}{=\!=\!=} -\infty.$

3. $\lim\limits_{x \to 3} \dfrac{1}{x - 3}$ does not exist ($1/0^\pm$).

4. $\lim\limits_{x \to 4} \dfrac{x}{(x-4)^2} \stackrel{4/0^+}{=\!=\!=} \infty.$

In the last limit, the denominator is positive as $x \to 4$ from both sides because it is a perfect (polynomial) square, and nonzero. Notice also that, strictly speaking, it is of the form $4/0^+$, but that is also *determinate* (i.e., not indeterminate). In fact, it is just 4 times a limit, which is of the form $1/0^+$ if we factor out the 4 (and accept that $4 \cdot \infty = \infty$):

$$\lim_{x \to 4} \frac{x}{(x-4)^2} = 4 \lim_{x \to 4} \frac{x/4}{(x-4)^2} \stackrel{4(1/0^+)}{=\!=\!=\!=} 4 \cdot \infty = \infty. \tag{2.41}$$

Notice that limits allow us to—somewhat—extend our arithmetic when it is understood that it is really a statement about limit forms. For instance, we could say that

$$(\forall a > 0)[a \cdot \infty = \infty], \quad \text{and} \quad (\forall a > 0)[a \cdot (-\infty) = -\infty]. \tag{2.42}$$

These mean that if part of the function approaches $a > 0$, and the other "approaches" ∞, then so does the limit of the product approach ∞. Similarly, for $a > 0$ we have $a \cdot (-\infty)$ gives us $-\infty$. For (2.41), we used the fact that if a function grows positive without bound as we approach the limit point, then so will another function that is roughly 4 times that function (it will just "grow" roughly four times as fast). On the other hand,

$$(\forall a < 0)[a \cdot \infty = -\infty], \quad \text{and} \quad (\forall a < 0)[a \cdot (-\infty) = \infty]. \tag{2.43}$$

Multiplying a function that "blows up" (meaning the output grows without bound in size) as we approach the limit point, by another function which approaches a negative number, does not change the fact that the product will still blow up, but the blow up will occur with the opposite sign (+/−).

In terms of division, we can write[30]

$$(\forall a > 0)\left[\left(\frac{a}{0^+} = a \cdot \frac{1}{0^+} = a \cdot \infty = \infty\right) \wedge \left(\frac{a}{0^-} = a \cdot \frac{1}{0^-} = a \cdot (-\infty) = -\infty\right)\right], \tag{2.44}$$

$$(\forall a < 0)\left[\left(\frac{a}{0^+} = a \cdot \frac{1}{0^+} = a \cdot \infty = -\infty\right) \wedge \left(\frac{a}{0^-} = a \cdot \frac{1}{0^-} = a \cdot (-\infty) = \infty\right)\right]. \tag{2.45}$$

We take this opportunity to point out that knowing where these infinite limits occur is useful in sketching a graph of the function. Consider, for instance, the limit from the previous example:

$$\lim_{x \to 4} \frac{x}{(x-4)^2} = \infty.$$

If we were to construct a sign chart for this function and note its infinite limit as $x \to 4$, as well as its continuity except at $x = 4$, we can conclude it will have a vertical asymptote at $x = 4$. This gives us some idea of what its graph looks like near that vertical asymptote.

Function:	$f(x) = x/(x-4)^2$		
Test $x =$	-1	2	5
Sign Factors:	$\ominus/\ominus^2$	$\oplus/\ominus^2$	$\oplus/\oplus^2$
Sign $f(x)$:	$\ominus$ 0	$\oplus$	$\oplus$

In fact, this function also has horizontal asymptotes, a topic which will occur in a later section. A computer-generated graph is given in Figure 2.18, page 151, and so our sign chart at least accurately displays the sign and the behavior as $x \to 4$. (Note also the x-intercept at $x = 0$.)

---

[30] Another way to look at these limits is to first make note that they are blowing up—for instance NUM $\to a \ne 0$ but DEN $\to 0$—and then just take note of the sign of the fraction as a whole. If it is positive and blowing up, the limit must be ∞; if negative and blowing up, the limit must be $-\infty$; if both signs occur consistently as x approaches the limit point, *and* the fraction is blowing up in absolute size, then the limit must not exist (part of the function blows up towards ∞, and another part towards $-\infty$, as $x \to a$).

2.6. INFINITE LIMITS AT POINTS

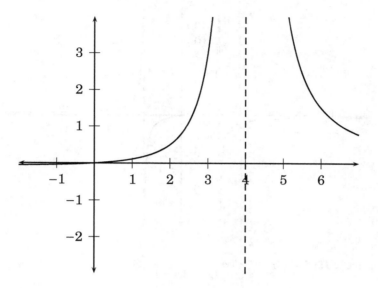

Figure 2.18: Partial graph of $f(x) = x/(x-4)^2$. Besides the vertical asymptote at $x = 4$, note the sign change at $x = 0$, the latter being not easily discerned due to the scale being true. (For instance, $f(-1) = -1/25 = -0.04$.)

In fact just knowing that the function "blows up" at $x = 4$, along with the sign chart, gives us the limits from *both* sides. For another example, we revisit the function $f(x) = x/[(x+1)(x-1)]$, which we encountered in Example 2.3.7, page 117.

Example 2.6.3

For the function $f(x) = \dfrac{x}{(x+1)(x-1)}$ from Example 2.3.7, page 117, we repeat the sign chart for convenience, though we add features to the chart to illustrate locations of the x-intercept and vertical asymptotes, the latter occurring when the denominator approaches zero but the numerator approaches another value:

Function: $\qquad f(x) = \dfrac{x}{(x+1)(x-1)}$

Test $x =$ $\qquad -2 \qquad -0.5 \qquad 0.5 \qquad 2$

Sign Factors: $\qquad \dfrac{\ominus}{\ominus\ominus} \qquad \dfrac{\ominus}{\oplus\ominus} \qquad \dfrac{\oplus}{\oplus\ominus} \qquad \dfrac{\oplus}{\oplus\oplus}$

Sign $f(x)$: $\qquad \ominus \quad -1 \quad \oplus \quad 0 \quad \ominus \quad 1 \quad \oplus$

The following limits can be found by considering the sign chart for the function $f(x) = \dfrac{x}{(x+1)(x-1)}$:

$$\lim_{x \to -1^-} \frac{x}{(x+1)(x-1)} = -\infty, \qquad \lim_{x \to 1^-} \frac{x}{(x+1)(x-1)} = -\infty,$$

$$\lim_{x \to -1^+} \frac{x}{(x+1)(x-1)} = \infty, \qquad \lim_{x \to 1^+} \frac{x}{(x+1)(x-1)} = \infty.$$

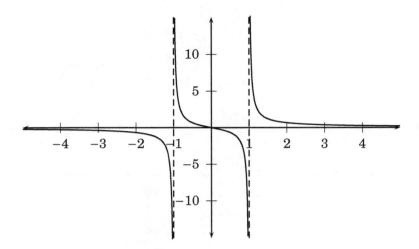

Figure 2.19: Graph of $f(x) = \dfrac{x}{(x+1)(x-1)}$. See Example 2.6.3, page 151.

All of these are of forms $1/0^+$, $1/0^-$, $-1/0^+$, or $-1/0^-$, since the numerators are approaching ± 1 and the denominators are approaching zero from left or right. Thus we have the function "blowing up" at ± 1, so we just need to know what its signs are as we approach ± 1 from either side.

The graph of $f(x)$ is given in Figure 2.19.[31]

These limits can also be computed more directly, without reference to the sign chart of the function, though some care must be used to correctly determine the sign of the function in the limiting behavior. For instance, we might write[32]

$$\lim_{x \to -1^-} \frac{x}{(x+1)(x-1)} \overset{-1/0}{\underset{\text{ALG}}{=\!=\!=}} \lim_{x \to -1^-} \frac{x}{x^2 - 1} \overset{-1/0^+}{=\!=\!=} -\infty.$$

While it was possible to note the sign of the function as $x \to -1^-$ with the denominator factored, it might appear easier to do so with the denominator multiplied as $x^2 - 1$. For $x < -1$, we would have to notice that $x^2 > 1$, and so the denominator is positive as it shrinks to zero. Similarly, we would compute the following, noting that for $x > -1$ but very close to -1 (in particular for $x \in (-1, 1)$), we have $x^2 < 1$ and so $x^2 - 1 < 0$, and thus

$$\lim_{x \to -1^+} \frac{x}{(x+1)(x-1)} \overset{-1/0}{\underset{\text{ALG}}{=\!=\!=}} \lim_{x \to -1^+} \frac{x}{x^2 - 1} \overset{-1/0^-}{=\!=\!=} \infty.$$

And so the labor needed to produce the sign chart in computing a single limit can be balanced against the technical difficulty of computing the limit without reference to the complete sign chart. If we are already constructing the sign chart for some other reason (e.g.,

---

[31]The graph also reflects that $f(x)$ shrinks in size when x is large, which we will develop further in the next section.

[32]Knowing that the limit is of the form $-1/0$ is not enough to know its value or if it exists, because these things depend upon the nature of the denominator's approach to zero. One could write $-1/0^?$ to express this ambiguity in casual presentations.

2.6. INFINITE LIMITS AT POINTS

to aid in graphing the entire function), we can use it to also compute our limits at the vertical asymptotes. But first we need to know that there is in fact "blow up" as we approach those input values, and so we need an eye to what a computation without a sign chart would look like. The following limits are computed without reference to full sign charts of their respective functions.

Example 2.6.4

Consider the following limits, which we compute without graphing in any way.

1. $\lim\limits_{x \to 9^+} \dfrac{x}{3-\sqrt{x}} \overset{9/0^-}{=\!=\!=\!=} -\infty.$ $\qquad \lim\limits_{x \to 9^-} \dfrac{x}{3-\sqrt{x}} \overset{9/0^+}{=\!=\!=\!=} \infty.$ $\qquad \lim\limits_{x \to 9} \dfrac{x}{3-\sqrt{x}} \overset{9/0^\pm}{=\!=\!=\!=}$ DNE.

2. $\lim\limits_{x \to 6^-} \dfrac{3-2x}{x^2+2x-48} = \lim\limits_{x \to 6^-} \dfrac{3-2x}{(x+8)(x-6)} \; \overset{\frac{-9}{(14)(0^-)}}{\underset{\frac{-9}{14} \cdot \frac{1}{0^-}}{=\!=\!=\!=}} \; \infty.$

Computations such as these are fairly routine, but did use some subtlety to get the correct signs. In the first three limits, it was important to notice that the denominator $3-\sqrt{x}$ is negative for $x > 9$ and positive for $x < 9$. The last limit in the example can be seen as a form $\frac{-9}{14} \cdot (-\infty) = \infty$. In that case, it helped to have the denominator factored. Note how we rewrote the limit form in the space below the elongated = symbol to extract a more obvious form $a \cdot \infty$ or $a \cdot (-\infty)$. Another legitimate rewriting would have $\frac{-9}{14 \cdot 0^-} = -9 \cdot \frac{1}{0^-} = -9(-\infty) = \infty$, since as a form, $14 \cdot 0^-$ still represents a term approaching zero from the left (because if we multiply such a term by 14, it still shrinks to zero but remains negative). In other words, as a form, $14 \cdot 0^- = 0^-$. Similarly, $-9(-\infty) = 9 \cdot \infty$ represents a limit form of a function that is growing positive and without bound. We will continue to develop this arithmetic of forms as we proceed.

It is possible that a 0/0 form can simplify to one of these determinate forms, as in the following example. If it is clear the function "blows up," then we need only check its sign to see if the limit is ∞, $-\infty$, or nonexistent (DNE).

Example 2.6.5

Consider the following limits (the algebra is the same for each so we only show details for the first limit):

1. $\lim\limits_{x \to 3} \dfrac{81-x^4}{(x^2-6x+9)^2} \overset{0/0}{\underset{\text{ALG}}{=\!=\!=}} \lim\limits_{x \to 3} \dfrac{(9-x^2)(9+x^2)}{[(x-3)^2]^2} \overset{0/0}{\underset{\text{ALG}}{=\!=\!=}} \lim\limits_{x \to 3} \dfrac{(3-x)(3+x)(9+x^2)}{(x-3)^4}$

   $\overset{0/0}{\underset{\text{ALG}}{=\!=\!=}} \lim\limits_{x \to 3} \dfrac{-(x-3)(3+x)(9+x^2)}{(x-3)^4} \overset{0/0}{\underset{\text{ALG}}{=\!=\!=}} \lim\limits_{x \to 3} \dfrac{-(3+x)(9+x^2)}{(x-3)^3} \overset{\frac{-6 \cdot 18}{0^\pm}}{=\!=\!=\!=}$ DNE.

2. $\lim\limits_{x \to 3^+} \dfrac{81-x^4}{(x^2-6x+9)^2} \overset{0/0}{\underset{\text{ALG}}{=\!=\!=}} \lim\limits_{x \to 3^+} \dfrac{-(3+x)(9+x^2)}{(x-3)^3} \overset{\frac{-6 \cdot 18}{0^+}}{=\!=\!=\!=} -\infty.$

3. $\lim\limits_{x \to 3^-} \dfrac{81-x^4}{(x^2-6x+9)^2} \overset{0/0}{\underset{\text{ALG}}{=\!=\!=}} \lim\limits_{x \to 3^-} \dfrac{-(3+x)(9+x^2)}{(x-3)^3} \overset{\frac{-6 \cdot 18}{0^-}}{=\!=\!=\!=} \infty.$

Rational exponents within limits can pose their own challenges. Recall that if a/b is a simplified fraction (so we do not allow 4/6 but do allow 2/3, for example), then

$$x^{a/b} = \left(x^a\right)^{1/b} = \left(x^{1/b}\right)^a. \tag{2.46}$$

Also recall that $x^{1/b} = \sqrt[b]{x}$. Finally, $(x^2)^{1/2} = \sqrt{x^2} = |x|$, for instance, while $(x^3)^{1/3} = \sqrt[3]{x^3} = x$. As usual, the odd roots have fewer complications and are thus simpler to work with than even roots, at least for abstract computations.

Example 2.6.6

Consider the following limits:

1. $\displaystyle\lim_{x \to 0^-} \frac{1}{x^{2/3}} = \lim_{x \to 0^-} \frac{1}{\sqrt[3]{x^2}} \stackrel{1/0^+}{=\!=\!=} \infty,$

2. $\displaystyle\lim_{x \to 0^-} x^{-5/3} = \lim_{x \to 0^-} \frac{1}{\sqrt[3]{x^5}} \stackrel{1/0^-}{=\!=\!=} -\infty,$

3. $\displaystyle\lim_{x \to -4} \frac{x}{(x+4)^{4/3}} = \lim_{x \to -4} \frac{x}{\left[(x+4)^4\right]^{1/3}} \stackrel{-4/0^+}{=\!=\!=} -\infty,$

4. $\displaystyle\lim_{x \to -4^-} \frac{x}{(x+4)^{1/3}} \stackrel{-4/0^-}{=\!=\!=} \infty,$

5. $\displaystyle\lim_{x \to -4^-} \frac{x}{\left[(x+4)^2\right]^{1/2}} = \lim_{x \to -4^-} \frac{x}{|x+4|} \stackrel{-4/0^+}{=\!=\!=} -\infty.$

There are applications where it is interesting to know what occurs in the "extreme," or limit, case. We now consider two of these.

Example 2.6.7

According to electrostatic theory, if two protons are brought to within the distance $d > 0$ of each other, the magnitude of the repelling force they would exert on each other will be given by $F = k/d^2$, where $k \approx 2.3 \times 10^{-28}$ N-m^2 is constant. So, we have that F is a function of d and we could write $F(d)$ instead of F in what follows, but we will use the physics style of notation where we "suppress" the independent variable d.

Then since $k > 0$, we can write

$$\lim_{d \to 0^+} F = \lim_{d \to 0^+} \frac{k}{d^2} \stackrel{k/0^+}{=\!=\!=} \infty.$$

Thus, according to electrostatic theory, it would require infinite force to bring two protons together such that the distance between them approaches zero.[33]

---

[33] The electrostatic theory mentioned here is far from complete, because it does not take into account other forces and quantum mechanical effects, nor is it clear what it means for the distance to be zero. One should ask if d is only the distances between their "centers," for instance, or if it is the distances between their edges, and is the "edge" of a proton well-defined? Still, it is interesting to see what even the incomplete theory says about what occurs to the force "in the limit."

2.6. INFINITE LIMITS AT POINTS

Example 2.6.8

Suppose an object lies along a line running through the center of a thin double convex lens, and further suppose that the line is perpendicular to the plane containing the outer edges of the lens. Also suppose the lens has a focal length of f, and d_o is the distance from the object to the center of the lens, where $d_o > f > 0$. If d_i is the distance from the lens center at which the resulting image of the object would be located on the opposite side of the lens, then the **thin lens equation** states that

$$\frac{1}{d_o} + \frac{1}{d_i} = \frac{1}{f}. \tag{2.47}$$

Find the trend in the distance d_i as the object is placed closer and closer to the focal length. Assume $d_o > f > 0$ throughout. Note that $d_o, d_i, f > 0$ are distances and not positions.

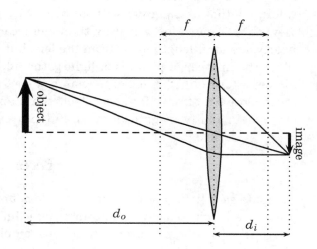

Solution: This situation is illustrated on the right. The focal length $f > 0$ is characteristic of the lens and should be considered constant. We assume the object to be movable horizontally relative to the lens as shown, with the base centered along a line running through the lens and perpendicular to the plane in which the outer edge of lens resides.

We wish to compute $\lim_{d_o \to f^+} d_i$, so first we solve for d_i as a function of d_o:

$$\frac{1}{d_i} = \frac{1}{f} - \frac{1}{d_o}$$

$$\implies d_i = \frac{1}{\frac{1}{f} - \frac{1}{d_o}} \cdot \frac{d_o f}{d_o f} = \frac{d_o f}{d_o - f}$$

$$\implies \lim_{d_o \to f^+} d_i = \lim_{d_o \to f^+} \frac{d_o f}{d_o - f} \xrightarrow[\text{ALG}]{d_o f / 0^+} \infty.$$

Thus, as the object moves to the right and approaches the focal length position, its image on the opposite side of the lens appears to move farther and farther (and inevitably more rapidly) away and to the right. One might conclude that the image of an object placed at the focal length will never be seen on the opposite side of the lens. This effect can be observed in a laboratory to the extent we can approach the ideal of having an object lie precisely and only at the focal length, and cam be observed numerically using some sample values. Suppose $f = 20\,\text{cm}$. Testing several relevant values of d_o, using $d_i = d_o f / (d_o - f)$, we get

$$d_o = 25\,\text{cm} \implies d_i = \frac{(25\,\text{cm})(20\,\text{cm})}{25\,\text{cm} - 20\,\text{cm}} = \frac{500\,\text{cm}^2}{5\,\text{cm}} = 100\,\text{cm},$$

$$d_o = 22\,\text{cm} \implies d_i = \frac{(22\,\text{cm})(20\,\text{cm})}{22\,\text{cm} - 20\,\text{cm}} = \frac{440\,\text{cm}^2}{2\,\text{cm}} = 220\,\text{cm},$$

$$d_o = 21\,\text{cm} \implies d_i = \frac{(21\,\text{cm})(20\,\text{cm})}{21\,\text{cm} - 20\,\text{cm}} = \frac{420\,\text{cm}^2}{1\,\text{cm}} = 420\,\text{cm},$$

$$d_o = 20.5\,\text{cm} \implies d_i = \frac{(20.5\,\text{cm})(20\,\text{cm})}{20.5\,\text{cm} - 20\,\text{cm}} = \frac{410\,\text{cm}^2}{0.5\,\text{cm}} = 820\,\text{cm}.$$

Finally, we would again see (for the case $f = 20\,\text{cm}$, but suppressing units) that

$$d_o \to 20^+ \implies d_i = \frac{d_o \cdot 20}{d_o - 20} = \frac{20}{1 - \frac{20}{d_o}} \xrightarrow[20/0^+]{20/(1-1^-)} \infty.$$

These last two examples illustrate that such limits do have analytical value in the physical sciences. In fact, the case with the two repellant protons is interesting in showing how there are other factors besides the electrostatic force governing interactions among protons and electrons, or the universe would otherwise consist of hydrogen atoms only (as any nuclei with more than one proton would be poised to explosively fly apart). The fact that we have many more elements than hydrogen is explained in part by the "strong nuclear force," or simply "nuclear force," attracting both protons and neutrons together more strongly than the electrostatic force can cause protons to repel each other, but only when they are in extremely close proximity (roughly 10^{-15} m).

In a future section, we will consider limits as a variable "approaches infinity," where we analyze the long-term trend in a function as the input variable grows "without bound." We would see that as d_0 grows without bound, $d_i \to f^+$. This allows us to design telescopes, to view objects much farther away than one focal length, and to expect the image to be very nearly one focal length away from the lens but on the other side of the lens from the object. Since the number of photons of light gathered by such a lens increases as the square of its diameter (or radius), much of the value of such a telescope lies in the fact that it can gather much more light from a far-away object than can our eyes. This ability renders the telescope's other main ability, magnification, useful in resolving images of far-off objects.

Exercises

Compute each limit (stating which ones do not exist) without reference to a sign chart, unless otherwise instructed. In computing these limits, write the form that allows you to make the conclusion where appropriate. (See examples throughout this section. For some limits, factoring is useful but it is not always necessary.)

1. $\lim_{x \to -5} \dfrac{1}{(x+5)^2}$

2. $\lim_{x \to 1^+} \dfrac{x}{x^2 - 1}$

3. $\lim_{x \to 1^-} \dfrac{x}{x^2 - 1}$

4. $\lim_{x \to 1} \dfrac{x}{x^2 - 1}$

5. $\lim_{x \to -1^+} \dfrac{x}{x^2 - 1}$

6. $\lim_{x \to -1^-} \dfrac{x}{x^2 - 1}$

7. $\lim_{x \to -1} \dfrac{x}{x^2 - 1}$

8. $\lim_{x \to 2^+} \dfrac{x^2 - 4x + 4}{x^2 - 4}$

9. $\lim_{x \to 2^-} \dfrac{x^2 - 4x + 4}{x^2 - 4}$

10. $\lim_{x \to 2} \dfrac{x^2 - 4x + 4}{x^2 - 4}$

11. $\lim_{x \to -2^+} \dfrac{x^2 - 4x + 4}{x^2 - 4}$

12. $\lim_{x \to -2^-} \dfrac{x^2 - 4x + 4}{x^2 - 4}$

13. $\lim_{x \to -2} \dfrac{x^2 - 4x + 4}{x^2 - 4}$

14. $\lim_{x \to -3} \dfrac{x}{|x - 3|}$

15. $\lim_{x \to 3} \dfrac{x}{|x - 3|}$

16. $\lim_{x \to 3^+} \dfrac{|x - 3|}{x^2 - 9}$

2.6. INFINITE LIMITS AT POINTS

17. $\lim\limits_{x \to 3^-} \dfrac{|x-3|}{x^2-9}$

18. $\lim\limits_{x \to 3} \dfrac{|x-3|}{x^2-9}$

19. $\lim\limits_{x \to -3^+} \dfrac{|x-3|}{x^2-9}$

20. $\lim\limits_{x \to -3^-} \dfrac{|x-3|}{x^2-9}$

21. $\lim\limits_{x \to -3} \dfrac{|x-3|}{x^2-9}$

22. Suppose $f(x) = \dfrac{x}{\sqrt{1-x^2}}$.

    (a) Discuss all possible points $a \in \mathbb{R}$ at which $f(x)$ may have infinite limits as x approaches a from one side or both sides. List all such limits and their values.

    (b) Draw a sign chart for this function. (First, note its domain; where is it defined?)

    (c) Use this information to sketch a rough graph of the function.

23. Suppose $f(x) = \dfrac{1}{x^4 - 9}$.

    (a) Make a sign chart for $f(x)$.

    (b) Discuss all possible points $a \in \mathbb{R}$ at which $f(x)$ may have infinite limits as x approaches a from one side or both sides. List all such limits and their values, using the sign chart.

    (c) Use all of this information to sketch a rough graph of the function.

24. According to Einstein's Special Relativity theory, the mass of an object with resting mass m and velocity v is given by

$$M(v) = \dfrac{m}{\sqrt{1 - \dfrac{v^2}{c^2}}}$$

where c is the speed of light. Assume $m > 0$ and $0 \le v < c$. Compute the following (for (b) and (c) write your final answer as some decimal number times m, using three significant digits):

    (a) $M(0)$

    (b) $M(c/2)$

    (c) $M(0.9c)$

    (d) $M(0.999c)$

    (e) $\lim\limits_{v \to c^-} M(v)$

    (f) What physical insight(s) should this limit provide us?

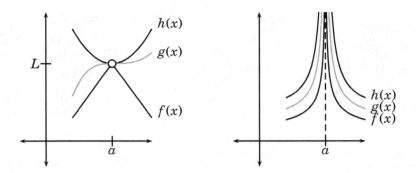

Figure 2.20: Examples illustrating the Sandwich Theorem. In both cases, $f(x) \leq g(x) \leq h(x)$ near $x = a$, i.e., for some set $0 < |x-a| < d$, some $d > 0$. Since the limits, as $x \to a$, of $f(x)$ and $h(x)$ are the same (L and $+\infty$ for the respective graphs), the function $g(x)$ must also have the same limit as $x \to a$.

2.7 Sandwich, Composition, and Trigonometric Theorems

In this section we will state the Sandwich Theorem and use it for computing several limits, including those that prove that the trigonometric functions are continuous where defined.[34]

2.7.1 Sandwich Theorem

Theorem 2.7.1 *(Sandwich Theorem) Suppose that there exists some $d > 0$ such that for every $x \in (a-d, a) \cup (a, a+d)$, i.e., for $0 < |x - a| < d$, we have*

$$f(x) \leq g(x) \leq h(x). \tag{2.48}$$

Then

$$\left(\lim_{x \to a} f(x) = L\right) \wedge \left(\lim_{x \to a} h(x) = L\right) \implies \lim_{x \to a} g(x) = L.$$

The idea is that f and h "sandwich" g between them, and if f and h both approach L, then g is forced to approach L as well. This is graphed for two cases in Figure 2.20, first where L is a finite real number and then where $L = \infty$. The functions $f(x)$ and $h(x)$ can be thought of as variable lower and upper bounds for the function $g(x)$ by (2.48). Thus the behavior of $f(x)$ and $h(x)$ can, in some circumstances (as in the theorem) force certain behavior from $g(x)$. The logic of the argument for the theorem is often graphed in various ways. We will employ the style of Figure 2.21, page 159, to illustrate many of our arguments, except that we will not continue to include the labels "(Hypothesis)" and "(Conclusion)," as they will become apparent in context.

There are several variations of the Sandwich Theorem, in which behavior of one or more *bounding functions* $f(x)$ and $h(x)$ can force behavior upon a (variably) *bounded* function $g(x)$. These variations are perhaps most clearly seen by graphing their respective situations. For

---

[34]The Sandwich Theorem is also called the Squeeze Theorem and the Pinching Theorem in other texts.

2.7. SANDWICH, COMPOSITION, AND TRIGONOMETRIC THEOREMS

$$\text{(Hypothesis) As } x \to a: \quad \underbrace{f(x)}_{\downarrow \, L} \leq g(x) \leq \underbrace{h(x)}_{\downarrow \, L}$$

$$\text{(Conclusion):} \quad \therefore g(x) \longrightarrow L$$

Figure 2.21: Figure illustrating the argument for the Sandwich Theorem.

instance, it is easily seen that we can replace $f(x) \leq g(x) \leq h(x)$ with $f(x) < g(x) < h(x)$ (see again Figure 2.20 at the beginning of this section).[35]

One-sided versions of the theorem also hold as in, for instance, the left-sided limit version:

$$\Big((\exists d > 0)(\forall x)\Big[x \in (a-d, a) \longrightarrow f(x) \leq g(x) \leq h(x)\Big]\Big) \wedge \Big(\lim_{x \to a^-} f(x) = L = \lim_{x \to a^-} h(x)\Big)$$
$$\implies \lim_{x \to a^-} g(x) = L.$$

The following limit is a very traditional example for the original Sandwich Theorem. Note that it relies on the fact that $\sin\theta$ is defined for every $\theta \in \mathbb{R}$, and that $-1 \leq \sin\theta \leq 1$.

Example 2.7.1

Compute the limit $\lim_{x \to 0} x \sin \frac{1}{x}$.

<u>Solution:</u> Note that $x \sin \frac{1}{x}$ is always between $x \cdot 1$ and $x \cdot (-1)$, but these switch roles as top and bottom bounding functions depending on the sign of x. However, we can always write

$$-|x| \leq x \sin \frac{1}{x} \leq |x|.^{36} \tag{2.49}$$

By continuity of $|x|$ and $-|x|$, we have

$$\lim_{x \to 0}(-|x|) \stackrel{\text{CONT}}{=\!=\!=} -|0| = 0, \quad \text{and} \quad \lim_{x \to 0}|x| \stackrel{\text{CONT}}{=\!=\!=} |0| = 0,$$

and so by the Sandwich Theorem we must conclude as well that $\lim_{x \to 0} x \sin \frac{1}{x} = 0$. See also the summary explanation that follows.

---

[35] Note that $f(x) < g(x) < h(x) \implies f(x) \leq g(x) \leq h(x)$, so the fact that we can replace the latter with the former follows quickly from logic; if we have the strict inequalities, then we also have the non-strict inequalities and that hypothesis of the Sandwich Theorem will still hold.

[36] We can make the argument leading to (2.49) more precise. For $x \neq 0$, we can say

$$\left|x \sin \frac{1}{x}\right| = |x| \cdot \left|\sin \frac{1}{x}\right| \implies \left|x \sin \frac{1}{x}\right| \leq |x| \iff -|x| \leq x \sin \frac{1}{x} \leq |x|,$$

where the "$\iff$" comes from the general rule that $|z| \leq K \iff -K \leq z \leq K$.

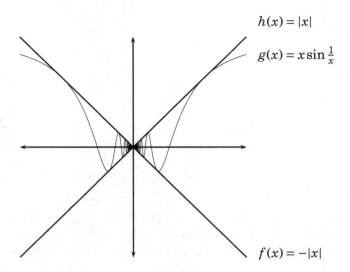

Figure 2.22: Partial graph of $g(x) = x \sin \frac{1}{x}$, which is bounded from above by $h(x) = |x|$ and from below by $f(x) = -|x|$. It oscillates wildly, running through infinitely many periods in the argument $1/x$ of the sine function as $x \to 0$, but is bounded in amplitude by functions that shrink to zero as $x \to 0$. ($g(x)$ is undefined at $x = 0$ but that fact is not apparent in the given graph.)

The Sandwich Theorem argument for this limit can be summarized graphically as follows:

$$\text{As } x \to 0: \quad \underbrace{-|x|}_{\downarrow \atop 0} \leq x \sin \tfrac{1}{x} \leq \underbrace{|x|}_{\downarrow \atop 0}$$

$$\therefore \ x \sin \tfrac{1}{x} \longrightarrow 0.$$

The function $g(x) = x \sin \frac{1}{x}$ is graphed in Figure 2.22, together with the bounding functions $-|x|$ and $|x|$. It has some interesting features, which make it very valuable for later examples that clarify some limit principles. We note how the argument $1/x$ of the sine function here runs through infinitely many periods of sine as $x \to 0$, so the function oscillates in sign with infinitely increasing rapidity as $x \to 0$. However, the "amplitude" $|x|$ is variable and shrinking to zero.

One use of this function is as an example of a rather general theorem, based on the Sandwich Theorem, regarding limits of products where one factor approaches zero while the other factor, however else it is ill-behaved, is at least bounded and defined as we approach the limit point.

Theorem 2.7.2 *Suppose that $f(x)$ is defined for $0 < |x - a| < d$ for some $d > 0$, and that for such x, $f(x)$ is of the form $f(x) = g(x)h(x)$ where $|g(x)| \leq M$ for some $M > 0$, and $h(x) \longrightarrow 0$ as $x \to a$. Then $\lim_{x \to a} f(x) = 0$.*

2.7. SANDWICH, COMPOSITION, AND TRIGONOMETRIC THEOREMS

The proof consists of noting that $-M|h(x)| \leq f(x) \leq M|h(x)|$, so $\pm M|h(x)| \longrightarrow 0$ as $x \to a$ implies $f(x) \longrightarrow 0$ as $x \to a$ as well:

$$\text{As } x \to a: \quad \underbrace{-M|h(x)|}_{\downarrow \atop 0} \leq g(x)h(x) \leq \underbrace{M|h(x)|}_{\downarrow \atop 0}$$

$$\therefore f(x) = g(x)h(x) \longrightarrow 0.$$

This theorem could have been used in the previous example to compute that limit immediately. For convenience, we will use "B" to refer to a function that is both defined and *bounded* as we approach the limit point. Then "$B \cdot 0$" is in fact a *determinate* limit form that yields zero in the limit. For the previous example, we would note the fact that for $x \neq 0$ we have $\left|\sin\frac{1}{x}\right| \leq 1$ and so, aside from being defined, $\sin\frac{1}{x}$ is bounded as $x \to 0$, and we can write

$$\lim_{x \to 0} x \sin\frac{1}{x} \overset{0 \cdot B}{=\!=\!=} 0.$$

While based on the Sandwich Theorem, this argument is somewhat intuitive and certainly less verbose.

2.7.2 Visualizing "Approaches" for Independent versus Dependent Variables

This was discussed briefly in Subsection 2.4.2, page 133, and we further clarify it here.

As with any visualization, the rigorous definitions ultimately determine the meanings of terms such as "approaches" in various contexts, and so those rigorous definitions should always be kept in mind. The shorthand diction used here is typical of how professional mathematicians might express limit ideas in a professional talk or abstract to an audience well aware of the underlying definitions.

The point we clarify in this subsection is that when we visualize the *independent* variable x "approaching" some point, say $x \to a$, we should visualize x as gradually and continuously (not skipping any values) getting closer to that point a—as close as we like and then even closer—but never actually achieving the value $x = a$, and approaching a from both directions (for a two-sided limit). That is built into our first definition of limit in the antecedent $0 < |x - a| < \delta$ of the defining implication. On the other hand, we have more flexibility with the *dependent* variable $f(x)$ in the implication's consequent $|f(x) - L| < \varepsilon$, though we still write $f(x) \longrightarrow L$.

And so in our latest example where we could write $x \to 0 \implies x \sin\frac{1}{x} \longrightarrow 0$, the value of x is supposed to approach zero from both sides but never reach zero, while the function $x \sin\frac{1}{x}$ not only gets closer to zero consistently (in the sense of being sandwiched between $-|x|$ and $|x|$ which both approach zero), but in this case, $x \sin\frac{1}{x}$ also *achieves* the value zero repeatedly (infinitely many times!) as $x \to 0$. So the independent variable x is forced to approach *but avoid* its limiting value, while the dependent variable need not actually avoid its limiting value when we write $x \to a \implies f(x) \to L$.

For another, rather trivial example, consider a constant function $g(x) = 0$ and see how $g(x) = 0 \to 0$ as $x \to \frac{\pi}{2}$. This function's output not only approaches zero; it is never anything but zero. Still we use the notation $x \to \frac{\pi}{2} \implies g(x) \to 0$. The manner in which $g(x) \to 0$ does indeed differ from what we mean by $x \to 0$ for the input variable x, for which we exclude the value $x = 0$.

One-sided limits would have the input variable such as x approaching the limit point continuously but from only one direction.

Infinite limits also use the symbols for "approaching," with $f(x) \to \infty$ or $f(x) \to -\infty$. When we write $f(x) \to \infty$, it obviously cannot get close to the value ∞ but should eventually be greater than any previously fixed number M, as per Definition 2.6.2, page 148.

2.7.3 "Sandwiching" from One Side, or "Pushing"

For the infinite limit case, we can have a Sandwich Theorem-type of argument using only one bounding function if it "bounds" from the appropriate side (above or below). In (2.50), f is the bounding function pushing h, while in (2.51), h is the bounding function pushing f.

Theorem 2.7.3 *Suppose $f(x) \leq h(x)$ on $0 < |x - a| < d$ for some $d > 0$. Then (separately) we have the following:*

$$\lim_{x \to a} f(x) = \infty \implies \lim_{x \to a} h(x) = \infty \tag{2.50}$$

$$\lim_{x \to a} h(x) = -\infty \implies \lim_{x \to a} f(x) = -\infty. \tag{2.51}$$

In other words, if the lesser (lower) function "blows up" towards ∞, it will push the same behavior onto the greater (upper) function; if the greater function "blows up" but towards $-\infty$, it will push the same behavior onto the lower function. Note that if the upper function has limit ∞, this does not force any behavior onto the lower function, and if the lower function has limit $-\infty$, it will not imply any behavior for the upper function.

Such arguments can be verified easily by graphing the situations and seeing how the "blowup" of one function can force a similar behavior of another. (See, for instance, Figure 2.20, page 158.) It is also useful to see the following, also somewhat visual style of argument in support of (2.50) and (2.51). Note that both diagrams are, for now, hypothetical; instead of $\therefore$ we could instead write $\implies$. The two cases (2.50) and (2.51) are separate:

$$\text{As } x \to a: \quad \underset{\underset{\infty}{\downarrow}}{f(x)} \leq h(x) \qquad\qquad \text{As } x \to a: \quad f(x) \leq \underset{\underset{-\infty}{\downarrow}}{h(x)}$$

$$\therefore h(x) \longrightarrow \infty \qquad\qquad\qquad\qquad \therefore f(x) \longrightarrow -\infty$$

Again, we always have to be careful forming conclusions from a diagram. For instance, suppose $f(x) \leq h(x)$ and $h(x) \longrightarrow \infty$. It is not necessarily the case that $f(x) \longrightarrow \infty$ as well, since $h(x)$ is above $f(x)$ and thus unable to "push" $f(x)$ up towards ∞ with it.

2.7. SANDWICH, COMPOSITION, AND TRIGONOMETRIC THEOREMS

Example 2.7.2

Compute $\lim_{x \to 2^+} \dfrac{x^2 + \sin x}{x - 2}$.

<u>Solution</u>: Since $-1 \leq \sin x \leq 1$, we have

$$x^2 - 1 \leq x^2 + \sin x \leq x^2 + 1.$$

Furthermore, as $x \to 2^+$, we have $x > 2$, and so $x - 2 > 0$ and we can therefore divide the inequality above by $x - 2$ without changing the directions of any inequalities:

$$\frac{x^2 - 1}{x - 2} \leq \frac{x^2 + \sin x}{x - 2} \leq \frac{x^2 + 1}{x - 2}. \tag{2.52}$$

In other words, the least that $\frac{x^2 + \sin x}{x-2}$ can be as $x \to 2^+$ is $\frac{x^2-1}{x-2}$, i.e., where $\sin x = -1$, and the greatest it can be is $\frac{x^2+1}{x-2}$, the case where $\sin x = 1$. Next we notice that

$$\lim_{x \to 2^+} \frac{x^2 - 1}{x - 2} \overset{3/0^+}{=\!=\!=} \infty \quad \Longrightarrow \quad \lim_{x \to 2^+} \frac{x^2 + \sin x}{x - 2} = \infty.$$

As noted before, the first inequality in (2.52) is in fact enough to "push" the desired limit to be ∞:

$$\text{As } x \to 2^+: \qquad \underbrace{\frac{x^2 - 1}{x - 2}}_{\substack{\downarrow 3/0^+ \\ \infty}} \leq \frac{x^2 + \sin x}{x - 2}$$

$$\therefore \frac{x^2 + \sin x}{x - 2} \longrightarrow \infty$$

We can compute this limit without a Sandwich Theorem argument if we note that $\sin x$ is continuous at $x = 2$ and $\sin 2 \approx 0.909297426$, and so the limit form is $(4 + \sin 2)/0^+$, which would output ∞ since $4 + \sin 2$ is a positive number (and we know that last fact even if we only know $\sin 2 \in [-1, 1]$).

Alternatively, we can break the fraction into the sum of two fractions and, again noting $\sin 2 > 0$, we can write

$$\lim_{x \to 2^+} \frac{x^2 + \sin x}{x - 2} = \lim_{x \to 2^+} \left[\frac{x^2}{x - 2} + \frac{\sin x}{x - 2} \right] \overset{\frac{4}{0^+} + \frac{\sin 2}{0^+}}{=\!=\!=\!=} \infty.$$
$$ \infty + \infty$$

This used the fact that, as a form, $\infty + \infty$ yields (reasonably enough) limits that are also ∞, as we will note in later sections. That does not mean we can avoid using sandwich arguments altogether, as in our first example where we showed that $x \sin \frac{1}{x} \longrightarrow 0$ as $x \to 0$.

Note too that a small change in the previous example adds an interesting complication. Consider a similar limit but where the numerator of the function is now $x^2 - \sin x$. Then we have

$$\lim_{x \to 2^+} \frac{x^2 - \sin x}{x - 2} = \lim_{x \to 2^+} \left[\frac{x^2}{x-2} - \frac{\sin x}{x-2} \right] \underset{\infty - \infty}{\overset{\frac{4}{0^+} - \frac{\sin 2}{0^+}}{=}} ?$$

The problem here is that the form $\infty - \infty$ is indeterminate because we do not know which of the two functions "blows up" faster. The form $\infty - \infty$ is discussed properly in Section 2.8, and like 0/0, requires other techniques to rewrite it. In fact, if we were given the limit as a difference of fractions that both grow to infinity, the usual methods require us to combine them into one fraction and work from there, using one of the previous methods. For instance, applying a Sandwich Theorem-type arguments we could write

$$\text{As } x \to 2^+: \qquad \underbrace{\frac{x^2 - 1}{x - 2}}_{\downarrow \atop \infty} \leq \frac{x^2 - \sin x}{x - 2}$$

$$\therefore \frac{x^2 - \sin x}{x - 2} \longrightarrow \infty$$

A non-sandwiching argument could note that $x^2 - \sin x \longrightarrow 4 - 0.909297426 > 3 > 0$, and so the limit is of form $(4 - \sin 2)/0^+$, or $a/0^+$ where $a > 0$, and so the limit will again be ∞.

2.7.4 Limits with Compositions of Functions

Theorem 2.7.4 *Suppose that, for some limiting behavior of x, we have $g(x) \longrightarrow L$, and $f(x)$ continuous at $x = L$. Then for the same limiting behavior of x we have $f(g(x)) \longrightarrow f(L)$.*

So for instance, if $f(x)$ is continuous at L and $\lim_{x \to a} g(x) = L$, then

$$\lim_{x \to a} f(g(x)) = f(L);$$

that is, *assuming that $\lim_{x \to a} g(x) = L$ is a point of continuity of f,* then

$$\lim_{x \to a} f(g(x)) = f\left(\lim_{x \to a} g(x)\right) = f(L). \tag{2.53}$$

Rephrased again, if the limit of the "inside" function is a point of continuity of the "outside" function, then we can "move the limit notation (lim) inside."[37]

In the interest of avoiding errors, students are often discouraged from using (2.53) as it is written because it is tempting to attempt to use it when the hypotheses are not met, in which case it can be false.

We will concentrate on the theorem itself, and the limit forms that arise from its proper application.

The theorem is generalizable in terms of the type of limit (basic, left, right or the "at infinity" variety we will have in Section 2.8). We will give a proof for the most basic case at the end of this section. That proof will be a simple modification of the proof of Theorem 2.2.5, page 93. The theorem can be illustrated graphically as follows:

---

[37]The reader is invited to consider this idea in light of a function diagram for $f(g(x))$. It can look like the diagram on page 94, with $g(a)$ replaced everywhere by L, $f(g(a))$ replaced by $f(L)$, and $|x - a| < \delta$ replaced by $0 < |x - a| < \delta$.

2.7. SANDWICH, COMPOSITION, AND TRIGONOMETRIC THEOREMS

$$\text{As } x \to a: \quad f(\underbrace{g(x)}_{\downarrow}) \xrightarrow{\therefore \text{ (by continuity)}} f(L)$$
$$L$$

Again, this diagram is not necessarily true if $f(x)$ is not continuous at $x = L$. When f is, we might not always give the justification "(by continuity)" within the diagram.

Example 2.7.3

Suppose that $\lim_{x \to a} g(x) = 12$, while $\lim_{x \to b} g(x) = 0$. Then

- $\lim_{x \to a} \sqrt[3]{g(x)} \xrightarrow{\sqrt[3]{12}} \sqrt[3]{12}$,

- $\lim_{x \to b} \sqrt[3]{g(x)} \xrightarrow{\sqrt[3]{0}} \sqrt[3]{0} = 0$,

- $\lim_{x \to a} \sqrt{g(x)} \xrightarrow{\sqrt{12}} \sqrt{12} = 2\sqrt{3}$,

- $\lim_{x \to b} \sqrt{g(x)}$ cannot be determined from the given information. This is discussed next.

In these computations, some other *determinate* limit forms "$f(L)$" emerged, where f is continuous at L, as in Theorem 2.7.4, page 164. For instance, the *form* $\sqrt[3]{12}$ describes a limit where the input of the cube root function is approaching 12. Such a limit will necessarily be equal to the *number* $\sqrt[3]{12}$ because the cube root function is continuous at 12, since it is continuous at each real number. However, the square root function is only continuous at positive input values, so the final limit in the example could not be computed. Now $\sqrt{(\ )}$ is right-continuous at zero, so $\sqrt{0^+}$ is a form that returns the value zero for such a limit. But for our example, as $x \to b$ we have the following possible scenarios:

- If $g(x) \to 0^+$, and so $g(x) \to 0$ but $g(x) > 0$ as $x \to b$, then $\lim_{x \to b} \sqrt{g(x)} \xrightarrow{\sqrt{0^+}} \sqrt{0} = 0$.

- In fact, if we just have $g(x) \geq 0$ as $x \to b$, we have the limit being $\sqrt{0} = 0$, since the square root is not fed negative numbers to process as $x \to b$.

- However, if $g(x) \to 0^-$ or $g(x) \to 0^\pm$, and so $g(x)$ is at least sometimes negative as $x \to b$, then $\lim_{x \to b} \sqrt{g(x)} \xrightarrow{\sqrt{0^\pm}}$ does not exist.

This theorem will prove to be more useful—and in fact will be crucial—in later sections. For now it is simply offered as a means to more efficiently dispatch certain limits.

2.7.5 Continuity Considerations for Trigonometric Functions

Our theorem is as follows, with the proof left for Subsection 2.7.6:

Theorem 2.7.5 *The six basic trigonometric functions,* $\sin x$, $\cos x$, $\tan x$, $\cot x$, $\sec x$, *and* $\csc x$ *are continuous everywhere they are defined. Thus,*

1. $\sin x$ *and* $\cos x$ *are continuous for* $x \in \mathbb{R}$;

2. $\tan x$ *and* $\sec x$ *are continuous* **except** *where* $\cos x = 0$, *and are thus continuous for all*
$$x \neq \frac{\pm\pi}{2}, \frac{\pm 3\pi}{2}, \frac{\pm 5\pi}{2}, \ldots;$$

3. $\cot x$ *and* $\csc x$ *are continuous* **except** *where* $\sin x = 0$, *and are thus continuous for all*
$$x \neq 0, \pm\pi, \pm 2\pi, \pm 3\pi, \ldots.$$

Considering the unit circle definitions of $\sin\theta$ and $\cos\theta$, it is reasonable that these two functions are continuous in θ, i.e., that they change continuously with θ. The continuity conclusions for the other trigonometric functions—which are quotients of the everywhere continuous functions 1, $\sin\theta$, and $\cos\theta$—then follow immediately.

We will more often use the variable x for limit computations, but the results are the same. For instance, $\sin x$ and $\cos x$ are continuous for all $x \in \mathbb{R}$, and so therefore are $\sec x = \frac{1}{\cos x}$ and $\tan x = \frac{\sin x}{\cos x}$, except where $\cos x = 0$, which is at odd multiples of $\pi/2$, as stated in the theorem. Similarly, $\csc x = \frac{1}{\sin x}$, and $\cot x = \frac{\cos x}{\sin x}$ are continuous except where $\sin x = 0$, which is at the integer multiples of π (that is, at the even multiples of $\pi/2$).

Because the results listed in Theorem 2.7.5 are intuitive, we will defer the proof until the end of the section. The proofs that $\sin\theta$ and $\cos\theta$ are continuous (as functions of θ) use the sandwich theorem and a geometric argument, and are interesting in their own rights, but for now we will concentrate our efforts in *applications* of the theorem.

Example 2.7.4

The following limits follow directly from continuity of the trigonometric functions (where defined), and of other functions:

- $\lim\limits_{x \to \pi} \sin x \xRightarrow{\text{CONT}} \sin \pi = 0.$

- $\lim\limits_{x \to 1} \sec x \xRightarrow{\text{CONT}} \sec 1 \approx 1.8508..$

- $\lim\limits_{x \to \pi/4} \tan x \xRightarrow{\text{CONT}} \tan \frac{\pi}{4} = 1.$

- $\lim\limits_{x \to \sqrt{\pi}} \cos x^2 \xRightarrow{\text{CONT}} \cos\left(\sqrt{\pi}\right)^2 = \cos \pi = -1.$

The last limit was computable as shown since x^2 is continuous on all of $\mathbb{R}$, and so is $\cos x$, and so the composition $\cos x^2$ is continuous on all of $\mathbb{R}$, which includes $x = \sqrt{\pi}$. Put another way, the function $\cos(\ (\ )^2\ )$ is continuous with respect to its input because $(\ )^2$ is continuous for all real inputs, and so is $\cos(\ )$. (Here we applied Theorem 2.2.5, page 93, on compositions of continuous functions.)

In the last example, the cosine function was the "outer" function, but the trigonometric functions can be combined with other functions in a variety of ways. For limits and continuity of functions involving trigonometric functions, we need only take care nothing "goes wrong," such as the function or combination thereof being undefined or otherwise discontinuous.

2.7. SANDWICH, COMPOSITION, AND TRIGONOMETRIC THEOREMS

Example 2.7.5

Consider the following limit computations:

- $\lim\limits_{x \to \frac{\pi}{2}} \dfrac{1 - \sin^2 x}{\cos x} \underset{\text{ALG}}{\overset{0/0}{=\!=\!=}} \lim\limits_{x \to \frac{\pi}{2}} \dfrac{\cos^2 x}{\cos x} \underset{\text{ALG}}{\overset{0/0}{=\!=\!=}} \lim\limits_{x \to \frac{\pi}{2}} \cos x \overset{\text{CONT}}{=\!=\!=} \cos \dfrac{\pi}{2} = 0.$

- $\lim\limits_{x \to 0} \sqrt{1 - \underbrace{\cos^2 x}_{<1}} \overset{\sqrt{0^+}}{=\!=\!=} 0.$ Alternatively, $\lim\limits_{x \to 0} \sqrt{\underbrace{1 - \cos^2 x}_{\geq 0}} \overset{\text{CONT}}{=\!=\!=} \sqrt{1 - \cos^2 0} = \sqrt{1 - 1} = 0.$

The first was a standard 0/0-form simplification using the trigonometric identity $\cos^2 x + \sin^2 x = 1$. The second limit was first computed by relying on the fact that $\cos^2 \theta < 1$ as $x \to 0$. But it was enough that $\cos x \in [-1, 1]$, so $\cos^2 x \in [0, 1]$ and so—though we are only interested in behavior as we approach zero—the expression inside the square root is never negative, and so continuity is never "broken." (We could have written "> 0" in our last limit comment instead, since that limit describes behavior for x near zero, but not at zero.)

Example 2.7.6

Consider the following limit computation. Note that since 0/0 is indeterminate, so is $\sin \frac{0}{0}$. As usual, all angles not otherwise specified are assumed to be in units of radians.

$$\lim\limits_{x \to 3} \sin\left(\dfrac{x^2 - 9}{x^2 - 5x + 6}\right) \underset{\text{ALG}}{\overset{\sin \frac{0}{0}}{=\!=\!=}} \lim\limits_{x \to 3} \sin\left(\dfrac{(x+3)(x-3)}{(x-2)(x-3)}\right) \underset{\text{ALG}}{\overset{\sin \frac{0}{0}}{=\!=\!=}} \lim\limits_{x \to 3} \sin\left(\dfrac{x+3}{x-2}\right)$$

$$\overset{\text{CONT}}{=\!=\!=} \sin 6 \approx -0.279415498.$$

Since 0/0-form is indeterminate, any "function" of it is also, so we began our analysis with the input (or "argument") of the sine function. Of course the *exact* value of that limit is $\sin 6$, but one may naturally be curious about its numerical value, which was given as well, as a nine-digit approximation.

In Theorem 2.7.4, page 164, we had an alternative method for analyzing the previous example's limit: We could instead exploit the everywhere-continuity of the sine function to allow manipulations such as

$$\lim\limits_{x \to 3} \sin\left(\dfrac{x^2 - 9}{x^2 - 5x + 6}\right) = \sin\left(\lim\limits_{x \to 3} \dfrac{x^2 - 9}{x^2 - 5x + 6}\right) = \sin\left[\lim\limits_{x \to 3}\left(\dfrac{(x+3)(x-3)}{(x-2)(x-3)}\right)\right] = \sin\left[\lim\limits_{x \to 3} \dfrac{x+3}{x-2}\right] = \sin 6.$$

One problem with this second method is that the first "=" is a "provisional =," since it assumes that the limit inside of the sine function actually exists. Later in the text we will have to string together such "provisional computations," meaning that we have to be aware that there may be a problem farther down the chain of computations revealing that an earlier computation in the chain was false, for instance if that limit which we were trying to

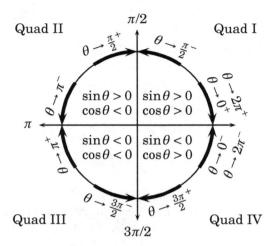

Figure 2.23: Unit circle graph showing the signs of $\sin\theta$ and $\cos\theta$ as θ approaches various axial angles from the right and left (i.e., from above and below the values at the axes). For many limit problems, the "angle" will be represented by the variable x. For instance, as $x \to \frac{\pi}{2}^+$, we have $\cos x \to 0^-$, since in such a case x is approaching the angle $\pi/2$ from angles within the second quadrant (Quad II), in which the cosine is negative. As a result, $x \to \frac{\pi}{2}^+ \implies \sec x = \frac{1}{\cos x} \to -\infty$, the limit form being $1/0^-$.

compute did not actually exist.[38] We will usually opt for the first method—as is written in Example 2.7.6—wherever possible for reasons stated previously.[39]

Trigonometric functions can also give rise to infinite-valued limits. In such cases, it is crucial to determine from within which quadrant the argument of (i.e., the angle inputted to) the function is approaching the limit point, and thus determine the sign of the trigonometric function. If the argument is x, then $x \to 0^+$ means the "angle" x approaches zero from within Quadrant I (where x is the measure of an angle in standard position, i.e., with the initial side being the ray which is the "positive horizontal axis"). On the other hand, an angle x for which $x \to 0^-$ is approaching zero from within Quadrant IV. For $x \to \pi^+$, we have x approaching π from angles with measure slightly greater than π, i.e., from the third quadrant. For $x \to \frac{\pi}{2}^+$, we are in the second quadrant, and so on. See Figure 2.23, where the part of "x" is played by

---

[38] This is a bit like what can occur in logic, where $P_1 \implies P_2, P_2 \implies P_3$, and so on, perhaps until we have, say, $P_8 \implies P_9$, so that $P_1 \implies P_2 \implies P_3 \implies P_4 \implies P_5 \implies P_6 \implies P_7 \implies P_8 \implies P_9$, and then we find out that P_9 is false, so we would have to conclude that P_1 (and everything in between) is also false.

[39] Some textbooks prescribe exactly this second method (not preferred here) of moving the limit inside the continuous function, for such a problem. Instead, we solved it by our standard method, which is to replace (if possible) the original function in its entirety with one that was continuous at the limit point $x = 3$ (and, technically, calling on Theorem 2.4.3, page 125), or, more generally, find a substitute function with the same limiting behavior so we can read the limit off the substitute function (which, technically, means using Theorem 2.4.3, page 125, or its one-sided versions).

We will require this alternative kind of manipulation later, particularly with limits "at infinity," but we will otherwise usually avoid it when possible because it requires very specific hypotheses: that the "inner function" has a limit, say L, and that the outer function is continuous at that point L. But in some sense we already use it when we get some of our new forms, such as $\sqrt{0^+}$ or even $\sin 6$ (as a "form," meaning the limit is of a function of the form $\sin(g(x))$ where $g(x) \to 6$).

2.7. SANDWICH, COMPOSITION, AND TRIGONOMETRIC THEOREMS

the angle θ.

Example 2.7.7

Consider the following trigonometric limits:

1. $\lim\limits_{x \to 0^+} \csc x = \lim\limits_{x \to 0^+} \dfrac{1}{\sin x} \overset{1/0^+}{=\!=\!=} \infty$

2. $\lim\limits_{x \to \pi^+} \csc x = \lim\limits_{x \to \pi^+} \dfrac{1}{\sin x} \overset{1/0^-}{=\!=\!=} -\infty$

3. $\lim\limits_{x \to \frac{\pi}{2}^-} \tan x = \lim\limits_{x \to \frac{\pi}{2}^-} \dfrac{\sin x}{\cos x} \overset{1/0^+}{=\!=\!=} +\infty$

4. $\lim\limits_{x \to \frac{\pi}{2}^+} \tan x = \lim\limits_{x \to \frac{\pi}{2}^+} \dfrac{\sin x}{\cos x} \overset{1/0^-}{=\!=\!=} -\infty$

5. $\lim\limits_{x \to \pi^-} \cot x = \lim\limits_{x \to \pi^-} \dfrac{\cos x}{\sin x} \overset{-1/0^+}{=\!=\!=} -\infty$

6. $\lim\limits_{x \to \frac{\pi}{2}} \tan x = \lim\limits_{x \to \frac{\pi}{2}} \dfrac{\sin x}{\cos x} \overset{1/0^{\pm}}{=\!=\!=}$ DNE

2.7.6 Proofs

First we prove Theorem 2.7.4, page 164, which states that if $g(x) \longrightarrow L$, and $f(x)$ is continuous at $x = L$, then $f(g(x)) \longrightarrow f(L)$. Here we will prove the basic case where the limiting behavior is as $x \to a \in \mathbb{R}$ (that is, where a is finite).

Proof: We will prove that if $a \in \mathbb{R}$ (so that a is not infinite), we have

$$\left(\lim_{x \to a} g(x) = L\right) \wedge (f(x) \text{ continuous at } x = L) \implies \left(\lim_{x \to a} f(g(x)) = f(L)\right).$$

To show this, we have to show that for any $\varepsilon > 0$, we can find a $\delta > 0$ such that

$$0 < |x - a| < \delta \implies |f(g(x)) - f(L)| < \varepsilon.$$

By our continuity and limit assumptions, respectively, we know that

$$(\forall \varepsilon_1 > 0)(\exists \delta_1 > 0)(\forall x)(|x - L| < \delta_1 \longrightarrow |f(x) - f(L)| < \varepsilon_1), \qquad (2.54)$$

$$(\forall \varepsilon_2 > 0)(\exists \delta_2 > 0)(\forall x)(0 < |x - a| < \delta_2 \longrightarrow |g(x) - L| < \varepsilon_2). \qquad (2.55)$$

So for this ε, choose $\varepsilon_1 = \varepsilon$, which gives a $\delta_1 > 0$ so that

$$|x - L| < \delta_1 \implies |f(x) - f(L)| < \varepsilon.$$

Next set $\varepsilon_2 = \delta_1 > 0$. This gives a $\delta_2 > 0$ so that

$$0 < |x - a| < \delta_2 \implies |g(x) - g(a)| < \varepsilon_2 = \delta_1.$$

Finally, let $\delta = \delta_2$, corresponding to ε_2 in the limit requirement for $g(x) \longrightarrow L$. This gives (with the part of "x" in (2.54) played by $g(x)$ in the third and fourth lines):

$$0 < |x - a| < \delta \iff 0 < |x - a| < \delta_2$$
$$\implies |g(x) - L| < \varepsilon_2$$
$$\iff |g(x) - L| < \delta_1$$
$$\implies |f(g(x)) - f(L)| < \varepsilon_1 = \varepsilon, \qquad \text{q.e.d.}$$

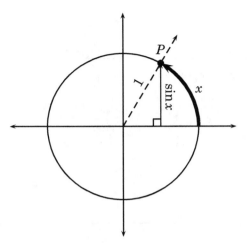

Figure 2.24: Unit circle graph showing the relative sizes of $\sin x$ and x, where x is the angle measure in radians, which coincides with the directed length of the arc because the radius is 1. (Recall that $\theta = s/r$ if s is the directed arc length and r is the radius.) More generally, the distance from the horizontal axis to P on the terminal side is $|\sin x|$, and the arc length distance from $(1,0)$ to P is given by $|x|$.

Next we prove Theorem 2.7.5, page 166, that the trigonometric functions $\sin x$, $\cos x$, $\tan x$, $\cot x$, $\sec x$ and $\csc x$ are continuous wherever they are defined.

Proof: Our "proof" will be in four parts, and will be cut somewhat shorter than a proof from "first principles" would be by using an observation about the geometry of the unit circle. In this abbreviated proof, we will see the Sandwich Theorem in action, in particular as applied to a useful inequality, (2.56), which will be our "observation."

The order in which we will prove our results is as follows: (1) continuity of $\sin x$ at $x = 0$, implying (2) continuity of $\cos x$ at $x = 0$, together implying (3) continuity of $\sin x$ and $\cos x$ at every $x \in \mathbb{R}$, which implies (4) continuity of the other trigonometric functions wherever they are defined.

(1) $\sin x$ is continuous at $x = 0$.

Consider the unit circle graphed in Figure 2.24, page 170. Now $|\sin x|$ is the distance from the horizontal axis to a point P where the terminal side of the angle intersects the unit circle. The arc in the figure is another, albeit non-straight path of length $|x|$ from the horizontal axis to P. Thus

$$|\sin x| \leq |x|, \tag{2.56}$$

which is the same as $-|x| \leq \sin x \leq |x|$. Letting $x \to 0$, we get the following:

$$\underbrace{-|x|}_{\downarrow \atop 0} \leq \sin x \leq \underbrace{|x|}_{\downarrow \atop 0}$$

2.7. SANDWICH, COMPOSITION, AND TRIGONOMETRIC THEOREMS

The Sandwich Theorem then gives us $\lim_{x\to 0} \sin x = 0$. Since $\sin 0 = 0$ as well, we have $\sin x$ is continuous at $x = 0$, q.e.d.[40]

(2) $\cos x$ is continuous at $x = 0$. This follows immediately, since near $x = 0$ (so the "angle" x terminates in the first or fourth quadrants) we have $\cos x > 0$ and thus (again, near $x = 0$) $\cos x = \sqrt{1 - \sin^2 x}$, and so we can replace $\cos x$ with that expression (according to Theorem 2.4.3):

$$\lim_{x\to 0} \cos x = \lim_{x\to 0} \sqrt{1 - \sin^2 x} = \sqrt{1 - \sin^2 0} = \sqrt{1} = 1 = \cos 0, \text{ q.e.d.}$$

We will take a moment here to explain why we could compute this limit as we did. Because $\sin x$ is continuous at $x = 0$, so is $1 - \sin^2 x$, and since that function approaches $1 > 0$ as $x \to 0$, its square root is also continuous at $x = 0$.

(3) $\sin x$ and $\cos x$ are continuous for all $x \in \mathbb{R}$. These follow from the two previous results and the trigonometric identities that follow:

$$\sin(\alpha + \beta) = \sin\alpha\cos\beta + \cos\alpha\sin\beta; \qquad (2.57)$$
$$\cos(\alpha + \beta) = \cos\alpha\cos\beta - \sin\alpha\sin\beta. \qquad (2.58)$$

From these we can write:

$$\lim_{x\to a}\sin x = \lim_{x\to a}\sin(a + (x-a))$$
$$= \lim_{x\to a}(\sin a \cos\underbrace{(x-a)}_{0} + \cos a \sin\underbrace{(x-a)}_{0}) = \sin a \cos 0 + \cos a \sin 0$$
$$= (\sin a)(1) + (\cos a)(0) = \sin a,$$

$$\lim_{x\to a}\cos x = \lim_{x\to a}\cos(a + (x-a))$$
$$= \lim_{x\to a}(\cos a \cos\underbrace{(x-a)}_{0} - \sin a \sin\underbrace{(x-a)}_{0}) = \cos a \cos 0 - \sin a \sin 0$$
$$= (\cos a)(1) - (\sin a)(0) = \cos a, \text{ q.e.d.}$$

Here we used what we will later call a *substitution argument*, which will be introduced properly in Section 2.9 (though we could instead invoke Theorem 2.7.4, page 164, directly). The idea is, roughly, that $x \to a \iff x - a \to 0$ in the sense of limit (where x is never actually equal to a, and $x - a$ is never equal to zero).

(4) All six trigonometric functions are continuous where they are defined.

---

[40] Recall that $f(x)$ is continuous at $x = a$ if and only if $\lim_{x\to a} f(x) = f(a)$. See Theorem 2.4.2, page 124.

Of course $\sin x$ and $\cos x$ were already shown continuous for all $x \in \mathbb{R}$, i.e., where defined, earlier. The other functions are defined by quotients where the numerators are either $\sin x$, $\cos x$, or 1, which are continuous everywhere, while the denominators are either $\sin x$ or $\cos x$, again continuous everywhere. Since a ratio of two functions is continuous if both numerator and denominator are continuous and the denominator is nonzero, the functions $\tan x$ and $\sec x$ are continuous except where $\cos x = 0$, and $\cot x$ and $\csc x$ are continuous except where $\sin x = 0$. Summarizing, all trigonometric functions are continuous where defined, q.e.d.

Simply knowing that the trigonometric functions are continuous wherever they are defined adds whole classes of functions whose limits and continuity questions we can now answer, as demonstrated in this section. In Section 2.9 we introduce two new, somewhat deep limits involving the sine and cosine functions. Those new limits will expand still further our classes of limits we can analyze, ultimately opening the differential calculus to trigonometric functions in Chapter 3.

Exercises

1. Compute $\lim_{x \to 0^+} \left[\sqrt{x} \sin\left(\frac{1}{x}\right)\right]$.

2. Compute $\lim_{x \to 1^+} \frac{\sin x}{x - 1}$. (Hint: $\sin 1 \approx 0.841470985$.)

3. Compute using a Sandwich Theorem-type argument $\lim_{x \to 5^+} \frac{x + \cos x}{x^2 - 25}$.

4. Compute $\lim_{x \to 2} \cos\left(\frac{x^2 - 4x + 4}{x^2 - 4}\right)$.

5. Compute the following limits.

   (a) $\lim_{x \to \frac{\pi}{2}^+} \sec x$

   (b) $\lim_{x \to \frac{\pi}{2}^-} \sec x$

   (c) $\lim_{x \to \frac{3\pi}{2}^+} \sec x$

   (d) $\lim_{x \to \frac{3\pi}{2}^-} \sec x$

6. Compute the following limits.

   (a) $\lim_{x \to 0^+} \cot x$

   (b) $\lim_{x \to 0^-} \cot x$

   (c) $\lim_{x \to \pi^+} \cot x$

   (d) $\lim_{x \to \pi^-} \cot x$

7. Suppose $f(x) \le h(x)$ for $0 < |x - a| < d$, for some $d > 0$, and that $\lim_{x \to a} h(x) = \infty$. By drawing several graphs, show that $\lim_{x \to a} f(x)$ can be anything: finite, ∞, $-\infty$, or nonexistent.

8. Suppose that $-x^3 + 2x^2 - x + 2 < f(x) < x^2 - 2x + 3$ for all $x \in [0, 2]$, except for $x = 1$. Find $\lim_{x \to 1} f(x)$ if possible.

9. Suppose for all $x \ne 2$ we have $4 \le f(x) \le (x - 2)^2 + 4$. Find $\lim_{x \to 2} f(x)$ if possible.

For the following, compute each limit that exists, and state those that do not. Be sure to show work justifying your answers.

10. $\lim_{x \to 0} \sqrt[3]{x \sin \frac{1}{x}}$

11. $\lim_{x \to 0} \sqrt{x \sin \frac{1}{x}}$

12. $\lim_{x \to 0} \sqrt{\left|x \sin \frac{1}{x}\right|}$

13. $\lim_{x \to 0} \sqrt{x^2 \sin^2 \frac{1}{x}}$

14. $\lim_{x \to 0^+} \left[\sqrt{x} \sin(\csc x)\right]$

15. $\lim_{x \to 0^+} \left[\sqrt{x} \sin\left(\csc\left(\frac{1}{x}\right)\right)\right]$

16. $\lim_{x \to 0} x^2 \cos \frac{1}{\sqrt[3]{x}}$

17. $\lim_{x \to 0} \sqrt[3]{x} \sin\left(\frac{1}{x}\right)$

18. $\lim_{x \to 0} \frac{\cos \frac{1}{x}}{x^2}$

19. $\lim_{x \to 0} \sin x \csc x$

20. $\lim_{x \to 0} \cot x \csc x$

21. $\lim_{x \to 0^+} \frac{\cos x}{x}$

22. $\lim_{x \to 0^-} \frac{\cos x}{x}$

23. $\lim_{x \to 0} \frac{\cos x}{x}$

24. $\lim_{x \to 0} \frac{\cos x}{x^2}$

25. $\lim_{x \to 0} \frac{\cos x}{|x|}$

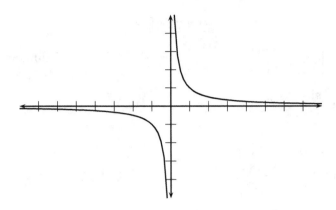

Figure 2.25: Partial graph of $f(x) = 1/x$. We see here that $f(x) \longrightarrow 0$ as $x \to \infty$ and as $x \to -\infty$. To be more precise, $x \to \infty \implies \frac{1}{x} \to 0^+$, while $x \to -\infty \implies \frac{1}{x} \to 0^-$.

2.8 Limits "At Infinity"

The limits we introduce here differ from previous limits in that here we are interested in the "long-term" behavior of functions $f(x)$, meaning as x grows without bound rather than as x approaches a finite point. There are new "forms" we will come across here, such as $1/\infty$, $\infty \cdot \infty$, ∞/∞, $0 \cdot \infty$, and $\infty - \infty$. (Only the first two are *determinate*.)

The first forms we will look at are $1/\infty$ and $1/(-\infty)$. For these, we look again to the function $f(x) = 1/x$, produced again here in Figure 2.25. Numerically we see that as the input moves to the right through values such as $x = 1, 2, 3, 10, 100, 1000$, and 10^6, the function takes on respective values $f(x) = 1, 1/2, 1/3, 1/10, 1/100, 1/1000$, and 10^{-6}. This trend continues, and as x grows "more positive" without bound, the output $f(x)$ shrinks towards (though is never equal to) zero. A similar phenomenon occurs when we take negative x-values $x = -1, -2, -3, -10, -100, -1000$, and -10^6, except the values of $f(x)$ are then $f(x) = -1, -1/2, -1/3, -1/10, -1/100, -1/1000$, and -10^{-6}. So as x becomes "increasingly negative" without bound, the function values are negative numbers shrinking in absolute size. In both cases, we can get as close to zero in the values of $f(x)$ as we could like (without necessarily *achieving* the value zero) by choosing x large enough, as is reflected in the statements

$$\lim_{x \to \infty} \frac{1}{x} \stackrel{\frac{1}{\infty}}{=\!=} 0, \tag{2.59}$$

$$\lim_{x \to -\infty} \frac{1}{x} \stackrel{\frac{1}{(-\infty)}}{=\!=} 0. \tag{2.60}$$

The forms $1/\infty$ and $1/(-\infty)$ are determinate, both yielding zero limits. Recall that a growing denominator tends to shrink a fraction, and in particular a large number will have a small reciprocal. From earlier discussions of the graph of $y = 1/x$, we can see how, as x gets arbitrarily large, $1/x$ gets arbitrarily small (though never quite zero) in absolute size.[41]

It is common to read the left-hand side of (2.59) as, "the limit, as x approaches infinity, of $1/x$." Of course, x does not approach ∞ in the sense of "getting close" to ∞, because ∞ is not

---

[41] In the presence of a growing denominator, unless there is another influence to counteract its effect—such as a growing numerator—the overall fraction will shrink. However, we will soon see that ∞/∞ is indeterminate.

2.8. LIMITS "AT INFINITY"

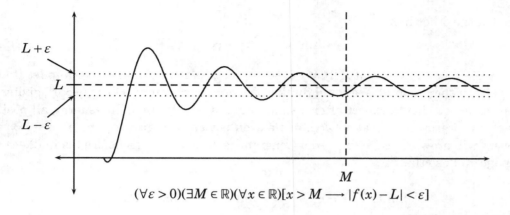

Figure 2.26: Illustration of the definition (2.61) of a finite limit L of a function as $x \to \infty$.

a point on the line x can approach, but the notation means that we are computing what the behavior of $1/x$ will be as x grows consistently positive and without bound. It is similar for $x \to -\infty$. To make these precise, we give the following definitions.

Definition 2.8.1 *For a finite number $L \in \mathbb{R}$, we say*

$$\lim_{x \to \infty} f(x) = L \iff (\forall \varepsilon > 0)(\exists M \in \mathbb{R})(\forall x \in \mathbb{R})[x > M \longrightarrow |f(x) - L| < \varepsilon], \qquad (2.61)$$

$$\lim_{x \to -\infty} f(x) = L \iff (\forall \varepsilon > 0)(\exists N \in \mathbb{R})(\forall x \in \mathbb{R})[x < N \longrightarrow |f(x) - L| < \varepsilon]. \qquad (2.62)$$

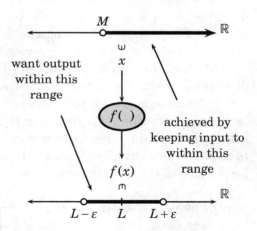

In (2.61), we could also write $f((M, \infty)) \subseteq (L - \varepsilon, L + \varepsilon)$, while in (2.62), we could write $f((-\infty, N)) \subseteq (L - \varepsilon, L + \varepsilon)$. A case of (2.61) for a particular ε is illustrated in Figure 2.26. We will leave the illustrations of (2.62) to the reader.

The figure on the right is another illustration of the definition of $x \longrightarrow \infty \implies f(x) \longrightarrow L$. It is likely that, as ε shrinks to be smaller and smaller, the number M will tend to move farther to the right on the number line representing $\mathbb{R}$.

Next we point out that it is natural to have a notion of an *infinite* limit as $x \to \infty$ or $x \to -\infty$. For instance,

$$\lim_{x \to \infty} x = \infty \qquad (2.63)$$

seems quite reasonable, as does

$$\lim_{x \to -\infty} x^2 \stackrel{(-\infty)^2}{=\!=\!=} \infty. \qquad (2.64)$$

There are many common functions that grow without bound as x grows without bound. Note that (2.64) can be thought of as a form $(-\infty) \cdot (-\infty)$ or $(-\infty)^2$, which reasonably yields the limit ∞. On the other hand,[42]

$$\lim_{x\to -\infty} x^3 \overset{(-\infty)^3}{=\!=\!=} -\infty, \tag{2.65}$$

since for $x \to -\infty$ we have eventually $x < 0$, so $x^3 < 0$. It is clear that $x \to -\infty$ implies that x^3 grows in size without bound. We could think of the previous limit as a form $(-\infty)^3$, giving the limit as $-\infty$ as we should expect after a small amount of reflection. In general, all positive powers of x will grow to $+\infty$ as $x \to \infty$, while even powers will grow to $+\infty$ as $x \to -\infty$ and odd powers will grow to $-\infty$ as $x \to -\infty$. Constant factors behave as before (as in (2.44) and (2.45), page 150), so that[43]

$$\lim_{x\to\infty} 5x \overset{5\cdot\infty}{=\!=\!=} \infty, \qquad \lim_{x\to -\infty}(-3x) \overset{-3\cdot(-\infty)}{=\!=\!=} \infty.$$

The definition of $\lim_{x\to\infty} f(x) = \infty$ is given next:

Definition 2.8.2 *We make the following definition:*

$$\lim_{x\to\infty} f(x) = \infty \iff (\forall M)(\exists N)(\forall x)(x > N \longrightarrow f(x) > M). \tag{2.66}$$

In other words, for any fixed M, we can force $f(x)$ to be greater than M by taking $x > N$, so that $f((N,\infty)) \subseteq (M,\infty)$.

The diagram on the right illustrates the definition for a limit in which $x \to \infty \implies f(x) \to \infty$. Note that as M is increased, we expect N to also be eventually increased.

Note that the part of M in the previous diagram is played by N here. This choice is only because, besides ε and δ, we tend to introduce Latin alphabet symbols in alphabetic order, absent some historical reason to do otherwise. Usually M and N are used for bounds, in this case lower bounds.

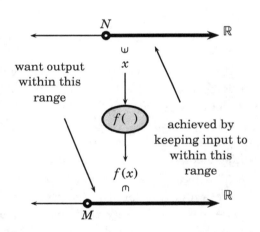

The definition of $f(x) \longrightarrow -\infty$ as $x \to \infty$, and similar definitions, are left as exercises.

To see (2.66) in action, consider how we might prove $\lim_{x\to\infty} x^2 = \infty$: Let $M \in \mathbb{R}$, and take $N = \sqrt{|M|}$, so that $x > N = \sqrt{|M|} \implies f(x) = x^2 > N^2 = \left(\sqrt{|M|}\right)^2 = |M| \geq M$. (We needed $N \geq 0$ so that $x > N \implies x^2 > N^2$. That implication is not necessarily true if we do not have $N \geq 0$.)

---

[42] Of course $(-\infty)\cdot(-\infty)$ is a particular *form* representing a product of two functions that are both negative and growing without bound. The product is naturally positive and also growing without bound, the resulting limit then being ∞. Similarly $(-\infty)^3 = -\infty$, as an equality of limit forms.

[43] Noninteger powers of x are more complicated for $x \to -\infty$. For instance, if $m/n > 0$ is a reduced fraction, then $x^{m/n} = (x^m)^{1/n} = \sqrt[n]{x^m}$. As $x \to -\infty$, eventually $x < 0$, and $x^{1/3} \to -\infty$, $x^{2/3} \to \infty$, $x^{1/2}$ is undefined, as is $x^{\sqrt{2}}$.

2.8. LIMITS "AT INFINITY"

Some relevant limit forms that occur in this and other contexts, and which are not indeterminate, include the following:

1. $\infty + a = \infty$ for any fixed $a \in \mathbb{R}$;

2. $\infty + \infty = \infty$;

3. $a \cdot \infty = \infty$ if $a > 0$, while $a \cdot \infty = -\infty$ if $a < 0$.

As before, we can perform *some* "arithmetic" of limit forms, though we always have to be careful not to make definite conclusions about indeterminate forms (see Example 2.8.1).

The cases mentioned in item 3 were also mentioned in previous sections (first on page 149). Note again that $a \cdot \infty$ as a limit form means that we have a limit where one function is approaching some constant $a \in \mathbb{R}$, and the other function is "approaching ∞" (positive and growing without bound in the limit), and so their product approaches ∞ if $a > 0$, and $-\infty$ if $a < 0$. If $a = 0$, then the form is indeterminate, and we have to attempt to rewrite it algebraically to see if it can be written into a determinate form.

These forms are relatively intuitive. The following are more subtle, and in fact are indeterminate:

$$\infty - \infty, \qquad 0 \cdot \infty, \qquad \infty/\infty, \qquad 0/0.$$

To see the first is indeterminate, consider, for instance, the following $\infty - \infty$-form limits. *Note how we often use factoring, as with 0/0 forms, to produce determinate forms, though we may need a different kind of cleverness here, namely removing the highest-degree term (which grows the fastest) as a factor. We will return to this theme repeatedly in what follows.*[44]

Example 2.8.1

$$\lim_{x \to \infty} \left[(x^2 + 1) - (x^2)\right] \overset{\infty - \infty}{\underset{\text{ALG}}{=\!=\!=}} \lim_{x \to \infty} (1) = 1,$$

$$\lim_{x \to \infty} \left(x^3 - x^2\right) \overset{\infty - \infty}{\underset{\text{ALG}}{=\!=\!=}} \lim_{x \to \infty} x^2(x - 1) \overset{\infty \cdot \infty}{=\!=\!=} \infty,$$

$$\lim_{x \to \infty} \left(x^4 - x^6\right) \overset{\infty - \infty}{\underset{\text{ALG}}{=\!=\!=}} \lim_{x \to \infty} x^4(1 - x^2) \overset{\infty \cdot (-\infty)}{=\!=\!=} -\infty.$$

The questions for $\infty - \infty$ form become "Which 'infinity' is larger, i.e., which function grows faster when we have a difference $f(x) - g(x)$ of functions f and g which both grow without bound? Or is there ultimately a compromise?" Similar examples can be found for forms $0 \cdot \infty$ (or $\infty \cdot 0$) and ∞/∞. The former we look at next, with a few examples to show that it is in fact indeterminate.

---

[44]Scrutiny of the first limit shows that it is possible to come up with a limit of the form $\infty - \infty$ which when evaluated gives any predetermined real value we would like (just replace the number 1 with the desired value). The second and third show we can, furthermore, find limits of form $\infty - \infty$ which return infinite limits as well.

Example 2.8.2

Consider the following limits of form $\infty \cdot 0$:

$$\lim_{x\to\infty}\left[x\cdot\frac{1}{x^2}\right] \overset{\infty\cdot 0}{\underset{\text{ALG}}{=\!=\!=}} \lim_{x\to\infty}\frac{1}{x} \overset{1/\infty}{=\!=} 0,$$

$$\lim_{x\to\infty}\left[x^2\cdot\frac{1}{x}\right] \overset{\infty\cdot 0}{\underset{\text{ALG}}{=\!=\!=}} \lim_{x\to\infty} x = \infty,$$

$$\lim_{x\to\infty}\left[x\cdot\frac{5}{x}\right] \overset{\infty\cdot 0}{\underset{\text{ALG}}{=\!=\!=}} \lim_{x\to\infty} 5 = 5.$$

So clearly $\infty \cdot 0$ is indeterminate; knowing we are multiplying two functions, one of which grows without bound and the other of which shrinks to zero, does not tell us the limiting behavior of the product. Some algebraic rewriting was necessary to cancel what competing influences we could, as we did previously with 0/0-form limits.

Now let us turn to polynomial and rational functions. Our first theorem is the following:

Theorem 2.8.1 *For a polynomial function $p(x) = a_n x^n + a_{n-1} x^{n-1} + \cdots + a_1 x + a_0$, where $a_n \ne 0$ (so the polynomial is indeed of degree n), we have*

$$\lim_{x\to\infty} p(x) = \lim_{x\to\infty} a_n x^n, \tag{2.67}$$

$$\lim_{x\to-\infty} p(x) = \lim_{x\to-\infty} a_n x^n. \tag{2.68}$$

In other words, for $x \to \infty$ and $x \to -\infty$, a polynomial function's growth is ultimately dictated by its leading (highest-degree) term. Rather than prove this in general, we can see the essence of a proof in the following examples and leave the actual proof as an exercise.

Example 2.8.3

Consider the following limits. Forms are first given above the "=" and then simplified underneath.

$$\lim_{x\to\infty}(3x^2 - 5x + 11) = \lim_{x\to\infty} x^2\left(3 - \frac{5}{x} + \frac{11}{x^2}\right) \overset{\infty\cdot(3-0+0)}{\underset{\infty\cdot 3}{=\!=\!=\!=\!=}} \infty,$$

$$\lim_{x\to-\infty}(x^3 + 95x^2 - 15x + 1000) = \lim_{x\to-\infty} x^3\left(1 + \frac{95}{x} - \frac{15}{x^2} + \frac{1000}{x^3}\right) \overset{-\infty(1+0-0+0)}{\underset{-\infty\cdot 1}{=\!=\!=\!=\!=}} -\infty.$$

While it might be considered "bad form," clearly we could have ignored the lower-order terms and instead computed the limits, respectively, of $3x^2$ and of x^3, though we will usually perform the full computations as shown. Still, it is useful to predict the outcome from the start, as we notice that when we factor out the highest power, the lower-order terms we are left with have negative powers of x, which then shrink to zero as $x \to \pm\infty$, leaving only the coefficient of the highest-order term, and its power of x, as relevant factors in the limit. This phenomenon is also very useful when we look at rational limits as $x \to \pm\infty$. Such limits are often of the form ∞/∞, $(-\infty)/(-\infty)$, and so on. Some examples follow.

2.8. LIMITS "AT INFINITY"

Example 2.8.4

Consider the following limits. Note how the largest degree is removed as a factor in both numerator and denominator.

$$\lim_{x\to\infty}\frac{3x^2+5x-9}{6x+11} \xrightarrow{\infty/\infty} \lim_{x\to\infty}\frac{x^2\left(3+\frac{5}{x}-\frac{9}{x^2}\right)}{x\left(6+\frac{11}{x}\right)} \xrightarrow{\infty/\infty}_{\text{ALG}} \lim_{x\to\infty} x\cdot\frac{3+\frac{5}{x}-\frac{9}{x^2}}{6+\frac{11}{x}} \xrightarrow{\infty\cdot\frac{3}{6}} \infty,$$

$$\lim_{x\to\infty}\frac{9x^2+2x+1}{16x^2+3x-100} \xrightarrow{\infty/\infty} \lim_{x\to\infty}\frac{x^2\left(9+\frac{2}{x}+\frac{1}{x^2}\right)}{x^2\left(16+\frac{3}{x}-\frac{100}{x^2}\right)} \xrightarrow{\infty/\infty}_{\text{ALG}} \lim_{x\to\infty}\frac{9+\frac{2}{x}+\frac{1}{x^2}}{16+\frac{3}{x}-\frac{100}{x^2}} = \frac{9+0+0}{16+0-0} = \frac{9}{16},$$

$$\lim_{x\to-\infty}\frac{5-3x}{2x^2+x+1} \xrightarrow{\infty/\infty} \lim_{x\to-\infty}\frac{x\cdot\left(\frac{5}{x}-3\right)}{x^2\left(2+\frac{1}{x}+\frac{1}{x^2}\right)} \xrightarrow{\infty/\infty}_{\text{ALG}} \lim_{x\to-\infty}\left[\frac{1}{x}\cdot\frac{\frac{5}{x}-3}{2+\frac{1}{x}+\frac{1}{x^2}}\right] \xrightarrow{\frac{1}{-\infty}\cdot\frac{-3}{2}}_{0\cdot\frac{-3}{2}} 0.$$

This indicates the likelihood of a theorem similar to Theorem 2.8.1 but for rational functions, and indeed we have the following (only valid as $x \to \pm\infty$):

Theorem 2.8.2 *For any rational function* $f(x) = \frac{p(x)}{q(x)}$, *where* $p(x) = a_n x^n + \cdots + a_1 x + a_0$ *and* $q(x) = b_m x^m + \cdots + b_1 x + b_0$, *with* $a_n, b_m \neq 0$, *we have*

$$\lim_{x\to\infty}\frac{p(x)}{q(x)} = \lim_{x\to\infty}\frac{a_n x^n}{b_m x^m}, \tag{2.69}$$

$$\lim_{x\to-\infty}\frac{p(x)}{q(x)} = \lim_{x\to-\infty}\frac{a_n x^n}{b_m x^m}. \tag{2.70}$$

So as we take $x \to \infty$ or $x \to -\infty$, the limiting behavior of a rational function is governed by the leading terms of the numerator and denominator. We will use this theorem for anticipating results, but will work the actual limits as in Example 2.8.4.[45]

---

[45]Note that the "leading term" means the nonzero term of the highest degree, not necessarily the first term appearing. For instance, in the polynomial $6-5x^2$, the leading term is $-5x^2$.

We will continue to compute the limits longhand for three reasons. First, it is good reinforcement of the underlying principles. Second, it is not entirely standard to write, for instance,

$$\lim_{x\to\infty}\frac{5x^2+3x-11}{7x^2-9x+1{,}000} = \lim_{x\to\infty}\frac{5x^2}{7x^2} = \lim_{x\to\infty}\frac{5}{7} = 5/7.$$

A reader might be confused about the whereabouts of the terms that were dropped and generally lose confidence that the writer's understanding is correct. Finally, the theorem requires that $x \to \infty$ or $x \to -\infty$, so if we reflexively drop terms, we may be tempted to do so for a limit at a finite point, where the theorem does not hold. That said, it is not uncommon for a trained mathematician to simply drop all steps above and write

$$\lim_{x\to\infty}\frac{5x^2+3x-11}{7x^2-9x+1{,}000} = 5/7.$$

If the limits for $x \to \infty$ and $x \to -\infty$ are the same, as they often are, we might write, for instance,

$$\lim_{x\to\pm\infty}\frac{5x^2+3x-11}{7x^2-9x+1{,}000} = 5/7.$$

It is common for trigonometric limits, and variations of the Sandwich Theorem (originally Theorem 2.7.1, page 158) to appear with limits "at infinity."

Example 2.8.5

Consider the limit $\lim\limits_{x\to\infty} \dfrac{\cos x}{x}$. This yields to the Sandwich Theorem quickly:

$$\text{As } x \to \infty: \quad \underbrace{\dfrac{-1}{x}}_{\downarrow\ 0} \leq \dfrac{\cos x}{x} \leq \underbrace{\dfrac{1}{x}}_{\downarrow\ 0} \qquad \therefore \dfrac{\cos x}{x} \longrightarrow 0.$$

One would usually then summarize: $\lim\limits_{x\to\infty} \dfrac{\cos x}{x} = 0$. Alternatively, one could write $\lim\limits_{x\to\infty} \dfrac{\cos x}{x} \stackrel{B/\infty}{=\!=} 0$, where B represents a function that is defined but bounded as $x \to \infty$, as discussed next.

We could also look at the previous limit as the limit of a product of two functions, one of which is bounded ($\cos x$) but defined as $x \to \infty$, and the other of which simultaneously approaches zero ($1/x$), yielding $B \cdot 0$ form, which is a determinate form giving zero in the limit (by a Sandwich Theorem argument mentioned in the previous section). Alternatively, we could define a form "B/∞" which will always yield zero since the denominator grows without bound (shrinking the fraction) while the numerator is unable to compensate (by growing the fraction) since it is bounded (but assumed to be defined). We could also write $B/\infty = B \cdot \frac{1}{\infty} = B \cdot 0$. The "algebra" of forms is interesting and intuitive, but one needs to be careful to identify legitimate underlying mechanisms—such as the Sandwich Theorem—to make such arguments regarding limit forms.

Example 2.8.6

Consider the limit $\lim\limits_{x\to\infty}(x + \sin x)$. Here we have a sum of functions, the first growing without bound and the second being bounded. Intuitively this sum should grow without bound since the function $\sin x$ is unable to check the growth of x. We can again use the Sandwich Theorem (though only the first inequality is necessary here, unlike in the previous example):

$$\text{As } x \to \infty: \quad \underbrace{x - 1}_{\downarrow\ \infty} \leq x + \sin x \leq \underbrace{x + 1}_{\downarrow\ \infty} \qquad \therefore (x + \sin x) \longrightarrow \infty.$$

In fact, recall that in such a case we only need the first inequality to form our conclusion.

We could look at this limit as an example of a form we could define as "$\infty + B$," which will always give us the actual limit being ∞. To see this, note that for such a case we are looking

2.8. LIMITS "AT INFINITY"

at sums $f(x) + g(x)$ where $g(x)$ is defined and bounded, i.e., $|g(x)| \leq M$ for some finite fixed M, and $f(x) \longrightarrow \infty$. By the boundedness of $g(x)$, we get

$$f(x) - M \leq f(x) + g(x) \leq f(x) + M.$$

Since $f(x) - M \longrightarrow \infty$, and (though not necessary for the conclusion) $f(x) + M \longrightarrow \infty$ also, we would conclude $f(x) + g(x) \longrightarrow \infty$ as well.

It should be pointed out that the limits $\lim_{x \to \infty} \sin x$ and $\lim_{x \to \infty} \cos x$ both do not exist. This is because these functions oscillate between -1 and 1 and do not approach any particular value to the exclusion of others (recall that a limit must be unique). However, the previous example shows that such functions can still be involved in limits "at infinity," especially when their (bounded) oscillations can be checked by, or dominated by, the influences of other functions in the limits.

The methods of the previous two examples are important and should be mastered, but we can use observations about forms (proved the same ways) and have more abbreviated computations:

$$\lim_{x \to \infty} \frac{\cos x}{x} \stackrel{B/\infty}{=\!=\!=} 0,$$

$$\lim_{x \to \infty} (x + \sin x) \stackrel{\infty + B}{=\!=\!=} \infty.$$

Here B stands for any bounded function, including constants. In both cases, the "B" cannot check the growth of the other ("∞") function, and so the other function's influence ultimately prevails in the limit. Note that B/∞ and $\infty + B$ are *determinate* forms. We list these and some others next. Note that the left sides are *forms* while the right sides are final limit values.

$$B/\infty = 0, \tag{2.71}$$
$$B/(-\infty) = 0, \tag{2.72}$$
$$B + \infty = \infty, \tag{2.73}$$
$$B - \infty = -\infty. \tag{2.74}$$

All these are intuitive and provable using the Sandwich Theorem and its variations. As always, it is important that we are aware of technicalities. For instance,

$$\lim_{x \to \infty} \frac{\tan x}{x \sec x} \quad \text{DNE}.$$

Though $(\tan x)/(x \sec x) = (\sin x/\cos x)/(x/\cos x) = (\sin x)/x$, this simplification is only valid if $\tan x$, $\sec x$ are defined, i.e., when $\cos x \neq 0$. But there are infinitely many times $\cos x = 0$ as $x \to \infty$, and for that matter within any interval (M, ∞), so none of our definitions for limits as $x \to \infty$ can hold true. This is despite the fact that if we (naively) simplify the function within the limit we would get $\lim_{x \to \infty} \frac{\sin x}{x} = 0$. We cannot say the same about the original limit because the function is undefined infinitely many times as $x \to \infty$, so its limit cannot exist. (Most novice, and some experienced, calculus students would miss that subtle point and declare the limit to be the same as that of $(\sin x)/x \to 0$.)

Also, knowing we have a bounded function combined with one that blows up does not always tell us the limit, depending upon the form. For instance, we cannot really say anything about a limit of form $B \cdot \infty$ without more information. If the first function is $f(x)$ defined by $(\forall x)(f(x) = 0)$, then we have a zero limit. If $f(x) \to 0$, we definitely need more information.[46]

If we instead know $1 \leq f(x) \leq 2$, and $g(x) \longrightarrow \infty$, multiplying by $g(x)$ (which is eventually positive), we have $1 \cdot g(x) \leq f(x)g(x) \leq 2g(x)$, $1 \cdot g(x) \to \infty$, and $2g(x) \longrightarrow \infty$, so we can say in this case that $f(x)g(x) \longrightarrow \infty$ (though we only needed $1 \cdot g(x) \to \infty$ for our conclusion).[47]

The upshot of all this is the fact that we sometimes *do* need to refer to the Sandwich Theorem-type arguments for these, unless the form gives us an obvious answer.

The continuity of the trigonometric functions (where they are defined) can also come into play with these limits, for instance in light of Theorem 2.7.4, page 164, on the compositions of functions, namely $(f(x)$ continuous at $x = L) \wedge (g(x) \to L) \implies f(g(x)) \to f(L)$:

Example 2.8.7

Consider $\lim_{x \to \infty} \sec\left(\dfrac{x}{x^2+1}\right)$. From what we know of rational functions, the input of the secant function here is approaching zero. Since the secant (=1/cosine) is continuous at zero, our answer should be $\sec 0 = 1/\cos 0 = 1/1 = 1$. For a more computational argument we might write

$$\lim_{x \to \infty} \sec\left(\frac{x}{x^2+1}\right) \stackrel[\text{ALG}]{\sec(\infty/\infty)}{=\!=\!=} \lim_{x \to \infty} \sec\left(\frac{x(1)}{x\left(x+\frac{1}{x}\right)}\right) = \lim_{x \to \infty} \sec\left(\frac{1}{x+\frac{1}{x}}\right) \stackrel[\sec 0]{\sec\left(\frac{1}{\infty+0}\right)}{=\!=\!=} 1.$$

(We used the fact that $1/x \longrightarrow 0$ as $x \to \infty$ in the denominator of the input of the secant function.) Again, the last step utilized the fact that $\sec x$ is continuous at $x = 0$. One well-versed in how the input $x/(x^2 + 1)$ of the secant function behaves as $x \to \infty$ could more quickly write

$$\lim_{x \to \infty} \sec\left(\frac{x}{x^2+1}\right) \stackrel{\sec 0}{=\!=\!=} \sec 0 = 1.$$

For our next example, we will return to a limit of form $\infty - \infty$. When problematic terms do not cancel from subtraction, a rewriting of the expression as a quotient will often achieve some useful cancellation and a determinate form. (This theme will return several times in the text.) In the case that follows, we use a conjugate multiplication step because (as the reader is invited to verify) no factoring of the growth factors will yield a determinate form from the beginning.

Example 2.8.8

Consider the limit $\lim_{x \to \infty}\left(\sqrt{x^2+x+1}-x\right)$. Clearly $x^2 + x + 1 \longrightarrow \infty$, and so we are taking square roots of numbers as large as we like. In fact, since $x > 0$ we can write $\sqrt{x^2+x+1} > \sqrt{x^2} = |x| = x \longrightarrow \infty$

---

[46] For example, consider $f(x) = K/x$, $g(x) = x$ and $x \to \infty$. This gives a limit of K, i.e., $f(x)g(x) = (K/x)\cdot x = K \longrightarrow K$, and we can choose K to be any real number we like.

[47] For example, this is exactly what occurs with the limit $\lim_{x \to \infty}\left[\left(\dfrac{3+\sin x}{2}\right)\cdot x\right] = \infty$. This is because for $x > 0$,

$$\left(\frac{3-1}{2}\right)\cdot x \leq \left(\frac{3+\sin x}{2}\right)\cdot x \leq \left(\frac{3+1}{2}\right)\cdot x,$$

and so the function is sandwiched between $x \to \infty$ and (redundantly, because the first is sufficient) $2x \to \infty$.

2.8. LIMITS "AT INFINITY"

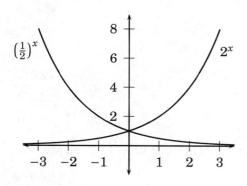

Figure 2.27a. 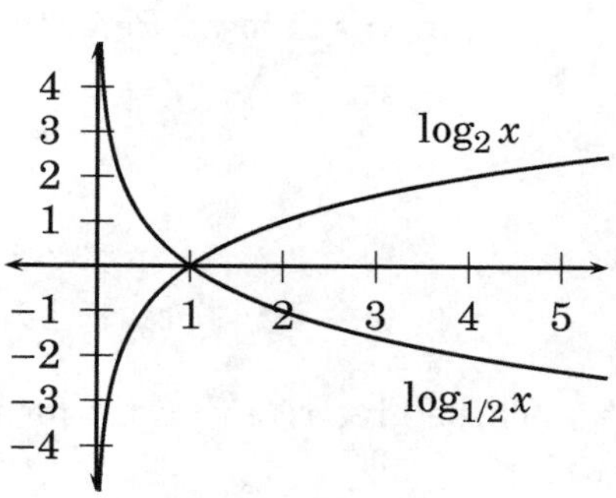   Figure 2.27b.

Figure 2.27: Partial graphs of exponential functions 2^x and $(1/2)^x$, along with logarithmic functions $\log_2 x$ and $\log_{1/2} x$, showing continuity and limiting behaviors.

as $x \to \infty$, and so this limit is of the form $\infty - \infty$.[48]

We solve this using the following method:

$$\lim_{x \to \infty} \left(\sqrt{x^2+x+1} - x\right) \xrightarrow[\text{ALG}]{\infty-\infty} \lim_{x \to \infty} \left[\left(\sqrt{x^2+x+1} - x\right) \cdot \frac{\sqrt{x^2+x+1}+x}{\sqrt{x^2+x+1}+x}\right]$$

$$= \lim_{x \to \infty} \frac{x^2+x+1-x^2}{\sqrt{x^2+x+1}+x} = \lim_{x \to \infty} \frac{x+1}{\sqrt{x^2+x+1}+x}$$

$$\xrightarrow[\text{ALG}]{\infty/\infty} \lim_{x \to \infty} \frac{x\left(1+\frac{1}{x}\right)}{x\left[\sqrt{1+\frac{1}{x}+\frac{1}{x^2}}+1\right]}$$

$$= \lim_{x \to \infty} \frac{1+\frac{1}{x}}{\sqrt{1+\frac{1}{x}+\frac{1}{x^2}}+1} = \frac{1+0}{\sqrt{1+0+0}+1} = 1/2.$$

This result is correct but probably not at all obvious from the original form. Only after finding a useful fractional form could we use our earlier techniques to compute its value. Limits that are writable as ratios are often easier to compute than other forms. Here it allowed us to compare the powers of x in the numerator and denominator. In our first limit section, namely Section 2.4, we had many 0/0 forms that we could easily simplify to get determinate forms.

There are applications, both conceptual and practical, for limits as the input variable "blows up." Many interesting applications involve exponential functions $f(x) = a^x$, or their variants such as $f(x) = C \cdot a^{kx}$. These are continuous for $x \in \mathbb{R}$, and their limits as $x \to \infty$ or

---

[48]It is also valid to define a limit form $\sqrt{\infty}$, which will return limits of ∞. (See Exercise 22, page 187, for some idea of a proof for this fact.)

$x \to -\infty$ are as follows. Refer to Figure 2.27a, page 183.

$$a > 1: \qquad x \to \infty \implies a^x \to \infty, \tag{2.75}$$
$$x \to -\infty \implies a^x \to 0^+, \tag{2.76}$$
$$a \in (0,1): \qquad x \to \infty \implies a^x \to 0^+, \tag{2.77}$$
$$x \to -\infty \implies a^x \to \infty. \tag{2.78}$$

These give rise to limit forms, so for examples (recalling that $e \approx 2.71828 > 1$), we can write

$$\lim_{x \to \infty} 2^x \overset{2^\infty}{=\!=\!=} \infty, \qquad\qquad \lim_{x \to \frac{\pi}{2}^-} e^{\tan x} \overset{e^\infty}{=\!=\!=} \infty,$$

$$\lim_{x \to \infty} 1.5^{-x} \overset{1.5^{-\infty}}{=\!=\!=} 0, \qquad\qquad \lim_{x \to \frac{\pi}{2}^+} e^{\tan x} \overset{e^{-\infty}}{=\!=\!=} 0,$$

$$\lim_{x \to \infty} \frac{e^x}{e^{-x}+1} \overset{\frac{e^\infty}{e^{-\infty}+1}}{=\!=\!=} \infty, \qquad\qquad \lim_{x \to \infty} \frac{2^{x+1}}{3^x} \overset{\infty/\infty}{\underset{\text{ALG}}{=\!=\!=}} \lim_{x \to \infty} \frac{2^x \cdot 2}{3^x}$$

$$= \lim_{x \to \infty} 2 \cdot \left(\frac{2}{3}\right)^x \overset{2 \cdot (\frac{2}{3})^\infty}{=\!=\!=} 0.$$

Related to the behaviors of the exponential functions are those of the logarithmic functions. Recall

$$\log_a x = y \iff a^y = x,$$

so looking at $y = \log_a x$ is the same as looking at $x = a^y$, or $y = a^x$ but with x and y trading roles. We can see from the graphs in Figure 2.27b that

$$a > 1: \qquad x \to \infty \implies \log_a x \to \infty, \tag{2.79}$$
$$x \to 0^+ \implies \log_a x \to -\infty, \tag{2.80}$$
$$a \in (0,1) \qquad x \to \infty \implies \log_a x \to -\infty, \tag{2.81}$$
$$x \to 0^+ \implies \log_a x \to \infty. \tag{2.82}$$

These are the logarithmic analogs of (2.75)–(2.78). In fact, it is not immediately clear from the figure that $\log_2 x \to \infty$ as $x \to \infty$, but we can go back to our definition in (2.66), page 176, and so for $M > 0$ we can take $N = 2^M$ and get $x > N = 2^M \implies \log_2 x > \log_2 2^M = M$. So the logarithmic graphs do "blow up" for $(a > 0) \wedge (a \ne 1)$, though they do so very slowly (for instance for $\log_2 x > 10$ we need $x > 2^{10} = 1024$). We will thus get limit forms such as $\log_2(\infty)$ yielding a limit of ∞, $\log_2(0^+)$ yielding $-\infty$, and others. Recall that $\ln x = \log_e x$, with $e \approx 2.71828 > 1$ and so $\ln x$ has a similar shape and asymptotics as $\log_2 x$, which is shown in Figure 2.27b, page 183. (Logarithms with bases $a \in (0,1)$ are rare, but do appear on occasion.)

We can now quickly compute some limits involving logarithms:

$$\lim_{x \to 0^+} \ln(\sin x) \overset{\ln 0^+}{=\!=\!=} -\infty, \qquad\qquad \lim_{x \to \infty} \ln\left(\frac{x^2 + 5x - 9}{3x^2 - 8x + 27}\right) \overset{\ln \frac{1}{3}}{=\!=\!=} \ln \frac{1}{3},$$

$$\lim_{x \to \infty} \ln(x^2 + 5x - 9) \overset{\ln \infty}{=\!=\!=} \infty, \qquad\qquad \lim_{x \to \frac{\pi}{2}^-} \ln(\tan x) \overset{\ln \infty}{=\!=\!=} \infty,$$

$$\lim_{x \to \infty} \frac{1}{\ln x} \overset{1/\ln \infty}{\underset{1/\infty}{=\!=\!=}} 0, \qquad\qquad \lim_{x \to 0^+} \sqrt{\ln \frac{1}{x}} \overset{\sqrt{\ln \infty}}{\underset{\sqrt{\infty}}{=\!=\!=}} \infty.$$

2.8. LIMITS "AT INFINITY"

Note that $\lim\limits_{x \to \frac{\pi}{2}^+} \ln(\tan x)$ does not exist because $\tan x < 0$ as $x \to \pi/2^+$, i.e., when x is in the second quadrant; recall that logarithms can only process positive inputs.

Next we consider an application of such limits. Limits as the input variable grows towards $+\infty$ are particularly valuable in the analysis of expected long-term behaviors of different systems. It is interesting because it can describe the state of a system as it seems to mostly "settle down." For many systems, it does not take unreasonably long for the state of the system to be near its limit (sometimes also called "equilibrium" if that limit is finite). Put another way, if $x \to \infty \implies f(x) \to L$, then for large enough x we should have $f(x) \approx L$. Thus the limit point L is interesting even though we cannot, in fact, "travel to infinity" in x (or whatever we call the input variable) to experience the limit, but may be able to experience the state of the system where x is large enough that $f(x) \approx L$ satisfactorily.

Example 2.8.9

In Section 3.3 we will consider electrical circuits that contain a resistor and an inductor in series, as seen to the right.

With a circuit having voltage V, a resistor with resistance R, and an inductor with inductance L, and a switch that is first "closed" at time $t = 0$, the current $I(t)$ flowing through the circuit at time t will be given by the following, for $t \geq 0$:

$$I(t) = \frac{V}{R}\left(1 - e^{-tR/L}\right).$$

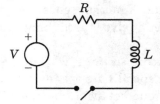

(a) What is the current at time $t = 0$?

(b) What are the current values at times $t = \frac{L}{R}, \frac{2L}{R}, \frac{3L}{R}$?

(c) As $t \to \infty$, what value does I approach?

Solution:

(a) The current at $t = 0$ is $I(0) = \frac{V}{R}\left(1 - e^0\right) = \frac{V}{R}(1-1) = 0$.

(b) For the other times we get

$$I\left(\frac{L}{R}\right) = \frac{V}{R}\left(1 - e^{-\left(\frac{L}{R}\right)\cdot R/L}\right) = \frac{V}{R}\left(1 - e^{-1}\right) \approx 0.63\left(\frac{V}{R}\right),$$

$$I\left(\frac{2L}{R}\right) = \frac{V}{R}\left(1 - e^{-\left(\frac{2L}{R}\right)\cdot R/L}\right) = \frac{V}{R}\left(1 - e^{-2}\right) \approx 0.86\left(\frac{V}{R}\right),$$

$$I\left(\frac{3L}{R}\right) = \frac{V}{R}\left(1 - e^{-\left(\frac{3L}{R}\right)\cdot R/L}\right) = \frac{V}{R}\left(1 - e^{-3}\right) \approx 0.95\left(\frac{V}{R}\right),$$

(c) Here we compute $\lim\limits_{t \to \infty} I(t)$:

$$\lim_{t \to \infty} I(t) = \lim_{t \to \infty} \frac{V}{R}\left(1 - e^{-tR/L}\right) \xrightarrow{\frac{V}{R}(1-e^{-\infty})} \frac{V}{R}(1-0) = \frac{V}{R}.$$

In Chapter 3 we will introduce Ohm's Law, which can be written $V = IR$. Note that in our example, as $t \to \infty$ we have $I \to \frac{V}{R}$, i.e., $I = V/R$ "in the limit" which is equivalent to Ohm's Law. An inductor will resist any sudden voltage change, in fact countering that change with a back voltage of its own, but in the presence of a steady voltage, an inductor will behave like a conductor. When $t = 0$ and the switch is first thrown, the voltage change felt by the inductor is most sudden, and for that instant no current flows as the inductor completely counters the voltage source. However, its capacity to resist (L) is not unlimited, and the voltage change it experiences (and its reactance as well) fades until the inductor behaves more and more like a conductor, so that nearly all (and in fact all, in the limit for the ideal case) of the resistance in the circuit eventually comes from the resistor.

Sometimes computing the limit as the input approaches infinity is also useful simply to indicate what could theoretically occur if a variable was allowed to get large enough. For instance, if some process's output is logarithmic, with a base greater than 1, even though growth may be very slow, it is theoretically possible for the output to be as large as we like. Just that observation is at times valuable.

In the meantime, the limit computations in the exercises help us to gain further "number sense" and "function sense" as we explore more aspects of the behaviors of functions so we can better analyze these things in theory and use them in the practice of analyzing real-world problems.

In future sections, we will make much use of limits "at infinity" in several contexts. For instance such limits are important in graphing a function when we wish to accurately represent its "end behavior," i.e., how the function behaves as $x \to \pm\infty$. If $f(x) \longrightarrow L \in \mathbb{R}$ as $x \to \infty$ or as $x \to -\infty$, then the function will have a horizontal asymptote $y = L$, which the function's graph "approaches" as x grows large. If the function grows without bound as $x \to \pm\infty$, that is also important to represent with the graph. Other very important examples occur later in the text, in which these limits are crucial for developing some of our most important calculus tools, but discussing these requires more background development.

Exercises

For problems 1–15, compute the limits where they exist (and if not, state so), showing all steps. You may wish to use Theorem 2.8.1 (page 178) or Theorem 2.8.2 (page 179) to **anticipate** an answer, but perform all the computations as in Examples 2.8.3 and 2.8.4 (starting on page 178).

1. $\lim\limits_{x \to \infty} x^5$

2. $\lim\limits_{x \to -\infty} x^5$

3. $\lim\limits_{x \to -\infty} x^4$

4. $\lim\limits_{x \to \infty} (x^4 - 5x^5)$

5. $\lim\limits_{x \to -\infty} (x^4 - 5x^5)$

6. $\lim\limits_{x \to \infty} (x^4 - 5x^6)$

7. $\lim\limits_{x \to -\infty} (x^4 - 5x^6)$

8. $\lim\limits_{x \to \infty} \dfrac{1}{x^3}$

9. $\lim\limits_{x \to -\infty} \dfrac{1}{x^3}$

10. $\lim\limits_{x \to \infty} \dfrac{x^2}{x^4 + 1}$

11. $\lim\limits_{x \to -\infty} \dfrac{1 - 2x^2}{x^3 + x^2 + x + 9}$

2.8. LIMITS "AT INFINITY"

12. $\lim_{x\to\infty} \dfrac{3x^2+5x-11}{2x^2-27x+100}$

13. $\lim_{x\to-\infty} \dfrac{3x^2+5x-11}{2x^2-27x+100}$

14. $\lim_{x\to\infty} \dfrac{x^2+3x-7}{x+5}$

15. $\lim_{x\to-\infty} \dfrac{x^2+3x-7}{x+5}$

16. Compute the limits, showing logical steps to justify answers.

    (a) $\lim_{x\to\infty}(x+\cos x)$

    (b) $\lim_{x\to\infty}(x-\cos x)$

    (c) $\lim_{x\to-\infty}(x+\cos x)$

    (d) $\lim_{x\to-\infty}(x^2+\cos x)$

17. Compute the limits, showing logical steps to justify answers.

    (a) $\lim_{x\to\infty} \dfrac{\sin x}{x^2}$

    (b) $\lim_{x\to\infty} \dfrac{x+\sin x}{x^2+1}$

    (c) $\lim_{x\to\infty} \dfrac{x^2}{x-\sin x}$

    (d) $\lim_{x\to\infty} \dfrac{x^2+2x+1-\sin x}{3x^2+2x-1}$

    (e) $\lim_{x\to\infty} x\sin x$

18. Compute the limit
$$\lim_{x\to\infty}\left(\sqrt{x^2+3x+9}-x\right).$$

19. Compute the limit
$$\lim_{x\to-\infty}\left(\sqrt{x^2-5x+9}-x\right).$$

    It is actually simpler than the previous limit (one line!).

20. Compute the following.

    (a) $\lim_{x\to\infty} \sin\dfrac{1}{x}$

    (b) $\lim_{x\to\infty} \cos\dfrac{1}{x}$

    (c) $\lim_{x\to\infty} \sin\left[\dfrac{\pi x^2+2x+9}{6x^2-11x+45}\right]$

21. Write definitions for the following (see (2.66), page 176). It may help to graph situations where these are true.

    (a) $\lim_{x\to\infty} f(x)=\infty$.

    (b) $\lim_{x\to-\infty} f(x)=\infty$.

    (c) $\lim_{x\to\infty} f(x)=-\infty$.

    (d) $\lim_{x\to-\infty} f(x)=-\infty$.

22. Prove that $\lim_{x\to\infty}\sqrt{x}=\infty$ using the definition found in (2.66), page 176. (For a hint, see the example in the paragraph immediately following that definition.)

23. A 10Ω resistor and a variable resistor R are placed in parallel in a circuit. The equivalent resistance R_p is related by the equation
$$\dfrac{1}{R_p}=\dfrac{1}{10}+\dfrac{1}{R},$$
where all resistances are in ohms (Ω):

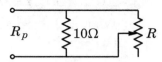

    (a) Solve algebraically for R_p.

    (b) Compute $\lim_{R\to\infty} R_p$. Is this answer physically reasonable? Why?

    (c) Compute $\lim_{R\to 0^+} R_p$. Is this answer physically reasonable? Why?

24. An employee can produce approximately
$$N(x) = \frac{50x+7}{2x+5}$$
items per day on the production line after x days on the job. In Chapter 3 we will be able to show that this is an increasing function. Find the maximum number of items that can be produced per day by computing $\lim_{x\to\infty} N(x)$.

25. A series circuit consisting of a voltage source, a resistor, a capacitor (initially discharged), and a switch is diagrammed next.

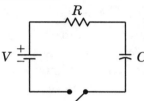

If the switch is first closed ("on") at $t = 0$ the charge on the capacitor is given by
$$q(t) = CV\left(1 - e^{-t/RC}\right).$$

   (a) What is the charge at $t = 0$?
   (b) Expressed as a percentage of CV, what is the charge at $t = RC, t = 2RC, t = 3RC$?
   (c) What is the trend in the charge as $t \to \infty$?

26. Suppose the population of a city grows at an annual rate of 2 percent per year. Then after t years the population will be given by
$$P(t) = P_0(1.02)^t.$$

   (a) What is the population at $t = 0$?
   (b) What is the population at $t = 1$?
   (c) What is the population at $t = 2$?
   (d) Assuming $P_0 > 0$, find $\lim_{t\to\infty} P(t)$.
   (e) Is this model for the city's population likely to be valid in the long term?

27. Compute the following (note that $\exp(x) = e^x$):

   (a) $\lim_{x\to\infty} 2^{1-\sqrt{x}}$
   (b) $\lim_{x\to 0^+} e^{1/x}$
   (c) $\lim_{x\to 0^-} e^{1/x}$
   (d) $\lim_{x\to\infty} e^{1/x}$
   (e) $\lim_{x\to\infty} \exp\left(\frac{2x^2+6x-5}{-2x^2+9x+56}\right)$

28. Consider $f(x) = \ln(e^x)$.

   (a) Without simplifying $f(x)$, compute $\lim_{x\to\infty} f(x)$.
   (b) Without simplifying $f(x)$, compute $\lim_{x\to-\infty} f(x)$.
   (c) Compute both limits again, this time by first simplifying $f(x)$.

29. Compute the following:

   (a) $\lim_{x\to 0^+} \ln(\csc x)$
   (b) $\lim_{x\to 0^+} \frac{1}{\ln x}$
   (c) $\lim_{x\to 5^-} \ln\left(\frac{1}{5-x}\right)$
   (d) $\lim_{x\to\infty} \ln(\ln x)$
   (e) $\lim_{x\to\infty} \ln(\ln(\ln x))$
   (f) $\lim_{x\to\infty} x \ln\left(\frac{1}{x}\right)$

30. Prove Theorem 2.8.1, page 178. Note that $p(x) = x^n\left(a_n + \frac{a_{n-1}}{x} + \cdots + \frac{a_0}{x^n}\right)$.

31. Prove Theorem 2.8.2, page 179.

2.9 Further Limit Theorems and Trigonometric Limits

In this section, we wrap up our discussion of fundamental methods for computing limits of functions. We also look at some very general theorems on limits, ranging from very intuitive to rather sophisticated. Into the mix we introduce and prove an interesting trigonometric limit (2.83), which will be the basis for another fundamental trigonometric limit (2.88), and many consequent trigonometric limits we could not have proved with previous methods. Finally, we will develop our most sophisticated (so far) limit method, which is substitution.[49] The aforementioned trigonometric limits will give rise to many other interesting limits, which require these new methods as well as the methods of previous sections.

2.9.1 Simple Limit Theorems

Many of the limit computations that we have already performed in this textbook were based on what we knew from continuity arguments. We took that approach because it is arguably more intuitive than the usual treatment found in most calculus textbooks (see Footnote 23, page 124.) In fact, the more common treatment is to instead rely on theorems about limits independent of possible underlying continuity (or continuity of replacement functions). Those basic limit theorems we combine below into one theorem, the different parts of which can be proved in ways very similar to the proofs of corresponding continuity theorems (see especially Section 2.2).

Theorem 2.9.1 *Suppose that, as x approaches some value, we have $f(x) \to L$ and $g(x) \to M$, where $L, M \in \mathbb{R}$. Then as x approaches that same value, we have*

(i) $f(x) + g(x) \longrightarrow L + M$,

(ii) $f(x) - g(x) \longrightarrow L - M$,

(iii) $Cf(x) \longrightarrow CL$, *for any fixed constant* $C \in \mathbb{R}$,

(iv) $f(x)g(x) \longrightarrow LM$,

(v) *and if* $M \neq 0$, *then* $f(x)/g(x) \longrightarrow L/M$.

(vi) *Also,* $\lim_{x \to a} C = C$, *where* $C \in \mathbb{R}$ *is any fixed constant.*

The theorem assumes that L and M exist and are finite. The theorem can be extended to include several (but not all!) cases where L or M do not exist or are infinite. Furthermore, the theorem extends in the obvious ways to numerous *forms*, so for instance if $f(x) \xrightarrow{\infty} \infty$ and $g(x) \xrightarrow{\infty} \infty$, then $f(x) + g(x) \xrightarrow{\infty + \infty} \infty$. We still have to be careful: For such f and g we have $f(x) - g(x)$ is of form $\infty - \infty$, which is indeterminate, as demonstrated in Example 2.8.1, page 177.

---

[49] Actually, we will add other methods after we develop derivatives, and particularly in the sequel. Still, what we finish in this section are the *foundational* methods for limits of functions. Even with the methods we will introduce after derivatives, we cannot avoid the methods of this section or this chapter. Indeed, the later methods are quite powerful where applicable, but have very limited scopes and therefore cannot replace what we develop here.

Textbooks include part (vi) for theoretical reasons we will consider momentarily, but also give it special emphasis because it is so simple it sometimes confuses. For an example illustrating how to interpret (vi) properly, consider the statement that $\lim_{x\to 5} 10 = 10$. What this means is that a function $h(x)$, where $h(x) = 10$ for all $x \in \mathbb{R}$, will give $\lim_{x\to a} h(x) = 10$ regardless of a (finite or infinite). This $h(x)$ is a "constant" function, which always returns the value 10 regardless of the input; its graph is the horizontal line $y = 10$. When graphed, it is clear that $h(x) \longrightarrow 10$ as $x \to a$ (regardless of a, chosen in advance). Rather than explicitly defining such an h, it is customary to simply write, for example, $\lim_{x\to a} 10 = 10$.

A common method, found in most of today's calculus textbooks, for obtaining a preliminary theory of finite limits—and by extension continuity—is based on this Theorem 2.9.1 using the following scheme:

1. State, and possibly prove, all parts of Theorem 2.9.1.

2. Prove (with ε-δ or by graphical demonstration) that $\lim_{x\to a} x = a$.

3. Use (i)–(iv) and (vi) from the theorem repeatedly to show that polynomials $p(x)$ have the property that $\lim_{x\to a} p(x) = p(a)$, using the limit version of the argument given in the proof of Theorem 2.2.4, page 92, but without reference to continuity (defined later in that approach).

4. From there, (v) gives that rational functions $p(x)/q(x)$, i.e., where p and q are polynomials, are continuous where defined, that being where $q(x) \ne 0$.

5. Then one looks at rational cases with 0/0 form, mentioning Theorem 2.4.3, page 125, on replacing functions with other functions that agree near the limit point and are continuous there.

6. One then progresses through the more sophisticated cases (radicals, infinite limits, limits at infinity, Sandwich Theorem, etc.).

7. Finally, *define* continuity at $x = a$ by the criterion that $\lim_{x\to a} f(x) = f(a)$ (our standalone ε-δ definition of continuity being logically equivalent, according to Theorem 2.4.2, page 124). It is hoped that the student has a good grasp of limits before continuity is introduced in this way.

This approach is mentioned so the reader will be aware of this common alternative treatment: defining limits, stating limit theorems, and defining continuity in terms of limits.

Our treatment in this textbook instead looked at continuity in its own right, and then introduced limits when either (1) continuity is "broken," or (2) we wish to describe other behaviors of functions, such as asymptotes and limits "at infinity," which are not so tightly connected with continuity.

Though we did not follow that logical scheme (instead opting for the advanced calculus and real analysis style of continuity before limits), we will occasionally have use for Theorem 2.9.1 in the rest of the text. Now we will look at an (admittedly) abstract example.

2.9. FURTHER LIMIT THEOREMS AND TRIGONOMETRIC LIMITS

Example 2.9.1

Suppose that $\lim_{x \to 3} f(x) = 5$ and $\lim_{x \to 3} g(x) = 7$. Then

$$\lim_{x \to 3}[f(x) + g(x)] \stackrel{5+7}{=\!=} 5 + 7 = 12,$$

$$\lim_{x \to 3}[f(x) - g(x)] \stackrel{5-7}{=\!=} 5 - 7 = -2,$$

$$\lim_{x \to 3}[4f(x)] \stackrel{4 \cdot 5}{=\!=} 4 \cdot 5 = 20,$$

$$\lim_{x \to 3}[f(x)g(x)] \stackrel{5 \cdot 7}{=\!=} 5 \cdot 7 = 35,$$

$$\lim_{x \to 3} \frac{f(x)}{g(x)} \stackrel{5/7}{=\!=} \frac{5}{7},$$

$$\lim_{x \to 3} 191 = 191.$$

Here we note limit "forms" such as $5 + 7$, which represents the limit of the sum of two functions, one whose output approaches 5 and the other whose output approaches 7. Such a limit's value must then be $5 + 7 = 12$, by our theorem. The last limit, of course, has nothing to do with the functions f or g. For a more specific example, consider the following:

Example 2.9.2

Compute $\lim_{x \to \infty}\left[17 + \dfrac{x}{x^2 + 1} + \dfrac{3x^2 + 5x + 9}{2 - x^2} \cdot \dfrac{2x - 3}{6x - 5}\right].$

Solution: There are several subexpressions here whose limits exist. In this particular example we are lucky that we can partition the whole expression into such well-behaved subexpressions:

$$\lim_{x \to \infty}\left[\underbrace{17}_{} + \underbrace{\frac{x}{x^2+1}}_{} + \underbrace{\frac{3x^2+5x+9}{2-x^2}}_{} \cdot \underbrace{\frac{2x-3}{6x-5}}_{}\right] = 17 + 0 - 3 \cdot \frac{1}{3} = 17 - 1 = 16.$$

$$\downarrow \downarrow \downarrow \downarrow$$
$$17 + 0 + (-3) \cdot 1/3$$

Note that we relied on our previous experience with limits at infinity, as outlined in Section 2.8, especially for rational expressions, where only the highest powers were ultimately relevant, as $x \to \infty$ (or as $x \to -\infty$).

This limit could be computed as it was because all the limits of subexpressions, as we organized them, existed and so our general Theorem 2.9.1, page 189, allows us to so combine them. Clearly this argument is easier than combining the subexpressions into a single rational expression. Sometimes recombining cannot be avoided, but often the complete computation can be avoided, as in what follows.

Example 2.9.3

Compute $\lim_{x \to \infty}\left[\dfrac{x - 2}{x^3 - 9x + 5} \cdot \dfrac{x^4 + 10x^3 - 9x^2 + 11x + 5}{2x^2 + x - 9}\right].$

Solution: As it stands, the form of this limit is $0 \cdot \infty$ (by Theorem 2.8.2, page 179, and the rationale for that result), which is indeterminate. One brute-force method of computing this limit is to combine the two fractions into one, but this requires some lengthy multiplication calculations. Instead we offer the two methods that follow, which work well because it is a limit at infinity.

1. We can factor the largest power of x which appears in each term and cancel:

$$\lim_{x\to\infty}\left[\frac{x-2}{x^3-9x+5}\cdot\frac{x^4+10x^3-9x^2+11x+5}{2x^2+x-9}\right]$$

$$=\lim_{x\to\infty}\left[\frac{x\left(1-\frac{2}{x}\right)}{x^3\left(1-\frac{9}{x^2}+\frac{5}{x^3}\right)}\cdot\frac{x^4\left(1+\frac{10}{x}-\frac{9}{x^2}+\frac{11}{x^3}+\frac{5}{x^4}\right)}{x^2\left(1+\frac{1}{x}-\frac{9}{x^2}\right)}\right]$$

$$=\lim_{x\to\infty}\left[\frac{x^5\left(1-\frac{2}{x}\right)\left(1+\frac{10}{x}-\frac{9}{x^2}+\frac{11}{x^3}+\frac{5}{x^4}\right)}{x^5\left(1-\frac{9}{x^2}+\frac{5}{x^3}\right)\left(2+\frac{1}{x}-\frac{9}{x^2}\right)}\right]$$

$$=\lim_{x\to\infty}\left[\frac{\left(1-\frac{2}{x}\right)\left(1+\frac{10}{x}-\frac{9}{x^2}+\frac{11}{x^3}+\frac{5}{x^4}\right)}{\left(1-\frac{9}{x^2}+\frac{5}{x^3}\right)\left(2+\frac{1}{x}-\frac{9}{x^2}\right)}\right]=\frac{1\cdot 1}{1\cdot 2}=\frac{1}{2}.$$

2. Another method is to observe what the leading terms of the numerator and denominator polynomials would be if we were to multiply and simplify them. For a limit at infinity, as we know, we need only look at the highest-order terms in the numerator and denominator:

$$\lim_{x\to\infty}\left[\frac{x-2}{x^3-9x+5}\cdot\frac{x^4+10x^3-9x^2+11x+5}{2x^2+x-9}\right]\quad=\quad\lim_{x\to\infty}\frac{x^5+\cdots-10}{2x^5+\cdots-45}\quad=\quad\frac{1}{2}.$$

The highest- and lowest-order terms are the easiest to compute for a polynomial product (while the intermediate-order terms may be complicated sums). Here it is only the highest-order terms which are relevant, again, *because the limit is at infinity*.

Next we consider what to do if our partition of the limit's function yields subexpressions whose limits do not necessarily exist. It is true that knowing that one of the component limits does not exist can sometimes allow us to conclude that the entire limit does not. However this is not always the case. In the next example we give an argument where the nonexistence of a component limit can, in that context, imply nonexistence of the full limit. We follow that example with one in which nonexistence of a component limit does not wreck the full limit. Both types should become intuitive, but again we will see the perennial lesson that we must never be too cavalier with our limit arguments.

Example 2.9.4 _____

Suppose that $\lim_{x\to a} f(x) = 5$ and $\lim_{x\to a} g(x)$ DNE. Then

$$\lim_{x\to a}[f(x)+g(x)]\;\overset{\text{5+DNE}}{=\!=\!=\!=\!=}\;DNE.$$

To see this, we can argue that since $g(x)$ is not approaching a well-defined limit value for whatever reason (perhaps being undefined near $x = a$, or oscillating, or having different left and right limits), then adding $f(x)$, which **is** approaching a number, will not compensate for the behavior of $g(x)$, and the final limit cannot exist. A more rigorous argument is given next.

Proof: Suppose again $\lim_{x\to a} f(x) = 5$ and $\lim_{x\to a} g(x)$ DNE. We will prove by contradiction that $\lim_{x\to a}[f(x)+g(x)]$ cannot exist. Suppose that it does exist. If this limit is some

2.9. FURTHER LIMIT THEOREMS AND TRIGONOMETRIC LIMITS

finite number $L \in \mathbb{R}$, then according to our general Theorem 2.9.1, page 189, we would have

$$\begin{aligned}
f(x)+g(x) &\longrightarrow L \implies -f(x)+(f(x)+g(x)) \longrightarrow -5+L &&\text{(by Theorem 2.9.1(i))}\\
&\implies (-f(x)+f(x))+g(x) \longrightarrow -5+L &&\text{(algebra)}\\
&\implies (0+g(x)) \longrightarrow -5+L &&\text{(algebra near limit point)}^{50}\\
&\iff g(x) \longrightarrow -5+L \text{ exists} &&\text{(algebra)}.
\end{aligned}$$

But that contradicts our original information that $\lim_{x\to a} g(x)$ does not exist. Thus we have to conclude that our assumption $\lim_{x\to a}[f(x)+g(x)]$ exists—which leads to a contradiction—must be false, and so $\lim_{x\to a}[f(x)+g(x)]$ does not exist, q.e.d. If instead, $f(x)+g(x) \longrightarrow \infty$ or $-\infty$, while $f(x) \longrightarrow 5$, it is not difficult to see from a similar argument that $g(x) \longrightarrow \pm\infty$, depending on "which infinity" is involved in the previous limit.[51]

In this example, the fact that the limit of $f(x)$ was finite meant that it could not—through simple addition—compensate for "bad behavior" of $g(x)$ in the limit. It is possible, however, for f to still compensate, in the sense that $f(x)$ can also simply dominate $g(x)$ in the expression $f(x)+g(x)$ if $g(x)$ is defined but bounded and $f(x)$ blows up. (The reader may wish to revisit (2.71)–(2.74), page 181, as we work the next example.)

Example 2.9.5

Suppose that $f(x) \longrightarrow \infty$ and $|g(x)| \leq M$ for some $M > 0$ (but real and therefore finite) as $x \to a$. Then we employ a Sandwich Theorem argument, with $-M+f(x) \leq f(x)+g(x) \leq M+f(x)$, with the first and third expressions approaching ∞ (though the first doing so is enough), carrying $f(x)+g(x)$ to an infinite limit as well. We could write

$$\lim_{x\to a}[f(x)+g(x)] \stackrel{\infty+B}{=\!=\!=} \infty.$$

For a specific example, consider: $\lim_{x\to\infty}(x+\sin x) \stackrel{\infty+B}{=\!=\!=} \infty.$

Actually, we worked exactly this example during the discussion of the form $B+\infty$, introduced on page 181.

The lesson to be gleaned from these examples is that if one part of the function "inside the limit" has a nonexistent limit (for the prescribed approach in the independent variable), sometimes we can conclude the same about the whole limit and sometimes we cannot. We usually need to examine more deeply the behavior of the other parts of the function and take into account how their influences combine. Sometimes the form is enough to determine the actual limit (or its nonexistence). With experience, the various cases become intuitive. (Recall also Example 2.7.1, page 159, and its associated Figure 2.22.)

---

[50]At $x = a$ we may have that $f(x)$ is undefined, so there it may be incorrect to say $-f(x)+f(x) = 0$. However, near (but not at) the limit point we know $f(x)$ is defined, from the assumption $f(x) \to 5$ as $x \to a$. Implied in our definition of limits ((2.22), page 123) is that $f(x)$ exists for $0 < |x-a| < \delta$, for some $\delta > 0$.

[51]Note how we once again used $P \longrightarrow (\sim P) \implies \sim P$.

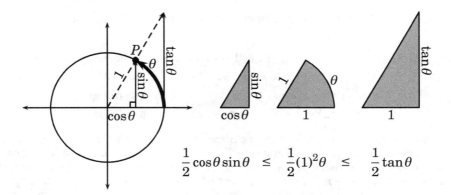

Figure 2.28: Illustration of relative sizes of $\frac{1}{2}\cos\theta\sin\theta$, $\frac{1}{2}\theta$ and $\frac{1}{2}\tan\theta$ for an angle $\theta \in [0, \pi/2)$. Note that each area is a superset of any preceding area illustrated, as most clearly illustrated in the far-left figure. Recall also that $\theta = s/r$, where s is the directed arc length and r the radius, so on the unit circle $\theta = s/1 = s$. Also recall that the area of a circular wedge is given by $A = \frac{1}{2}r^2\theta$. (To help remember this formula, think of what this means if $\theta = 2\pi$: Area $= \frac{1}{2}r^2 \cdot 2\pi = \pi r^2$ if θ sweeps the whole circle.)

2.9.2 A Trigonometric Limit

An interesting limit, which is surprisingly useful for future results, is the following:

Theorem 2.9.2 *With θ given in radians, we have the following limit:*

$$\boxed{\lim_{\theta \to 0} \frac{\sin\theta}{\theta} = 1.} \tag{2.83}$$

Proof: We include a proof here for completeness. The argument is quite interesting, but the result is far more important than the proof.

The proof relies on the Sandwich Theorem and a geometric observation, which is given in Figure 2.28. In that figure, θ is the radian measure of the angle, which terminates in the first quadrant. The observation involves three areas defined by this angle θ: a right triangle, contained within a circular wedge, which is in turn contained in another right triangle. The smaller triangle has "base" $\cos\theta$ and "height" $\sin\theta$, and thus has area $\frac{1}{2}\cos\theta\sin\theta$. For the circular wedge, recall that a wedge with radius r and radian-measure angle θ has area $\frac{1}{2}r^2\theta$. (The reader is invited to test this formula for the cases $\theta = 0$ and $\theta = 2\pi$, the latter representing an area formula for the entire circle.) The other triangle has base 1 and height $\tan\theta$. To see this, note that it is similar to the smaller triangle, and so we have the proportion of sides: $\sin\theta/\cos\theta = h/1$, where h is the height of the larger triangle. It follows that $h = \tan\theta$. Thus the larger triangle has area $\frac{1}{2} \cdot 1 \cdot \tan\theta$. Now for $\theta \in [0, \pi/2)$, we have

$$\frac{1}{2}\cos\theta\sin\theta \leq \frac{1}{2}\theta \leq \frac{1}{2}\tan\theta.$$

2.9. FURTHER LIMIT THEOREMS AND TRIGONOMETRIC LIMITS

As $\theta \to 0^+$, we have $\sin\theta > 0$ so not only can we multiply by 2, but we can also then divide by $\sin\theta$, giving us

$$\cos\theta \le \frac{\theta}{\sin\theta} \le \frac{1}{\cos\theta}. \qquad (2.84)$$

This gives us a Sandwich Theorem-type argument as $\theta \to 0^+$:

$$\text{As } \theta \to 0^+: \quad \underbrace{\cos\theta}_{\downarrow \atop 1} \le \frac{\theta}{\sin\theta} \le \underbrace{\frac{1}{\cos\theta}}_{\downarrow \atop 1}$$

giving us $\frac{\theta}{\sin\theta} \longrightarrow 1$ as $\theta \to 0^+$, i.e.,

$$\lim_{\theta \to 0^+} \frac{\theta}{\sin\theta} = 1. \qquad (2.85)$$

Next we dispatch the left-side limit. In fact, the same inequality (2.84) holds as $\theta \to 0^-$, because all three expressions are the same if we replace θ with $-\theta$. This follows because all three functions are "even," i.e., $\cos(-\theta) = \cos\theta$, $(-\theta)/\sin(-\theta) = \theta/\sin\theta$, $1/\cos(-\theta) = 1/\cos\theta$, and furthermore $\theta \to 0^+ \iff (-\theta) \to 0^-$. With these one can perform the above computations with $(-\theta) \to 0^+$, and thus $\theta \to 0^-$.

Perhaps a less convoluted approach is to more explicitly borrow the substitution method from upcoming Subsection 2.9.3. Here we let $\phi = -\theta$ so that $\theta \to 0^- \iff \phi \to 0^+$. Thus

$$\lim_{\theta \to 0^-} \frac{\theta}{\sin\theta} = \lim_{\phi \to 0^+} \frac{-\phi}{\sin(-\phi)} = \lim_{\phi \to 0^+} \frac{-\phi}{-\sin\phi} = \lim_{\phi \to 0^+} \frac{\phi}{\sin\phi} = 1, \qquad (2.86)$$

the last limit being, of course, (2.85) with the variable renamed. Putting (2.85) together with (2.86), of course, gives us

$$\lim_{\theta \to 0} \frac{\theta}{\sin\theta} = 1. \qquad (2.87)$$

Finally, we can then get our result based on the previous limit, looking at the reciprocal function:

$$\lim_{\theta \to 0} \frac{\sin\theta}{\theta} = \lim_{\theta \to 0} \frac{1}{\left(\frac{\theta}{\sin\theta}\right)} \stackrel{1/1}{=\!=\!=} 1, \quad \text{q.e.d.}$$

Notice that the limit we proved, (2.83), says something about how fast $\sin\theta$ approaches zero as θ approaches zero: Eventually, $\sin\theta$ and θ approach zero at approximately the same rate. In fact, many physics problems use the approximation $\sin\theta \approx \theta$, which follows from $(\sin\theta)/\theta \approx 1$ for $|\theta|$ small and in radians. This approximation is graphed in Figure 2.29, page 196, using x instead of θ as the independent (domain) variable. We will see how this

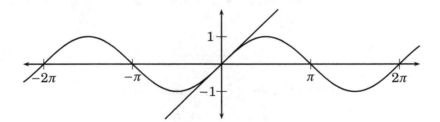

Figure 2.29: Partial graphs of $y = x$ and $y = \sin x$. Since $(\sin x)/x \longrightarrow 1$ as $x \to 0$, we get $(\sin x)/x \approx 1$ for x near zero, and thus $\sin x \approx 1 \cdot x$ for small x. In the graph we see how quickly $\sin x$ and x become close enough that the limited resolution of the graphic rendering device makes it difficult to distinguish the two functions for small x. (However, it should be noted that the two functions x and $\sin x$ only agree at $x = 0$. Indeed, a closer examination of Figure 2.28, page 194, shows that for $\theta > 0$ we have $\theta > \sin\theta$, and for $\theta < 0$ we have $\theta < \sin\theta$.)

very important approximation, and the underlying limit, arise from other calculus techniques in later chapters.[52]

Many other interesting limits follow from this limit (2.83), as we can see in the following example. Note that the second limit computation there utilizes the trigonometric identities $(1 - \cos x)(1 + \cos x) = 1 - \cos^2 x = \sin^2 x$.

Example 2.9.6

Consider the following limits, which require both our basic trigonometric limit (2.83) and our general limit theorem, Theorem 2.9.1 with which we began the section (on page 189).

- $\lim\limits_{x \to 0} \dfrac{\tan x}{x} = \lim\limits_{x \to 0} \dfrac{\sin x}{x \cos x} = \lim\limits_{x \to 0} \left[\dfrac{\sin x}{x} \cdot \dfrac{1}{\cos x} \right] \overset{1 \cdot \frac{1}{1}}{=\!=\!=} 1.$

- $\lim\limits_{x \to 0} \dfrac{1 - \cos x}{x} \overset{0/0}{=\!=\!=} \lim\limits_{x \to 0} \left[\dfrac{1 - \cos x}{x} \cdot \dfrac{1 + \cos x}{1 + \cos x} \right] = \lim\limits_{x \to 0} \dfrac{1 - \cos^2 x}{x(1 + \cos x)} = \lim\limits_{x \to 0} \dfrac{\sin^2 x}{x(1 + \cos x)}$

  $= \lim\limits_{x \to 0} \left[\dfrac{\sin x}{x} \cdot \dfrac{\sin x}{1 + \cos x} \right] \overset{1 \cdot \frac{0}{2}}{=\!=\!=} 1 \cdot \dfrac{0}{2} = 0.$

The first limit is also used in physics in the form $\tan\theta \approx \theta$ for $|\theta|$ small. The second limit occurs enough to warrant its own theorem, though as we demonstrated it is relatively quickly derived—using a standard trick from trigonometry—from the more basic $\lim_{\theta \to 0} \frac{\sin\theta}{\theta} = 1$.

---

[52]While (2.87) is the immediate result of our analysis, and is interesting in its own right, most texts (as here) go ahead and present the reciprocal limit (2.83). The reason is that $\lim_{\theta \to 0}((\sin\theta)/\theta)$ gives a (perhaps) more intuitive comparison of the behavior of $\sin\theta$ *versus* that of the independent variable θ. We have already had many limits of ratios of functions $f(x)/g(x)$, which compare the two functions' behaviors in the sense of limits. A very useful way to analyze a function is to compare it to its input variable by ratios $f(x)/x$, $f(x)/(x - a)$, or $(f(x) - f(a))/(x - a)$. Though this can often be accomplished instead by looking at the reciprocals of these, it is arguably less intuitive to do so.

2.9. FURTHER LIMIT THEOREMS AND TRIGONOMETRIC LIMITS

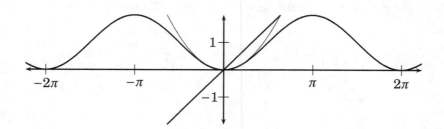

Figure 2.30: Partial graphs of $y = x$ and $y = 1 - \cos x$, showing how $1 - \cos x$ approaches zero faster than x as $x \to 0$—so much so that $(1-\cos x)/x \longrightarrow 0$ (see (2.88)). The graph of $y = \frac{1}{2}x^2$ is also given (thinner curve), illustrating how $(1-\cos x)/x^2 \approx \frac{1}{2}$, i.e., $1 - \cos x \approx \frac{1}{2}x^2$, or $\cos x \approx 1 - \frac{1}{2}x^2$ for x small, as derived in Example 2.9.7.

Theorem 2.9.3 *The following limit (proved previously) holds:*

$$\boxed{\lim_{\theta \to 0} \frac{1 - \cos \theta}{\theta} = 0.} \qquad (2.88)$$

In other words, $1 - \cos \theta$ shrinks to zero faster than θ does, as $\theta \to 0$. This is reasonable when we see the graphs of $y = 1 - \cos x$ and $y = x$, given (as darker curves) in Figure 2.30.[53]

For the rest of this section, we will assume (2.83) and (2.88) and compute other trigonometric limits based on these, the work from previous sections, and Theorem 2.9.1. As with all limits, it is important to be careful and not jump to incorrect conclusions; our basic trigonometric limits (2.83) and (2.88)—which the reader should commit to memory—given here again as

$$\lim_{\theta \to 0} \frac{\sin \theta}{\theta} = 1, \qquad \lim_{\theta \to 0} \frac{1 - \cos \theta}{\theta} = 0,$$

are very specific in their scopes. Moreover, with trigonometric limits it is sometimes still necessary to exploit the algebraic identities among those functions. Consider the following (perhaps surprising) trigonometric limit calculation:

Example 2.9.7

Compute $\lim\limits_{x \to 0} \dfrac{1 - \cos x}{x^2}$.

Solution: A first, perhaps more obvious attempt is quickly seen to be a dead end:

$$\lim_{x \to 0} \frac{1 - \cos x}{x^2} = \lim_{x \to 0} \left[\frac{1 - \cos x}{x} \cdot \frac{1}{x} \right] \xrightarrow{0 \cdot (\pm \infty)} ?,$$

leading to the indeterminate form $0 \cdot (\pm \infty)$. As happens so frequently with limits, we look for some other way of rewriting the function. The usual method of computing this limit is to again exploit the fact that

$$(1 - \cos x)(1 + \cos x) = 1 - \cos^2 x = \sin^2 x.$$

---

[53] One could also say that $\cos \theta \to 1$ very rapidly as $\theta \to 0$, but of course neither description is as precise as (2.88), i.e., that $(1 - \cos \theta)/\theta \to 0$ as $\theta \to 0$.

By multiplying the function inside the limit by $(1+\cos x)/(1+\cos x)$, we can compute the limit as follows:

$$\lim_{x\to 0}\frac{1-\cos x}{x^2} = \lim_{x\to 0}\left[\frac{1-\cos x}{x^2}\cdot\frac{1+\cos x}{1+\cos x}\right] = \lim_{x\to 0}\frac{1-\cos^2 x}{x^2(1+\cos x)} = \lim_{x\to 0}\frac{\sin^2 x}{x^2(1+\cos x)}$$
$$= \lim_{x\to 0}\left[\frac{\sin x}{x}\cdot\frac{\sin x}{x}\cdot\frac{1}{1+\cos x}\right] = 1\cdot 1\cdot\frac{1}{1+1} = \frac{1}{2}.$$

In the last step of above example, we used Theorem 2.9.1 (the limit of a product being the product of the limits, when the "factor" limits exist), and the fact that the factor $1/(1+\cos x)$ is continuous at $x = 0$.

One could argue using this limit that, for $|x|$ small, $1-\cos x \approx \frac{1}{2}x^2$, which is also illustrated in Figure 2.30, page 197. This fact we will derive later in the form $\cos x \approx 1 - \frac{1}{2}x^2$, which is sometimes used in applications. It is much more accurate than the earlier approximation that $\cos x \approx 1$ for x small (though they coincide at $x = 0$).

We will compute many more trigonometric limits in this section, but first we need a new limit technique, which we introduce next.

2.9.3 Limits by Substitution

To bring our analytical methods of computing limits to the next level, we now develop some substitution techniques. These techniques require delicacy, but with care they are quite powerful and sophisticated. The next two examples are offered to motivate the techniques.

Example 2.9.8

Compute $\lim_{x\to 0}\dfrac{\sin 5x}{x}$.

<u>Solution</u>: The usual method—justified in the subsequent discussion—is to multiply by $\frac{5}{5}$:

$$\lim_{x\to 0}\frac{\sin 5x}{x} = \lim_{x\to 0}\frac{5\sin 5x}{5x} = \lim_{x\to 0}\left[5\cdot\frac{\sin 5x}{5x}\right] = 5\cdot 1 = 5.$$

Notice that the original limit was of the form "$(\sin 0)/0$" (i.e., $0/0$) which is indeterminate. It is important in our basic trigonometric limit (2.83), i.e., $\theta \to 0 \implies (\sin\theta)/\theta \to 1$, that the two θ variables are "zeros" approaching zero at the same rate (though that is not always quite enough, as we will eventually see). By multiplying the fraction by $5/5$, we were able to get a form "$5\cdot[(\sin 0)/0]$," but where the rates of the "zeros" (i.e., the $5x$'s) approaching zero were exactly the same. Most presentations of the computation of this limit are exactly as given in Example 2.9.8, but it is to be understood that we are transforming this limit by way of a substitution. For instance, one might instead write

$$\lim_{x\to 0}\frac{\sin 5x}{x} = \lim_{x\to 0}\left[5\cdot\frac{\sin 5x}{5x}\right] = \lim_{(5x)\to 0}\left[5\cdot\frac{\sin 5x}{5x}\right] = 5\cdot 1 = 5,$$

2.9. FURTHER LIMIT THEOREMS AND TRIGONOMETRIC LIMITS

or

$$\lim_{x \to 0} \frac{\sin 5x}{x} = \lim_{x \to 0} \left[5 \cdot \frac{\sin 5x}{5x} \right] = \lim_{\theta \to 0} \left[5 \cdot \frac{\sin \theta}{\theta} \right] = 5 \cdot 1 = 5,$$

where $\theta = 5x$, and $x \to 0 \iff \theta \to 0$. The nature of the mechanism whereby $\theta \to 0 \iff x \to 0$ is crucial, as we will explain in due course. The next example displays a slightly more sophisticated argument.

Example 2.9.9

Compute $\lim\limits_{x \to 0} \dfrac{\cos x^2 - 1}{x^4}$. (Compare to Example 2.9.7, page 197.)

<u>Solution</u>: Here we will make a substitution $\theta = x^2$, so that $x \to 0 \implies \theta = x^2 \to 0^+$. Then we can write

$$\lim_{x \to 0} \frac{\cos x^2 - 1}{x^4} \overset{0/0}{=\!=} \lim_{\theta \to 0^+} \frac{\cos \theta - 1}{\theta^2} = \lim_{\theta \to 0^+} \left[\frac{\cos \theta - 1}{\theta^2} \cdot \frac{\cos \theta + 1}{\cos \theta + 1} \right] = \lim_{\theta \to 0^+} \frac{\cos^2 \theta - 1}{\theta^2 (\cos \theta + 1)}$$

$$= \lim_{\theta \to 0^+} \frac{-\sin^2 \theta}{\theta^2 (\cos \theta + 1)} = \lim_{\theta \to 0^+} \left[-\frac{\sin \theta}{\theta} \cdot \frac{\sin \theta}{\theta} \cdot \frac{1}{\cos \theta + 1} \right] = -1 \cdot 1 \cdot \frac{1}{1+1} = -\frac{1}{2}.$$

Notice that in this example we used the fact that full limits imply one-sided limits. Specifically, we used that

$$\lim_{\theta \to 0} \frac{\sin \theta}{\theta} = 1 \implies \lim_{\theta \to 0^+} \frac{\sin \theta}{\theta} = 1.$$

Now we consider circumstances where we can make a substitution to compute a limit. First we take another look at what it means for $x \to a$. (Subsection 2.7.2, page 161, began this discussion.) When we write $x \to a$ under "lim," we visualize x getting arbitrarily close to **but not equal to** the value a. But we also assume that x does so through a continuum of values, so that x does not "skip over" any values as it approaches a, as we see from the definition of $\lim_{x \to a} f(x)$, as for example in the case this limit is a finite number L:

$$(\forall \varepsilon > 0)(\exists \delta > 0)(\forall x)[\ \underbrace{0 < |x - a| < \delta}_{x \in (a-\delta, a) \cup (a, a+\delta)} \longrightarrow |f(x) - L| < \varepsilon].$$

In order to substitute algebraically to produce another limit, say $\lim_{u \to \beta} F(u)$, and claim it is the same as $\lim_{x \to a} f(x)$, we would like to be sure that $x \to a \implies u \to \beta$, and that the latter "approach" is still somehow a **proper** approach. Usually what is obvious is only that $x \to a \implies u \to \beta$ in the sense that $\lim_{x \to a} u = \beta$, which allows for the approach of u to β to be quite sloppy or narrow. It may be that the approach u takes to β is one-sided, or has gaps as $x \to a$, or actually *achieves* $u = \beta$ while we *require* $x \neq a$. Indeed, when we developed the convention that $\lim_{x \to a} u = \beta$ would also be written $x \to a \implies u \to \beta$, we visualized the approach of u to β quite loosely, while that of x to a was more strict. To force $u \to \beta$ to occur in the same way that $x \to a$ requires much more structure in the relationship between x and u than we usually wish to have to accommodate. One notable example where this does happen is when $u = mx + b$, where $m \neq 0$, giving that $x \to a \iff u \to (ma + b)$ and both approaches are proper in every way. Another example is $x \to 0 \implies x^2 \to 0^+$ (note the absence of the converse $\impliedby$). Other cases can be more or less complicated.

What we can do to produce a useful theorem is to not make a strong statement about the relationship between x and u, but rather qualify the result in a different way (which also makes a proof easier to formulate, were we to include it), still yielding a useful result. What we settle on here is the following, the proof of which is similar to that of Theorem 2.7.4, page 164, and is left as an exercise for the interested reader.

Theorem 2.9.4 (Limit Substitution Theorem) *Suppose that the variables x and u are related in such a way that*

(a) $\lim\limits_{x \to a} u = \beta$,

(b) $(\exists d > 0)(\forall x)[0 < |x - a| < d \longrightarrow (u \neq \beta)]$,

(c) $(\exists d > 0)(\forall x)[0 < |x - a| < d \longrightarrow (f(x) = F(u))]$.

Then

$$\lim_{x \to a} f(x) = \lim_{u \to \beta} F(u) \qquad \text{if this second limit exists.} \qquad (2.89)$$

There are many theorems we will come across where we have an equality like (2.89), which is qualified by the criterion that the second quantity exists, or otherwise makes sense in the context. To remove that criterion would require a much more complicated set of conditions than (a)–(c). That the second limit exists is key, as is that $u \to \beta$ be an "approach" of a similar kind in the sense that not only should $u \to \beta$, but $u \neq \beta$ as $x \to a$ according to (b).[54] In fact, the validity of (a) and (c) usually is clear from the actual algebraic substitution; it is (b) that requires a bit more scrutiny but is also usually not difficult to see.

Example 2.9.10

Compute $\lim\limits_{x \to \frac{\pi}{2}} \dfrac{\sin\left(x - \frac{\pi}{2}\right)}{x - \frac{\pi}{2}}$.

Here we make the substitution $u = x - \frac{\pi}{2}$, so that $x \to \frac{\pi}{2} \implies u \to 0$ in the sense of the hypotheses (a)–(c) of the theorem. Thus we can write (as long as the final limit exists!)

$$\lim_{x \to \frac{\pi}{2}} \frac{\sin\left(x - \frac{\pi}{2}\right)}{x - \frac{\pi}{2}} = \lim_{u \to 0} \frac{\sin u}{u} = 1.$$

It would be reasonable to use the notation $\underset{u = x - \pi/2}{=\!=\!=}$ in a limit computation as in the example, but explanations of *substitutions* are usually done within the paragraphs, and even if not, the nature of the substitution is usually pretty clear from how the new variable appears in the new limit.

---

[54]Later in the subsection we will have an example to show why (b) is necessary. In that example, $x \to a \implies u \to \beta$, but u oscillates, passing through the value β infinitely many times as $x \to a$. In that example the naive substitution is invalid: The new limit exists but the original does not, so obviously they are not equal.

2.9. FURTHER LIMIT THEOREMS AND TRIGONOMETRIC LIMITS

In the earlier Examples 2.9.8 and 2.9.9 we used θ to play the role of u in the theorem, but that is mostly a matter of taste. We will usually use u because it is the traditional variable of substitution later in the calculus.[55]

The theorem can be extended in obvious ways to cases such as

$$x \to a \implies u \to \infty,$$
$$x \to a \implies u \to \beta^+,$$
$$x \to a^+ \implies u \to \beta^+,$$

and so on.

It should be pointed out that if $x \to a$ implies a "proper" one-sided approach, for instance $u \to \beta^+$, then we can perform a substitution based on that, except then we would compute the one-sided limit $\lim_{u \to \beta^+} F(u)$. This was the case in Example 2.9.9, page 199. Furthermore, a one-sided approach in x to a may yield a proper approach in a variable u of some kind. Infinite "approaches" can also arise from substitutions, or give rise to substitutions, as we see next.

Example 2.9.11

Consider $\lim_{x \to \frac{\pi}{2}^+} \dfrac{2\tan^2 x + 3\tan x + 7}{\tan^2 x - 6\tan x + 30}$.

Here we let $u = \tan x$. Then $x \to \frac{\pi}{2}^+ \implies u \to -\infty$. (Because we will never have $u = -\infty$, we do not need an analog to the criterion (b) in the Limit Substitution Theorem, Theorem 2.9.4, page 200.) Thus

$$\lim_{x \to \frac{\pi}{2}^+} \frac{2\tan^2 x + 3\tan x + 7}{\tan^2 x - 6\tan x + 30} = \lim_{u \to -\infty} \frac{2u^2 + 3u + 7}{u^2 - 6u + 30} = \lim_{u \to -\infty} \frac{u^2\left(2 + \frac{3}{u} + \frac{7}{u^2}\right)}{u^2\left(1 - \frac{6}{u} + \frac{30}{u^2}\right)} = \frac{2+0+0}{1-6+0} = 2.$$

In fact this example can also be computed by multiplying numerator and denominator by $\cos^2 x$. A similar computation is left to the exercises.

The next example illustrates that if one type of indeterminate form is not easily dealt with, an algebraic manipulation can likely give rise to another that is more easily computed.

Example 2.9.12

Compute $\lim_{x \to \infty} x \sin \dfrac{1}{x}$.

<u>Solution:</u> Note that $\frac{1}{x} \longrightarrow 0^+$ as $x \to \infty$, so the form here is essentially $\infty \cdot \sin 0 = \infty \cdot 0$ (more precisely, $\infty \cdot \sin 0^+ = \infty \cdot 0^+$), which is indeterminate (as shown in Example 2.8.2, page 178). However, we can rewrite the limit with a power of x in the denominator, instead of having x as a multiplicative factor. Then we will perform a substitution and use (2.83), page 194, for the final computation.

$$\lim_{x \to \infty} x \sin \frac{1}{x} \xuparrow{\infty \cdot 0}{\text{ALG}} \lim_{x \to \infty} \frac{\sin \frac{1}{x}}{\frac{1}{x}} \quad \text{(form } \sin 0^+/0^+\text{)}.$$

---

[55] Actually, θ is becoming increasingly common as a variable of substitution, and we will have occasion to use it as we did in our first substitution examples.

Now we let $u = 1/x$, so that $x \to \infty \implies u \to 0^+$ properly (see page 199, and Theorem 2.9.4, page 200), and so

$$\lim_{x\to\infty} x \sin\frac{1}{x} \underset{\text{ALG}}{=\!=\!=} \lim_{x\to\infty} \frac{\sin\frac{1}{x}}{\frac{1}{x}} = \lim_{u\to 0^+} \frac{\sin u}{u} = 1.$$

Substitution is a very powerful method, but sometimes the mechanics of it become needlessly complicated and a certain amount of "hand-waving" becomes appropriate. For instance, depending on the author and the audience, the u-limit above might be omitted, but the substitution principle should be understood. When we think of the limits

$$\lim_{\theta\to 0} \frac{\sin\theta}{\theta} = 1, \qquad \lim_{\theta\to 0} \frac{1-\cos\theta}{\theta} = 0,$$

what is important is that the "θ" inside the sine and cosine functions approaches (but never achieves!) zero at the same rate as the "θ" in the denominator. Thus one may just write

$$\lim_{x\to 0} \frac{\sin(\sin x)}{\sin x} = 1$$

since the substitution $\theta = \sin x$ gives us $x \to 0 \implies \theta \to 0$ properly, and $\lim_{\theta\to 0}\frac{\sin\theta}{\theta} = 1$ is known. To be cautious, one can explicitly write this substitution step:

$$\lim_{x\to 0} \frac{\sin(\sin x)}{\sin x} = \lim_{\theta\to 0} \frac{\sin\theta}{\theta} = 1,$$

where $\theta = \sin x \to 0$ in a manner consistent with our theorem (as $x \to 0$), as an examination of the graph (or just the nature) of the sine curve indicates.

The next example shows how we can save some effort by simply noting the approaches to zero are at the same rate. Still, we should have the limit substitution theorem and its criteria (a)–(c) in mind. We will show the careful substitution method, and then the more summary (or terse or "hand-waving") argument.

Example 2.9.13

Consider the limit $\lim\limits_{x\to 0} \dfrac{1-\cos x^2}{x}$. Again, we know what this limit would be if we had x^2 in the denominator, so we will rewrite the function in a form where that is the case and compensate in the numerator:

$$\lim_{x\to 0} \frac{1-\cos x^2}{x} = \lim_{x\to 0} \frac{x(1-\cos x^2)}{x^2} = \lim_{x\to 0} \left[x \cdot \frac{1-\cos x^2}{x^2}\right].$$

At this point we could use a substitution, say $u = x^2$, and so $x \to 0 \implies u \to 0^+$, but we have different expressions for our function as $x \to 0^+$ and $x \to 0^-$:

$$\lim_{x\to 0^+} \left[x \cdot \frac{1-\cos x^2}{x^2}\right] = \lim_{u\to 0^+} \left[\sqrt{u} \cdot \frac{1-\cos u}{u}\right] \underset{}{=\!=\!=} 0^+ \cdot 0 \ \ 0,$$

$$\lim_{x\to 0^-} \left[x \cdot \frac{1-\cos x^2}{x^2}\right] = \lim_{u\to 0^+} \left[-\sqrt{u} \cdot \frac{1-\cos u}{u}\right] \underset{}{=\!=\!=} -0^+ \cdot 0 \ \ 0.$$

2.9. FURTHER LIMIT THEOREMS AND TRIGONOMETRIC LIMITS

This is all correct, but instead we will simply rewrite the function so that we get the same rate of approach to zero inside the cosine and in the denominator:

$$\lim_{x \to 0} \frac{1-\cos x^2}{x} = \lim_{x \to 0}\left[x \cdot \frac{1-\cos x^2}{x^2} \right] \stackrel{0 \cdot 0}{=\!=\!=} 0.$$

Rather than making the substitution, we rewrote the function and noticed we have matching rates of approach to zero in the $(1-\cos x^2)/x^2$ term: We see that $x^2 \to 0^+$ *properly*, as in our limit substitution theorem, so we can just cite the result $(1-\cos\theta)/\theta \to 0$ as $\theta \to 0$ (which includes both left and right limits).

As we saw in this example, sometimes substitution requires us to consider cases, where instead we could do some hand-waving (based on anticipating what would happen if we did perform the substitution). The next example perhaps makes the case for selective, "informed hand-waving" more strongly:

Example 2.9.14

Consider $\lim\limits_{x \to 0} \dfrac{\sin 2x}{\sin 5x}$.

Solution: We need a $2x$ to oppose the $\sin 2x$, and $5x$ to oppose the $\sin 5x$, all the while not actually changing the value of the function. We do this by multiplying the numerator by $2x/2x$, and the denominator by $5x/5x$, with factors arranged strategically:

$$\lim_{x \to 0} \frac{\sin 2x}{\sin 5x} = \lim_{x \to 0} \frac{2x \cdot \frac{\sin 2x}{2x}}{5x \cdot \frac{\sin 5x}{5x}} = \lim_{x \to 0} \frac{2 \cdot \frac{\sin 2x}{2x}}{5 \cdot \frac{\sin 5x}{5x}} \stackrel{\frac{2 \cdot 1}{5 \cdot 1}}{=\!=\!=} 2/5.$$

To actually use a verbose substitution method, we would have to factor the function first into three factors, $\frac{2}{5} \cdot \frac{\sin 2x}{2x} \cdot \frac{5x}{\sin 5x}$ and invoke two separate substitutions, one each for the second and third factors, and use our general limit Theorem 2.9.1, page 189. Such an approach would be correct, but quite cumbersome.

When we have $x \to a \implies u \to \beta$, there are exceptional cases where the u-variable approach to β is problematic for limit substitution methods. We always need to be aware that we require $u \ne \beta$ (and over a continuum of values near β) as $x \to a$, but again it is rare that there is a problem when the substitutions are routine. The next two examples give some idea of one kind of substitution to avoid.[56]

Example 2.9.15

We claim that the limit $\lim\limits_{x \to 0} \dfrac{\sin\left[x \sin \frac{1}{x}\right]}{\left[x \sin \frac{1}{x}\right]}$ does not exist.

Now the function $x \sin \frac{1}{x}$ is plotted in Figure 2.22, page 160, and has two relevant features for our discussion here:

---

[56] In fact, it is not necessary for $u \ne \beta$ if f is continuous at $x = a$ and $u = \beta$ when $x = a$, but then the value of the limit is just $f(a)$ and this analysis is unnecessary. These more advanced limit techniques, such as substitution, are for dealing with the cases that continuity is "broken," or the quantities are not all finite.

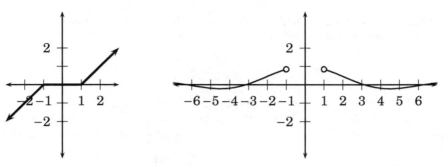

Figure 2.31.a. Graph of $y = g(x)$.   Figure 2.31.b. Graph of $y = \frac{\sin(g(x))}{g(x)}$.

Figure 2.31: Graph of $y = g(x)$ and $y = (\sin(g(x)))/(g(x))$ from Example 2.9.16. The function in Figure 2.31 is undefined for $x \in [-1, 1]$ due to a zero denominator there, the limit at $x \to 0$ does not exist.

1. $x \sin \dfrac{1}{x} \longrightarrow 0$ as $x \to 0$, which was proved by the Sandwich Theorem; however,

2. $x \sin \dfrac{1}{x}$ has infinitely many zeros (that is, points x where the function is zero) as $x \to 0$.

From the first point, we see that if we let $\theta = x \sin \frac{1}{x}$, then $x \to 0 \implies \theta \to 0$. Unfortunately, $\theta = 0$ infinitely many times as $x \to 0$, and so θ is not approaching zero *properly*. Thus, even though $\lim_{\theta \to 0} \frac{\sin \theta}{\theta} = 1$, we cannot say that the original limit is the same value. Indeed, the original limit does not exist, because as we have $x \to 0$ there are infinitely many points where the original function is not even defined, forcing us to conclude (since we cannot possibly satisfy the ε-δ definition of the limit, as given on page 123) that

$$\lim_{x \to 0} \frac{\sin \left[x \sin \frac{1}{x} \right]}{\left[x \sin \frac{1}{x} \right]} \quad \text{does not exist.}$$

The reason this limit does not exist compares to the reason $\lim_{x \to 5} \sqrt{x^2 - 25}$ does not exist, though the latter is more obvious: that $\sqrt{x^2 - 25}$ is *undefined* for part of the path of approach, in particular for $x \in (-5, 5)$ so that $\sqrt{x^2 - 25}$ is undefined as $x \to 5^-$. With our example's limit, the function does not exist (that is, is undefined) everywhere that the denominator, $x \sin \frac{1}{x}$ is zero, i.e., wherever $1/x = n\pi$, or $x = 1/(n\pi)$ where $n \in \mathbb{Z} - \{0\}$, and this happens infinitely many times inside of any interval $(-\delta, 0)$ or $(0, \delta)$ if $\delta > 0$. As with $\lim_{x \to 5} \sqrt{x^2 - 25}$, we cannot *approach* the limit point on a continuum (from both sides in the example $\sqrt{x^2 - 25}$) and be sure to stay in the domain of the function, and thus we are forced to conclude the limit does not exist. For a less subtle example, consider the following.

Example 2.9.16

Suppose $g(x) = \begin{cases} x - 1 & \text{if} & x \in (1, \infty) \\ 0 & \text{if} & x \in [-1, 1] \\ x + 1 & \text{if} & x \in (-\infty, -1) \end{cases}$ and consider $\lim_{x \to 0} \dfrac{\sin(g(x))}{g(x)}$.

This is graphed in Figure 2.31.a. Next, consider the limit $\lim_{x \to 0} \dfrac{\sin(g(x))}{g(x)}$. If we let $u = g(x)$, then one could say $x \to 0 \implies u \to 0$, but the latter ($u \to 0$) will not be a proper approach, because $u = 0$

for all $x \in [-1, 1]$ (not just at the limit point $x = 0$). In other words, u is not simply "approaching" zero for x near zero, but rather u is the **constant** zero for such x sufficiently close to zero. Clearly $(\sin g(x))/g(x)$ is undefined for $x \in [-1, 1]$ (because $g(x) = 0$ there), and $[-1, 1]$ contains the values of x on the eventual path as $x \to 0$. Thus, we **cannot** say $\lim_{x \to 0}(\sin g(x))/g(x)$ is equal to $\lim_{u \to 0}(\sin u)/u$ because the former limit does not exist, while the latter limit is known to be 1.

2.9.4 Epilogue

The point of this section is that there are limits of complicated functions for which we can rightfully summarize how some of the substructures of a function are behaving, and synthesize that into a statement about the overall limit. We do have to be very clear to make our arguments in the correct spirit of limits: We see what values are approached by this or that part of the function, sometimes making sure that those values do not actually occur if that is problematic, and sometimes allowing that they do without forcing the overall limit not to exist.

So we sometimes have to consider how the values are approached (including for our approaches "to infinity"), for instance, along a continuum of values, or in fits and spurts, skipping some values or staying constant for part or all of the approach. At times one kind of "approach" is necessary for the limit to exist, and at other times in a computation any kind of approach suffices. This depends mainly on what is used as an input variable and what is used as an output variable, and if there are variables "in between" (as in the case of substitution), how they are behaving as the input variable makes its approach to the value specified in the limit.

With training and depth of understanding, many complicated limits can be dispatched fairly quickly, but attempts do so without the requisite understanding can and likely will cause predictable (but avoidable) errors in limit computations. With practice, these become more and more rare, until they are completely vanquished.

Our development of limits of functions ends here temporarily and resumes in the sequel. There, these techniques will be revisited and expanded, and new techniques based on the calculus developed in between will also be introduced.

We will use the limit techniques developed here in the meantime, and the concept of limit will be embedded in much of what we do throughout the text. Both differential and integral calculus are *defined* in terms of limits, so limits indeed are foundational to all of calculus.

Exercises

For Exercises 1–24, compute the given limit or explain why it does not exist.

1. $\lim\limits_{x \to 0} \dfrac{\sin 9x}{x}$

2. $\lim\limits_{x \to 0} \dfrac{x}{\sin 9x}$

3. $\lim\limits_{x \to 0} \dfrac{\sin 9x}{\sin 7x}$

4. $\lim\limits_{x \to 0} \dfrac{\sin x^2}{x}$

5. $\lim\limits_{x \to 0^+} \dfrac{\sin x}{x^2}$

6. $\lim\limits_{x \to 0} \dfrac{\sin^3 5x}{x^3}$

7. $\lim\limits_{x \to 0} \sqrt[3]{\dfrac{\sin x}{x}}$

8. $\lim\limits_{x \to 0} \sqrt{\dfrac{\sin x}{x}}$

9. $\lim\limits_{x \to 0^+} \dfrac{\tan \sqrt{x}}{\sqrt{x}}$

10. $\lim\limits_{x \to 0} \dfrac{\tan 2x}{x}$

11. $\lim\limits_{x \to 0} \dfrac{1 - \cos 2x}{x}$

12. $\lim\limits_{x \to 0} \dfrac{1 - \cos 2x}{x^2}$

13. $\lim\limits_{x \to \frac{\pi}{2}} \dfrac{1 - \cos 2x}{x^2}$

14. $\lim\limits_{x \to 0} \dfrac{1 - \cos x}{x^4}$

15. $\lim\limits_{x \to 0} \dfrac{\sin(\tan x)}{\tan x}$

16. $\lim\limits_{x \to \frac{\pi}{2}^+} \dfrac{\sin(\tan x)}{\tan x}$

17. $\lim\limits_{x \to 0} \dfrac{\sin |x|}{|x|}$

18. $\lim\limits_{x \to 0} \dfrac{\sin |x|}{x}$

19. $\lim\limits_{x \to 0} \dfrac{x}{\sin x}$

20. $\lim\limits_{x \to 0} \dfrac{x^2}{\sin x}$

21. $\lim\limits_{x \to -\infty} \dfrac{\sin e^x}{e^x}$

22. $\lim\limits_{x \to \infty} \dfrac{\sin e^x}{e^x}$

23. $\lim\limits_{x \to -\infty} \dfrac{1 - \cos e^x}{e^x}$

24. $\lim\limits_{x \to -\infty} \dfrac{\cos e^x}{e^x}$

25. Compute or state that the following limit does not exist and why:
$$\lim\limits_{x \to 0} \dfrac{1 - \cos\left(x \sin \frac{1}{x}\right)}{x \sin \frac{1}{x}}.$$

26. Recompute the limit of Example 2.9.11, page 201, this time by first multiplying the numerator and denominator of the function by $\cos^2 x$.

27. Compute $\lim\limits_{x \to 0^+} \dfrac{2\csc^2 x + 3\csc x + 11}{5\csc^2 x + 4\csc x + 7}$ two ways:

    (a) using a substitution argument;

    (b) by first rewriting the function, and not using a substitution argument.

28. Compute $\lim\limits_{x \to \infty} \left[\dfrac{2x - 9}{3x + 5} \cdot \dfrac{6x^2 - 9x + 10}{x^2 - 7x + 8} + \dfrac{7x^2 + 6}{3x^3 + 4x - 3} \cdot \cos\left(\dfrac{1}{x}\right) \right]$.

29. Compute $\lim\limits_{x \to -\infty} \left[\dfrac{x^2 + 8x - 9}{5x^2 - 6x + 42} + \sin e^x \right]$.

30. Compute $\lim\limits_{x \to \infty} \left[\dfrac{x}{x^2 + 1} + \cos x \right]$.

31. Compute $\lim\limits_{x \to \infty} \left[\dfrac{x^2 + 1}{x - 1} + \cos x \right]$.

Chapter 3

The Derivative

The earlier chapters are the analytical preludes to calculus. This chapter begins the study of calculus proper, starting with the study of *differential calculus*, also known as the calculus of derivatives. We will develop all of the fundamental derivative computations in this chapter.

Once we complete our development of derivatives and see many of their most immediate applications in this chapter and in Chapter 4, we will then look towards *integral calculus* in Chapter 5, where roughly speaking we see how to reverse what we do here.[1]

The integral calculus of Chapter 5 is expanded in the sequel. Subsequent chapters there develop several topics, which are either offshoots of differential and integral calculus or are greatly extended by these.

As we will see, differential calculus addresses rates at which quantities change with respect to each other, while integral calculus addresses how quantities (or the changes in quantities) accumulate. Once we lay the foundations of both differential and integral calculus, we will further develop and apply both in very diverse circumstances for the remainder of the text.

3.1 Introduction to Derivatives

In this section we look at perhaps the main breakthrough concept of calculus, namely the *derivative*, which allows us to measure how quickly a quantity is changing. Its significance cannot be overstated. It will have implications for many types of situations, both theoretical and applied, each with its own inherent intuition. The derivative itself owes its existence to the concept of limit, which is therefore even more foundational.

3.1.1 An Example: Position Gives Rise to Velocity

Suppose an object is in one-dimensional motion, which we can therefore assume is along a number line, so we can take its *position* along that line to be a function $s(t)$, where t is in units of time.

---

[1] In this chapter, we take a function and find its derivative. In the final chapter, we are given the derivative and are tasked with determining the underlying function. This reverse process is often a more difficult task. However, a thorough understanding of the material in this chapter greatly simplifies the learning of integral calculus, just as learning multiplication well makes learning division much easier.

As a philosophical matter, upon reflection it seems reasonable that if we know the position $s(t)$ of an object at all times t, then we should also know its *velocity* $v(t)$ at all times t. If, for instance, we are driving a car and wish to know the velocity of another car, we can pull our own car alongside of the other car, match its every move, and read our own speedometer. If our position along our own axis matches that of the other car along its axis at all times, how could the velocities be different? It seems that they cannot. In the example of this subsection we describe a theoretical argument and mechanism for determining velocity $v(t)$ from knowledge of position $s(t)$.[2]

There are different ways in which we may be able to determine the position $s(t)$ of an object at any given time t that we choose. We can simply observe where the object is at that time t (perhaps using a timed camera), measure it with some other device such as a GPS (Global Positioning System) receiver, or we may have sufficient theory to give us an actual equation for $s(t)$, which is the ideal case. In the case that follows, we have an object in free-fall whose vertical position is given by

$$s(t) = -4.9t^2 + 40t,$$

where s is in meters and t is in seconds. Much of what follows does not rely on such a function being explicitly known, but only that there is such a function in theory.

Suppose we wish to find $v(3)$, the velocity at time 3 sec. To explore this for our case, let us first construct a table of $s(t)$ versus t, for regularly occurring times, centered at $t = 3$:

$t =$	1.5	2	2.5	**3**	3.5	4	4.5
$s(t) =$	48.975	60.4	69.375	**75.9**	79.975	81.6	80.775

If this motion is along an axis oriented so that the positive numbers are to the right of zero, then the object seems to be rightward-moving at $t = 3$; the values of the positions are increasing on the chart until $t \in [4, 4.5]$. If motion is instead along a standard vertical axis, the object seems to be rising until that interval.

From the chart we have several ways to approximate $v(3)$. For instance, we can compute the *average velocity* for any time interval with one endpoint being $t = 3$ and the other listed above. We will usually use Δt for "change in time," in this case the change from 3 to $3 + \Delta t$, and write this average velocity as

$$\frac{s(3 + \Delta t) - s(3)}{\Delta t},$$

where the interval under consideration is either $[3 + \Delta t, 3]$ if $\Delta t < 0$, or $[3, 3 + \Delta t]$ if $\Delta t > 0$.[3] So for $[2.5, 3]$ and $[3, 4.5]$ these can be computed (where units would then be meters/second):

- Average velocity for $[2.5, 3]$ is $\frac{s(2.5) - s(3)}{-0.5} = \frac{69.375 - 75.9}{-0.5} = 13.05$.

- Average velocity for $[3, 4.5]$ is $\frac{s(4.5) - s(3)}{1.5} = \frac{80.775 - 75.9}{1.5} = 3.25$.

---

[2]Traditional physics textbooks tend to use $s(t)$ for one-dimensional motion. Depending upon the context, other possibilities include $x(t)$, $y(t)$, $r(t)$, and still others. In one-dimensional motion *position* can be any real number—positive, negative, or zero—unlike *distance*, which must be nonnegative.

[3]While it is natural and proper to take the differences in a "final minus initial" order, with the ratio of such differences it does not matter as long as we are consistent in the orders of both numerator and denominator. Note that $\frac{a-b}{c-d} = \frac{a-b}{c-d} \cdot \frac{-1}{-1} = \frac{b-a}{d-c}$. If we change the order of the difference in the numerator, we must also change the order in the denominator, and the ratio will be the same.

3.1. INTRODUCTION TO DERIVATIVES

We can similarly fill the following table of average velocities over the given intervals, using data from the previous table (here "Av." means "Average," and "Undef." means "Undefined"):

Interval t	[1.5,3]	[2,3]	[2.5,3]	[3,3]	[3,3.5]	[3,4]	[3,4.5]
Av. Velocity	$\frac{48.975-75.9}{-1.5}$	$\frac{60.4-75.9}{-1.0}$	$\frac{69.375-75.9}{-0.5}$	$\frac{75.9-75.9}{0}$	$\frac{79.975-75.9}{0.5}$	$\frac{81.6-75.9}{1.0}$	$\frac{80.775-75.9}{1.5}$
Decimal	17.95	15.5	13.05	Undef.	8.15	5.7	3.25

As we see, there is still much variation in these average velocities over these intervals where one endpoint is $t = 3$, leading one to believe that none of the intervals up for consideration is likely to be a very good approximation of the actual velocity at time $t = 3$, i.e., for $v(3)$.

However, it seems reasonable that most objects in motion, when not encountering some radical event, will at least have *continuous* velocities, and perhaps our intervals are simply too long.

Assuming continuity of $v(t)$, then given a desired tolerance for $v(3)$, namely $\varepsilon > 0$, we should be able to find a tolerance $\delta > 0$ for t such that on the interval $|t - 3| < \delta$ we have $|v(t) - v(3)| < \varepsilon$. Not knowing the exact velocity function (yet), we do not know the level of smallness required of $\delta > 0$, which will give us the desired accuracy, nor do we know $v(3)$, except that it makes sense that it exists (barring some unusual physics of its motion).

Reasoning further (though not rigorously proven here), if the velocity does not change much on some interval $|t - 3| < \delta$, or even for $|t - 3| \leq \delta$, then the velocity at $t = 3$ should also be within ε of the average velocity on the interval $[3 - \delta, 3 + \delta]$. (If no value $v(t)$ for t in the interval $[t - \delta, t + \delta]$ is more than ε from $v(3)$, then neither should the average velocity for the interval be more than ε from $v(3)$.) We can then go further to remark that this should also be true of the even smaller subintervals $[3 - \delta, 3]$ and $[3, 3 + \delta]$.[4]

So for instance, in the previous table our various interval endpoints were incremented by 0.5. If instead we increment by 0.1, we see, unsurprisingly, that the average velocity values vary less:

Interval t	[2.7,3]	[2.8,3]	[2.9,3]	[3,3]	[3,3.1]	[3,3.2]	[3,3.3]
Av. Velocity	$\frac{72.279-75.9}{-0.3}$	$\frac{73.584-75.9}{-0.2}$	$\frac{74.791-75.9}{-0.1}$	$\frac{75.9-75.9}{0}$	$\frac{76.911-75.9}{0.1}$	$\frac{77.824-75.9}{0.2}$	$\frac{78.639}{0.3}$
Decimal	12.07	11.58	11.09	Undef.	10.11	9.62	9.13

Here we see the values of the average velocity stabilizing somewhat, and we should expect that average velocities over yet smaller intervals to give still better approximations of the actual velocity $v(3)$.

It becomes more convenient to have a chart where we give shrinking values for $\Delta t \neq 0$ and compute $[s(3 + \Delta t) - s(3)]/[\Delta t]$ to see if there is a more apparent trend. The following are reported to within 0.0000001, i.e., 10^{-7}. Note that units would be in meter/second:

$$\Delta t = 0.01 \implies \frac{s(3+\Delta t) - s(3)}{\Delta t} = 10.551, \qquad \Delta t = -0.01 \implies \frac{s(3+\Delta t) - s(3)}{\Delta t} = 10.649,$$

$$\Delta t = 0.001 \implies \frac{s(3+\Delta t) - s(3)}{\Delta t} = 10.5951, \qquad \Delta t = -0.001 \implies \frac{s(3+\Delta t) - s(3)}{\Delta t} = 10.6049,$$

$$\Delta t = 0.0001 \implies \frac{s(3+\Delta t) - s(3)}{\Delta t} = 10.59951, \qquad \Delta t = -0.0001 \implies \frac{s(3+\Delta t) - s(3)}{\Delta t} = 10.60049.$$

---

[4] While the continuity definition calls for intervals $|t - 3| < \delta$, i.e., $(t - \delta, t + \delta)$, arguments used to prove many of the limit and continuity theorems allow us to, if necessary, take slightly smaller values of δ for the given ε so that we can instead use the closed intervals $|t - 3| \leq \delta$, i.e., $[t - \delta, t + \delta]$ (for the new, smaller δ).

It seems the trend is towards $v(3) \approx 10.60$ m/sec. It is reasonable that we will be more confident that we are approaching the actual value of $v(3)$ when we look at the average velocity over yet shorter time intervals. Of course we cannot allow $\Delta t = 0$, because we would be dividing by zero in the respective quotient $(s(3+\Delta t) - s(3))/(\Delta t)$, which would read 0/0 if we let $\Delta t = 0$.

The obvious punchline is that we should use a tool from Chapter 2 to break through the analysis, using a limit as $\Delta t \to 0$ to deduce $v(3)$. For our particular example, since we know the form of $s(t)$, namely $s(t) = -4.9t^2 + 40t$, we can compute the following, noting that there are incidents where we can cancel products without actually computing them (such as $-4.9 \cdot 3^2$ and $40 \cdot 3$):

$$\lim_{\Delta t \to 0} \frac{s(3+\Delta t) - s(3)}{\Delta t} \overset{0/0}{\underset{\text{ALG}}{=\!=}} \lim_{\Delta t \to 0} \frac{[-4.9(3+\Delta t)^2 + 40(3+\Delta t)] - [-4.9(3)^2 + 40(3)]}{\Delta t}$$

$$\overset{0/0}{\underset{\text{ALG}}{=\!=}} \lim_{\Delta t \to 0} \frac{-4.9\left(3^2 + 6\Delta t + (\Delta t)^2\right) + 40 \cdot 3 + 40\Delta t + 4.9(3)^2 - 40 \cdot 3}{\Delta t}$$

$$\overset{0/0}{\underset{\text{ALG}}{=\!=}} \lim_{\Delta t \to 0} \frac{-4.9 \cdot 6\Delta t - 4.9(\Delta t)^2 + 40\Delta t}{\Delta t}$$

$$\overset{0/0}{\underset{\text{ALG}}{=\!=}} \lim_{\Delta t \to 0} \frac{(\Delta t)(-29.4 - 4.9\Delta t + 40)}{\Delta t}$$

$$\overset{0/0}{\underset{\text{ALG}}{=\!=}} \lim_{\Delta t \to 0} (-29.4 - 4.9\Delta t + 40) \overset{\text{CONT}}{=\!=\!=} -29.4 + 0 + 40 = 10.6,$$

as we suspected. It is reasonable then to conclude that the actual velocity of the object was, in fact, 10.6 m/sec, for that was the value that the average velocities approached as the lengths of the time intervals from $t = 3$ shortened to zero: Since we expected the *change* in velocity to become "closer and closer to zero" as those intervals shrank to length zero, we therefore expected these average velocities on the shrinking intervals to approach the actual velocity at $t = 3$.

Note that all limits here were of the form 0/0 until the common factor Δt cancelled from the numerator and denominator. We usually will not bother to note the 0/0 form in this context, nor note the step that used the continuity of a function of Δt (when the other terms were constant with respect to the limit), because such limits generally do have the 0/0 form until near the end of the computation, when the continuity (in the variable Δt) is invoked as we let $\Delta t \longrightarrow 0$.

In a situation where the velocity is not constant, it does not make sense to use a simple formula such as "(directed) distance divided by time" for velocity at a given time t. From the development and discussion so far, though, it seems reasonable to *define* velocity by such limits, as we do next.

Definition 3.1.1 *Given a position function $s(t)$, define the* **velocity** *(or* **instantaneous velocity**, *to distinguish it from average velocity) at a time t to be the function given by the limit*

$$v(t) = \lim_{\Delta t \to 0} \frac{\Delta s}{\Delta t} = \lim_{\Delta t \to 0} \frac{s(t + \Delta t) - s(t)}{\Delta t}, \tag{3.1}$$

for each t for which the limit (3.1) exists and is finite, and where we also define

$$\Delta s = s(t + \Delta t) - s(t). \tag{3.2}$$

3.1. INTRODUCTION TO DERIVATIVES

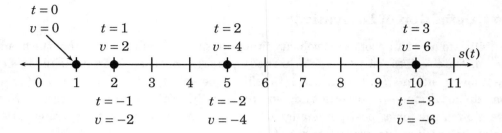

Figure 3.1: Here we trace the one-dimensional motion $s(t) = t^2 + 1$ as a position on a number line for times $t = -3, -2, -1, 0, 1, 2, 3$ (note the progression of those positions in the figure). The velocities $v = 2t$ are also given. The graph reflects how the particle comes in from the right for negative t, stops at $t = 0$ ($s = 1$, $v = 0$), and moves back out toward the right for positive t (faster and faster as $t > 0$ increases). Note that no "t-axis" appears explicitly. See Example 3.1.1.

If we are fortunate enough to know an algebraic formula for $s(t)$ as a function, then we can use the limit (3.1) to calculate a formula for $v(t)$. This describes a situation known to physicists as *one-dimensional motion*. Note how (3.1) allows for $\Delta t \to 0^-$ as well as $\Delta t \to 0^+$. Note also that for continuous $s(t)$—a reasonable assumption in classical physics—limits of form (3.1) will be of 0/0 form.

Next we compute $v(t)$ for a slightly simpler example, but for all $t \in \mathbb{R}$ (and not just a specific value of t), and make some attempt to graphically represent the motion in Figure 3.1.

Example 3.1.1

Suppose that position is given by $s(t) = t^2 + 1$. We can use (3.1) to calculate the velocity function for any fixed t as follows. As is often the case with derivative computations, this limit will be of 0/0 form until the last step, as we perform algebra to attempt to can cancel the Δt factor in the denominator.[5]

$$\begin{aligned}
v(t) &= \lim_{\Delta t \to 0} \frac{s(t + \Delta t) - s(t)}{\Delta t} \\
&= \lim_{\Delta t \to 0} \frac{\left((t + \Delta t)^2 + 1\right) - \left(t^2 + 1\right)}{\Delta t} \\
&= \lim_{\Delta t \to 0} \frac{(t^2 + 2t\Delta t + (\Delta t)^2 + 1) - (t^2 + 1)}{\Delta t} \\
&= \lim_{\Delta t \to 0} \frac{t^2 + 2t\Delta t + (\Delta t)^2 + 1 - t^2 - 1}{\Delta t} \\
&= \lim_{\Delta t \to 0} \frac{2t\Delta t + (\Delta t)^2}{\Delta t} \\
&= \lim_{\Delta t \to 0} (2t + \Delta t) \\
&= 2t.
\end{aligned}$$

We showed that $s(t) = t^2 + 1 \implies v(t) = 2t$. Some position and velocity data are given for various times t in Figure 3.1. Note that when $s > 0$ the position is to the right of our "origin" $s = 0$ (as is always the case here), and when $s < 0$ the position is to the left of the origin. Also, when $v > 0$ the motion is rightward, and when $v < 0$ the motion is towards the left. For example, at time $t = 2$ we have the position $s(2) = 2^2 + 1 = 5$, and velocity $v(2) = 2(2) = 4$. If s is measured in meters, and t in seconds, then the units in the limit (3.1) are meters/second, as we would expect.

---

[5]Here we treat t as a constant in the calculation of $v(t)$; t is fixed, while the variable $\Delta t \to 0$.

3.1.2 Definition of Derivative

The ability to find a nonconstant velocity function is a tremendous leap from the grade school notion of "rate = distance ÷ time." As argued earlier in this section, in principle it seems that if we know position $s(t)$ at all times t, we should be able to find velocity $v(t)$. Having limits at our disposal made it possible to do so analytically. Note that $v(t)$ is really a *limit* of an expression of the form "position change divided by time change," that is, $(\Delta s)/(\Delta t)$, so it has some of the spirit of the grade school notion.

Such limits are useful in more than just "position knowledge $\Longrightarrow$ velocity knowledge" problems; we will have use for them throughout the text in numerous contexts. Because they are ubiquitous, we generalize the notation and call the functions that arise from these limits *derivatives*. (The following definition should be committed to memory.)

Definition 3.1.2 *Given any quantity Q that is a function of the variable x, i.e., $Q = Q(x)$:*

- *The* **derivative** *of Q* **with respect to** *x is the function $Q'(x)$, read* **"Q-prime of x,"** *defined by*[6]

$$Q'(x) = \lim_{\Delta x \to 0} \frac{Q(x + \Delta x) - Q(x)}{\Delta x} \qquad (3.3)$$

  wherever that limit exists and is finite.

- *If this limit does not exist* **or** *is infinite at a given x_0, we say $Q'(x_0)$* **does not exist**. *If the limit does exist as a finite number at $x = x_0$, we say $Q(x)$ is* **differentiable** *at x_0.*

- *$Q'(x)$ is also called the* **instantaneous rate of change of $Q(x)$ with respect to** *x.*[7]

To define the derivative $Q'(x)$ at a given value x, we require not only that the limit (3.3) exists, but also that it is finite (i.e., exists as a real number). We will make more use of the term *differentiable* in later sections where its justification is clearer. For now, it should suffice to note (Q differentiable at x_0) $\Longleftrightarrow$ ($Q'(x_0)$ exists).

In most cases, we expect the limit (3.3) which defines the derivative to be of 0/0 form, requiring the usual techniques (e.g., algebraic simplification) to compute.

Definition 3.1.3 *We also define the* average rate of change *over an interval as before: If the initial value of x is x_0 (pronounced "x-naught" or "x sub(script) zero"), and the final value is x_f, then the* **average rate of change of $Q(x)$ with respect to** *x for the interval $[x_0, x_f]$ or $[x_f, x_0]$ (depending on whether $x_0 < x_f$ or $x_0 > x_f$) is given by the* **difference quotient**

$$\frac{Q(x_f) - Q(x_0)}{x_f - x_0} = \frac{Q(x_0 + \Delta x) - Q(x_0)}{\Delta x} = \frac{\Delta Q}{\Delta x} \qquad (3.4)$$

where

$$\Delta x = x_f - x_0, \quad \text{and} \qquad (3.5)$$
$$\Delta Q = Q(x_f) - Q(x_0) = Q(x_0 + \Delta x) - Q(x_0). \qquad (3.6)$$

---

[6] This "prime" notation was first introduced by Italian mathematician and astronomer Joseph-Louis Lagrange (1736–1813).

[7] "Instantaneous" rate of change means the rate of change "at that instant" (as opposed to an average rate of change of the output variable as it is measured against an entire interval's length of values of the input variable).

3.1. INTRODUCTION TO DERIVATIVES

So we see that the derivative (3.3) is just the limit of the average rate of change in Q given in (3.4) on an interval with endpoints x and $x + \Delta x$, assuming that limit as $\Delta x \to 0$ exists and is finite.

With this notation, we can rewrite the (instantaneous) velocity function for a given $s(t)$ as the following, which is an equation well known to physics students:

$$v(t) = s'(t). \tag{3.7}$$

Because there are so many contexts, there are several different relevant notations, such as those in Equations (3.5)–(3.6). Each has its value and all are worth knowing.[8]

Example 3.1.2

Suppose a car has a very precise fuel gauge and a very precise odometer, perhaps GPS-enabled. Let $V(D)$ be the volume of fuel in the tank after traveling a distance D, as read from the odometer. Then whenever this derivative exists (i.e., the limit exists and is finite),[9]

$$V'(D) = \lim_{\Delta D \to 0} \frac{\Delta V}{\Delta D} = \lim_{\Delta D \to 0} \frac{V(D + \Delta D) - V(D)}{\Delta D}$$

represents the instantaneous rate of fuel volume change with respect to distance traveled. If D is in units of miles and V is in units of gallons, this rate would have units of gallons per mile. (To compute the more standard miles/gallon, we can take the reciprocal.) Furthermore, this derivative of volume can also reflect the flow of a fluid, almost. There is one minor technicality, as we explain next.

Notice that the fuel should be leaving the tank whenever the engine is running, so V should be decreasing as we drive. This gives $\Delta V < 0$ when $\Delta D > 0$, giving $\Delta V/\Delta D < 0$, and thus $V'(D) \leq 0$.[10]

However, the fuel running through the engine is exactly the fuel leaving the tank, and so the actual flow rate out of the tank we would report would be the opposite, $-V'(D)$, a nonnegative quantity, for any particular position D.[11]

There are countless other applications of the derivative. All we need is a quantity Q as a function of another quantity x, to measure the rate that Q changes as x changes. The average rate of change of Q with respect to x is again $(\Delta Q)/(\Delta x)$, and the instantaneous rate is the number we get when we consider $(\Delta Q)/(\Delta x)$ and let $\Delta x \to 0$, giving what we take to be the rate "at that instant."

---

[8]In the next section we will introduce the very powerful Leibniz notation for the derivative $Q'(x)$, which we will then write dQ/dx (notice the resemblance to $\Delta Q/\Delta x$).

[9]Technically, odometers display total distance D traveled and not position s. By its essence, D must be nonnegative, while s can have different signs, or even several coordinates, in the context of vectors, which appear much later in the textbook.

[10]Recall that a negative function or sequence can have a limit that is either negative (including $-\infty$), zero (approached from below), or nonexistent. This is why knowing $\Delta V/\Delta D < 0$ only allows us to conclude that its limit is nonpositive, or nonexistent.

[11]One interesting aspect of this relationship between D and V is that, in practice, V will not always be a function of D, because there will be times that V is changing but D is not, for instance when the car is stopped and the motor is idling, for which there would be a continuum of values of V for that particular D. However, when the car is moving we will have V as a function of D and can consider the derivative $V'(D)$. Later we will look closely at calculus for cases that might not be functions "globally" (where we look at the entire graph of the relation between two variables), but may be "locally" (where the graph does pass the vertical line test).

3.1.3 Slope of a Curve

All these applications have their own interpretations. Interestingly enough, the *analytic geometric* interpretation of the derivative of a function unifies them for many analytical purposes in one graphical setting. For the bulk of the remainder of this section, we will concentrate on the significance of the derivative $f'(x)$ to the graph of $y = f(x)$.

First we will consider a very simple case. Suppose the graph of $y = f(x)$ is linear, so that

$$f(x) = mx + b,$$

where $m, b \in \mathbb{R}$ are fixed constants. Then, noticing when various terms "cancel," we compute:

$$f'(x) = \lim_{\Delta x \to 0} \frac{f(x + \Delta x) - f(x)}{\Delta x} = \lim_{\Delta x \to 0} \frac{[m(x + \Delta x) + b] - [mx + b]}{\Delta x}$$
$$= \lim_{\Delta x \to 0} \frac{mx + m\Delta x + b - mx - b}{\Delta x} = \lim_{\Delta x \to 0} \frac{m\Delta x}{\Delta x} = \lim_{\Delta x \to 0} m = m.$$

Thus, when $y = f(x)$ is the line $y = mx + b$, we get $f'(x) = m$; *if the function's graph is a line, then the derivative is its slope*.

Recall that the slope of a line measures how rapidly that line rises or falls as we move, for instance along the line and to the right. In other words, slope measures the rate of change in y with respect to x (also known as "the rise over the run," in the case of lines). That rate is constant on a line, but varies along most curves. Still, if we look closely at a point $(a, f(a))$ on the graph of $y = f(x)$, we can often associate a slope with the curve there.[12]

To measure this slope, again we would in effect measure the way $y = f(x)$ changes (instantaneously) with respect to x at $x = a$. With this motivation, we make the following definition:

Definition 3.1.4 *Given a function $f(x)$, the **slope** of the graph of $y = f(x)$ at any point $(a, f(a))$ on the graph is given by $f'(a)$, assuming this derivative exists there.*

It is important to note that a function and its derivative give two types of information about the graph of the function:

---

[12]We can imagine instances where a naive and perhaps nearsighted observer standing on a curve at $(a, f(a))$, and very focused on the curve at and around that point, might believe it has a constant slope. What we take to be the actual slope of the curve at that point is what we get when we see "the rise divided by the run" between two points $(a, f(a))$ and $(a + \Delta x, f(a + \Delta x))$, and then let $\Delta x \to 0$. Consider the illustration that follows:

The (dotted) line connecting the two curve points $(a, f(a))$ and $(a + \Delta x, f(a + \Delta x))$ is called a *secant line* (from the Latin, meaning "to cut"). Its slope is the difference quotient whose limit is $f'(a)$:

$$\Delta x \longrightarrow 0 \implies \frac{f(a + \Delta x) - f(a)}{\Delta x} \longrightarrow f'(a).$$

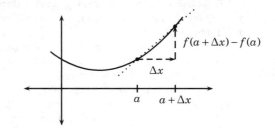

This can be visualized as varying the point $(a + \Delta x, f(a + \Delta x))$ closer and closer to $(a, f(a))$, and "in the limit" achieving the *tangent line* to the curve at $x = a$.

3.1. INTRODUCTION TO DERIVATIVES

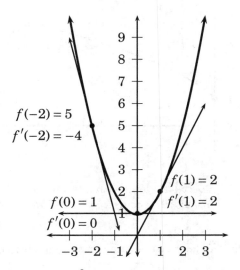

Figure 3.2: The graph of $f(x) = x^2 + 1$, along with the tangent lines to the graph at $x = -2, 0, 1$. A *tangent line* is a line through a point $(a, f(a))$ on the curve that has the same slope as the curve at that point. The height at each x is given by $f(x) = x^2 + 1$, while the slope is given by $f'(x) = 2x$. The tangent line at $(a, f(a))$ can be given by $y = f(a) + f'(a)(x - a)$.

- $f(x)$ gives the *height* of the graph for a particular x-value;
- $f'(x)$ gives the *slope* of the graph at that x-value.

For instance, we saw in Example 3.1.1, page 211 (albeit using different variables), that $f(x) = x^2 + 1 \implies f'(x) = 2x$. When we graph $y = x^2 + 1$, i.e., when we graph the function $f(x) = x^2 + 1$, the function $f(x)$ gives the height at each x, and the derivative $f'(x) = 2x$ gives the slope there. This is illustrated in Figure 3.2.

Also illustrated in that figure, and of geometric interest, is the *tangent line* to the graph of $y = f(x)$ at a given point $(a, f(a))$ on the curve. This is just the line through $(a, f(a))$ with the same slope as the curve, i.e., with slope $f'(a)$.

Definition 3.1.5 *The line through $(a, f(a))$ with slope $f'(a)$, i.e., the same slope as the function at $x = a$, is the* **tangent line** *to the graph of $y = f(x)$ through $(a, f(a))$.*

Three separate tangent lines are drawn in Figure 3.2. A formula for the tangent line through $(a, f(a))$ presents itself readily, since we have a point $(a, f(a))$, and a slope $f'(a)$, the modified point-slope form, $y = y_1 + m(x - x_1)$, gives us

$$y = f(a) + f'(a)(x - a). \tag{3.8}$$

For the function $f(x) = x^2 + 1$ in Figure 3.2, (3.8) gives us the following tangent lines, among others:

- At $x = 1$: $(a, f(a)) = (1, 2)$, $f'(a) = f'(1) = 2$, so the tangent line is $y = 2 + 2(x - 1)$, or $y = 2x$.

- At $x = -2$: $(a, f(a)) = (-2, 5)$, $f'(a) = f'(-2) = -4$, so the tangent line is $y = 5 - 4(x + 2)$, or $y = -4x - 3$.

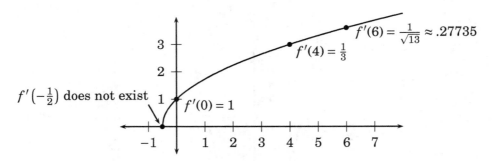

Figure 3.3: The graph of $f(x) = \sqrt{2x+1}$, along with a few slopes. Notice the behavior of the slope, in particular how $f'(x) = 1/\sqrt{2x+1} \to \infty$ as $x \to -1/2^+$, and how $f'(x) \to 0^+$ as $x \to \infty$.

- At $x=0$: $(a, f(a)) = (0, 1)$, $f'(a) = f'(0) = 0$, so the tangent line is $y = 1 + 0(x-0)$, or simply $y = 1$.

Example 3.1.3

Consider the function $f(x) = \sqrt{2x+1}$. Then, using a conjugate multiplication (third line) we can compute $f'(x)$, noting all limits are of form 0/0 until the final step:

$$\begin{aligned} f'(x) &= \lim_{\Delta x \to 0} \frac{f(x + \Delta x) - f(x)}{\Delta x} \\ &= \lim_{\Delta x \to 0} \frac{\sqrt{2(x + \Delta x) + 1} - \sqrt{2x+1}}{\Delta x} \\ &= \lim_{\Delta x \to 0} \frac{\sqrt{2(x + \Delta x) + 1} - \sqrt{2x+1}}{\Delta x} \cdot \frac{\sqrt{2(x + \Delta x) + 1} + \sqrt{2x+1}}{\sqrt{2(x + \Delta x) + 1} + \sqrt{2x+1}} \\ &= \lim_{\Delta x \to 0} \frac{(2(x+\Delta x)+1) - (2x+1)}{\Delta x \left(\sqrt{2(x+\Delta x)+1} + \sqrt{2x+1} \right)} \\ &= \lim_{\Delta x \to 0} \frac{2x + 2\Delta x + 1 - 2x - 1}{\Delta x \left(\sqrt{2(x+\Delta x)+1} + \sqrt{2x+1} \right)} \\ &= \lim_{\Delta x \to 0} \frac{2\Delta x}{\Delta x \left(\sqrt{2(x+\Delta x)+1} + \sqrt{2x+1} \right)} \\ &= \lim_{\Delta x \to 0} \frac{2}{\left(\sqrt{2(x+\Delta x)+1} + \sqrt{2x+1} \right)} = \frac{2}{2\sqrt{2x+1}} = \frac{1}{\sqrt{2x+1}}. \end{aligned}$$

To summarize, $f(x) = \sqrt{2x+1} \implies f'(x) = \dfrac{1}{\sqrt{2x+1}}$.

We can make several observations about the form of this derivative. First, note that $f(-1/2) = 0$ exists, but $f'(-1/2)$ does not. Of course, for $x = -1/2$ we cannot take $\Delta x \to 0^-$ and avoid expressions with square roots of negative numbers. Still, it is interesting to notice that $f'(x) \to \infty$ as $x \to (-1/2)^+$, and to note how this is borne out on the graph, as the slope becomes increasingly vertical as $x \to -\frac{1}{2}^+$. At the other extreme, as $x \to \infty$ we have $f'(x) \to 0^+$, so the function becomes less sloped as we move x farther to the right. This is all reflected in the graph, as illustrated in Figure 3.3.

As mentioned previously, we will mostly refrain from writing $\overset{0/0}{=\!=\!=}$ when computing these limits. By the nature of derivatives, it is expected that the limits involved will, under ordi-

3.1. INTRODUCTION TO DERIVATIVES

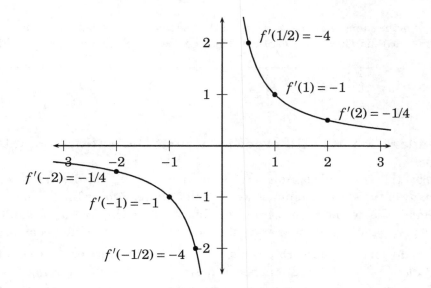

Figure 3.4: Illustration of the graph of $f(x) = 1/x$, from Example 3.1.4, showing also data from its derivative $f'(x) = -1/x^2$. From the graph or from the formula for $f'(x)$, we can notice that $f'(x) < 0$ for all $x \neq 0$, and observe the behavior of $f'(x)$ as $x \to \infty$, as $x \to -\infty$, and as $x \to 0$.

nary circumstances, be of the form 0/0 until the final computation where we have a determinate form, allowing us to finally compute the limit $\lim_{\Delta x \to 0} \frac{\Delta Q}{\Delta x}$. In fact, if there are problems at some x-values, such as nonexistence, including for disagreement of left and right limit, they tend to become apparent in the formula we get for the derivative, so our care in computing the actual derivatives should be more focused upon the algebra of computing the limits that define the derivatives. For instance, the fact that the derivative of $f(x) = \sqrt{2x+1}$ does not exist at $x = -1/2$, or for that matter anywhere to the left of $x = -1/2$, is apparent from the formula $f'(x) = 1/\sqrt{2x+1}$. To repeat, for the most part if there is a "problem" at some point with the function's differentiability, it will be apparent in the derivative formula as well.

Example 3.1.4 _____

Consider next the function $f(x) = \frac{1}{x}$. We will find its slope everywhere that it is defined. The method involves multiplying the numerator and denominator by a factor—namely the least common multiple of both denominators (that appear in the numerator)—which will remove the fractions in the numerator. Note how $x(x + \Delta x)$ is distributed across the numerator in the second line, separately canceling the denominators found in that numerator. (That is the whole point of that multiplication.)

$$f'(x) = \lim_{\Delta x \to 0} \frac{f(x + \Delta x) - f(x)}{\Delta x} = \lim_{\Delta x \to 0} \frac{\frac{1}{x+\Delta x} - \frac{1}{x}}{\Delta x} = \lim_{\Delta x \to 0} \frac{\frac{1}{x+\Delta x} - \frac{1}{x}}{\Delta x} \cdot \frac{x(x+\Delta x)}{x(x+\Delta x)}$$
$$= \lim_{\Delta x \to 0} \frac{x - (x + \Delta x)}{(\Delta x)(x)(x + \Delta x)} = \lim_{\Delta x \to 0} \frac{-\Delta x}{(\Delta x)(x)(x + \Delta x)} = \lim_{\Delta x \to 0} \frac{-1}{x(x + \Delta x)} = \frac{-1}{x^2}.$$

Summarizing, $f(x) = \dfrac{1}{x} \implies f'(x) = -\dfrac{1}{x^2}$.

We see some interesting features of this derivative as well. For instance, it is always negative, so the graph is always sloping downwards, where the derivative exists (and the function is defined).

Furthermore, it is the same at $x = a$ as $x = -a$, which is a result of the symmetry. Finally, we note that $f'(x) \to 0$ as $x \to \pm\infty$, and $f'(x) \to -\infty$ as $x \to 0$. This is indeed reflected in the graph in Figure 3.4.

3.1.4 Standard Types of Difference-Quotient Derivative Computations

Since the derivative is usually a limit of 0/0 form, in computing the derivative we can expect to require many of the same techniques we used in Chapter 2 for 0/0-form limits, though the term that makes the denominator zero in the limit will be the Δx, Δt, or similar variable, and the variable x, t, or similar will be treated as a *constant* in the limit computations.

In an introductory section such as this, the types of derivatives we compute will tend to be from functions of a limited number of basic types, though the computations can be somewhat complicated. The types we will encounter in this section tend to be *algebraic* in nature, meaning they involve polynomials, radicals, and fractions thereof. It is useful to take note of the typical techniques (some would call them "tricks") and when to use them. They do indeed mirror the techniques used for 0/0-form limits from the previous chapter. We show some of the more involved examples of standard types here.

Polynomials: For each of these, we simply multiply ("expand") and simplify the polynomials in the numerator of the difference quotient, after which a factor Δx can be extracted to cancel with the denominator:

$$f(x) = x^3 - 9x \implies f'(x) = \lim_{\Delta x \to 0} \frac{f(x + \Delta x) - f(x)}{\Delta x}$$
$$= \lim_{\Delta x \to 0} \frac{\left[(x + \Delta x)^3 - 9(x + \Delta x)\right] - \left[x^3 - 9x\right]}{\Delta x}$$
$$= \lim_{\Delta x \to 0} \frac{x^3 + 3x^2 \Delta x + 3x(\Delta x)^2 + (\Delta x)^3 - 9x - 9\Delta x - x^3 + 9x}{\Delta x}$$
$$= \lim_{\Delta x \to 0} \frac{3x^2 \Delta x + 3x(\Delta x)^2 + (\Delta x)^3 - 9\Delta x}{\Delta x}$$
$$= \lim_{\Delta x \to 0} \frac{(\Delta x)\left[3x^2 + 3x \Delta x + (\Delta x)^2 - 9\right]}{\Delta x}$$
$$= \lim_{\Delta x \to 0} \left[3x^2 + 3x \Delta x + (\Delta x)^2 - 9\right]$$
$$= 3x^2 - 9.$$

Rational functions: For these, our difference quotients will have "fractions within fractions," which we simplify by multiplying by a common denominator (preferably the least common multiple) of every denominator within the overall fraction (as seen in

3.1. INTRODUCTION TO DERIVATIVES

the second line):

$$f(x) = \frac{2x}{3x-5} \implies f'(x) = \lim_{\Delta x \to 0} \frac{f(x+\Delta x) - f(x)}{\Delta x}$$

$$= \lim_{\Delta x \to 0} \frac{\frac{2(x+\Delta x)}{3(x+\Delta x)-5} - \frac{2x}{3x-5}}{\Delta x} \cdot \frac{[3(x+\Delta x)-5][3x-5]}{[3(x+\Delta x)-5][3x-5]}$$

$$= \lim_{\Delta x \to 0} \frac{2(x+\Delta x)(3x-5) - 2x[3(x+\Delta x)-5]}{(\Delta x)[3(x+\Delta x)-5][3x-5]}$$

$$= \lim_{\Delta x \to 0} \frac{2[3x^2 - 5x + 3x\Delta x - 5\Delta x] - 2x(3x + 3\Delta x - 5)}{(\Delta x)[3x + 3\Delta x) - 5][3x-5]}$$

$$= \lim_{\Delta x \to 0} \frac{6x^2 - 10x + 6x\Delta x - 10\Delta x - 6x^2 - 6x\Delta x + 10x}{(\Delta x)[3(x+\Delta x)-5][3x-5]}$$

$$= \lim_{\Delta x \to 0} \frac{-10\Delta x}{(\Delta x)[3(x+\Delta x)-5][3x-5]}$$

$$= \lim_{\Delta x \to 0} \frac{-10}{[3(x+\Delta x)-5][3x-5]} = \frac{-10}{(3x-5)^2}.$$

Functions with radicals: If our difference quotient has a difference of radicals, multiplying by the conjugate—or whatever missing factor will remove the radicals—will usually allow us, after simplification, to find a factor Δx to cancel with the same factor in the denominator. That step is included in the second line, where we use the fact that $(\alpha - \beta)(\alpha + \beta) = \alpha^2 - \beta^2$, or more specifically, $\left(\sqrt{a} - \sqrt{b}\right)\left(\sqrt{a} + \sqrt{b}\right) = a - b$:

$$f(x) = \sqrt{1-2x} \implies f'(x) = \lim_{\Delta x \to 0} \frac{f(x+\Delta x) - f(x)}{\Delta x}$$

$$= \lim_{\Delta x \to 0} \frac{\sqrt{1-2(x+\Delta x)} - \sqrt{1-2x}}{\Delta x} \cdot \frac{\sqrt{1-2(x+\Delta x)} + \sqrt{1-2x}}{\sqrt{1-2(x+\Delta x)} + \sqrt{1-2x}}$$

$$= \lim_{\Delta x \to 0} \frac{[1-2(x+\Delta x)] - [1-2x]}{(\Delta x)\left[\sqrt{1-2(x+\Delta x)} + \sqrt{1-2x}\right]}$$

$$= \lim_{\Delta x \to 0} \frac{1 - 2x - 2\Delta x - 1 + 2x}{(\Delta x)\left[\sqrt{1-2(x+\Delta x)} + \sqrt{1-2x}\right]}$$

$$= \lim_{\Delta x \to 0} \frac{-2\Delta x}{(\Delta x)\left[\sqrt{1-2(x+\Delta x)} + \sqrt{1-2x}\right]}$$

$$= \lim_{\Delta x \to 0} \frac{-2}{\sqrt{1-2(x+\Delta x)} + \sqrt{1-2x}} = \frac{-2}{2\sqrt{1-2x}} = \frac{-1}{\sqrt{1-2x}}.$$

Other derivative computations from the definition may use combinations of these techniques, or somewhat more involved but similar techniques, such as using the factorization $\alpha^3 - \beta^3 = (\alpha - \beta)(\alpha^2 + \alpha\beta + \beta^2)$, but more in the form of

$$a - b = \left(\sqrt[3]{a} - \sqrt[3]{b}\right)\left(\left(\sqrt[3]{a}\right)^2 + \left(\sqrt[3]{a}\right)\left(\sqrt[3]{b}\right) + \left(\sqrt[3]{b}\right)^2\right),$$

and thus

$$\sqrt[3]{a} - \sqrt[3]{b} = \left(\sqrt[3]{a} - \sqrt[3]{b}\right) \cdot \frac{\left(\sqrt[3]{a}\right)^2 + \left(\sqrt[3]{a}\right)\left(\sqrt[3]{b}\right) + \left(\sqrt[3]{b}\right)^2}{\left(\sqrt[3]{a}\right)^2 + \left(\sqrt[3]{a}\right)\left(\sqrt[3]{b}\right) + \left(\sqrt[3]{b}\right)^2} = \frac{a-b}{\left(\sqrt[3]{a}\right)^2 + \left(\sqrt[3]{a}\right)\left(\sqrt[3]{b}\right) + \left(\sqrt[3]{b}\right)^2},$$

to compute $f'(x)$ if $f(x) = \sqrt[3]{x}$.

While these derivative computations may seem very involved, fortunately in later sections we will see how we can use some very general results for quickly computing the derivative $f'(x)$ from the form of the function $f(x)$, so that we rarely have to return to the definition of the derivative, as in Equation (3.3), page 212, in order to compute it. However, there are indeed times when this exact nature of the derivative as a particular limit is crucial, and so (1) its limit definition should be memorized, and (2) its exact nature should always be kept in mind, for those cases where the shortcuts we learn later might not apply. Just as being aware of the deeper nature of limits could keep one from making naive mistakes in that context, we should expect the same with derivatives, especially since derivatives are, by definition, special cases of limits.

3.1.5 Marginal Cost

As we will see, the derivative has many applications. While geometrically it represents the slope of the curve of a function, we have already seen it can represent any instantaneous rate of change of any quantity $Q(x)$ with respect to its input variable x. A very good illustration is an object's velocity $v(t) = s'(t)$ when the function $s(t)$ is its position and t is time, but another very different example was $V'(D)$ measuring the change of volume V in a vehicle's fuel tank as its total distance traveled D varies. Nearly any field that deals with numerical quantities has some use for the derivative, which measures rates at which those quantities change with respect to each other. One such field is business economics.

For example, in economics the precise definition of *marginal cost* is the cost of the *next* item, or $(x+1)$st item, after x items are already produced and their costs for production already paid. In function notation, if $C(x)$ is the total cost of producing x items, then (on reflection) it is clear that the marginal cost at that level—that is, the cost of the $(x+1)$st item—would be $C(x+1) - C(x)$. One might assume that with the infrastructure required to produce the first x items already in place, this next item would be relatively inexpensive—though probably not free—to produce. However, there are exceptions to this, where it could be much more expensive than the previous items if more infrastructure is suddenly needed to produce that next item. For instance, if one jetliner can carry a maximum of 400 passengers, the 400th may be nearly free for the airline to seat and fly (after the previous 399 are provided for), but the 401st would likely be very expensive since another plane is required (even if ultimately the 401 passengers would be divided more evenly between the planes). That would be a case where $C(401) - C(400)$ is quite large.

In practice, rather than computing the actual marginal cost $C(x+1) - C(x)$, a proxy (substitute) for this is often used, usually the instantaneous rate of change of total cost $C(x)$ with respect to the number of units x, that is, $C'(x)$, where $C(x)$ is a function whose formula seems to make computational sense for x on actual intervals (and not just at nonnegative integer values). Thus, many textbooks instead make the following definition (without the parenthetical):[13]

---

[13] In many fields, there will be a precise definition and a different, "working definition," which itself may change from context to context. For example, in statistics a numerical datum said (as a working definition) to be in the 75th percentile is often understood to be one for which 75 percent of the other data are below it. Elsewhere being in the 75th percentile is described as meaning 25 percent of the other data are above it. However, these cannot

3.1. INTRODUCTION TO DERIVATIVES

Definition 3.1.6 *If $C(x)$ is the total cost function for producing x items, then the (proxy)* **marginal cost function** *is given by $C'(x)$.*

This assumes $C'(x)$ makes sense as a function, usually requiring some formula for $C(x)$. Then $C'(x)$ will have dimensions which are units of money per units of items produced; if C is in dollars and x is in items produced, then $C'(x)$ will be in dollars per item. This is because the limit inherits the units from the quotient in the definition

$$C'(x) = \lim_{\Delta x \to 0} \frac{C(x + \Delta x) - C(x)}{\Delta x}.$$

In contrast, the units of the actual (non-proxy) marginal cost $C(x+1) - C(x)$ will be in dollars only, though since it is understood to be for one item (or one unit of items), it can be still considered a dollar amount "per (the next) item."

Example 3.1.5

Consider the cost function $C(x) = -3x^2 + 90x + 500$, representing the cost (in hundreds of dollars) of manufacturing x cases of a particular product. Assume this function is valid for $0 \le x \le 25$. Find the average rate of change of cost per case as x ranges from 10 to 15 cases. Also find the instantaneous rate of change in cost when 10 cases are manufactured, i.e., the (proxy) marginal cost when $x = 10$.

<u>Solution</u>: The average rate of change of cost per case on $[10, 15]$ will be given by the difference quotient $(C(15) - C(10))/(15 - 10)$ (see Equation (3.4) on page 212). We might first compute

$$C(15) = -3(15)^2 + 90(15) + 500 = 1175,$$
$$C(10) = -3(10)^2 + 90(10) + 500 = 1100,$$

and so the average rate of change of cost per case for $10 \le x \le 15$ is

$$\frac{C(15) - C(10)}{15 - 10} = \frac{1175 - 1100}{5} = \frac{75}{5} = 15.$$

both be exact because if 75 percent are below and 25 percent are above, when we include the datum in question, we would have over 100 percent of our data, which is impossible. If the data set is very large, so that one datum is much less than 1 percent of the total observations, the discrepancy may be minor enough to ignore. For a much smaller sample the discrepancy is large. (The third highest of four data could be described as being in the 50th percentile by one definition, and 75th in the other.)

Often, one measurement is used as a proxy for the measurement that matches the exact definition, and if that proxy is used more and more, it is sometimes given as *the* definition. This is indeed the case with marginal cost, as there are textbooks that give only the proxy, derivative definition. The two definitions are connected by the approximation that follows, in which the precise definition of marginal cost at level x is on the left-hand side, and its usual proxy, which is often given as *the* definition, is on the right in the following:

$$\underbrace{C(x+1) - C(x)}_{\text{exact definition}} = \frac{C(x+1) - C(x)}{1} \approx \underbrace{C'(x)}_{\text{proxy}}.$$

While this may be somewhat confusing (or simply unsatisfying), it is similar to the idea that, if $s'(t_1) = 5$ ft/sec, we expect to displace by approximately 5 ft in the next second after t_1, though the more exact answer would be $s(t_1 + 1 \text{ sec}) - s(t_1)$. In a factory producing 10,000 cars, it is reasonable to assume the approximation using $C'(x)$ is quite good, assuming a reasonable formula for $C(x)$ that makes sense algebraically on intervals, while a defense contractor producing a small number of stealth fighter jets would likely use the exact definition $C(x+1) - C(x)$.

With the derivative rules we will develop as this chapter unfolds, we will see that it is often easier and more efficient to compute $s'(t_1)$ than to compute the difference between $s(t)$-values at the end and start of the interval $t \in [t_1, t_1 + 1 \text{ sec}]$. That is usually not the case here, where we find derivatives from the limit definition.

Since C is in hundreds of dollars, and the quantities in the denominator are in cases of product, it follows that the units of this final quantity will be in hundreds of dollars per case. We could report our answer as 15 hundred dollars per case, or \$1,500/case.

To compute the instantaneous rate at $x = 10$, we will first compute $C'(x)$ in the abstract and then input $x = 10$. As usual, there will be much cancellation in the numerator until we can extract a factor Δx to cancel with that term in the denominator. Until that term is canceled, all limits are of the form 0/0, as is usual in derivative computations that use the limit definition of the derivative.

$$\begin{aligned} C'(x) &= \lim_{\Delta x \to 0} \frac{C(x+\Delta x) - C(x)}{\Delta x} \\ &= \lim_{\Delta x \to 0} \frac{\left[-3(x+\Delta x)^2 + 90(x+\Delta x) + 500\right] - \left[-3x^2 + 90x + 500\right]}{\Delta x} \\ &= \lim_{\Delta x \to 0} \frac{\left[-3\left(x^2 + 2x\Delta x + (\Delta x)^2\right) + 90x + 90\Delta x + 500\right] + 3x^2 - 90x - 500}{\Delta x} \\ &= \lim_{\Delta x \to 0} \frac{-3x^2 - 6x\Delta x - 3(\Delta x)^2 + 90x + 90\Delta x + 500 + 3x^2 - 90x - 500}{\Delta x} \\ &= \lim_{\Delta x \to 0} \frac{-6x\Delta x - 3(\Delta x)^2 + 90\Delta x}{\Delta x} \\ &= \lim_{\Delta x \to 0} \frac{(\Delta x)[-6x - 3\Delta x + 90]}{\Delta x} \\ &= \lim_{\Delta x \to 0} [-6x - 3\Delta x + 90] = -6x + 90 \end{aligned}$$

Thus $C'(10) = -6(10) + 90 = 30$. Because of the form of the limit, since $C(x + \Delta x)$ and $C(x)$ are in hundreds of dollars and Δx is in cases (of product), these quotients are in units of hundreds of dollars per case, and so therefore is the limit $C'(x)$. Thus $C'(10) = 30$ hundreds of dollars/case, and so we can instead report that the instantaneous rate of change at $x = 10$ cases is \$3000/case.

This derivative computation may seem tedious, but fortunately techniques of later sections will make the derivation

$$C(x) = -3x^2 + 90x + 500 \implies C'(x) = -6x + 90$$

nearly as simple as writing just that. In Exercise 16 of this section it is noted that there is a general formula for the derivative of a function of the form $f(x) = ax^2 + bx + c$, depending only on the coefficients. That and other rules for derivatives will make computations such as this one almost trivial.

We should note that with the exact definition of marginal cost, at $x = 10$, we would compute the marginal cost to be (in hundreds of dollars per unit, i.e., hundreds of dollars per case) given by

$$\begin{aligned} C(11) - C(10) &= \left[-3(11)^2 + 90(11) + 500\right] - \left[-3(10)^2 + 90(10) + 500\right] \\ &= 1127 - 1100 \\ &= 27. \end{aligned}$$

This is reasonably well approximated by our computation of $C'(10) = 30$. Economics being rather far from an exact, predictive science anyhow, using $C'(10) = -6(10) + 90 = 30$ seems

3.1. INTRODUCTION TO DERIVATIVES

like a reasonable proxy for the computation of the exact marginal cost $C(11) - C(10) = 27$ given. From time to time we will make some use out of the idea that for small Δx,

$$\frac{f(x + \Delta x) - f(x)}{\Delta x} \approx f'(x),$$

depending on how "small" Δx is, and on how quickly $[f(x + \Delta x) - f(x)]/\Delta x \to f'(x)$ as $\Delta x \to 0$. While the notation will change, this idea is present in many contexts in this text. A special case of this, then, is with the marginal cost but with $\Delta x = 1$ item:

$$\frac{C(x+1) - C(x)}{1} \approx C'(x).$$

This computation would automatically contain the dimensions of money/item, which is understood when the question is asked, "How much would the next item of product cost?" but flows better from the approximation given (by the presence of the denominator, even if it is simply 1 unit).

It should also be pointed out that there are similar, related quantities in economics, such as total revenue versus marginal revenue, and total profit versus marginal profit. Whether total or marginal, it is assumed that profit is the difference between revenue and cost. Hence one often sees the equations $P = R - C$ and $P' = R' - C'$, relating the total profit, revenue and cost, and also the derivative proxies for the marginal profit, revenue, and cost.

3.1.6 Some Further Applications

In Definition 3.1.2, page 212, we saw that we can consider $Q'(x)$ as the derivative of $Q(x)$ with respect to x, whatever Q and x represent, and $Q'(x)$ is given by (3.3):

$$Q'(x) = \lim_{\Delta x \to 0} \frac{Q(x + \Delta x) - Q(x)}{\Delta x}.$$

The real power of the definition lies in the interpretation and intuition in the meaning of the derivative $Q'(x)$, as the instantaneous rate of change of $Q(x)$ with respect to x. So far, we have had four applications, namley:

- velocity $v(t) = s'(t)$, the instantaneous rate of change in position $s(t)$ with respect to the time t;

- the rate $V'(D)$ of fuel volume change with respect to distance traveled D;

- the slope $f'(x)$ of the graph of $y = f(x)$ at a given value of x, which also represents the change in the height of the function $f(x)$ with respect to the horizontal position x;

- the (proxy) marginal cost $C'(x)$, measuring the rate of change in cost with respect to the number of items produced.

Throughout this chapter and the next we will consider further contexts for the derivative. Indeed, the derivative arguably offers a unique insight into each setting, though all of these insights are ultimately unified mathematically in the limit formula for $Q'(x)$, as the instantaneous rate of change of $Q(x)$ with respect to x.

As with any application problem, some care must be taken to be sure that the natural quantities considered are the same quantities that are usually considered in the common conversation. For instance, recall that if the volume of fuel in a tank is $V(D)$, what we usually call the fuel flow rate is actually $-V'(D)$. (See Example 3.1.2, page 213.)

Example 3.1.6

Suppose a liquid is stored in an inverted conical container where the height of the cone is twice its radius. Find a formula for the instantaneous rate of change of the volume V with respect to the height h of the liquid in the cone.

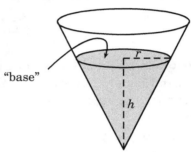

Solution: What we seek here is $V'(h)$, and so we first need to find a formula for $V(h)$. In general, the volume of a cone (such as that represented by the liquid in the tank) is given by $V = \frac{1}{3}\pi r^2 h$, where r is the radius of the "base" of the cone and h is the "height" of the cone. For the whole tank, we would have $h = 2r$, and so by similar triangles this will be the case regardless of the height of the fluid in the cone.

We can use $V = \frac{1}{3}\pi r^2 h$, but substitute for r using $r = \frac{1}{2}h$. Putting these together, we get V as a function of h, namely $V = \frac{1}{3}\pi r^2 h = \frac{1}{3}\pi(h/2)^2 h$, or

$$V(h) = \frac{1}{12}\pi h^3.$$

Now we compute $V'(h)$ in the usual way:

$$V'(h) = \lim_{\Delta h \to 0} \frac{V(h+\Delta h) - V(h)}{\Delta h} = \lim_{\Delta h \to 0} \frac{\frac{1}{12}\pi(h+\Delta h)^3 - \frac{1}{12}\pi h^3}{\Delta h}$$

$$= \lim_{\Delta h \to 0} \left[\frac{\pi}{12} \cdot \frac{h^3 + 3h^2 \Delta h + 3h(\Delta h)^2 + (\Delta h)^3 - h^3}{\Delta h} \right] = \lim_{\Delta h \to 0} \left[\frac{\pi}{12} \cdot \frac{(\Delta h)(3h^2 + 3h(\Delta h) + (\Delta h)^2)}{\Delta h} \right]$$

$$= \lim_{\Delta h \to 0} \left[\frac{\pi}{12}(3h^2 + 3h\Delta h + (\Delta h)^2) \right] = \frac{\pi}{12}(3h^2) = \frac{\pi}{4}h^2.$$

Thus $V'(h) = \frac{1}{4}\pi h^2$.

It is interesting to examine some of the values of $V'(h)$ for various values of h, such as the following:

$$V'(5\,\text{cm}) = \frac{\pi}{4}(5\,\text{cm})^2 = \frac{25\pi}{4}\frac{\text{cm}^3}{\text{cm}} \approx 19.6\,\frac{\text{cm}^3}{\text{cm}};$$

$$V'(2\,\text{cm}) = \frac{\pi}{4}(2\,\text{cm})^2 = \pi\,\frac{\text{cm}^3}{\text{cm}} \approx 3.14\,\frac{\text{cm}^3}{\text{cm}};$$

$$V'\left(\frac{1}{2}\,\text{cm}\right) = \frac{\pi}{4}\left(\frac{1}{2}\,\text{cm}\right)^2 = \frac{\pi}{16}\,\frac{\text{cm}^3}{\text{cm}} \approx 0.196\,\frac{\text{cm}^3}{\text{cm}}.$$

The units first appear as cm², but we write the units as cm³/cm because the rate should be in units of volume change per unit of height change. Note that the volume changes more

3.1. INTRODUCTION TO DERIVATIVES

slowly with respect to height when the height is smaller, but increases faster for larger h. It is also somewhat interesting to note that $V'(h) = \frac{1}{4}\pi h^2 \longrightarrow 0^+$ as $h \to 0^+$, and also that $V'(h) = \frac{1}{4}\pi h^2 \longrightarrow \infty$ as $h \to \infty$, both illustrating this rate of change in the "extreme" cases. Indeed, for larger h, V changes rapidly with h, while for smaller h, V changes more slowly with h. Put another way, when h is large we require a larger change in V to produce the same change in h than when h is small.

Example 3.1.7

Suppose a right triangle has a vertical leg of height 4, and a horizontal leg of variable length x. Let l be the length of the hypotenuse. We compute the derivative of l with respect to x and note some trends.

From the figure on the right it is clear that
$$l^2 = x^2 + 4^2.$$

Writing $l = l(x)$, since $l \geq 0$ we have
$$l(x) = \sqrt{x^2 + 16}.$$

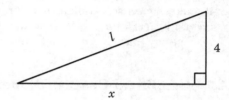

Now we compute $l'(x)$ as usual, with a conjugate multiplication required due to the square root:

$$\begin{aligned}
l'(x) &= \lim_{\Delta x \to 0} \frac{l(x+\Delta x) - l(x)}{\Delta x} = \lim_{\Delta x \to 0} \frac{\sqrt{(x+\Delta x)^2 + 16} - \sqrt{x^2 + 16}}{\Delta x} \\
&= \lim_{\Delta x \to 0} \frac{\sqrt{(x+\Delta x)^2 + 16} - \sqrt{x^2 + 16}}{\Delta x} \cdot \frac{\sqrt{(x+\Delta x)^2 + 16} + \sqrt{x^2 + 16}}{\sqrt{(x+\Delta x)^2 + 16} + \sqrt{x^2 + 16}} \\
&= \lim_{\Delta x \to 0} \frac{[(x+\Delta x)^2 + 16] - [x^2 + 16]}{\Delta x \left[\sqrt{(x+\Delta x)^2 + 16} + \sqrt{x^2 + 16}\right]} = \lim_{\Delta x \to 0} \frac{x^2 + 2x\Delta x + (\Delta x)^2 + 16 - x^2 - 16}{\Delta x \left[\sqrt{(x+\Delta x)^2 + 16} + \sqrt{x^2 + 16}\right]} \\
&= \lim_{\Delta x \to 0} \frac{2x\Delta x + (\Delta x)^2}{\Delta x \left[\sqrt{(x+\Delta x)^2 + 16} + \sqrt{x^2 + 16}\right]} = \lim_{\Delta x \to 0} \frac{(\Delta x)(2x + \Delta x)}{\Delta x \left[\sqrt{(x+\Delta x)^2 + 16} + \sqrt{x^2 + 16}\right]} \\
&= \lim_{\Delta x \to 0} \frac{2x + \Delta x}{\sqrt{(x+\Delta x)^2 + 16} + \sqrt{x^2 + 16}} = \frac{2x}{2\sqrt{x^2 + 16}} \\
\Longrightarrow l'(x) &= \frac{x}{\sqrt{x^2 + 16}}.
\end{aligned}$$

We next compute this for some values of x and note some trends. Let us suppose these lengths are all in feet. Note that the derivatives will be in units of ft/ft, technically dimensionless, though there is meaning, as we will see. (The reader should visualize each case and the meaning of the derivative values or trends for each case.)

- $l'(3) = 3/\sqrt{3^2 + 16} = 3/5$, meaning that when $x = 3$ ft, $l(x)$ is changing at a rate of 3/5 foot per foot of change in x (but only at that exact position of x). In fact, l' will be dimensionless because the difference quotients are all in units of ft/ft.

- $l'(10) = 10/\sqrt{10^2 + 16} = 10/\sqrt{116} \approx 0.9285$.

- As $x \to \infty$, we see $l'(x) = x/\sqrt{x^2 + 16} \to 1$, meaning that when x is large it requires roughly one foot of length enlargement in l for each foot enlargement in x.

- On the other hand, as $x \to 0^+$ we see $l'(x) = x/\sqrt{x^2 + 16} \xrightarrow{0^+/16} 0$, and so when x is small l will change relatively little (as a matter of proportion) as x changes.

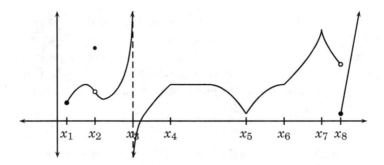

Figure 3.5: Graph of a function for which the derivative does not exist at $x_1, x_2, \ldots, x_7$, as it makes no sense to talk about a finite slope at any of those points. Recall that the slope is given by $f'(x) = \lim_{\Delta x \to 0}[(f(x+\Delta x)-f(x))/(\Delta x)]$, and in that limit we have $\Delta x \to 0^{\pm}$. Recall also we say that $f'(x)$ does not exist if this limit is not finite.

3.1.7 Graphical Cases of Nonexistent Derivatives

Before we leave this section, a few graphical examples where derivatives do not exist at particular points can be illustrative. Recall that the derivative $f'(x)$ of a function $f(x)$ gives the value of the *slope* of the graph at the point $(x, f(x))$. Also recall that for us to say $f'(x)$ exists, we require $f'(x) = \lim_{\Delta x \to 0}[(f(x+\Delta x)-f(x))/(\Delta x)]$ to exist as a finite number. In such a case, there is a tangent line to the graph at $(x, f(x))$.

In the examples in Figure 3.5, we have points x_i where it would be impossible to define a finite slope at that point, and so there would be no equation of a tangent line there with finite slope. At x_1 we cannot have $\Delta x \to 0^-$. At x_2, the left and right limits would not be finite. At x_3, we have $f(x_3)$ undefined. At x_4, the slope from the left is positive (as $\Delta x \to 0^-$) but the slope from the right (as $\Delta x \to 0^+$) is zero, and so $f'(x)$ DNE. At x_5, x_6, and x_7, we similarly have disagreement from the two sides, and at x_8 we have that and more, in that the left-side limit is not finite.

In all cases of the points x_1 through x_8 in the figure, we would be hard-pressed to draw reasonable tangent lines at any of those eight indexed points. In the next section and beyond we will see further implications of the existence of the derivative.

This concludes this introductory section on the nature of the derivative as a measure of how a function changes compared to its input. In later sections we will see more applications, as well as rules for greatly simplifying the process of computing the derivative.

Exercises

For problems 1–14, use the definition to compute $f'(x)$ for the given function:

$$f'(x) = \lim_{\Delta x \to 0} \frac{f(x+\Delta x)-f(x)}{\Delta x}.$$

1. $f(x) = 5 - 2x$.

2. $f(x) = 10$.

3. $f(x) = 2x^2 + 3$.

3.1. INTRODUCTION TO DERIVATIVES

4. $f(x) = 3x^2 - 5x + 9$.

5. $f(x) = \sqrt{x}$.

6. $f(x) = \dfrac{3}{x+2}$.

7. $f(x) = \sqrt{9-5x}$.

8. $f(x) = \dfrac{1}{x^2}$.

9. $f(x) = \dfrac{2}{\sqrt{x}}$.

10. $f(x) = 2x^3$. (Recall $(a+b)^3 = a^3 + 3a^2b + 3ab^2 + b^3$.)

11. $f(x) = \sqrt[3]{x+1}$. (Hint: We made use of the difference of two squares in Example 3.1.3, page 216. Here you will need to use the difference of two cubes in a similar manner, i.e., writing $a^3 - b^3 = (a-b)(a^2 + ab + b^2)$.)

12. $f(x) = x^4$. (Hint: What is the general formula for $(a+b)^4$?)

13. $f(x) = \dfrac{x}{x+1}$. (Hint: This is easier if $f(x)$ is rewritten first using long division.)

14. $f(x) = \dfrac{x+1}{x-1}$.

15. Suppose $s(t) = -16t^2 + 15t + 20$ describes the height of a projectile in free fall. (Here t is in seconds and s is in feet.)

    a. Find the velocity function $v(t)$, using (3.1), page 210.
    b. What is the projectile's velocity when $t=0$? $t=10$?
    c. Find t so that the projectile is stationary (i.e., $v=0$).
    d. How high is the projectile when it is stationary?

16. Consider the general quadratic function $f(x) = ax^2 + bx + c$.

    a. Use the definition of derivative to find a formula for the derivative of the general quadratic function $f(x) = ax^2 + bx + c$.
    b. Assuming $a \neq 0$, this represents a parabola. Assuming also that the slope is zero at the vertex, find a general formula for the x-coordinate of the vertex.
    c. Find a general formula for the y-coordinate of the vertex, and thus a formula for the point (x,y) at the vertex.

For Exercises 17–20, find the tangent line to the graph at the given point $x=a$ for the given function. (See (3.8), page 215, and the examples following.)

17. $f(x) = x^2 - 9$, $a = 4$.

18. $f(x) = x^3$, $a = -1$.

19. $f(x) = \sqrt{x}$, $a = 9$.

20. $f(x) = \dfrac{1}{x}$, $a = \dfrac{1}{10}$.

21. The cost (in hundreds of dollars) from manufacturing x cases of a product is $C(x) = 8x - 0.2x^2$, for $x \in [0, 20]$.

    a. Find the average rate of change of cost between when 10 and 20 cases are manufactured.
    b. Find the marginal cost when 15 cases are manufactured.
    c. Do the same for 18 cases.

22. Find $f'(x)$ if $f(x) = x^{3/2}$. (Hint: Rewrite as $f(x) = \sqrt{x^3}$.)

23. Find $f'(x)$ if $f(x) = x^{2/3}$.

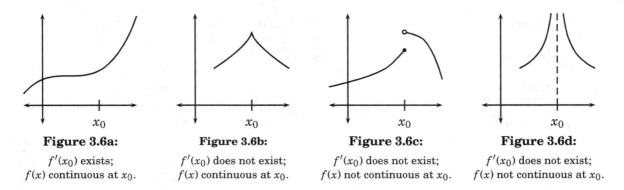

Figure 3.6a:
$f'(x_0)$ exists;
$f(x)$ continuous at x_0.

Figure 3.6b:
$f'(x_0)$ does not exist;
$f(x)$ continuous at x_0.

Figure 3.6c:
$f'(x_0)$ does not exist;
$f(x)$ not continuous at x_0.

Figure 3.6d:
$f'(x_0)$ does not exist;
$f(x)$ not continuous at x_0.

Figure 3.6: Graphs illustrating Theorem 3.2.1, page 229, namely ($f'(x_0)$ exists) $\implies$ ($f(x)$ is continuous at x_0). In Figure 3.6a, we have differentiability at x_0, and the continuity at x_0 follows. In Figure 3.6b, our theorem is vacuously true, also illustrating that a function can be continuous at x_0 but not have a derivative there. Figure 3.6c and Figure 3.6d are similar logical cases, not only being again vacuous cases of the theorem as stated, but also being useful illustrations of its contrapositive: *If $f(x)$ is not continuous at $x = x_0$, then $f'(x_0)$ does not exist.* (That is also stated as (3.9), page 229.)

3.2 First Differentiation Rules; Leibniz Notation

In this section, we derive rules that let us easily *differentiate*, i.e., compute the derivative function $f'(x)$ of, any polynomial function $f(x)$, and do the same with the trigonometric functions $\sin x$ and $\cos x$. Along the way we will derive a few general (though not comprehensive) rules for derivatives. We will also introduce the very powerful *Leibniz notation* for derivatives, and show how knowing the derivative helps us to further analyze a function. One consequence is that we can more accurately predict and visualize a function's behavior, and better sketch its graph by hand if we like. But first we note an important theorem.

3.2.1 Differentiability Implies Continuity

It should seem reasonable that in order to be able to discuss the slope of the graph of $y = f(x)$ at $x = a$, the function should at least be continuous there. Indeed, it is difficult to imagine a graph of a function which has a discontinuity at $x = a$ but nonetheless has a definable slope there, because slope should give some idea of the function's "direction" both left and right of the point on the curve. Try as we might, reconciling slope with any kind of discontinuity (e.g., jump discontinuities, vertical asymptotes, points where a function is undefined) seems doomed to fail, particularly if we wish to imagine a meaningful tangent line $y = f(a) + f'(a)(x - a)$ with finite slope $f'(a)$ at that point. (Recall $f'(a)$ is said to exist if and only if the limit by which it is defined both exists and is finite.) Fortunately it is a provable fact that a function must be continuous in order for it to have a finite slope:[14]

---

[14]There are notions of left and right derivatives, just as there are notions of left and right continuity, and left and right limits. For left and right derivatives, we simply take $\Delta x \to 0^-$ and $\Delta x \to 0^+$, respectively. As one might suspect, the (usual) derivative exists exactly when both left and right derivatives exist and are equal. Furthermore, a left derivative's existence implies left continuity, and the same is true with "right" replacing "left." We will see left and right derivatives later in the section and briefly in later chapters, particularly Chapter 5.

3.2. FIRST DIFFERENTIATION RULES; LEIBNIZ NOTATION

Theorem 3.2.1 $\boxed{(f'(a)\ exists) \implies (f(x)\ is\ continuous\ at\ x = a)}$.

Proof: Suppose $f'(a)$ exists, meaning that $f'(a) = \lim\limits_{\Delta x \to 0} \dfrac{f(a + \Delta x) - f(a)}{\Delta x} \in \mathbb{R}$, i.e., the limit exists as a (finite) real number. We need to show that this implies $f(x)$ is continuous at $x = a$, which we do by proving instead the equivalent statement that $\lim\limits_{x \to a} f(x) = f(a)$ (Theorem 2.4.2, page 124).

First we re-write this limit using the substitution $x = a + \Delta x$, which gives $x \to a \iff \Delta x \to 0$ properly. Then we perform a somewhat clever algebraic expansion of the expression within the limit by subtracting and adding $f(a)$ (which exists and is real or this previous limit defining $f'(a)$ could not exist and be finite), and eventually divide and multiply by Δx, to get the following (one "key" being $f'(a)$ is a real, and therefore finite, number by hypothesis):

$$\lim_{x \to a} f(x) = \lim_{\Delta x \to 0} f(a + \Delta x) = \lim_{\Delta x \to 0} [f(a + \Delta x) - f(a) + f(a)]$$

$$= \lim_{\Delta x \to 0} \left[\underbrace{\frac{f(a + \Delta x) - f(a)}{\Delta x} \cdot \Delta x + f(a)}_{\text{"key"}} \right] = f'(a) \cdot 0 + f(a) = f(a), \qquad \text{q.e.d.}$$

The upshot of the theorem is that a function must be continuous in order to have a derivative, as the contrapositive also makes clear:

$$\boxed{(f(x)\ discontinuous\ at\ x_0) \implies (f'(x_0)\ does\ not\ exist)}. \tag{3.9}$$

However, it is possible to have a function be continuous at x_0 but not have a derivative there, as shown in Figure 3.6b, page 228.

The theorem's importance makes itself known on occasion. In coming sections, we will be more focused on differentiability, i.e., whether a function's derivative exists. If the function is discontinuous, there is no need to look for its derivative. Also, if we know a function's derivative, then we know the function is continuous. It is a useful insight that differentiability is a stronger condition than continuity, just as continuity is a stronger condition than the function simply being defined. These facts are stated in the implications that follow, along with their contrapositive forms:

$$f'(x_0)\ exists \implies f(x)\ is\ continuous\ at\ x = x_0 \implies f(x_0)\ is\ defined;$$
$$f(x_0)\ undefined \implies f(x)\ discontinuous\ at\ x = x_0 \implies f'(x_0)\ does\ not\ exist.[15]$$

It is also useful to consider a position function $s(t)$ and its velocity function $v(t) = s'(t)$. It is perhaps difficult to visualize a discontinuous position function, absent such phenomena as teleportation or imagination-inspiring quantum mechanical effects. In such cases, we do not expect to have a finite-valued velocity either (and we only consider finite-valued derivatives anyway). Or, as is logically equivalent, it seems that if there is a (finite and) definable velocity, the position should only vary continuously.

---

[15] Recall when we write $\mathscr{P} \implies \mathscr{Q} \implies \mathscr{R}$, this is a shorthand, which means that we have essentially three valid implications: $\mathscr{P} \implies \mathscr{Q}$, and $\mathscr{Q} \implies \mathscr{R}$, and (thus) $\mathscr{P} \implies \mathscr{R}$.

3.2.2 Positive Integer Power Rule

We will often be interested in finding derivatives of functions $f(x) = x^n$. Fortunately, there is a simple rule that covers all such functions. It is usually called the *Power Rule*, as stated next, for the case $n \in \mathbb{N} = \{1, 2, 3, 4, 5, 6, \ldots\}$ (that is, *natural numbers n*). Here we interpret x^0 to be the function 1, though for $x \leq 0$ that has some theoretical difficulties. In this context, though, we get a true statement if we interpret $x^0 = 1$.

Theorem 3.2.2 $(f(x) = x^n) \wedge (n \in \mathbb{N}) \implies f'(x) = n \cdot x^{n-1}$.

Note that implicit in this theorem is that the derivative of x^n exists for every $x \in \mathbb{R}$—i.e., "exists everywhere"—since it is equal to nx^{n-1}, defined everywhere. Before giving the proof, we list some quick results.

Example 3.2.1

Here we list the derivatives of some of the positive integer powers of x. The first case listed below ($n = 1$) does follow from the proof, though we would be reading the statement of the theorem for that case $f(x) = x^1 \implies f'(x) = 1x^0 = 1$. Again, we do not really wish to say $x^0 = 1$ regardless of x, for several technical reasons (though it is fine as long as $x > 0$), but we see how the formula naively gives us what we want for $n = 1$. Note also that $x^1 = x$. The rest of the table is more straightforward:

If $f(x) =$	x	x^2	x^3	x^4	$\cdots$	x^{100}	$\cdots$
then $f'(x) =$	1	$2x$	$3x^2$	$4x^3$	$\cdots$	$100x^{99}$	$\cdots$

Next we give the proof of the (Positive Integer) Power Rule.

Proof: The proof we give here depends on the binomial expansion (which we do here), or Pascal's Triangle (which gives the same numbers as coefficients of the powers of x). It is important to remember that x is a fixed number in the limit, and Δx is the variable approaching zero as far as the limit is concerned. With that, and the binomial expansion or Pascal's Triangle in mind, we compute (for $n \in \mathbb{N}$) that $f(x) = x^n$ implies that

$$f'(x) = \lim_{\Delta x \to 0} \frac{f(x + \Delta x) - f(x)}{\Delta x} = \lim_{\Delta x \to 0} \frac{(x + \Delta x)^n - x^n}{\Delta x}$$

$$= \lim_{\Delta x \to 0} \frac{\left(x^n + nx^{n-1}\Delta x + \frac{n(n-1)x^{n-2}(\Delta x)^2}{1 \cdot 2} + \cdots + nx(\Delta x)^{n-1} + (\Delta x)^n\right) - x^n}{\Delta x}$$

$$= \lim_{\Delta x \to 0} \frac{nx^{n-1}\Delta x + \frac{(n)(n-1)}{1 \cdot 2}x^{n-2}(\Delta x)^2 + \cdots + nx(\Delta x)^{n-1} + (\Delta x)^n}{\Delta x}$$

$$= \lim_{\Delta x \to 0} \left(nx^{n-1} + \frac{(n)(n-1)}{1 \cdot 2}x^{n-2}(\Delta x)^1 + \cdots + nx(\Delta x)^{n-2} + (\Delta x)^{n-1}\right)$$

$$= nx^{n-1} + 0 + \cdots + 0 = nx^{n-1}, \quad \text{q.e.d.}$$

The only term that survives in the limit in the fourth line is the nx^{n-1} term because the others have positive integer powers of Δx, which is approaching zero. We will see later that this Power Rule is actually much more general. In fact, it can be used for $n \in \mathbb{R}$ but we need some more advanced methods to prove such generality. For now, we will apply it only to $n \in \mathbb{N}$.

3.2. FIRST DIFFERENTIATION RULES; LEIBNIZ NOTATION

3.2.3 Leibniz Notation

We will find that other derivative rules can be unwieldy and sometimes ambiguous to write with our present notation. Thus we will introduce the very powerful Leibniz notation and use it, except in a few settings where our present (prime) notation is simpler.[16]

Definition 3.2.1 $\boxed{\dfrac{d}{dx}f(x) = f'(x).}$

Mimicking actual fractions, this is also written $\dfrac{df(x)}{dx}$ and sometimes shortened to df/dx when it is clearly understood that f is a function of x.

The symbol $\dfrac{d}{dx}$ is a *differential operator*, which takes a function of x and returns the function's derivative *with respect to x*. Thus $\dfrac{d}{dx}[\ ]$ takes a function as its input, and returns a (derivative) function as its output. When we are interested in position and velocity, we can write $s'(t) = \dfrac{ds(t)}{dt}$, or when it is understood that $s = s(t)$, we might simply write

$$v = \frac{ds}{dt}. \tag{3.10}$$

The notation naturally resembles difference quotients. With our limit definition of derivatives, where $\Delta f(x) = f(x+\Delta x) - f(x)$ and $\Delta s = s(t+\Delta t) - s(t)$, when $\Delta x \to 0$ and $\Delta t \to 0$, respectively, we have in the limits that

$$\frac{\Delta f(x)}{\Delta x} \longrightarrow \frac{df(x)}{dx}, \qquad \text{and} \qquad \frac{\Delta s(t)}{\Delta t} \longrightarrow \frac{ds(t)}{dt}.$$

It is similar for any such related quantities. With Leibniz notation, our Power Rule becomes:

$$\boxed{\frac{d}{dx}(x^n) = nx^{n-1}.} \tag{3.11}$$

If we would like to compute $f'(a)$, i.e., the derivative at a particular point, in the Leibniz notation and elsewhere, we often use a vertical line which is read, "evaluated at," as in

$$\boxed{f'(a) = \left.\frac{d}{dx}f(x)\right|_{x=a}.}$$

So, for example, $\dfrac{dx^5}{dx} = 5x^4$, and the slope at $x = 1$ of the function $f(x) = x^5$ is given by[17]

$$f'(1) = 5 \cdot 1^4 = 5, \qquad \text{or} \qquad \left.\frac{dx^5}{dx}\right|_{x=1} = 5x^4\Big|_{x=1} = 5 \cdot 1^4 = 5.$$

---

[16] Named for Gottfried Wilhelm Leibniz (July 1, 1646–November 14, 1716), a German mathematician and philosopher. Most credit him and English philosopher, mathematician, physicist, and theologian Sir Isaac Newton (December 25, 1642–March 20, 1727) with independently discovering calculus. Much is written about rivalries between the "Newton camp" and the "Leibniz camp," regarding who discovered what first. Newton's notation for derivative used a dot above the function, as in $ds/dt = \dot{s}(t)$, a notation still used in some physics textbooks.

[17] Often, the "$x =$" is omitted when the variable is obvious, as in $\left.\dfrac{dx^5}{dx}\right|_1 = 5x^4\Big|_1 = 5 \cdot 1^4 = 5.$

The Leibniz notation is often written as if it behaves like a fraction, e.g., $\frac{d}{dx}\left(x^5\right) = \frac{dx^5}{dx}$, but we should be clear that the "d" in the numerator, and (separately) the "dx" in the denominator are treated as indivisible; we do not break those terms up further. In fact, for now we should consider "$\frac{d}{dx}$" to be a single, indivisible operator.

That said, note the flexibility of the Leibniz notation in the following, using operators $\frac{d}{dx}$, $\frac{d}{du}$ and $\frac{d}{dt}$, respectively:

$$\frac{dx^3}{dx} = 3x^2, \qquad \frac{du^3}{du} = 3u^2, \qquad \frac{dt^3}{dt} = 3t^2.$$

These are actually the same rule (with different variables): that the cube of a quantity changes with respect to that quantity at the (instantaneous) rate of 3 times the square of the quantity. For instance, if the horizontal axis is given by t and we graph the height t^3 on the vertical axis, then the slope of the curve at any point (t, t^3) on the curve is always $3t^2$. Similarly if the variable is, x, u, or any other variable.

The Leibniz notation also keeps us from making the mistake of trying to use the derivative rules (such as the Power Rule) to compute, for example, $\frac{du^3}{dx}$. Since the variables (u and x) do not match, the Power Rule cannot be used directly.[18]

With the Power Rule (3.11) and a few other results we can quickly calculate the derivatives of polynomials. Much of this chapter will be devoted to calculating derivatives using known rules, which save an enormous amount of time when compared to calculating derivatives using limits of difference quotients as in the previous section.

3.2.4 Sum and Constant Derivative Rules

Theorem 3.2.3 (Sum Rule) *Suppose* $\dfrac{d}{dx}f(x)$ *and* $\dfrac{d}{dx}g(x)$ *exist. Then*

$$\boxed{\frac{d}{dx}(f(x) + g(x)) = \frac{d}{dx}f(x) + \frac{d}{dx}g(x).} \qquad (3.12)$$

In other words the derivative of a sum is the sum of the respective derivatives. Before we give the proof, it is worth mentioning that some texts write this using the prime notation:

$$(f + g)' = f' + g'.$$

---

[18]Later in the text we will have the *Chain Rule*, which helps us get around the problem of computing $\frac{du^3}{dx}$, and similar derivatives where the variable in the numerator does not match the variable of the denominator. There we will see some of the true power of the Leibniz notation, as we compute for instance

$$\frac{du^3}{dx} = \frac{du^3}{du} \cdot \frac{du}{dx} = 3u^2 \cdot \frac{du}{dx}.$$

Notice how we apparently multiplied and divided by du to achieve the second expression. While it is nontrivial to prove the validity of such seemingly algebraic manipulations, their beauty and analytical power are features of the Leibniz notation.

3.2. FIRST DIFFERENTIATION RULES; LEIBNIZ NOTATION

Proof: Assume that $\frac{d}{dx}f(x)$ and $\frac{d}{dx}g(x)$ exist at a particular value x. Then from the definition of derivative and some rearrangement of terms, we can compute

$$\begin{aligned}\frac{d}{dx}(f(x)+g(x)) &= \lim_{\Delta x\to 0}\frac{(f(x+\Delta x)+g(x+\Delta x))-(f(x)+g(x))}{\Delta x}\\ &= \lim_{\Delta x\to 0}\frac{f(x+\Delta x)-f(x)+g(x+\Delta x)-g(x)}{\Delta x}\\ &= \lim_{\Delta x\to 0}\left(\frac{f(x+\Delta x)-f(x)}{\Delta x}+\frac{g(x+\Delta x)-g(x)}{\Delta x}\right)\\ &= \lim_{\Delta x\to 0}\frac{f(x+\Delta x)-f(x)}{\Delta x}+\lim_{\Delta x\to 0}\frac{g(x+\Delta x)-g(x)}{\Delta x}\\ &= \frac{d\,f(x)}{dx}+\frac{d\,g(x)}{dx}, \qquad \text{q.e.d.}\end{aligned}$$

The reason that we could break this into two limits legitimately is that the two new limits both existed and were finite by assumption. (That is Theorem 2.9.1, page 189.)

Example 3.2.2

$$\frac{d}{dx}\left(x^3+x^2+x\right) = \frac{dx^3}{dx}+\frac{dx^2}{dx}+\frac{dx}{dx}=3x^2+2x+1.$$

This can also be written $f(x)=x^3+x^2+x \implies f'(x)=3x^2+2x+1$. With practice, one learns to skip the first step in this example. Note how $\frac{dx}{dx}=1$, as one might hope. This reflects the (trivial?) fact that x and x change at the same rate (i.e., the ratio of their rates of change is always 1). Put another way, the slope of the line $y=x$ is always 1.

It is not so obvious geometrically why "slopes are additive." Indeed, if we were to graph three functions (with the same domain), namely $f(x)$, $g(x)$, and their sum $h(x)=f(x)+g(x)$, then the height of the function $h(x)$ at a given point x is clearly the sum of the heights of $f(x)$ and $g(x)$ at that value. (After all, when we graph a function, its height for a given input x-value is its output value at x, and the output $h(x)$ is the sum of outputs $f(x)$ and $g(x)$.) The (now proved) theorem states that the slope of h at x, namely $h'(x)$, is indeed the sum of the slopes of the functions f and g there: $h'(x)=f'(x)+g'(x)$. If we consider a physical example, it might be more obvious:

Example 3.2.3

Suppose we are producing a mixture of three liquids, which we dub liquid 1, liquid 2, and liquid 3, by pumping them into a vat. The liquids are pumped into the vat at individual rates r_1, r_2, and r_3. If the volumes of each are V_1, V_2, and V_3, respectively, and the total volume is $V = V_1 + V_2 + V_3$, which is produced at rate R, then it should seem reasonable that $R = r_1 + r_2 + r_3$. In other words,

$$\frac{dV}{dt}=\frac{d}{dt}[V_1+V_2+V_3]=\frac{dV_1}{dt}+\frac{dV_2}{dt}+\frac{dV_3}{dt}.$$

For instance, if liquid 1 is pumped into the vat at 20 gal/min, liquid 2 is pumped at 30 gal/min and liquid 3 is pumped at 50 gal/min, then the mixture is being produced at $(20+30+50)$ gal/min = 100 gal/min.

So if we are to find the additive rule for derivatives believable for our physical example (mixing liquids), we should accept it is true for our geometric example (slopes).

The next theorem is usually given separately for emphasis.

Theorem 3.2.4 *The derivative of a constant is zero; if a function is defined by $f(x) = C$ for all $x \in \mathbb{R}$, where C is some fixed constant, then $f'(x) = 0$ for all $x \in \mathbb{R}$. Written two different ways, we have $f(x) = C \implies f'(x) = 0$, i.e.,*

$$\boxed{\frac{d}{dx} C = 0.} \tag{3.13}$$

There are several ways to see this. From the limit definition of the derivative (3.3), page 212, regardless of $\Delta x \neq 0$, the difference quotient is $[f(x + \Delta x) - f(x)]/\Delta x = [C - C]/\Delta x = 0/\Delta x = 0$, so it remains zero in the limit as $\Delta x \to 0$. From another perspective regarding what we know about lines we have that the graph of $f(x) = C$ is a line of slope $m = 0$. From a qualitative standpoint this theorem is reasonable since constants have rate of change zero (hence the term *constant*) with respect to x. Some texts write $(C)' = 0$. With this theorem and its predecessors we can write, for example,[19]

$$\frac{d}{dx}\left(x^3 + 28\right) = \frac{d}{dx}\left(x^3\right) + \frac{d}{dx}(28) = 3x^2 + 0 = 3x^2.$$

With very little practice one learns to write simply $\frac{d}{dx}\left(x^3 + 28\right) = 3x^2$.

With one more result we can easily compute derivatives of arbitrary polynomials. All that is left is to answer the question of what to do with the multiplicative constants (coefficients) of the various terms of a polynomial. The following applies to multiplicative constants in general.

Theorem 3.2.5 *Multiplicative constants are preserved in the derivative. In other words,*

$$\boxed{\frac{d}{dx}(C \cdot f(x)) = C \cdot \frac{d}{dx} f(x).} \tag{3.14}$$

The proof is left as an exercise. It follows from the fact that multiplicative constants "go along for the ride" in limits as well. (See again Theorem 2.9.1, page 189.)[20]

---

[19] One weakness of Lagrange's "prime" notation is that we do not know what variable we are taking the derivative with respect to. For instance, in an earlier example we have fuel volume V as a function of total distance traveled D, and so dV/dD measured the flow rate of fuel per mile. However, since $D = D(t)$, we have $V = V(D(t))$, so ultimately $V = V(t)$, i.e., V can be written as a (algebraically different) function of t instead, in which case we can calculate dV/dt, measuring the flow rate with respect to time. So when asked to calculate V', or even $V'(5)$, there is this ambiguity that is not present in the Leibniz notation.

If one wrote $V'(D)$, it would probably be understood to mean dV/dD and not dV/dt. Similarly, $V'(5 \text{ seconds})$ would be understood to mean dV/dt evaluated at $t = 5$ seconds.

[20] From another perspective, multiplicative constants "amplify" the outputs of functions, but consistently, and thus "amplify" their rates of change consistently as well. However, variable factors do not amplify consistently, and themselves vary, so they cannot be "factored" from a derivative computation as if they were constant multipliers. By their nature, variables vary and the derivative operators (e.g., $\frac{d}{dx}$) measure that variation. To treat a variable as a constant multiplier, and remove it from being measured by the derivative operator, is to ignore this fact. (By contrast, constants do not vary.)

3.2. FIRST DIFFERENTIATION RULES; LEIBNIZ NOTATION

Example 3.2.4

Consider the following derivatives:

- $\frac{d}{dx}(5x^7) = 5 \cdot \frac{d}{dx}(x^7) = 5 \cdot 7x^6 = 35x^6$,
- $\frac{d}{dx}(-x^4) = -1 \cdot \frac{dx^4}{dx} = -4x^3$,
- $\frac{d}{dx}\left(\frac{x^6}{5}\right) = \frac{1}{5} \cdot \frac{dx^6}{dx} = \frac{1}{5} \cdot 6x^5 = \frac{6}{5}x^5$,
- $\frac{d}{dx}\left(\frac{1}{4}x^{100}\right) = \frac{1}{4} \cdot \frac{dx^{100}}{dx} = \frac{100}{4}x^{99} = 25x^{99}$,
- $\frac{d}{dx}(5x) = 5 \cdot \frac{dx}{dx} = 5 \cdot 1 = 5$,
- $\frac{d}{dx}(x^2 \cdot x^5) = \frac{d}{dx}(x^7) = 7x^6$.

With very little practice, one learns to compute many derivatives in one line, though theoretically there are several steps involved. While learning the process of computing derivatives, there are usually some clarifying moments when one learns what can and cannot be done. For instance, the first five of those "bulleted" computations were clearly of the form $f(x) = Cx^n \implies f'(x) = C \cdot nx^{n-1}$, though the placement of the constant "C" may have used some cleverness. The final derivative, $\frac{d}{dx}(x^2 \cdot x^5)$ was not explicitly written as an unchanging *constant* multiplying some term x^n, since x^2 is not a constant. We had to *rewrite* the function $x^2 x^5 = x^7$, from which the Power Rule gave us its derivative immediately. Similarly, we have to be aware when asked to compute, for instance, $\frac{d}{dx}(x^2)^3 = \frac{dx^6}{dx} = 6x^5$. It would have been tempting to "differentiate" $(\ )^3$ and get $3(\ )^2$, regardless of the contents of the parentheses (as we do with function evaluation), but for differentiation (computing derivatives), the nature of the terms in the parentheses is crucial. (And how we deal with the different situations is, in fact, part of what makes calculus interesting!) We will see many more examples of what we can and cannot do with the differentiation rules as we continue to develop them. Fortunately, once the scopes of the rules are well understood, differentiation is rather straightforward.[21]

Now we combine the Power Rule, the rule on the derivative of a sum, and rules on the behaviors of (especially multiplicative) constants to derive how to compute derivatives of polynomials. It is important to recall that $\frac{d}{dx}$ treats additive constants (which do not "survive" the differentiation process) differently from multiplicative constants (which do survive). For any given polynomial (where the $a_k \in \mathbb{R}$ and each $k \in \{0, 1, 2, \ldots, n\}$), we have:

$$\frac{d}{dx}\left(a_n x^n + a_{n-1}x^{n-1} + \cdots + a_2 x^2 + a_1 x + a_0\right) \tag{3.15}$$
$$= a_n \cdot nx^{n-1} + a_{n-1} \cdot (n-1)x^{n-2} + \cdots + a_2 \cdot 2x + a_1.$$

At first glance, this formula may seem confusing. To be clear on the logic, note that we first use the Sum Rule to break this into a sum of derivatives of the $a_k x^k$, $k = 1, \ldots, n$ and a_0 (i.e., "distribute" $\frac{d}{dx}$ over the sum), calculating the derivatives of each $a_k x^k$ term in turn, each time using the fact that the multiplicative constants a_k are preserved (as multiplicative constants) in what are otherwise simple Power Rules: $\frac{d}{dx}(a_k x^k) = a_k \cdot kx^{k-1}$. The final term a_0 is an additive constant with derivative zero and thus does not appear on the right-hand side of (3.15).

---

[21] Derivatives tend to be more straightforward to compute than limits, at least for the examples one encounters in typical calculus textbooks, though the algebra of simplifying derivatives can be more involved. Especially for transcendental functions, sometimes half of the effort lies in convincing oneself that two ostensible expressions for a derivative are indeed equal. (In other words, whether the student's answer matches the one given in the answer key in the back of the book.)

Example 3.2.5

To see how (3.15) can be carried out quickly, we list a couple of brief examples:

$$\frac{d}{dx}\left(5x^4 + 9x^2 + 13x + 47\right) = 5 \cdot 4x^3 + 9 \cdot 2x^1 + 13 \cdot 1 + 0 = 20x^3 + 18x + 13,$$

$$\frac{d}{dx}\left(9 - 6x + 5x^{11}\right) = 0 + (-6) \cdot 1 + 5 \cdot 11x^{10} = -6 + 55x^{10} = 55x^{10} - 6.$$

One learns quickly to "think" but not necessarily write the first computational step in such problems, writing instead for instance $\frac{d}{dx}\left(5x^4 + 9x^2 + 13x + 47\right) = 20x^3 + 18x + 13$, and similarly $\frac{d}{dx}\left(9 - 6x + 5x^{11}\right) = -6 + 55x^{10}$. Note that the negative sign also "goes along for the ride," in the latter computation since it is just a multiplicative factor of -1: $\frac{d}{dx}\left(9 - 6x + 5x^{11}\right) = \frac{d}{dx}\left(9 + (-1 \cdot 6)x + 5x^{11}\right)$. In fact, we could list a *Difference Rule*,

$$\frac{d}{dx}(f(x) - g(x)) = \frac{d}{dx}f(x) - \frac{d}{dx}g(x),$$

but that would be redundant given the Sum Rule and how multiplicative constants, even if negative, are preserved in the derivative.

We should also point out that to use (3.15), page 235, we need to have the function written in the form of the left-hand side of that equation, i.e., expanded and not left factored. Consider, for instance, the following example.

Example 3.2.6

$$\frac{d}{dx}[(x^2+1)^2] = \frac{d}{dx}[x^4 + 2x^2 + 1] = 4x^3 + 4x. \quad \text{Note that } \frac{d}{dx}[(x^2+1)^2] \neq 2(x^2+1).$$

Here we needed first to expand the polynomial by performing the multiplication (squaring). Thus $\frac{d}{dx}[(x^2+1)^2] \neq 2(x^2+1)^1$, that is, $\frac{d}{dx}[(x^2+1)^2] \neq 2x^2+2$, as the correct result is noted in the example. This is because we wish to apply $\frac{d}{dx}$, which asks for derivative with respect to x, and not (x^2+1).[22]

We again note that derivatives do not allow *variable* quantities to "go along for the ride." Thus $\frac{d}{dx}\left[x \cdot x^3\right] \neq x \cdot \frac{d}{dx}\left[x^3\right]$. Indeed, $x \cdot \frac{dx^3}{dx} = x \cdot 3x^2 = 3x^3$, while $\frac{d(x \cdot x^3)}{dx} = \frac{dx^4}{dx} = 4x^3$.

Finally, we point out again that it should be clear from the previous examples that using the algorithm (3.15), page 235, for such computations is much simpler than using the original definition of the derivative (letting $\Delta x \to 0$ in a limit of difference quotients as in (3.3), page 212) to calculate derivatives of polynomials. However, the function must first be written as a sum of terms $a_k x^k$, where a_k can be any fixed real number and the variable of differentiation is x. Once we have that form, assuming our function is in fact a polynomial, we can quickly compute the derivative.

---

[22]In fact, $\dfrac{d(x^2+1)^2}{d(x^2+1)} = 2(x^2+1)$, because the variable being squared, (x^2+1), is also the variable of differentiation, i.e., the variable with respect to which the derivative is computed. This will be discussed further in later sections.

3.2. FIRST DIFFERENTIATION RULES; LEIBNIZ NOTATION

Example 3.2.7

Consider the following derivative computations.[23]

- $\frac{d}{dx}[(x+2)(x-3)] = \frac{d}{dx}(x^2 - x - 6) = 2x - 1$,
- $\frac{d}{dx}[x^2(3x-9)] = \frac{d}{dx}[3x^3 - 9x^2] = 9x^2 - 18x$,
- $\frac{d}{dt}[(3t-5)^2] = \frac{d}{dt}[9t^2 - 30t + 25] = 18t - 30$,
- $\frac{d}{du}\left[\frac{u^2-9u}{3u}\right] = \frac{d}{du}\left[\frac{1}{3}u - 3\right] = \frac{1}{3}, \qquad u \neq 0.$

3.2.5 Increasing and Decreasing Functions; Graphing Polynomials

Recall that while $f(x)$ gives the height of the graph of $y = f(x)$ at a particular value of x, the derivative $f'(x)$ gives the slope there. If the slope is positive for all x on an interval (a, b), then the graph is "sloping upwards" and therefore rising on that interval; if the slope is negative for all such x, then the graph is "sloping downwards" and falling. Another way to speak of graphs "rising" and "falling" is to discuss functions that are *increasing* or *decreasing* on an interval (a, b).

Definition 3.2.2 *Consider a function $f(x)$ with an interval (a, b) contained within its domain.*

1. *We say $f(x)$ is* **increasing** *on (a, b) if and only if $(\forall x, y \in (a, b))[x < y \longleftrightarrow f(x) < f(y)]$.*

2. *We say $f(x)$ is* **decreasing** *on (a, b) if and only if $(\forall x, y \in (a, b))[x < y \longleftrightarrow f(x) > f(y)]$.*

Note that it is possible that a function is not consistently increasing or consistently decreasing on a given interval.

Clearly, for an increasing function on (a, b), the height increases as x increases through the values on the interval. Similarly, for a decreasing function on (a, b), the height decreases as x increases through the interval. If we know exactly where a function is increasing, and where it is decreasing, that information can be of great help in plotting or otherwise analyzing the function. To see what this has to do with derivatives, we state the following theorem. Its proof relies on the Mean Value Theorem, which will be introduced in a later chapter (Section 4.3). However, this important theorem should already have the ring of truth given what we know of derivatives and slopes.

---

[23] The last computation in Example 3.2.7, technically excludes $u = 0$ because the original function is undefined there (and is technically not a polynomial because of that), but it is a removable discontinuity, and that single-point discontinuity will not affect the slope elsewhere, which is, geometrically, what the derivative measures. More precisely, we can always choose Δu so that $|\Delta u| < |u|$ in our limit definition of the derivative (so that $u + \Delta u \neq 0$), and it will be the same computation for the function $\frac{u^2-9u}{3u}$ as for $\frac{1}{3}u - 3$. On open intervals where the original function coincides with the polynomial $\frac{1}{3}u - 9$, we can take the derivative of the latter to deduce the derivative of the former, just as we could with limits, but in both cases we do require open intervals on which these two functions coincide in their actions and therefore graphs. It is not enough to know that two functions are equal at a point in order to declare their derivatives equal because their slopes through the points can be very different. However, if the two functions coincide on an open *interval*, then their derivatives must coincide there as well.

Theorem 3.2.6 *Suppose $f(x)$ is defined for $x \in (a,b)$, and $f'(x)$ exists for $x \in (a,b)$. Then*

1. *$(\forall x \in (a,b))[f'(x) > 0] \implies f(x)$ is increasing on (a,b);*

2. *$(\forall x \in (a,b))[f'(x) < 0] \implies f(x)$ is decreasing on (a,b).*

If $f'(x)$ changes sign on (a,b), then $f(x)$ is neither consistently increasing nor consistently decreasing on (a,b). Note that by the hypotheses, $f(x)$ is continuous on (a,b).

Typically we apply the theorem by constructing a sign chart for f', to mark where f is increasing and where it is decreasing, assuming f' exists and we can utilize the theorem. For the introductory example we will be more verbose in our analysis, with added explanation, though we will streamline the process for later examples.

Example 3.2.8

Use the theorem, and previous techniques, to graph $f(x) = x^3 - 3x$.

<u>Solution</u>: This function is clearly continuous on all of $\mathbb{R} = (-\infty, \infty)$. We might also notice the following asymptotic behavior, which we compute here "from scratch," but also recall that it is enough to look at the highest-order term in the polynomial. Either way we come to the same conclusions:

$$\lim_{x\to\infty} f(x) = \lim_{x\to\infty} (x^3 - 3x) \xrightarrow[\text{ALG}]{\infty-\infty} \lim_{x\to\infty}\left[x^3\left(1 - \frac{3}{x^2}\right)\right] \xrightarrow{\infty\cdot 1} \infty,$$

$$\lim_{x\to-\infty} f(x) = \lim_{x\to-\infty} (x^3 - 3x) \xrightarrow[\text{ALG}]{-\infty+\infty} \lim_{x\to-\infty}\left[x^3\left(1 - \frac{3}{x^2}\right)\right] \xrightarrow{-\infty\cdot 1} -\infty.$$

If we draw a sign chart for $f(x)$, showing where the function is positive and where it is negative, we can get some idea of what the graph looks like. To construct a sign chart for any function, we look at all the possible points where the function can change signs. Recall that the Intermediate Value Theorem (Corollary 2.3.2, page 108) implies a function $f(x)$ can only change signs, as we increase the input x, by either passing through a point of zero output (or zero height on the graph) or having a discontinuity. Since our particular $f(x)$ here is continuous on all of $\mathbb{R}$, we look to where $f(x) = 0$ to divide $\mathbb{R}$ into intervals of constant sign. Now $f(x) = x^3 - 3x = x(x^2 - 3)$ is zero for $x = 0, \pm\sqrt{3}$. This gives us four intervals on which $f(x)$ does not change signs. We can test for the sign of $f(x)$ at a single point in each interval to get the sign of $f(x)$ on that interval. Doing so, as we did in Section 2.3, we construct the sign chart for $f(x)$ (noting $\sqrt{3} \approx 1.73205$):

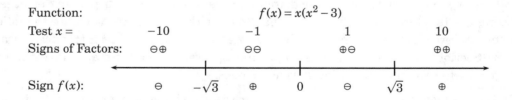

From the sign chart and the behavior as $x \to \pm\infty$, we can get some idea of what the graph of $f(x)$ looks like. That information is reflected, however imprecisely, in Figure 3.7, page 239.

A serious drawback to such a graph is that we know from the Extreme Value Theorem (Corollary 2.3.1, page 107) that there will be a value in $[-\sqrt{3}, 0]$ which is a **local maximum**, and another in $[0, \sqrt{3}]$ which is a **local minimum** (we will formally define these boldface terms shortly), but we do not know exactly where these are—or how many there are—from the sign chart of the function. However, a sign chart for the *derivative* of $f(x)$ can give us this information.

Since $f(x) = x^3 - 3x$, it follows quickly that $f'(x) = 3x^2 - 3$. Recall that on intervals where $f' > 0$, the function f is increasing, while on those intervals on which $f' < 0$, the function is decreasing. Since

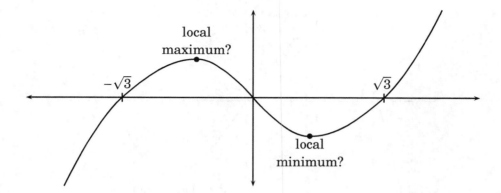

Figure 3.7: Rough graph of $f(x) = x^3 - 3x$ based upon its sign chart and behavior as $x \to \pm\infty$. In particular, we do not know the exact locations of any local maxima or minima without investigating the derivative of $f(x)$.

$f'(x)$ is also an easily factored—and continuous—polynomial, constructing its sign chart is easy. Note $f'(x) = 3x^2 - 3 = 3(x^2 - 1) = 3(x+1)(x-1)$ is zero exactly where $x = \pm 1$, the only points where $f'(x)$ changes sign.

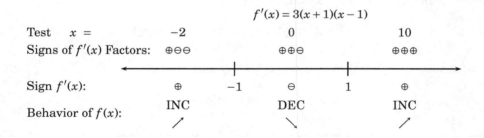

Here we used "INC" to abbreviate increasing, which we also signified by the arrow pointing upwards ($\nearrow$), and we used "DEC" and ($\searrow$) to signify decreasing. From this, we deduce that we get a local maximum at $(-1, f(-1)) = (-1, 2)$, and a local minimum at $(1, f(1)) = (1, -2)$. These two bits of information allow us to draw a more accurate sketch of the graph of $f(x) = x^3 - 3x$, as illustrated in Figure 3.8, page 240. That graph is computer-generated, but we can get a very accurate picture of the function's general behavior by plotting the information we have gathered: the sign of f, including the x-**intercepts** (where $f(x) = 0$), the limiting behavior of $f(x)$ as $x \to \pm\infty$, and where $f(x)$ is increasing/decreasing, including any local maximum and minimum points.[24]

It is important to distinguish the meanings of a sign chart for $f(x)$ and one for $f'(x)$. The former just tells us where the function is below or above the x-axis; the latter tells us where the function is increasing and where the function is decreasing.

In this example we used the following terms, which we now define. Here, $\exists (a,b) \ni x_0$ is read "there exists an open interval (a,b) containing x_0."

---

[24] It is also worth noticing that $f'(x) = 3(x+1)(x-1) \longrightarrow \infty$ for both $x \to \infty$ and $x \to -\infty$, and so the slope of $f(x)$ grows larger (as a positive number) as $x \to \pm\infty$. This is not the case with all graphs (see Figure 3.3, page 216, for example), but it is a nice feature to include when plotting a graph such as in Figure 3.8, page 240.

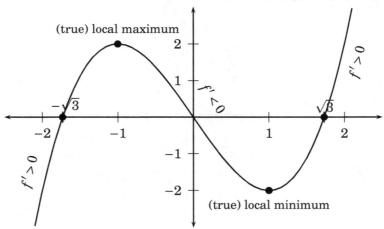

Figure 3.8: Partial graph of $f(x) = x^3 - 3x$ showing the sign of $f(x)$, the limiting behavior as $x \to \pm\infty$, and the sign of $f'(x)$ (which indicates also the locations of local extrema). The x-intercepts (where $f(x) = 0$), the local maximum and local minimum points are also illustrated, as are the facts that $x \to \infty \implies f(x) \to \infty$, and $x \to -\infty \implies f(x) \to -\infty$.

Definition 3.2.3 *Given a function $f(x)$.*

1. *We call a point x_0 a **local maximum** of $f(x)$ if and only if*

$$(\exists (a,b) \ni x_0)(\forall x \in (a,b))[f(x) \leq f(x_0)]. \tag{3.16}$$

2. *We call a point x_0 a **local minimum** of $f(x)$ if and only if*

$$(\exists (a,b) \ni x_0)(\forall x \in (a,b))[f(x) \geq f(x_0)]. \tag{3.17}$$

In other words, x_0 is a local maximum of $f(x)$ if there is an open interval containing x_0 in which the function is never greater than $f(x_0)$ on that interval. Local minima are defined analogously. If $f(x)$ is continuous in an open interval around x_0, and f' exists in that interval, then a change of signs of f' at x_0 indicates one of these *local extrema*. If, for instance, $f' > 0$ to the left of x_0 and $f' < 0$ to the right, then f increases before and decreases after x_0, making x_0 a local maximum. This can be seen in the derivative sign chart and graph (Figure 3.8).

Example 3.2.9

Use the derivative to determine where the graph of $f(x) = x^4 - 6x^2 + 8x$ is increasing, and where it is decreasing. Use this information to sketch a graph of $y = f(x)$.

<u>Solution</u>: We wish to know where $f'(x) > 0$ and $f'(x) < 0$, so we compute this derivative and construct its sign chart.

$$f'(x) = \frac{d}{dx}\left[x^4 - 6x^2 + 8x\right]$$
$$\implies f'(x) = 4x^3 - 12x + 8$$
$$\implies f'(x) = 4(x^3 - 3x + 2).$$

3.2. FIRST DIFFERENTIATION RULES; LEIBNIZ NOTATION

To construct the sign chart we thus need to solve $x^3 - 3x + 2 = 0$. While solving a third-degree polynomial equation can be quite difficult, in this case we are somewhat fortunate that $x = 1$ is one solution: $(1)^3 - 3(1) + 2 = 1 - 3 + 2 = 0$, and so $(x - 1)$ is one factor of the polynomial. From this we can use either synthetic division or polynomial long division to find the quotient when $x^3 - 3x + 2$ is divided by $x - 1$. Long division is shown on the right.

$$\begin{array}{r} x^2 + x -2 \\ x-1 \overline{\smash{\big)}\, x^3 -3x +2} \\ \underline{x^3 - x^2} \\ x^2 - 3x \\ \underline{x^2 - x} \\ -2x +2 \\ \underline{-2x +2} \\ 0 \end{array}$$

At this point it is relatively simple to complete our factorization of $f'(x)$:

$$f'(x) = 4(x-1)(x^2 + x - 2) = 4(x-1)(x+2)(x-1) = 4(x-1)^2(x+2).$$

From this we get $f'(x) = 0 \iff x \in \{-2, 1\}$, and can construct our sign chart for $f'(x)$:

Derivative: $\qquad\qquad\qquad f'(x) = 4(x-1)^2(x+2)$

Test $x =$	-10		0		10
Signs $f'(x)$ Factors:	$\oplus\oplus\ominus$		$\oplus\oplus\oplus$		$\oplus\oplus\oplus$
Sign $f'(x)$:	$\ominus$	-2	$\oplus$	1	$\oplus$
Behavior of $f(x)$:	DEC $\searrow$		INC $\nearrow$		INC $\nearrow$

The graph of $f(x)$ will clearly have a local minimum at $x = -2$. At $x = 1$ we have a curious situation where the graph is increasing on the intervals $(-2, 1)$ and $(1, \infty)$, but has slope zero at $x = 1$. This is illustrated in Figure 3.9, page 242, showing that the curve momentarily "levels off" at $x = 1$.

In fact, $f(x)$ here can be said to be increasing on $[-2, \infty)$ and decreasing on $(-\infty, -2]$. Whether a function is increasing or decreasing (or neither) is technically a description of its behavior on an interval, not at a particular point. (See Definition 3.2.2, page 237.) In graphing contexts, we usually list the intervals where the function is increasing or decreasing as open.

Note also that

$$\lim_{x \to \pm\infty} f(x) = \lim_{x \to \pm\infty} [x^4 - 6x^2 + 8x] = \lim_{x \to \pm\infty} \left[x^4 \left(1 - \frac{6}{x^2} + \frac{8}{x^3} \right) \right] \xrightarrow{\infty \cdot 1} \infty.$$

We can easily see one x-intercept occurs at $x = 0$ (since $f(0) = 0$ clearly), and from our computer-generated graph or from experimentation and the Intermediate Value Theorem, another occurs just to the right of $x = -3$. Computing also[25]

$$f(-2) = (-2)^4 - 6(-2)^2 + 8(-2) = 16 - 24 - 16 = -24,$$
$$f(1) = (1)^4 - 6(1)^2 + 8(1) = 1 - 6 + 8 = 3,$$

we get the graph in Figure 3.9, page 242.

It is important to note that having $f'(x) = 0$ does not imply that there is a local maximum or minimum point there. We should always consider the full sign chart of $f'(x)$. Local extrema do exist at points of continuity $x = a$ on the graph if $f'(x)$ changes sign at $x = a$. In the

---

[25] In Section 4.5 we will see how to find x-intercepts of most functions to as much accuracy as we desire.

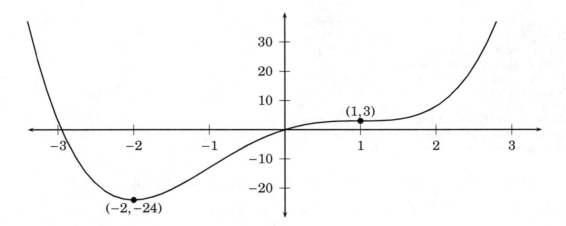

Figure 3.9: Partial graph of $f(x) = x^4 - 6x^2 + 8x$. From the derivative's sign chart and form $f'(x) = 4x^3 - 12x + 8 = 4(x-1)^2(x+2)$, it is clear that $f'(x) = 0$ at $x = -2, 1$. While the function has a local (indeed "global") minimum at $x = -2$, by contrast at $x = 1$ the graph only "levels out" momentarily, and then continues to increase from then on. Thus $f(x)$ is in fact increasing on all of $(-2, \infty)$. (Technically, one can even assert $f(x)$ is increasing on $[-2, \infty)$ and decreasing on $(-\infty, -2]$, but we usually list only open intervals in these graphing contexts.)

previous example, $x = 1$ is a zero of $f'(x)$ of multiplicity 2, which is even, and so $f'(x)$ does not change sign there, and there is no local extremum at $x = 1$. At $x = -2$ we have a zero of $f'(x)$ of multiplicity 1, which is odd, and so $f'(x)$ does change sign there, giving us a local extremum (a minimum in that case).

3.2.6 Derivatives of Sine and Cosine

In this subsection we show how $\sin x$ and $\cos x$ are both differentiable, compute their derivatives, and apply them to functions involving these in simple ways. (More complicated combinations of these will have to await later sections.) In particular we will prove the following theorem, which states that the derivative of the sine function is the cosine function, and the derivative of the cosine function is the opposite of the sine function.

Theorem 3.2.7 *The functions $\sin x$ and $\cos x$ are differentiable for all $x \in \mathbb{R}$, and—when x is measured in* **radians**—*their derivatives are given by:*

$$\boxed{\frac{d \sin x}{dx} = \cos x,} \tag{3.18}$$

$$\boxed{\frac{d \cos x}{dx} = -\sin x.} \tag{3.19}$$

These should be memorized. They are reasonable given their respective graphs as in Figure 3.10, page 243. For instance, for the sine and cosine curves we have the following data:

3.2. FIRST DIFFERENTIATION RULES; LEIBNIZ NOTATION

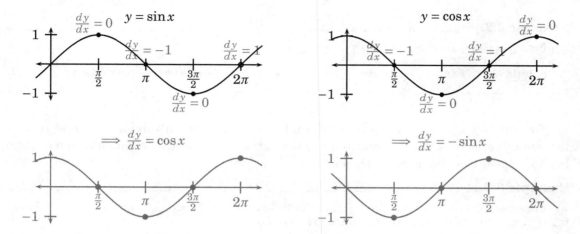

Figure 3.10: Partial graphs of $y = \sin x$, $y = \cos x$, and their respective derivative functions graphed below them. Note the points labeled on the (top) functions, and their derivative values highlighted on the derivative functions below them.

$$\begin{array}{rccccc} x = & 0, & \frac{\pi}{2}, & \pi, & \frac{3\pi}{2}, & 2\pi. \\ y = \sin x = & 0, & 1, & 0, & -1, & 0. \\ \frac{dy}{dx} = \cos x = & 1, & 0, & -1, & 0, & 1. \end{array} \qquad \begin{array}{rccccc} x = & 0, & \frac{\pi}{2}, & \pi, & \frac{3\pi}{2}, & 2\pi. \\ y = \cos x = & 1, & 0, & -1, & 0, & 1. \\ \frac{dy}{dx} = -\sin x = & 0, & -1, & 0, & 1, & 0. \end{array}$$

Looking at the graph of $\sin x$ as drawn in Figure 3.10, the slopes at these points and the values for $\cos x$ seem at least compatible. It is similar for $\cos x$ and $-\sin x$. We will prove the derivative formula for $\sin x$, and leave the derivative of $\cos x$ as an exercise. (The two computations are very similar.)

Proof: (3.18): The proof is based upon the following:

$$\lim_{\theta \to 0} \frac{\sin \theta}{\theta} = 1, \qquad \lim_{\theta \to 0} \frac{1 - \cos \theta}{\theta} = 0, \qquad \sin(\alpha + \beta) = \sin \alpha \cos \beta + \cos \alpha \sin \beta,$$

which are, respectively, (2.83) from page 194, (2.88) from page 197, and a trigonometric identity (usually proved geometrically). We will use the limit-definition of the derivative, expand using the formula for $\sin(\alpha + \beta)$, and rearrange the terms so we can use these trigonometric limits. If $f(x) = \sin x$, then

$$f'(x) = \lim_{\Delta x \to 0} \frac{f(x + \Delta x) - f(x)}{\Delta x} = \lim_{\Delta x \to 0} \frac{\sin(x + \Delta x) - \sin x}{\Delta x}$$
$$= \lim_{\Delta x \to 0} \frac{\sin x \cos \Delta x + \cos x \sin \Delta x - \sin x}{\Delta x} = \lim_{\Delta x \to 0} \frac{\sin x (\cos \Delta x - 1) + \cos x \sin \Delta x}{\Delta x}$$
$$= \lim_{\Delta x \to 0} \left[\sin x \cdot \frac{\cos \Delta x - 1}{\Delta x} + \cos x \cdot \frac{\sin \Delta x}{\Delta x} \right] = \sin x \cdot 0 + \cos x \cdot 1 = \cos x, \qquad \text{q.e.d.}$$

This proves $\frac{d}{dx} \sin x = \cos x$. The proof that $\frac{d}{dx} \cos x = -\sin x$ is similar and left as an exercise. Now we use these alongside our previous rules.

Example 3.2.10

Find $f'(x)$ if $f(x) = x^2 + \sin x - 3\cos x$.

Solution: $f'(x) = \dfrac{d}{dx}\left[x^2 + \sin x - 3\cos x\right] = 2x + \cos x - 3(-\sin x) = 2x + \cos x + 3\sin x.$

We can also use these derivatives to find where functions involving $\sin x$ and $\cos x$ are increasing or decreasing, and thus find any local extrema (that is, local maxima and minima), but the algebra can be more involved.

Example 3.2.11

Consider the function $f(x) = \sin x - \cos x$. Find where $f(x)$ is increasing and where $f(x)$ is decreasing, and use this information to sketch the graph of $f(x)$.

Solution: Here $f'(x) = \cos x - (-\sin x) = \cos x + \sin x$. Since this is defined and continuous everywhere, we will check where it is zero to detect where it ($f'(x)$ here) possibly changes signs. The following technique works anytime we are interested in solving $a\sin x + b\cos x = 0$, where $a, b \neq 0$:

$$f'(x) = 0 \iff \cos x + \sin x = 0 \iff \sin x = -\cos x \iff \frac{\sin x}{\cos x} = -1 \iff \tan x = -1.$$

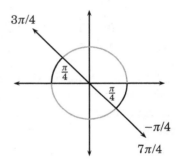

The reason we can divide by $\cos x$ is because there are no solutions where $\cos x = 0$, because such solutions would require also $\sin x = 0$, and these cannot be zero simultaneously because $\sin^2 x + \cos^2 x = 1$. So we are looking for $x \in \mathbb{R}$ such that $\tan x = -1$. This occurs with angles x in the second quadrant and in the fourth quadrant, with reference angles $\pi/4$:

Thus we are looking for angles $x = \frac{3\pi}{4} + n\pi$, where $n = 0, \pm 1, \pm 2, \pm 3, \ldots$. Now the function $f(x) = \cos x + \sin x$ is (at most) 2π-periodic, so we can analyze one interval of length 2π to see what the entire graph should look like. We will use the points $x = -\pi/4, 3\pi/4, 7\pi/4$ (and so on) to partition our sign chart, and declare the pattern from there.[26]

$$f'(x) = \cos x + \sin x$$

Test $x =$	0	π	2π	3π
$f'(x) =$	$1 + 0 = 1$	$-1 + 0 = -1$	$1 + 0 = 1$	$-1 + 0 = -1$
	$\frac{-\pi}{4}$	$\frac{3\pi}{4}$	$\frac{7\pi}{4}$	$\frac{11\pi}{4}$ $\quad\quad \frac{15\pi}{4}$
Sign f':	$\oplus$	$\ominus$	$\oplus$	$\ominus$
Behavior of f:	↗	↘	↗	↘

---

[26]In fact, $f(x) = \sin x - \cos x$ is exactly 2π-periodic, because if we use $\sin(\alpha - \beta) = \sin\alpha\cos\beta - \cos\alpha\sin\beta$, we can rewrite $f(x) = \sqrt{2}\left(\sin x \cdot \frac{1}{\sqrt{2}} - \cos x \cdot \frac{1}{\sqrt{2}}\right) = \sqrt{2}\sin\left(x - \frac{\pi}{4}\right)$. That also makes graphing this function a simple trigonometric problem not requiring calculus, but such workarounds are not always available or obvious. And calculus is more fun.

3.2. FIRST DIFFERENTIATION RULES; LEIBNIZ NOTATION

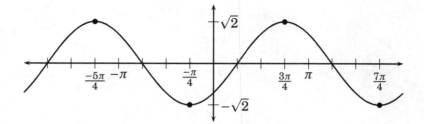

Figure 3.11: Partial graph of $f(x) = \sin x - \cos x$, showing for instance the local minima at $x = 7\pi/4$ and $x = -\pi/4$, and the local maxima at $x = 3\pi/4$ and $x = -5\pi/4$. Each local extremum is repeated every 2π. See Example 3.2.11.

We see a local maximum at $\left(\frac{3\pi}{4}, f\left(\frac{3\pi}{4}\right)\right) = \left(\frac{3\pi}{4}, \sqrt{2}\right)$, since

$$f(3\pi/4) = \sin\frac{3\pi}{4} - \cos\frac{3\pi}{4} = \frac{\sqrt{2}}{2} - \frac{-\sqrt{2}}{2} = \frac{\sqrt{2} + \sqrt{2}}{2} = \frac{2\sqrt{2}}{2} = \sqrt{2}.$$

Because $f(x)$ is (at most, and as it will emerge, exactly) 2π-periodic, this local maximum height then repeats every 2π in both left and right directions. Similarly, because of the sign chart and the fact that this function (and its derivative and its derivative's sign chart) repeats every 2π, we have a local minimum at, for instance (using a computation similar to the previous one), $\left(\frac{7\pi}{4}, f\left(\frac{7\pi}{4}\right)\right) = \left(\frac{7\pi}{4}, -\sqrt{2}\right)$, which also repeats every 2π in both directions. This function is graphed in Figure 3.11.

The previous examples are good illustrations of how we can have emergent properties of combinations of functions, and how these properties (such as extrema at each point $\frac{3\pi}{4} + n\pi$ for the previous function) might not be immediately obvious from the form of the function $f(x)$ alone, but they can become more clear from the form of the derivative $f'(x)$. The next function's behavior is a bit more predictable in some ways, but some subtleties emerge from the derivative as well.

Example 3.2.12

Let $f(x) = x + \sin x$. Find where $f(x)$ is increasing and where $f(x)$ is decreasing.

<u>Solution</u>: Here, $f'(x) = 1 + \cos x = 0$ when $\cos x = -1$, which is at $x = \pm\pi, \pm 3\pi, \pm 5\pi, \ldots$. A partial sign chart is given next:

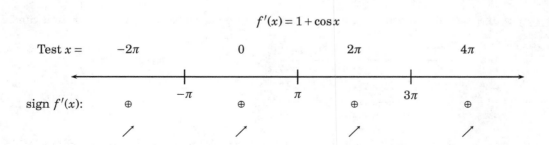

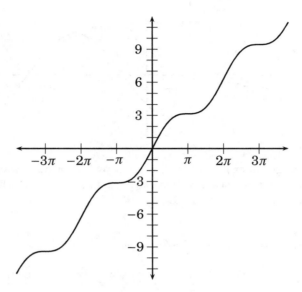

Figure 3.12: Partial graph of $f(x) = x + \sin x$. The derivative being $f'(x) = 1 + \cos x$, which is positive except at $x = \pm\pi, \pm 3\pi, \pm 5\pi, \ldots$, the function is always increasing, momentarily "leveling off" at these points where $f'(x) = 0$.

From the chart we see that this function is always increasing, only momentarily having zero slope at the odd multiples of π. Note that these points occur at (π,π), $(3\pi,3\pi)$, $(7\pi,7\pi)$, etc., and $(-\pi,-\pi)$, $(-3\pi,-3\pi)$, $(-7\pi,-7\pi)$, etc. This function is graphed in Figure 3.12 showing this behavior.

3.2.7 Limits That Are Derivative Forms

In this subsection we note that there are 0/0-form limits which are in fact derivatives by definition, which we can see if we properly rewrite the limits. Some are more subtle than others, of course, and so we start with some that are less subtle.

Example 3.2.13

Compute $\displaystyle\lim_{\Delta x \to 0} \frac{\sin(\pi/4 + \Delta x) - \frac{1}{\sqrt{2}}}{\Delta x}$.

Solution: It is useful to define the function $f(x) = \sin x$, in which case we can see our limit can be written

$$\lim_{\Delta x \to 0} \frac{\sin(\pi/4 + \Delta x) - \frac{1}{\sqrt{2}}}{\Delta x} = \lim_{\Delta x \to 0} \frac{f(\pi/4 + \Delta x) - f(\pi/4)}{\Delta x} = f'(\pi/4) = \cos(\pi/4) = \frac{1}{\sqrt{2}}.$$

Of course we might have been alerted to the nature of that limit due to the choice of Δx

3.2. FIRST DIFFERENTIATION RULES; LEIBNIZ NOTATION

as our limiting variable, and that we might be able to write the limit into some form

$$\lim_{\Delta x \to 0} \frac{f(x+\Delta x)-f(x)}{\Delta x} = f'(x).$$

If we know the appropriate function f and its derivative f', then the limit amounts to a simple evaluation (albeit of the *derivative* function).

Then again, there is nothing special about any variable, except for the meaning we give it, so another variable might also be used. A popular one is h, which is often used for horizontal shifts as well.

Example 3.2.14 _____

Compute $\lim_{h \to 0} \dfrac{\sin h - 0}{h}$.

<u>Solution:</u> In fact, we have seen this one before, using θ and omitting the redundant "-0." Nonetheless, we might consider $f(x) = \sin x$ again and note that[27]

$$\lim_{h \to 0} \frac{\sin h - 0}{h} = \lim_{h \to 0} \frac{\sin(0+h) - \sin 0}{h} = f'(0) = \cos 0 = 1.$$

Next we point out that there are three common definitions for $f'(x)$, all of which (of course) yield the same result. Each has a context in which it is most appropriate. That the first two are equal is obvious, and the proof of the third is left for the exercises:

$$f'(a) = \lim_{\Delta x \to 0} \frac{f(a+\Delta x)-f(a)}{\Delta x}, \tag{3.20}$$

$$f'(a) = \lim_{h \to 0} \frac{f(a+h)-f(a)}{h}, \tag{3.21}$$

$$f'(a) = \lim_{x \to a} \frac{f(x)-f(a)}{x-a}. \tag{3.22}$$

A simple substitution of $h = \Delta x$ shows the equality of the first two. That the first is also equal to the third is clear geometrically, where the slope of the secant line connecting $(a, f(a))$ with $(x, f(x))$ is given, and then $x \to a$ would be the same limit of secant line slopes as we get from the limit of secant line slopes connecting $(a, f(a))$ and $(a + \Delta x, f(a + \Delta x))$, and letting $\Delta x \to 0$. The proof by substitution is left for the exercises.

This third form (3.22) can be useful on occasion, such as in the following:

Example 3.2.15 _____

Compute $\lim_{x \to 2\pi} \dfrac{\cos x - 1}{x - 2\pi}$.

---

[27] Recall again our special limit $\lim_{\theta \to 0} \dfrac{\sin \theta}{\theta} = 1$. Its sister limit was $\lim_{\theta \to 0} \dfrac{1 - \cos \theta}{\theta} = 0$.

248   CHAPTER 3. THE DERIVATIVE

Solution: This time we let $f(x) = \cos x$, and compute
$$\lim_{x \to 2\pi} \frac{\cos x - 1}{x - 2\pi} = \lim_{x \to 2\pi} \frac{f(x) - f(2\pi)}{x - 2\pi} = f'(2\pi) = -\sin 2\pi = -0 = 0.$$

It would have been very difficult to compute this limit using previous techniques, though not impossible. The next limit is somewhat involved if using previous techniques, but it is also interesting to see it as a derivative-form limit.

Example 3.2.16 _____

Compute $\lim_{x \to 2} \dfrac{x^3 - 8}{x - 2}$.

Solution: Here we let $f(x) = x^3$ and note that $f(2) = 8$. Also note that $f'(x) = 3x^2$. Thus we can write
$$\lim_{x \to 2} \frac{x^3 - 8}{x - 2} = \lim_{x \to 2} \frac{f(x) - f(2)}{x - 2} = f'(2) = 3(2)^2 = 12.$$
Without utilizing the derivative nature of this limit, we would have resorted to previous techniques:
$$\lim_{x \to 2} \frac{x^3 - 8}{x - 2} \underset{\text{ALG}}{\overset{0/0}{=\!=\!=}} \lim_{x \to 2} \frac{(x-2)(x^2 + 2x + 4)}{x - 2} \underset{\text{ALG}}{\overset{0/0}{=\!=\!=}} \lim_{x \to 2} (x^2 + 2x + 4) = 4 + 4 + 4 = 12.$$

In this example, the two techniques may seem equally involved, even if we use a third alternative which is to use polynomial long division (or synthetic division) to simplify the quotient at the end of the example. However, if the power were higher than 3, the derivative method is likely to be significantly simpler (but more clever!).

3.2.8 Absolute Value and Other Piecewise-Defined Functions

Here we begin by exploring how one would compute derivatives for piecewise-defined functions such as $\frac{d}{dx}|x|$. This is indeed a piecewise-defined (and continuous) function, as we have

$$|x| = \begin{cases} x & \text{if } x \geq 0, \text{ i.e., } x \in [0, \infty) \\ -x & \text{if } x < 0, \text{ i.e., } x \in (-\infty, 0). \end{cases} \implies \frac{d|x|}{dx} = \begin{cases} 1 & \text{if } x > 0, \text{ i.e., } x \in (0, \infty) \\ -1 & \text{if } x < 0, \text{ i.e., } x \in (-\infty, 0). \end{cases}$$

If x is safely within one of the pieces, i.e, x is within an open interval in which the formula for $f(x)$ is simple and differentiable, then the derivative is easily computed using other rules. For instance, if $f(x) = |x|$, then as the graph helps to demonstrate, we have

$$f'(5) = \frac{d}{dx}(x)\bigg|_{x=5} = 1\bigg|_{x=5} = 1,$$

$$f'(-3) = \frac{d}{dx}(-x)\bigg|_{x=-3} = -1\bigg|_{x=-3} = -1.$$

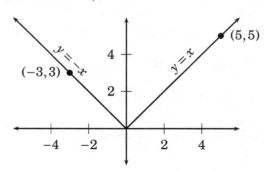

3.2. FIRST DIFFERENTIATION RULES; LEIBNIZ NOTATION 249

We can do this, because when we look at the definition of the derivative, with $|\Delta x|$ small enough, $x + \Delta x$ will still be within the interior of the interval $[0, \infty)$ (i.e., within $(0, \infty)$) or within $(-\infty, 0)$. However, the case $x = 0$ is problematic. In fact, $f'(0)$ will not exist:

$$\left. \begin{array}{l} \lim_{\Delta x \to 0^+} \dfrac{f(0+\Delta x) - f(0)}{\Delta x} = \lim_{\Delta x \to 0^+} \dfrac{|\Delta x|}{\Delta x} \overset{\Delta x > 0}{=\!=\!=} \lim_{\Delta x \to 0^+} \dfrac{\Delta x}{\Delta x} = \lim_{\Delta x \to 0^+} 1 = 1 \\ \lim_{\Delta x \to 0^-} \dfrac{f(0+\Delta x) - f(0)}{\Delta x} = \lim_{\Delta x \to 0^-} \dfrac{|\Delta x|}{\Delta x} \overset{\Delta x < 0}{=\!=\!=} \lim_{\Delta x \to 0^-} \dfrac{-\Delta x}{\Delta x} = \lim_{\Delta x \to 0^-} (-1) = -1 \end{array} \right\} \Longrightarrow f'(0) \text{ DNE}.$$

For $f'(0)$ to exist, these two limits would have to be equal (and thus equal to $f'(0)$), but these limits are different and so $f'(0)$ does not exist.

However, this first method is somewhat more involved than is often necessary. We can simply note that the left-side limit ($\Delta x \to 0^-$) would yield the same as $\frac{d}{dx}(-x) = -1$ and the right-side limit would yield the same as $\frac{d}{dx}(x) = 1$, which are not equal and so the two-sided limit $\frac{d}{dx}|x|$ does not exist at $x = 0$.

We next extend this analysis to other, piecewise-defined functions.

Example 3.2.17 _____

Find $f'(x)$ if $f(x) = \begin{cases} 1, & \text{if} & x \geq \pi/2, \\ \sin x, & \text{if} & -\pi/2 < x < \pi/2, \\ -x, & \text{if} & x \leq -\pi/2. \end{cases}$

Solution: Note first that $f(x)$ is not continuous at $x = -\pi/2$, since the left and right limits are not equal: $x \to (-\pi/2)^- \implies f(x) = -x \to \pi/2$, while $x \to (-\pi/2)^+ \implies f(x) = \sin x \to -1$. Therefore the limit of $f(x)$ as $x \to -\pi/2$ cannot exist, implying that the function must not be continuous there, and therefore not differentiable there either. (Recall for continuity the limit must equal the function value, and differentiability requires continuity.) However, $f(x)$ is continuous at $x = \pi/2$ ($x \to (\pi/2)^\pm \implies f(x) \to 1 = f(\pi/2)$), so there is a possibility of a derivative there, which we will explore momentarily.

But first we notice that the other x-values in the domain are safely within the interiors of intervals where f is given by a differentiable function, so we **at least** have

$$x \in (\pi/2, \infty) \implies f'(x) = \frac{d}{dx}(1) = 0,$$
$$x \in (-\pi/2, \pi/2) \implies f'(x) = \frac{d}{dx}(\sin x) = \cos x,$$
$$x \in (-\infty, -\pi/2) \implies f'(x) = \frac{d}{dx}(-x) = -1.$$

Since $f(x)$ is not even continuous at $x = -\pi/2$ (and therefore cannot have a derivative there), all that is left to check is $f'(\pi/2)$. This is simpler than one might think at first. When we look at the definition of $f'(\pi/2)$ using limits, computing this with $\Delta x \to 0^+$ would be exactly the same computation as if the function in question were the (constant) function 1 (with derivative 0), and so that limit would be 0. When taking $\Delta x \to 0^-$, that limit computation would be exactly the same as if the function were $\sin x$, and we know its derivative is $\cos x$, and $\cos(\pi/2) = 0$. Thus the left and right limits in the definition of $f'(\pi/2)$ can be found using the functions in the branches that define $f(x)$—assuming that the function is continuous there, which it is—and these limits match, both having value 0. We can conclude $f'(\pi/2) = 0$.

Summarizing,
$$f'(x) = \begin{cases} 0, & \text{if} & x \geq \pi/2, \\ \cos x, & \text{if} & -\pi/2 < x < \pi/2, \\ -1, & \text{if} & x < -\pi/2, \end{cases}$$

and $f'(x)$ does not exist otherwise (i.e., at $x = -\pi/2$). The function is drawn in Figure 3.13.

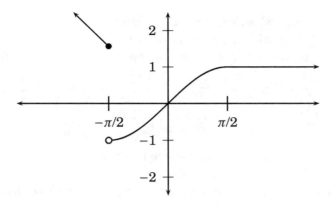

Figure 3.13: Figure for Example 3.2.17, page 249 and the discussion that follows.

Computing derivatives for functions such as these indicates one of many reasons it is important to recall the original definition of $f'(x)$, namely $f'(x) = \lim_{\Delta x \to 0} \frac{f(x+\Delta x)-f(x)}{\Delta x}$. To make the argument in the previous example more formally, one could write the following. Note in the first we perform the limit directly, and in the second we interpret it as a derivative.

$$\left.\begin{array}{l} \lim_{\Delta x \to 0^+} \dfrac{f\left(\frac{\pi}{2}+\Delta x\right)-f\left(\frac{\pi}{2}\right)}{\Delta x} = \lim_{\Delta x \to 0^+} \dfrac{1-1}{\Delta x} = \lim_{\Delta x \to 0^+} 0 = 0, \\ \lim_{\Delta x \to 0^-} \dfrac{f\left(\frac{\pi}{2}+\Delta x\right)-f\left(\frac{\pi}{2}\right)}{\Delta x} = \lim_{\Delta x \to 0^-} \dfrac{\sin\left(\frac{\pi}{2}+\Delta x\right)-1}{\Delta x} = \dfrac{d\sin x}{dx}\bigg|_{\frac{\pi}{2}} = \cos\dfrac{\pi}{2} = 0. \end{array}\right\} \Longrightarrow f'\left(\dfrac{\pi}{2}\right) = 0$$

A similar argument shows $f'(-\pi/2)$ does not exist (DNE), though we knew that already due to the discontinuity of $f(x)$ at $x = -\pi/2$. Ignoring that for a moment we note the following:

$$\left.\begin{array}{l} \lim_{\Delta x \to 0^+} \dfrac{f\left(\frac{-\pi}{2}+\Delta x\right)-f\left(\frac{-\pi}{2}\right)}{\Delta x} = \lim_{\Delta x \to 0^+} \dfrac{\sin\left(\frac{-\pi}{2}+\Delta x\right)-\left(\frac{\pi}{2}\right)}{\Delta x} \xrightarrow{\frac{-1-\pi/2}{0^+}} -\infty \\ \lim_{\Delta x \to 0^-} \dfrac{f\left(\frac{-\pi}{2}+\Delta x\right)-f\left(\frac{-\pi}{2}\right)}{\Delta x} = \lim_{\Delta x \to 0^-} \dfrac{[-(x+\Delta x)]-(-x)}{\Delta x} = \dfrac{d(-x)}{dx}\bigg|_{\frac{-\pi}{2}} = -1 \end{array}\right\} \Longrightarrow f'\left(\dfrac{-\pi}{2}\right) \text{ DNE.}^{28}$$

We could also have noticed that slope formulas for the pieces on $(-\infty, -\pi/2]$ and $(-\pi/2, \pi/2)$ are themselves continuous but would have values, namely $\frac{dy}{dx} = -1$ and $\frac{dy}{dx} = \cos x$, which do not coincide at $x = -\pi/2$. That observation alone can also lead us to conclude that $f'(-\pi/2)$ DNE, as can our previous observation that $f(x)$ is discontinuous at $x = -\pi/2$.

For computing derivatives of a piecewise-defined function, we make the following three observations:

- *For a point x on the interior of a piece on which f is defined by a formula, we can use the derivative rules that apply to that formula at that interior point to find $f'(x)$.*

- *If x is a boundary point of two adjacent interval pieces given in the definition of f, and if f is not continuous at x, then $f'(x)$ DNE.*

---

[28]In this second analysis to determine $f'(-\pi/2)$ DNE, even if both limits were $-\infty$ we would conclude the derivative does not exist at $-\pi/2$ because, by definition, a derivative only exists if its defining limit exists and is finite.

3.2. FIRST DIFFERENTIATION RULES; LEIBNIZ NOTATION

- However, if continuity is established at a boundary point x of the two interval pieces on which f is defined by separate formulas, if there is a useful formula for each of the derivatives on the two adjacent pieces, and if those two formulas agree numerically at the boundary point of those pieces, we can declare the function differentiable there as well, with the derivative's value $f'(x)$ there being the same as what those formulas would have given.

Though the most direct explanation for a particular differentiation of a piecewise-defined function would resemble the limit computations occurring after the previous example, we can often make do with an argument such as the one listed in these bullet points.

Example 3.2.18

Find a formula for $f'(x)$ if $f(x) = \begin{cases} \cos x & \text{if} \quad x \geq \pi, \\ -\frac{1}{\pi}x & \text{if} \quad -\pi < x < \pi, \\ 1 + \frac{1}{\pi}\sin x & \text{if} \quad x \leq -\pi \end{cases}$

<u>Solution</u>: For the interior points of these intervals we can compute the derivatives immediately:

$$x \in (\pi, \infty) \implies f'(x) = \frac{d}{dx}\cos x = -\sin x,$$

$$x \in (-\pi, \pi) \implies f'(x) = \frac{d}{dx}\frac{-1}{\pi}x = \frac{-1}{\pi},$$

$$x \in (-\infty, -\pi) \implies f'(x) = \frac{d}{dx}\left(1 + \frac{1}{\pi}\sin x\right) = \frac{1}{\pi}\cos x.$$

Next we check the continuity at the boundary points of the interval pieces on which $f(x)$ is defined.

$$\lim_{x \to \pi^+} f(x) = \lim_{x \to \pi^+} \cos x = -1,$$
$$\lim_{x \to \pi^-} f(x) = \lim_{x \to \pi^-} \left[\frac{-1}{\pi}x\right] = -1,$$
$$f(\pi) = \cos \pi = -1.$$

$$\lim_{x \to -\pi^+} f(x) = \lim_{x \to -\pi^+} \left[-\frac{1}{\pi}x\right] = 1,$$
$$\lim_{x \to -\pi^-} f(x) = \lim_{x \to -\pi^-} \left[1 + \frac{1}{\pi}\sin x\right] = 1,$$
$$f(-\pi) = 1 + \frac{1}{\pi}\sin(-\pi) = 1.$$

From these we see that $f(x)$ is continuous on all of $\mathbb{R}$, because (1) it is clearly continuous on the open intervals $(-\infty, -\pi)$, $(-\pi, \pi)$ and (π, ∞) since on those intervals it coincides with functions continuous on all of $\mathbb{R}$, and (2) it is also continuous at the boundary points of these intervals since the function values coincide with the limits at those two points $x = \pm \pi$. (Recall the alternative, limit-definition of continuity, given by Theorem 2.4.2, page 124.)

Furthermore, we can see that $f'(x) = -\sin x$ on (π, ∞), while $f'(x) = -1/\pi$ on $(-\pi, \pi)$, and $f'(x) = \frac{1}{\pi}\cos x$ on $(-\infty, -\pi)$. Next we check to see if the derivative formulas on the adjacent pieces match at the boundary points.

$$\frac{d\cos x}{dx}\bigg|_\pi = -\sin x\bigg|_\pi = 0,$$
$$\frac{d\left[\frac{-1}{\pi}x\right]}{dx}\bigg|_\pi = \frac{-1}{\pi}\bigg|_\pi = -\frac{1}{\pi}.$$

$$\frac{d\left[\frac{-1}{\pi}x\right]}{dx}\bigg|_{-\pi} = \frac{-1}{\pi}\bigg|_{-\pi} = -\frac{1}{\pi},$$
$$\frac{d\left[1 + \frac{1}{\pi}\sin x\right]}{dx}\bigg|_{-\pi} = \frac{1}{\pi}\cos x\bigg|_{-\pi} = -\frac{1}{\pi}.$$

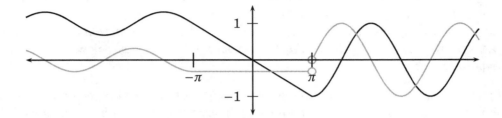

Figure 3.14: Figure for Example 3.2.18. The function $f(x)$ is graphed in black, and its derivative $f'(x)$ is graphed in gray. Note that $f'(-\pi)$ exists but $f'(\pi)$ does not. It is worth noting where f' is zero, positive, negative, constant, and nonexistent and noting the behavior of f at those points.

From these we see that $f'(\pi)$ does not exist, while $f'(-\pi) = -1/\pi$. (The latter required also that $f(x)$ be continuous at $x = -\pi$.) Putting all of this together we get

$$f(x) = \begin{cases} \cos x & \text{if} \quad x \geq \pi, \\ -\frac{1}{\pi}x & \text{if} \quad -\pi < x < \pi, \\ 1 + \frac{1}{\pi}\sin x & \text{if} \quad x \leq -\pi \end{cases} \implies f'(x) = \begin{cases} -\sin x & \text{if} \quad x > \pi, \\ -1/\pi & \text{if} \quad -\pi < x < \pi, \\ \frac{1}{\pi}\cos x & \text{if} \quad x \leq -\pi. \end{cases}$$

Understood from this formula is that $f'(\pi)$ does not exist. Also note that we could have included the point $x = -\pi$ in either the leftmost interval or the middle interval, just as we could have in the definition of $f(x)$.

The function is plotted in Figure 3.14 above.

In the example above the function was continuous at both π and $-\pi$, but the derivatives on adjacent pieces only matched at $-\pi$ but did not match at π. In the previous Example 3.2.17, page 249, we had

$$f(x) = \begin{cases} 1, & \text{if} \quad x \geq \pi/2, \\ \sin x, & \text{if} \quad -\pi/2 < x < \pi/2, \\ -x, & \text{if} \quad x \leq -\pi/2. \end{cases} \implies f'(x) = \begin{cases} 0, & \text{if} \quad x \geq \pi/2, \\ \cos x, & \text{if} \quad -\pi/2 < x < \pi/2, \\ -1, & \text{if} \quad x < -\pi/2. \end{cases}$$

In that example we have continuity at $\pi/2$ but not at $-\pi/2$, and so in the derivative formula we have to omit $-\pi/2$. Since the derivative (and function) formulas agree on the adjacent pieces at $\pi/2$, we can include that point in the derivative formula.

3.2.9 Velocities and Slopes

In this section, we have begun to occasionally graph the derivative function along with the original function. As we will see in the next section and beyond, there are as many specific

3.2. FIRST DIFFERENTIATION RULES; LEIBNIZ NOTATION

interpretations of the derivative as there are quantities which the original functions, and their inputs, represent. We will repeatedly mention that the derivative as represented by slope can somewhat unify these for purposes of geometrically analyzing the changes in the quantities themselves.

For instance, consider $v(t) = \frac{ds(t)}{dt}$, i.e., $v(t) = s'(t)$. It is occasionally interesting to consider the graph of $v(t)$, particularly when juxtaposed with that of $s(t)$ (just as we sometimes compare the graphs of f' and f). This we do in the next example. Note the sign of v when s is increasing, decreasing or briefly stationary.

Example 3.2.19

Consider a position function $s(t) = 2 + \sin t$, for $t \in \mathbb{R}$. If we were to graph $s(t)$ versus t, then the slope of the curve at each point $t \in \mathbb{R}$ would be the velocity $v(t) = s'(t) = \cos t$:

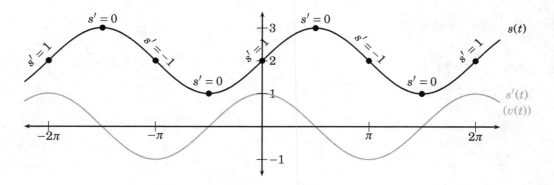

To repeat: If the one-dimensional position $s(t)$ is graphed on the vertical axis with t on the horizontal, then **the slope of $s(t)$ gives** $v(t)$; that is, the slope of the position function is the velocity function. From this we can see where the position is increasing ($v > 0$), decreasing ($v < 0$), and at least briefly stationary ($v = 0$).

Exercises

For Exercises 1–11, compute the given derivatives.

1. $\dfrac{d}{dx}\left[x^2 - 199x + 27\right]$

2. $\dfrac{d}{dx}\left[\dfrac{1}{2}x^2 + 2x\right]$

3. $\dfrac{d}{dt}\left[t^7 - 19t + 10^6\right]$

4. $\dfrac{d}{dx}\left[(x+9)(x-3)\right]$

5. $\dfrac{d}{dy}\left[10 - 9y^8\right]$

6. $\dfrac{d}{dx}(2x+5)^2$

7. $\dfrac{d}{dx}\left[\dfrac{5x^3 - 6x^2 - 8x + 100}{3}\right]$

8. $\dfrac{d}{dx}\left[\dfrac{x^4 - 16}{x^2 + 4}\right]$

9. $\dfrac{d}{dx}[5\sin x + 3\cos x]$

10. $\dfrac{d}{d\theta}[2\cos\theta - \sin\theta]$

11. $\dfrac{d}{dy}\left[10y^2 - 3y - 4\sin y\right]$

12. In a previous section we derived $f(x) = \sqrt{x} \implies f'(x) = \dfrac{1}{2\sqrt{x}}$. Show that this is consistent with the Power Rule formula from this section, even though $\sqrt{x}$ is not a positive integer power of x.

13. Repeat the previous exercise for $f(x) = \dfrac{1}{x} \implies f'(x) = -1/x^2$.

14. Graph $f(x) = x^2$, including all integer values for $x \in [-3, 3]$. On the graph label $f(x)$ and $f'(x)$ for $x = -2, -1, 0, 1, 2$, and draw tangent lines at these points. Note how the values for $f'(x)$ make sense for these values.

15. Repeat the previous exercise for $f(x) = x^3$. (Feel free to use a different scale for x and y axes. However, it may be useful to note immediately that $f'(0) = 0$ when sketching the graph.)

16. Show that

$$\dfrac{d}{dx}\left(x^2 \cdot x^3\right) \neq \left(\dfrac{d}{dx}\left(x^2\right)\right)\left(\dfrac{d}{dx}\left(x^3\right)\right).$$

Why does this not violate Theorem 3.2.5 (i.e., (3.14)), page 234?

17. Give an alternate proof of the integer Power Rule, Theorem 3.11, by using

$$a^n - b^n = (a-b)\left(a^{n-1} + a^{n-2}b \right.$$
$$\left. + a^{n-3}b^2 + \cdots + ab^{n-2} + b^{n-1}\right).$$

(Hint: a^n will be your $f(x + \Delta x)$ term, and b^n will be $f(x)$.)

18. Suppose $s(t) = 3t^2 - 2t + 19$. Find $v(t)$. Also find when the particle is moving to the right ($v > 0$) and when it is moving left ($v < 0$).

19. Graph the function $f(x) = x^4 - 4x^2$, showing all x-intercepts, all local maxima and minima. (See Example 3.2.8, page 238.)

20. Use a substitution argument to show that (3.22) is equivalent to (3.21), page 247. In other words, show that

$$\lim_{x \to a} \dfrac{f(x) - f(a)}{x - a} = \lim_{\Delta x \to 0} \dfrac{f(a + \Delta x) - f(a)}{\Delta x},$$

and thus both are equal to $f'(a)$. (Hint: A substitution that has $x = a + \Delta x$ will accomplish this, though some explanation should be included.)

For Exercises 21–26, use the fact that $\lim_{x \to a} \frac{f(x)-f(a)}{x-a} = f'(a)$ to compute the following limits. (See Examples 3.2.13–3.2.16, starting on page 246.)

21. $\lim_{x \to 2} \frac{x^4 - 16}{x - 2}$.

22. $\lim_{x \to -1} \frac{3x^9 + 3}{x + 1}$.

23. $\lim_{x \to 1} \frac{\sin x - \sin 1}{x - 1}$.

24. $\lim_{x \to \frac{\pi}{2}} \frac{\cos x}{x - \pi/2}$.

25. $\lim_{x \to 0} \frac{\cos x - 1}{x}$.

26. $\lim_{x \to 1} \frac{x^{100} - 1}{x - 1}$.

27. Use $\cos(\alpha + \beta) = \cos\alpha \cos\beta - \sin\alpha \sin\beta$ to prove (3.19), page 242:

$$\frac{d\cos x}{dx} = -\sin x.$$

It may be helpful to see the proof for the derivative of $\sin x$, and to recall $\lim_{\theta \to 0} \frac{\cos\theta - 1}{\theta} = 0$, which is basically Equation (2.88), page 197.

28. Graph $f(x) = \sin x + \cos x$ for $x \in [-2\pi, 2\pi]$, showing where this function is increasing and where it is decreasing.

29. Graph $f(x) = x + 2\cos x$ over a reasonable interval, showing where this function is increasing and where it is decreasing. Also show its behavior as $x \to \pm\infty$.

30. Consider the function

$$f(x) = \begin{cases} 2x & \text{if } x \geq 1, \\ x^2 & \text{if } -1 < x < 1, \\ -2x - 1 & \text{if } x \leq -1. \end{cases}$$

(a) For what values of x is $f(x)$ continuous?

(b) Find a (piecewise-defined) formula for $f'(x)$.

31. Repeat the instructions in the previous exercise for

$$f(x) = \begin{cases} 0 & \text{if } x \geq \pi, \\ \sin x & \text{if } 0 \leq x < \pi, \\ x & \text{if } x < 0. \end{cases}$$

32. Consider $f(x) = |x^2 - x - 6|$.

(a) Construct a sign chart for $y = x^2 - x - 6$.

(b) Define $f(x)$ piece-wise.

(c) Find a (piecewise-defined) formula for $f'(x)$.

33. Exercise 6 above must be expanded ("multiplied out") before using the Power Rule. The answer is $8x + 20$. Compute that derivative using instead the technique in Footnote 18, page 232. (There $u = 2x + 5$).

34. The voltage V across a resistor in an electrical circuit is the product of the current I and the resistance R, that is, $V = IR$. (This is Ohm's Law, discussed at length in Section 3.3.) If both the current and resistance vary with time t and are given by

$$I = 3 + 2t + 0.1t^2,$$
$$R = 20 - 0.2t,$$

then answer the following, where I is in amps (A), R is in ohms (Ω), V is in volts (V) and t is in seconds.

(a) Find dI/dt. What are the units of this derivative?

(b) Do the same for dR/dt.

(c) Do the same for dV/dt. (First simplify $V = IR$.)

(d) Show that $\frac{dV}{dt} \neq \frac{dI}{dt} \cdot \frac{dR}{dt}$.

35. Find the general rate of change dV/ds of the volume of a cube with respect to the length s of one side, given $V = s^3$. Then find the specific rate when $s = 10$ cm. What are its units?

36. Find the rate of change dA/dr of the area of a circle with respect to its radius, recalling that $A = \pi r^2$. Does the formula for dA/dr look familiar? Find both exact and approximate values fo the following.

   (a) dA/dr when $r = 10$ cm
   (b) dA/dr when $r = 100$ cm
   (c) $\lim\limits_{r_0 \to \infty} \dfrac{dA}{dr}\bigg|_{r_0}$
   (d) $\lim\limits_{r_0 \to 0^+} \dfrac{dA}{dr}\bigg|_{r_0}$

For Exercises 37–40, sketch a reproduction of the graph of $s(t)$ versus t, and match it to the graph of its derivative $s'(t) = v(t)$ from (A)–(D), drawing the appropriate derivative graph on the same grid.

37.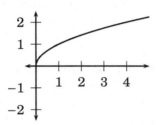

(A)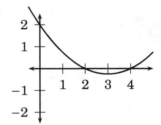

38.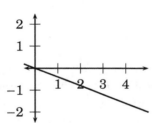

(B)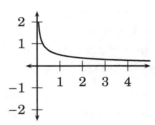

39.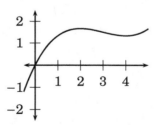

(C)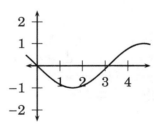

40.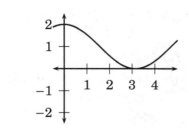

(D)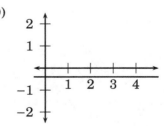

3.3. FIRST APPLICATIONS OF DERIVATIVES

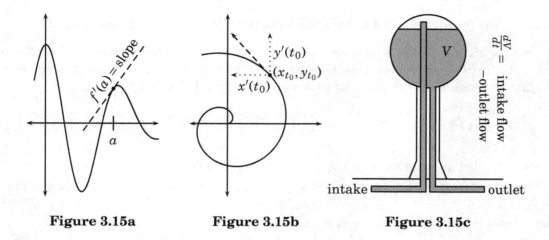

Figure 3.15a   **Figure 3.15b**   **Figure 3.15c**

Figure 3.15: Illustrations of three applications of derivatives. In Figure 3.15a we have the geometric interpretation of the derivative of the function $f(x)$: namely that $f'(a)$ is the slope of the curve at $(a, f(a))$ or, equivalently, the slope of the tangent line to the curve at that point. In Figure 3.15b we have a path where the horizontal position $x(t)$ and vertical position $y(t)$ are both functions of time t. Then there will be a horizontal velocity $x'(t)$ and a vertical velocity $y'(t)$. Both can be evaluated for a particularly time $t = t_0$. Because of the particular path and the way it is traversed here, in this case $x'(t_0) < 0$ and $y'(t_0) > 0$, so the object is moving left and rising at time t_0. (The student versed in vectors may note that the position vector is $\langle x(t), y(t) \rangle$, and then the velocity vector is $\langle x'(t), y'(t) \rangle$, also denoted $\frac{d}{dt} \langle x(t), y(t) \rangle$.) Figure 3.15c shows a water tower with a spherical tank containing liquid of volume V, whose net instantaneous fluid flow is the difference between the intake flow and the outlet flow, and this difference is the same as $\frac{dV}{dt}$.

3.3 First Applications of Derivatives

Geometrically, the derivative $f'(x)$ represents the slope of the curve $y = f(x)$ at various points $(x, f(x))$ on the curve. This graphical significance has many uses but is only one application. The simplest physics example of the derivative comes from one-dimensional motion, where if $s(t)$ is the position of an object along some number line or axis, then its derivative represents the object's velocity along that axis, and is variously denoted by $s'(t)$, $v(t)$, or ds/dt.

These two interpretations of the derivative—slope and velocity—have very different natural visualizations. (See for instance Figure 3.15a for the slope application, and Figure 3.1, page 211, for an illustration of one-dimensional velocity.)

In fact, the derivative adds much sophistication to our understanding in a variety of settings, depending on the nature of the variables involved. Fortunately, all are eventually intuitive. In this section we consider several such settings and the physical interpretations of the derivatives therein.

All physical interpretations of the derivative, regardless of setting, are unified by the theme of instantaneous rate of change of one quantity with respect to another on which it depends. (See for instance Definition 3.1.2, page 212, from the first section of this chapter.)

3.3.1 One- and Two-Dimensional Velocity: Vertical and Horizontal

We can revisit one-dimensional position and velocity in light of recently introduced differentiation (derivative computing) rules. For instance, if we have a particle with a one-dimensional position function $s(t) = 6t^2 - 9t + 15$, we immediately get the velocity function:

$$s(t) = 6t^2 - 9t + 15 \implies v(t) = \frac{ds(t)}{dt} = \frac{d}{dt}\left[6t^2 - 9t + 15\right] = 12t - 9.$$

If s is in meters and t in seconds, then $v = \frac{ds}{dt}$ is in meters/second.

If instead, $s(t) = \sin t$, then $v(t) = \frac{ds(t)}{dt} = \frac{d}{dt}\sin t = \cos t$, so a sinusoidal motion gives a similar periodic velocity. In short, anytime we know a formula for position $s(t)$, we can in theory compute velocity $v(t) = \frac{ds(t)}{dt}$, usually with relative ease.[29, 30]

So far, all velocities we have looked at were one-dimensional. If we care to consider two-dimensional velocity—velocity in the xy-plane for instance—we would look at cases where position would be given by $(x(t), y(t))$, and then we can look separately at how horizontal position changes with time, and how vertical position (height) changes with time, to have the horizontal and vertical velocities:

$$\begin{cases} x(t) = \text{horizontal position at time } t \\ y(t) = \text{vertical position at time } t \end{cases} \implies \begin{cases} x'(t) = \text{horizontal velocity at time } t \\ y'(t) = \text{vertical velocity at time } t. \end{cases}$$

These can indeed be analyzed separately, though the total motion is of course important. Where a motion is only horizontal we may opt to use $x(t)$ instead of $s(t)$, and when only vertical we may opt for $y(t)$. If motion is a combination of both, we will usually use $x(t)$ and $y(t)$, respectively. We will usually write $\frac{dx}{dt}$ and $\frac{dy}{dt}$ for these velocities $x'(t)$ and $y'(t)$, but both notations will appear. See Figure 3.15b, page 257.

Example 3.3.1 _____

From ground level, an object thrown straight up at 96 ft/sec at time $t = 0$ (in seconds) will have its vertical position in feet given by $y(t) = 96t - 16t^2$, until it hits the ground.

(a) Find how long it will be in the air.

(b) Starting at $t = 0$, find its velocity at each subsequent second until it hits the ground.

Solution: Part (a) is an algebra problem, where we wish to find all t such that $y(t) > 0$. To that end, we first solve for t when $y(t) = 0$:

$$0 = 96t - 16t^2 \iff 0 = 16t(6 - t) \iff t \in \{0, 6\}.$$

---

[29]Any curve whose graph is a transformation of a sine curve through translations, reflections, contractions, or dilations (or any combination of these) is referred to as "sinusoidal." This includes cosine curves. It is rare to come across the term "cosinusoidal," though the meaning would be clear enough.

[30]If $s(t) = \sin t$ is in meters (m) and t in seconds (sec), then a more explicit formula for s would be $s(t) = (\sin(t/\text{sec}))$ meter, and then $\frac{ds(t)}{dt} = (\cos(t/\text{sec}))$ meter/second, though many texts would simply write $s(t) = \sin t$, explain s is in meters and t in seconds, and assume the reader understands where the units would appear in the more explicit formula, and how they carry over in the derivative. Note t/sec would be dimensionless, or interpreted to be in units of radians. We can see how we get meter/second from difference quotients, but it will also be clear after the next sections where the Chain Rule is introduced, without resorting to difference quotients.

3.3. FIRST APPLICATIONS OF DERIVATIVES

We can see that the object is in the air (i.e., $y > 0$) for $0 < t < 6$. (We could use a sign chart for $y = 16t(6-t)$ to see this if we like.) Thus the object will be in the air for 6 seconds.[31]

For Part (b) we first compute $\frac{dy}{dt} = \frac{d}{dt}\left(96t - 16t^2\right) = 96 - 32t$. From this we can easily make a chart:

t	0^*	1	2	3	4	5	6^*
dy/dt	96	64	32	0	-32	-64	-96

*We include velocities at $t = 0, 6$ as if the formula for motion continued for $t < 0$ and $t > 6$. Alternatively, we can allow a right-side derivative at $t = 0$ and a left-side derivative at $t = 6$, meaning only having $\Delta t \to 0^+$ in the difference quotient at $t = 0$, and only $\Delta t \to 0^-$ in the difference quotient at $t = 6$.

Note that when $dy/dt > 0$ the object is rising (y is increasing with t), and when $dy/dt < 0$ the object is falling. Also note that we can deduce the maximum height of the object from the velocity, if we observe from experience that it will reach its maximum height at the moment $t \in (0, 6)$ it is neither rising nor falling, i.e., that moment when $dy/dt = 0$, which is at $t = 3$ (which we can see from the chart but which we should check by solving $dy/dt = 96 - 32t = 0$), at which time the height is $y(3) = 96(3) - 16(3)^2 = 288 - 144 = 144$, i.e., $y = 144$ ft.

Example 3.3.2

An object is thrown horizontally at time $t = 0$ off of the top of a 200-ft building at 50 ft/sec, arcing to the ground, which is flat and horizontal. How far from the base of the building will it land, and at what vertical velocity will it hit the ground? The equations for its position before it strikes the ground will be the following, with x and y in feet and $t \geq 0$ in seconds: [32]

$$x(t) = 50t,$$
$$y(t) = 200 - 16t^2.$$

Solution: The object will be in flight until $y(t) = 0$, which we solve for t:

$$200 - 16t^2 = 0 \iff 16t^2 = 200 \iff t = \pm\sqrt{200/16} = \pm\sqrt{2 \cdot 100/16} = \pm\frac{10}{4}\sqrt{2} = \pm\frac{5}{2}\sqrt{2}.$$

Since we do not know anything about the period $t < 0$, and in any event are not interested in such times here, we will conclude that the only relevant solution is $t = \frac{5}{2}\sqrt{2} \approx 3.54$. At that time,

$$x\left(\frac{5}{2}\sqrt{2}\right) = 50 \cdot \frac{5}{2}\sqrt{2} = 75\sqrt{2} \approx 106.$$

We conclude that the object will land approximately 106 ft from the base of the building. We also compute $dy/dt = -32t$, so

$$y'\left(\frac{5}{2}\sqrt{2}\right) = -32 \cdot \frac{5}{2}\sqrt{2} = -80\sqrt{2} \approx -113,$$

that is, the vertical speed at which the object hits the ground will be approximately 113 ft/sec, though $dy/dt < 0$ when the object hits the ground, so it is (clearly) falling.

---

[31]One could argue it is not in the air for a "full six seconds," but it is customary to assign the length of $b - a$ to any interval with endpoints a, b with $a < b$, be it $[a, b]$, (a, b), $[a, b)$, or $(a, b]$. Thus the length of the time interval (0, 6 sec) is 6 seconds; there really is no other length we can logically assign to it.

[32]Note that $y(t) = 200 - 16t^2$, if we include all the units, would actually read $y(t) = 200$ ft $- \frac{16\text{ft}}{\text{sec}^2}t^2$. This would imply also that dy/dt will have units of ft/sec. Units are often omitted from (or "suppressed in") the equation when it is well understood what they should be.

These two examples use formulas, which will be developed later, for motion which assume that air resistance is negligible, and that the ground is horizontal and flat.

3.3.2 Volume and Fluid Flow in Liquids

The relationship between total volume of a liquid in a container, such as a tank, and the net fluid flow into or out of the tank, is another which can be described using derivatives. The volume V of liquid in the tank can be given as a function of time t. For the volume to change (barring evaporation, condensation, and the like), fluid must flow into or out of the tank. The net fluid flow rate into or out of the tank would be the same, except for perhaps a sign $+/-$.[33]

We will take the *net flow rate into the container* to be the difference between the rate at which any liquid is flowing *into* the container, and the rate at which any liquid is flowing *out of* the container. We then conclude that, if V is the volume of liquid in the container, it changes only by having liquid pumped into or out of the container. Upon reflection, it should seem reasonable that

$$\frac{dV}{dt} = \text{(flow rate of liquid into the container)} \qquad (3.23)$$
$$- \text{(flow rate of liquid out of the container)}.$$

If $dV/dt > 0$, then more fluid is flowing into the container than out of it, while if $dV/dt < 0$ more fluid is flowing out of the container than into it. If we have fluid flowing into the container at 5 ℓ/sec but also flowing out of the container at 3 ℓ/sec (perhaps from a leak or open spigot), the net flow rate would be 2 ℓ/sec into the container, and we would write either $dV/dt = 5\ \ell/\text{sec} - 3\ell/\text{sec}$ or simply $dV/dt = 2\ \ell/\text{sec}$. If, on the other hand, fluid is flowing out at 10 ℓ/sec and in at 4 ℓ/sec, then the net flow is 6 ℓ/sec *out of* the container and $dV/dt = -6\ \ell/\text{sec}$.

Example 3.3.3

Suppose the volume of water in a tank is given by $V(t) = t(10-t)$, $0 \leq t \leq 10$. Then the volume of liquid in the tank is changing at a rate (for $0 < t < 10$) of

$$\frac{dV}{dt} = \frac{d}{dt}\left[10t - t^2\right] = 10 - 2t = 2(5-t).$$

If V is in gallons and t is in minutes, then $\frac{dV}{dt}$ is in gallons/minute. The maximum volume of the tank occurs at $t = 5$, since before then we have $\frac{dV}{dt} > 0$, while after $t = 5$ we have $\frac{dV}{dt} < 0$. The actual maximum volume is then $V(5) = 5(10-5) = 25$. If we were to graph $V(t)$ for this situation, we would produce a picture as in Figure 3.16, page 261.

Note from the graph in Figure 3.16, page 261, how the volume increases when $dV/dt > 0$, i.e., when $t \in (0,5)$, and decreases when $dV/dt < 0$, on $t \in (5,10)$. In fact, one often looks where a derivative is zero to find where a function might be maximized, or for that matter minimized. This will be discussed at length as we continue to discuss derivatives.

---

[33] Again, this assumes no gains or losses in the volume of liquid from such things as evaporation, condensation or chemical reactions. We also assume that the liquid is "incompressible," so we cannot "squeeze" it into a smaller volume. In reality, we may also need to assume the liquid is of constant temperature, because liquids can change volumes when heated or cooled, a phenomenon useful in constructing mercury thermometers.

3.3. FIRST APPLICATIONS OF DERIVATIVES

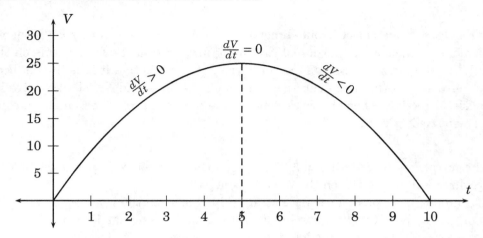

Figure 3.16: Graph of volume versus time from Example 3.3.3, page 260, where $V = t(10-t) = 10t - t^2$, so $\frac{dV}{dt} = 10 - 2t$. The maximum volume occurs at time $t = 5$. It is common for a maximum or minimum of a quantity to occur where its derivative is zero.

3.3.3 Zero-Order Chemical Reactions

Most chemical reactions occur at a rate dependent on the concentration of the reactants, i.e., the chemicals that react and are consumed in the course of a chemical reaction. This is not the case with zero-order chemical reactions, which can occur under certain circumstances. If the chemical formula for a reactant is A, then the concentration of A is often given by the notation $[A]$.

For notational simplicity, we will let $X = [A]$. In a zero-order reaction, we will have

$$X = X_0 - kt,$$

where $k > 0$ is a rate constant, X_0 is the initial concentration of A, and t is time. Then the rate of change of the concentration X with respect to time is given by

$$\frac{dX}{dt} = \frac{d}{dt}\left[X_0 - kt\right] = -k,$$

which is constant (the defining characteristic of zero-order chemical reactions). This rate is negative because the reactant is being consumed and so its concentration is decreasing.

In practice, these equations hold true until one of the reactants is completely consumed and the reaction ceases, or until some other mechanism causes this rate to change.[34]

3.3.4 Energy, Work, and Power

Work is first defined as a force (push or pull) acting on an object as it displaces in the direction of the force. We can extend this idea to include both positive and negative work. Assuming

---

[34]There are also, for instance, first-order ($dX/dt = -kX$) and second-order ($dX/dt = -kX^2$) reactions, but we will need some more derivative rules to be able to analyze these higher-order reactions, which we will do in Section 3.11.

that the motion of the object is one-dimensional, we can choose an axis collinear with this displacement for measuring purposes. Assuming further that the force acting on the object is parallel to this axis, we can define the force F to be positive if it is in the direction of the ray representing the nonnegative numbers on the axis, and F to be negative if it is in the opposite direction. Then the work done by the constant force F on an object through its displacement Δs is given by

$$W = F\Delta s. \qquad (3.24)$$

If the force F is in the direction of Δs, then the product W is positive. If F is in the opposite direction from Δs, then the work W is negative.

For example, if we push a car up a straight incline, then the work we do is positive (the motion of the car is in the direction in which we push it), whereas if we push against a car that is rolling down the same incline in order to attempt to slow its motion, the work we do to the car is negative because the motion of the car is in the opposite direction as the force we exert. (Worth noting is that the work done to us by the car is then positive.)

In the English system, force is given in units of pounds (lbs) and displacement in feet (ft), and so work is given in foot-pounds (ft-lbs).

In the metric system of units, force is defined in newtons (N) and distance in meters (m). The work accomplished in pushing an object using 25 N of force over a distance of 4 m would then be 25 N · 4 m = 100 N-m.[35]

While the N-m is an acceptable unit of work, for convenience this was renamed as the joule, written J:[36]

$$J = N \cdot m. \qquad (3.25)$$

Thus we could rework our original example and claim that the work performed by using 25N over 4m would be

$$25 \, N \cdot 4 \, m = 100 \, J.$$

Energy is defined as the ability to do work, and quantitatively is considered equal to the work that can be performed by the energy. Similarly, work done is the same as the energy spent performing the work. Thus we could also say that 100 J of energy was expended in using 100 J of work to move this object.

It is well known that under everyday circumstances—not involving, say, nuclear energy where energy is produced by converting mass to energy—energy will be "conserved," meaning that it is neither created nor destroyed, although it can be converted to different forms or passed from one object to another. Energy forms include heat, light, electrical, magnetic, kinetic (energy due to motion), sound (acoustic), chemical, and nuclear. Some of these are

---

[35]To give some perspective, one newton (lower-case when not a proper name) is equal in force to approximately 0.2248089 pounds (or lbs), or roughly the weight of two large chicken eggs, or (as some authors point out) one medium apple. One pound of force is thus approximately 4.4482216 newtons.

If a reader familiar with the length units of the English system of measures can remember that one inch is now *defined* to be exactly 2.54 cm, then converting meters to feet is simple enough:

$$1 \, m = 1 \, m \times \frac{100 \, cm}{m} \times \frac{in}{2.54 \, cm} \times \frac{ft}{12 \, in} \approx 3.280839895 \, ft.$$

[36]Named for James Prescott Joule (1818–1889), an English physicist who discovered the relationship between heat and work.

3.3. FIRST APPLICATIONS OF DERIVATIVES

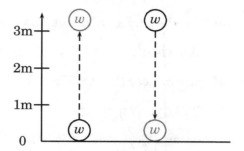

Figure 3.17: Potential and kinetic energy example. When an object with weight w in newtons N is at height 0 and not moving, it has zero potential and kinetic energy. If work is performed to lift it 3m where it stops, the total work required (force times displacement) is $w \cdot 3$ m, with units of N-m or joules (J). That work (energy) is then described as "stored" in the object as potential energy, also equal to $w \cdot 3$ m. If the object is then stopped at that height, it has zero kinetic energy, because kinetic energy is the energy it possesses due to its motion. If the object is then allowed to drop in free-fall, it will release this potential energy in the form of kinetic energy, as gravity performs work on it, and it will have increasing kinetic energy until it reaches the ground, at which time it will have given up all its potential energy to kinetic energy $w \cdot 3$ m, and then on impact it will release the kinetic energy (usually as heat, work performed on the ground, sound, and perhaps energy required to deform the weight and the earth). This does assume losses due to air resistance are negligible.

just reinterpretations of each other, as for instance light is considered to be a form of electromagnetic energy.

For conceptual reasons, mechanical energy is often broadly categorized into two types:

Potential energy: Energy an object is capable of making available to do work. This is sometimes called stored energy.

Kinetic energy: Energy an object possesses due to its motion.

We will show in a later chapter (Footnote 50, page 600) that kinetic energy can be given by

$$E_k = \frac{1}{2}mv^2 \qquad (3.26)$$

where m is the mass of the object and v is its velocity. In Figure 3.17 we can solve for v when the object hits the ground (or any height during free-fall) because the energy (work) expended on the object to lift it will be wh, where w is its weight, h is the height to which it is raised, and the kinetic energy it has at the bottom of its path will be $\frac{1}{2}mv^2$.

Sometimes we will have use for Newton's Second Law of Motion, given in (3.27), which states that an object with mass m subject to net force F will experience an acceleration a relate by

$$F = ma. \qquad (3.27)$$

Then the weight w of an object is a measure of the force of gravity on the object, and so if there are no other forces on the object, i.e., if the object were in "free-fall," then to know its weight we would need to know how gravity would accelerate the object. The acceleration due to gravity is given by g, which on the surface of the earth we take to be[37]

$$g \approx 9.8 \text{ m/sec}^2, \quad \text{or} \quad g \approx 32 \text{ ft/sec}^2. \tag{3.28}$$

The weight w of an object of mass m near the surface of the earth is then

$$w = mg. \tag{3.29}$$

With these, we can find v when an object lands if it is lifted a height h and then dropped (assuming negligible air resistance), because the kinetic energy ($\frac{1}{2}mv^2$) as it hits the ground will be equal to the potential (stored) energy it had at its maximum height, which is equal to the work ($wh = mgh$) done to lift it there, and so

$$\underbrace{mgh}_{\text{top}} = \underbrace{\frac{1}{2}mv^2}_{\text{bottom}}. \tag{3.30}$$

Example 3.3.4

Suppose the object in Figure 3.17, page 263, has mass 10 kg and has velocity zero at height 3 m.

1. What is its velocity as it hits the ground?
2. What is its velocity in free-fall at height 1.5 m?

Solution: (Note the difference between symbols for mass m and meters m.)

1. We can use (3.30) to find its velocity as it hits the ground, since at that point it has released all of its potential energy it possessed from being lifted to the height of 3 m as kinetic energy:

$$mgh = \frac{1}{2}mv^2 \implies m(9.8\,\text{m/sec}^2)(3\,\text{m}) = \frac{1}{2}mv^2$$
$$\implies 2(9.8\,\text{m/sec}^2)(3\,\text{m}) = v^2$$
$$\implies \pm\sqrt{2(9.8\,\text{m/sec}^2)(3\,\text{m})} = v$$
$$\implies \pm 7.67\,\text{m/sec} \approx v.$$

One could argue that since the motion is downward, we should use $v \approx -7.67\,\text{m/sec}$.

2. As the object descends from height 3 m to height 1.5 m, the potential energy difference is

$$mg(3\,\text{m}) - mg(1.5\,\text{m}) = mg(1.5\,\text{m}) = 10\,kg \times 9.8\,\text{m/sec}^2 \times 1.5\,\text{m} = 147\,\text{J}.$$

That potential energy lost became kinetic energy, added to the object that had zero kinetic energy at height 3 m. And so at height 1.5 m we have

$$\frac{1}{2}(10\,\text{kg})v^2 = 147\,\text{J} \implies v = \pm\sqrt{\frac{2 \times 147\,\text{J}}{10\,\text{kg}}} = \pm\sqrt{29.4}\,\text{m/sec} \approx \pm 5.42\,\text{m/sec}.$$

Recalling that $1\,\text{J} = 1\,\text{kg-m}^2/\text{sec}^2$, we see that the correct units do emerge. We report $v \approx -5.42\,\text{m/sec}$.

---

[37]The value $9.8\,\text{m/sec}^2$ is denoted g and is the approximate acceleration due to gravity near the surface of the earth assuming zero wind resistance. It is also often taken to be $32\,\text{ft/sec}^2$. At a latitude near $45°$ this value was measured to be closer to $9.80665\,\text{m/sec}^2$ or approximately $32.1740\,\text{ft/sec}^2$. For our purposes in this textbook, we will use the 9.8 and 32 as approximations, respectively, as one finds in typical physics texts.

3.3. FIRST APPLICATIONS OF DERIVATIVES

Power is a measure of how quickly energy is available (or spent). If the total energy produced in a process as a function of t is given by $E = E(t)$, then power $P = P(t)$ is given by

$$P = \frac{dE}{dt}. \tag{3.31}$$

Average power over a time interval $[t_0, t_f]$ would then be $\frac{\Delta E}{\Delta t} = \frac{E(t_f) - E(t_0)}{t_f - t_0}$. One could describe power as a rate of "flow" of energy. The unit of power is the joule/second, also known as the watt, written W.[38]

If 100 J of work is performed in 2 seconds, then the average power used during those seconds was (100 J)/(2 sec) = 50 J/sec = 50 W. If the same work is accomplished in 1 second, the power required is (100 J)/(1 sec) = 100 W.

Example 3.3.5

A bullet weighing 50 g (grams) exits a rifle barrel at 750 m/s.

1. What is its kinetic energy in Joules?

2. If the total time the bullet is accelerated within the rifle barrel is 0.0015 sec, what was the average power required to accelerate the bullet within the barrel? (Ignore other energy consumed to heat the bullet, spin the bullet, recoil the gun, or produce the sound from firing the rifle.)

3. If instead a gunmaker wishes a 25 g bullet to have the same kinetic energy, what must its velocity be?

Solution: Recall the kinetic energy of the bullet will be given by the equation $\frac{1}{2}mv^2$, where v is the velocity.

(1) The mass of the bullet is $50 \text{ g} = 50 \text{ g} \times \frac{1 \text{ kg}}{1000 \text{ g}} = 0.05$ kg. Thus the kinetic energy is

$$\frac{1}{2}mv^2 = \frac{1}{2}(0.05 \text{ kg})(750 \text{ m/s})^2 \approx 14{,}060 \text{ J}.$$

(2) To find the average power, we take the kinetic energy of the bullet divided by the time required to produce the energy:

$$\text{Power} = \frac{14{,}060 \text{ J}}{0.0015 \text{ sec}} \approx 9.4 \times 10^6 \, \frac{\text{J}}{\text{sec}} = 9.4 \times 10^6 \text{ W}.$$

(3) In order to have a bullet with half of the mass still possess that same amount of kinetic energy, we would need its velocity to satisfy $\frac{1}{2}(0.025 \text{ kg})v^2 \approx 14{,}060$ J. Solving:

$$\frac{1}{2}(0.025 \text{kg})v^2 \approx 14{,}060 \text{kg-}\frac{\text{m}^2}{\text{s}^2} \iff 0.025 \text{kg} v^2 \approx 28{,}120 \text{kg-}\frac{\text{m}^2}{\text{s}^2}$$

$$\iff v^2 \approx 1{,}124{,}800 \, \frac{\text{m}^2}{\text{s}^2} \iff v \approx \sqrt{1{,}124{,}800} \, \frac{\text{m}}{\text{s}}.$$

Note that if we solved $\frac{1}{2}(50\text{g})(750\text{m/sec}^2) = \frac{1}{2}(25\text{g})v^2$ exactly, we would arrive at $v = 750\sqrt{2} \, \frac{\text{m}}{\text{s}} \approx 1060 \, \frac{\text{m}}{\text{s}}$.

---

[38] Named for James Watt (1736–1819), a Scottish engineer and inventor, known especially for the unit that bears his name, and for his work in greatly improving steam engines which considerably hastened the Industrial Revolution, as his work unlocked the potential power of steam-powered engines. While studying power, he developed the concept of horsepower as a unit which, depending on the underlying definition, varies between 735 W and 750 W.

Thus if we halve the mass of the bullet, we can compensate to have the same kinetic energy by increasing the bullet's velocity by a factor of $\sqrt{2} \approx 1.414$. This is also intuitive from the form of the expression for kinetic energy, namely $\frac{1}{2}mv^2$.

3.3.5 Electrical Charge and Current

Analogous to the volume-fluid flow relationship is the relationship between total electrical charge within, and current flowing out of, say, a capacitor or battery. Differences arise because there are signs associated with the charges that are "flowing," and positive and negative charges within the object "cancel" each other, so that only the net charge is relevant in the computation of Q. It is complicated even more by conventions according to which the relevant charged particles are called positive or negative, and by two different ways of looking at the flow of current. Either way, the relevant equation is

$$I = -\frac{dQ}{dt}, \qquad (3.32)$$

where I is the current (rate of charges flowing and their signs), and Q is the total net charge usually given in units of coulombs.[39]

As usual, t is in seconds. Then dQ/dt is given in coulombs/second, also known as amperes, or amps, denoted A.[40]

One coulomb of charge is the charge of approximately 6.24×10^{18} protons. If a one-amp current flows for exactly one second, there will be "one coulomb" of electrons flowing, though technically it would be -1 coulomb of charge since electrons carry negative charges.[41]

The fundamental carriers of charge in electrical currents are electrons which are negatively charged, and protons whose charge is positive. Atoms, of which all everyday matter in solid, liquid, or gas form is constituted, are formed by positive nuclei consisting of positively charged protons and neutrally charged neutrons, surrounded by clouds of electrons in various energy "shells." Since like, non-neutral charges (+/+ or −/−) repel and unlike,

---

[39]Named for Charles-Augustin de Coulomb (1736–1806), a French physicist who discovered Coulomb's Law, which describes the attraction or repulsion between charged particles, depending on their charges and the distance between them. He also did some important theoretical work on friction. There are many "giants" of electrical theory, with units named for them, as we will see periodically.

[40]Named for André-Marie Ampère (1775–1836), a French physicist and mathematician, who expounded greatly on the discovery by Hans Christian Oersted (Danish physicist and chemist, 1777–1851) that an electrical current can move a compass needle, that is, create a magnetic field. Oersted eventually mapped the magnetic field surrounding the wire, and Ampère's research netted many more principles of how electrical currents in conductors cause them to either attract or repel, depending on the directions of the currents and the placement of the wires. He called this line of investigation *electrodynamics*, but today it is considered a subset of *electromagnetism*. Electric motors are examples of applications of Ampère's principles.

[41]It is for historical reasons that the proton was assigned a positive charge, and the electron's charge therefore assigned to be negative. It was the American scientist and statesman Dr. Benjamin Franklin (1706–1790) who noted the existence of opposite natures of charges and gave them signs (+/−), which ultimately implied which respective signs would be given to the proton and electron. Many physicists have opined that it would have been better to have the names of the charges reversed, since one "adds" to electrical current by adding more electrons to the current. Instead, students of physics and electronics learn to look at the movement of charges (electricity) two ways: as movement of electrons, and as movement of the positions of positive charges (see Figure 3.18, page 268.)

3.3. FIRST APPLICATIONS OF DERIVATIVES

non-neutral charges (+/−) attract, total charge of a mass of matter tends to be zero, as any atom with, for instance, a net positive charge will attract electrons or other atoms with net negative charges, and repel those with similar positive charges. Similarly, an atom with a net negative charge will be easily coaxed into giving up its extra electrons for an atom with a net positive charge. There are other, nonelectric forces that can pry an electron from one atom to another, but then the two will usually stay within each other's proximity due to the attraction because they have opposite net charges (and sometimes for other chemical bonding reasons as well). And there are nuclear forces keeping the protons from ejecting each other from the nucleus, but we will not discuss them here.

By expending energy, one can force masses of atoms to have a net positive or net negative charge. This occurs in batteries and capacitors, where positive and negative charges are segregated, at least temporarily (total charge still being zero). When these two reservoirs of charges are connected by a conductor or other path (perhaps with some resistance), charges can flow to attempt to reestablish the normal state of things where both reservoirs have a zero net charge. Similarly, two objects experiencing friction between each other can each develop nonzero net charge—one with a net positive charge and the other a net negative charge—and then if the objects come back together, or in contact with a charge "sink" such as a grounded metal object, the charges can come back into equilibrium quickly, as attestable by anyone who has received a high-power (even if low-energy) "static electricity" shock.

Electricity refers to the flow of charges. This is almost always accomplished by the movement of electrons, the negatively charged particles, which are much lighter (9.11×10^{-31} kg) and move much more freely than the relatively heavy atoms or even a proton (1.67×10^{-27} kg), which is usually bound to an atomic nucleus. So if we follow actual moving particles, we should say that electrical current flows from negative to positive.

However, a casual reader of electrical or electronic literature is likely to note that sometimes "current" is said to flow from positive to negative, as if protons were moving that direction. Indeed, for many purposes engineers and technicians find it useful to follow the "flow" of positive charges, though in fact they are following the "holes" left by vacating electrons. Note that when an electron leaves an otherwise neutral atom, leaving a "hole" in the atom's electron shells, the atom is rendered positive, and attractive to other electrons. This flow of "holes" in the opposite direction as the flow of electrons is illustrated in Figure 3.18, page 268.

A battery is a device for storing electrical energy and consists of a *cathode*, which is capable of holding a positive charge, an *anode*, which is capable of holding a negative charge, and a separator.[42]

If a battery is placed in a *circuit* allowing electrons to flow externally from the anode to the cathode, these charges may be able to perform useful work, and in the process will somewhat neutralize the charges of the anode and cathode, thus partially *discharging* the battery. Battery theory is somewhat complicated by the fact that the chemical composition of components of the anode and cathode can sometimes change as their charges change. Depending on the chemistry, the battery may be rechargeable, reversing the process and thus letting the battery again store energy. Charging occurs when an external supply of

---

[42] The names "cathode" and "anode" correspond to names given to the different charges ions can carry: positive ions are also called *cations*, and negative ions are called *anions*. Incidentally, the term *ion* was coined by Michael Faraday, whose name will appear in a later footnote.

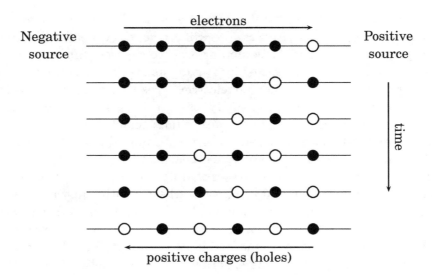

Figure 3.18: Illustration of the motivation for two types of current paradigms: electronic current, which follows the movement of electrons; and conventional current, which follows the movement of positive charges. As electrons move to the right, positive charges (holes) left by the vacating electrons move to the left. Electronic current is movement of electrons (− to +), where conventional current is movement of the positive charge locations (+ to −).

electrical energy is applied to the battery in such a way that a positive wire from the charger (or similar source) is attached to the cathode's terminal and the negative wire of the charger is attached to the anode's terminal.

Example 3.3.6

A battery is connected to a charger in such a way that the total charge of the battery's cathode is given by $Q = t + \sin t$, where Q is in coulombs and t is in seconds, until the battery can no longer accept new charge. What is the current I coming from the battery's cathode as a function of time t during this charging?

Solution: Current in coulombs/second, i.e., amps, will be given by

$$I = -\frac{dQ}{dt} = -\frac{d}{dt}(t + \sin t) = -(1 + \cos t).$$

For this example, the following graph in Figure 3.19, page 269, shows some early values for Q, $\frac{dQ}{dt}$ and $I = -\frac{dQ}{dt}$.

In this example, note that the battery is, over time, accumulating (positive) charge Q at the cathode, and $-Q$ at the anode, as the charger is sending electrons into the anode and pulling them from the cathode. The accumulation is not occurring at a constant rate, so for instance at $t = \pi$ the charger is sending no charges, and the total charge in the battery's anode levels off momentarily. At $t = 2\pi$ the charger is pulling electrons at the cathode at a rate of 1A (so they come out of the battery with a negative charge, hence $I = -1$A), and we can see the charge Q building more rapidly at that time, as the slope of Q is steeper there.

3.3. FIRST APPLICATIONS OF DERIVATIVES

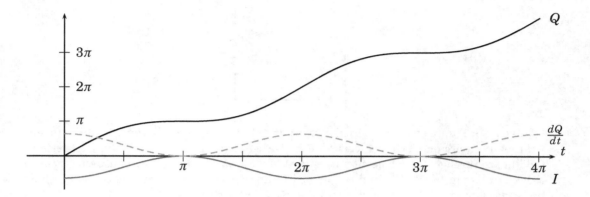

Figure 3.19: Graph for early values of t for Example 3.3.6, page 268

3.3.6 Ohm's Law

One of the most basic laws in electrical theory is Ohm's Law.[43] It describes how much current I that a voltage of V will force to pass through a material that measurably resists the passage of the current. There are many reasons to employ standard *resistors* in a circuit, one reason being to regulate voltages and currents passing through various components. Incandescent light bulbs and heating elements in electrical heaters are basically resistors that rob the flowing charges of some of their energy and in the process give off energy as light and heat.[44]

Resistance is measured in standard units of ohms which are conveniently given the symbol Ω. The usual form given for Ohm's law is

$$V = IR. \tag{3.33}$$

This can describe a simple circuit such as that drawn in Figure 3.20, page 270. Note that a resistor will be shown as a zig-zagging line, signifying the difficulty with which charges pass,

---

[43] Named for Georg Simon Ohm (1789–1854), a German physicist and teacher, mostly self-taught, with initial home-schooling from his also self-taught father. The law that bears his name was first found in his 1827 treatise on his complete theory of electricity.

[44] While it is true that resistors rob the flowing charges of energy and power, the amounts can at first be counterintuitive. For instance, a light bulb designed to consume 60 W of power when wired to a 120 V outlet will have a resistance of $240\,\Omega$, so that $I = (120\,\text{volt})/(240\,\Omega) = 0.5\,\text{A}$, which gives us its power (described later in this section) consumption of $P = IV = 0.5\,\text{A} \times 120\,\text{volt} = 60\,\text{W}$. However, a similar computation shows that a 30 W bulb will have a resistance of $480\,\Omega$, giving half of the current but at the same 120 volt. So the bulb with higher resistance (counterintuitively?) robs the current of less power. This is because the higher resistance will let fewer electrons (less current) pass in a given amount of time, and will therefore extract less power from them (and less energy from their source).

On the other hand, if we had two unequal resistors in series, while the total resistance is higher than each one alone and so the circuit consumes less power, the resistor with the higher resistance will consume more power than the resistor with the lower resistance, because (by Ohm's Law and the fact that the same current passes through both resistors) the resistor with the higher resistance will have a larger "voltage drop" across it. (Voltage is the measure of energy per coulomb. Higher voltage implies the particles are, on average, individually more energetic.)

While the basic ideas of electricity are simple enough, analyzing actual circuits often requires very careful mathematical computations. The results afterwards seem intuitive, but often only after they are carefully, mathematically established.

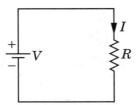

Figure 3.20: Figure showing a circuit with voltage V coming from a battery, sending current I through the circuit which has resistance R. The current that can pass through the resistor (and hence through the whole circuit) is given by $I = V/R$. Ohm's Law is usually given in the equivalent form $V = IR$. In the figure, current is drawn as "conventional current," which proceeds clockwise, while actual electrons move counter-clockwise.

compared to the straight-line conductors, which are assumed to have negligible resistance. Also note the symbol for a battery on the left of the circuit.

From the equation, we can see that if a voltage V is placed across a circuit with resistance R, the circuit will allow $I = V/R$ current to flow. For a fixed voltage, doubling the resistance would halve the current allowed to flow. On the other hand, for a fixed resistance, doubling the voltage would double the current flow.

Note that (3.33) clearly holds when $V, I, R = 1$, i.e., when $V = 1\text{volt}$, $I = 1\text{A}$ and $R = 1\Omega$. Since we can also write $R = V/I$, we can use this example of a circuit with 1Ω of resistance and 1 volt of voltage causing $1A$ of current to flow, and note therefore that

$$1\Omega = \frac{1 \text{ volt}}{1\text{A}}, \qquad (3.34)$$

and so to overcome 1 ohm of resistance requires one volt per amp of current flowing.

Example 3.3.7 _____

If we have a circuit that always has 5Ω of resistance, and variable current $I = (2\cos t)\text{A}$ where t is in seconds, then find

1. the voltage V in the circuit, as a function of time;

2. the rate of change of the current with respect to time;

3. the rate of change of the voltage with respect to time.

Solution: When we know we are using standard units, so $V = IR$ here, we have the option of carrying units through our computation, or suppressing them and interpreting the results in their proper units. Here we will include units, though to do so can clutter the computations somewhat.

1. $V = IR = (2\cos t)\text{A} \cdot 5\Omega = 10(\cos t)\text{A}\Omega$, which by (3.34) can be rewritten $V = 10(\cos t)\text{volt}$.

2. Since $I = (2\cos t)\text{A}$, we can write $\frac{dI}{dt} = \frac{d}{dt}[(2\cos t)\text{A}] = -2(\sin t)\text{A/sec}$.

3. Similarly, $\frac{dV}{dt} = \frac{d}{dt}[10(\cos t)\text{volt}] = -10(\sin t)\text{volt/sec}$.

3.3. FIRST APPLICATIONS OF DERIVATIVES

Example 3.3.8

Find the time rate of change of voltage V across a resistor (as in Figure 3.20, page 270) at time $t = 1\,\text{sec}$ if $R = 20 - 0.1t^2$ and $I = 3 + 0.2t^3$.

Solution: Here we again use $V = IR$ (suppressing units), and compute

$$V = IR = (3 + 0.2t^3)(20 - 0.1t^2) = -0.02t^5 + 4t^3 - 0.3t^2 + 60$$

$$\implies \frac{dV}{dt} = -0.1t^4 + 12t^2 - 0.6t$$

$$\implies \left.\frac{dV}{dt}\right|_{t=1} = -0.1 + 12 - 0.6 = 11.3$$

with the final units being understood to be volt/second.

Exercises

1. Assume position s is given in feet, and time is given in seconds.

    (a) What are the units of velocity $v = \frac{ds}{dt}$?

    (b) What are the units of acceleration $a = \frac{dv}{dt}$?

    (c) What are the units of the jerk $\frac{da}{dt}$? (See Footnote 88, page 391.)

2. An object travels in a circular path given by

    $$x(t) = 2\cos t,$$
    $$y(t) = 2\sin t.$$

    (a) Graph the positions of the object at times $t = 0, \frac{\pi}{2}, \pi, \frac{3\pi}{2}, 2\pi$, and label those points on the graph as $t = 0, \frac{\pi}{2}, \pi, \frac{3\pi}{2}, 2\pi$, respectively. Be sure the graph is large enough to accommodate the information in the next part.

    (b) Compute $\frac{dx}{dt}$ and $\frac{dy}{dt}$.

    (c) Find the vertical and horizontal velocities dx/dt and dy/dt at each of the times listed, and list these velocities "$dx/dt = \_$," "$dy/dt = \_$."

    (d) Do the velocities make sense for each given position?

3. Repeat Exercise 2 but for the times $t = \frac{\pi}{4}, \frac{3\pi}{4}, \frac{5\pi}{4}, \frac{7\pi}{4}$.

4. Repeat Exercise 2 but with

    $$x(t) = 2\sin t,$$
    $$y(t) = 2\cos t.$$

5. A *cube* is a solid object bounded by six identical squares (faces of the cube) of uniform side length s. The point where three faces meet is a *vertex* of the cube, consisting of three right angles. The volume of a cube of side length s is therefore $V = s^3$. Find the rate of change of the volume of a cube with respect to side length s when $s = 10$ cm.

6. A *circle* is the set of all points in a plane, which are some fixed distance $r > 0$, called the circle's *radius*, from a fixed point in the plane, called the circle's *center*. The area bounded by

the circle, usually called the "area of the circle," is given by $A = \pi r^2$. Find the rate of change in A with respect to r when $r = 15$ cm.

7. A *sphere* is the set of all points in space which are the same distance $r > 0$, called the sphere's *radius*, from a fixed point called the *center* of the sphere. A sphere will form the boundary of a volume $V = \frac{4}{3}\pi r^3$, this V usually referred to as the "volume of the sphere."

   (a) Find the rate of change of the volume of a sphere with respect to its radius.

   (b) At what radius will $\frac{dV}{dr} = 20\pi$ cm^3/cm?

8. If we graph $s(t)$ on a vertical axis, versus t on a horizontal axis, what is the geometrical significance of $v(t)$?

9. The total charge passing a point in a circuit is given by the equation

   $$Q(t) = 2t^3 + 3t.$$

   (a) Find the current as a function of time t.

   (b) Find the current when $t = 4$ sec.

10. Suppose that a circuit has a resistance of $10\,\Omega$, and the current is given as a function of time t by the equation $I = \left(2t^3 - 3t^2 + 11\right)$ A, where t is in seconds. (Here A is the unit of amperes.)

    (a) Find the voltage V as a function of time t.

    (b) Calculate dI/dt.

    (c) Calculate dV/dt at $t = 2$ sec.

11. Given that power is the rate of change of work with respect to time, suppose that the work done on an object is given as a function of time by the equation

    $$W(t) = 50 - 2t^2 + 2t^3$$

    with units of Joules. Find the power at time $t = 2$ sec.

12. Suppose a large water balloon containing only water is able to keep a spherical shape regardless of the volume of water it contains. Assume that the radius is given in units of feet by $r = 9 - t$ for $t \in [1, 8]$, where t is in units of minutes.

    (a) Find the flow rate of water from the balloon for $t \in (1, 8)$. (Hint: Find V as a function of t. Note $V = \frac{4}{3}\pi r^3$.)

    (b) Find the flow rate of water from the balloon for time values $t = 2, 4, 7$. What is the trend?

13. Suppose a 10 kg weight is lifted 100 m high and dropped. What is the velocity with which it will hit the ground, if air resistance is negligible?

14. A baseball weighs approximately 145 g. A typical baseball pitcher uses a path of approximately 1.3 m to accelerate the baseball from 0 to approximately 90 mi/hr. What is the average power used to accomplish this? (Note that 2.54 cm = 1 in, 12 in = 1 ft, and 5280 ft = 1 mi.)

3.4. CHAIN RULE I

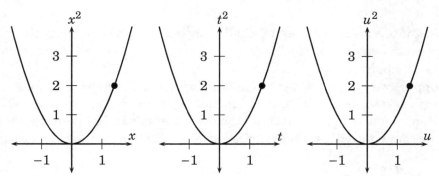

Figure 3.21: Three identical graphs: x^2 versus x, t^2 versus t, and u^2 versus u. All have the same slope—though these are dubbed $\frac{dx^2}{dx} = 2x$, $\frac{dt^2}{dt} = 2t$ and $\frac{du^2}{du} = 2u$, respectively—at each fixed horizontal axis value. One such value is represented by a black dot on each of the three graphs (at $(\sqrt{2}, 2)$), where the input and output variables agree, as do the slopes ($2\sqrt{2}$).

3.4 Chain Rule I

The Chain Rule is perhaps the most important and recurring differentiation rule. Immensely rich in application, it can be very easily and elegantly stated when notation is chosen wisely. It also typically requires much practice to execute with consistent accuracy.

In this section, we begin by looking at an introductory version of the computational mechanics of using the Chain Rule—mechanics which, at least "formally," make the rule almost self-explanatory—and then transition to more efficient statements of the rule as used in practice. We also examine how it is analogous to other computational phenomena whose intuitions might be more easily visualized, and along the way give some idea of what its proof might look like.

In its simplest form, the Chain Rule dictates how we must calculate derivatives of compositions of functions, i.e., functions of the form $h(x) = f(g(x))$, especially when we know how to calculate f' and g'. (In fact in a Chain Rule context, the functions are more often written $h(x) = f(u(x))$, but the principle is the same.) With the Chain Rule we will be able to calculate derivatives for a much wider class of functions, to find slopes on "implicit curves," and to determine so-called related rates relationships among variables. In later sections we will see how it also helps us to compute derivatives of inverse functions.

3.4.1 Leibniz Notation to Explain the Chain Rule

Before we see how the Chain Rule is stated with the very powerful Leibniz notation, first we will make some observations about that notation. For example, the following three formulas express the same thing—the instantaneous rate of change of the square of a quantity with respect to that quantity—albeit written below with three different variables (recalling for instance that $\frac{dx^2}{dx} = \frac{d}{dx}[x^2]$):

$$\frac{dx^2}{dx} = 2x, \qquad \frac{dt^2}{dt} = 2t, \qquad \frac{du^2}{du} = 2u.$$

See Figure 3.21 above for a graphical interpretation of this fact. It is important that in each equation the variables in the numerators matched those in the denominators, be they x, t or

u. So for instance $du^2/dx \neq 2u$, as we shall soon see, the problem being that the variables involved do not match, so the original Power Rule, namely $\frac{dx^n}{dx} = nx^{n-1}$ or equivalently $\frac{du^n}{du} = nu^{n-1}$, does not apply.

Now suppose we have a differentiable function $u = u(x)$ and want to find the derivative of $(u(x))^2$ with respect to x. The "outer" function is $f(x) = x^2$ (i.e., squaring what is inside), while the "inner" function is $u = u(x)$. If we are open-minded a bit, perhaps we would accept the following computation, as if these were algebraic fractions subject to cancellation:

$$\underbrace{\frac{d(u(x))^2}{dx}}_{\text{variables } u(x), x \text{ do not match}} = \underbrace{\frac{d(u(x))^2}{d\,u(x)}}_{\text{variables } u(x) \text{ match}} \cdot \frac{d\,u(x)}{dx} = 2(u(x))^1 \cdot \frac{d\,u(x)}{dx}.$$

This is a special case of the Chain Rule. Note that this same computation could also be written with Lagrange's prime notation, though it is not as intuitive, particularly if we "suppress" the x-variable in the Leibniz notation:

$$\text{Prime (Lagrange):} \quad \big((u(x))^2\big)' = 2(u(x))^1 \cdot u'(x) \tag{3.35}$$

$$\text{Leibniz:} \quad \frac{d\,u^2}{dx} = \frac{d\,u^2}{du} \cdot \frac{du}{dx} = 2u \cdot \frac{du}{dx}. \tag{3.36}$$

Let us consider for a moment this Equation (3.36). Note that we are hoping to measure how u^2 changes with x. To do so, we measure how u^2 changes with u, and multiply by how u changes with x. That the logic parallels the following may be useful:

Suppose we wish to know the cost of fueling a passenger car as a function of miles driven. This will vary as the efficiency of the car varies due to load, speed and other conditions. But at any given moment, the rate of cost per mile can be determined from multiplying the cost per gallon by the fueling rate of gallons per mile, which in units would read:

$$\frac{\text{dollar}}{\text{mile}} = \frac{\text{dollar}}{\text{gallon}} \cdot \frac{\text{gallon}}{\text{mile}}.$$

To utilize the Leibniz strategy in (3.36) above requires a modicum of cleverness in breaking the original Leibniz derivative notational "fraction" into other useful Leibniz notation "fractions," such that our new derivative expressions invoke known rules.[45]

Example 3.4.1 _____

Consider the following derivative computations, where we differentiate short "chains" of (two) functions:

- $\dfrac{d}{dx}\left(x^3 + 8x\right)^5 = \dfrac{d(x^3 + 8x)^5}{d(x^3 + 8x)} \cdot \dfrac{d(x^3 + 8x)}{dx} = 5\left(x^3 + 8x\right)^4 \left(3x^2 + 8\right).$

---

[45] Some calculus textbooks emphasize instead the prime notation for the Chain Rule. We discuss it and its reasonableness towards the end of this section, and use it for a partial proof of the Chain Rule. Some mathematicians have reservations about the Leibniz factorization such as (3.36) which appears to express the Chain Rule as a multiplication of fractions (after all, they are actually short-hand for some *limits* of fractions), though if one gets farther along one sees that this is a case where "it all works out." For that reason, and its usefulness for pedagogy and exploration, and because we ultimately will use a *slightly* different form before long which is less objectionable, we do include it here.

3.4. CHAIN RULE I

- $\dfrac{d}{dx}\sin x^5 = \dfrac{d(\sin x^5)}{dx^5}\cdot\dfrac{dx^5}{dx} = \cos x^5 \cdot 5x^4 = 5x^4\cos x^5.$

  It is customary in such cases to put the $5x^4$ before the trigonometric function in the final form, so that it is clear that $5x^4$ is not a factor *inside* of the trigonometric function.

- $\dfrac{d\sin^5 x}{dx} = \dfrac{d(\sin x)^5}{dx} = \dfrac{d(\sin x)^5}{d\sin x}\cdot\dfrac{d\sin x}{dx} = 5(\sin x)^4\cos x = 5\sin^4 x\cos x.$

  Another convention is to write $(\sin x)^n$ as $\sin^n x$, as long as $n \neq -1$ (recall $\sin^{-1} x = \arcsin x$). For computational purposes it is easier to work with $(\sin x)^n$.

- $\dfrac{d(x^2+3\cos x)^8}{dx} = \dfrac{d(x^2+3\cos x)^8}{d(x^2+3\cos x)}\cdot\dfrac{d(x^2+3\cos x)}{dx} = 8(x^2+3\cos x)^7(2x^2-3\sin x).$

- $\dfrac{d\cos(\sin x)}{dx} = \dfrac{d\cos(\sin x)}{d\sin x}\cdot\dfrac{d\sin x}{dx} = -\sin(\sin x)\cdot\cos x.$

It is also possible to extend this to a "chain" of three or more functions.

Example 3.4.2 _____

$$\dfrac{d(\sin 5x^3)^9}{dx} = \dfrac{d(\sin 5x^3)^9}{d(\sin 5x^3)}\cdot\dfrac{d(\sin 5x^3)}{d(5x^3)}\cdot\dfrac{d(5x^3)}{dx}$$
$$= 9(\sin 5x^3)^8(\cos 5x^3)(15x^2)$$
$$= 135x^2\sin^8 5x^3 \cos 5x^3.$$

To be clear, the presentation style above is very useful for first understanding the Chain Rule, and for referring back to when uncertain. However, one usually does not write out such things after becoming better versed in differentiation techniques, instead opting for a slight shorthand as described in the next subsection.

Worth noting is that, typical of how we apply differentiation rules, we work "outside" to "inside," in this case noting the overall structure of the functions $(\ )^9$, $\sin(\ )$ and finally $5(\ )^3$, and their implications for the derivative. Also worth mentioning is that the above derivative would require seemingly Herculean effort to derive using the limit definition of the derivative.

3.4.2 Chain Rule Versions of the Power, Sine, and Cosine Rules

Here we use the Chain Rule to give generalized Power, Sine and Cosine differentiation rules, whose "proofs" are given in Leibniz notation:

Theorem 3.4.1 *Assuming $u(x)$ is differentiable, we have the following specific formulas:*

$$\textbf{Power Rule:} \quad \frac{d(u(x))^n}{dx} = n(u(x))^{n-1} \cdot \frac{du(x)}{dx}, \quad i.e., \quad \boxed{\frac{du^n}{dx} = n\, u^{n-1} \cdot \frac{du}{dx},} \quad (3.37)$$

$$\textbf{Sine Rule:} \quad \frac{d\sin(u(x))}{dx} = \cos(u(x)) \cdot \frac{du(x)}{dx}, \quad i.e., \quad \boxed{\frac{d\sin u}{dx} = \cos u \cdot \frac{du}{dx},} \quad (3.38)$$

$$\textbf{Cosine Rule:} \quad \frac{d\cos(u(x))}{dx} = -\sin(u(x)) \cdot \frac{du(x)}{dx}, \quad i.e., \quad \boxed{\frac{d\cos u}{dx} = -\sin u \cdot \frac{du}{dx}.} \quad (3.39)$$

Any of these can be demonstrated as before, as for instance $\frac{d\sin u}{dx} = \frac{d\sin u}{du} \cdot \frac{du}{dx} = \cos u \frac{du}{dx}$. When it is understood that u is a function of x, sometimes written $u = u(x)$, the formula is often then written with the (x) "suppressed," for simplicity. The three results (3.37)–(3.39) are relatively simple in action:

- $\dfrac{d(\sin x)^4}{dx} = 4(\sin x)^3 \cdot \dfrac{d\sin x}{dx} = 4(\sin x)^3 \cos x,$

- $\dfrac{d\sin x^4}{dx} = \cos x^4 \cdot \dfrac{dx^4}{dx} = \cos x^4 \cdot 4x^3 = 4x^3 \cos x^4,$

- $\dfrac{d\cos(1-2x)}{dx} = -\sin(1-2x) \cdot \dfrac{d(1-2x)}{dx} = -\sin(1-2x) \cdot (-2) = 2\sin(1-2x).$

A more general statement of the chain rule is perhaps best given with a mix of Leibniz and prime notation, written here with $u(x)$ written as simply u:[46]

$$\textbf{General Chain Rule:} \quad \frac{df(u(x))}{dx} = \frac{df(u(x))}{du(x)} \cdot \frac{du(x)}{dx}, \quad i.e., \quad \boxed{\frac{df(u)}{dx} = f'(u) \cdot \frac{du}{dx}.} \quad (3.40)$$

We will return to this form later, but for now will concentrate on these Power, Sine and Cosine rules.

One of the primary goals of this chapter is to demonstrate how to *parse* the various rules as they are called for, by each other. In the next subsection we show more such scenarios.

3.4.3 More Complicated Functions: Rules Calling Other Rules

Computing the derivative of a more complicated function will usually require us to employ several differentiation rules, and it is best that each such step is written out, in turn, exactly as the rule dictates: if a rule directs us to take the derivative of some component part of the function, *we should write that we are to compute that derivative, before actually computing the derivative.* We do this as we work, from the outer structure of the function to the more inner structures as the differentiation rules direct. This extra verbosity allows us (1) to avoid many errors, and (2) to easily check our work.

---

[46]Equation (3.40) is best considered after the more specific examples like we have here, because the execution is somewhat confusing until the logic of the Chain Rule is better internalized for several examples of functions f. At that point, (3.40) is arguably the best general formulaic statement of the Chain Rule for use in practice.

3.4. CHAIN RULE I

Example 3.4.3

Compute $f'(x)$ if $f(x) = \sin^2 x + \cos^2 x$.

Solution: The overall structure of this function is that of a sum, so (according to the Sum Rule, (3.12), page 232) we are to compute the derivatives of $\sin^2 x$ and $\cos^2 x$ and add them. Here we will have the Sum Rule "calling" the Chain Rule twice:[47]

$$f'(x) = \frac{d}{dx}\left[\sin^2 x + \cos^2 x\right] = \frac{d(\sin x)^2}{dx} + \frac{d(\cos x)^2}{dx} = 2(\sin x)\cdot\frac{d\sin x}{dx} + 2(\cos x)\cdot\frac{d\cos x}{dx}$$
$$= 2\sin x \cos x + 2\cos x(-\sin x) = 2\sin x \cos x - 2\cos x \sin x = 0.$$

In fact the final answer is not surprising if we noticed $f(x) = \sin^2 x + \cos^2 x = 1$, for all $x \in \mathbb{R}$, and so

$$f'(x) = \frac{d}{dx}\left[\sin^2 x + \cos^2 x\right] \underset{\text{ALG}}{=\!=\!=} \frac{d}{dx}[1] = 0.$$

Again, as a matter of careful and correct application of the various differentiation rules as they apply, it is useful to write out each rule as it is invoked, *and to do so using the literal statements of the rule as they apply to the problem at hand*, and not to attempt to consolidate too many steps. With such care, very complicated functions can have their derivatives computed accurately and expeditiously.

Unlike, say, the algebraic rules for exponents which are ultimately about counting the total number of factors of some quantity, it is less clear at first why the differentiation rules are internally consistent. The reader should be assured that they are consistent, when correctly applied, and so we should not be surprised at all that when we compute the derivative of $f(x) = \sin^2 x + \cos^2 x$, we get the same answer (zero) as if we had computed the derivative of the function rewritten as $f(x) = 1$. In fact the internal consistency of the differentiation rules eventually can be a help in both understanding them, and in finding creative ways to rewrite a function to exploit the differences in such rules. In particular if a function appears complicated but has an easy simplification, it is usually better to compute the derivative of the simplified function.

Looking again at the example above we see also the important lesson that, when computing a function's derivative, we have to look at the outer structure of the function to see which is the first differentiation rule to apply (in that case the Sum Rule), and then apply the others as subsequent rules call upon them. For instance, we can redo a previous example, Example 3.4.2, page 275, by "parsing" through the rules:

Example 3.4.4

Compute $f'(x)$ if $f(x) = \sin^9 5x^3$.

---

[47]There are some conventions whereby mathematicians and others in related disciplines have many types of "short-hand," which can be confusing to students. With trigonometric functions in particular, what is written can be ambiguous at times, so for instance $\sin x \cos x = (\sin x)(\cos x)$ is understood, while $\sin 2x = \sin(2x)$ seems to follow a different convention. As already pointed out, $\sin^2 x = (\sin x)^2$ is understood, though $\sin^{-1} x = \arcsin(x)$ while $(\sin x)^{-1} = \csc x$. Like all conventions, one gets accustomed to these eventually, but the reader should be aware that some initial confusion is normal.

Solution: First note that $f(x) = \left(\sin 5x^3\right)^9$. This requires a Power Rule, calling a Sine Rule, both in Chain Rule versions.

$$f'(x) = 9(\sin 5x^3)^8 \cdot \frac{d(\sin 5x^3)}{dx} = 9(\sin 5x^3)^8 \cdot \cos 5x^3 \cdot \frac{d\, 5x^3}{dx}$$
$$= 9(\sin 5x^3)^8 \cos 5x^3 \cdot 15x^2 = 135x^2 \sin^8 5x^3 \cos 5x^3.$$

The actual calculus was finished before the last line, and then it is a matter of style how the final answer should be written.

Compare the computation above to the manner of computation in Example 3.4.2, page 275:

$$f'(x) = \frac{d(\sin^9 5x^3)}{dx} = \frac{d(\sin 5x^3)^9}{d(\sin 5x^3)} \cdot \frac{d(\sin 5x^3)}{d(5x^3)} \cdot \frac{d(5x^3)}{dx} = 9(\sin 5x^3)^8 (\cos 5x^3)(15x^2).$$

In the example above it seemed in some ways faster to use the older method, and that is arguably true when we have a pure chain of functions, where the input variable is fed into one function, whose output is fed into another, whose output is fed into another, and so on. However, (1) it is currently not standard procedure to use such a "decomposition," with "denominators" such as $d(\sin 5x^3)$ and $d(5x^3)$ in our operators even though it shows clearly how the factors in the final answer appear in turn, and (2) when the Chain Rule is mixed in with other rules (such as the Sum Rule in the previous example, or Product and Quotient Rules in future sections), it is usually simpler and less cluttered to write out what the Chain Rule says for the particular "outer function," before computing the derivative of the "inner function" contained in the Chain Rule.

Example 3.4.5 _____

Compute $f'(x)$ if $f(x) = \cos\left(\sin x^2 + 9x^3\right)$.

Solution: This is a Chain Rule (more specifically the Cosine Rule for chains of functions), calling a Sum Rule, calling another Chain Rule (Sine version) and a Power Rule:

$$f'(x) = \frac{d}{dx}\left[\cos\left(\sin x^2 + 9x^3\right)\right]$$
$$= -\sin\left(\sin x^2 + 9x^3\right) \cdot \frac{d}{dx}\left(\sin x^2 + 9x^3\right)$$
$$= -\sin\left(\sin x^2 + 9x^3\right) \cdot \left[\frac{d\left(\sin x^2\right)}{dx} + \frac{d\left(9x^3\right)}{dx}\right]$$
$$= -\sin\left(\sin x^2 + 9x^3\right) \cdot \left[\cos x^2 \cdot \frac{dx^2}{dx} + 9 \cdot 3x^2\right]$$
$$= -\sin\left(\sin x^2 + 9x^3\right) \cdot \left[\cos x^2 \cdot 2x + 27x^2\right]$$
$$= -\sin\left(\sin x^2 + 9x^3\right) \cdot \left[2x\cos x^2 + 27x^2\right]$$
$$= -\left(2x\cos x^2 + 27x^2\right)\sin\left(\sin x^2 + 9x^3\right).$$

The calculus was finished once the differential operator $\frac{d}{dx}$ ceased appearing (in the last three lines), and the rest was algebraic simplifying or rewriting according to style.

3.4. CHAIN RULE I

It is important to understand the example above completely: the outer structure of the original function (as cosine), the order in which the differentiation rules are called in turn, the necessity of parentheses and brackets, and the general organization of the various terms in the derivative. There are several opportunities for errors in the above computation, though if one is fluent in the rules and careful in writing all of the steps, and makes liberal use of grouping symbols ( ) and [ ], it is actually straightforward.

3.4.4 More on "Matching" of Variables

We will repeatedly come back to this theme of how best to write our parsings of the various rules invoked in the differentiation process for a given function. For now we will revisit the notion of what to do when the variable in a function does, or does not, match the variable of differentiation.

Example 3.4.6

Compute $\frac{df(z)}{dz}$ if $f(z) = \cos(z^3 + \sin z)$.

<u>Solution</u>: Here the names of the variables have changed, but the principle of the Chain Rule is the same. Again we will compute this using the Cosine Rule (3.39), page 276, except with z in place of x:

$$f'(z) = \frac{d}{dz}\cos(z^3 + \sin z) = \underbrace{-\sin(z^3 + \sin z)}_{\substack{\text{"first factor"} \\ \frac{d\cos(z^3+\sin z)}{d(z^3+\sin z)}}} \cdot \underbrace{\frac{d(z^3 + \sin z)}{dz}}_{\text{"second factor"}} = -\sin(z^3 + \sin z) \cdot (3z^2 + \cos z).$$

We remind the reader that the reason we can use that rule is that the first factor in the Chain Rule was the derivative of a cosine with respect to its input, though we only wrote that fact in the comments under the brace $\underbrace{}$. While we move towards writing such derivative computations as we did in the example above but without such comments, it is still useful to remember that it "worked" because the first factor in the above computation was actually of that form written under the brace.

The most common practice for computing the derivative above is as it would read without the braces. Indeed many instructors teach that *the derivative of cosine is minus sine..., multiplied by the derivative of what is inside the cosine.* That is perhaps an over-simplification, in particular because it does not specify the variables involved, but is useful if informed by awareness of what the Leibniz-style expansion would yield.

To be clear on what the Sine and Cosine Rules say, and why these should hold, consider the following abstract equations, which are in fact restatements of (3.38) with different variables, attempting to understand but omit the first rewriting:[48]

$$\frac{d\sin u}{dw} = \frac{d\sin u}{du} \cdot \frac{du}{dw} = \cos u \cdot \frac{du}{dw}, \qquad \frac{d\cos u}{dw} = \frac{d\cos u}{du} \cdot \frac{du}{dw} = -\sin u \cdot \frac{du}{dw}$$

$$\frac{d\sin\theta}{d\xi} = \frac{d\sin\theta}{d\theta} \cdot \frac{d\theta}{d\xi} = \cos\theta \cdot \frac{d\theta}{d\xi}, \qquad \frac{d\cos\theta}{d\xi} = \frac{d\cos\theta}{d\theta} \cdot \frac{d\theta}{d\xi} = -\sin\theta \cdot \frac{d\theta}{d\xi},$$

$$\frac{d\sin x}{dt} = \frac{d\sin x}{dx} \cdot \frac{dx}{dt} = \cos x \cdot \frac{dx}{dt}, \qquad \frac{d\cos x}{dt} = \frac{d\cos x}{dx} \cdot \frac{dx}{dt} = -\sin x \cdot \frac{dx}{dt}.$$

Note that in all the cases, the decomposition's first factor let us use the known, simple derivative formula for sine or cosine—because the variable inside of the sine or cosine matched the variable of differentiation—and then we compensated for introducing the new variable's derivative (as a fraction of sorts) with the second factor. This is consistent with the original sine and cosine rules:

$$\frac{d}{dx}\sin u = \cos u \cdot \frac{du}{dx}, \qquad \frac{d}{dx}\cos u = -\sin u \cdot \frac{du}{dx}.$$

Thus when the variables do not match, one expects to have the extra factor which is the derivative of the "inside" function, or variable, to compensate for the disparity, so for instance

$$\frac{d}{d\xi}\sin\theta = \cos\theta \cdot \frac{d\theta}{d\xi}.$$

Thus to see how $\sin\theta$ changes with ξ we first see how $\sin\theta$ changes with θ, and multiply by how θ changes with ξ. The more one reflects upon this, (hopefully) the more natural it seems.

Now when variables *do* match, the Chain Rule can still be invoked correctly, though needlessly, adding only a trivial factor which will not change the outcome:

$$\frac{d}{dx}\sin x = \frac{d}{dx}\sin x \cdot \frac{dx}{dx} = \cos x \cdot 1 = \cos x.$$

So again, the Chain Rule is there for us when the variable within the ("outer") function does not match the variable with respect to which we are computing the derivative, but the Chain Rule is also consistent in any situation in which it is invoked correctly (even if unnecessarily). It is unnecessary to invoke the Chain Rule when the variables match (though there is no harm if we do), but it is necessary to invoke when they do not.

3.4.5 Power Rule for Rational Powers

With the Chain Rule we have enough theoretical development to show that the Power Rule actually holds for any constant power which is a rational number p/q (where $p, q \in \mathbb{Z}$, and of course $q \neq 0$). Recall that the set of all rational numbers was denoted $\mathbb{Q}$, for "quotients," i.e., fractions of integers. We already proved the Power Rule for powers $n \in \{0, 1, 2, 3, \ldots\}$, and that result is used in the proof for rational powers, which we leave until the end of this section, so we can expeditiously come to examples. But first we state the theorem.

---

[48]Just to be clear, it should be repeated that when we write for instance $\cos x \cdot \frac{dx}{dt}$, we mean that the $\frac{dx}{dt}$ is outside of the cosine function, i.e., we mean $(\cos x) \cdot \frac{dx}{dt}$. Note that many texts assume this meaning without making it explicit with the dot ".," and simply write $\cos x \frac{dx}{dt}$. As a matter of style, it is assumed the derivative factor $\frac{dx}{dt}$ is not part of the argument (input) of the cosine function in such a case.

3.4. CHAIN RULE I

Theorem 3.4.2 (Power Rule for Rational Powers) *For any $r \in \mathbb{Q} - \{0\}$ (i.e., nonzero rational numbers),*[49]

$$\boxed{\frac{d x^r}{dx} = rx^{r-1},} \qquad (3.41)$$

$$\boxed{\frac{d u^r}{dx} = ru^{r-1} \cdot \frac{du}{dx}.} \qquad (3.42)$$

Example 3.4.7

The following, which were nontrivial exercises requiring limits of difference quotients in Section 3.1, yield quickly to this extended Power Rule:

$$\frac{d\sqrt{x}}{dx} = \frac{dx^{1/2}}{dx} = \frac{1}{2}x^{1/2-1} = \frac{1}{2}x^{-1/2} = \frac{1}{2\sqrt{x}},$$

$$\frac{d}{dx}\left[\frac{1}{x}\right] = \frac{dx^{-1}}{dx} = -1x^{-2} = \frac{-1}{x^2}.$$

In fact, these particular derivatives occur often enough that they, along with their Chain Rule versions, deserve special attention (and should be committed to memory):

$$\boxed{\frac{d\sqrt{x}}{dx} = \frac{1}{2\sqrt{x}},} \qquad \boxed{\frac{d\sqrt{u}}{dx} = \frac{1}{2\sqrt{u}} \cdot \frac{du}{dx},} \qquad (3.43)$$

$$\boxed{\frac{d}{dx}\left[\frac{1}{x}\right] = \frac{-1}{x^2},} \qquad \boxed{\frac{d}{dx}\left[\frac{1}{u}\right] = \frac{-1}{u^2} \cdot \frac{du}{dx}.} \qquad (3.44)$$

Example 3.4.8

To find $f'(x)$ for $f(x) = \sqrt{x^2 + 1}$, we compute

$$f'(x) = \frac{d}{dx}\sqrt{x^2 + 1} = \frac{1}{2\sqrt{x^2+1}} \cdot \frac{d}{dx}(x^2+1) = \frac{1}{2\sqrt{x^2+1}} \cdot (2x) = \frac{x}{\sqrt{x^2+1}}.$$

Note that in the above example the "outer" function was the square root, or $(\ )^{1/2}$, while the "inner" function is $(\ )^2 + 1$. One could write (though again, it is not standard practice):

$$\frac{d}{dx}\sqrt{x^2+1} = \frac{d\sqrt{x^2+1}}{d(x^2+1)} \cdot \frac{d(x^2+1)}{dx} = \frac{1}{2\sqrt{x^2+1}} \cdot 2x = \frac{x}{\sqrt{x^2+1}}.$$

---

[49] Even in the case $r = 0$, this is still "formally true," in that $\frac{dx^0}{dx} = 0x^{-1}$ if we allow this to be zero for all x-values (even $x = 0$), *and* assume $x^0 = 1$ for all x-values (which is also problematic). This is the same as writing $\frac{d1}{dx} = 0$. So this Power Rule, even when naively applied, is quite robust, as long as r is a fixed constant.

We could have also done either computation using $\sqrt{(\ )} = (\ )^{1/2}$ and used the Power Rule directly, rather than our formulas for the special case.

Example 3.4.9

Suppose $f(x) = \sqrt{x + \sqrt{x}}$. Then the Leibniz decomposition to compute its derivative would look like

$$\frac{df(x)}{dx} = \frac{d\sqrt{x+\sqrt{x}}}{d\left(x+\sqrt{x}\right)} \cdot \frac{d\left(x+\sqrt{x}\right)}{dx} = \frac{1}{2\sqrt{x+\sqrt{x}}} \cdot \left(1 + \frac{1}{2\sqrt{x}}\right).$$

Again—and especially with practice—one would usually not write the Leibniz decomposition in the first step, but should instead write

$$\frac{d}{dx}\sqrt{x+\sqrt{x}} = \frac{1}{2\sqrt{x+\sqrt{x}}} \cdot \frac{d\left(x+\sqrt{x}\right)}{dx} = \frac{1}{2\sqrt{x+\sqrt{x}}} \cdot \left(1 + \frac{1}{2\sqrt{x}}\right).$$

A common mistake in the above example is to think of $\sqrt{x}$ as the inner function, since geometrically it somehow appears to be innermost. In fact the inner function is actually the whole of $x + \sqrt{x}$, hence the derivative factor $\left(1 + \frac{1}{2\sqrt{x}}\right)$. This mistake is indeed quite common, particularly among students who previously learned a small amount of calculus but failed to grasp the "whys" along with the "hows."

Example 3.4.10

Suppose $f(x) = \dfrac{2}{(x^3 - 9x + 7)^7}$. Then, noting the factor of 2 "going along for the ride," we compute

$$f'(x) = \frac{d}{dx}\left[2(x^3 - 9x + 7)^{-7}\right] = 2 \cdot (-7)(x^3 - 9x + 7)^{-8} \frac{d}{dx}(x^3 - 9x + 7)$$

$$= -14(x^3 - 9x + 7)^{-8}(3x^2 - 9) = \frac{-14(3x^2 - 9)}{(x^3 - 9x + 7)^8}.$$

Example 3.4.11

Suppose $f(x) = \sqrt[3]{\dfrac{1}{x + \sqrt{x^3 + 9}}}$. Note how we rewrite this in computing $f'(x)$:

$$f'(x) = \frac{d}{dx}\left[\sqrt[3]{\frac{1}{x+\sqrt{x^3+9}}}\right] = \frac{d}{dx}\left(x+\sqrt{x^3+9}\right)^{-1/3}$$

$$= -\frac{1}{3}\left(x+\sqrt{x^3+9}\right)^{-4/3} \cdot \frac{d}{dx}\left(x+\sqrt{x^3+9}\right)$$

$$= -\frac{1}{3}\left(x+\sqrt{x^3+9}\right)^{-4/3} \cdot \left[1 + \frac{1}{2\sqrt{x^3+9}} \cdot \frac{d}{dx}(x^3+9)\right]$$

$$= -\frac{1}{3}\left(x+\sqrt{x^3+9}\right)^{-4/3} \cdot \left[1 + \frac{3x^2}{2\sqrt{x^3+9}}\right].$$

3.4. CHAIN RULE I

In the calculation above, we first rewrote the expression as a $-1/3$ power, then used the Chain Rule version of the Power Rule, and used the Chain Rule *again* in calculating the derivative of that "inner" function. It is very common to rewrite a function so that the derivative rules are more easily applied. We do so again in the next example.

Example 3.4.12

Compute $\frac{d}{dx}\sec x$.

<u>Solution:</u> Here we will use that $\sec x = (\cos x)^{-1}$. We will also use Equation (3.44) rather than the Power Rule directly.

$$\frac{d\sec x}{dx} = \frac{d(\cos x)^{-1}}{dx} = \frac{-1}{(\cos x)^2} \cdot \frac{d\cos x}{dx} = \frac{-1}{\cos^2 x} \cdot (-\sin x) = \frac{\sin x}{\cos^2 x} = \frac{1}{\cos x} \cdot \frac{\sin x}{\cos x} = \sec x \tan x.$$

The calculus in this example was finished after the third "=" symbol, and the rest was simplification and rearrangement for style. In a later section, where the derivative of the secant function is listed as one to be memorized, the final form will be $\frac{d}{dx}\sec x = \sec x \tan x$.

It should be emphasized that, while it is useful to memorize the derivatives of the square root and the reciprocal, they are, after all, simple powers and so those formulas can (and should occasionally for memory reinforcement) be re-derived from the Power Rule.

3.4.6 Chain Rule Proof of the Power Rule for Rational Powers

We now divert temporarily to prove the Power Rule for rational numbers

$$r \in \mathbb{Q} \implies \frac{dx^r}{dx} = rx^{r-1},$$

from which the Chain Rule version also follows. The technique used will recur several times throughout the text. Fortunately it is somewhat self-explanatory, assuming knowledge of the general Chain Rule.

The proof is in two steps, the first being a proof in the case of negative integer powers, from which we can eventually recover all rational power cases.

<u>Proof:</u> We will use the Chain Rule to show that the Power Rule, $\frac{d}{dx}x^r = rx^{r-1}$, holds also for any rational power $r = p/q$, with p,q nonzero integers. (The case $p = 0$ is trivial and the case $q = 0$ is meaningless.)

First we will show that the Power Rule holds for $y = x^n$ for any negative integer exponents n. In such cases we can write $y = x^{-m}$ for a positive integer exponent m (namely $m = -n$). But then $y^{-1} = x^m$. Furthermore we already showed in Section 3.1 (Example 3.1.4, page 217) that the derivative definition gives us $\frac{dy^{-1}}{dy} = -1/y^2$ (though the variable used in the proof there was x). Using this, the Chain Rule and (eventually) the fact that $y = x^n$ we get the following, again using $m = -n \in \{1, 2, 3, \ldots\}$:

$$y = x^n \implies y^{-1} = x^m$$
$$\implies \frac{d}{dx}[y^{-1}] = \frac{d}{dx}[x^m]$$
$$\implies -\frac{1}{y^2} \cdot \frac{dy}{dx} = mx^{m-1}$$
$$\implies \frac{dy}{dx} = -y^2 mx^{m-1}$$
$$= -(x^n)^2(-n)x^{-n-1}$$
$$= nx^{2n-n-1} = nx^{n-1}, \quad \text{q.e.d.}^{50}$$

Now we will use the Chain Rule in a similar way to compute the derivatives of rational powers of x. Suppose $y = x^{p/q}$, where $p, q \in \mathbb{Z} - \{0\}$ are nonzero integers, and that $r = p/q$ is in simplified form. Then we can raise both sides of $y = x^{p/q}$ to the power q to get[51]

$$y^q = x^p$$
$$\implies \frac{dy^q}{dx} = \frac{dx^p}{dx}$$
$$\implies qy^{q-1}\frac{dy}{dx} = px^{p-1}$$
$$\implies \frac{dy}{dx} = \frac{p}{q}y^{1-q}x^{p-1} = \frac{p}{q}\left(x^{p/q}\right)^{1-q}x^{p-1} = \frac{p}{q} \cdot x^{\frac{p}{q}-p}x^{p-1} = \frac{p}{q} \cdot x^{\frac{p}{q}-1},$$

which, since $p/q = r$, can be rewritten $\frac{dy}{dx} = rx^{r-1}$, q.e.d.[52]

3.4.7 Applied Examples

Here we introduce a few further examples showing the derivative's usefulness in applications.

---

[50] It is important to that we interpret the first implication correctly. Recall that $y = x^n$, so y is a function of x. But then so is y^{-1} and, in fact, y^{-1} and x^m are *the same functions of x*. (So for instance, if y^{-1} and x^m were graphed versus x, the graphs would be the same, so the slopes at each x-value would be the same.) Therefore y^{-1} and x^m have the same derivative with respect to x.

[51] Note that p or q (but not both, since p/q is simplified) could be negative, but what we are about to do is justified by the previous result that the Power Rule also works for negative integer exponents.

[52] In fact, once we have logarithms we can define $y = x^r$ for all $r \in \mathbb{R}$ and prove that $\frac{dx^r}{dx} = rx^{r-1}$ still holds, at least for $x > 0$, since x^r may not be defined for negative values of x, but then neither will be the derivative rx^{r-1} for those values of x. And so for $x > 0$ we have $\frac{dx^{\sqrt{2}}}{dx} = \sqrt{2} \cdot x^{\sqrt{2}-1}$, and $\frac{dx^\pi}{dx} = \pi x^{\pi-1}$, for two examples. Finally it should be recalled that the rule is even more robust if we write "formally" $\frac{dx^0}{dx} = 0x^{-1} = 0$, if we (1) take $x^0 = 1$ for all $x \in \mathbb{R}$ (which we will see is somewhat problematic in later chapters), and (2) $0x^{-1} = 0$ for all $x \in \mathbb{R}$, which is problematic when $x = 0$. In other words, with these formalities, the Power Rule allows for $\frac{d1}{dx} = 0$ to be a special case of the Power Rule.

3.4. CHAIN RULE I

Example 3.4.13

Suppose charge q_1 is fixed and charge q_2 is moving away from charge q_1, and r is the distance between them. According to Coulomb's Law, the force F which each charge imposes upon the other is given by $F = k\frac{q_1 q_2}{r^2}$. Suppose further that when $r = 0.1$m, we have $F = 0.6$N. Find $\frac{dF}{dr}$ when $r = 0.1$m and when $r = 0.3$m.

Solution: Here we can start with Coulomb's Law to find collectively the constant multiplier kq_1q_2:

$$0.6 = \frac{kq_1q_2}{(0.1)^2} \iff 0.6(0.1)^2 = kq_1q_2 \iff kq_1q_2 = 0.006$$

$$\implies F = \frac{kq_1q_2}{r^2} = \frac{0.006}{r^2}.$$

(Recall that k, q_1 and q_2 are all constants.) From this we can use the Power Rule to deduce

$$\frac{dF}{dr} = \frac{d}{dr}\left[\frac{0.006}{r^2}\right] = \frac{d}{dr}\left[0.006 r^{-2}\right] = -2(0.006)r^{-3} = -0.012 r^{-3}$$

$$\implies \frac{dF}{dr} = -\frac{0.012}{r^3}.$$

With that, we have

$$\left.\frac{dF}{dr}\right|_{r=0.1} = -\frac{0.012}{0.1^3} = -12,$$

$$\left.\frac{dF}{dr}\right|_{r=0.3} = -\frac{0.012}{0.3^3} \approx -0.44.$$

Note that for $r > 0$ we have $\frac{dF}{dr} < 0$, so F decreases as r increases. Furthermore, the rate of decrease in F itself slows quickly.

Example 3.4.14

Supply, demand and price for an item are usually related. The number q demanded for a certain item will usually be higher if the price P is lower, and vice-versa. The **price elasticity of demand** measures how "elastic" is the demand relative to changes of price, and is given by the formula[53]

$$\varepsilon = \left(\frac{P}{q}\right)\left|\frac{dq}{dP}\right|, \tag{3.45}$$

where P is the price and q is the quantity for a given demand equation which relates P and q. Demand is said to be **elastic** if $\varepsilon > 1$, **inelastic** if $\varepsilon < 1$, and **unitary** if $\varepsilon = 1$. Describe the elasticity of the following demand curve equations:

---

[53]This formula is the derivative version of an alternative definition, in which ε is defined as

$$\frac{(\Delta q)/q}{(\Delta P)/P} = \frac{100\% \cdot (\Delta q)/q}{100\% \cdot (\Delta P)/P} = \frac{\text{percent change in }q}{\text{percent change in }P}.$$

Using the first expression in the line above, letting $\Delta P \to 0$ we see that for small ΔP this is approximately $(P/q) \cdot (dq/dP)$. However, this quantity being almost always negative (why?), economists tend to verbalize its absolute value instead. Since we expect P and q to be nonnegative, using absolute values with the derivative's output generally leaves a nonnegative value for ε.

Note that if $\varepsilon = 1$ in the non-calculus definition (after absolute values), then a 1% reduction in price would yield a 1% increase in demand. If $\varepsilon > 1$ then a 1% decrease in price would yield an increase of more than 1% in demand. We leave the interpretation of $\varepsilon < 1$ to the reader.

(a) $q = 100 - \frac{1}{3}P$ (b) $q = \sqrt{1200 - P}$

Solution: We take these in turn.

(a) For $q = 100 - \frac{1}{3}P$, we have $\frac{dq}{dP} = -\frac{1}{3}$, and so

$$\varepsilon = \left(\frac{P}{q}\right)\left|\frac{dq}{dP}\right| = \left(\frac{P}{100 - \frac{1}{3}P}\right)\left|-\frac{1}{3}\right| = \frac{P}{100 - \frac{1}{3}P} \cdot \frac{1}{3} = \frac{P}{300 - P}.$$

We can either solve directly for the three cases $P/(300-P) < 1, > 1, = 1$ or we can introduce $F(P) = \frac{P}{300-P} - 1$ and see where this is negative, positive and zero—in short, construct a sign chart for $F(P)$:
$F(P) = \frac{P}{300-P} - 1 = \frac{P-(300-P)}{300-P} = \frac{2P-300}{300-P}$. We can see this is zero when $2P - 300 = 0 \iff 2P = 300 \iff P = 150$. This corresponds to $\varepsilon = 1$, so demand is unitary ($\varepsilon = 1$) when $P = 150$.

Before constructing the sign chart, we notice that $q = 100 - \frac{1}{3}P$ requires that $P < 300$, lest q be negative. Furthermore, we can assume $P > 0$ as well. With these constraints in mind, we produce our sign chart:

Function: $F(P) = \frac{2(P-150)}{300-P}$

Test $P =$ 50   200

Sign Factors: $\ominus/\oplus$ $\oplus/\oplus$

Sign $F(P)$:   0 $\ominus$ 150 $\oplus$ 300

From the chart we get $F(P) < 0$ on $0 < P < 150$, $F(P) = 0$ at $P = 150$ and $F(P) > 0$ on $150 < P < 300$. Equivalently, we get

(inelastic) $\varepsilon < 0$ $\iff$ $0 < P < 150$,
(unitary) $\varepsilon = 0$ $\iff$ $P = 150$,
(elastic) $\varepsilon > 0$ $\iff$ $150 < P < 300$.

(b) For the case $q = \sqrt{1200 - P}$, we have

$$\varepsilon = \left(\frac{P}{q}\right)\left|\frac{dq}{dP}\right| = \left(\frac{P}{\sqrt{1200-P}}\right)\left|\frac{1}{2\sqrt{1200-P}} \cdot \frac{d}{dP}(1200-P)\right|$$

$$= \frac{P}{\sqrt{1200-P}} \cdot \left|\frac{-1}{2\sqrt{1200-P}}\right| = \frac{P}{\sqrt{1200-P}} \cdot \frac{1}{2\sqrt{1200-P}}$$

$$= \frac{P}{2(1200-P)}.$$

We can use the same technique as we did in (a) to find where $\varepsilon < 1, > 1, = 1$, though this time we will opt for solving these directly. Note that we must have $0 < P < 1200$ due to the definition of q. Under that assumption, we have (showing the entire solution process for the first two cases):

$$\varepsilon = 1 \iff \frac{P}{2400-2P} = 1$$
$$\iff P = 2400 - 2P$$
$$\iff 3P = 2400 \iff P = 800,$$

$$\varepsilon < 1 \iff \frac{P}{2400-2P} < 1$$
$$\iff P < 2400 - 2P$$
$$\iff 3P < 2400 \iff P < 800,$$

$$\varepsilon > 1 \iff \frac{P}{2400-2P} > 1 \iff P > 800.$$

3.4. CHAIN RULE I

Remembering our constraint $0 < P < 1200$, we can write in fact that

$$\begin{array}{lcll}\text{(inelastic)} & \varepsilon < 0 & \Longleftrightarrow & 0 < P < 800, \\ \text{(unitary)} & \varepsilon = 0 & \Longleftrightarrow & P = 800, \\ \text{(elastic)} & \varepsilon > 0 & \Longleftrightarrow & 800 < P < 1200.\end{array}$$

It should be noted that we could multiply both sides of the inequalities by $2400 - 2P$ because we were operating under the assumption that $0 < P < 1200$, and this implies $2400 - 2P > 0$. If we are unsure of the sign of an expression we wish to multiply by to manipulate an inequality, it is safer to use a sign chart method, which avoids multiplying both sides of an inequality by a variable quantity (of unknown sign). Also, an observant student might combine both ideas, finding where $\varepsilon = 1$ and simply testing the other regions in simpler cases such as the above.

Example 3.4.15

Suppose an object slides on a straight, frictionless track. Suppose further that one end of a spring is attached to a fixed point at the end of the track, and that the other end of the spring is attached to the sliding object. If the object is otherwise free to slide, its motion will be periodic, of a type known as simple harmonic motion.

Suppose that the equilibrium point for the object attached to the spring is at position $x = 0$, A is the maximum positive displacement of the object, $x(t)$ is the position of the object at any given time t, and $x(0) = A$, we can give the position by

$$x = A \cos\left(\frac{2\pi}{T} t\right),$$

where T is the period of the motion, that is, the time required for one complete cycle before the motion repeats. This equation would hold, for instance, if the object were pulled to position A and then released at time $t = 0$.

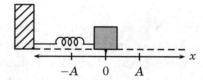

Find the velocity for any time $t > 0$.

<u>Solution</u>: As usual we will define velocity $v(t) = \frac{d}{dt} x(t)$. From the Chain Rule we get the following computations (note $2\pi/T$ is a constant):

$$v(t) = \frac{d}{dt}\left[A \cos\left(\frac{2\pi}{T} t\right)\right] = -A \sin\left(\frac{2\pi}{T} t\right) \cdot \frac{d}{dt}\left[\frac{2\pi}{T} t\right] = -A \sin\left(\frac{2\pi}{T} t\right) \cdot \frac{2\pi}{T} = -A \left(\frac{2\pi}{T}\right) \sin\left(\frac{2\pi}{T} t\right).$$

It is common in the study of oscillating functions to define an angular velocity by the lower-case Greek letter (omega) $\omega = 2\pi/T$, so that the above computation can be summarized:

$$x = A \cos(\omega t) \stackrel{\frac{d}{dt}}{\Longrightarrow} v = -A \omega \sin(\omega t).$$

Note that when $|x(t)|$ is maximized (where $\cos(\omega t) = \pm 1$, in this case at $t = n \cdot \pi/\omega$ for $n \in \mathbb{N}$, we also have $v = 0$ (since $\sin(\omega t) = 0$ there). So the object is "at rest" when it is at the extremes of its position. That should seem intuitive enough if we picture its motion. Also, vice-versa, in that $|v(t)|$ is maximized when $x(t) = 0$. Finally note that, while position is bounded within $[-A, A]$, the velocity's range is $[-A\omega, A\omega]$ (assuming $\omega > 0$), and so a larger ω would imply a proportionately larger v. These things should all seem intuitive upon reflection.

We also note that the period is given by $T = 2\pi/\omega$. This is the case with most simple springs which (until fully stretched or compressed) conform to Hooke's Law $F = -kx$, where $k > 0$ is a constant, x is the position relative to the equilibrium (rest) position of the spring and F is the force it exerts. Recall that $F = ma$ where m is the mass of the accelerated object. In our example above we could have further computed $a = \frac{dv}{dt} = -A\omega^2 x$, where $x = \cos(\omega t)$. Then multiplying by m gives us $F = -Am\omega^2 x$, which is Hooke's Law with $k = Am\omega^2$ for our constant.[54]

3.4.8 Chain Rule in Prime Notation

Many texts introduce the Chain Rule first using the prime notation, though here we opted for the more self-explanatory Leibniz notation for introductory purposes. We include this section here which re-introduces the Chain Rule using the other, prime notation approach. It is very useful to be fluent in both, because of the ubiquity of both approaches. Both have their conveniences. For instance, the Chain Rule's proof, and many special cases of the rule, are easier to write efficiently using prime notation.

Theorem 3.4.3 (Chain Rule) *Suppose that $h(x) = f(g(x))$. Then, wherever the following exist, we have (in Lagrange's "prime" notation):*

$$h'(x) = (f(g(x)))' = f'(g(x))g'(x). \tag{3.46}$$

Note that there is an "outer" function, namely f, and an "inner function" g, and (3.46) describes how to compute the derivative of the "chain" of functions $f \circ g(x) = f(g(x))$. The rule is sometimes stated that we compute the derivative of the outer function *with respect to the inner function*, that is $f'(g(x))$, and then multiply by the derivative of the inner function, i.e., by $g'(x)$. It should be mentioned that "multiplying by the derivative of the inner function" is the step that is most commonly forgotten or performed incorrectly in such derivative computations.

Recall Equation (3.40), page 276, which stated $\frac{d}{dx}f(u) = f'(u) \cdot \frac{du}{dx}$; with the names changed somewhat we can see how we can write Equation (3.46) above, where $h(x) = f(g(x))$, as

$$h'(x) = \frac{d}{dx}f(g(x)) = f'(g(x)) \cdot \frac{dg(x)}{dx} = f'(g(x)) \cdot g'(x),$$

---

[54] Named for Robert Hooke (1635–1703), an English inventor, professor, natural philosopher and architect. His 1678 statement of his law in words read, *Ut tensio, sic vis* (Latin), meaning, "As the extension, so the force."

Hooke was a contemporary of Newton's, and was very much involved in many of the significant discussions of his time. For instance, Newton gave him some credit for reviving Newton's interest in astronomical mechanics, leading to Newton's development of the inverse square law of gravitational attraction, which mirrors Coulomb's electrostatic law describing how charged particles attract or repel, depending in part upon their distance of separation. Of course with charged particles, "like charges repel" though there are no cases of gravitational repulsion in Newton's law (only attraction).

3.4. CHAIN RULE I

as stated in Equation (3.46). Or we can bring even more Leibniz notation to bear:

$$h'(x) = \frac{d}{dx}f(g(x)) = \frac{d f(g(x))}{d g(x)} \cdot \frac{d g(x)}{dx},$$

and get the same result. But for this subsection we will demonstrate both how to make Chain Rule computations using only the prime notation, and how it corresponds to the Leibniz-style computations from before.

Example 3.4.16

Compute $h'(x)$ if $h(x) = (x^2 + 3x)^2$. Use prime notation.

Solution: Besides the unwieldy "limit definition" using difference quotients, thus far there are two possible methods for computing $h'(x)$ here:

- Expand the function first, and then compute the derivative, as we would need to do if we had little more than the Power Rule to rely upon:

$$h(x) = x^4 + 6x^3 + 9x^2 \implies h'(x) = 4x^3 + 18x^2 + 18x.$$

- Instead, use the Chain Rule, the "outer function" being $f(x) = x^2$ (squaring the input), and "inner function" $g(x) = x^2 + 3x$. Since $f'(x) = 2x^1$ while $g'(x) = 2x + 3$, by (3.46) we have

$$h'(x) = [f'(g(x))]g'(x) = \left[2(g(x))^1\right]g'(x) = \left[2(x^2+3x)^1\right](2x+3).$$

Note that multiplying these factors gives us $h'(x) = 2(2x^3 + 9x^2 + 9x) = 4x^3 + 18x^2 + 18x$ as before.

In some contexts with functions, "empty parentheses" can illustrate the actions of functions. For the above derivative, we can look at the following:

$$\begin{array}{rlcl} & \text{outer function} & \text{is} & (\ )^2 \\ \implies & \text{its derivative} & \text{is} & 2(\ ) \\ \hline & \text{inner function} & \text{is} & (\ )^2 + 3(\ ) \\ \implies & \text{its derivative} & \text{is} & 2(\ ) + 3 \end{array}$$

The Chain Rule for this problem could then read:

$$\left[((x)^2 + 3(x))^2\right]' = 2((x)^2 + 3(x)) \cdot [2(x) + 3] = 2(x^2 + 3x) \cdot (2x+3).$$

Example 3.4.17

Find $h'(x)$ if $h(x) = (x^3 + 27x + 9)^{55}$. Use prime notation.

Solution: Certainly we do not want to multiply this out to use earlier rules. Instead we just notice that this is a composition of two functions, with the "outer" function being $f(x) = x^{55}$ and the "inner" function being $g(x) = x^3 + 27x + 9$. Now $f'(x) = 55x^{54}$, while $g'(x) = 3x^2 + 27$. Thus

$$h'(x) = [f'(g(x))]g'(x) = 55(g(x))^{54} \cdot g'(x) = 55(x^3 + 27x + 9)^{54} \cdot (3x^2 + 27).$$

Even if we had somehow expanded the original, 165-degree polynomial first, and then calculated the derivative, it is unlikely we would have noticed that our resulting 164-degree polynomial answer factors so nicely.

Of course the Chain Rule has much to say about derivatives of functions which contain trigonometric functions in their structures. Below are two examples where we use the prime notation, and where the "outer function" and "inner function" are squaring and sine functions, respectively, and then vice-versa.

Example 3.4.18

Find $h'(x)$ if $h(x) = \sin^2 x$. Use prime notation.

Solution: Note that $h(x) = (\sin x)^2$, so the "outer function" is $f(x) = x^2$, while the "inner function" is $\sin x$. Next note that $f'(x) = 2x$ and $g'(x) = \cos x$. Thus

$$h'(x) = [f'(g(x))]g'(x) = [2(g(x))]g'(x) = 2\sin x \cdot \cos x.$$

Compare to the Leibniz-style computation $\frac{d}{dx}(\sin x)^2 = 2(\sin x)\frac{d\sin x}{dx} = 2\sin x \cos x$.

Example 3.4.19

Suppose instead $h(x) = \sin x^2$. Here $f(x) = \sin x$, $g(x) = x^2$, $f'(x) = \cos x$, $g'(x) = 2x$. Hence

$$h'(x) = [f'(g(x))]g'(x) = \cos g(x) \cdot g'(x) = \cos x^2 \cdot 2x = 2x\cos x^2.$$

Compare to the Leibniz-style computation $\frac{d}{dx}\sin(x^2) = \cos(x^2) \cdot \frac{dx^2}{dx} = \cos(x^2) \cdot 2x = 2x\cos x^2$.

It is crucial to identify the outer function and the inner function. It is also important that the inner function $g(x)$ is inputted into the derivative f' of the outer function:

$$(f(g(x)))' = f'\left(\underline{g(x)}\right) \cdot g'(x).$$

While we identified the outer and inner functions by name, in fact naming an "$f(x)$" and "$g(x)$" is unwieldy and unnecessary. Before long (though usually not immediately) the pattern of these computations does become natural enough. Throughout the text we will usually opt for the Leibniz-style derivations because their logic is more clear as we compute derivatives of more complicated functions.

3.4.9 Further Examples

Here we consider several examples to review this new differentiation technique and the Power Rule for rational powers. We will use the Leibniz notation, but in the form that one who is practiced and careful in using the Chain Rule would likely use, namely that $\frac{df(u)}{dx} = f'(u) \cdot \frac{du}{dx}$. We will assume that the forms of f and f' are already understood.

- $\frac{d}{dx}\left[x\sqrt{x}\right] = \frac{dx^{3/2}}{dx} = \frac{3}{2}x^{1/2} = \frac{3}{2}\sqrt{x}$.

- $\frac{d}{dx}\left[\frac{-5}{(x^2+9)^2}\right] = \frac{d}{dx}\left[-5(x^2+9)^{-2}\right] = -5(-2)(x^2+9)^{-3}\frac{d(x^2+9)}{dx} = \frac{10}{(x^2+9)^3} \cdot 2x = \frac{20x}{(x^2+9)^3}$.

3.4. CHAIN RULE I

- $\frac{d}{dx}\left[\sin^2(x+\sin 2x)\right] = \frac{d}{dx}[\sin(x+\sin 2x)]^2 = 2[\sin(x+\sin 2x)]\frac{d}{dx}[\sin(x+\sin 2x)]$

    $= 2\sin(x+\sin 2x)\cos(x+\sin 2x)\frac{d}{dx}[x+\sin 2x]$

    $= 2\sin(x+\sin 2x)\cos(x+\sin 2x)\left[1+\cos 2x \cdot \frac{d\, 2x}{dx}\right]$

    $= 2\sin(x+\sin 2x)\cos(x+\sin 2x)[1+2\cos 2x].$

- $\frac{d\sec^3 x^3}{dx} = \frac{d}{dx}\left[(\cos x^3)^{-3}\right] = (-3)(\cos x^3)^{-4}\frac{d}{dx}[\cos x^3] = -3\sec^4 x^3 \cdot \left(-\sin x^3 \cdot \frac{dx^3}{dx}\right)$

    $= 3\sec^4 x^3 \sin x^3 \cdot 3x^2 = \frac{9x^2 \sin x^3}{\cos^4 x^3}.$

- $\frac{d\sin x}{dt} = \frac{d\sin x}{dx} \cdot \frac{dx}{dt} = \cos x \cdot \frac{dx}{dt}$. (The first step is often not shown.)

- Suppose $h(x) = f(g(x))$, and $f(3) = 5$, $f'(9) = 8$, $g(6) = 9$ and $g'(6) = 7$. Then what is $h'(6)$?

    <u>Solution</u>: Using prime notation, $h'(x) = f'(g(x))g'(x)$, and so
    $$h'(6) = f'(g(6))g'(6) = f'(9)g'(6) = 8 \cdot 7 = 56.$$

3.4.10 Partial Proof of Chain Rule

To show that the Chain Rule makes sense from the limit-definition standpoint as a derivative rule, we next offer a partial proof of the Chain Rule. Note how the way the limit is re-written reflects the ultimate statement of the Chain Rule. It also reflects much of the intuition of the rule.

We will not completely prove the Chain Rule in this context because of some technical difficulties which arise when proving the rule in its most general form. However, we will look at a proof in perhaps the most common case, which is the case that $g(x+\Delta x) - g(x) \longrightarrow 0$ "properly" (so that in particular we also have $(\exists \delta > 0)[0 < |\Delta x| < \delta \longrightarrow g(x+\Delta x) - g(x) \neq 0]$, as described in Subsection 2.9.3, starting on page 198). In such a case we can safely divide and multiply by $g(x+\Delta x) - g(x)$ in the limit definition of the derivative to get

$$\frac{d}{dx}[f(g(x))] = \lim_{\Delta x \to 0} \frac{f(g(x+\Delta x)) - f(g(x))}{\Delta x}$$

$$= \lim_{\Delta x \to 0} \underbrace{\frac{f(g(x+\Delta x)) - f(g(x))}{g(x+\Delta x) - g(x)}}_{(I)} \cdot \underbrace{\frac{g(x+\Delta x) - g(x)}{\Delta x}}_{(II)} = \lim_{\Delta x \to 0} (I) \cdot \lim_{\Delta x \to 0} (II),$$

if these last two limits exist. Now we claim that they do, and will give us that this overall limit is $f'(g(x))g'(x)$, at least under the assumption that $g(x+\Delta x) - g(x) \longrightarrow 0$ "properly."

The second term (II) clearly has limit $g'(x)$, by definition:

$$\lim_{\Delta x \to 0} (II) = \lim_{\Delta x \to 0} \frac{g(x+\Delta x) - g(x)}{\Delta x} = g'(x).$$

Under the assumption that $g(x+\Delta x) - g(x) \longrightarrow 0$ properly as $\Delta x \to 0$, we can substitute $\Delta g(x) = g(x+\Delta x) - g(x) \longrightarrow 0$, and rewrite the limit (I)[55]

$$\lim_{\Delta x \to 0}(I) = \lim_{\Delta x \to 0} \frac{f(g(x+\Delta x)) - f(g(x))}{g(x+\Delta x) - g(x)} = \lim_{\Delta g(x) \to 0} \frac{f(g(x) + \Delta g(x)) - f(g(x))}{\Delta g(x)} = f'(g(x)).$$

Note that the computation above is also correct even if $\Delta g(x) \longrightarrow 0^+$ or $\Delta g(x) \longrightarrow 0^-$ properly, because the existence of the two-sided limit represented by $f'(g(x))$ is assumed in our hypotheses of the Chain Rule theorem. Thus $\frac{d}{dx} f(g(x)) = \lim_{\Delta x \to 0}(I) \cdot \lim_{\Delta x \to 0}(II) = f'(g(x)) \cdot g'(x)$, q.e.d.

3.4.11 Simple Graphical Examples of Chain Rule

Here we offer an admittedly very simple example of the Chain Rule's implications for the graph of a composite function. In Figure 3.22, page 293, we have graphs of the functions $y_1 = \sin x$ and $y_2 = \sin 2x$, for which we have (using the Chain Rule to find y_2')

$$y_1 = \sin x \quad \Longrightarrow \quad y_1' = \cos x,$$
$$y_2 = \sin 2x \quad \Longrightarrow \quad y_2' = 2\cos 2x.$$

Three points each on the graphs of y_1 and y_2 are shown in particular in Figure 3.22, those being corresponding points in the periods of their respective functions.

function	point	height	slope
y_1	$(0,0)$	$y_1(0) = 0$	$y_1'(0) = 1$
y_2	$(0,0)$	$y_2(0) = 0$	$y_2'(0) = 2$
y_1	$(\pi, 0)$	$y_1(\pi) = 0$	$y_1'(\pi) = -1$
y_2	$(\frac{\pi}{2}, 0)$	$y_2(\frac{\pi}{2}) = 0$	$y_2'(\frac{\pi}{2}) = -2$
y_1	$(\frac{3\pi}{2}, -1)$	$y_1(\frac{3\pi}{2}) = -1$	$y_1'(\frac{3\pi}{2}) = 0$
y_2	$(\frac{3\pi}{4}, -1)$	$y_2(\frac{3\pi}{4}) = 0$	$y_2'(\frac{3\pi}{4}) = 0$

From the graph and from the chart above, we see how the *horizontal variable* transformation (horizontal contraction of the graph by a factor of 2) from y_1 to y_2 also caused a change in the rate at which the vertical variable changed with respect to the horizontal variable. In this case, that rate of change was doubled as we transformed from y_1 to y_2, because x was replaced with $2x$, and the velocity at which $2x$ passes through values is twice that of x.

Nonlinear transformations would be even more complicated. Consider

$$y_3 = \sin x^2 \quad \Longrightarrow \quad y_3' = 2x \cos x^2.$$

---

[55] When we rewrite $f(g(x+\Delta x)) = f(g(x) + \Delta g(x))$, we were justified because

$$g(x + \Delta x) = (g(x+\Delta x) - g(x)) + g(x) = \Delta g(x) + g(x) = g(x) + \Delta g(x).$$

3.4. CHAIN RULE I

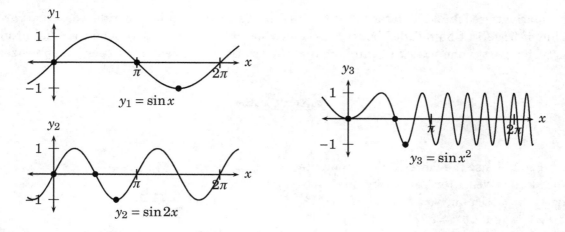

Figure 3.22: Graphs of the sine function with three different inputs: x, $2x$ and x^2. Also highlighted are three points which correspond to the same input values for the sine function. Because of the multiplicative nature of the Chain Rule derivatives, the slopes at the three points can change from graph to graph. (See the discussion within the text.)

function	point	height	slope
y_1	$(0,0)$	$y_1(0) = 0$	$y_1'(0) = 1$
y_3	$(0,0)$	$y_3(0) = 0$	$y_3'(0) = 0$
y_1	$(\pi, 0)$	$y_1(\pi) = 0$	$y_1'(\pi) = -1$
y_3	$(\sqrt{\pi}, 0)$	$y_3(\sqrt{\pi}) = 0$	$y_3'(\sqrt{\pi}) = 2\sqrt{\pi}(-1) \approx -3.54$
y_1	$(\frac{3\pi}{2}, -1)$	$y_1(\frac{3\pi}{2}) = -1$	$y_1'(\frac{3\pi}{2}) = 0$
y_3	$(\sqrt{3\pi/2}, -1)$	$y_3(\sqrt{3\pi/2}) = -1$	$y_3'(\sqrt{3\pi/2}) = 2\sqrt{3\pi/2}\cos(3\pi/2) = 0$

We see from the table that changing the input of the sine function from x to x^2 actually flattens the slope at $x = 0$ due to the extra factor of $2x$ in the derivative $y_3' = 2x\cos x^2$, but it amplifies the slope for the same reason at $x = \sqrt{\pi}$ (corresponding to $x = \pi$ in the pure sine function y_1). At $x = \sqrt{3\pi/2}$ the extra factor of $2x$ is "amplifying" a zero slope, which still yields a zero slope.

While such intuitions are interesting, fortunately we do not have to have them first in order to compute Chain Rule derivatives. Indeed, we will often rely on the somewhat formulaic derivative techniques to flush out such observations in order to improve our intuitions. For that reason we will spend much time on the mechanics of computing derivatives, and usually let the insights follow, though at times we will switch that order.

We will have use for the Chain Rule in nearly every section of the text from here on, and will develop it further as we proceed. Two later sections (Section 3.6 and Section 4.6) are dedicated to insights it offers in special and important contexts, and so those sections have "Chain Rule" in their titles. Other sections (starting with Section 5.8) dealing with antiderivatives found still later in the text are dedicated to what is sometimes described as a kind of "reverse Chain Rule," but it has its own special insights and by then the Chain Rule is such a part of one's calculus thinking that the relationship need not be overly emphasized.

Exercises

For 1–6, use the Power Rule to compute the derivative, after re-writing the problem. In particular, you can use $\frac{d}{dx}(ax^n) = a \cdot nx^{n-1}$, and $(ab)^n = a^n b^n$ anytime $a, b \geq 0$.

1. $\dfrac{d}{dx}\left[\dfrac{1}{x^{11}}\right]$

2. $\dfrac{d}{dx}\left[\dfrac{1}{\sqrt{x}}\right]$

3. $\dfrac{d}{dx}\left[\sqrt[3]{x^4}\right]$

4. $\dfrac{d}{dx}\left[\dfrac{6}{x}\right]$

5. $\dfrac{d}{dt}\left[\dfrac{1}{2t^2}\right]$

6. $\dfrac{d}{dy}\left[\sqrt{9y}\right]$

For 7–10, compute the given derivative.

7. $\dfrac{d}{dx}\left[(1-9x)^{11}\right]$

8. $\dfrac{d}{dx}\left[27(3x^2 - 10x + 55)^2\right]$

9. $\dfrac{d}{dx}\left[\sqrt{2x^5 - 1}\right]$

10. $\dfrac{d}{dx}(3x+1)^2$. Do this two ways: using the Chain Rule, and by first expanding the square. Show that the answers are the same.

For Exercises 11–17, compute $f'(x)$.

11. $f(x) = (x+5)^{100}$

12. $f(x) = (2x+5)^{100}$

13. $f(x) = (3x^2 + 5x - 9)^{100}$

14. $f(x) = \sqrt[3]{x^2 + 2x + 9}$

15. $f(x) = \dfrac{1}{(x^4 - x + 1)^3}$

16. $f(x) = \sin(x + \sin(x + \sin x))$

17. $f(x) = \sqrt{x + \sqrt{x + \sqrt{x}}}$

For Exercises 18–21, compute the given derivatives.

18. $\dfrac{d \sin z}{dx}$

19. $\dfrac{d \cos \theta}{dt}$

20. $\dfrac{d x^7}{dt}$

21. $\dfrac{d \sin(\cos x)}{d \cos x}$

For Exercises 22–26, compute the given derivative.

22. $\dfrac{d \sin(\cos x)}{dx}$

23. $\dfrac{d \sin \sqrt{x}}{dx}$

24. $\dfrac{d}{dx}\sqrt{\sin x}$

25. $\dfrac{d \cos^3 x}{dx}$

26. $\dfrac{d}{dx}\cos\left(\dfrac{1}{x}\right)$

27. Find $h'(9)$ if $h(x) = f(g(x))$, $g(9) = 5$, $g'(9) = 2$, and $f'(5) = 7$.

28. Since $\pi/4$ is a constant, so is $\sin\frac{\pi}{4}$, so the derivative of $\sin\frac{\pi}{4}$ should be zero. Show that we could also use the Chain Rule to verify this fact that $\frac{d}{dx}\sin\frac{\pi}{4} = 0$.

3.4. CHAIN RULE I

29. On the unit circle, $y^2 = 1 - x^2$. If we take either the upper semicircle or the lower semicircle, then y is also a function of x. Find the tangent line to the graph at the point $(3/5, 4/5)$ by finding $\frac{dy}{dx}$ two ways:

    (a) Using $y = \sqrt{1-x^2}$ for the upper semicircle, and the Chain Rule.

    (b) By applying $\frac{d}{dx}$ to both sides of $y^2 = 1 - x^2$, as we did in the proof of Theorem 3.4.2, and then solving for $\frac{dy}{dx}$, and plugging into that expression $(x, y) = (3/5, 4/5)$.

30. Using $\sec x = (\cos x)^{-1}$,

    (a) derive $\dfrac{d}{dx} \sec x = \sec x \tan x$.

    (b) Use (a) and the Chain Rule to compute $\dfrac{d}{dx} \sec \sqrt{x^2 + 1}$.

31. Show that if k, q_1, q_2 are constant in Coulomb's law, $F = k\frac{q_1 q_2}{r^2}$, then
    $$\frac{dF}{dr} = -2\frac{F}{r}.$$

32. As a charge q_2 moves away from a stationary charge q_1, the instantaneous rate of change of the Coulomb force F with respect to r is measured to be 5N/m.

    (a) Find the magnitude of the force between the charges when they are 0.3m apart. (See Example 3.4.13, page 285.)

    (b) If we are given that $k = 9 \times 10^9$ N$\cdot$m^2/C, and $q_2 = 3 \times 10^{-6}$C, find the equation for the force between the two particles when separated by a distance r.

    (c) Find $\frac{dF}{dr}$ from (b) when $r = 0.25$m.

33. The gravitational force of attraction between any two masses is directly proportional to the product of the masses and inversely proportional to the square of the distance between them. In particular,
    $$F = G\frac{m_1 m_2}{r^2}, \quad (3.47)$$
    where $G \approx 6.67 \times 10^{-11}$N(m)^2/(kg)^2 is the universal gravitational constant, m_1, m_2 are in kg, and r is in m. Find the rate of change of F with respect to distance r (i.e., find dF/dr) when $r = 0.5$m.

34. An object moves along the x-axis according to the equation $s(t) = t + 2\sin t$. Find those values of t when the object is moving to the left, i.e., find where $s'(t) < 0$, within the interval $t \in [0, 2\pi]$.

35. Suppose that the cost in thousands of dollars to manufacture x thousands of items is given by
    $$C(x) = 0.003x^3 + 0.2x^2 + 0.3x + 70,$$
    where $x \in [0, 10]$.

    (a) Find the (proxy) marginal cost function, $C'(x)$.

    (b) Determine the marginal cost at $x = 4$ (in units of thousands of dollars per thousands of items, which simplifies to dollars per item).[56]

    (c) Find the average rate of change of cost in going from $x = 3.9$ to $x = 4.1$.

    (d) Compare the answers to parts (b) and (c).

36. Consider the following alternating current circuit:

---
[56]Here marginal cost would represent the cost of the next thousand items.

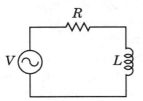

The current I where the resistor R and an inductor L are present is given by
$$I = \frac{V}{\sqrt{R^2 + (2\pi f L)^2}},$$
where V is the voltage and f is the current's frequency of alternation (in cycles/second, or Hertz, also written Hz). Assuming that V, R and f are constants, find an equation for the instantaneous rate of change of current I with respect to inductance L. The units of L will be henries H.[57]

---

[57]Named for Joseph Henry (December 17, 1797–May 13, 1878), an American scientist who discovered electromagnetic inductance independently of British Scientist Michael Faraday, (September 22, 1791–August 25, 1867) who discovered and published his work on the phenomenon first. Henry also discovered self-inductance by which a change in the current in a winding of wire in electromagnet or "inductor" (as in our diagram) can induce reactive voltages in other windings (just as changes in current in one coil of an alternating current transformer can cause voltages in other coils, which is the basic mechanism by which such transformers operate). His work on relays helped to lead to the invention of the telegraph.

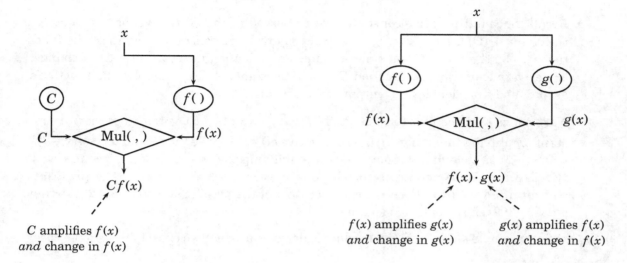

Figure 3.23: Idea of the Product Rule. The instantaneous rate of change of $f(x)$ is amplified (multiplied) by C in the first diagram. In the second diagram there are two contributions to the change in $f(x)g(x)$: the change in $g(x)$, as multiplied (or amplified) by $f(x)$, and the change in $f(x)$, multiplied by $g(x)$.

3.5 Product, Quotient, and Remaining Trigonometric Rules

In this section we first introduce the rule for the differentiation of a product of two functions. From that and the Chain Rule, we will derive a rule for differentiating a quotient of two functions. With a Quotient Rule we will be able to use the rules for $\sin x$ and $\cos x$ to derive rules for $\tan x$ and $\cot x$. For completeness we will also compute the rules for $\sec x$ and $\csc x$ (the Chain Rule suffices) and thus finish our rules for the six basic trigonometric functions.

The rules for calculating the derivative of a product or a quotient are not as simple as for a sum or difference. However they are straightforward when applied correctly.

3.5.1 Product Rule Stated and First Applied

We begin with the statement and some discussion of the Product Rule, followed by several examples demonstrating its mechanics. The actual proof we leave until the next subsection.

Theorem 3.5.1 (Product Rule) *At each value of x for which $\frac{d}{dx}f(x)$ and $\frac{d}{dx}g(x)$ exist, so does the derivative $\frac{d}{dx}(f(x) \cdot g(x))$ exist, and it is given by*

$$\boxed{\frac{d}{dx}[f(x) \cdot g(x)] = f(x)\frac{d}{dx}g(x) + g(x)\frac{d}{dx}f(x).} \tag{3.48}$$

Though we defer the proof, we can make a couple of observations.

- The derivative of $f(x)g(x)$ is **not** simply the product $f'(x)g'(x)$ of the two derivatives. (See for example Exercise 16 in Section 3.2, page 254.) It is more interesting than that.

- Recall that multiplicative constants are preserved in the derivative, or "go along for the ride": $\frac{d}{dx}(Cf(x)) = C\frac{d}{dx}f(x)$. One could say that the constant "amplifies" the function $f(x)$ by the factor C, causing the same amplification factor in the rate of change, or derivative, of the new function $Cf(x)$. (For example, $C = 2$ doubles the function's output, and thus doubles the output's rate of change.)

- Next notice that the first term of (3.48), $f(x)\frac{dg(x)}{dx}$ treats $f(x)$ as though it were a constant amplifying the change in (i.e., derivative of) $g(x)$, while the second term $g(x)\frac{df(x)}{dx}$ treats $g(x)$ as though it were a constant amplifying the change in $f(x)$. **In this way the Product Rule accounts for the changes in each function in the product, as amplified by the other.** A close scrutiny of the proof (Subsection 3.5.2, starting on page 302) shows how this emerges.

Our first example shows how the Product Rule gives us what we expect for a simple case.

Example 3.5.1

Let $f(x) = x^5$. Then $f'(x) = 5x^4$ from the Power Rule. But we can also write $f(x) = x^3 \cdot x^2$, from which the Product Rule gives

$$\frac{d}{dx}\left[x^3 \cdot x^2\right] = x^3 \cdot \frac{d}{dx}(x^2) + x^2 \cdot \frac{d}{dx}(x^3) = x^3 \cdot 2x + x^2 \cdot 3x^2 = 2x^4 + 3x^4 = 5x^4.$$

Of course the Product Rule will be of much more use than proving things we already knew. The next example requires the Product Rule (or some *very* clever tricks with difference quotients!):

Example 3.5.2

Suppose $f(x) = x^2 \sin x$. This is a product of two differentiable functions. Its derivative is given by[58]

$$f'(x) = \frac{d(x^2 \sin x)}{dx} = x^2 \cdot \frac{d\sin x}{dx} + \sin x \cdot \frac{d(x^2)}{dx} = x^2 \cos x + \sin x \cdot 2x = x(x\cos x + 2\sin x)$$

The last step was just an algebraic one, factoring the final answer as much as possible.

Now we list several simple examples to illustrate the basic mechanics of the Product Rule.

- $\dfrac{d}{dx}\left[(3x^2 + 5x - 9)(5x^3 + 7x^2 + 27x - 4)\right]$

$= (3x^2 + 5x - 9)\dfrac{d}{dx}(5x^3 + 7x^2 + 27x - 4) + (5x^3 + 7x^2 + 27x - 4)\dfrac{d}{dx}(3x^2 + 5x - 9)$

$= (3x^2 + 5x - 9)(15x^2 + 14x + 27) + (5x^3 + 7x^2 + 27x - 4)(6x + 5).$

---

[58] If the functions are not differentiable, this fact appears as we take the derivatives on the right-hand side of the Product Rule statement (3.48). Thus we usually just naively apply the rule—instead of checking differentiability first—and if the result makes sense, the original function satisfied the hypotheses of the Product Rule.

3.5. PRODUCT, QUOTIENT, AND REMAINING TRIGONOMETRIC RULES

- $\frac{d}{dx}(x\cos x) = x \cdot \frac{d}{dx}\cos x + \cos x \cdot \frac{d}{dx}x = x(-\sin x) + \cos x \cdot 1 = -x\sin x + \cos x.$

- $\frac{d}{dt}[PV] = P \cdot \frac{dV}{dt} + V \cdot \frac{dP}{dt}.$

One of the interesting aspects of the calculus is the various ways that the consistency of differentiation (derivative-taking) rules can be seen by using different strategies for particular derivatives. Earlier we showed how to use the Product Rule to compute $\frac{d}{dx}(x^2 \cdot x^3) = \cdots = 5x^4$, which we can also compute with the Power Rule $\frac{d}{dx}(x^5) = 5x^4$. We can also see how the Product Rule gives us the behavior of multiplicative constants in derivative computations. For example, instead of writing $\frac{d}{dx}(2\sin x) = 2\frac{d}{dx}\sin x = 2\cos x$, we can consider $2\sin x$ to be the product of two functions, and compute the derivative using the Product Rule if we care to:

$$\frac{d}{dx}(2\sin x) = 2 \cdot \frac{d}{dx}\sin x + \sin x \cdot \frac{d}{dx}(2) = 2\cos x + \sin x \cdot 0 = 2\cos x + 0 = 2\cos x.$$

Of course the rule on multiplicative constants (page 234) is faster: $\frac{d\,2\sin x}{dx} = 2 \cdot \frac{d\,\sin x}{dx} = 2\cos x.$

Because the Product Rule calls for the computation of derivatives of the factors, it often "calls" upon other rules to compute these component derivatives. Conversely, other rules may call upon the Product Rule. As we saw with the Chain Rule, it is crucial that we look at the overall structure of a function to see which rule to apply first, and then work our way in towards the inner structures as the differentiation rules require in their turns; we compute derivatives "from the outside to the inside." The next two examples are Product Rules first, which then call the Chain Rule.

Example 3.5.3

Suppose $f(x) = \sin x^2 \cos x^3$. This is foremost a product of two functions, so we need the Product Rule first.[59]

$$\begin{aligned}
f'(x) &= \frac{d}{dx}\left[\sin x^2 \cos x^3\right] \\
&= \sin x^2 \cdot \frac{d}{dx}\cos x^3 + \cos x^3 \cdot \frac{d}{dx}\sin x^2 &&\text{(Product Rule)} \\
&= \sin x^2 \cdot \left(-\sin x^3 \cdot \frac{d}{dx}x^3\right) + \cos x^3 \cdot \left(\cos x^2 \cdot \frac{d}{dx}x^2\right) &&\text{(Chain Rule, twice)} \\
&= (\sin x^2)(-\sin x^3)(3x^2) + \cos x^3 \cos x^2 \cdot 2x \\
&= -3x^2 \sin x^2 \sin x^3 + 2x\cos x^3 \cos x^2. &&\text{(Rearrangement)}
\end{aligned}$$

Thus, when we took the derivatives called for by the Product Rule, these required the Chain Rule. (We could have factored the final computation but it is not necessary.)

---

[59]Note that $\sin x^2 \cos x^3$ is taken to be a product. Indeed, it is understood that the sine and cosine functions here are separate factors. Also note that x^2 is the input of the sine, and x^3 the input of the cosine. The convention is to understand this function, as written, in the following way:

$$\sin x^2 \cos x^3 = (\sin x^2)(\cos x^3) = \left[\sin(x^2)\right]\left[\cos(x^3)\right].$$

For a polynomial example, consider the following:

Example 3.5.4

$f(x) = (x^2 + 2x + 3)^2(x^2 + 1)^3$. Without the product rule we would be forced to carry out the multiplications, but since this is written as a product of two functions, we can instead use the Product Rule. (The calculus is finished before the dividing line, and the rest is algebra, which is optional, but the factoring is interesting.)

$$\begin{aligned}
f'(x) &= \frac{d}{dx}\left[(x^2+2x+3)^2(x^2+1)^3\right] \\
&= (x^2+2x+3)^2 \cdot \frac{d}{dx}(x^2+1)^3 + (x^2+1)^3 \cdot \frac{d}{dx}(x^2+2x+3)^2 \\
&= (x^2+2x+3)^2 \cdot 3(x^2+1)^2 \cdot \frac{d}{dx}(x^2+1) + (x^2+1)^3 \cdot 2(x^2+2x+3)^1 \frac{d}{dx}(x^2+2x+3) \\
&= (x^2+2x+3)^2 \cdot 3(x^2+1)^2(2x) + (x^2+1)^3 \cdot 2(x^2+2x+3)(2x+2) \\
\hline
&= 6x(x^2+2x+3)^2(x^2+1)^2 + (4x+4)(x^2+1)^3(x^2+2x+3) \\
&= (x^2+2x+3)(x^2+1)^2\left[6x(x^2+2x+3) + (4x+4)(x^2+1)\right] \\
&= (x^2+2x+3)(x^2+1)^2\left[6x^3 + 12x^2 + 18x + 4x^3 + 4x + 4x^2 + 4\right] \\
&= (x^2+2x+3)(x^2+1)^2\left[10x^3 + 16x^2 + 22x + 4\right].
\end{aligned}$$

Again, the statement of the Product Rule here called for derivatives of the factors, and each of those required a Chain Rule. Note how one factor of (x^2+2x+3) and two factors of (x^2+1) could be extracted from each term to allow for factoring. Such algebraic manipulations are useful, for instance, if a sign chart for the derivative is desired, for instance for graphing purposes.

It is also possible that a Product Rule can occur within a Chain Rule, as in the following.

Example 3.5.5

Suppose $f(x) = \sqrt{\sin x \cos x}$. Then

$$\begin{aligned}
f'(x) &= \frac{d}{dx}\sqrt{\sin x \cos x} \\
&= \frac{1}{2\sqrt{\sin x \cos x}} \cdot \frac{d}{dx}(\sin x \cos x) \\
&= \frac{1}{2\sqrt{\sin x \cos x}} \cdot \left[\sin x \frac{d}{dx}\cos x + \cos x \frac{d}{dx}\sin x\right] \\
&= \frac{1}{2\sqrt{\sin x \cos x}} \cdot [\sin x(-\sin x) + \cos x \cos x] \quad = \quad \frac{-\sin^2 x + \cos^2 x}{2\sqrt{\sin x \cos x}}.
\end{aligned}$$

This is not the only method for solving this problem, but it is perhaps the most straightforward.[60]

---

[60]Actually, with trigonometry we can rewrite the problem and the answer, using $\sin 2\theta = 2\sin\theta\cos\theta$ and $\cos 2\theta = \cos^2\theta - \sin^2\theta$. Below, "$\Longrightarrow$" represents another Chain Rule problem (calling yet another Chain Rule).

$$f(x) = \sqrt{\tfrac{1}{2}\sin 2x} \quad \Longrightarrow \quad f'(x) = \cdots = \frac{\cos 2x}{2\sqrt{\tfrac{1}{2}\sin 2x}}.$$

3.5. PRODUCT, QUOTIENT, AND REMAINING TRIGONOMETRIC RULES

We can also use these Product-Rule–derived derivatives to help graph functions.

Example 3.5.6

Suppose $f(x) = x\sqrt{1-x^2}$. This function we will differentiate and then graph. But even before delving into the derivative, we can first notice the domain of $f(x)$ is $-1 \le x \le 1$, and the function (height) itself is zero at $x = 0, \pm 1$ (x-intercepts). Next we consider its derivative:

$$f'(x) = x \cdot \frac{d}{dx}\sqrt{1-x^2} + \sqrt{1-x^2} \cdot \frac{d}{dx}(x) = x \cdot \frac{1}{2\sqrt{1-x^2}} \cdot \frac{d}{dx}(1-x^2) + \sqrt{1-x^2} \cdot 1$$

$$= x \cdot \frac{1}{2\sqrt{1-x^2}} \cdot (-2x) + \sqrt{1-x^2} = \frac{-x^2}{\sqrt{1-x^2}} + \sqrt{1-x^2}.$$

To graph this function $f(x)$ we would like to know where it is increasing and where it is decreasing and thus locate local extrema. To those ends we next proceed to see where $f' > 0$ and $f' < 0$. For such a task, it is best if the derivative is written as a single fraction:

$$f'(x) = \frac{-x^2}{\sqrt{1-x^2}} + \sqrt{1-x^2} \cdot \frac{\sqrt{1-x^2}}{\sqrt{1-x^2}} = \frac{-x^2 + 1 - x^2}{\sqrt{1-x^2}} = \frac{1 - 2x^2}{\sqrt{1-x^2}} \xrightarrow{\text{optional}} \frac{(1 - \sqrt{2}x)(1 + \sqrt{2}x)}{\sqrt{1-x^2}}.$$

Now we see that this is undefined (f' DNE) except for $-1 < x < 1$. The fraction which is f' is zero exactly where the numerator is zero and the denominator is not. Thus

$$f'(x) = 0 \iff 1 - 2x^2 = 0 \iff 1 = 2x^2 \iff \frac{1}{2} = x^2 \iff x = \pm\frac{1}{\sqrt{2}} \approx \pm 0.7071.$$

From this we can make a sign chart for f' to see where f is increasing/decreasing.

$$f'(x) = \frac{1 - 2x^2}{\sqrt{1-x^2}}$$

Test $x =$		-0.9		0		0.9	
Factors of $f'(x)$:		$\ominus/\oplus$		$\oplus/\oplus$		$\ominus/\oplus$	
Sign $f'(x)$:	-1	$\ominus$	$-\frac{1}{\sqrt{2}}$	$\oplus$	$\frac{1}{\sqrt{2}}$	$\ominus$	1
Behavior of $f(x)$:		DEC ↘		INC ↗		DEC ↘	

We see a local minimum at $x = -1/\sqrt{2}$, and a local maximum at $x = 1/\sqrt{2}$. The actual points are as follow, noting that $\sqrt{1/2} = 1/\sqrt{2}$:

$$\left(-\frac{1}{\sqrt{2}}, f\left(-\frac{1}{\sqrt{2}}\right)\right) = \left(-\frac{1}{\sqrt{2}}, -\frac{1}{\sqrt{2}} \cdot \sqrt{1/2}\right) = \left(-\frac{1}{\sqrt{2}}, -\frac{1}{2}\right) \approx (-0.7071, -0.5),$$

$$\left(\frac{1}{\sqrt{2}}, f\left(\frac{1}{\sqrt{2}}\right)\right) = \left(\frac{1}{\sqrt{2}}, \frac{1}{\sqrt{2}} \cdot \sqrt{1/2}\right) = \left(\frac{1}{\sqrt{2}}, \frac{1}{2}\right) \approx (0.7071, 0.5).$$

All this behavior leads us to the graph, which is given in Figure 3.24, page 302. Notice the (computer-generated) graph there also reflects that $f'(x) = \frac{1 - 2x^2}{\sqrt{1-x^2}} \longrightarrow -\infty$ as $x \to -1^+$ or as $x \to 1^-$. Also somewhat useful is that $f'(0) = 1$. Finally worth noting is that $f(-x) = -f(x)$, meaning f is an *odd function*, and so its graph is symmetric with respect to the origin.

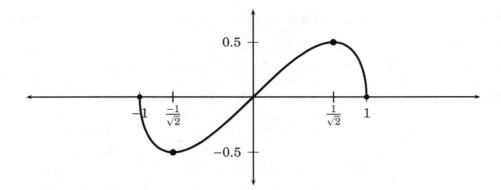

Figure 3.24: Complete graph of $f(x) = x\sqrt{1-x^2}$, with local extrema marked at $x = \pm 1/\sqrt{2}$.

3.5.2 Product Rule Proof

For completeness we include here a proof of the Product Rule. The proof is not accomplished so much by "brute force," but instead utilizes some clever rewriting.

It is normal to wonder *why* one would think of the "trick" used to produce a proof, but this should not immediately distract from the fact that a proof "works." Many proofs used today are quite condensed or even very different from the original discoverers' proofs, as broader knowledge uncovers other proof strategies which may be more efficient, or proofs are more condensed. These changes often happen over the decades or even centuries since the first discovered valid proofs. As a result, proofs often look less like the natural paths of discovery and more like terse arguments which may feel less satisfying than a natural path towards discovery, but if the logic is valid then the conclusions have to be respected.

Furthermore, there is knowledge to be gained from even these short or clever proofs—for instance, because the "trick" may be useful in other contexts—and so they are worth reading and understanding, though again we will almost always just quote the results, without reference to their proofs, when solving problems.

Proof: (Product Rule) Suppose $f(x)$ and $g(x)$ are both differentiable at a given x, i.e., $f'(x)$ and $g'(x)$ both exist. Then f and g are both continuous at x, and

$$\begin{aligned}
\frac{d}{dx}\left[f(x)g(x)\right] &= \lim_{\Delta x \to 0} \frac{f(x+\Delta x)g(x+\Delta x) - f(x)g(x)}{\Delta x} \\
&= \lim_{\Delta x \to 0} \frac{f(x+\Delta x)g(x+\Delta x) - f(x+\Delta x)g(x) + f(x+\Delta x)g(x) - f(x)g(x)}{\Delta x} \\
&= \lim_{\Delta x \to 0} \frac{f(x+\Delta x)\left[g(x+\Delta x) - g(x)\right] + g(x)\left[f(x+\Delta x) - f(x)\right]}{\Delta x} \\
&= \lim_{\Delta x \to 0} \left[f(x+\Delta x) \cdot \frac{g(x+\Delta x) - g(x)}{\Delta x} + g(x) \cdot \frac{f(x+\Delta x) - f(x)}{\Delta x}\right] \\
&= f(x)g'(x) + g(x)f'(x), \qquad \text{q.e.d.}
\end{aligned}$$

3.5. PRODUCT, QUOTIENT, AND REMAINING TRIGONOMETRIC RULES

The last line of the proof follows because by continuity at x (implied by differentiability there, according to Theorem 3.2.1, page 229) $\Delta x \to 0 \implies f(x+\Delta x) \to f(x)$. The two difference quotients in the proof approach $f'(x)$ and $g'(x)$ respectively, while $g(x)$ is constant in the limit (which is in Δx, not x). The middle two lines were simply algebra, with the "clever trick" in the second line, where we subtract and add the same quantity (thus not changing the value of our expression), in such a way that we can have two factored expressions. The basic form is $AB - CD = AB - AD + AD - CD = A(B-D) + D(A-C)$, except here these terms were $\underbrace{f(x+\Delta x)}_{A}\underbrace{g(x+\Delta x)}_{B} - \underbrace{f(x)}_{C}\underbrace{g(x)}_{D}$.

3.5.3 Quotient Rule

We often need to find derivatives of functions of the form $h(x) = f(x)/g(x)$. We can rewrite these as $h(x) = f(x)(g(x))^{-1}$, and use the Product Rule, which will then call the Chain Rule, to get

$$h'(x) = \frac{d}{dx}\left[f(x)(g(x))^{-1}\right] = f(x)\frac{d}{dx}\left[(g(x))^{-1}\right] + (g(x))^{-1}\frac{d}{dx}f(x)$$

$$= f(x)\left[(-1)(g(x))^{-2}\frac{d}{dx}g(x)\right] + (g(x))^{-1}\frac{d}{dx}f(x)$$

$$= \frac{-f(x)\frac{d}{dx}g(x)}{(g(x))^2} + \frac{\frac{d}{dx}f(x)}{g(x)}$$

$$= \frac{-f(x)\frac{d}{dx}g(x)}{(g(x))^2} + \frac{g(x)\frac{d}{dx}f(x)}{(g(x))^2} = \frac{g(x)\frac{d}{dx}f(x) - f(x)\frac{d}{dx}g(x)}{(g(x))^2}.$$

This is conveniently summarized in the following theorem (which should be memorized):

Theorem 3.5.2 (Quotient Rule) *If f and g are differentiable at input value x, and $g(x) \neq 0$, then*

$$\boxed{\frac{d}{dx}\left[\frac{f(x)}{g(x)}\right] = \frac{g(x)\frac{d}{dx}f(x) - f(x)\frac{d}{dx}g(x)}{(g(x))^2}.} \tag{3.49}$$

As noted in the derivation, this rule is actually redundant given the availability of the Product and Chain Rules. However, it is useful and efficient, especially because the resulting derivative emerges as an already combined fraction, useful for sign charts, graphing and other applications.

Example 3.5.7

Find $f'(x)$ if $f(x) = \dfrac{\sin x}{x}$.

Solution: Using the Quotient Rule we have

$$f'(x) = \frac{x\frac{d}{dx}\sin x - \sin x \frac{d}{dx}(x)}{(x)^2} = \frac{x\cos x - \sin x}{x^2}.$$

Note that this derivative makes sense exactly when $x \neq 0$.

The Quotient Rule is especially useful for rational functions, i.e., ratios of polynomials.

Example 3.5.8

Suppose $f(x) = \dfrac{x^3-8}{x^2-9}$. Then

$$f'(x) = \frac{(x^2-9)\frac{d}{dx}(x^3-8) - (x^3-8)\frac{d}{dx}(x^2-9)}{(x^2-9)^2} = \frac{(x^2-9)(3x^2) - (x^3-8)(2x)}{(x^2-9)^2}$$

$$= \frac{3x^4 - 27x^2 - 2x^4 + 16x}{(x^2-9)^2} = \frac{x^4 - 27x^2 + 16x}{(x^2-9)^2}.$$

The calculus was finished by the end of the first line; the remaining work was algebraic. The final form shows where $f(x)$ is differentiable ($x \neq \pm 3$).

Example 3.5.9

Suppose $f(x) = \dfrac{\sqrt{x}}{\cos x}$. Then

$$f'(x) = \frac{\cos x \cdot \frac{d\sqrt{x}}{dx} - \sqrt{x} \cdot \frac{d\cos x}{dx}}{(\cos x)^2} = \frac{\cos x \cdot \frac{1}{2\sqrt{x}} - \sqrt{x}(-\sin x)}{\cos^2 x} = \frac{\frac{\cos x}{2\sqrt{x}} + \sqrt{x}\cdot \sin x}{\cos^2 x} \cdot \frac{2\sqrt{x}}{2\sqrt{x}} = \frac{\cos x + 2x\sin x}{2\sqrt{x}\cdot \cos^2 x}.$$

Later when we have derivative formulas for all of the trigonometric functions, we could instead use the product rule after rewriting $f(x) = \sqrt{x}\sec x$. In fact in Exercise 30a, page 295, the differentiation rule $\frac{d}{dx}\sec x = \sec x \tan x$ is derived. That rule is also found later in this section, among other trigonometric derivatives, on page 305. We use it in the next example.

Example 3.5.10

Suppose $f(x) = \sec\left(\dfrac{x}{x-1}\right)$. Then

$$f'(x) = \frac{d}{dx}\sec\left(\frac{x}{x-1}\right)$$

$$= \sec\left(\frac{x}{x-1}\right)\tan\left(\frac{x}{x-1}\right)\cdot \frac{d}{dx}\left(\frac{x}{x-1}\right)$$

$$= \sec\left(\frac{x}{x-1}\right)\tan\left(\frac{x}{x-1}\right)\cdot \left(\frac{(x-1)\cdot \frac{d}{dx}(x) - x\cdot \frac{d}{dx}(x-1)}{(x-1)^2}\right)$$

$$= \sec\left(\frac{x}{x-1}\right)\tan\left(\frac{x}{x-1}\right)\cdot \frac{(x-1)(1) - (x)(1)}{(x-1)^2}$$

$$= \sec\left(\frac{x}{x-1}\right)\tan\left(\frac{x}{x-1}\right)\cdot \frac{x-1-x}{(x-1)^2}$$

$$= \frac{-1}{(x-1)^2}\sec\left(\frac{x}{x-1}\right)\tan\left(\frac{x}{x-1}\right).$$

3.5. PRODUCT, QUOTIENT, AND REMAINING TRIGONOMETRIC RULES

In fact the Quotient Rule computation for this particular problem could have been avoided through long division, giving $\frac{d}{dx}\left(\frac{x}{x-1}\right) = \frac{d}{dx}\left(1 + \frac{1}{x-1}\right)$, making for a simple Power/Chain Rule, but the Quotient Rule was straightforward and left that factor as a single fraction. For an example of Chain Rules inside a Quotient Rule, consider the next example. Recall $\frac{d}{dx}\sqrt{x} = \frac{d}{dx}x^{1/2} = \frac{1}{2}x^{-1/2} = \frac{1}{2\sqrt{x}}$.

Example 3.5.11

Suppose $f(x) = \dfrac{\cos 2x}{\sqrt{x^2-1}}$. Then

$$f'(x) = \frac{\sqrt{x^2-1}\frac{d}{dx}\cos 2x - \cos 2x \cdot \frac{d}{dx}\sqrt{x^2-1}}{\left(\sqrt{x^2-1}\right)^2}$$

$$= \frac{\sqrt{x^2-1}\cdot\left(-\sin 2x \cdot \frac{d}{dx}(2x)\right) - \cos 2x \cdot \frac{1}{2\sqrt{x^2-1}} \cdot \frac{d}{dx}(x^2-1)}{x^2-1}$$

$$= \frac{\sqrt{x^2-1}\cdot(\sin 2x \cdot (-2)) - \dfrac{\cos 2x \cdot 2x}{2\sqrt{x^2-1}}}{x^2-1} \cdot \underbrace{\frac{\sqrt{x^2-1}}{\sqrt{x^2-1}}}_{\text{To Simplify}} = \frac{-2(x^2-1)\sin 2x - x\cos 2x}{(x^2-1)^{3/2}}.$$

It is an interesting exercise in both calculus and algebraic simplification to derive the same conclusion using $f(x) = \cos 2x \cdot (x^2-1)^{-1/2}$ and the Product Rule (which would also call the Chain Rule twice).

3.5.4 Tangent, Cotangent, Secant, and Cosecant Rules

The following are derivative rules for the remaining trigonometric functions. These rules are given in both simple ("matching variable") and Chain Rule versions.

$$\boxed{\frac{d\tan x}{dx} = \sec^2 x,} \qquad \boxed{\frac{d\tan u}{dx} = \sec^2 u \cdot \frac{du}{dx}.} \qquad (3.50)$$

$$\boxed{\frac{d\cot x}{dx} = -\csc^2 x,} \qquad \boxed{\frac{d\cot u}{dx} = -\csc^2 u \cdot \frac{du}{dx}.} \qquad (3.51)$$

$$\boxed{\frac{d\sec x}{dx} = \sec x \tan x,} \qquad \boxed{\frac{d\sec u}{dx} = \sec u \tan u \cdot \frac{du}{dx}.} \qquad (3.52)$$

$$\boxed{\frac{d\csc x}{dx} = -\csc x \cot x,} \qquad \boxed{\frac{d\csc u}{dx} = -\csc u \cot u \cdot \frac{du}{dx}.} \qquad (3.53)$$

Though each derivative formula above can be derived with relative ease, these should all be memorized. It may help to notice patterns when comparing the derivatives of tangent and cotangent, secant and cosecant, and how these are similar to the comparison of sine and cosine derivatives. In short, these formulas come in function/cofunction pairs.

We will prove the derivative of $\tan x$ is $\sec^2 x$ and leave the rest as exercises. The Chain Rule versions then follow. To see the formula for the tangent, we rewrite it as the quotient of sine and cosine, and use the quotient rule.

$$\frac{d}{dx}\tan x = \frac{d}{dx}\left[\frac{\sin x}{\cos x}\right] = \frac{\cos x \cdot \frac{d\sin x}{dx} - \sin x \cdot \frac{d\cos x}{dx}}{(\cos x)^2}$$

$$= \frac{\cos x \cos x - (\sin x)(-\sin x)}{\cos^2 x} = \frac{\cos^2 x + \sin^2 x}{\cos^2 x} = \frac{1}{\cos^2 x} = \sec^2 x,$$

$$\therefore \quad \frac{d}{dx}\tan u = \frac{d}{du}\tan u \cdot \frac{du}{dx} = \sec^2 u \cdot \frac{du}{dx}.$$

We used the trigonometric identities ($\sin^2\theta + \cos^2\theta = 1$, $1/\cos\theta = \sec\theta$). The last line was our usual Chain Rule argument, given that $\frac{d\tan x}{dx} = \sec^2 x$ was already proved. With (3.50)–(3.53), and derivatives of sine and cosine from earlier ((3.18), (3.19), page 242), we finally have derivatives of all six trigonometric functions.

Every student of calculus should memorize the derivatives of the trigonometric functions, *and be able to derive these four new ones* from knowing the derivatives of $\sin x$ and $\cos x$. Now we can apply these. First we look at some of the simpler examples.

Example 3.5.12

- $\dfrac{d\cot\sqrt{x}}{dx} = -\csc^2\sqrt{x} \cdot \dfrac{d\sqrt{x}}{dx} = -\csc^2\sqrt{x} \cdot \dfrac{1}{2\sqrt{x}} = \dfrac{-\csc^2\sqrt{x}}{2\sqrt{x}}.$

  (Note: the expressions $\sqrt{x}$ cannot "cancel" here.)

- $\dfrac{d\sqrt{\cot x}}{dx} = \dfrac{1}{2\sqrt{\cot x}} \cdot \dfrac{d\cot x}{dx} = \dfrac{1}{2\sqrt{\cot x}} \cdot (-\csc^2 x) = \dfrac{-\csc^2 x}{2\sqrt{\cot x}}.$

- $\dfrac{d}{dx}\sec x^9 = \sec x^9 \tan x^9 \cdot \dfrac{d}{dx}(x^9) = \sec x^9 \tan x^9 \cdot 9x^8 = 9x^8 \sec x^9 \tan x^9.$

- $\dfrac{d}{dx}x^2\tan x = x^2 \cdot \dfrac{d}{dx}\tan x + \tan x \cdot \dfrac{d}{dx}(x^2) = x^2\sec^2 x + 2x\tan x.$

- $\dfrac{d}{dx}\left[\dfrac{x}{\tan x}\right] = \dfrac{d}{dx}(x\cot x) = x \cdot \dfrac{d}{dx}(\cot x) + \cot x \cdot \dfrac{d}{dx}(x) = (x)(-\csc^2 x) + \cot x = -x\csc^2 x + \cot x.$

Note that we opted to rewrite the last derivative as a Product Rule problem. If we preferred to have a single fraction as our immediate answer, we could have left it as a Quotient Rule problem. (As it is, we can rewrite it as a fraction if we wish, using the definitions of cosecant and cotangent.)

Before continuing with more complicated examples, we should briefly consider why it makes sense graphically that $\frac{d}{dx}\tan x = \sec^2 x$. The graph of $f(x) = \tan x$ is given in Figure 3.25, page 307, and so we can consider its derivative formula in light of that graph. Note that this derivative is always positive where defined; $\sec^2 x > 0$, and in fact $\sec^2 x = 1/\cos^2 x \geq 1$, since $|\cos x| \leq 1$. Thus $\tan x$ is always increasing in any interval on which it is defined.

3.5. PRODUCT, QUOTIENT, AND REMAINING TRIGONOMETRIC RULES

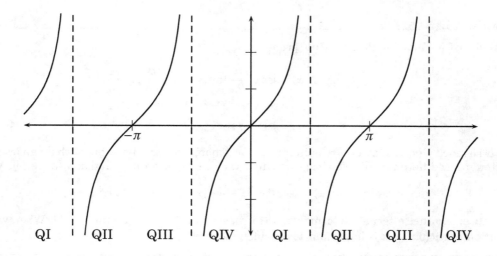

Figure 3.25: Partial graph of $y = \tan x$. Since $y = \sin x/\cos x$, it is continuous except where undefined, i.e., except where there are vertical asymptotes, at each x-value where $\cos x = 0$ (and thus $\sin\theta = \pm 1$). Recall that $\tan x$ is positive if x represents an angle terminating in the first or third quadrants, and negative if terminating in the second or fourth quadrants, so the quadrants represented by the x-values are labeled QI–QIV. Note that $\frac{d}{dx}\tan x = \sec^2 x$, which is positive where defined (coinciding with where $\tan x$ is defined), and thus $\tan x$ is an increasing function on any interval on which it is defined.

Furthermore $\sec^2 x \to \infty$ as $x \to \pm\frac{\pi}{2}, \pm\frac{3\pi}{2}, \pm\frac{5\pi}{2}$, etc., and that has implications for the graph. Of course $\tan x = \sin x/\cos x$ has vertical asymptotes at each of those x-values (where $\cos x = 0$ and $\sin x = \pm 1$). Finally, note that $\frac{d\tan x}{dx}\big|_{x=0} = \sec^2 x\big|_{x=0} = \frac{1}{\cos^2 0} = 1$, for instance, so the slope through the $(0, \tan 0) = (0,0)$ is 1. That slope repeats every π units in both directions, due to the π-periodic nature of the tangent.

Example 3.5.13

Here is a typical Chain Rule problem involving the tangent.

$$\frac{d}{dx}\tan\sqrt{x^2-1} = \sec^2\sqrt{x^2-1} \cdot \frac{d}{dx}\sqrt{x^2-1} = \sec^2\sqrt{x^2-1} \cdot \frac{1}{2\sqrt{x^2-1}} \cdot \frac{d(x^2-1)}{dx}$$

$$= \frac{\sec^2\sqrt{x^2-1}}{2\sqrt{x^2-1}} \cdot 2x = \frac{x\sec^2\sqrt{x^2-1}}{\sqrt{x^2-1}}.$$

Note that, tempting as it may be, the radicals above cannot be combined or canceled in any of the steps; one is outside the secant-squared function, and the other is safely quarantined inside. Note also that squaring the secant function does not alter its input $\sqrt{x^2-1}$.

Example 3.5.14

Another Chain Rule problem is the following. Note $\cot^3 x = (\cot x)^3$.

$$\frac{d}{dx}\cot^3 x = 3(\cot x)^2 \frac{d}{dx}\cot x = 3\cot^2 x(-\csc^2 x) = -3\cot^2 x\csc^2 x.$$

Example 3.5.15

Consider the following Product Rule problem:

$$\frac{d}{dx}(\sec x \tan x) = \sec x \cdot \frac{d}{dx}\tan x + \tan x \frac{d}{dx}\sec x$$
$$= \sec x \sec^2 x + \tan x \sec x \tan x$$
$$= \sec^3 x + \sec x \tan^2 x.$$

There is so much algebraic structure built into the trigonometric functions that such an answer can be rewritten many different ways. For instance, since $\tan^2 x + 1 = \sec^2 x$ our final answer can be written

$$\sec x(\sec^2 x + \tan^2 x) = \sec x(\sec^2 x + \sec^2 x - 1) = \sec x(2\sec^2 x - 1),$$

for instance. Another alternative form is $\sec x(\tan^2 x + 1 + \tan^2 x) = \sec x(2\tan^2 x + 1)$. When we study integration particularly, it is important to consider such options.

3.5.5 More on Putting Rules Together—Carefully

This subsection is just a reminder that there are times when several differentiation rules are called when computing the derivative of a complicated function. In such cases we need to recognize which rules apply, and then consider exactly how to invoke them. And it is always better to write out the rules, literally, as they apply.

It is easy to lapse into intellectual laziness by skipping steps, but this is an error-prone habit which does not save any time in the long run. Some steps can be combined into other steps with little risk, especially with long-term practice where we can easily anticipate how a step will play out (the rules for sums and additive and multiplicative constants come to mind). However, the Quotient, Product, Power, trigonometric, and all forms of the Chain Rule should be written out in their own steps *before* we compute the derivatives called for by the internals of these rules. For instance, it is tempting to compute all at once:

$$\frac{d}{dx}\left[(x^3 + 9x^2 + \sin 2x)(\tan x^5)\right]$$
$$= (x^3 + 9x^2 + \sin 2x)(\sec^2 x^5 \cdot 5x^4) + \tan x^5 \cdot (3x^2 + 18x + \cos 2x \cdot 2).$$

However, this approach has a couple of disadvantages which become more important as functions become increasingly complicated. First, we have to mentally keep track of what rule applies where, without the benefit of breaking it into steps and so noting each rule. Second, if we (or anyone else) would like to check our work we can try to re-read what we wrote, but we will likely find ourselves again performing the same mental gymnastics we did the first time, and likely repeating any mistakes we made that first time. We can go a long way towards avoiding these difficulties by writing out all the steps. Since each step invokes a single differentiation rule (though we may apply different rules to different terms in the same "step"), much of our work is re-copying the line above, which takes very little time. Care in "bookkeeping" will translate into clearer thinking and less error (and easier error correction). Consider the following approach to the problem above, and imagine a "momentum" to writing out these steps fluidly:

3.5. PRODUCT, QUOTIENT, AND REMAINING TRIGONOMETRIC RULES

$$\frac{d}{dx}\left[(x^3+9x^2+\sin 2x)(\tan x^5)\right]$$

$$=\left(x^3+9x^2+\sin 2x\right)\frac{d}{dx}\left(\tan x^5\right)+\left(\tan x^5\right)\cdot\frac{d}{dx}\left(x^3+9x^2+\sin 2x\right)$$

$$=\left(x^3+9x^2+\sin 2x\right)\sec^2 x^5\cdot\frac{dx^5}{dx}+\tan x^5\cdot\left(3x^2+18x+\cos 2x\cdot\frac{d(2x)}{dx}\right)$$

$$=(x^3+9x^2+\sin 2x)(\sec^2 x^5\cdot 5x^4)+\tan x^5\cdot(3x^2+18x+\cos 2x\cdot 2)$$

$$=5x^4(x^3+9x^2+\sin 2x)\sec^2 x^5+(3x^2+18x+2\cos 2x)\tan x^5.$$

Ultimately the original function is a product, so the first step written here was exactly the statement of the Product Rule for the particular factors. In the next line, we begin to take the derivatives demanded within the Product Rule, and find some Power Rules (with the multiplicative constants along for the ride—i.e., with the rule that multiplicative constants are preserved), and a couple of Chain Rules which we write out exactly. Next we compute the derivatives of the "inside" functions demanded by the Chain Rule, and finally we do some algebra to make the result more presentable. Note how we (or another reader) can re-read this with great assurance that it is correct, since each step follows differentiation rules in obvious (though the terms themselves may be complicated) ways. Note that when applying derivative rules to complicated functions, we work from the outside (large structure) of the function inwards.

Example 3.5.16

Find $f'(x)$ if $f(x)=2\sin^3 5x+\csc\sqrt{x^2+1}$.

Solution: This is first a sum, and then there are several Chain Rules which come into play.

$$f'(x)=\frac{d}{dx}\left(2\sin^3 5x\right)+\frac{d}{dx}\left(\csc\sqrt{x^2+1}\right)$$

$$=2\frac{d}{dx}\sin^3 5x+\left(-\csc\sqrt{x^2+1}\cot\sqrt{x^2+1}\cdot\frac{d\sqrt{x^2+1}}{dx}\right)$$

$$=2\cdot 3\sin^2 5x\cdot\frac{d\sin 5x}{dx}-\csc\sqrt{x^2+1}\cot\sqrt{x^2+1}\cdot\frac{1}{2\sqrt{x^2+1}}\frac{d(x^2+1)}{dx}$$

$$=6\sin^2 5x\cos 5x\cdot\frac{d(5x)}{dx}-\frac{\csc\sqrt{x^2+1}\cot\sqrt{x^2+1}}{2\sqrt{x^2+1}}\cdot 2x$$

$$=6\sin^2 5x\cos 5x\cdot 5-\frac{x\csc\sqrt{x^2+1}\cot\sqrt{x^2+1}}{\sqrt{x^2+1}}$$

$$=30\sin^2 5x\cos 5x-\frac{x\csc\sqrt{x^2+1}\cot\sqrt{x^2+1}}{\sqrt{x^2+1}}.$$

Example 3.5.17

Find $f'(x)$ if $f(x)=\sin^3\left(x^2\tan x\right)$.

This is first a Power and Chain Rule problem, since $f(x)=\left[\sin\left(x^2\tan x\right)\right]^3$. After a second, trigonometric Chain Rule we will have a Product Rule. It is not necessary to notice all this structure from the beginning; it all becomes apparent as we note the function's structure from the outside inwards, applying the appropriate differentiation rules as we go:

$$\begin{aligned}f'(x) &= 3\left[\sin\left(x^2\tan x\right)\right]^2 \cdot \frac{d}{dx}\left[\sin\left(x^2\tan x\right)\right]\\ &= 3\sin^2\left(x^2\tan x\right)\cdot\cos(x^2\tan x)\cdot\frac{d}{dx}\left(x^2\tan x\right)\\ &= 3\sin^2\left(x^2\tan x\right)\cos(x^2\tan x)\cdot\left(x^2\frac{d\tan x}{dx}+\tan x\cdot\frac{dx^2}{dx}\right)\\ &= 3\sin^2\left(x^2\tan x\right)\cos\left(x^2\tan x\right)\cdot\left(x^2\sec^2 x+\tan x\cdot 2x\right) \qquad \text{(Now Rearrange)}\\ &= 3\left(x^2\sec^2 x+2x\tan x\right)\sin^2\left(x^2\tan x\right)\cos\left(x^2\tan x\right).\end{aligned}$$

In some cases it is best to simplify algebraically before applying differentiation rules. For an obvious illustration of this, consider the following.

Example 3.5.18 _____

Compute $\frac{d}{dx}(\cos x\sec x)$.

Solution: We have two methods that come to mind immediately:

- $\dfrac{d(\cos x\sec x)}{dx} = \cos x\cdot\dfrac{d\sec x}{dx}+\sec x\cdot\dfrac{d\cos x}{dx} = \cos x\sec x\tan x+\sec x(-\sin x)$
  $= 1\cdot\tan x-\tan x = 0.$

- $\dfrac{d(\cos x\sec x)}{dx} = \dfrac{d}{dx}(1) = 0.$

As mentioned before, sometimes there can be more than one "natural" method of computing a derivative. The rules are consistent (they are theorems after all!), and if correctly applied different methods will yield the same result. Of course in the above example, the second method is much faster. The next example is perhaps not so obvious.

Example 3.5.19 _____

Here we compute $\dfrac{d}{dz}\left[\dfrac{1+\frac{z+1}{z-1}}{1-\frac{z+1}{z-1}}\right]$.

A method of brute force would be to perform the Quotient Rule and continue from there:

$$= \frac{\left(1-\frac{z+1}{z-1}\right)\frac{d}{dz}\left(1+\frac{z+1}{z-1}\right)-\left(1+\frac{z+1}{z-1}\right)\frac{d}{dz}\left(1-\frac{z+1}{z-1}\right)}{\left(1-\frac{z+1}{z-1}\right)^2},$$

from which we need to compute two more quotient rules. However, if we instead simplify the function from the beginning, our work is greatly simplified:

3.5. PRODUCT, QUOTIENT, AND REMAINING TRIGONOMETRIC RULES

$$\frac{d}{dz}\left[\frac{1+\frac{z+1}{z-1}}{1-\frac{z+1}{z-1}}\right] = \frac{d}{dz}\left[\frac{1+\frac{z+1}{z-1}}{1-\frac{z+1}{z-1}}\cdot\frac{z-1}{z-1}\right] = \frac{d}{dz}\left[\frac{z-1+(z+1)}{z-1-(z+1)}\right] = \frac{d}{dz}\left[\frac{2z}{-2}\right] = \frac{d}{dz}(-z) = -1.$$

We should be a bit careful here about the domain, in particular that no denominator has value zero. And so the original expression of this function requires $z \neq 1$, due to the two denominators equal to $z - 1$.

In fact, with a short exercise one can show that the "large denominator" $1 - \frac{z+1}{z-1}$ is never zero (for if it were, that would imply $z - 1 = z + 1$, which is impossible, for it then implies $-1 = 1$).

In applications especially, we are often led to complicated expressions for functions, for which we then need to compute the derivatives. It is always useful to look out for such cases in which algebraic simplification from the beginning will simplify our calculus tasks, as well as give us a look at a simpler form of the original function. As complicated as the above function first appeared, it was simply the function $-z$ (where both the original and the simplified function are defined, which for this problem means $z \neq 1$). Note that even if we computed this derivative the longer way, we might not recognize our answer to be simply -1.

3.5.6 Extended Product Rule

The Extended Product Rule is redundant, in that it simply comes from applying the product rule multiple times, but it can be useful in some circumstances. We will use the next example to illustrate its logic.

Example 3.5.20

Find $f'(x)$ if $f(x) = x^2 \sin 2x \cos 3x$.

Solution: If we would like, we can consider $f(x)$ to first be the product of two functions, those being x^2 and $(\sin 2x \cos 3x)$. Finding $f'(x)$ is then a matter for the Product Rule. When we get into the derivatives the Product Rule tells us to call for the second such factor, namely $(\sin 2x \cos 3x)$, that will in turn require another Product Rule. Hence

$$f'(x) = \frac{d}{dx}\left[x^2(\sin 2x \cos 3x)\right] = x^2 \frac{d}{dx}(\sin 2x \cos 3x) + (\sin 2x \cos 3x)\frac{dx^2}{dx}$$

$$= x^2\left[\sin 2x \cdot \frac{d\cos 3x}{dx} + \cos 3x \cdot \frac{d\sin 2x}{dx}\right] + \sin 2x \cos 3x \cdot 2x$$

$$= x^2\left[\sin 2x\left(-\sin 3x \cdot \frac{d(3x)}{dx}\right) + \cos 3x\left(\cos 2x \cdot \frac{d(2x)}{dx}\right)\right] + 2x \sin 2x \cos 3x$$

$$= x^2 \sin 2x(-\sin 3x)(3) + x^2 \cos 3x \cos 2x(2) + 2x \sin 2x \cos 3x$$

$$= -3x^2 \sin 2x \sin 3x + 2x^2 \cos 3x \cos 2x + 2x \sin 2x \cos 3x.$$

If we expand the second line in our solution computation, we can see a pattern in which

$$\frac{d}{dx}\left[f(x)g(x)h(x)\right] = f(x)g(x)\frac{d\,h(x)}{dx} + f(x)h(x)\frac{d\,g(x)}{dx} + g(x)h(x)\frac{d\,f(x)}{dx}.$$

The obvious pattern extends to four or more functions as well: the derivative of a product of n functions $u_1 u_2 \cdots u_n$ will be the sum of n products $u_1 u_2 \cdots \left(\frac{du_k}{dx}\right) \cdots u_n$ for $k = 1, 2, \ldots, n$. Written in "prime" notation, for instance,

$$(f_1 f_2 f_3 f_4)' = f_1 f_2 f_3 f_4' + f_1 f_2 f_3' f_4 + f_1 f_2' f_3 f_4 + f_1' f_2 f_3 f_4,$$

or as some texts would write,

$$(f_1 f_2 f_3 f_4)' = f_1' f_2 f_3 f_4 + f_1 f_2' f_3 f_4 + f_1 f_2 f_3' f_4 + f_1 f_2 f_3 f_4'.$$

Whichever form, the idea is that each function $f_k(x)$ has an instantaneous rate of change which contributes to the change in the product, and its contribution is $f_k'(x)$ but multiplied (or "amplified") by the other functions multiplying it.

Example 3.5.21

Find $f'(x)$ if $f(x) = x^2 \sec x \tan x$.

Solution:

$$\begin{aligned} f'(x) &= (x^2)'(\sec x)(\tan x) + x^2 (\sec x)' \tan x + x^2 (\sec x)(\tan x)' \\ &= 2x \sec x \tan x + x^2 \sec x \tan x \tan x + x^2 \sec x \sec^2 x \\ &= x \sec x \left(2 \tan x + x \tan^2 x + x \sec^2 x\right). \end{aligned}$$

One could further rewrite the final expression using $\sec^2 x = \tan^2 x + 1$ if so desired.

Exercises

For Exercises 1–8, compute the derivatives two different ways:

(a) by using the rule called upon by the way the function is originally written; and

(b) by first simplifying the function, and then computing the derivative.

(c) Show that the answers are the same.

For instance, in 1, first use the Product Rule, and then compute $\frac{d}{dx}(x^4)$ with the Power Rule, and finally show that the answers are the same.

1. $\dfrac{d}{dx}\left[x^2 \cdot x^2\right]$

2. $\dfrac{d}{dx}\left[\dfrac{x^9}{x^3}\right]$

3. $\dfrac{d}{dx}[\cos x \sec x]$

4. $\dfrac{d}{dx}[\cos x \tan x]$

5. $\dfrac{d}{dx}\left[\dfrac{1}{\cos x}\right]$

6. $\dfrac{d}{dx}[\tan x \cot x]$

7. $\dfrac{d}{dx}[\sin x \sec x]$

8. $\dfrac{d}{dx}\left[\dfrac{1}{\sin x}\right]$

3.5. PRODUCT, QUOTIENT, AND REMAINING TRIGONOMETRIC RULES

For Exercises 9–24, compute the given derivatives.

9. $\dfrac{d}{dx}[\tan^2 x]$

10. $\dfrac{d}{dx}\left[\dfrac{x^2+1}{\sin x-1}\right]$

11. $\dfrac{d}{dx}\left[\sqrt{1-\csc 2x}\right]$

12. $\dfrac{d}{dx}[x\sin x \cos x]$

13. $\dfrac{d}{dx}\left[\dfrac{x^3-7x+5}{x^2-3}\right]$

14. $\dfrac{d}{dx}\left[\dfrac{1+\frac{1}{x+1}}{1-\frac{1}{x-1}}\right]$

15. $\dfrac{d}{dx}[\sin^2 x \cos^3 x]$

16. $\dfrac{d}{dx}[x^2 \cos x^2]$

17. $\dfrac{d}{dx}\left[\dfrac{x^4+3}{x-1}\right]$

18. $\dfrac{d}{dx}\left[\sec^4(x^3+2x)\right]$

19. $\dfrac{d}{dx}\left[\dfrac{\sin x+3}{\cos x+2}\right]$

20. $\dfrac{d}{dx}[\sec 3x \cot 5x]$

21. $\dfrac{d}{dx}\left[\dfrac{x}{\sqrt{x^2+1}}\right]$

22. $\dfrac{d}{dx}\left[\dfrac{(x+5)^3}{(x-4)^5}\right]$. Factor the numerator in your answer to simplify.

23. $\dfrac{d}{dx}\left[3\cot^2 9x - \sqrt{\cos 6x+1}\right]$

24. $\dfrac{d}{dx}[\tan(x+\tan(x+\tan x))]$

25. Find a formula for $(f_1 f_2 f_3 f_4 f_5)'$.

26. Use the Quotient Rule to show that $\dfrac{d}{dx}\left[\dfrac{f(x)}{C}\right] = \dfrac{f'(x)}{C}$, assuming $C \neq 0$. (One can also use the rule for multiplicative constants.)

27. Show that $\dfrac{d}{dx}\left[\dfrac{1}{f(x)}\right] = \dfrac{-f'(x)}{[f(x)]^2}$.

28. Consider $f(x) = \cot x = \dfrac{\cos x}{\sin x}$.

    (a) Where are the vertical asymptotes of $f(x)$?

    (b) What is the sign of $f'(x)$ where it exists?

    (c) Use this information, along with the x-intercepts (points $(x,0)$ where $f(x) = 0$) to graph this function. (It will be helpful to mark the relevant terminating quadrants if x represents an angle.)

29. A simple, Earth-bound pendulum of length L in meters has a period of oscillation of T seconds. That is, every T (seconds) the pendulum completes one cycle. The period varies with the length by the equation $T = 2\pi\sqrt{L/g}$, where $g = 9.8\text{m/sec}^2$. Find the rate of change of T with respect to L when $T = 3\text{sec}$.

30. If the cost to manufacture x items is $C(x)$, then we define the *average cost per item* to be

$$\overline{C}(x) = \dfrac{C(x)}{x}, \qquad (3.54)$$

and from this define the *marginal average cost* by $\overline{C}'(x)$. Find the average cost per item function, and the marginal average cost function, if

$$C(x) = \dfrac{x^2+3x}{x+4} + 100.$$

How do you interpret marginal average cost?

31. The intensity I of a 100 watt (100W) light bulb is given by

$$I(x) = \dfrac{7.92W}{x^2},$$

where x is the distance in meters from the bulb. (The W in the equation above can be suppressed in actual computations.)

    (a) What are the units of I?

    (b) Find $\dfrac{dI}{dx}$. What are its units?

    (c) At what distance (accurate to the nearest cm=0.01m) will the rate of change of intensity be equal to -3W/m?

32. A variable resistance R is given in a circuit and the voltage is found by the formula
$$V = \frac{5R+10}{R+3}.$$
Find the instantaneous rate of change of V with respect to R when $R = 6\Omega$. (V will be measured in volts.)

33. A double convex converging lens will focus an object p distance in front of the lens to an image a distance q from the lens on the opposite side of the lens. The lens has a focal length of f. (Here all distances are in cm.) A well-known formula in optics gives
$$\frac{1}{p} + \frac{1}{q} = \frac{1}{f}.$$
Find an equation which represents the rate of change of q with respect to p, assuming f is constant. (Hint: Solve the given equation for q and then differentiate—that is, take its derivative—while assuming that f is constant. From this you can determine dq/dp.)

34. On any interval on which $f(x) = \cot x$ is defined, what can you say about whether it is increasing or decreasing? Explain.

3.6 Chain Rule II: Implicit Differentiation

In this section we apply the Chain Rule in the setting of so-called *implicit functions*, which are more general than the *explicit functions* we have dealt with so far. In a later chapter we will use the Chain Rule in still a third context, *related rates*.

First we recall all differentiation rules considered so far, including a reexamination of the Chain Rule. We then perform some simple differentiation problems of the type typically encountered in a first study of implicit functions. Then we proceed to the topic of this section, namely finding the slope dy/dx for *curves*—in our case given by equations—which might not define functions in the purest sense, but which can represent functions locally ("implicitly"), and for which it is still reasonable to ask about slopes.

3.6.1 Review of Chain Rule and Other Differentiation Rules

At this point we have many differentiation rules. If we look at all the rules so far, they can be summed up as follows. First we had the very general rules, regardless of $f(x)$ and $g(x)$, and *fixed constants* $C \in \mathbb{R}$, as long as the expressions on the right existed:

- $\dfrac{d[Cf(x)]}{dx} = C \cdot \dfrac{df(x)}{dx}$
- $\dfrac{d[f(x) \pm g(x)]}{dx} = \dfrac{df(x)}{dx} \pm \dfrac{dg(x)}{dx}$
- $\dfrac{d[f(x) \cdot g(x)]}{dx} = f(x) \cdot \dfrac{dg(x)}{dx} + g(x) \cdot \dfrac{df(x)}{dx}$
- $\dfrac{d}{dx}\left[\dfrac{f(x)}{g(x)}\right] = \dfrac{g(x) \cdot \frac{df(x)}{dx} - f(x) \cdot \frac{dg(x)}{dx}}{[g(x)]^2}$
- $\dfrac{dC}{dx} = 0$

Then we had rules for specific functions, and their Chain Rule versions, illustrated below using empty parentheses instead of the variable u:[61]

- $\frac{d}{dx}[x^n] = n \cdot x^{n-1}$
- $\frac{d}{dx}\sin x = \cos x$
- $\frac{d}{dx}\cos x = -\sin x$
- $\frac{d}{dx}\tan x = \sec^2 x$
- $\frac{d}{dx}\cot x = -\csc^2 x$
- $\frac{d}{dx}\sec x = \sec x \tan x$
- $\frac{d}{dx}\csc x = -\csc x \cot x$
- $\frac{d}{dx}\sqrt{x} = \frac{1}{2\sqrt{x}}$
- $\frac{d}{dx}\left[\frac{1}{x}\right] = \frac{-1}{x^2}$

- $\frac{d}{dx}[(\ )^n] = n \cdot (\ )^{n-1} \cdot \frac{d(\ )}{dx}$
- $\frac{d}{dx}\sin(\ ) = \cos(\ ) \cdot \frac{d(\ )}{dx}$
- $\frac{d}{dx}\cos(\ ) = -\sin(\ ) \cdot \frac{d(\ )}{dx}$
- $\frac{d}{dx}\tan(\ ) = \sec^2(\ ) \cdot \frac{d(\ )}{dx}$
- $\frac{d}{dx}\cot(\ ) = -\csc^2(\ ) \cdot \frac{d(\ )}{dx}$
- $\frac{d}{dx}\sec(\ ) = \sec(\ )\tan(\ ) \cdot \frac{d(\ )}{dx}$
- $\frac{d}{dx}\csc(\ ) = -\csc(\ )\cot(\ ) \cdot \frac{d(\ )}{dx}$
- $\frac{d}{dx}\sqrt{(\ )} = \frac{1}{2\sqrt{(\ )}} \cdot \frac{d(\ )}{dx}$
- $\frac{d}{dx}\left[\frac{1}{(\ )}\right] = \frac{-1}{(\ )^2} \cdot \frac{d(\ )}{dx}$

---

[61] The formulas in the right column are usually given with the variable u inside the empty parentheses, assuming u is a function of x. If that is not the case—that is, if u is not a function of x—then both sides of those equations on the right do not make sense. It is therefore customary to proceed to write these formulas because a problem in the left-hand side of one of these equations will manifest in the right-hand side of that equation.

(Note that the last few rules were just special cases of the Power Rule.)

Recall that when the variable of differentiation matches the input variable of the function, we can use the simple derivative rule for that particular function. And when these variables do not match, we must compensate by multiplying by the derivative of that input variable with respect to the variable of differentiation. These two cases can be stated thusly:

$$\text{variables match:} \qquad \frac{df(x)}{dx} = f'(x), \tag{3.55}$$

$$\text{variables do not match:} \qquad \frac{df(u)}{dx} = f'(u) \cdot \frac{du}{dx}. \tag{3.56}$$

In fact the second generalizes the first, since we could have included a factor $\frac{dx}{dx} = 1$ in (3.55). As an example of (3.56), further justified with Leibniz notation, we can have $f(x) = \sin x$ and $u = u(x)$ yielding the following:

$$\frac{d\sin u}{dx} = \frac{d\sin u}{du} \cdot \frac{du}{dx} = \cos u \cdot \frac{du}{dx}.$$

The middle expression is simply the Leibniz decomposition, allowing us to use the simple sine derivative formula for $\frac{d\sin u}{du}$ because the variable in the sine—u—matched the differential operator $\frac{d}{du}$. Though writing that middle expression seems to justify the Chain Rule "formally," it becomes more burdensome if the input of sine is more complicated, and so we will usually skip (but not forget!) that step in most future computations. For a few other illustrative examples, consider the following, noting that the previous rules still apply if we assume that y is a function of x (else Footnote 61, page 315 applies):

Example 3.6.1

Consider the following four derivative computations.

- $\frac{d}{dx}(y^2) = 2y \cdot \frac{dy}{dx}.$

- $\frac{d}{dx}\sqrt{y} = \frac{1}{2\sqrt{y}} \cdot \frac{dy}{dx}.$

- $\frac{d}{dx}(\csc y^2) = -\csc y^2 \cot y^2 \cdot \frac{dy^2}{dx} = -\csc y^2 \cot y^2 \cdot 2y \cdot \frac{dy}{dx} = -2y\csc y^2 \cot y^2 \cdot \frac{dy}{dx}.$

- $\frac{d}{dx}(y\sin y) = y \cdot \frac{d(\sin y)}{dx} + \sin y \cdot \frac{d(y)}{dx} = y \cdot \cos y \cdot \frac{dy}{dx} + \sin y \cdot \frac{dy}{dx}.$

Notice that the last example required first the Product Rule, and so the first step was to write—exactly—the statement of the Product Rule (from the previous page, and originally as Equation (3.48), page 297) where the product is $y \cdot \sin y$. This is ultimately a product of two functions of x, since we are assuming y is a function of x, and thus $y\sin y$ is itself also a function of x. As with earlier derivatives, it is crucial to write out *precisely* what the general rules dictate for the given function. This becomes even more critical in the following:

3.6. CHAIN RULE II: IMPLICIT DIFFERENTIATION

Example 3.6.2

Consider the following derivative computations:

- $\frac{d}{dx}(xy^2) = x\frac{dy^2}{dx} + y^2\frac{dx}{dx} = x \cdot 2y\frac{dy}{dx} + y^2(1) = 2xy\frac{dy}{dx} + y^2.$

- $\frac{d}{dx}\left[\frac{x}{y}\right] = \frac{y\frac{dx}{dx} - x\frac{dy}{dx}}{y^2} = \frac{y - x\frac{dy}{dx}}{y^2}.$

- $\frac{d}{dx}(x^2+y^2)^2 = 2(x^2+y^2)^1 \frac{d}{dx}(x^2+y^2) = 2(x^2+y^2)\left(2x + 2y \cdot \frac{dy}{dx}\right) = 4(x^2+y^2)\left(x + y\frac{dy}{dx}\right).$

Notice that we skipped an explicit writing of the "Sum Rule" step in the last computation above:

$$\frac{d}{dx}(x^2 + y^2) = \frac{d}{dx}(x^2) + \frac{d}{dx}(y^2) = 2x + 2y \cdot \frac{dy}{dx}.$$

The first term, $\frac{d}{dx}(x^2)$ was a simple Power Rule because the variables matched. For the second term they do not match, so we need the Chain Rule to compute $\frac{d}{dx}(y^2) = 2y\frac{dy}{dx}$.[62]

Example 3.6.3

Consider the following derivative computation, noting that $\sin xy = \sin(xy)$:

$$\frac{d}{dx}\sin xy = \cos xy \cdot \frac{d(xy)}{dx} = \cos xy \cdot \left(x\frac{dy}{dx} + y\frac{dx}{dx}\right) = \cos xy \cdot \left(x\frac{dy}{dx} + y\right) = x\cos xy \cdot \frac{dy}{dx} + y\cos xy.$$

The above example used the Chain Rule first, and then the product rule "called" by the Chain Rule. Note that the first step could have been written

$$\frac{d\sin xy}{dx} = \frac{d\sin xy}{d(xy)} \cdot \frac{d(xy)}{dx}. \tag{3.57}$$

Of course it is important to note that, in the Leibniz notation, $d(xy)$ is taken to be one encapsulated quantity (and not, for instance, a product of d and xy).

Two more points should be emphasized regarding the Product Rule computation above, which we repeat here:

$$\frac{d(xy)}{dx} = x\frac{dy}{dx} + y\frac{dx}{dx}. \tag{3.58}$$

First, note that (3.58) is *exactly what the Product Rule says* for this product xy. Thus it should be established true from the explicit result of the Product Rule. Next, when computing $\frac{d}{dx}(y)$,

---

[62]Again, to give a more consistent feel to our process we could write $\frac{d(x^2)}{dx} = 2x \cdot \frac{dx}{dx}$ but it is not necessary since the "inner function's" derivative is just $\frac{dx}{dx} = 1$. One therefore usually gets into the habit of using the simpler derivative rule when the Chain Rule is unnecessary. Furthermore one can think of $\frac{dy}{dx}$ as $\frac{d(y^1)}{dx} = 1y^0\frac{dy}{dx} = 1 \cdot \frac{dy}{dx} = \frac{dy}{dx}$, or just realize that no more work is needed on that term once the factor $\frac{dy}{dx}$ emerges, however that occurred.

we get exactly what we would from the Chain Rule, or from the Power Rule. Thinking of y as a function "raised to the power 1," we might write:

$$\frac{d(y)}{dx} = \frac{d(y)}{dy} \cdot \frac{dy}{dx} = 1 \cdot \frac{dy}{dx} = \frac{dy}{dx}. \tag{3.59}$$

Thus once again the derivative rules are self-consistent, as is Leibniz notation, properly interpreted.

3.6.2 Implicit Functions and Their Derivatives

In this subsection we discuss the context and concepts of implicit differentiation, following one example through the discussion as an illustration. In subsequent subsections we will discuss more numerous examples.

In previous sections we were interested in functions $y = f(x)$, i.e., where y was given *explicitly* as a function of x. However, there are relationships and graphs of interest where y is related to x *implicitly*, for instance, by an equation which contains both y and x as variables, except that y is not solved for as an explicit function of x. Indeed, the graph of the equation (i.e., the graph of all (x, y) which satisfy the equation) might not represent a function at all. However, it is quite possible for y to be a function of x *locally* near a point (x_0, y_0) on the graph, and to speak of "slope" there, found through so-called *implicit differentiation* with the equation describing the curve.

Definition 3.6.1 *We will say y is **locally** (or **implicitly**) a function of x near (x_0, y_0) if and only if there exist $\delta, \varepsilon > 0$ and an open rectangle*

$$\left\{ (x, y) \, \Big| \, (|x - x_0| < \delta) \wedge (|y - y_0| < \varepsilon) \right\}$$

such that, within that open rectangle, the graph represents y as a function of x.

A simple example is a circle such as $x^2 + y^2 = 25$. Except at $(5, 0)$ and $(-5, 0)$, we can find a small enough, open rectangle around each point (x_0, y_0) on the graph so that y is a function of x *inside that rectangle*. Note that it is not necessary that there exists a vertical line touching the graph inside the rectangle, but if such a line does touch the graph inside the rectangle it can only do so at one point.

Now suppose we would like to find the slope of the graph of $x^2 + y^2 = 25$, without solving for y first.[63] If we are at a point (x_0, y_0) satisfying the equation $x^2 + y^2 = 25$, and for which we can (in principle) find an open rectangle centered at (x_0, y_0) in which $y = y(x)$, i.e., in which y is a function of x, then *inside such a rectangle*, we have that $y^2 = (y(x))^2$ is a (composite) function of x, and $x^2 + y^2$ is therefore also a function of x. Furthermore, 25 is a (rather trivial) function of x and, inside the rectangles, $x^2 + y^2$ and 25 are in fact the *same* function of x:

$$\underbrace{x^2 + y^2}_{\substack{\text{function of } x \\ \text{in each rectangle}}} = \underbrace{25.}_{\substack{\text{same function of } x \\ \text{in the rectangles}}}$$

---

[63]For this one we can solve for y, almost: $y = \pm\sqrt{25 - x^2}$, with the "plus" case being the upper semicircle and the "minus" case being the lower semicircle. For many curves we cannot isolate y, or cannot do so easily. Even when we can, it is sometimes easier to use the equation as it stands than in its "solved" form.

3.6. CHAIN RULE II: IMPLICIT DIFFERENTIATION

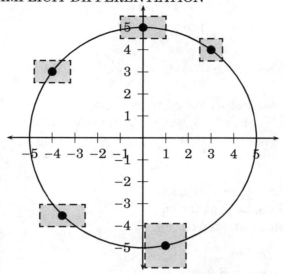

Figure 3.26: The graph of $x^2 + y^2 = 25$ gives y as a function of x locally except at $(-5,0)$ and $(5,0)$. Some possible open rectangles centered at various (x_0, y_0) on the graph are drawn, within each of which (separately) y is a function of x.

Because these are the same functions of x in an open rectangle, they have the same derivatives with respect to x there. Thus we can state

$$\frac{d}{dx}\left(x^2 + y^2\right) = \frac{d}{dx}\left(25\right). \tag{3.60}$$

Now suppose that we restrict our analysis to just such a rectangle in which y is a function of x. In the above we are taking the derivative with respect to x. The first term is a simple Power Rule, but the second is a Chain Rule version of the Power Rule since the variables do not match, while the other side of the equation is a constant. Thus, (3.60) becomes

$$2x + 2y\frac{dy}{dx} = 0.$$

The process above is called *implicit differentiation*, since we are differentiating both sides of an equation which represents a curve in the xy-plane, on which y is *locally* a function of x, given "implicitly" (not explicitly) by the equation. The differentiation process is simply a manifestation of the Chain Rule, though the name "implicit differentiation" does give the process context. It is $\frac{dy}{dx}$ that we want, i.e., the slope on the original curve. Fortunately we can now solve for it. Starting again with the original equation, we get:[64]

---

[64]The term *implicit differentiation* unfortunately is sometimes used to refer to any computation of a derivative of an expression with respect to one variable, where the expression contains other variables. For instance sometimes the term is used when the variable of differentiation is "hidden" in the original equation, as is t in $PV = C \implies \frac{d}{dt}(PV) = \frac{d}{dt}(C) \implies P\frac{dV}{dt} + V\frac{dP}{dt} = 0$ (assuming C is constant). In all cases it is *implied* that the other variables are functions of the variable of differentiation (or the differential operator with respect to that variable cannot be meaningfully applied), though technically "implicit differentiation" should only refer to differentiation on curves on which one variable is "implicitly" (i.e., not explicitly) given as a function (at least "locally") of the variable of differentiation.

In fact in all of these cases, we are simply applying the Chain Rule in different contexts.

$$x^2 + y^2 = 25 \implies 2x + 2y\frac{dy}{dx} = 0 \iff 2y\frac{dy}{dx} = -2x \iff \frac{dy}{dx} = \frac{-2x}{2y} = -\frac{x}{y}.$$

Some more notation is now needed for when we wish to evaluate this at particular points on the curve. The value of $\frac{dy}{dx}$ at such (x_0, y_0) on the curve is written

$$\left.\frac{dy}{dx}\right|_{(x_0,y_0)}.$$

Out loud, this is said, "$\frac{dy}{dx}$ evaluated at the point (x_0, y_0)." Looking back at the circle, we can find the slope at any (x_0, y_0) which is on the graph:

$$\left.\frac{dy}{dx}\right|_{(3,4)} = -\left.\frac{x}{y}\right|_{(3,4)} = -\frac{3}{4},$$

$$\left.\frac{dy}{dx}\right|_{(0,-5)} = -\left.\frac{x}{y}\right|_{(0,-5)} = -\frac{0}{-5} = 0,$$

$$\left.\frac{dy}{dx}\right|_{(-4,3)} = -\left.\frac{x}{y}\right|_{(-4,3)} = -\frac{-4}{3} = \frac{4}{3},$$

$$\left.\frac{dy}{dx}\right|_{(5/\sqrt{2},-5/\sqrt{2})} = -\frac{5/\sqrt{2}}{-5/\sqrt{2}} = 1.$$

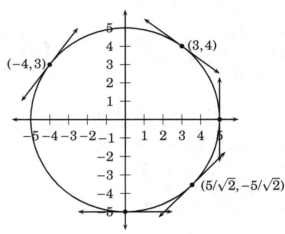

A quick check of the graph above shows that these slope calculation results seem reasonable. In fact, it is interesting to notice what happens if we attempt to evaluate $\left.\frac{dy}{dx}\right|_{(5,0)}$. If we naively insert $(5,0)$ into $\frac{dy}{dx} = -\frac{x}{y}$, we would get $-\frac{5}{0}$, which is undefined. Thus the derivative there does not exist, which is born out by the graph in the sense that there is no defined slope there. (Note that in the Euclidean geometry sense the tangent to the circle is a vertical line there.) Similarly for $(-5, 0)$.

We might also be interested in equations of tangent lines to this graph. For instance, at $(x, y) = (-4, 3)$, we have the slope

$$\left.\frac{dy}{dx}\right|_{(-4,3)} = \left.\frac{-x}{y}\right|_{(-4,3)} = \frac{-(-4)}{3} = 4/3,$$

and so with the point $(-4, 3)$, and the slope $dy/dx = 4/3$ we have the equation for the tangent line given by

$$y = 3 + \frac{4}{3}(x + 4).$$

Finding the slope at points on the circle above does not absolutely require the *implicit differentiation* technique, because we can write y as an *explicit* function of x and compute the derivative, near any (x_0, y_0) on the curve except at $(\pm 5, 0)$, which have undefined slope anyhow. Indeed, on the upper semicircle we have $y = \sqrt{25 - x^2}$, and so

$$\frac{dy}{dx} = \frac{d}{dx}\sqrt{25 - x^2} = \frac{1}{2\sqrt{25-x^2}} \cdot \frac{d(25-x^2)}{dx} = \frac{-2x}{2\sqrt{25-x^2}} = \frac{-x}{\sqrt{25-x^2}}.$$

3.6. CHAIN RULE II: IMPLICIT DIFFERENTIATION

Notice that (on the upper semicircle) this is the same as the derivative acquired through implicit differentiation: $\frac{dy}{dx} = \frac{-x}{\sqrt{25-x^2}} = \frac{-x}{y}$. A similar analysis shows that on the lower semicircle we also get two equivalent forms of the derivative:

$$\frac{dy}{dx} = \frac{d}{dx}\left[-\sqrt{25-x^2}\right] = \frac{-1}{2\sqrt{25-x^2}} \cdot \frac{d(25-x^2)}{dx} = \frac{-(-2x)}{2\sqrt{25-x^2}} = \frac{x}{\sqrt{25-x^2}},$$

which also can be written on the lower semicircle as $\frac{dy}{dx} = \frac{x}{-y} = -\frac{x}{y}$. However it is arguably easier to compute the derivative with the implicit form of the curve as we did originally.

Next we consider other examples where it is clearly not possible to solve for y as an explicit function of x, though like the previous example, "pieces" of the curve can be written $y = f(x)$ (using the arcsine function, whose derivative we will derive in a later section).

Example 3.6.4 _____

Consider the equation $x = \sin y$, pictured below and to the right.

The slope $\frac{dy}{dx}$ is then computed as follows:

$$x = \sin y$$
$$\implies \frac{d}{dx}[x] = \frac{d}{dx}[\sin y]$$
$$\implies 1 = \cos y \frac{dy}{dx}$$
$$\implies \frac{1}{\cos y} = \frac{dy}{dx}.$$

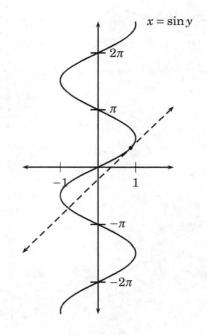

So for instance the slope at $\left(\frac{\sqrt{3}}{2}, \frac{\pi}{3}\right)$ is given by $\frac{dy}{dx} = 1/\cos\frac{\pi}{3} = \frac{1}{1/2} = 2$, and the equation of the tangent line there (shown on the graph) is then

$$y = \frac{\pi}{3} + 2\left(x - \frac{\sqrt{3}}{2}\right).$$

At those points on the graph where we have $\cos y = 0$, namely $y = \pm\frac{\pi}{2}, \pm\frac{3\pi}{2}, \pm\frac{5\pi}{2}, \ldots$, one could say we have vertical tangent lines, and this is reflected in the slope equation $\frac{dy}{dx} = \frac{1}{\cos y}$.

___

We can always solve algebraically for $\frac{dy}{dx}$ in these problems. This is because $\frac{dy}{dx}$ will always be a *factor* in the terms in which it appears, and will only appear to the first degree, so we are basically solving a *linear* (first-degree) equation in the "variable" $\frac{dy}{dx}$, albeit with nonconstant coefficients. So in fact it is no different fundamentally from solving $Ax + By + C = Dx + Ey + F$ for y (or for x, for that matter). One moves all terms containing that variable to one side of the equation, moves the other terms to the other side, factors the variable from the side which then contains it, and divides by the other factor.

Of course our implicit differentiation technique begins with calculus steps and ends with algebra steps. A slightly more complete description of the process than our summary above can be given by the following five-step process, assuming our curve is given by an equation:

(1) Apply $\frac{d}{dx}$ to both sides of the equation describing the curve.

(2) Complete all differentiation steps, flushing out all terms with a factor $\frac{dy}{dx}$.

(3) Put all terms with $\frac{dy}{dx}$ factors on one side, other terms on the other side of the equation.

(4) Factor the $\frac{dy}{dx}$ from the side which contains it.

(5) Finally, divide by the remaining factor, leaving $\frac{dy}{dx}$ on one side by itself. (Simplify as justified.)

Example 3.6.5 _____

Find dy/dx on the curve $x^2 + 3xy - 5y^2 = 9$.

<u>Solution</u>: We compute $\frac{dy}{dx}$, labeling which steps in our computation correspond to steps in our algorithm above.[65]

$$x^2 + 3xy - 5y^2 = 9 \quad \overset{(1)}{\Longrightarrow} \quad \frac{d}{dx}[x^2 + 3xy - 5y^2] = \frac{d}{dx}[9]$$

$$\overset{(2)}{\Longrightarrow} \quad \frac{dx^2}{dx} + 3\frac{d(xy)}{dx} - 5\frac{dy^2}{dx} = 0$$

$$\overset{(2)}{\Longrightarrow} \quad 2x + 3\left(x\frac{dy}{dx} + y\frac{dx}{dx}\right) - 5 \cdot 2y\frac{dy}{dx} = 0$$

$$\overset{(2)}{\Longrightarrow} \quad 2x + 3x\frac{dy}{dx} + 3y - 10y\frac{dy}{dx} = 0$$

$$\overset{(3)}{\Longrightarrow} \quad 3x\frac{dy}{dx} - 10y\frac{dy}{dx} = -2x - 3y$$

$$\overset{(4)}{\Longrightarrow} \quad (3x - 10y)\frac{dy}{dx} = -2x - 3y$$

$$\overset{(5)}{\Longrightarrow} \quad \frac{dy}{dx} = \frac{-2x - 3y}{3x - 10y} = \frac{2x + 3y}{10y - 3x}.$$

Example 3.6.6 _____

Consider the curve $(x^2 + y^2)^{3/2} = 2xy$, which is graphed in Figure 3.27, page 324. (Much later in

---

[65] In the computations leading to finding $\frac{dy}{dx}$, all but the first implication $\Longrightarrow$ can be replaced by an equivalence $\Longleftrightarrow$. We do lose some information when we compute a derivative, because functions which differ by a constant will have the same derivative. Also, the last step may have had cases in which we have $0 = 0$ before the division, but those tend to be isolated points on the curve.

3.6. CHAIN RULE II: IMPLICIT DIFFERENTIATION

the text we will see how this graph came about.) We find $\frac{dy}{dx}$ using the steps outlined above.

$$(x^2+y^2)^{3/2} = 2xy$$
$$\implies \frac{d}{dx}\left[(x^2+y^2)^{3/2}\right] = \frac{d}{dx}[2xy]$$
$$\implies \frac{3}{2}(x^2+y^2)^{1/2} \cdot \frac{d}{dx}(x^2+y^2) = 2\left[x\frac{dy}{dx} + y\frac{dx}{dx}\right]$$
$$\implies \frac{3}{2}(x^2+y^2)^{1/2} \cdot \left(2x + 2y\frac{dy}{dx}\right) = 2\left(x\frac{dy}{dx} + y\cdot 1\right)$$
$$\implies \frac{3}{2}(x^2+y^2)^{1/2}(2x) + \frac{3}{2}(x^2+y^2)^{1/2}\left(2y\frac{dy}{dx}\right) = 2x\frac{dy}{dx} + 2y$$
$$\implies 3x\sqrt{x^2+y^2} + 3y\sqrt{x^2+y^2}\cdot\frac{dy}{dx} = 2x\frac{dy}{dx} + 2y$$
$$\implies 3y\sqrt{x^2+y^2}\cdot\frac{dy}{dx} - 2x\frac{dy}{dx} = 2y - 3x\sqrt{x^2+y^2}$$
$$\implies \left[3y\sqrt{x^2+y^2} - 2x\right]\frac{dy}{dx} = 2y - 3x\sqrt{x^2+y^2}$$
$$\implies \frac{dy}{dx} = \frac{2y - 3x\sqrt{x^2+y^2}}{3y\sqrt{x^2+y^2} - 2x}.$$

Now we will compute the slope for two points on the curve, $\left(\frac{1}{\sqrt{2}}, \frac{1}{\sqrt{2}}\right)$ and $\left(\frac{3}{4}, \frac{\sqrt{3}}{4}\right)$, which are labeled in the aforementioned figure (Figure 3.27, page 324). (The reader should verify that these points are on the original curve.)

- $\left.\frac{dy}{dx}\right|_{\left(\frac{1}{\sqrt{2}}, \frac{1}{\sqrt{2}}\right)} = \left.\frac{2y - 3x\sqrt{x^2+y^2}}{3y\sqrt{x^2+y^2} - 2x}\right|_{\left(\frac{1}{\sqrt{2}}, \frac{1}{\sqrt{2}}\right)} = \frac{2\cdot\frac{1}{\sqrt{2}} - 3\cdot\frac{1}{\sqrt{2}}\sqrt{\left(\frac{1}{\sqrt{2}}\right)^2 + \left(\frac{1}{\sqrt{2}}\right)^2}}{3\cdot\frac{1}{\sqrt{2}}\sqrt{\left(\frac{1}{\sqrt{2}}\right)^2 + \left(\frac{1}{\sqrt{2}}\right)^2} - 2\cdot\frac{1}{\sqrt{2}}}$

$= \frac{\sqrt{2} - \frac{3}{\sqrt{2}}\cdot 1}{\frac{3}{\sqrt{2}}\cdot 1 - \sqrt{2}} = -\frac{\left(\frac{3}{\sqrt{2}} - \sqrt{2}\right)}{\frac{3}{\sqrt{2}} - \sqrt{2}} = -1.$

This should seem reasonable given the position of this point on the graph. The tangent line at that point would then be $y = \frac{1}{\sqrt{2}} - \left(x - \frac{1}{\sqrt{2}}\right)$, i.e., $y = -x + \frac{2}{\sqrt{2}}$, or $y = -x + \sqrt{2}$. For the computation of $\frac{dy}{dx}$ at $\left(\frac{3}{4}, \frac{\sqrt{3}}{4}\right)$ we will be more brief.

- $\left.\frac{dy}{dx}\right|_{\left(\frac{3}{4}, \frac{\sqrt{3}}{4}\right)} = \frac{2\cdot\frac{\sqrt{3}}{4} - 3\cdot\frac{3}{4}\sqrt{\frac{12}{16}}}{3\cdot\frac{\sqrt{3}}{4}\sqrt{\frac{12}{16}} - 2\cdot\frac{3}{4}} = \frac{\frac{\sqrt{3}}{2} - \frac{9}{4}\cdot\sqrt{\frac{3}{4}}}{3\cdot\frac{\sqrt{3}}{4}\sqrt{\frac{3}{4}} - \frac{3}{2}} = \frac{\frac{\sqrt{3}}{2} - \frac{9}{4}\cdot\frac{\sqrt{3}}{2}}{3\cdot\frac{\sqrt{3}}{4}\cdot\frac{\sqrt{3}}{2} - \frac{3}{2}} \cdot \frac{8}{8} = \frac{4\sqrt{3} - 9\sqrt{3}}{9 - 12}$

$= \frac{-5\sqrt{3}}{-3} = \frac{5}{3}\sqrt{3} \approx 2.88675135.$

This should also seem reasonable from the graph in Figure 3.27, page 324.

The argument in the proof of the Power Rule for rational powers of x, Theorem 3.4.2 on page 281, used this implicit differentiation technique. As in that proof, we can compute $\frac{dy}{dx}$ without worrying about actually finding the open rectangles which give y locally as a

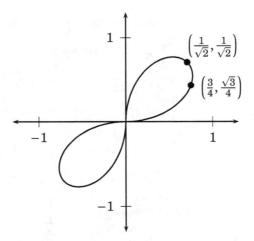

Figure 3.27: Graph of $(x^2+y^2)^{3/2}=2xy$. The slope dy/dx at each point (x,y) on the curve is calculated in Example 3.6.6, page 322. The points $(x,y)=\left(\frac{1}{\sqrt{2}},\frac{1}{\sqrt{2}}\right)$ and $(x,y)=\left(\frac{3}{4},\frac{\sqrt{3}}{4}\right)$ are plotted as well, for which slopes were computed in that example.

function of x. The rectangles justify the technique, but if the open rectangle hypothesis is false, that problem will usually manifest in the final form of the computed derivative.

Example 3.6.7

Find $\frac{dy}{dx}$ for the graph $5x+x^2+y^2+xy=\tan y$. This one will require two Chain Rule calls, for the y^2 and $\tan y$ terms, and a Product Rule for the xy term. As before, if we are careful in writing out the Product Rule we are less likely to have errors.

$$\frac{d}{dx}\left[5x+x^2+y^2+xy\right]=\frac{d\tan y}{dx}$$

$$\implies 5+2x+2y\cdot\frac{dy}{dx}+\left[x\frac{dy}{dx}+y\frac{dx}{dx}\right]=\sec^2 y\cdot\frac{dy}{dx}$$

$$\implies 5+2x+2y\cdot\frac{dy}{dx}+x\frac{dy}{dx}+y=\sec^2 y\cdot\frac{dy}{dx}$$

$$\implies 5+2x+y=\sec^2 y\frac{dy}{dx}-2y\frac{dy}{dx}-x\frac{dy}{dx}$$

$$\implies 5+2x+y=\left(\sec^2 y-2y-x\right)\frac{dy}{dx}$$

$$\implies \frac{5+2x+y}{\sec^2 y-2y-x}=\frac{dy}{dx}.$$

To find the tangent line through $(0,0)$, which is on the graph, we then compute

$$\left.\frac{dy}{dx}\right|_{(0,0)}=\left.\frac{5+2x+y}{\sec^2 y-2y-x}\right|_{(0,0)}=\frac{5+0+0}{1-0-0}=5,$$

and so the tangent line through $(0,0)$ has equation $y=5x$.

The next example is quite long, but with moderate persistence is also quite doable.

3.6. CHAIN RULE II: IMPLICIT DIFFERENTIATION

Example 3.6.8

Find $\frac{dy}{dx}$ on the graph of $y^3 \sec\sqrt{x^2+y^2} = \cos 2x$.

$$\frac{d}{dx}\left[y^3 \sec\sqrt{x^2+y^2}\right] = \frac{d\cos 2x}{dx}$$

$$\implies y^3 \frac{d}{dx}\left[\sec\sqrt{x^2+y^2}\right] + \sec\sqrt{x^2+y^2} \cdot \frac{dy^3}{dx} = -\sin 2x \cdot \frac{d\,2x}{dx}$$

$$\implies y^3 \sec\sqrt{x^2+y^2} \tan\sqrt{x^2+y^2} \cdot \frac{d}{dx}\sqrt{x^2+y^2} + \sec\sqrt{x^2+y^2} \cdot 3y^2 \cdot \frac{dy}{dx} = -\sin 2x \cdot 2$$

$$\implies y^3 \sec\sqrt{x^2+y^2} \tan\sqrt{x^2+y^2} \cdot \frac{1}{2\sqrt{x^2+y^2}} \cdot \frac{d}{dx}(x^2+y^2) + 3y^2 \sec\sqrt{x^2+y^2} \cdot \frac{dy}{dx} = -2\sin 2x$$

$$\implies \frac{y^3 \sec\sqrt{x^2+y^2} \tan\sqrt{x^2+y^2}}{2\sqrt{x^2+y^2}} \cdot \left(2x + 2y\frac{dy}{dx}\right) + 3y^2 \sec\sqrt{x^2+y^2} \cdot \frac{dy}{dx} = -2\sin 2x.$$

Next we apply the distributive law in the first term to flush out the $\frac{dy}{dx}$ terms and get

$$\frac{xy^3 \sec\sqrt{x^2+y^2} \tan\sqrt{x^2+y^2}}{\sqrt{x^2+y^2}} + \frac{y^4 \sec\sqrt{x^2+y^2} \tan\sqrt{x^2+y^2}}{\sqrt{x^2+y^2}} \cdot \frac{dy}{dx} + 3y^2 \sec\sqrt{x^2+y^2} \cdot \frac{dy}{dx} = -2\sin 2x.$$

Now we put all terms with the factor $\frac{dy}{dx}$ on one side (here, the left side), and the others on the opposite side:

$$\frac{y^4 \sec\sqrt{x^2+y^2} \tan\sqrt{x^2+y^2}}{\sqrt{x^2+y^2}} \cdot \frac{dy}{dx} + 3y^2 \sec\sqrt{x^2+y^2} \cdot \frac{dy}{dx} = -2\sin 2x - \frac{xy^3 \sec\sqrt{x^2+y^2} \tan\sqrt{x^2+y^2}}{\sqrt{x^2+y^2}}$$

Next we factor the $\frac{dy}{dx}$:

$$\left(\frac{y^4 \sec\sqrt{x^2+y^2} \tan\sqrt{x^2+y^2}}{\sqrt{x^2+y^2}} + 3y^2 \sec\sqrt{x^2+y^2}\right)\frac{dy}{dx} = -2\sin 2x - \frac{xy^3 \sec\sqrt{x^2+y^2} \tan\sqrt{x^2+y^2}}{\sqrt{x^2+y^2}}.$$

Finally, we divide to solve for $\frac{dy}{dx}$:

$$\frac{dy}{dx} = \frac{-2\sin 2x - \dfrac{xy^3 \sec\sqrt{x^2+y^2} \tan\sqrt{x^2+y^2}}{\sqrt{x^2+y^2}}}{\dfrac{y^4 \sec\sqrt{x^2+y^2} \tan\sqrt{x^2+y^2}}{\sqrt{x^2+y^2}} + 3y^2 \sec\sqrt{x^2+y^2}}.$$

If we would like, we can multiply the numerator and denominator by $\sqrt{x^2+y^2}$ to get:

$$\frac{dy}{dx} = \frac{-2\sqrt{x^2+y^2}\sin 2x - xy^3 \sec\sqrt{x^2+y^2} \tan\sqrt{x^2+y^2}}{y^4 \sec\sqrt{x^2+y^2} \tan\sqrt{x^2+y^2} + 3y^2 \sqrt{x^2+y^2} \sec\sqrt{x^2+y^2}}.$$

Although this latest example may seem tedious, no particular step is conceptually difficult. Success in such a project is nearly as much dependent upon our bookkeeping skills as

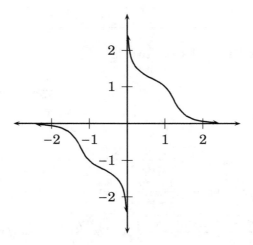

Figure 3.28: Graph of the curve $xy^5 + x^5y = 2$ from Example 3.6.9. The curve can have no horizontal tangents, though we can see that $\frac{dy}{dx} \to 0$ as $x \to \pm\infty$. It is also interesting to note that $\left|\frac{dy}{dx}\right| \to \infty$ as $x \to 0$. We did not prove this fact, though we can see that y must get large quickly as $x \to 0$ if the equation $xy^5 + x^5y = 2$ is to hold, since $x \to 0$ alone would shrink xy^5 and x^5y, and so y must grow to compensate if the sum is to be fixed at 2.

upon understanding of derivative rules. The next example requires some further analysis of the derivative as computed.

Example 3.6.9

Show that there are no horizontal tangent lines for the curve $xy^5 + x^5y = 2$.

Solution: As before, we apply $\frac{d}{dx}$ to both sides:

$$xy^5 + x^5y = 2 \implies \frac{d}{dx}\left[xy^5 + x^5y\right] = \frac{d}{dx}(2)$$

$$\implies x\frac{dy^5}{dx} + y^5\frac{dx}{dx} + x^5\frac{dy}{dx} + y\frac{dx^5}{dx} = 0$$

$$\implies x \cdot 5y^4\frac{dy}{dx} + y^5 + x^5\frac{dy}{dx} + y \cdot 5x^4 = 0$$

$$\implies \left[5xy^4 + x^5\right]\frac{dy}{dx} = -y^5 - 5x^4y$$

$$\implies \frac{dy}{dx} = -\frac{y^5 + 5x^4y}{5xy^4 + x^5} = -\frac{y(y^4 + 5x^4)}{x(5y^4 + x^4)}.$$

The curve is graphed in Figure 3.28 above. To have a horizontal tangent, we would need the numerator above to be zero, i.e., $y(y^4 + 5x^4) = 0$. This would imply $y = 0$ or $y^4 + 5x^4 = 0$. We will see that neither is possible on this curve.

1. Case $y = 0$: This never happens on the curve $x^5y + xy^5 = 2$, because it would imply $0 = 2$.

2. Case $y^4 + 5x^4 = 0$: This would require $x, y = 0$, since $(y^4 \geq 0) \wedge (5x^4 \geq 0) \implies y^4 + 5x^4 \geq 0$, and we get equality only when both (nonnegative) terms in the sum are zero, thus requiring $x, y = 0$, which is impossible again due to the fact that we require (x, y) to be on the curve $xy^5 + x^5y = 2$.

3.6. CHAIN RULE II: IMPLICIT DIFFERENTIATION

The above example demonstrates the importance of applying the derived formula for dy/dx only to points (x,y) which are on the original curve, even if the formula for dy/dx on its face "makes sense" for nearly any pair (x,y). The actual curve is the setting in which the formula for dy/dx is meaningful. See again Figure 3.28, page 326. A similar example, but with a very different looking graph, is the following:

Example 3.6.10

Find all points of horizontal tangency for the curve $xy^3 - x^2y^2 = 4$.

Solution: We proceed as before, applying $\frac{d}{dx}$ to both sides, to first find the form of $\frac{dy}{dx}$:

$$\frac{d}{dx}[xy^3 - x^2y^2] = \frac{d}{dx}[4]$$

$$\implies x\frac{dy^3}{dx} + y^3\frac{dx}{dx} - \left(x^2\frac{dy^2}{dx} + y^2\frac{dx^2}{dx}\right) = 0$$

$$\implies x \cdot 3y^2\frac{dy}{dx} + y^3 - x^2 \cdot 2y\frac{dy}{dx} - y^2 \cdot 2x = 0$$

$$\implies [3xy^2 - 2x^2y]\frac{dy}{dx} = 2xy^2 - y^3$$

$$\implies \frac{dy}{dx} = \frac{2xy^2 - y^3}{3xy^2 - 2x^2y}.$$

If a point on the graph has a horizontal tangent, then at that point $\frac{dy}{dx} = 0$, which would require $2xy^2 - y^3 = 0$. Solving for y would yield

$$2xy^2 - y^3 = 0 \implies y^2(2x - y) = 0$$
$$\implies (y = 0) \lor (2x - y = 0).$$

We can disregard the $y = 0$ solution of this equation since no such point exists on the graph of our curve $xy^3 - x^2y^2 = 4$.

We are then left with solutions in which $2x - y = 0$, i.e., $y = 2x$. Because we are only interested in points on the original curve, we can solve

$$(y = 2x) \land (xy^3 - x^2y^2 = 4) \implies x(2x)^3 - x^2(2x)^2 = 4$$
$$\implies 8x^4 - 4x^4 = 4$$
$$\implies 4x^4 = 4$$
$$\implies x^4 = 1$$
$$\implies x = \pm\sqrt[4]{1} = \pm 1.$$

Since $y = 2x$ and $x = \pm 1$ at such points, we conclude that if such points on the graph at which the slope is zero exist, they should be in the set $\{(1,2),(-1,-2)\}$. To be sure our implication is not vacuously true, we check that these points are indeed on the graph, and we check that the slope is zero at these points.

At the point $(1,2)$, the equation of the curve would read $1 \cdot 2^3 - 1^2 2^2 = 4$, or $8 - 4 = 4$, which is true.

At the point $(-1,-2)$, the equation of the curve would read $(-1)(-2)^3 - (-1)^2(-2)^2 = 4$, or $8 - 4 = 4$ as before.

The slopes at those two points would be given by the equation $\frac{dy}{dx} = \frac{2xy^2 - y^3}{3xy^2 - 2x^2y}$:

$$\left.\frac{dy}{dx}\right|_{(1,2)} = \frac{2(1)(2)^2 - 2^3}{3(1)(2^2) - 2(1)^2(2)} = \frac{0}{8} = 0, \qquad \left.\frac{dy}{dx}\right|_{(-1,-2)} = \frac{2(-1)(-2)^2 - (-2)^3}{3(-1)(-2)^2 - 2(-1)^2(-2)} = \frac{0}{-8} = 0.$$

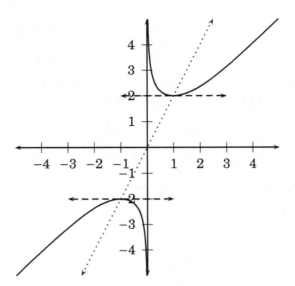

Figure 3.29: The graph of $xy^3 - x^2y^2 = 4$ for Example 3.6.10, page 327. The two points at which the tangent lines are horizontal (i.e., where the slope of the curve is zero) occur at $(1,2)$ and $(-1,-2)$. The relevant tangent lines are shown (dashed). Also shown is the line $y = 2x$ (dotted). In the process of finding where the tangent lines were horizontal, the system $xy^3 - x^2y^2 = 4$ (the curve) and $y = 2x$ described where the points of horizontal tangency on the curve were located.

This verifies that the points on the curve $xy^3 - x^2y^2 = 4$ at which the tangent lines are horizontal are exactly the points $(1,2)$ and $(-1,-2)$.

The graph for Example 3.6.10 is given in Figure 3.29 above, from which it is clearly reasonable that the slope is zero at $(1,2)$ and $(-1,-2)$. Also included in that graph is the line $y = 2x$, shown to illustrate that part of the derivation in which we solved the system

$$\begin{cases} xy^3 - x^2y^2 &= 4 \\ y &= 2x, \end{cases}$$

the solution of which is $(x,y) \in \{(1,2),(-1,-2)\}$.

The graphs of implicit curves can be very difficult to produce, and are often not at all obvious. We included several here to offer some visualization of the phenomena we observed from the derivatives. Fortunately the analyses here did not require the visual representations (graphs), and that is a testament to the power of our analytical tools developed here.[66]

---

[66] For the interested reader, we mention here that the code used to create graphs for this document allows us to easily plot curves where y is a function of x given by a formula, or both x and y are functions of another variable, say t. In Figure 3.29 above we had a curve in which we could, using the quadratic formula, solve for x in terms of y, then let $y = t$ and let $x = g(y) = g(t)$, although there were two solutions for x in terms of y (due to the nature of the quadratic formula) and so both had to be plotted separately. For the curve in Figure 3.28, page 326, an external package was used to generate data which were then plotted and connected smoothly.

The reader should not be worried if it is far from obvious why these curves have the graphs represented here.

3.6. CHAIN RULE II: IMPLICIT DIFFERENTIATION

3.6.3 A Mistake to Avoid

Before finishing this section, a remark is in order: $f(a) = g(a) \not\Rightarrow f'(a) = g'(a)$. In a graphical sense, the first equation is about heights of functions and the second is about their slopes.

It is tempting to try to simplify an algebraic equation by taking derivatives of both sides if the new equations are indeed simpler (e.g., with many polynomial equations). However, in such cases we are looking for *points* where one expression (or function) equals the other, which is very different from a scenario where two sides of an equation are the same *as functions* of, say, x.

Example 3.6.11

Consider the equation
$$x^2 - 2x + 1 = 5 - 2x. \tag{3.61}$$
This succumbs easily to the earlier methods of solution:
$$x^2 - 2x + 1 = 5 - 2x$$
$$\iff x^2 = 4$$
$$\iff x = -2, 2$$
so the solution is simply $x = \pm 2$. Now suppose instead we first take derivatives of both sides of (3.61):
$$2x - 2 = -2 \tag{3.62}$$
$$\iff 2x = 0$$
$$\iff x = 0.$$

We see that we get a value for x which is not a solution of the original equation (3.61): if we let $x = 0$ in (3.61), we get $0^2 - 2 \cdot 0 + 1 = 5 - 2 \cdot 0$, i.e., $1 = 5$, which is clearly false. We can think of (3.61) as an equation where we solve $f(x) = g(x)$, with $f(x) = x^2 - 2x + 1$ and $g(x) = 5 - 2x$, which has solutions $x = -2, 2$. Then (3.62) is the equation $f'(x) = g'(x)$, with solution $x = 0$. Clearly (3.61) does not imply (3.62) (i.e., $x \in \{-2, 2\}$ does not imply $x \in \{0\}$).

The upshot is that we cannot solve an algebraic equation by first taking derivatives of both sides, and expect the new equation to have the same solution as the original.

By contrast, when we perform implicit differentiation we are not intending to solve equations, but to derive slopes (or more generally, rates of change) on implicit curves.

For implicit differentiation, recall the idea is that in some open rectangles in the plane the curve defines y as a function of x, and so when we have an equation for a curve such as $xy^3 - x^2y^2 = 4$ from Example 3.6.10, page 327, the left-hand side and right-hand side of that equation are both ultimately functions of x inside such a rectangle, and moreover are *the same* function of x there, and therefore have the same derivative there. By applying $\frac{d}{dx}$ to both sides we are left with a new equation relating x, y and $\frac{dy}{dx}$, in which we can solve for $\frac{dy}{dx}$ in terms of x and y, though that expression for $\frac{dy}{dx}$ is only valid at points on the implicit curve.

For further explanation see Figure 3.30, page 330.

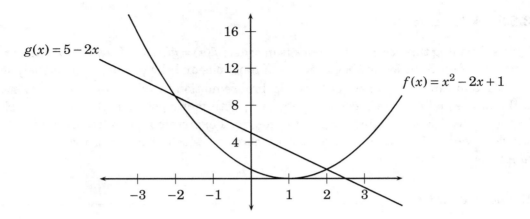

Figure 3.30: The graphs of $y = f(x)$ and $y = g(x)$ intersect at $x = \pm 2$, which is the solution to $f(x) = g(x)$. However, it is clear from the picture that, while $f(x) = g(x)$ at $x = \pm 2$, the derivatives (slopes) $f'(x)$ and $g'(x)$ at those two points are not the same. In fact, $f'(x) = g'(x)$ at $x = 0$ only. Thus $f(a) = g(a) \not\Rightarrow f'(a) = g'(a)$.

3.6.4 An Application

Example 3.6.12

When two resistors with values R_1 and R_2 are connected in parallel, the resulting resistance R is given by the equation

$$\frac{1}{R} = \frac{1}{R_1} + \frac{1}{R_2}. \tag{3.63}$$

Suppose we always have $R_2 = R_1 + 4\Omega$.

1. Find an equation for R in terms of $r = R_1$.
2. Show that $r^2 = 2rR + 4R - 4r$.
3. Find dR/dr in terms of r using the equation you found in part 1 directly.
4. Find dR/dr in terms of r by differentiating the equation in part 2.
5. Use part 1 to show this is the same as the answer in part 3.

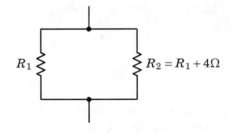

Solution: We take these in turn.

1. After making the substitution $R_2 = R_1 + 4$ we have

$$\frac{1}{R} = \frac{1}{R_1} + \frac{1}{R_1 + 4}.$$

Using $r = R_1$ for simplicity this then becomes

$$\frac{1}{R} = \frac{1}{r} + \frac{1}{r+4}.$$

3.6. CHAIN RULE II: IMPLICIT DIFFERENTIATION

Taking reciprocals of both sides gives us

$$R = \frac{1}{\frac{1}{r} + \frac{1}{r+4}} = \frac{1}{\frac{1}{r} + \frac{1}{r+4}} \cdot \frac{r(r+4)}{r(r+4)} = \frac{r^2 + 4r}{(r+4) + r}$$

$$\implies R = \frac{r^2 + 4r}{2r + 4}. \tag{3.64}$$

2. Multiplying the above equation by $(2r+4)$ gives us

$$2rR + 4R = r^2 + 4r$$
$$\iff 2rR + 4R - 4r = r^2$$
$$\iff r^2 = 2rR + 4R - 4r, \text{ q.e.d.} \tag{3.65}$$

3. We find dR/dr using the Quotient Rule applied to (3.64):

$$\frac{dR}{dr} = \frac{(2r+4)\frac{d}{dr}(r^2+4r) - (r^2+4r)\frac{d}{dr}(2r+4)}{(2r+4)^2} = \frac{(2r+4)(2r+4) - (r^2+4r)(2)}{[2(r+2)]^2}$$
$$= \frac{4r^2 + 16r + 16 - 2r^2 - 8r}{4(r+2)^2} = \frac{2r^2 + 8r + 16}{4(r+2)^2} = \frac{2(r^2 + 4r + 8)}{4(r+2)^2}$$
$$\implies \frac{dR}{dr} = \frac{r^2 + 4r + 8}{2(r+2)^2}. \tag{3.66}$$

4. Using part 2, i.e., (3.65), we apply $\frac{d}{dr}$ to both sides:

$$\frac{d}{dr}[r^2] = \frac{d}{dr}[2rR + 4R - 4r]$$
$$\implies 2r = 2r\frac{d}{dr}(R) + R\frac{d}{dr}(2r) + 4\frac{dR}{dr} - 4$$
$$\implies 2r = 2r\frac{dR}{dr} + 2R + 4\frac{dR}{dr} - 4$$
$$\implies 2r + 4 - 2R = (2r+4)\frac{dR}{dr}$$
$$\implies \frac{dR}{dr} = \frac{2r + 4 - 2R}{2r + 4}.$$

5. Now, using $R = (r^2 + 4r)/(2r + 4)$ we get

$$\frac{dR}{dr} = \frac{2r + 4 - 2 \cdot \frac{r^2+4r}{2r+4}}{2r+4} \cdot \frac{2r+4}{2r+4} = \frac{(2r+4)^2 - 2(r^2+4r)}{(2r+4)^2},$$

which was one of our intermediate results in the computations in part 3 (specifically, the second line there), and so continued simplification would give us the same as the final simplification in part 3 as well, q.e.d.

Exercises

For Exercises 1–10 compute $\frac{dy}{dx}$ by implicit differentiation.

1. $Ax + By = C$, where A, B and C are constants, and $B \neq 0$.
2. $\sin x + \cos y = \frac{1}{2}$.
3. $x^2 - y^2 = 9$. Find also the tangent line at $(5, -4)$.
4. $x^2 + 3x + y^2 = 16 + 7y$.
5. $x + \tan x = y^3 + \sec y$.
6. $x^2 + 2xy + y^2 = \cos x$.
7. $x \sin y + y \sin x = 2$.
8. $\sin x^2 = \cos y^3$.
9. $\tan \frac{x}{y} = 9$.
10. $\sin(xy) + y = \csc x - 11$
11. Consider the curve graphed below given by $(x^2 + y^2)^3 = (x^2 - y^2)^2$.

    (a) Find $\frac{dy}{dx}$.
    (b) Find the equation of the tangent line at $\left(\frac{1}{4}, \frac{\sqrt{3}}{4}\right)$.

    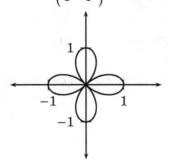

    (c) From the graph, what can you say about the slope $\frac{dy}{dx}$ as $y \to 1^-$ and as $y \to -1^+$?

    (d) From the graph, what can you say about the slope $\frac{dy}{dx}$ as $x \to 1^-$ and $x \to -1^+$?

    (e) Note from the graph that, for instance, $y \to 1^- \implies x \to 0$ on the given curve. Using that fact and your equation for dy/dx, find
    $$\lim_{y \to 1^-} \frac{dy}{dx}.$$

12. Consider the curve $xy = 1$.

    (a) Use implicit differentiation to find $\frac{dy}{dx}$.
    (b) Use $y = \frac{1}{x}$ to find $\frac{dy}{dx}$.
    (c) Show algebraically how these are the same answers.

13. Consider the algebraic equation
    $$x^2 = 9.$$

    (a) Define functions $f(x) = x^2$ and $g(x) = 9$ and graph them together on one grid.
    (b) What is its solution of the original equation, i.e., of $f(x) = g(x)$?
    (c) What is the graphical significance of the solution of $f(x) = g(x)$?
    (d) Now consider the equation $f'(x) = g'(x)$ for these two functions f and g. What is the solution of this new equation $f'(x) = g'(x)$?
    (e) Explain the significance of the solution of $\frac{d}{dx}f(x) = \frac{d}{dx}g(x)$ for this particular example.
    (f) Explain why, if a particular x satisfies $f(x) = g(x)$, we cannot expect that it also satisfies $f'(x) = g'(x)$.

14. Show that there are no horizontal tangent lines to the curve
$$x^2 y^2 - xy^3 = 4.$$

15. Show that there is one horizontal tangent line for the curve
$$xy^3 - x^2 y = 8$$
and determine the point of horizontal tangency.

16. Consider the curve
$$\sin x \cos y = \frac{1}{2},$$
where $x, y \in [0, \pi/2]$.

   (a) Use implicit differentiation to find dy/dx.

   (b) Find dy/dx when $y = \pi/4$.

17. Suppose $y = f(x)/g(x)$. Derive the quotient rule for dy/dx by multiplying by $g(x)$ and using implicit differentiation. (Hint: Replace y with $f(x)/g(x)$ at some point after the differentiation step.)

3.7 Arctrigonometric Functions and Their Derivatives

In this and the next three sections, we will explore derivatives of the remaining standard classes of functions. Presently we will look at the arctrigonometric functions, while in the next three sections we will look respectively at exponential functions, logarithmic functions, and related techniques. While we will need to be mindful of the exact natures of all these functions, with their derivative formulas we will usually be able to simply apply the new formulas together with the previous rules, and in so doing nearly finish our study of computing derivatives. However, their algebraic properties (or some would say, "quirks") and precise definitions cannot be ignored. Indeed, they are often exploited to simplify our calculus computations.

The arctrigonometric functions are also called the *inverse trigonometric functions*, which is somewhat problematic in that the trigonometric functions are not invertible *per se*. But that does not keep us from defining inverse functions in specific local contexts, in ways which are still useful in other contexts. It is analogous to how we define a square root, and while it does not "undo" the square in all circumstances (since $\sqrt{x^2} = |x|$ and not simply x), it is obviously still quite useful.

3.7.1 One-to-One Functions and Inverses, Reviewed

Recall what it means for a function $y = f(x)$ to be invertible, or *one-to-one*, meaning that for $f : S \longrightarrow \mathbb{R}$, i.e., where S is the domain of f, we have

$$(\forall x_1, x_2 \in S)[(x_1 = x_2) \longleftrightarrow (f(x_1) = f(x_2))] \tag{3.67}$$

Note that for f to be a function we already have $(\forall x_1, x_2 \in S)[(x_1 = x_2) \longrightarrow (f(x_1) = f(x_2))]$, i.e., if we know the input we (in principle) know the output. For the function to be one-to-one means we also have $\longleftarrow$, i.e., if we know the output we can (in principle) deduce the input. When we have such an f, we call its *inverse* f^{-1}, the function defined by the property

$$(\forall x \in S)(\forall y \in f(S))[f(x) = y \longleftrightarrow f^{-1}(y) = x]. \tag{3.68}$$

Recall that here the superscript "-1" is not an exponent (power), but an indication that we are looking at the "reverse" of the process that takes x to y.

Recall also that just as f^{-1} reverses f, so does f reverse f^{-1}, as shown in the diagrams below left:

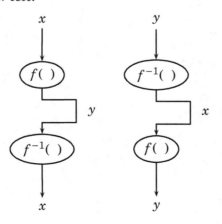

Thus, if f is one-to-one, we have

$$(\forall x \in S) \quad [f^{-1}(f(x)) = x], \tag{3.69}$$
$$(\forall y \in f(S))[f(f^{-1}(y)) = y]. \tag{3.70}$$

In other words, the functions f and f^{-1} "undo" each other; for a one-to-one function f with inverse f^{-1}, we have that their map-

3.7. ARCTRIGONOMETRIC FUNCTIONS AND THEIR DERIVATIVES

pings are reversible: $x \xmapsto{f} \underbrace{y}_{f(x)} \xmapsto{f^{-1}} \underbrace{x}_{f^{-1}(y)}$, which as mappings of sets would look like $S \xrightarrow{f} f(S) \xrightarrow{f^{-1}} S$.

The usual algebraic way to attempt to invert a function f, if possible, is to solve the equation $y = f(x)$ for x, producing an equation of the form $x = f^{-1}(y)$. Often in the process of attempting to calculate $f^{-1}(y)$ in this way we will discover if it in fact exists, i.e., if f is one-to-one. There can be many technicalities, depending upon the original function, but many computations are straightforward.

Recall also that there is often confusion regarding the naming of input and output variables. If x is the input of f and y the output, then it is useful to let y be the input of f^{-1} and x the output, hence $y = f(x) \iff x = f^{-1}(y)$ when f is one-to-one. However the inverse is a function in its own right so the usual convention is to discuss $f^{-1}(x)$ just as we discuss $f(x)$.

3.7.2 The Arctrigonometric Functions and Their Derivatives

In Table 3.1, page 338, we summarize the most important conclusions of what a careful development of the arctrigonometric functions would yield, including their domains, their ranges, and their derivatives. We save the actual derivations for later subsections, since at this stage we are most interested in rules for computing derivatives involving these functions. However, the detailed derivations are ultimately quite informative and should themselves be studied for a couple of reasons. First, the technicalities involved there will reappear many times in later chapters. Second, the derivations are also quite interesting—and useful as exercises—because, as occurs in many applied problems, the techniques used form an interesting mix of algebra, trigonometry and our calculus techniques, particularly implicit differentiation.[67]

However in this subsection we concentrate on their derivatives, though we keep an eye towards their actual definitions, including their domains and ranges. Armed with the derivative formulas, we will be able to greatly expand the class of functions we can differentiate.

Since the trigonometric functions are periodic, i.e., repeat periodically across their domains, there is no guarantee for instance that $\sin x_1 = \sin x_2$ will imply $x_1 = x_2$. In other words, *the trigonometric functions are not invertible!*

The way we nonetheless endeavor to "invert" the trigonometric functions is to temporarily restrict their domains so that they are forced to be one-to-one within these restricted domains. For instance, $\sin x$ is one-to-one if we restrict x so that $x \in [-\pi/2, \pi/2]$ only:

$$\left(\forall x_1, x_2 \in \left[-\frac{\pi}{2}, \frac{\pi}{2}\right]\right)[(\sin x_1 = \sin x_2) \longleftrightarrow (x_1 = x_2)].$$

This kind of domain restriction is explained in subsequent subsections, where for each of the six standard trigonometric functions we look for a subset of its domain on which (1) the

---

[67]Indeed, the further along one gets in mathematical or applied studies, the more one has to borrow from a growing diversity of subjects. It is very often the "technicalities"—themselves often first found in derivations, footnotes or otherwise parenthetically—which play key roles in solving interesting problems, in both pure and applied mathematics.

function is one-to-one, and (2) the set of all outputs covers the whole range of the original function.[68]

There are some other issues which turn out to be easier to accommodate, such as consistency and compatibility with the other six trigonometric functions, and which will be explained as we develop the theory.

Of course the simplest uses for the arctrigonometric functions arise from solving trigonometric equations. For instance, it often happens in applications that we need to solve an equation such as $\tan\theta = x$ for the variable θ. For this and other reasons, arctrigonometric functions become of interest. These are functions which take a *number*, such as x, and return an *angle* such as θ (also represented by a number, but the distinction is important) for which x is some given trigonometric function of that angle. So if we are interested in knowing an angle θ such that $\tan\theta = x$, there is an arctrigonometric function $\arctan x$ which will give us such an angle θ—even if it does not give us all such angles θ—so that $\tan\theta = x$. Three arctrigonometric functions, namely arcsine, arccosine and arctangent are built into scientific calculators, along with the original trigonometric functions sine, cosine and tangent. These functions allow us to move from angle to trigonometric function of the angle, and back, almost.

The trouble is that there is important ambiguity about the angle θ in such a problem, namely that if some angle θ solves $\tan\theta = x$, so do all angles of the form $\theta + n\pi$, where $n \in \mathbb{Z}$ (i.e., $\theta \pm \pi, \theta \pm 2\pi, \theta \pm 3\pi, \ldots$). Presently calculators only output a single angle for such a problem. Still, knowing one solution is quite useful, as it indirectly gives us such information as the reference angle of the other solutions. For instance, if we use a calculator for a solution to $\tan\theta = 5$, one solution will be given by the calculator to be $\theta = \tan^{-1} 5 \approx 1.373400767$ (or approximately $78.69006753°$). If we need an angle in the third quadrant (not practical in most right-triangle trigonometry, but useful in many applications nonetheless) we can instead use, for one example, $\tan^{-1} 5 + \pi \approx 4.514993421$ (or $\tan^{-1} 5 + 180° \approx 258.6900675°$ if we want to work in degrees).

Leaving the derivations for later, we now list derivatives of arctrigonometric functions in basic and Chain Rule forms. Note that $\sin^{-1} x$ is also denoted $\arcsin x$, $\cos^{-1} x = \arccos x$, and so on, if more verbose notations are desired.[69]

The derivatives of the (as yet undefined) inverse (or arc) trigonometric functions are as follow:

---

[68] It is akin to the problem of trying to invert the function $f(x) = x^2$, which is not one-to-one on all of $\mathbb{R} = (-\infty, \infty)$, since for instance $f(-5) = f(5)$, while $-5 \ne 5$. Since it is impossible to truly invert $f(x) = x^2$, instead we define the "principal square root" $g(y) = \sqrt{y} \ge 0$, which does give us an inverse to $f(x)$ on the set $x \ge 0$, i.e., $x \in [0, \infty)$:

$$(\forall x_1, x_2 \in [0, \infty)) \left[(x_1 = x_2) \longleftrightarrow \left(x_1^2 = x_2^2 \right) \right] \quad \text{and so} \quad (\forall x_1, x_2 \in [0, \infty)) \left[(x_1 = x_2) \longleftrightarrow \left(\sqrt{x_1} = \sqrt{x_2} \right) \right].$$

While these quantified statements above are not true if we replace $[0, \infty)$ with the whole domain $\mathbb{R}$ of f, it is still useful to define such a square root function g, or $\sqrt{}$. For instance, knowing $f(x) = K$ gives us $x = \pm\sqrt{K} = \pm g(K)$, so the value of $g(K)$ is still useful in finding a particular x, regardless of whether we ultimately need $x = \sqrt{K}$ or $x = -\sqrt{K}$. Which one is chosen usually follows from further information contained within the problem.

[69] Again note that the superscript -1 in $\sin^{-1} x$ is not an exponent in the sense of "power," but is rather a notation borrowed from the study of *inverse functions*. Indeed, for the -1 "power" we have another name for $(\sin x)^{-1}$, specifically $\csc x$.

3.7. ARCTRIGONOMETRIC FUNCTIONS AND THEIR DERIVATIVES

$$\frac{d\sin^{-1}x}{dx} = \frac{1}{\sqrt{1-x^2}}, \qquad \frac{d\sin^{-1}u}{dx} = \frac{1}{\sqrt{1-u^2}} \cdot \frac{du}{dx}, \qquad (3.71)$$

$$\frac{d\cos^{-1}x}{dx} = \frac{-1}{\sqrt{1-x^2}}, \qquad \frac{d\cos^{-1}u}{dx} = \frac{-1}{\sqrt{1-u^2}} \cdot \frac{du}{dx}, \qquad (3.72)$$

$$\frac{d\tan^{-1}x}{dx} = \frac{1}{x^2+1}, \qquad \frac{d\tan^{-1}u}{dx} = \frac{1}{u^2+1} \cdot \frac{du}{dx}, \qquad (3.73)$$

$$\frac{d\cot^{-1}x}{dx} = \frac{-1}{x^2+1}, \qquad \frac{d\cot^{-1}u}{dx} = \frac{-1}{u^2+1} \cdot \frac{du}{dx}, \qquad (3.74)$$

$$\frac{d\sec^{-1}x}{dx} = \frac{1}{|x|\sqrt{x^2-1}}, \qquad \frac{d\sec^{-1}u}{dx} = \frac{1}{|u|\sqrt{u^2-1}} \cdot \frac{du}{dx}, \qquad (3.75)$$

$$\frac{d\csc^{-1}x}{dx} = \frac{-1}{|x|\sqrt{x^2-1}}, \qquad \frac{d\csc^{-1}u}{dx} = \frac{-1}{|u|\sqrt{u^2-1}} \cdot \frac{du}{dx}. \qquad (3.76)$$

Remarkably all these derivatives are "algebraic" in nature, meaning that they involve only multiplication, division, polynomials and radicals. These emerging derivative forms, developed next and summarized in Table 3.1, page 338, will prove crucial in the later development of antiderivatives.[70]

While somewhat complicated, clearly these derivatives come in pairs, the difference being in the sign ($\pm$), so for instance $\frac{d}{dx}\cos^{-1}x = -\frac{d}{dx}\sin^{-1}x$, and similarly for other cofunction pairs. Table 3.1, page 338, contains the forms, along with their domains and ranges.

For our first examples we observe the following simple, similar Chain Rule computations. Note how $\sin^{-1}x^2$ is interpreted to be $\sin^{-1}(x^2)$, and similarly with the other forms.

Example 3.7.1

Here we compute examples of each form where the "inner function" is x^2. Note that $|x^2| = |x|^2 = x^2$.

$$\frac{d}{dx}\sin^{-1}x^2 = \frac{1}{\sqrt{1-(x^2)^2}} \cdot \frac{dx^2}{dx} \qquad = \frac{1}{\sqrt{1-x^4}} \cdot 2x \qquad = \frac{2x}{\sqrt{1-x^4}},$$

$$\frac{d}{dx}\cos^{-1}x^2 = \frac{-1}{\sqrt{1-(x^2)^2}} \cdot \frac{dx^2}{dx} \qquad = \frac{-1}{\sqrt{1-x^4}} \cdot 2x \qquad = \frac{-2x}{\sqrt{1-x^4}},$$

$$\frac{d}{dx}\tan^{-1}x^2 = \frac{1}{(x^2)^2+1} \cdot \frac{dx^2}{dx} \qquad = \frac{1}{x^4+1} \cdot 2x \qquad = \frac{2x}{x^4+1},$$

$$\frac{d}{dx}\cot^{-1}x^2 = \frac{-1}{(x^2)^2+1} \cdot \frac{dx^2}{dx} \qquad = \frac{-1}{x^4+1} \cdot 2x \qquad = \frac{-2x}{x^4+1},$$

$$\frac{d}{dx}\sec^{-1}x^2 = \frac{1}{|x^2|\sqrt{(x^2)^2-1}} \cdot \frac{dx^2}{dx} = \frac{1}{x^2\sqrt{x^4-1}} \cdot 2x = \frac{2x}{x^2\sqrt{x^4-1}} = \frac{2}{x\sqrt{x^4-1}},$$

$$\frac{d}{dx}\csc^{-1}x^2 = \frac{-1}{|x^2|\sqrt{(x^2)^2-1}} \cdot \frac{dx^2}{dx} = \frac{-1}{x^2\sqrt{x^4-1}} \cdot 2x = \frac{-2x}{x^2\sqrt{x^4-1}} = \frac{-2}{x\sqrt{x^4-1}}.$$

---

[70]"Algebraic" is in contrast to "transcendental," the latter referring to trigonometric, arctrigonometric, logarithmic and exponential functions, for instance. Note that the absolute value can be considered algebraic, since $|x| = \sqrt{x^2}$.

Function	Inputs (Domain)	Outputs (Range)	Outputs Graphed	Derivative
$\sin^{-1} x$	$x \in [-1, 1]$	$\theta \in \left[-\frac{\pi}{2}, \frac{\pi}{2}\right]$		$\dfrac{d \sin^{-1} x}{dx} = \dfrac{1}{\sqrt{1-x^2}}$
$\cos^{-1} x$	$x \in [-1, 1]$	$\theta \in [0, \pi]$		$\dfrac{d \cos^{-1} x}{dx} = \dfrac{-1}{\sqrt{1-x^2}}$
$\tan^{-1} x$	$x \in \mathbb{R}$	$\theta \in \left(-\frac{\pi}{2}, \frac{\pi}{2}\right)$		$\dfrac{d \tan^{-1} x}{dx} = \dfrac{1}{x^2+1}$
$\cot^{-1} x$	$x \in \mathbb{R} - \{0\}$	$\theta \in \left(-\frac{\pi}{2}, \frac{\pi}{2}\right) - \{0\}$		$\dfrac{d \cot^{-1} x}{dx} = \dfrac{-1}{x^2+1}$
$\sec^{-1} x$	$x \in (-\infty, -1] \cup [1, \infty)$	$\theta \in [0, \pi] - \left\{\frac{\pi}{2}\right\}$		$\dfrac{d \sec^{-1} x}{dx} = \dfrac{1}{|x|\sqrt{x^2-1}}$
$\csc^{-1} x$	$x \in (-\infty, -1] \cup [1, \infty)$	$\theta \in \left[-\frac{\pi}{2}, \frac{\pi}{2}\right] - \{0\}$		$\dfrac{d \csc^{-1} x}{dx} = \dfrac{-1}{|x|\sqrt{x^2-1}}$

Table 3.1: Summary of arctrigonometric functions, their domains and ranges (given in both interval notation and graphed as angles through the unit circle), and their derivatives. Note that all angles displayed in the "Outputs Graphed" column are assumed to be between $\frac{-\pi}{2}$ and π.

3.7. ARCTRIGONOMETRIC FUNCTIONS AND THEIR DERIVATIVES

Next we switch the "outer" and "inner" functions from above.

Example 3.7.2

Here we compute examples where the "outer function" is "squaring" in each.

$$\frac{d}{dx}\left(\sin^{-1}x\right)^2 = 2\left(\sin^{-1}x\right)^1 \cdot \frac{d\sin^{-1}x}{dx} = 2\sin^{-1}x \cdot \frac{1}{\sqrt{1-x^2}} = \frac{2\sin^{-1}x}{\sqrt{1-x^2}},$$

$$\frac{d}{dx}\left(\cos^{-1}x\right)^2 = 2\left(\cos^{-1}x\right)^1 \cdot \frac{d\cos^{-1}x}{dx} = 2\cos^{-1}x \cdot \frac{-1}{\sqrt{1-x^2}} = \frac{-2\cos^{-1}x}{\sqrt{1-x^2}},$$

$$\frac{d}{dx}\left(\tan^{-1}x\right)^2 = 2\left(\tan^{-1}x\right)^1 \cdot \frac{d\tan^{-1}x}{dx} = 2\tan^{-1}x \cdot \frac{1}{x^2+1} = \frac{2\tan^{-1}x}{x^2+1},$$

$$\frac{d}{dx}\left(\cot^{-1}x\right)^2 = 2\left(\cot^{-1}x\right)^1 \cdot \frac{d\cot^{-1}x}{dx} = 2\cot^{-1}x \cdot \frac{-1}{x^2+1} = \frac{-2\cot^{-1}x}{x^2+1},$$

$$\frac{d}{dx}\left(\sec^{-1}x\right)^2 = 2\left(\sec^{-1}x\right)^1 \cdot \frac{d\sec^{-1}x}{dx} = 2\sec^{-1}x \cdot \frac{1}{|x|\sqrt{x^2-1}} = \frac{2\sec^{-1}x}{|x|\sqrt{x^2-1}},$$

$$\frac{d}{dx}\left(\csc^{-1}x\right)^2 = 2\left(\csc^{-1}x\right)^1 \cdot \frac{d\csc^{-1}x}{dx} = 2\csc^{-1}x \cdot \frac{-1}{|x|\sqrt{x^2-1}} = \frac{-2\csc^{-1}x}{|x|\sqrt{x^2-1}}.$$

Example 3.7.3

Some miscellaneous examples of derivative computations using these follow:

- $\dfrac{d}{dx}\left[\dfrac{\cos^{-1}x}{x}\right] = \dfrac{x \cdot \frac{d\cos^{-1}x}{dx} - \cos^{-1}x \cdot \frac{dx}{dx}}{(x)^2} = \dfrac{\frac{-x}{\sqrt{1-x^2}} - \cos^{-1}x}{x^2} \cdot \dfrac{\sqrt{1-x^2}}{\sqrt{1-x^2}} = \dfrac{-x - \sqrt{1-x^2}\cos^{-1}x}{x^2\sqrt{1-x^2}}$

- $\dfrac{d}{dx}\left[x\sec^{-1}x\right] = x \cdot \dfrac{d}{dx}\left[\sec^{-1}x\right] + \sec^{-1}x \cdot \dfrac{dx}{dx} = x \cdot \dfrac{1}{|x|\sqrt{x^2-1}} + \sec^{-1}x.$

    If we happen to know $x > 0$, this simplifies to $\dfrac{x}{x\sqrt{x^2-1}} + \sec^{-1}x = \dfrac{1}{\sqrt{x^2-1}} + \sec^{-1}x.$

    If $x < 0$ we instead get $\dfrac{x}{-x\sqrt{x^2-1}} + \sec^{-1}x = \dfrac{-1}{\sqrt{x^2-1}} + \sec^{-1}x.$

- $\dfrac{d}{dx}\left[\tan^{-1}(\tan x)\right] = \dfrac{1}{(\tan x)^2 + 1} \cdot \dfrac{d}{dx}\tan x = \dfrac{1}{\tan^2 x + 1} \cdot \sec^2 x = \dfrac{1}{\sec^2 x}\sec^2 x = 1.$

    A careful look at this particular function, especially in light of our later development, would reveal that for each interval on which $\tan x$ is defined, there exists an integer n so that, $\tan^{-1}(\tan x) = x + n\pi$, so it is reasonable that this derivative above should be 1 (since $n\pi$ will be constant on such an interval). This sort of thing occurs on occasion when dealing with functions and their derivatives. Indeed sometimes it is the calculus considerations which first lead us to such simplifications of the functions.

- $\dfrac{d}{dx}\sqrt{x + \tan^{-1}2x} = \dfrac{1}{2\sqrt{x + \tan^{-1}2x}} \cdot \dfrac{d}{dx}\left[x + \tan^{-1}2x\right] = \dfrac{1}{2\sqrt{x + \tan^{-1}2x}} \cdot \left[1 + \dfrac{1}{(2x)^2 + 1} \cdot \dfrac{d}{dx}(2x)\right]$

    $= \dfrac{1}{2\sqrt{x + \tan^{-1}2x}} \cdot \left[1 + \dfrac{2}{4x^2 + 1}\right] = \dfrac{4x^2 + 1 + 2}{2(4x^2 + 1)\sqrt{x + \tan^{-1}2x}} = \dfrac{4x^2 + 3}{2(4x^2 + 1)\sqrt{x + \tan^{-1}2x}}.$

- $\dfrac{d}{dt}\sec^{-1}\left(\dfrac{1}{t}\right) = \dfrac{1}{\left|\frac{1}{t}\right|\sqrt{\left(\frac{1}{t}\right)^2 - 1}} \cdot \dfrac{d}{dt}\left(\dfrac{1}{t}\right) = \dfrac{1}{\left|\frac{1}{t}\right|\sqrt{\frac{1}{t^2} - 1}} \cdot \dfrac{-1}{t^2}.$

Now we will note that $t^2 = |t|^2$, and $|1/t| = 1/|t|$, and so we can simplify this somewhat, continuing:

$$= \frac{1}{\frac{1}{|t|}\sqrt{\left(\frac{1}{t}\right)^2 - 1}} \cdot \frac{-1}{|t|^2} = \frac{-1}{|t|\sqrt{\frac{1}{t^2} - 1}} = \frac{-1}{\sqrt{t^2}\sqrt{\frac{1}{t^2} - 1}} = \frac{-1}{\sqrt{1-t^2}}.$$

That derivative may seem familiar, because it is the same as the derivative of $\cos^{-1} t$ (with respect to t). This is no accident, because (as we will see) we define $\sec^{-1} x$ to be that angle $\theta \in [0, \pi]$ (excluding $\pi/2$ but that turns out not to be a problem here) such that $x = \sec\theta$. Then $1/x = \cos\theta$. We also define (later) $\cos^{-1} z$ to be that angle $\theta \in [0, \pi]$ such that $\cos\theta = z$, so these two arctrigonometric functions, arcsecant and arccosine, both return similar angles in their ranges. Moreover, if $\theta \in [0, \pi]$ and $\sec\theta = \frac{1}{t}$, then $\cos\theta = t$; that is, if $\theta = \sec^{-1}\left(\frac{1}{t}\right)$, then $\theta = \cos^{-1} t$. If we had noticed that previously, we could have made shorter work of this derivative:

$$\frac{d}{dt}\sec^{-1}\left(\frac{1}{t}\right) = \frac{d}{dt}\cos^{-1} t = \frac{-1}{\sqrt{1-t^2}}.$$

- $\frac{d}{dz}\left[\frac{1}{\cot^{-1} z}\right] = \frac{-1}{\left(\cot^{-1} z\right)^2} \cdot \frac{d}{dz}\cot^{-1} z = \frac{-1}{\left(\cot^{-1} z\right)^2} \cdot \frac{-1}{z^2 + 1} = \frac{1}{(z^2 + 1)\left(\cot^{-1} z\right)^2}.$

Compare the last two derivative examples above, where in the first we could take advantage of a reciprocal relationship, but not in the second. This is because with the arctrigonometric, i.e., inverse trigonometric, functions the reciprocal relationships occur with the inputs; with trigonometric functions the reciprocal relationships occur in the outputs. For instance, $\sec x = 1/\cos x$, while $\sec^{-1} x = \cos^{-1}(1/x)$. That is because the arcsecant inputs a secant of an angle, which is the reciprocal of what the arccosine inputs, namely the cosine of an angle (both arctrigonometric functions returning that same angle). We could use these facts to simplify $\frac{d}{dx}\left[\frac{1}{\sec x}\right] = \frac{d}{dx}\cos x = -\sin x$ instead of using a power or chain rule, while we used $\frac{d}{dt}\sec^{-1}\left(\frac{1}{t}\right) = \frac{d}{dt}\cos^{-1} t$ in one of the above examples here.

The previous general derivative rules (Chain, Power, Sum, Product, Quotient and trigonometric rules) mix easily with the rules for the arctrigonometric functions, and so one who knows the previous rules need only have a list of these new derivative rules to compute derivatives involving both sets of rules. Moreover, one can then begin to notice some peculiarities of these functions which emerge from the derivative computations. Many of these peculiarities are then predictable, at least in retrospect (tautologically!), when one completely understands the geometric and algebraic natures of the arctrigonometric functions. We now delve into those natures in turn, and note some of their peculiarities. With greater understanding of the functions come more opportunities to take advantage of them preemptively, just as one might notice immediately (before invoking four Chain Rule computations) that $\frac{d}{dx}\left(\cos^2 9x + \sin^2 9x\right) = 0$ because, after all, $\frac{d}{dx}\left(\cos^2 9x + \sin^2 9x\right) = \frac{d(1)}{dx} = 0$.

3.7.3 The Arcsine Function and Its Derivative

We begin our development with the inverse trigonometric functions which can be found on scientific calculators: arcsine, arccosine and arctangent (i.e., $\sin^{-1}$, $\cos^{-1}$ and $\tan^{-1}$ respectively). Though these are the most intuitive in their derivations, in fact calculus applications find the most use for the arcsine, arctangent and arcsecant, so the derivation of the arcsecant

3.7. ARCTRIGONOMETRIC FUNCTIONS AND THEIR DERIVATIVES

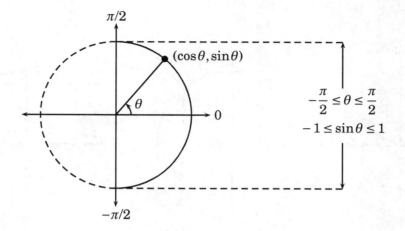

Figure 3.31: Diagram showing that part of the unit circle used to construct the arcsine function. For all $x \in [-1,1]$, there exists a unique $\theta \in [-\pi/2, \pi/2]$ so that $\sin\theta = x$.

will also be presented in full. For completeness, we will include results for arccosecant and arccotangent.

We begin with the arcsine function. We are thus interested in finding a function which we will call $\sin^{-1}x$, or $\arcsin x$, so that $\sin^{-1}x$ returns an angle θ so that $\sin\theta = x$. Now the range (of outputs) of $\sin\theta$ is $[-1,1]$, so our function $\sin^{-1}x$ should be able to input any such $x \in [-1,1]$ and return an angle θ whose sine is x. As shown in Figure 3.31, all such outputs $\sin\theta \in [-1,1]$ are achieved if we restrict the input of the sine function to $\theta \in \left[-\frac{\pi}{2}, \frac{\pi}{2}\right]$. Moreover, for each $x \in [-1,1]$ there exists a unique $\theta \in \left[-\frac{\pi}{2}, \frac{\pi}{2}\right]$ such that $\sin\theta = x$. Thus we make the following definition:

Definition 3.7.1 *For every $x \in [-1,1]$, define*

$$\sin^{-1}x = \text{"that } \theta \in \left[-\frac{\pi}{2}, \frac{\pi}{2}\right] \text{ such that } \sin\theta = x.\text{"} \tag{3.77}$$

Note that $\sin(\sin^{-1}x) = x$, because in that computation we are taking the sine of an angle—albeit inside of $\left[-\frac{\pi}{2}, \frac{\pi}{2}\right]$—whose sine is x, and so its sine is, naturally, x. Using this fact, we can note that $y = \sin^{-1}x \implies \sin y = \sin(\sin^{-1}x) \iff \sin y = x$, i.e.,[71]

$$y = \sin^{-1}x \implies \sin y = x. \tag{3.78}$$

The graph of $y = \sin^{-1}x$ is thus a subset of the graph of $\sin y = x$. This is shown in Figure 3.32a, page 342.

Using (3.78), we can now derive the derivative of the arcsine function. Eventually we will

---

[71]However, $\sin^{-1}(\sin x)$ is equal to x if (and only if) $x \in \left[-\frac{\pi}{2}, \frac{\pi}{2}\right]$, though $\sin^{-1}(\sin x)$ will at least share the same reference angle as x.

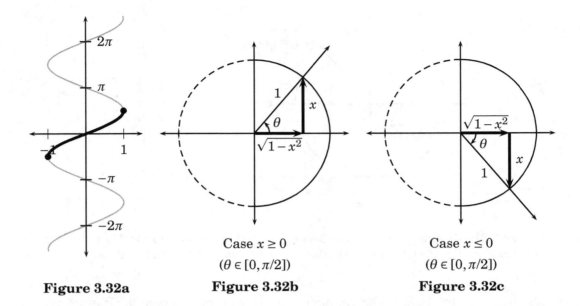

Figure 3.32: Figure 3.32a shows a partial graph of $x = \sin y$ in gray, with the graph of $y = \sin^{-1} x$, i.e., that of $\{(x,y) \mid x = \sin y, y \in \left[-\frac{\pi}{2}, \frac{\pi}{2}\right]\}$, in black. Figure 3.32b shows the triangle used in the final computation for $\frac{d}{dx}\sin^{-1} x$ for the case $x \geq 0$, which corresponds to the (angle-valued) output of the arcsine being in the first quadrant. Figure 3.32c shows the same but for $x \leq 0$ and the output of the arcsine being an angle in the fourth quadrant.

need to refer to Figure 3.32, but the initial computations are Chain Rules in nature:

$$\begin{aligned} y = \sin^{-1} x &\implies \sin y = \sin(\sin^{-1} x) \\ &\implies \sin y = x \\ &\implies \frac{d}{dx}(\sin y) = \frac{d}{dx}(x) \\ &\implies \cos y \cdot \frac{dy}{dx} = 1 \\ &\implies \frac{dy}{dx} = \frac{1}{\cos y}. \end{aligned}$$

At this point we need to rewrite this derivative in terms of x using $y = \sin^{-1} x$:

$$y = \sin^{-1} x \implies \frac{dy}{dx} = \frac{1}{\cos(\sin^{-1} x)}. \tag{3.79}$$

Now recall that $\sin^{-1} x \in \left[-\frac{\pi}{2}, \frac{\pi}{2}\right]$, and in fact $\sin^{-1} x$ is that angle $\theta \in \left[-\frac{\pi}{2}, \frac{\pi}{2}\right]$ so that $\sin\theta = x$. There are two basic cases for this θ: that θ is in Quadrant I or Quadrant IV. (If θ is axial, then the analysis of either case will still work.) These cases are given in Figure 3.32b,c.

It is important to construct angles θ with the proper representative triangles: Here the sine of θ must equal x, keeping in mind the signs of vertical (or horizontal) displacements—either positive or negative—represented by the "triangle legs." The hypotenuse must always be positive (since it is always a distance, or *radius*, and not a displacement), and

3.7. ARCTRIGONOMETRIC FUNCTIONS AND THEIR DERIVATIVES

the quadrants must be correct. The third side is constructed to be consistent with both the Pythagorean Theorem and the quadrant. In both cases, $x \geq 0$ and $x \leq 0$, we see that the third side of the representative triangle is $\sqrt{1-x^2}$ since it is a nonnegative—in fact usually rightward—displacement in the horizontal direction. From this we can read off $\cos(\sin^{-1} x) = \cos\theta = \sqrt{1-x^2}$.

Inserting this information into our derivative computation (3.79) gives us $y = \sin^{-1} x \implies \frac{dy}{dx} = 1/\cos(\sin^{-1} x) = 1/\sqrt{1-x^2}$. We give this result in summary form, and then give the Chain Rule version:

$$\boxed{\frac{d}{dx}\sin^{-1} x = \frac{1}{\sqrt{1-x^2}}}, \qquad \boxed{\frac{d}{dx}\sin^{-1} u = \frac{1}{\sqrt{1-u^2}} \cdot \frac{du}{dx}}. \qquad (3.80)$$

As usual, the latter can be decomposed into

$$\frac{d}{dx}\sin^{-1} u = \frac{d}{du}\sin^{-1} u \cdot \frac{du}{dx} = \frac{1}{\sqrt{1-u^2}} \cdot \frac{du}{dx}.$$

Note that (3.80) only makes sense for $x \in (-1, 1)$, and that $\frac{d}{dx}\sin^{-1} x = \frac{1}{\sqrt{1-x^2}} \longrightarrow \infty$ as $x \to 1^-$ and as $x \to -1^+$. This is borne out by the graph of $y = \sin^{-1} x$ given in Figure 3.32a, page 342. That graph also reflects how $\frac{d}{dx}\sin^{-1} x = \frac{1}{\sqrt{1-x^2}} > 0$ for all $x \in (-1, 1)$, that is, how $\sin^{-1} x$ is increasing in that interval (and in fact, in the whole domain $x \in [-1, 1]$).

3.7.4 The Arccosine and Its Derivative

The development for the arccosine function mirrors that of the arcsine, except that we will take the range of the arccosine function to contain angles in Quadrants I and II, specifically $\theta \in [0, \pi]$. That is because such angles form exactly the kind of set we need so that the cosine is a one-to-one function with outputs covering the whole range $[-1, 1]$ of the cosine function.

Definition 3.7.2 *For every $x \in [-1, 1]$, define*

$$\cos^{-1} x = \text{``that } \theta \in [0, \pi] \text{ such that } \cos\theta = x.\text{''} \qquad (3.81)$$

The arccosine function is given (in bold) in Figure 3.33a, page 344. The computation of the derivative of $\cos^{-1} x$ is similar to that for the arcsine, eventually referring to an illustration of two cases, namely $\theta = \cos^{-1} x$ terminating in Quadrant I and $\theta = \cos^{-1} x$ terminating in Quadrant II.

$$\begin{aligned}
y = \cos^{-1} x &\implies \cos y = x \\
&\implies \frac{d}{dx}(\cos y) = \frac{d}{dx}(x) \\
&\implies -\sin y \frac{dy}{dx} = 1 \\
&\implies \frac{dy}{dx} = \frac{1}{-\sin y} = \frac{1}{-\sin(\cos^{-1} x)}.
\end{aligned}$$

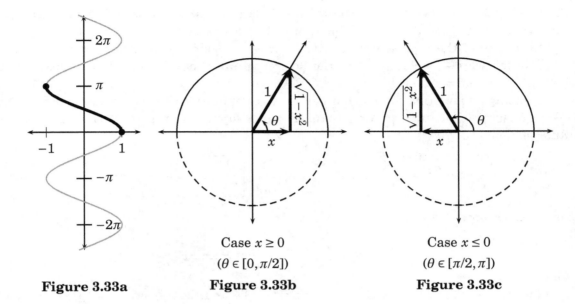

Figure 3.33a	Figure 3.33b	Figure 3.33c
	Case $x \geq 0$	Case $x \leq 0$
	($\theta \in [0, \pi/2]$)	($\theta \in [\pi/2, \pi]$)

Figure 3.33: Partial graph of $x = \cos y$ in gray, with the graph of $y = \cos^{-1} x$, i.e., that of the set $\{(x,y) \mid x = \cos y, \ y \in [0,\pi]\}$, in black.

From Figures 3.33b,c, we see that the sine of the angle $\theta = \cos^{-1} x$ is $\sqrt{1-x^2}$ in each of the two quadrants in which θ terminates. Continuing our earlier computation, we have

$$y = \cos^{-1} x \implies \frac{dy}{dx} = \frac{-1}{\sin(\cos^{-1} x)} = \frac{-1}{\sqrt{1-x^2}}.$$

Collecting this with its Chain Rule version, we have

$$\boxed{\frac{d}{dx} \cos^{-1} x = \frac{-1}{\sqrt{1-x^2}}}, \qquad \boxed{\frac{d}{dx} \cos^{-1} u = \frac{-1}{\sqrt{1-u^2}} \cdot \frac{du}{dx}}. \qquad (3.82)$$

The derivative of the arccosine function is negative for $x \in (-1, 1)$, and so the function itself is decreasing. Also, the derivative approaches $-\infty$ as $x \to 1^-$ and as $x \to -1^+$.

Another derivation for the arccosine function relies upon the fact that, with these definitions of $\sin^{-1} x$ and $\cos^{-1} x$, we have an identity

$$\sin^{-1} x + \cos^{-1} x = \frac{\pi}{2}. \qquad (3.83)$$

This is verified easily when $0 < x < 1$ because it reflects that the acute angles of a right triangle sum to $\pi/2$. Some checking (which we omit here) shows that (3.83) also holds for other cases in which $x \in [-1, 1]$. With (3.83), we can take derivatives of both sides and easily get that $\frac{d}{dx} \cos^{-1} x = -\frac{d}{dx} \sin^{-1} x$, which reflects why the derivatives of $\cos^{-1} x$ and $\sin^{-1} x$ are the same except for the sign.

3.7. ARCTRIGONOMETRIC FUNCTIONS AND THEIR DERIVATIVES

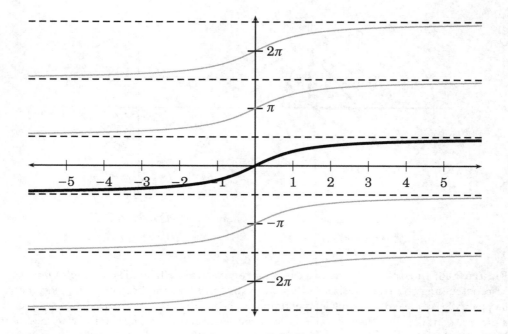

Figure 3.34: Partial graph of $x = \tan y$ in gray, with that part of the graph which represents $y = \tan^{-1} x$, i.e., that of $\{(x,y) \mid x = \tan y, y \in \left(-\frac{\pi}{2}, \frac{\pi}{2}\right)]\}$, in black.

3.7.5 The Arctangent Function and Its Derivative

Definition 3.7.3 *For every $x \in \mathbb{R}$, define*

$$y = \tan^{-1} x = \text{"that } \theta \in \left(-\frac{\pi}{2}, \frac{\pi}{2}\right) \text{ such that } \tan \theta = x.\text{"} \qquad (3.84)$$

Thus the range we will use for the arctangent function is $\theta \in \left(-\frac{\pi}{2}, \frac{\pi}{2}\right)$. The graph of $y = \tan^{-1} x$ is a subset of the graph of $x = \tan y$, as illustrated in Figure 3.34.

This function $y = \tan^{-1} x$ has some interesting features. For instance, the domain of $\tan^{-1} x$ is all of $\mathbb{R}$, so it makes sense to consider its behavior "at infinity." From the graph we can see that

$$x \longrightarrow \infty \implies \tan^{-1} x \longrightarrow \frac{\pi}{2}^{-}, \qquad (3.85)$$

$$x \longrightarrow -\infty \implies \tan^{-1} x \longrightarrow \left(\frac{-\pi}{2}\right)^{+}. \qquad (3.86)$$

These become important in later sections, as we consider more limits and discuss improper integrals.

In finding the derivative of the arctangent function, we proceed as in the earlier derivations, with Figure 3.35, page 346, giving the relevant triangles. Note how we construct the triangles, where $\theta = \tan^{-1} x$. In doing so, we must ensure that the quadrants and signs of the (vertical and horizontal) displacements are consistent with both the Pythagorean Theorem

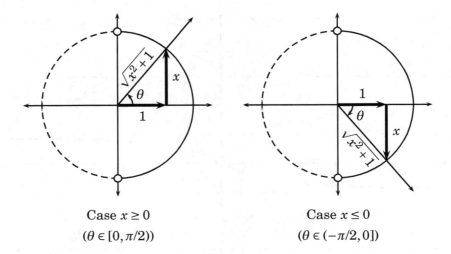

Case $x \geq 0$
($\theta \in [0, \pi/2)$)

Case $x \leq 0$
($\theta \in (-\pi/2, 0]$)

Figure 3.35: Illustration of the two cases for representative triangles for angles $\theta = \tan^{-1} x$. In constructing the triangles, note that we want $\tan \theta = x$, the horizontal displacement, to be positive (we picked 1), and the hypotenuse to be positive (as is *always* the case, since it is a *distance*, unlike the other two sides which represent horizontal or vertical *displacements*). Note also that these triangles reside within circles of radius $\sqrt{x^2+1}$, which are therefore not generally unit circles.

and the quadrant of the terminal side of $\theta = \tan^{-1} x$. Now we proceed with the derivative computation:

$$y = \tan^{-1} x \implies \tan y = x$$
$$\implies \frac{d}{dx}(\tan y) = \frac{d}{dx}(x)$$
$$\implies \sec^2 y \cdot \frac{dy}{dx} = 1$$
$$\implies \frac{dy}{dx} = \cos^2 y$$
$$\implies \frac{dy}{dx} = (\cos y)^2 = \left(\frac{1}{\sqrt{x^2+1}}\right)^2 = \frac{1}{x^2+1}.$$

Summarizing this, and the Chain Rule version, we have

$$\boxed{\frac{d}{dx}\tan^{-1} x = \frac{1}{x^2+1}}, \qquad \boxed{\frac{d}{dx}\tan^{-1} u = \frac{1}{u^2+1} \cdot \frac{du}{dx}}. \qquad (3.87)$$

Note that $\frac{d}{dx}\tan^{-1} x = \frac{1}{x^2+1} > 0$ for all $x \in \mathbb{R}$, and so the arctangent function is increasing everywhere (see Figure 3.34, page 345). Moreover, the slope of the graph becomes gentler as $|x|$ grows:

$$\lim_{x \to \pm\infty}\left[\frac{d}{dx}\tan^{-1} x\right] = \lim_{x \to \pm\infty} \frac{1}{x^2+1} \overset{1/\infty}{=\!=\!=} 0.$$

3.7. ARCTRIGONOMETRIC FUNCTIONS AND THEIR DERIVATIVES

This reflects the behavior of the slope of $y = \tan^{-1} x$ as the graph approaches its horizontal asymptotes (Figure 3.34, page 345). These become important in later sections, as we consider more limits and discuss improper integrals.

3.7.6 The Arcsecant Function and Its Derivative

We define the arcsecant to be consistent with the arccosine. We begin with the fact that
$$\sec\theta = x \iff \cos\theta = \frac{1}{x}.$$

Since the range of $\cos\theta$ is $|\cos\theta| \leq 1$, it follows that the range of secant is $|\sec\theta| = \left|\frac{1}{\cos\theta}\right| \geq 1$. We will define the arcsecant so that its domain (input) is the same as the output of $\sec\theta$, and that its range is the same as that of $\cos^{-1}\left(\frac{1}{x}\right)$ (almost that of arccosine, except that $\frac{1}{x}$ is never zero):

Definition 3.7.4 *For $x \in (-\infty, -1] \cup [1, \infty)$, define*
$$\sec^{-1} x = \text{"that } \theta \in \left[0, \frac{\pi}{2}\right) \cup \left(\frac{\pi}{2}, \pi\right] \text{ such that } \sec\theta = x\text{."} \tag{3.88}$$

There are some complications that arise in the derivation of the arcsecant function's derivative. That the derivation is not as straightforward as the others' is not surprising when we consider that the arcsecant function is not even continuous on its domain. Indeed, when we pick enough of the graph of $x = \sec y$ to cover all possible values for x but not so much that y is no longer a function of x, we are forced to take two separate branches of the graph $x = \sec y$. In Figure 3.36, page 348, that part of the graph of $x = \sec y$ which defines $y = \sec^{-1} x$ is highlighted. Now we derive $\frac{d}{dx}\sec^{-1} x$, beginning with the following computation:

$$\sec y = x \implies \frac{d}{dx}(\sec y) = \frac{d}{dx}(x)$$
$$\implies \sec y \tan y \frac{dy}{dx} = 1$$
$$\implies \frac{dy}{dx} = \cos y \cot y = \cos\left(\sec^{-1} x\right) \cot\left(\sec^{-1} x\right).$$

Referring to Figure 3.37, page 348, we see that the hypotenuse—which always must be positive—is x when x is positive, and $-x$ when x is negative. In both cases, we can summarize the hypotenuse as represented by the quantity $|x|$. Thus we can continue the implications to get the following (noting dy/dx DNE at $x = \pm 1$):

$$y = \sec^{-1} x \implies \frac{dy}{dx} = \cos(\sec^{-1} x)\cot(\sec^{-1} x) = \begin{cases} \dfrac{1}{x} \cdot \dfrac{1}{\sqrt{x^2-1}}, & \text{if } x > 1, \\ \dfrac{-1}{-x} \cdot \dfrac{-1}{\sqrt{x^2-1}}, & \text{if } x < -1. \end{cases}$$

Now we summarize these, using the fact that $x = |x|$ in the expression for $x \geq 1$, while $-x = |x|$ as well in the expression for $x \leq -1$. We also include the Chain Rule version:

$$\boxed{\frac{d}{dx}\sec^{-1} x = \frac{1}{|x|\sqrt{x^2-1}}}, \qquad \boxed{\frac{d}{dx}\sec^{-1} u = \frac{1}{|u|\sqrt{u^2-1}} \cdot \frac{du}{dx}}. \tag{3.89}$$

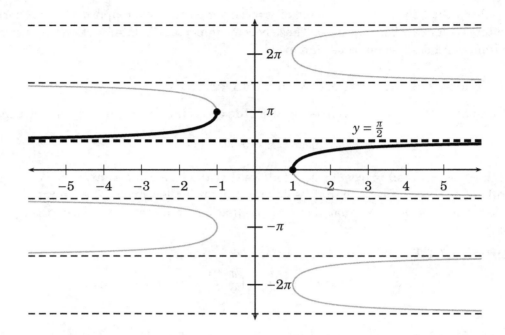

Figure 3.36: Partial graph of $x = \sec y$ in gray, with that part of the graph that represents $y = \sec^{-1} x$, i.e., that of $\{(x,y) \mid x = \sec y, \quad y \in [0, \frac{\pi}{2}) \cup (\frac{\pi}{2}, \pi]\}$, in black. Note $\sec^{-1} x$ is increasing on any *interval* on which it is defined.

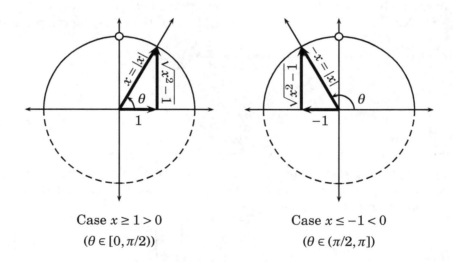

Case $x \geq 1 > 0$ Case $x \leq -1 < 0$

($\theta \in [0, \pi/2)$)   ($\theta \in (\pi/2, \pi]$)

Figure 3.37: Illustration of the two cases for representative triangles for angles $\theta = \sec^{-1} x \in [0, \pi] - \{\frac{\pi}{2}\}$. We draw the triangles so that $\sec\theta = x$, i.e., $\cos\theta = \frac{1}{x}$. In doing so, however, we must be sure that the hypotenuse is always positive, and that the other two sides have appropriate signs. In particular, we need the hypotenuse to be x when $x \geq 1$ and $-x$ when $x \leq -1$. In both cases the hypotenuse can be written $|x|$. Also, in both cases the vertical side represents a positive displacement of $\sqrt{x^2 - 1}$.

3.7. ARCTRIGONOMETRIC FUNCTIONS AND THEIR DERIVATIVES

Notice that (3.89) implies that $y = \sec^{-1} x$ is increasing wherever it is differentiable, since its derivative is positive wherever defined. Notice also the limiting behavior of $\sec^{-1} x$ as $x \to \infty$ and as $x \to -\infty$ (see Figure 3.36, page 348):

$$x \to \infty \implies \sec^{-1} x \to \frac{\pi^-}{2}, \quad \frac{d}{dx}\sec^{-1} x \to 0; \qquad (3.90)$$

$$x \to -\infty \implies \sec^{-1} x \to \frac{\pi^+}{2}, \quad \frac{d}{dx}\sec^{-1} x \to 0. \qquad (3.91)$$

A closely related derivative, which is left as an exercise, is the following (with the Chain Rule form included):

$$\boxed{\frac{d}{dx}\sec^{-1}|x| = \frac{1}{x\sqrt{x^2-1}},} \qquad \boxed{\frac{d}{dx}\sec^{-1}|u| = \frac{1}{u\sqrt{u^2-1}} \cdot \frac{du}{dx}.} \qquad (3.92)$$

Equation (3.92) can be proved using the Chain Rule and the fact that $|x|$ is the same as x for $x > 0$, and $-x$ for $x < 0$. Both cases, $x > 0$ and $x < 0$ (actually $x \geq 1$, $x \leq 1$ to be precise), should be proved separately. The Chain Rule then gives us the second equation in (3.92). These forms are preferred to (3.89) when we compute antiderivatives in later sections.

3.7.7 Geometric and Algebraic Manipulations Aiding Computations

As explained in the context of the derivative computation $\frac{d}{dt}\sec^{-1}(1/t) = \frac{d}{dt}\cos^{-1} t$, on page 339, and in light of the angles the various arctrigonometric functions output, the following should be clear:

$$\cos^{-1}\left(\frac{1}{x}\right) = \sec^{-1} x, \quad \sec^{-1}\left(\frac{1}{x}\right) = \cos^{-1} x, \qquad (3.93)$$

$$\sin^{-1}\left(\frac{1}{x}\right) = \csc^{-1} x, \quad \csc^{-1}\left(\frac{1}{x}\right) = \sin^{-1} x, \qquad (3.94)$$

$$\tan^{-1}\left(\frac{1}{x}\right) = \cot^{-1} x, \quad \cot^{-1}\left(\frac{1}{x}\right) = \tan^{-1} x. \qquad (3.95)$$

This can be useful if the argument of the arctrigonometric function is computationally more difficult to deal with than its reciprocal.

Example 3.7.4

Use a reciprocal relationship to compute $\frac{d}{dx}\tan^{-1}\frac{1}{x^2}$:

Solution: $\frac{d}{dx}\tan^{-1}\frac{1}{x^2} \stackrel{\text{ALG}}{=} \frac{d}{dx}\cot^{-1} x^2 = \frac{-1}{(x^2)^2 + 1} \cdot \frac{dx^2}{dx} = \frac{-2x}{x^4 + 1}$. Alternatively, we can draw in the plane the angle $\theta = \tan^{-1}\frac{1}{x^2}$ and observe that one of the other trigonometric functions of θ is simpler:

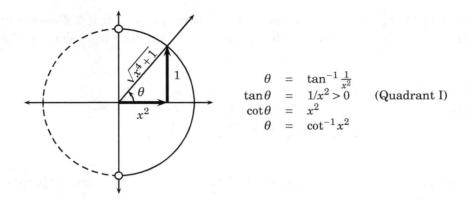

From the picture we see that $\tan^{-1}\frac{1}{x^2} = \theta = \cot^{-1} x^2$, and so we can use this reasoning, rather than the reciprocal relationships (3.93)–(3.95), to perform the "ALG" step in the derivative computation above. Note how we filled in the length of the hypotenuse based on the Pythagorean Theorem, though it was not necessary for this particular example (see remarks that follow). Note also that since $x \neq 0$, we have $1/x^2 > 0$, so the angle θ must be in the first quadrant.

Because the reciprocal relationships (3.93)–(3.95) are fairly straightforward, for the previous example the graphical method of deriving $\tan^{-1}\frac{1}{x^2} = \cot^{-1} x^2$ is unnecessarily involved. However, we present it here for this simpler problem to introduce the method in the context of actual derivative computations.[72]

When using a graphical method to approach such a problem, the basic approach is, more or less, the following:

(a) Draw representative angles θ (representing the arctrigonometric function in question).

(b) Draw triangles in the appropriate quadrants with the given trigonometric function value for the angle.

(c) Fill in the missing side length (if the hypotenuse) or displacement (if one of the other "legs") using the Pythagorean Theorem but taking care that all "legs" are labelled with the correct signs.

(d) Read off some other, computationally simpler, trigonometric function of the angle.

(e) Rewrite the angle as either the arctrigonometric function of the simpler expression in (d), or some related angle.

(f) Compute the derivative using this simpler expression.

Next we consider this approach for an example in which the reciprocal relations are insufficient.

---

[72]Essentially this technique of drawing representative triangles and angles was already used in our derivations of the derivative formulas for the arctrigonometric functions. We re-introduce the geometric perspective repeatedly because of its versatility in application.

Example 3.7.5

Compute $\dfrac{d}{dx}\tan^{-1}\left[\dfrac{x}{\sqrt{9-x^2}}\right]$.

<u>Solution</u>: Here we draw triangles for the two possible cases, $x > 0$ and $x < 0$, i.e., where θ has positive tangent and where θ has negative tangent, respectively. The zero tangent case can be absorbed into either. (To be more precise, the cases can be $x \in [0,3)$ and $x \in (-3,0)$, for example.) In both diagrams, we have $\theta = \tan^{-1}\left[\dfrac{x}{\sqrt{9-x^2}}\right]$:

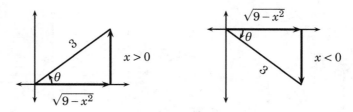

Note how the signs associated with the three sides of the triangle are consistent with the definition of θ, the sign of $\tan\theta$, and the geometry of the plane. Also in both cases we note that $\sin\theta = \dfrac{x}{3}$, and in fact these angles θ are in the correct intervals for the range (output) of arcsine. Thus we can write $\theta = \sin^{-1}\dfrac{x}{3}$. The differentiation problem then becomes

$$\dfrac{d}{dx}\tan^{-1}\left[\dfrac{x}{\sqrt{9-x^2}}\right] \stackrel{\text{GEOM}}{=\!=\!=} \dfrac{d}{dx}\sin^{-1}\left(\dfrac{x}{3}\right) = \dfrac{1}{\sqrt{1-\left(\frac{x}{3}\right)^2}} \cdot \dfrac{d}{dx}\left(\dfrac{x}{3}\right) = \dfrac{1}{\sqrt{1-\frac{x^2}{9}}} \cdot \dfrac{1}{3}$$

$$= \dfrac{1}{3\sqrt{1-\frac{x^2}{9}}} = \dfrac{1}{\sqrt{9}\sqrt{1-\frac{x^2}{9}}} = \dfrac{1}{\sqrt{9\left(1-\frac{x^2}{9}\right)}} = \dfrac{1}{\sqrt{9-x^2}}.$$

The calculus was finished by the end of the first line, but the algebraic simplifications are worthwhile. With little practice, one could easily have the final result one step after the end of the first line.

It should be pointed out that this geometric method is not always useful. For instance, it requires us to find the third side of the "triangle," and on inspection it is very possible that none of the trigonometric functions of the angle θ will be substantially simpler than the original. Indeed, it is usually the case the others will be more difficult. However, perhaps because these functions often arise from geometric arguments regarding some physical or abstract analysis, the technique is useful enough to be worth developing.

Even when the technique does yield a simpler expression for a trigonometric function for θ, there can still be complications from the quadrants. Most complications, in fact, turn out not to matter, as for instance two -1 factors cancel. But care must be taken to ensure proper representation of the angle and the sides of the triangles.

352   CHAPTER 3. THE DERIVATIVE

Example 3.7.6

Compute $\dfrac{d}{dx}\sin^{-1}\left[\dfrac{\sqrt{x^2-25}}{x}\right]$.

<u>Solution</u>: Here we draw angles $\theta \in [-\pi/2, \pi/2]$ such that $\sin\theta = \dfrac{\sqrt{x^2-25}}{x}$. Recall that the hypotenuse must be positive. For that reason, the second triangle will require us to add factors of -1 to two of the sides.

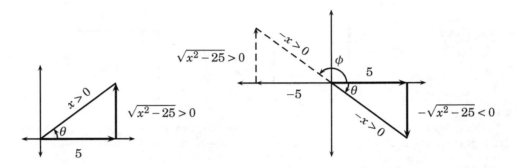

In both cases, the simplest trigonometric functions are in fact the secants of θ. Unfortunately, arcsecant has its range in the first and second quadrants. Note that in the second diagram, the angle $\phi = \sec^{-1}\dfrac{-x}{-5} = \sec^{-1}\dfrac{x}{5}$, and so $\theta = \phi - \pi$. This will be needed for the second computation:

$$x>0: \quad \frac{d}{dx}\sin^{-1}\left[\frac{\sqrt{x^2-25}}{x}\right] \stackrel{\text{GEOM}}{=\!=\!=} \frac{d}{dx}\sec^{-1}\left(\frac{x}{5}\right) = \frac{1}{\left|\frac{x}{5}\right|\sqrt{\left(\frac{x}{5}\right)^2-1}}\cdot\frac{d}{dx}\left(\frac{x}{5}\right)$$

$$= \frac{5}{|x|\sqrt{\frac{x^2}{25}-1}}\cdot\frac{1}{5} = \frac{5}{|x|\cdot 5\sqrt{\frac{x^2}{25}-1}} = \frac{5}{|x|\sqrt{x^2-25}}.$$

$$x<0: \quad \frac{d}{dx}\sin^{-1}\left[\frac{\sqrt{x^2-25}}{x}\right] \stackrel{\text{GEOM}}{=\!=\!=} \frac{d}{dx}\left[\sec^{-1}\left(\frac{x}{5}\right)-\pi\right] = \frac{1}{\left|\frac{x}{5}\right|\sqrt{\left(\frac{x}{5}\right)^2-1}}\cdot\frac{d}{dx}\left(\frac{x}{5}\right)-0$$

$$= \frac{5}{|x|\sqrt{\frac{x^2}{25}-1}}\cdot\frac{1}{5} = \frac{5}{|x|\cdot 5\sqrt{\frac{x^2}{25}-1}} = \frac{5}{|x|\sqrt{x^2-25}}.$$

$$\therefore \frac{d}{dx}\sin^{-1}\left[\frac{\sqrt{x^2-25}}{x}\right] = \frac{5}{|x|\sqrt{x^2-25}} \quad \text{(all cases)}.$$

We see both computations gave the same expression for the derivative, but it was not obvious from the outset. While there were geometric complications, and some algebraic cleverness needed, it is arguably easier (particularly with practice) than a direct approach, shown here for comparison:

$$\frac{d}{dx}\sin^{-1}\left[\frac{\sqrt{x^2-25}}{x}\right] = \frac{1}{\sqrt{1-\left[\frac{\sqrt{x^2-25}}{x}\right]^2}}\cdot\frac{d}{dx}\left[\frac{\sqrt{x^2-25}}{x}\right] = \frac{1}{\sqrt{1-\frac{x^2-25}{x^2}}}\cdot\frac{x\cdot\frac{d}{dx}\sqrt{x^2-25}-\sqrt{x^2-25}\cdot\frac{dx}{dx}}{(x)^2}$$

$$= \frac{1}{\sqrt{\frac{x^2-(x^2-25)}{x^2}}}\cdot\frac{x\cdot\frac{1}{2\sqrt{x^2-25}}\cdot\frac{d}{dx}(x^2-25)-\sqrt{x^2-25}}{x^2} = \frac{1}{\sqrt{\frac{25}{x^2}}}\cdot\frac{x\cdot\frac{2x}{2\sqrt{x^2-25}}-\sqrt{x^2-25}}{x^2}\cdot\frac{\sqrt{x^2-25}}{\sqrt{x^2-25}}$$

$$= \sqrt{\frac{x^2}{25}}\cdot\frac{x^2-(x^2-25)}{x^2\sqrt{x^2-25}} = \frac{|x|}{5}\cdot\frac{25}{|x|^2\sqrt{x^2-25}} = \frac{5}{|x|\sqrt{x^2-25}}.$$

3.7. ARCTRIGONOMETRIC FUNCTIONS AND THEIR DERIVATIVES

Note the repeated use of the Chain Rule and the common algebraic techniques, in particular using $\sqrt{x^2} = |x|$, $x^2 = |x|^2$, and the technique for simplifying the second fraction, which arose from the Chain Rule.

It should also be pointed out that, if we had made early substitutions for $|x|$, i.e., x when $x > 0$ and $-x$ when $x < 0$, we would have different expressions for the final answer in this example, but they can be reconciled by performing the reverse substitutions strategically. (We did that in our derivation of $\frac{d}{dx} \sec^{-1} x$.)

If one wishes to avoid the geometric approach and persevere through the derivative formulas, that is perfectly valid, and some would argue preferable. However, the geometric approach is closely related to some techniques required in the sequel. There the focus is with the sine, tangent, secant, and their respective arctrigonometric functions, and in fact a representative triangle in the first quadrant is enough, except for the important case in which the secant is used for the substitution. In that case, both first and second quadrant cases must be considered separately.

The geometric method can also give an efficient method in many applied problems in which an expression in one trigonometric function can be written in terms of another.

3.7.8 Alternative Definitions of Arcsecant, Arccosecant, Arccotangent

Here we point out, with little elaboration, that some texts define, for instance, $\sec^{-1} x$ to be that $\theta \in [0, \frac{\pi}{2}) \cup [\pi, \frac{3\pi}{2})$ so that $\sec \theta = x$. In other words, θ is chosen from the smallest positive angles found in Quadrants I and III. Such texts might similarly define the arccosecant to output values in Quadrants I and III, perhaps even represented by $\theta \in (-\pi, -\pi/2] \cup (0, \pi/2]$. For that matter, some texts define arccotangent to output values in Quadrants I and II, represented by $\theta \in (0, \pi)$.

While there is general agreement about how to define the arcsine, arccosine, and arctangent functions, there is indeed some disagreement regarding how best to define the arcsecant, arccosecant, and arccotangent. Fortunately these are rarely used, with the arcsecant being somewhat the exception, and so one simply needs to be clear which definition is being used if it becomes an issue.

It is understandable why someone would wish to define these functions arcsecant, arccosecant, and arccotangent these ways. For the first two, their derivatives will no longer contain absolute values.

Now arcsecant is defined in the earlier subsections of this section as outputting angles in Quadrants I and II, in such a way that we have $\sec^{-1} x = \cos^{-1}\left(\frac{1}{x}\right)$, which means we do not have to change the angles when switching between the two functions to describe the same angle (either by its cosine or by its secant, i.e., the reciprocal of its cosine). It is similar with $\csc^{-1}$ and $\sin^{-1}$, and again with $\tan^{-1}$ and $\cot^{-1}$. However, with our definition the derivatives of the arcsecant and arccosecant have the absolute value in the expression, which is a complication absent when one uses these alternative definitions of the arcsecant and arccosecant. Again, the trade-off is that these alternative definitions, as given here and in Exercise 30, have the less natural ranges of outputs—particularly considering one does at times need values consistent with what a calculator or computer programming language would output for the usual functions arcsine, arccosine, and arctangent—but simpler derivatives. At present our approach, consistent with calculators and simpler geometry, seems to be the more popular.

The approach in Problem 30 is, in fact, to then define the arccosecant and arccotangent by requiring that $\csc^{-1} x = \frac{\pi}{2} - \sec^{-1} x$, and $\cot^{-1} x = \frac{\pi}{2} - \tan^{-1} x$. This makes for derivatives consistent with ours, except for the lack of an absolute value in the expression in the arccosecant as well as the arcsecant.

The best justification for our approach, in fact, lies in the ability to use (3.93)–(3.95), page 349, where we can use, for instance, $\sec^{-1} x = \cos^{-1} \frac{1}{x}$. Indeed, this makes calculator exercises in computing these other inverse trigonometric functions simpler (by using the reciprocal function first, and then the appropriate arctrigonometric function). This is useful in part because most calculators do not have keys for arcsecant, arccosecant, or arccotangent (perhaps because of the lack of total agreement).

Even so, the reader should be aware that other authors and texts use different ranges, which are slightly awkward but admittedly do have simpler derivatives.

Exercises

For Exercises 1–12, compute and simplify the given derivative. (Note that $x \geq 0$ is implied for any derivative involving $\sqrt{x}$.)

1. $\frac{d}{dx} \sqrt{\sin^{-1} x}$
2. $\frac{d}{dx} \sqrt{\cos^{-1} x}$
3. $\frac{d}{dx} \sqrt{\tan^{-1} x}$
4. $\frac{d}{dx} \sqrt{\cot^{-1} x}$
5. $\frac{d}{dx} \sqrt{\sec^{-1} x}$
6. $\frac{d}{dx} \sqrt{\csc^{-1} x}$
7. $\frac{d}{dx} \sin^{-1} \sqrt{x}$
8. $\frac{d}{dx} \cos^{-1} \sqrt{x}$
9. $\frac{d}{dx} \tan^{-1} \sqrt{x}$
10. $\frac{d}{dx} \cot^{-1} \sqrt{x}$
11. $\frac{d}{dx} \sec^{-1} \sqrt{x}$
12. $\frac{d}{dx} \csc^{-1} \sqrt{x}$

For Exercises 13–19, compute and simplify the given derivative.

13. $\frac{d}{dx} \left[x \csc^{-1} x \right]$
14. $\frac{d}{dx} \left[\frac{\tan^{-1} x}{x} \right]$
15. $\frac{d}{dx} \left[(x^2 + 1) \tan^{-1} x - x \right]$
16. $\frac{d}{dx} \left[\sin^{-1} \left(\frac{x}{3} \right) \right]$
17. $\frac{d}{dx} \left[\frac{1}{3} \tan^{-1} \left(\frac{x}{3} \right) \right]$
18. $\frac{d}{dx} \left[\sin^{-1} x + \sin x^{-1} + (\sin x)^{-1} \right]$
19. $\frac{d}{dx} \left[25 \sec^{-1} \frac{x}{5} + x \sqrt{25 - x^2} \right]$
20. Compute $\frac{d}{dx} \sec^{-1} \left(\frac{1}{x} \right)$ two ways:

    (a) Directly, using the Chain Rule.
    (b) Rewriting the original function as an arccosine function.
    (c) Also show that the answers are in fact the same. (You may need to consider two cases, x positive and x negative.)

21. Compute $\frac{d}{dx} \left[\sin^{-1} \left(\frac{x}{\sqrt{x^2 + 1}} \right) \right]$ by drawing a representative triangle

3.7. ARCTRIGONOMETRIC FUNCTIONS AND THEIR DERIVATIVES

and re-writing the function. Be sure to consider both cases: $x \geq 0$ and $x < 0$.

Use the technique of Exercise 21 or similar to rewrite and then compute the derivatives in Exercises 22–26.

22. $\dfrac{d}{dx} \tan^{-1} \dfrac{1}{x}$

23. $\dfrac{d}{dx} \sin^{-1} \dfrac{1}{x^2}$

24. $\dfrac{d}{dx} \sec^{-1} \dfrac{5}{x}$

25. $\dfrac{d}{dx} \cos^{-1}\left(\dfrac{x}{\sqrt{25+x^2}}\right)$. Assume $x > 0$.

26. $\dfrac{d}{dx} \sin^{-1} \sqrt{\dfrac{x-1}{x}}$. Assume $x > 0$ first. Then modify for $x < 0$.

27. Prove (3.92), that
$$\dfrac{d}{dx} \sec^{-1} |x| = \dfrac{1}{x\sqrt{x^2-1}}.$$
To do so, use the previous formulas for the derivative of the arcsecant function and consider the cases $x > 0$ and $x < 0$ (or more precisely, $x > 1$ and $x < -1$) separately and show (3.92) holds for both cases.

28. Compute $\dfrac{d}{dx}\left[\sin^{-1} x + \cos^{-1} x\right]$. Explain why we should have known the answer would be simple given the algebraic relationship between the arcsine and arccosine functions. (See (3.83), page 344.)

29. Consider the general problem of $f(x)$ being one-to-one with inverse function $f^{-1}(x)$. Next consider that
$$y = f(x) \implies \dfrac{dy}{dx} = f'(x)$$
$$x = f^{-1}(y) \implies \dfrac{dx}{dy} = (f^{-1})'(y).$$

Next, using $x = f^{-1}(y)$, apply $\dfrac{d}{dx}$ to both sides to show that
$$\dfrac{dx}{dy} = \dfrac{1}{\left(\dfrac{dy}{dx}\right)}.$$

Compare to the version found in other texts: If $f(x)$ and $g(x)$ are inverse functions of each other, then
$$g'(x) = \dfrac{1}{f'(g(x))}.$$

30. Some texts define $\sec^{-1} x$ to be that $\theta \in \left[0, \dfrac{\pi}{2}\right) \cup \left[\pi, \dfrac{3\pi}{2}\right)$ so that $\sec \theta = x$. In other words, θ is chosen from the smallest positive angles in Quadrants I and III.

    (a) Produce a graph as in Figure 3.36, page 348, with the appropriate part of the curve highlighted, to show that the highlighted curve does indeed represent a one-to-one function.

    (b) Modify Figure 3.37, page 348, showing the same cases in which $\sec \theta = x$.

    (c) Use this modified graph to show that, with this definition, in fact we have
    $$\dfrac{d}{dx} \sec^{-1} x = \dfrac{1}{x\sqrt{x^2-1}}.$$

    (d) Explain why, with this definition, we will no longer always have $\sec^{-1} x = \cos^{-1}\left(\dfrac{1}{x}\right)$.

For Exercises 31–32, show the equation is valid by differentiation.

31. $\dfrac{d}{dx}\left[x \sin^{-1} x + \sqrt{1-x^2}\right] = \sin^{-1} x$.

32. $\dfrac{d}{dx}\left[x \cos^{-1} x - \sqrt{1-x^2}\right] = \cos^{-1} x$.

3.8 Exponential Functions

In this section, we look at a function $f(x) = e^x$, where e is a very important, irrational number, approximated by $e \approx 2.7182818$.[73]

What makes this function interesting, among other reasons, is that $\frac{d}{dx}(e^x) = e^x$. In fact, only functions that are constant multiples of e^x are their own derivatives. Before arguing that such a function exists, we will look briefly at exponential functions a^x in general, after which we will concentrate on e^x.

When we are finished with this section, we will then look at the inverse functions of such *exponential functions*. These inverses are better known as *logarithms*, and their derivatives fill an important gap in the theory. These functions also have many useful algebraic properties we can exploit before we compute the derivatives. In fact, in many problems that do not explicitly include logarithms, we can introduce them to exploit their properties, to make some derivative computations proceed much faster.

Much of what we do in this section relies on the fact that a^x is an everywhere continuous, differentiable function for any $a > 0$. To actually prove this requires integral calculus (the second part of this text), but this fact is believable through observations. For our purposes here, we will work from this assumption (a^x continuous and differentiable for $a > 0$, $x \in \mathbb{R}$), see how it is reasonable, but omit the proof.[74]

3.8.1 Exponential Functions

Of course, we will have use for the algebraic rules of exponential functions, which we quickly re-list here. Assuming $a, b > 0$, $r, s \in \mathbb{R}$, $m \in \{1, 2, 3, \ldots\}$ we have

$$a^r a^s = a^{r+s}, \qquad (ab)^r = a^r b^r, \qquad \frac{1}{a^{-r}} = a^r,$$

$$\frac{a^r}{a^s} = a^{r-s}, \qquad \left(\frac{a}{b}\right)^r = \frac{a^r}{b^r}, \qquad a^{1/m} = \sqrt[m]{a}$$

$$a^0 = 1, \qquad a^{-r} = \frac{1}{a^r}, \qquad (a^r)^s = a^{r \cdot s}.$$

In this subsection we look at functions $f(x) = a^x$. In order for such a function to have domain $x \in \mathbb{R}$, we restrict ourselves to $a > 0$. (Think of what $(-2)^x$ would be for $x = \frac{1}{2}, \frac{1}{4}, \frac{3}{2}, \pi$, etc.) We will generally avoid the case $a = 1$ as well, as the function 1^x is rather trivial.

We will look at two cases separately, namely $a > 1$ and $a \in (0, 1)$. For our prototypes, we will look at $a = 2$ and $a = \frac{1}{2}$ specifically, and argue that the same trends hold for similar a's in their respective cases.

---

[73] In fact, e is arguably at least equal in importance to π, though its importance is not as easily accessible.

[74] The proof that a^x is continuous and differentiable on all of $\mathbb{R}$ is interesting and worthwhile, but we omit it from this text. The proof is long and follows a path which is essentially backwards from how we most easily learn these functions. It defines logarithms first, using an alternative definition of logarithms based on integration theory (the subject of Chapter 5), proves all their algebraic properties still hold with that definition, and then considers the exponentials as inverses of the logarithms. It makes for more mathematically cohesive theory but is counter-intuitive as a path of discovery, where it seems more natural (as we do in this text) to develop exponential functions first, and logarithms as their inverses.

3.8. EXPONENTIAL FUNCTIONS

Example 3.8.1

Consider the function $f(x) = 2^x$. There are two trends—which are in fact reflections of each other—that we will observe with this function: what happens as $x \to \infty$ and what happens as $x \to -\infty$. We will do so by incrementing by 1 in each direction to observe the trends.

$$2^0 = 1 \qquad\qquad 2^{-1} = \frac{1}{2} = 0.5$$
$$2^1 = 2 \qquad\qquad 2^{-2} = \frac{1}{4} = 0.25$$
$$2^2 = 4 \qquad\qquad 2^{-3} = \frac{1}{8} = 0.125$$
$$2^3 = 8 \qquad\qquad 2^{-4} = \frac{1}{16} = 0.0625$$
$$2^4 = 16 \qquad\qquad 2^{-5} = \frac{1}{32} = 0.03125$$
$$2^5 = 32 \qquad\qquad 2^{-6} = \frac{1}{64} = 0.015625$$
$$2^6 = 64 \qquad\qquad 2^{-7} = \frac{1}{128} = 0.0078125$$
$$2^7 = 128 \qquad\qquad 2^{-8} = \frac{1}{256} = 0.00390625$$
$$2^8 = 256 \qquad\qquad \text{etc.}$$

These trends continue as $x \to \infty$ and $x \to -\infty$. They are also predictable since

$$2^{x+1} = 2^x 2^1 = 2^x \cdot 2, \qquad 2^{x-1} = 2^x 2^{-1} = 2^x \cdot \frac{1}{2}.$$

In other words, for every increment of one to the right, the height of the function is multiplied by a factor of 2; a movement of one to the left lowers the function's height by half.

To compute 2^r for rational numbers $r = \frac{p}{q}$, where $q \in \mathbb{N}$, is to compute $2^{p/q} = \sqrt[q]{2^p}$, or alternatively $\left(\sqrt[q]{2}\right)^p$. For an irrational number $s \in \mathbb{R} - \mathbb{Q}$, we simply take any sequence of rational numbers $r_1, r_2, \ldots$ so that $r_n \longrightarrow s$ and define $2^s = \lim_{n \to \infty} 2^{r_n}$. In doing so, we can eventually define 2^x for any $x \in \mathbb{R}$, thus achieving the first graph in Figure 3.38, page 358.

To be sure, this argument is not rigorous, but the graph should be somewhat convincing, at least in its behavior at the integers. For our purposes we will accept without proof that

- $f(x) = 2^x$ is continuous for all $x \in \mathbb{R}$,

- $f(x) = 2^x$ is one-to-one as a function $f : \mathbb{R} \longrightarrow (0, \infty)$.

Taking these facts and the graph for granted, we also notice the following limiting behavior:

$$x \longrightarrow \infty \implies 2^x \longrightarrow \infty, \tag{3.96}$$
$$x \longrightarrow -\infty \implies 2^x \longrightarrow 0^+. \tag{3.97}$$

We now contrast this behavior with that of the related function $g(x) = \left(\frac{1}{2}\right)^x$.

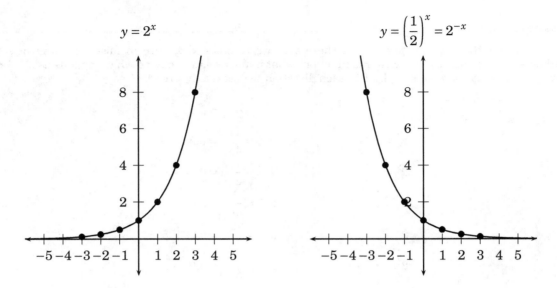

Figure 3.38: Partial graphs of $y = 2^x$ and $y = \left(\frac{1}{2}\right)^x = 2^{-x}$. In both, a move to the right or left by one unit causes a change in the height of the graph by a factor of 2. Such functions whose values change by a (positive) constant factor with each increment are called *exponential*. Increasing exponential functions are said to represent *exponential growth*, while decreasing exponential functions represent *exponential decay*.

Example 3.8.2

Consider $g(x) = \left(\frac{1}{2}\right)^x$. Some points on the graph of this function are indicated below:

$$\left(\frac{1}{2}\right)^0 = 1 \qquad \left(\frac{1}{2}\right)^{-1} = 2^1 = 2$$

$$\left(\frac{1}{2}\right)^1 = \frac{1}{2} = 0.5 \qquad \left(\frac{1}{2}\right)^{-2} = 2^2 = 4$$

$$\left(\frac{1}{2}\right)^2 = \frac{1}{4} = 0.25 \qquad \left(\frac{1}{2}\right)^{-3} = 2^3 = 8$$

$$\left(\frac{1}{2}\right)^3 = \frac{1}{8} = 0.125 \qquad \left(\frac{1}{2}\right)^{-4} = 2^4 = 16.$$

Such a function shrinks in height by a factor of 1/2 with each increment of one unit to the right in x, and increases by a factor 2 with each increment of one unit to the left. Thus the behavior of $g(x) = \left(\frac{1}{2}\right)^x$ is in some sense the opposite of that of $f(x) = 2^x$. This is not surprising, when we realize one is just the reflection of the other in the sense that $g(x) = f(-x)$:

$$g(x) = \left(\frac{1}{2}\right)^x = \frac{1}{2^x} = 2^{-x} = f(-x).$$

This function $g(x)$ is illustrated by the second graph in Figure 3.38.

Any function of the form $f(x) = a^x$ where $a > 1$ represents a function that increases by the factor $a > 1$ with every increment to the right. Such behavior will ultimately imply $a^x \longrightarrow \infty$

3.8. EXPONENTIAL FUNCTIONS

as $x \longrightarrow \infty$, and $a^x \longrightarrow 0^+$ as $x \longrightarrow -\infty$. On the other hand, we get the opposite if $a \in (0,1)$, for we can then write $a^x = \left(\frac{1}{a}\right)^{-x}$, which is of the form b^{-x} where $b = \frac{1}{a} > 1$. This limiting behavior is summarized next:

$$a > 1 \implies \begin{cases} a^x & \longrightarrow & \infty & \text{as} & x & \longrightarrow & \infty, \\ a^x & \longrightarrow & 0^+ & \text{as} & x & \longrightarrow & -\infty, \end{cases}$$

$$a \in (0,1) \implies \begin{cases} a^x & \longrightarrow & 0^+ & \text{as} & x & \longrightarrow & \infty, \\ a^x & \longrightarrow & \infty & \text{as} & x & \longrightarrow & -\infty. \end{cases}$$

The rate at which this limiting behavior occurs depends on a. For instance, 3^x increases faster than 2^x as x increases, and therefore decreases faster as x decreases. Since 2^x and 3^x agree at $x = 0$, and are positive everywhere, we thus have

$$x > 0 \implies 3^x > 2^x > 1,$$
$$x < 0 \implies 0 < 3^x < 2^x.$$

Similarly, $\left(\frac{1}{3}\right)^x$ shrinks faster than $\left(\frac{1}{2}\right)^x$ as x increases, with the opposite occurring as x decreases. Next we see how dramatic this difference in growth, of 2^x versus 3^x, is in two ways.

Example 3.8.3

Show that 3^x grows faster than 2^x as $x \longrightarrow \infty$.

Solution: Note first that 3^x and 2^x are both increasing as x increases, and have the same value at $x = 0$. That is, $y = 3^x$ and $y = 2^x$ both contain the point $(0,1)$. Furthermore,

$$\lim_{x \to \infty} (3^x - 2^x) = \lim_{x \to \infty} \left[2^x \left(\left(\frac{3}{2}\right)^x - 1\right)\right] \xrightarrow{\infty \cdot (\infty - 1)} \infty.$$

Thus 3^x is an increasing distance above 2^x, and in fact that distance increases without bound. Alternatively, we can show 3^x grows significantly faster than 2^x by noting that

$$\lim_{x \to \infty} \frac{3^x}{2^x} \xrightarrow[\text{ALG}]{\infty/\infty} \lim_{x \to \infty} \underbrace{(3/2)}_{>1}{}^x = \infty,$$

so 3^x is a nonconstant multiple of 2^x, and that multiple (namely $\left(\frac{3}{2}\right)^x$) is blowing up as $x \to \infty$.

3.8.2 Derivative of a Special Exponential Function

If we look back at the (computer-generated) graphs in Figure 3.38, page 358, it is not unreasonable to expect that slope can be defined along these curves. Indeed that is the case, though again we must wait to have more tools with which to prove it. Still, if we take for granted that $a > 0 \implies \frac{d}{dx}(a^x)$ exists, we can perform the following computation. Note that a^x is constant in the limit (which varies Δx, and not x itself).

$$f(x) = a^x \implies f'(x) = \lim_{\Delta x \to 0} \frac{f(x + \Delta x) - f(x)}{\Delta x} = \lim_{\Delta x \to 0} \frac{a^{x+\Delta x} - a^x}{\Delta x} = \lim_{\Delta x \to 0} \frac{a^x(a^{\Delta x} - a^0)}{\Delta x}$$
$$= a^x \cdot \lim_{\Delta x \to 0} \frac{a^{0+\Delta x} - a^0}{\Delta x} = a^x \cdot \lim_{\Delta x \to 0} \frac{f(0 + \Delta x) - f(0)}{\Delta x} = a^x \cdot f'(0).$$

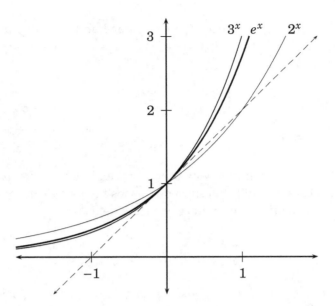

Figure 3.39: Graphs of 2^x, e^x, and 3^x. Only $y = e^x$ has slope 1 at $x = 0$, which implies ultimately that $\frac{d}{dx}(e^x) = e^x$. See the discussion leading to (3.98)–(3.100). The tangent line to $y = e^x$ through $(0,1)$ is also shown.

To get the last line from the one immediately prior is just to recognize the definition of $f'(0)$. In all cases $a > 0$ we see that

$$f(x) = a^x \implies f'(x) = a^x \cdot f'(0). \tag{3.98}$$

So if $f(x) = a^x$, then $f'(x)$ is a constant multiple of the original function a^x, that constant—namely $f'(0)$—depending on a.

Now, perhaps with the aid of a computer to approximate $f'(0)$ for various functions $f(x) = a^x$, it can be determined (for now without proof) that

$$f(x) = 2^x \implies f'(0) \approx 0.69314718,$$
$$f(x) = 3^x \implies f'(0) \approx 1.09861229.$$

Due to the nature of the functions, it is not hard to see that the larger the base a, the greater the slope of the graph of a^x at $x = 0$. So for $b \in (2,3)$, we get the slope of b^x at $x = 0$ should be between the slope of 2^x and the slope of 3^x at $x = 0$. It is then reasonable to believe that for some number $e \in (2,3)$, with $f(x) = e^x$ we will get $f'(0) = 1$, so that (3.98) becomes

$$f(x) = e^x \implies f'(x) = e^x \cdot 1 = e^x. \tag{3.99}$$

In fact, that number e does exist (and we will find other ways to *derive* it much later), and e^x is graphed along with 2^x and 3^x in Figure 3.39. We will refer to the function e^x as the *natural exponential function*. Thus we have a derivative formula, with the Chain Rule version following as always:

3.8. EXPONENTIAL FUNCTIONS

$$\boxed{\frac{d\,e^x}{dx} = e^x,} \qquad \boxed{\frac{d\,e^u}{dx} = e^u \cdot \frac{du}{dx}.} \qquad (3.100)$$

The number e is irrational, but can be given approximately by[75]

$$e \approx 2.71828\ 18284\ 59045\ 23536\ 02874\ 71352\ 66249\ 77572\ 47093\ 69996. \qquad (3.101)$$

The function e^x, together with 2^x and 3^x, is given in Figure 3.39, page 360. Later in the text we will show that only constant multiples of e^x are their own derivative functions. For now we can include such functions in derivative problems. In the next two examples we show some simple Chain Rule problems where the exponential is, alternatively, the "outer" and "inner" function.

Example 3.8.4

Here we compute derivatives of functions where the natural exponential function is the "outer" function.

$$\frac{d}{dx}\left[e^{\sin^{-1}x}\right] = e^{\sin^{-1}x} \cdot \frac{d}{dx}\left[\sin^{-1}x\right] = e^{\sin^{-1}x} \cdot \frac{1}{\sqrt{1-x^2}} = \frac{e^{\sin^{-1}x}}{\sqrt{1-x^2}},$$

$$\frac{d}{dx}\left[e^{\cos^{-1}x}\right] = e^{\cos^{-1}x} \cdot \frac{d}{dx}\left[\cos^{-1}x\right] = e^{\cos^{-1}x} \cdot \frac{-1}{\sqrt{1-x^2}} = \frac{-e^{\cos^{-1}x}}{\sqrt{1-x^2}},$$

$$\frac{d}{dx}\left[e^{\tan^{-1}x}\right] = e^{\tan^{-1}x} \cdot \frac{d}{dx}\left[\tan^{-1}x\right] = e^{\tan^{-1}x} \cdot \frac{1}{x^2+1} = \frac{e^{\tan^{-1}x}}{x^2+1},$$

$$\frac{d}{dx}\left[e^{\cot^{-1}x}\right] = e^{\cot^{-1}x} \cdot \frac{d}{dx}\left[\cot^{-1}x\right] = e^{\cot^{-1}x} \cdot \frac{-1}{x^2+1} = \frac{-e^{\cot^{-1}x}}{x^2+1},$$

$$\frac{d}{dx}\left[e^{\sec^{-1}x}\right] = e^{\sec^{-1}x} \cdot \frac{d}{dx}\left[\sec^{-1}x\right] = e^{\sec^{-1}x} \cdot \frac{1}{|x|\sqrt{x^2-1}} = \frac{e^{\sec^{-1}x}}{|x|\sqrt{x^2-1}},$$

$$\frac{d}{dx}\left[e^{\csc^{-1}x}\right] = e^{\csc^{-1}x} \cdot \frac{d}{dx}\left[\csc^{-1}x\right] = e^{\csc^{-1}x} \cdot \frac{-1}{|x|\sqrt{x^2-1}} = \frac{-e^{\csc^{-1}x}}{|x|\sqrt{x^2-1}}.$$

Example 3.8.5

Here we compute derivatives of functions where the natural exponential function is the "inner"

---

[75] We list 50 places after the decimal point because many students get the wrong impression when seeing the standard $e \approx 2.718281828$, leading them to leap to the conclusion that there is a pattern in the decimal representation. Of course, the number $2.7\overline{1828}$ is a "repeating decimal" and therefore rational, unlike e, which is irrational. (An interesting algebra exercise is to show that $2.7\overline{1828} = 271{,}801/99{,}990$, an obviously rational number.)

function. Note that $e^x > 0$ and so $|e^x| = e^x$.

$$\frac{d}{dx}\sin^{-1}e^x = \frac{1}{\sqrt{1-(e^x)^2}} \cdot \frac{de^x}{dx} = \frac{1}{\sqrt{1-e^{2x}}} \cdot e^x = \frac{e^x}{\sqrt{1-e^{2x}}},$$

$$\frac{d}{dx}\cos^{-1}e^x = \frac{-1}{\sqrt{1-(e^x)^2}} \cdot \frac{de^x}{dx} = \frac{-1}{\sqrt{1-e^{2x}}} \cdot e^x = \frac{-e^x}{\sqrt{1-e^{2x}}},$$

$$\frac{d}{dx}\tan^{-1}e^x = \frac{1}{(e^x)^2+1} \cdot \frac{de^x}{dx} = \frac{1}{e^{2x}+1} \cdot e^x = \frac{e^x}{e^{2x}+1},$$

$$\frac{d}{dx}\cot^{-1}e^x = \frac{-1}{(e^x)^2+1} \cdot \frac{de^x}{dx} = \frac{-1}{e^{2x}+1} \cdot e^x = \frac{-e^x}{e^{2x}+1},$$

$$\frac{d}{dx}\sec^{-1}e^x = \frac{1}{|e^x|\sqrt{(e^x)^2-1}} \cdot \frac{de^x}{dx} = \frac{1}{e^x\sqrt{e^{2x}-1}} \cdot e^x = \frac{1}{\sqrt{e^{2x}-1}},$$

$$\frac{d}{dx}\csc^{-1}e^x = \frac{-1}{|e^x|\sqrt{(e^x)^2-1}} \cdot \frac{de^x}{dx} = \frac{-1}{e^x\sqrt{e^{2x}-1}} \cdot e^x = \frac{-1}{\sqrt{e^{2x}-1}}.$$

Example 3.8.6

Here we compute $\frac{d}{dx}(e^x)^2$ two different ways: first by the Chain Rule in the obvious way, and then by instead rewriting the function using properties of exponents.

(a) $\dfrac{d(e^x)^2}{dx} = 2(e^x)\dfrac{de^x}{dx} = 2e^x \cdot e^x = 2e^{2x}.$

(b) $\dfrac{d(e^x)^2}{dx} = \dfrac{de^{2x}}{dx} = e^{2x} \cdot \dfrac{d\,2x}{dx} = e^{2x} \cdot 2 = 2e^{2x}.$

Of course, we expect to be able to rewrite a function algebraically, when convenient, before computing a derivative, and achieve the same result, perhaps in a different form.

Example 3.8.7

Here we compute $\dfrac{d}{dx}\left[\dfrac{e^x}{e^{8x}}\right] = \dfrac{d}{dx}e^{x-8x} = \dfrac{d}{dx}e^{-7x} = e^{-7x}\dfrac{d}{dx}(-7x) = -7e^{-7x}.$

This example could have been computed using the Quotient Rule (an interesting exercise), but the algebraic simplification made this easier. Now we list several further examples, all of which should be self-explanatory given previous rules and the derivative formula for the natural exponential function. Note where factors are usually placed in final expressions.

- $\dfrac{d}{dx}(5e^x) = 5 \cdot \dfrac{d}{dx}e^x = 5e^x.$

- $\dfrac{d}{dx}e^{x^2} = e^{x^2} \cdot \dfrac{dx^2}{dx} = e^{x^2} \cdot 2x = 2xe^{x^2}.$

3.8. EXPONENTIAL FUNCTIONS

- $\dfrac{d}{dx}\sin e^x = \cos e^x \cdot \dfrac{d\,e^x}{dx} = \cos e^x \cdot e^x = e^x \cos e^x.$

- $\dfrac{d}{dx}\left[\dfrac{e^x}{x^2}\right] = \dfrac{x^2 \frac{d}{dx}(e^x) - e^x \frac{d}{dx}(x^2)}{(x^2)^2} = \dfrac{x^2 e^x - e^x \cdot 2x}{x^4} = \dfrac{xe^x(x-2)}{x^4} = \dfrac{e^x(x-2)}{x^3}.$

- $\dfrac{d}{dx}(e^{\csc x}) = e^{\csc x}\dfrac{d}{dx}(\csc x) = e^{\csc x}(-\csc x \cot x) = -e^x \csc x \cot x.$

- $\dfrac{d}{dx}(e^{2x}\sin 3x) = e^{2x}\cdot\dfrac{d\sin 3x}{dx} + \sin 3x \cdot \dfrac{d\,e^{2x}}{dx} = e^{2x}\cos 3x\cdot\dfrac{d\,3x}{dx} + \sin 3x\cdot e^{2x}\cdot\dfrac{d\,2x}{dx}$
  $= 3e^{2x}\cos 3x + 2e^{2x}\sin 3x = e^{2x}(3\cos 3x + 2\sin 3x).$

- $\dfrac{d}{dx}\sqrt[3]{e^{5x}} = \dfrac{d}{dx}\left[(e^{5x})^{1/3}\right] = \dfrac{d}{dx}(e^{(5/3)x}) = e^{(5/3)x}\cdot\dfrac{d[(5/3)x]}{dx} = \dfrac{5}{3}e^{(5x/3)}.$

These are all exercises involving the old differentiation rules, combined with our newest derivative formula (3.100). Note also one often uses clever placement of factors to avoid ambiguity, though one must remember the "order of operations," so for instance $2e^{3x} = (2)\left(e^{(3x)}\right)$, the argument $(3x)$ of the exponential function automatically considered as "grouped."

In the last problem, we simplified first to avoid calling an extra Chain Rule.

3.8.3 A Note on Differences between Polynomials, Exponentials

It should be well noted that these functions a^x in general, and e^x in particular, are very different from any of the other functions we had previously. Even though they involve powers, in the past the x-variable was part of the **base** and **not the exponent**. Compare the behavior of x^2 and 2^x, for instance, as well as the natures of their derivatives. For a more dramatic example, consider

$$\dfrac{d}{dx}x^{20} = 20x^{19},$$
$$\dfrac{d}{dx}20^x = 20^x \cdot k, \qquad k = \left.\dfrac{d\,20^x}{dx}\right|_{x=0}.$$

Not only are the Power Rule and Exponential Rule formulas not the same, but the Power Rule in the polynomial example decreases the power for the derivative, which does not occur in the exponential problem.[76] The two functions x^{20} and 20^x share at most a vague resemblance in their behaviors. (Both increase without bound, but that is almost where the similarities end, for instance only 20^x increasing on all of the domain $\mathbb{R}$.) It is very different

---

[76]A common mistake among novice calculus students is to treat exponential functions as if they are similar to polynomials when, for instance, computing derivatives. It is important to notice, for example, that

$$\dfrac{d}{dx}20^x \ne x\cdot 20^{x-1}.$$

This derivative is not a Power Rule like $\frac{d\,x^a}{dx}$, but an Exponential Rule $\frac{d\,a^x}{dx}$, which we will derive later (essentially finding the formula for k in our already-derived formula that $\frac{d}{dx}a^x = k\cdot a^x$, assuming $a > 0$). Power Rules assume the reverse of the Exponential Rules: that the variable is in the base, not the exponent (and that the exponent is a constant).

to raise the (variable) x to a constant power, than to take a constant raised to the (variable) xth power. In fact, we will be able to show later that any exponential growth will always dominate any polynomial growth as $x \to \infty$:

$$\lim_{x \to \infty} \frac{a^x}{x^n} = \infty \quad \text{for } a > 1, \text{ and any fixed } n. \tag{3.102}$$

Thus, though the trend would not show itself until x is almost unimaginably large, we have for example

$$\lim_{x \to \infty} \frac{(1.00000000000000000001)^x}{x^{1000}} = \infty.$$

It would require special programming, with a very large number of significant digits allowed, to see this trend with the help of a computer, though in the sequel we will be able to prove this limit with relative ease.

Exercises

1. Compute the following derivatives (in pairs):

   (a) $\dfrac{d}{dx} e^{\sin x}$, $\quad \dfrac{d}{dx} \sin e^x$

   (b) $\dfrac{d}{dx} e^{\cos x}$, $\quad \dfrac{d}{dx} \cos e^x$

   (c) $\dfrac{d}{dx} e^{\tan x}$, $\quad \dfrac{d}{dx} \tan e^x$

   (d) $\dfrac{d}{dx} e^{\cot x}$, $\quad \dfrac{d}{dx} \cot e^x$

   (e) $\dfrac{d}{dx} e^{\sec x}$, $\quad \dfrac{d}{dx} \sec e^x$

   (f) $\dfrac{d}{dx} e^{\csc x}$, $\quad \dfrac{d}{dx} \csc e^x$

2. Compute the following derivatives by first rewriting the original function using properties of exponents. (These will still require Chain Rules.)

   (a) $\dfrac{d}{dx} \sqrt{e^x}$

   (b) $\dfrac{d}{dx} \left[\dfrac{1}{e^x} \right]$

   (c) $\dfrac{d}{dx} (e^{2x})^9$

   (d) $\dfrac{d}{dx} \left(e^x e^{3x} \right)$

   (e) $\dfrac{d}{dx} \left[\dfrac{e^{5x}}{e^{3x}} \right]$

For Exercises 3–13, compute the derivatives.

3. $\dfrac{d}{dx} \left(e^{-x} \right)$

4. $\dfrac{d}{dx} \left(e^{\sqrt{x}} \right)$

5. $\dfrac{d}{dx} \left(e^{-1/x^2} \right)$

6. $\dfrac{d}{dx} (2e^x + 9)^4$

7. $\dfrac{d}{dx} \sin^2 e^{x^3}$

8. $\dfrac{d}{dx} \left(x^3 e^x \right)$ (factor your answer)

9. $\dfrac{d}{dx} \left[\dfrac{e^{2x}}{x^2 + 1} \right]$

10. $\dfrac{d}{dx} \left[\dfrac{e^x}{5} \right]$ (rewrite as a multiplication first)

11. $\dfrac{d}{dx} \left[\dfrac{e^x - e^{-x}}{2} \right]$

12. $\dfrac{d}{dx} e^{e^x}$

3.8. EXPONENTIAL FUNCTIONS

13. $\dfrac{d}{dx}(xe^x - e^x)$ (simplify your answer)

14. Consider $f(x) = e^{-x^2}$. This curve is one example of a normal (also called bell) curve, slightly rescaled. Make a sign chart for $f'(x)$ (using the fact that $e^r > 0$ for all $r \in \mathbb{R}$). Use this to show that there is a maximum at $x = 0$. Draw this function, showing the limiting behavior of $f(x)$ as $x \to \pm\infty$.

15. For general $a > 1$, use the facts that $a^x > 0$ for all x, and that $x > 0 \implies a^x > 1$ to prove that a^x is increasing, i.e., $x_1 < x_2 \implies a^{x_1} < a^{x_2}$. (Hint: Consider $a^{x_2} - a^{x_1}$, factor, and show this is positive.)

16. Consider $f(x) = (-2)^x$. Clearly the domain *includes* the nonzero integers $\{\pm 1, \pm 2, \pm 3, \ldots\}$, and one might even allow for $(-2)^0$ to be 1. However, this function is not continuous, as we explore here.

    (a) What can you say about $f(x)$ at $x = 1/2, 3/2, 5/2, \ldots$?

    (b) What can you say about $f(x)$ at $x = 1/4, 3/4, 5/4, \ldots$?

    (c) What can you say about $f(x)$ at $x = 0.5, 0.1, 5.5, 9.1$? (Hint: What are these as fractions?)

    (d) What can you say about $f(x)$ at $x = 0.2$?

    (e) Does it make any sense to compute $f(x)$ for any *irrational* numbers x, such as $x = \pi = 3.141592653\ldots$?

    (f) Between any two points $a, b \in \mathbb{R}$, where $a < b$, is there some number x so that $f(x)$ does not exist?

    (g) What does this imply about the continuity of $f(x)$?

17. Consider $f(x) = 0^x$. What is the natural domain of this function?

18. Why do we not much discuss $f(x) = a^x$ for $a = 1$?

19. Why do we only consider $f(x) = a^x$ for $a \in (0,1) \cup (1, \infty)$?

3.9 The Natural Logarithm I

Since $\frac{de^x}{dx} = e^x > 0$ for all $x \in \mathbb{R}$, we know e^x is both continuous and always increasing on all of $\mathbb{R}$, and therefore one-to-one there. In this section, we introduce the function that is the inverse to e^x, namely the *natural logarithm* of x, denoted $\ln x$. Its derivative, as in the case with the arctrigonometric functions, is surprisingly simple and algebraic, and has nothing algebraically to do with logarithms, trigonometric, arctrigonometric, or exponential functions.

We will derive the derivative of the natural logarithm, and look at several examples of derivatives involving this function. Before exploring the calculus of the natural logarithm, we will review the algebra of general logarithm functions. In doing so, we will see examples where what would be difficult derivatives can be found more quickly when we rewrite the function to a more convenient form for calculus, using the algebraic properties of logarithms.

In the next section, we will use algebraic techniques to extend the results here to compute derivatives of more general logarithm and exponential functions, such as $\log_a x$ and a^x. We will also develop a technique known as *logarithmic differentiation*, which will allow faster differentiation in many cases, as well as differentiation of functions of the form $f(x)^{g(x)}$, where the base and the exponent are both allowed to vary.

3.9.1 Algebra of Logarithms

For $a \in (0,1) \cup (1,\infty)$ we next define the logarithm—with base a—of x, written $\log_a x$ (usually read, "log base a of x"), as described next:

Definition 3.9.1 *For $x > 0$, $\log_a x$ is that number y so that $a^y = x$. In other words, $\log_a x$ is that power of a which yields x.*

Example 3.9.1

Consider the following logarithm computations:

$$\log_2 8 = 3 \quad \text{since} \quad 2^3 = 8,$$
$$\log_3 9 = 2 \quad \text{since} \quad 3^2 = 9,$$
$$\log_{10} \frac{1}{100} = -2 \quad \text{since} \quad 10^{-2} = \frac{1}{100},$$
$$\log_{16} 2 = \frac{1}{4} \quad \text{since} \quad 16^{1/4} = 2,$$
$$\log_{27} 9 = \frac{2}{3} \quad \text{since} \quad 27^{2/3} = 9,$$
$$\log_4 \frac{1}{8} = -\frac{3}{2} \quad \text{since} \quad 4^{-3/2} = \frac{1}{8}.$$

An observation that follows very quickly from the definition of the logarithms is the following:

$$\log_a a^x = x. \tag{3.103}$$

We will make repeated use of that observation. For now we note that with (3.103) we can perform computations like these by instead rewriting the argument of the logarithm as a power of a:

3.9. THE NATURAL LOGARITHM I

Example 3.9.2

We compute the following examples using (3.103).

$$\log_2 8 = \log_2 2^3 = 3,$$
$$\log_{10} 1000 = \log_{10} 10^3 = 3,$$
$$\log_8 4 = \log_8 8^{2/3} = 2/3,$$
$$\log_3 \frac{1}{81} = \log_3 3^{-4} = -4,$$
$$\log_a a = \log_a a^1 = 1.$$

Now we list some properties of logarithms based on the definition. We also show how they mirror the related properties of exponents. In the table that follows, assume $M = a^m$ and $N = a^n$.

Logarithmic Property	Exponential Property
1. $\log_a(MN) = \log_a M + \log_a N$	1. $a^m a^n = a^{m+n}$
2. $\log_a \dfrac{M}{N} = \log_a M - \log_a N$	2. $\dfrac{a^m}{a^n} = a^{m-n}$
3. $\log_a M^p = p \cdot \log_a M$	3. $(a^m)^p = a^{mp}$
4. $\log_a 1 = 0$	4. $a^0 = 1$
5. $\log_a \dfrac{1}{M} = -\log_a M$	5. $a^{-m} = \dfrac{1}{a^m}$

The first logarithmic property reflects the first exponential property in the following way. When we say $M = a^m$ and $N = a^n$, we can think of these as stating that while M represents m factors of a, and N represents n factors of a, it follows that MN represents $m+n$ factors of a:

$$\log_a(\underbrace{\underbrace{M}_{a^m} \underbrace{N}_{a^n}}_{a^{m+n}}) = \log_a \underbrace{M}_{a^m} + \log_a \underbrace{N}_{a^n}, \text{ i.e.,}$$
$$\log_a a^{m+n} = \log_a a^m + \log_a a^n, \text{ i.e.,}$$
$$m + n = \quad m \quad + \quad n.$$

This makes perfect sense if $m, n \in \{0, 1, 2, 3, 4, \ldots\}$, but we can also make sense of having a half-factor of a, which is then $a^{1/2}$, i.e., the square root of a. We can also talk about having -3 factors of a, which is like removing 3 factors of a, or dividing by a^3, which is the same as having a factor of (or multiplying by) a^{-3}. Extending this to all real powers of a, we can say that a number representing m factors of a means that number a^m, as we would have computed in the previous section. (For that discussion, see page 357.)

The second property says, roughly, that if we have m factors of a, and we divide by n factors of a, then we are left with $m - n$ factors of a. (One could say n factors of a "cancel.") The third says that if M represents m factors of a, then M^p represents $p \cdot m$ factors of a.

The fourth can be interpreted as meaning, in the context of multiplication (which is arguably the context of exponents and therefore ultimately logarithms), having no factors of a is the same as being left with the factor 1 only. Of course, it also follows from our definition of logarithms, since 1 is the zeroth power of a.[77]

The fifth property can be achieved by the second (with help from the fourth) or from the third:

$$\log_a \frac{1}{M} = \log_a 1 - \log_a M = 0 - \log_a M = -\log_a M,$$
$$\log_a \frac{1}{M} = \log_a (M)^{-1} \quad\quad = -1\log_a M = -\log_a M.$$

For instance, $\log_2 \frac{1}{32} = -\log_2 32 = -5$. While not as obvious with negative or fractional powers, nonetheless the algebraic properties of logarithms follow because logarithms are basically about counting *factors* of the base a, and so products, quotients, and powers are more relevant to logarithms. On the other hand, for instance, there are no simple rules for expanding $\log_a(M + N)$, because logarithms are not about addition of numbers *per se*.

There is one last, crucial property we mention here, since we may be interested in computing approximations to numbers such as $\log_2 3$, though our calculating devices usually do not come equipped with the function $\log_2(\ )$. Thus we need a "change of base" formula, which follows. Here we assume $a, b \in (0, 1) \cup (1, \infty)$, which (as explored in the exercises) is what we require of "proper bases" for a logarithm. In the exercises of the next section, we outline the proof of the change of base formula, namely

$$\log_a x = \frac{\log_b x}{\log_b a}. \tag{3.104}$$

If $b = 10$, $\log_{10} x = \log x$, found on scientific calculators, with which we can compute

$$\log_2 3 = (\log 3)/(\log 2) \approx 1.584962501.$$

This is reasonable, since $2^1 = 1$ and $2^2 = 4$, and $f(x) = 2^x$ is continuous on all of $\mathbb{R}$, so for some $x \in (1, 2)$ we have $2^x = 3$. (See Figure 3.38, page 358.)

Most scientific calculators have two logarithmic keys: $\log_{10} x$, labeled "$\log x$" and $\log_e x$ given by "$\ln x$." Though many sciences use the logarithm with base 10, for calculus and its science applications it turns out that $\ln x$ is much more useful, as we will see.[78]

---

[77] In the context of addition, having "nothing," or subtracting everything in an expression from the expression, would mean being left with only zero added. In multiplication, we can divide everything leaving "nothing," in a sense meaning being left with the factor 1 only. In addition, we can say we begin with zero; in multiplication, we begin with the factor 1. Adding zero to a quantity does nothing (in the sense of changing the value to a new quantity); multiplying a quantity by one similarly changes nothing.

[78] One has to be careful when reading formulas which contain "$\log x$," because while most texts mean by this $\log_{10} x$, there are some which will mean $\log_e x$, i.e., $\ln x$. The problem is akin to knowing if $\sin x$ is a function of x in radians, or in degrees. We will always write $\ln x$ for $\log_e x$, and $\log x$ for $\log_{10} x$.

3.9. THE NATURAL LOGARITHM I

One interesting aspect of (3.104) is that every function $\log_a x$ is a constant multiple of every other such function. For instance, with $a \in (0,1) \cup (1,\infty)$ and $b = e$ in (3.104), we have

$$\log_a x = \frac{\log_e x}{\log_e a} = \frac{\ln x}{\ln a} = \frac{1}{\ln a} \cdot \ln x. \tag{3.105}$$

In this section, we will derive the derivative of the natural logarithm, and so with (3.105) we will have the derivatives of all logarithm functions.

We will see that algebraically the logarithm function in any base $a \in (0,1) \cup (1,\infty)$ is the inverse function of the exponential function with the same base. Thus the range of the exponential becomes the domain of the logarithm, and the domain of the exponential becomes the range of the logarithm. The range and domain of the exponential functions being $(0,\infty)$ and $\mathbb{R}$, respectively, we have the following for general $a \in (0,1) \cup (1,\infty)$:

$$\log_a a^x = x, \quad x \in \mathbb{R}, \tag{3.106}$$
$$a^{\log_a x} = x, \quad x > 0, \tag{3.107}$$
$$y = \log_a x \iff x = a^y. \tag{3.108}$$

We can now use these to prove our change of base formula, (3.104), page 368. Suppose $a, b \in (0,1) \cup (1,\infty)$ and $y = \log_a x$. Then $x = a^y = \left(b^{\log_b a}\right)^y = b^{y \log_b a}$. Taking this equation and applying $\log_b(\ )$ to both sides gives $\log_b x = y \log_b a$. Setting "$y = y$" from this and our original equation gives $\log_a x = y = \log_b x / \log_b a$, q.e.d.

3.9.2 Graph of the Natural Logarithm Function

Replacing a in (3.106)–(3.108) with e yields similar equations, listed next. We will use (3.111) for graphing the natural logarithm function.

$$\ln e^x = x, \quad x \in \mathbb{R}, \tag{3.109}$$
$$e^{\ln x} = x, \quad x > 0, \tag{3.110}$$
$$y = \ln x \iff x = e^y. \tag{3.111}$$

According to (3.111), we see that graphing $y = \ln x$ is the same as graphing $x = e^y$. Of course, graphing $x = e^y$ is the same as graphing $y = e^x$, except that x and y have changed roles, so for instance $(0,1)$ being in the graph of $y = e^x$ means that $(1,0)$ is in the graph of $x = e^y$, i.e., $y = \ln x$. In Figure 3.40, page 370, we have both $y = e^x$ (in gray), and $x = e^y$, i.e., $y = \ln x$ (in black).

Referring to the graph in Figure 3.40, we see also the following limiting behavior:

$$x \to 0^+ \iff \ln x \to -\infty, \tag{3.112}$$
$$x \to \infty \iff \ln x \to \infty. \tag{3.113}$$

These follow from $x \to -\infty \iff e^x \to 0^+$, and $x \to \infty \iff e^x \to \infty$, respectively. The growth in $\ln x$ shown in the graph is indeed unbounded, but it is a very slow type of growth. In fact, such slow growth is dubbed *logarithmic growth*. We will show in a later chapter that this growth is slower than any positive power of x, so for example $\lim_{x \to \infty} \frac{\ln x}{x^{0.00000000001}} = 0$. We

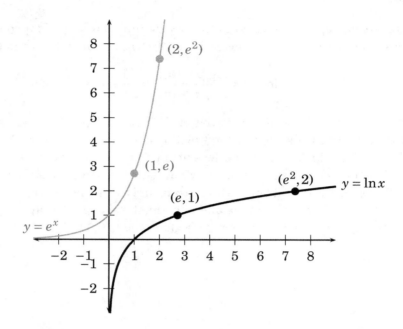

Figure 3.40: Partial graphs of $y = e^x$ in gray, and $y = \ln x$ (i.e., $x = e^y$) in black.

can replace 0.00000000001 with any other positive number and have the same limit. More generally, we have the following. Note that $\log_a x = -\log_{1/a} x$, since $a^y = x \iff \left(\frac{1}{a}\right)^{-y} = x$.

$$\lim_{x \to \infty} \frac{\log_a x}{x^s} = 0, \qquad \text{for all } s > 0 \text{ and } a \in (0,1) \cup (1,\infty). \tag{3.114}$$

This is the logarithmic version of (3.102), page 364.[79]

3.9.3 Derivative of the Natural Logarithm

We use (3.111), that is $y = \ln x \iff x = e^y$, to compute $\frac{d}{dx} \ln x = \frac{dy}{dx}$. We use implicit differentiation, recalling that $x = e^y$ again for the final step:

$$y = \ln x \iff e^y = x$$
$$\implies \frac{d}{dx}(e^y) = \frac{d}{dx}(x)$$
$$\implies e^y \cdot \frac{dy}{dx} = 1$$
$$\implies \frac{dy}{dx} = \frac{1}{e^y} = \frac{1}{x}.$$

---

[79] It may not be so obvious that $x \to \infty \implies \ln x \to \infty$ from the graph. Recall

$$\lim_{x \to \infty} f(x) = \infty \iff (\forall N)(\exists M)[x > M \longrightarrow f(x) > N].$$

Thus we need to show that for any N, we can force $\ln x > N$ by taking $x > M$ for some M. Later we will show that $\ln x$ is an increasing function on its domain $(0,\infty)$, so we can take $M = e^N$, so $x > M \implies \ln x > \ln M = \ln e^N = N$. So for example, if we want to show that eventually, as we move rightward on the x-axis, we have $\ln x > 10{,}000{,}000{,}000$, we just take $x > e^{10,000,000,000}$ $(= M)$, a very large number but certainly finite.

3.9. THE NATURAL LOGARITHM I

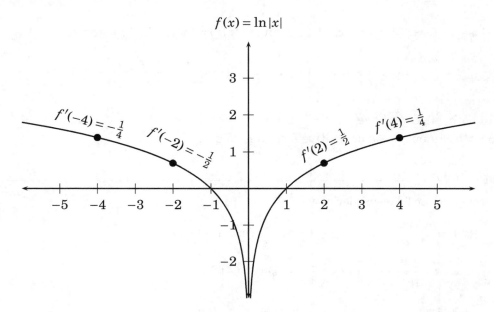

Figure 3.41: Partial graph of $f(x) = \ln|x|$ with slopes at some sample points.

We summarize this result and the Chain Rule version next:

$$\boxed{\frac{d\ln x}{dx} = \frac{1}{x}}, \qquad \boxed{\frac{d\ln u}{dx} = \frac{1}{u} \cdot \frac{du}{dx}}. \qquad (3.115)$$

Thus the derivative of the natural logarithm is in fact a simple power, albeit the -1 power.

As elegant as (3.115) might appear, it is incomplete, as one might suspect when noting that the derivative formulas seem to be defined as long as x (or u) is nonzero, while the logarithm requires positive x (or u). Thus we know what kind of function gives derivative $\frac{1}{x}$ as long as $x > 0$, but $\frac{1}{x}$ exists for $x \neq 0$, so we should like to know what function gives rise to derivative $\frac{1}{x}$ for $x < 0$ as well. (For example, what kind of position gives velocity $\frac{1}{t}$ even when $t < 0$?) The solution to this is to consider the functions $\ln|x|$, and more generally $\ln|u|$, which are defined as long as x and u are simply nonzero. As it happens, and as we will show next, our more general derivative formulas are the following:

$$\boxed{\frac{d}{dx}\ln|x| = \frac{1}{x}}, \qquad \boxed{\frac{d}{dx}\ln|u| = \frac{1}{u} \cdot \frac{du}{dx}}. \qquad (3.116)$$

If we know x or u, respectively, is positive then the absolute values are redundant. In considering (3.116), note that the graph of $y = \ln|x|$, given in Figure 3.41, shows how the derivative $\frac{1}{x}$ gives reasonable slopes at several points.

We now prove that (3.115)—which we derived—implies (3.116), that is (3.115) $\Longrightarrow$ (3.116). For convenience, we first define $f(x) = \ln|x|$, and so our task is to show $f'(x) = 1/x$ for all $x \in \mathbb{R} - \{0\}$.

1. Case $x \in (0, \infty)$: Here $|x| = x$ and so $f(x) = \ln|x| = \ln x$, implying $f'(x) = \frac{1}{x}$ as before.

2. Case $x \in (-\infty, 0)$: Here $|x| = -x$, giving us $f(x) = \ln|x| = \ln(-x)$, and so

$$x < 0 \implies f(x) = \ln(-x) \implies f'(x) = \frac{1}{-x} \cdot \frac{d(-x)}{dx} = \frac{1}{-x} \cdot (-1) = \frac{1}{x}.$$

Thus in both cases we have $f'(x) = \frac{1}{x}$, q.e.d.

Now we can put these derivative formulas to use.

Example 3.9.3 ───

We compute the following derivatives:

- $\dfrac{d}{dx}(x \ln x) = x \cdot \dfrac{d}{dx} \ln x + (\ln x) \dfrac{d}{dx}(x) = x \cdot \dfrac{1}{x} + \ln x \cdot 1 = 1 + \ln x.$

- $\dfrac{d}{dx} \ln \cos x = \dfrac{1}{\cos x} \cdot \dfrac{d}{dx} \cos x = \dfrac{1}{\cos x} \cdot (-\sin x) = -\tan x.$

- $\dfrac{d}{dx} \ln |\cos x| = \dfrac{1}{\cos x} \cdot \dfrac{d}{dx} \cos x = \dfrac{1}{\cos x} \cdot (-\sin x) = -\tan x.$

    Compare to the previous computation. Note that the domain for $\ln \cos x$ is $\{x \mid \cos x > 0\}$, while $\ln|\cos x|$ has domain $\{x \mid \cos x \neq 0\}$. Where $\cos x > 0$, these are the same function.

- $\dfrac{d}{dx} \sin(\ln x) = \cos(\ln x) \cdot \dfrac{d}{dx}(\ln x) = \cos(\ln x) \cdot \dfrac{1}{x} = \dfrac{\cos(\ln x)}{x}.$

- $\dfrac{d}{dx} \sin(\ln |x|) = \cos(\ln |x|) \cdot \dfrac{d}{dx}(\ln |x|) = \cos(\ln |x|) \cdot \dfrac{1}{x} = \dfrac{\cos(\ln |x|)}{x}.$

    Compare to the previous computation. Note that for this last function the domain is $x \neq 0$, while the previous function required $x > 0$. For $x > 0$ they are the same function.

───

Example 3.9.4 ───

Consider the following computation:

$$\frac{d}{dx} \ln \sqrt{x} = \frac{1}{\sqrt{x}} \cdot \frac{d}{dx} \sqrt{x} = \frac{1}{\sqrt{x}} \cdot \frac{1}{2\sqrt{x}} = \frac{1}{2x}.$$

Notice that we get one-half the answer we would have had if we had taken the derivative of $\ln x$. In fact, that this should be the case can be seen from one of the properties of logarithms:

$$\frac{d}{dx} \ln \sqrt{x} = \frac{d}{dx} \ln(x)^{1/2} = \frac{d}{dx} \left[\frac{1}{2} \ln x \right] = \frac{1}{2} \cdot \frac{1}{x} = \frac{1}{2x}.$$

───

Of course we had the same result using either method of computation. The lesson from this latest example is that we can at times use the properties of logarithms to algebraically rewrite the function in such a way that the derivative computation is simpler. For review,

3.9. THE NATURAL LOGARITHM I

and emphasis on the natural logarithm, we revisit the properties of logarithms as applied to the special case $a = e$. In what follows, $M, N > 0$ and $p \in \mathbb{R}$.

$$\ln(MN) = \ln M + \ln N, \tag{3.117}$$

$$\ln \frac{M}{N} = \ln M - \ln N, \tag{3.118}$$

$$\ln(M^p) = p \cdot \ln M, \tag{3.119}$$

$$\ln 1 = 0, \tag{3.120}$$

$$\ln \frac{1}{M} = -\ln M, \tag{3.121}$$

$$\log_a M = \frac{\ln M}{\ln a}. \tag{3.122}$$

The absolute value can be introduced easily into the arguments of the natural logarithm. In what follows we need only assume $M, N \ne 0$.[80]

$$\ln|MN| = \ln(|M| \cdot |N|) = \ln|M| + \ln|N|,$$
$$\ln|M/N| = \ln(|M|/|N|) = \ln|M| - \ln|N|,$$
$$\ln|M^p| = \ln|M|^p = p \cdot \ln|M|$$

Example 3.9.5

Find $f'(x)$ if $f(x) = \ln|x \sin x|$. <u>Solution</u>: We will compute $f'(x)$ using two methods:

1. $f'(x) = \dfrac{1}{x \sin x} \cdot \dfrac{d}{dx}(x \sin x) = \dfrac{1}{x \sin x}\left[x \cdot \dfrac{d \sin x}{dx} + \sin x \cdot \dfrac{dx}{dx}\right] = \dfrac{x \cos x + \sin x}{x \sin x}$
   $= \dfrac{x \cos x}{x \sin x} + \dfrac{\sin x}{x \sin x} = \cot x + \dfrac{1}{x}.$

2. If instead we first re-write $f(x) = \ln|x| + \ln|\sin x|$, we get
   $f'(x) = \dfrac{1}{x} + \dfrac{1}{\sin x} \cdot \dfrac{d}{dx} \sin x = \dfrac{1}{x} + \dfrac{1}{\sin x} \cdot \cos x = \dfrac{1}{x} + \cot x.$

The first method required a Chain Rule calling a Product Rule. The second required logarithm identity and a simple Chain Rule. As often occurs, algebraic re-writing of the function made the calculus easier.

Example 3.9.6

Find $\dfrac{d}{dx} \ln \left| \sin^4 x \cos^6(x^2) \right|$.

<u>Solution</u>: If we did not wish to use the algebraic properties of logarithms first, we would need a Chain Rule, calling a Product Rule, calling two Chain Rules, one of those calling yet another Chain

---

[80]Note that $|M^p| \ne |M|^{|p|}$ if $p < 0$, as for example

$$\left|2^{-3}\right| = \left|\frac{1}{8}\right| = \frac{1}{8}, \qquad |2|^{|-3|} = 2^3 = 8.$$

Rule. It is certainly doable, but not desirable if it can be avoided. Instead we will use the algebraic properties to expand the function to make for a simpler differentiation process. Consider the following:

$$\begin{aligned}\frac{d}{dx}\ln|\sin^4 x \cos^6(x^2)| &= \frac{d}{dx}\left[\ln|(\sin x)|^4 + \ln|(\cos(x^2))|^6\right] \\ &= \frac{d}{dx}\left[4\ln|\sin x| + 6\ln|\cos(x^2)|\right] \\ &= 4 \cdot \frac{1}{\sin x} \cdot \frac{d \sin x}{dx} + 6 \cdot \frac{1}{\cos(x^2)} \cdot \frac{d \cos(x^2)}{dx} \\ &= \frac{4\cos x}{\sin x} + \frac{6}{\cos(x^2)} \cdot \left(-\sin(x^2)\frac{dx^2}{dx}\right) \\ &= \frac{4\cos x}{\sin x} + \frac{6}{\cos(x^2)} \cdot (-\sin(x^2) \cdot 2x) \\ &= 4\cot x - 12x \tan(x^2).\end{aligned}$$

Note that the first two lines are algebraic in nature. (Note also that $\cos(x^2)$ is not a product.)

The natural logarithm transforms products to sums, quotients to differences, and powers to multiplying factors. Each of these transformations leaves us with simpler derivative rules. With some practice, the expansion of the logarithm becomes natural and quick (we will strive for a single step!), after which the differentiation steps are relatively easy.

Example 3.9.7 _____

Compute $\dfrac{d}{dx}\ln\left|\dfrac{x^3 \cos x}{\sqrt{1+x^2}}\right|$.

Solution: We will write all the steps in expanding the function, but it should become clearer with practice that the final rewriting of the function can be anticipated from the original form (see previous paragraph). The first three steps show the algebraic expansion, from which the calculus carries us to our final answer.

$$\begin{aligned}\frac{d}{dx}\ln\left|\frac{x^3 \cos x}{\sqrt{1+x^2}}\right| &= \frac{d}{dx}\left[\ln|x^3 \cos x| - \ln\sqrt{1+x^2}\right] \\ &= \frac{d}{dx}\left[\ln|x|^3 + \ln|\cos x| - \ln(1+x^2)^{1/2}\right] \\ &= \frac{d}{dx}\left[3\ln|x| + \ln|\cos x| - \frac{1}{2}\ln(1+x^2)\right] \\ &= 3 \cdot \frac{1}{x} + \frac{1}{\cos x} \cdot \frac{d \cos x}{dx} - \frac{1}{2} \cdot \frac{1}{1+x^2} \cdot \frac{d}{dx}(x^2+1) \\ &= \frac{3}{x} - \frac{\sin x}{\cos x} - \frac{1}{2(1+x^2)} \cdot (2x) \\ &= \frac{3}{x} - \tan x - \frac{x}{1+x^2}.\end{aligned}$$

Note how we did not use absolute values in the $\ln\sqrt{x^2+1}$ term (first line), since $x^2 + 1 \geq 1 > 0$.

3.9. THE NATURAL LOGARITHM I

It is interesting to note how the absolute values inside of the logarithms "disappear" in the derivative step. For that reason, some texts will skip them altogether, instead assuming that the quantity inside a logarithm is positive. But to be careful, even if the argument (input) of the natural logarithm in the original did not have absolute values, they still belong everywhere they appear on the right sides in the previous example, since it is possible that $x^3 \cos x > 0$ while the two factors are negative, making the subsequent logarithms undefined. Note how the following are all correct uses of the absolute values:

- $\dfrac{d}{dx}\left[\ln x^2\right] = \dfrac{d}{dx}[2\ln|x|] = \dfrac{2}{x}$. Note $x^2 \geq 0$, and $\ln x^2$ is defined for all $x \neq 0$. Also, $x^2 = |x|^2$.
  We could have been more verbose:
  $$\frac{d}{dx}\ln x^2 = \frac{d}{dx}\ln|x|^2 = \frac{d}{dx}(2\ln|x|) = 2 \cdot \frac{1}{x} = \frac{2}{x}.$$

- $\dfrac{d}{dx}\left[\ln x^3\right] = \dfrac{d}{dx}[3\ln x] = \dfrac{3}{x}$, but is undefined if $x < 0$.

- $\dfrac{d}{dx}\left[\ln|x|^3\right] = \dfrac{d}{dx}[3\ln|x|] = \dfrac{3}{x}$, defined for all $x \neq 0$.

While the algebraic properties are what often distinguish the natural and other logarithms, in a calculus setting it is also important to note the Chain Rule properties, particularly when the natural logarithm is the "outside" function, but it is on occasion the "inside" function as well, so we include the following examples to illustrate both patterns.[81]

Example 3.9.8

Note that -1 is not an exponent in $\sin^{-1}x$, $\cos^{-1}x$, and so on.

$$\frac{d}{dx}\ln\left|\sin^{-1}x\right| = \frac{1}{\sin^{-1}x} \cdot \frac{d\sin^{-1}x}{dx} = \frac{1}{\sin^{-1}x} \cdot \frac{1}{\sqrt{1-x^2}} = \frac{1}{\left(\sin^{-1}x\right)\sqrt{1-x^2}},$$

$$\frac{d}{dx}\ln\tan^{-1}x = \frac{1}{\tan^{-1}x} \cdot \frac{d\tan^{-1}x}{dx} = \frac{1}{\tan^{-1}x} \cdot \frac{1}{x^2+1} = \frac{1}{\left(\tan^{-1}x\right)\left(x^2+1\right)},$$

$$\frac{d}{dx}\ln\sec^{-1}x = \frac{1}{\sec^{-1}x} \cdot \frac{d\sec^{-1}x}{dx} = \frac{1}{\sec^{-1}x} \cdot \frac{1}{|x|\sqrt{x^2-1}} = \frac{1}{\left(\sec^{-1}x\right)|x|\sqrt{x^2-1}};$$

$$\frac{d}{dx}\sin^{-1}|\ln x| = \frac{1}{\sqrt{1-|\ln x|^2}} \cdot \frac{d\ln|x|}{dx} = \frac{1}{\sqrt{1-(\ln x)^2}} \cdot \frac{1}{x} = \frac{1}{x\sqrt{1-(\ln x)^2}},$$

$$\frac{d}{dx}\tan^{-1}\ln x = \frac{1}{(\ln x)^2+1} \cdot \frac{d\ln x}{dx} = \frac{1}{(\ln x)^2+1} \cdot \frac{1}{x} = \frac{1}{x\left[(\ln x)^2+1\right]},$$

$$\frac{d}{dx}\sec^{-1}\ln x = \frac{1}{|\ln x|\sqrt{(\ln x)^2-1}} \cdot \frac{d\ln x}{dx} = \frac{1}{|\ln x|\sqrt{(\ln x)^2-1}} \cdot \frac{1}{x} = \frac{1}{x|\ln x|\sqrt{(\ln x)^2-1}}.$$

---

[81] However, the most interesting and important patterns are left for the exercises. It is very important for the student to see the mechanics of differentiation with logarithmic functions.

A few notes are in order. For instance the first example would have looked very similar without absolute values, but the domain of the original function would be smaller. It also bears repeating that the "−1" superscript on the trigonometric functions is not an exponent, so for instance $\ln|\sin^{-1}x| \neq -1 \cdot \ln|\sin x|$. Finally, when we write for instance $\ln\tan^{-1}x$, this is taken to mean $\ln(\tan^{-1}x)$, and $\tan^{-1}\ln x$ means $\tan^{-1}(\ln x)$; these are not products.[82]

Exercises

For Exercises 1–7, compute the derivative two ways (See Example 3.9.4, page 372):

(a) By using the derivative rules called by the original form.

(b) By using properties of the natural logarithm to rewrite the function and then compute the derivative.

1. $\dfrac{d}{dx}\ln x^4$

2. $\dfrac{d}{dx}\ln x^5$

3. $\dfrac{d}{dx}\ln|x^5| = \dfrac{d}{dx}\ln|x|^5$

4. $\dfrac{d}{dx}\ln\left|\dfrac{1}{x}\right|$

5. $\dfrac{d}{dx}\ln\sqrt[3]{|x|} = \dfrac{d}{dx}\ln|\sqrt[3]{x}|$

6. $\dfrac{d}{dx}\ln e^x$

7. $\dfrac{d}{dx}\ln e^{x^2}$

8. Consider $f(x) = e^{\ln x}$. Note that $f(x) = x$ (though its domain is restricted to $x > 0$). Find $f'(x)$ using the original (unsimplified) form, and show that the form of the derivative can be simplified to $f'(x) = 1$ (as we should expect).

Compute and simplify the following derivatives (using (3.116), page 371):

9. $\dfrac{d}{dx}\ln|\sin x|$

10. $\dfrac{d}{dx}\ln|\cos x|$

11. $\dfrac{d}{dx}\ln|\tan x|$ (do not re-write first)

12. $\dfrac{d}{dx}\ln|\cot x|$ (do not re-write first)

13. $\dfrac{d}{dx}\ln|\sec x|$ (it is interesting both to use this form, and alternatively to rewrite the function first)

14. $\dfrac{d}{dx}\ln|\csc x|$ (see previous comment)

Compute the desired derivative for 15–22.

15. $\dfrac{d}{dx}(\ln x)^2$

16. $\dfrac{d}{dx}\sqrt{\ln x}$

17. $\dfrac{d}{dx}\sin(\ln x)$

18. $\dfrac{d}{dx}\tan(\ln x)$

19. $\dfrac{d}{dx}\sec(\ln|x|)$

20. $\dfrac{d}{dx}\ln|\ln x|$

21. $\dfrac{d}{dx}\ln(\ln(\ln x))$

22. $\dfrac{d}{dx}\ln|\ln|\ln|x|||$

---

[82]Students sometimes mistakenly interpret an expression like $\ln\tan^{-1}x$ to be a product. A moment's reflection shows that it cannot be a product, since ln( ) requires an input, even if the parentheses are absent.

3.9. THE NATURAL LOGARITHM I

For Exercises 23–27, compute $f'(x)$ for each of the following by first rewriting $f(x)$ by expanding the logarithms.

23. $f(x) = \ln\left|x^3(x^2 + 3x)^{20}\right|$

24. $f(x) = \ln\left|(2x+9)^3(3x^2+5x)^9(2-7x)^{10}\right|$

25. $f(x) = \ln\left|\dfrac{9x-1}{2x+4}\right|$

26. $f(x) = \ln\left|\dfrac{(3x+5)^7}{(7x^2+2)^5}\right|$

27. $f(x) = \ln\left|\dfrac{x^2\sin^3 x}{\cos^4 2x\sqrt[3]{x^2-9}}\right|$

28. Compute and simplify the following derivatives (which cannot take advantage of the properties of logarithms).

   (a) $\dfrac{d}{dx}(x\ln x - x)$

   (b) $\dfrac{d}{dx}\sin(\ln|\cos x|)$

   (c) $\dfrac{d}{dx}\ln(x^2+1)$. Why is $\ln|x^2+1|$ the same as $\ln(x^2+1)$?

29. Show that $\dfrac{d}{dx}\ln|\sec x + \tan x| = \sec x$. (Use the simplest derivative strategy. The key is in the simplification of the derivative.)

30. Compute the following two derivatives directly (using the Chain Rule formula for the derivative of the natural logarithm), show that they are the same, and explain why we should have expected they were the same:

   $\dfrac{d}{dx}\ln|\sec x|, \quad \dfrac{d}{dx}[-\ln|\cos x|].$

31. Compute the following derivatives and show that they simplify as described.

   (a) $\dfrac{d}{dx}\ln|\sec x + \tan x| = \sec x$.

   (b) $\dfrac{d}{dx}\ln|\csc x + \cot x| = -\csc x$.

32. For $f(x) = (g(x))^{h(x)}$, where $g(x) > 0$ derive the following formula for $f'(x)$:

$$f'(x) = h(x)(g(x))^{h(x)-1}g'(x)$$
$$+ (g(x))^{h(x)}h'(x)\ln(g(x)). \quad (3.123)$$

To do so, follow the following steps:

   (a) Use the idea that $a = e^{\ln a}$ (so what is a^x?) to show
   $$f(x) = e^{[\ln(g(x))\cdot h(x)]}.$$

   (b) Find $f'(x)$ using this form.

   (c) Simplify $f'(x)$ from the previous step to achieve (3.123).

33. Assume $h(x) = n$ is constant. Then rewrite $f(x) = (g(x))^{h(x)}$ and use (3.123) to compute $f'(x)$ for this case.

34. Repeat the previous problem supposing instead that $g(x) = a$ is a constant, and $h(x)$ is allowed to vary.

35. Now we consider the rationale for not allowing $a = 1$ to be the base of a logarithm. To do so, we consider how we would attempt to develop a function $\log_a x$.

   (a) First, recall we found the graph of $y = \ln x = \log_e x$ based upon the graph of $y = e^x$. What would be the graph of $y = 1^x$?

   (b) Separately, consider what would be the domain of $\log_1 x$.

   (c) Separately still, explain why (3.122) would preclude $a = 1$.

36. Prove that
$$\dfrac{d}{dx}\left[x\tan^{-1}x - \dfrac{1}{2}\ln(1+x^2)\right] = \tan^{-1}x.$$

37. Prove the following for the case that $x > 1$:
$$\dfrac{d}{dx}\left[x\sec^{-1}x - \ln\left|x+\sqrt{x^2-1}\right|\right]$$
$$= \sec^{-1}x.$$

3.10 The Natural Logarithm II: Further Results

In this section, we use the properties of the natural logarithm, and its relationship to exponential functions and other logarithms, to pursue further differentiation results. The first technique we will develop is called *logarithmic differentiation*, in which we actually introduce the natural logarithm into problems that can benefit from its presence. In later subsections, we use change-of-base techniques to rewrite problems into forms for which we have formulas. In doing so, some new and more general differentiation rules will emerge.

3.10.1 Logarithmic Differentiation

Because the natural logarithm takes products to sums, quotients to differences, and powers to multiplying factors, we have already seen that many derivative problems involving natural logarithms can be much reduced in complexity through algebraic expansion of a relevant logarithm. By properly *introducing* the natural logarithm into differentiation problems involving products, quotients, and powers, we can take advantage of the logarithm's algebraic expansion properties. Introducing the logarithm into the problem inserts its own complications, but these are relatively minor compared to previous, brute-force methods for many of these computations. We begin with an example to illustrate the method.

Example 3.10.1

Find $f'(x)$ if $f(x) = \dfrac{x \sin^3 x}{\sqrt{x^2+1}}$.

Solution: The technique that follows is similar to our implicit differentiation, except that first we apply the function $\ln |(\ )|$ to both sides in the following sense:

$$f(x) = \frac{x\sin^3 x}{\sqrt{x^2+1}} \quad \Longrightarrow \quad \ln|f(x)| = \ln\left|\frac{x\sin^3 x}{\sqrt{x^2+1}}\right|.$$

The whole point of doing so is to be able to take advantage of the algebraic properties of logarithms (so that the differentiation steps will be easier). Continuing where we left off,

$$\Longrightarrow \quad \ln|f(x)| = \ln|x| + \ln\left|\sin^3 x\right| - \ln\left|(x^2+1)^{1/2}\right|$$

$$\Longrightarrow \quad \ln|f(x)| = \ln|x| + 3\ln|\sin x| - \frac{1}{2}\ln|x^2+1|.$$

Next we differentiate, i.e., apply $\frac{d}{dx}$. Though we have complicated things by introducing a natural logarithm to both sides of the equation, the properties of the logarithm will help us to avoid the quotient and product rules, because the natural logarithm "takes products to sums, and quotients to differences," as well as taking exponents and turning them into multipliers. (Recall $\frac{d}{dx}\ln|u| = \frac{1}{u} \cdot \frac{du}{dx}$.)

$$\Longrightarrow \quad \frac{d}{dx}\ln|f(x)| = \frac{d}{dx}\left[\ln|x| + 3\ln|\sin x| - \frac{1}{2}\ln|x^2+1|\right]$$

$$\Longrightarrow \quad \frac{1}{f(x)} \cdot \frac{df(x)}{dx} = \frac{1}{x} + 3 \cdot \frac{1}{\sin x} \cdot \frac{d\sin x}{dx} - \frac{1}{2} \cdot \frac{1}{x^2+1} \cdot \frac{d(x^2+1)}{dx}$$

$$\Longrightarrow \quad \frac{f'(x)}{f(x)} = \frac{1}{x} + \frac{3\cos x}{\sin x} - \frac{2x}{2(x^2+1)}$$

$$\Longrightarrow \quad \frac{f'(x)}{f(x)} = \frac{1}{x} + 3\cot x - \frac{x}{x^2+1}$$

3.10. THE NATURAL LOGARITHM II: FURTHER RESULTS

We wanted $f'(x)$, so we next multiply both sides by $f(x)$, which gives $f'(x)$ in terms of both x and $f(x)$. We then finish by substituting the original form for $f(x)$:

$$f'(x) = f(x)\left[\frac{1}{x} + 3\cot x - \frac{x}{x^2+1}\right] = \frac{x\sin^3 x}{\sqrt{x^2+1}}\left[\frac{1}{x} + 3\cot x - \frac{x}{x^2+1}\right].$$

While this process did require several steps, those steps were arguably simpler than those in our earlier methods, which would have called for a Quotient Rule calling one Product Rule and two Chain Rules. Furthermore, with practice the process of logarithmic differentiation can be streamlined to be much faster. For instance, if we can anticipate the final logarithmic expansion, and also use the shortcut

$$\frac{d\ln|u(x)|}{dx} = \frac{u'(x)}{u(x)}, \tag{3.124}$$

particularly when $u'(x)$ is simple to compute, then we can consolidate steps from the previous example. For instance one practiced in the technique may skip the second line (in gray) in computing

$$f(x) = \frac{x\sin^3 x}{\sqrt{x^2+1}} \implies \ln|f(x)| = \ln|x| + 3\ln|\sin x| - \frac{1}{2}\ln|x^2+1|$$

$$\implies \frac{d}{dx}[\ln|f(x)|] = \frac{d}{dx}\left[\ln|x| + 3\ln|\sin x| - \frac{1}{2}\ln|x^2+1|\right]$$

$$\implies \frac{f'(x)}{f(x)} = \frac{1}{x} + 3\cdot\frac{\cos x}{\sin x} - \frac{1}{2}\cdot\frac{2x}{x^2+1}$$

$$\implies f'(x) = f(x)\left[\frac{1}{x} + 3\cot x - \frac{x}{x^2+1}\right]$$

$$\implies f'(x) = \frac{x\sin^3 x}{\sqrt{x^2+1}}\left[\frac{1}{x} + 3\cot x - \frac{x}{x^2+1}\right].$$

To be sure, more steps can be and arguably should be written when the technique is first learned, but it should appear true that this abbreviated process is much preferable to our earlier techniques *for this problem*, and that such efficiency is an achievable goal.

The process of logarithmic differentiation does not replace our earlier methods entirely. Indeed, it is only immediately useful for finding $f'(x)$ if $f(x)$ is the type of function whose natural logarithm (of its absolute value, to be more precise) can be expanded in a useful way. For instance, the previous function yielded nicely to the process because it consisted of powers of functions, combined through multiplication and division. However, it would not be advantageous, for instance, to attempt logarithmic differentiation on a function such as $f(x) = \sec x + \tan x + 9x^2 - \frac{1}{x}$, since this is a sum and there is no algebraic expansion for the natural log of a sum.[83]

The logarithmic differentiation process for finding $f'(x)$ for a given $f(x)$ is as follows:

---

[83]One could rewrite $f(x) = y_1 + y_2 + y_3 + y_4$, for instance, and perform logarithmic differentiation on each y_i separately to find each y'_i, and thus have $f'(x) = y'_1 + y'_2 + y'_3 + y'_4$, and indeed sometimes this is necessary. For the example to which this note refers, that would certainly not decrease the work required to find $f'(x)$.

1. Beginning with the equation that defines $f(x)$, apply $\ln |(\ )|$ to both sides.

2. Expand the logarithm on the right-hand side. (This may be consolidated into Step 1.)

3. Apply $\frac{d}{dx}$ to both sides. The left-hand side will then be $\frac{1}{f(x)} \cdot \frac{df(x)}{dx} = \frac{f'(x)}{f(x)}$.

4. Multiply the resulting equation by $f(x)$, thus solving for $f'(x)$ in terms of x and $f(x)$.

5. Substitute the original formula for $f(x)$ on the right-hand side. (This may be consolidated into Step 4.)

We will see in the sequel that there are other times where applying $\ln |(\ )|$ to both sides before differentiation is advantageous besides just for finding certain derivatives $f'(x)$, but for now another example of this first type is called for.

Example 3.10.2

Find $f'(x)$ if $f(x) = \dfrac{5\sin 2x \cos^3 4x}{\sqrt[3]{1+\tan 6x}}$.

Solution: We proceed as before, this time being more verbose in our Chain Rule computations.

$$f(x) = \frac{5\sin 2x \cos^3 4x}{\sqrt[3]{1+\tan 6x}}$$

$$\implies \ln |f(x)| = \ln \left| \frac{5\sin 2x \cos^3 4x}{(1+\tan 6x)^{1/3}} \right|$$

$$\implies \ln |f(x)| = \ln|5| + \ln|\sin 2x| + 3\ln|\cos 4x| - \frac{1}{3}\ln|1+\tan 6x|$$

$$\implies \frac{d}{dx}\ln |f(x)| = \frac{d}{dx}\left[\ln|5| + \ln|\sin 2x| + 3\ln|\cos 4x| - \frac{1}{3}\ln|1+\tan 6x|\right]$$

$$\implies \frac{1}{f(x)} \cdot \frac{df(x)}{dx} = 0 + \frac{1}{\sin 2x} \cdot \frac{d\sin 2x}{dx} + 3 \cdot \frac{1}{\cos 4x} \cdot \frac{d\cos 4x}{dx} - \frac{1}{3} \cdot \frac{1}{1+\tan 6x} \cdot \frac{d(1+\tan 6x)}{dx}$$

$$\implies \frac{f'(x)}{f(x)} = \frac{1}{\sin 2x} \cdot \cos 2x \cdot \frac{d 2x}{dx} + 3 \cdot \frac{1}{\cos 4x} \cdot (-\sin 4x) \cdot \frac{d 4x}{dx} - \frac{1}{3} \cdot \frac{1}{1+\tan 6x} \cdot \left(0 + \sec^2 6x \cdot \frac{d 6x}{dx}\right)$$

$$\implies \frac{f'(x)}{f(x)} = 2\cot 2x + 3(-\tan 4x)\cdot(4) - \frac{6\sec^2 6x}{3(1+\tan 6x)}$$

$$\implies f'(x) = f(x)\left[2\cot 2x - 12\tan 4x - \frac{2\sec^2 6x}{1+\tan 6x}\right]$$

$$\implies f'(x) = \frac{5\sin 2x \cos^3 4x}{\sqrt[3]{1+\tan 6x}}\left[2\cot 2x - 12\tan 4x - \frac{2\sec^2 6x}{1+\tan 6x}\right].$$

A few more notes about the process are in order.

(a) The absolute values introduced with the natural logarithm vanish in the derivative step. For this reason, some textbooks do not include them, but technically they should be included. One nice feature of the process is that the correct answer can be found even when absolute values are (naively?) omitted.

3.10. THE NATURAL LOGARITHM II: FURTHER RESULTS

(b) The answer this process delivers is of a different form from that produced by our earlier methods, but the two are algebraically the same, **except** that the answer here may need to be expanded and simplified to be, technically, completely correct. For instance, if $\cos 4x = 0$, this answer appears undefined because of the $\tan 4x$ term, though $\cos 4x = 0$ does not necessarily break the differentiability. When we distribute the $f(x)$ factor across the brackets in our final answer, a factor of $\cos 4x$ will cancel the denominator in the $\tan 4x$ term. Algebraically that is not valid reasoning if $\cos 4x = 0$, but in fact the naively simplified form—with the $\cos 4x$ term (as well as the $\sin 2x$ in the $f(x)$ and $\cot 2x$ terms) canceled—ultimately gives the correct derivative $f'(x)$.

(c) Related to the previous item, note that we cannot technically compute $\ln|f(x)|$ wherever $f(x) = 0$ (such as when $\sin 2x = 0$ or $\cos 4x = 0$ in our first example, namely Example 3.10.2), but this too gets glossed over in the differentiation process, especially in the final answer if $f(x)$ is distributed across the brackets and the offending terms canceled.

In some ways, logarithmic differentiation is better than expected, in that even when certain steps technically are problematic (such as presence of points where the input of a logarithm would be zero), in the end—at least in the simplified answer—we get the correct result for all points of differentiability. The proof of the following theorem of a generalized Product Rule includes this phenomenon, as the theorem holds even where one of the $g_i(x) = 0$.[84]

Theorem 3.10.1 *For* $f(x) = g_1(x)g_2(x)g_3(x)\cdots g_n(x)$, *we have*

$$f'(x) = [g'_1(x)g_2(x)g_3(x)\cdots g_n(x)] + [g_1(x)g'_2(x)g_3(x)\cdots g_n(x)]$$
$$+ [g_1(x)g_2(x)g'_3(x)\cdots g_n(x)] + \cdots + [g_1(x)g_2(x)g_3(x)\cdots g'_n(x)]. \quad (3.125)$$

A proof in the case $f(x) = g_1(x)g_2(x)g_3(x)$ shows the pattern of argument for the general case.

$$f(x) = g_1(x)g_2(x)g_3(x)$$
$$\implies \ln|f(x)| = \ln|g_1(x)g_2(x)g_3(x)|$$
$$\iff \ln|f(x)| = \ln|g_1(x)| + \ln|g_2(x)| + \ln|g_3(x)|$$
$$\implies \frac{d}{dx}[\ln|f(x)|] = \frac{d}{dx}[\ln|g_1(x)| + \ln|g_2(x)| + \ln|g_3(x)|]$$
$$\implies \frac{f'(x)}{f(x)} = \frac{g'_1(x)}{g_1(x)} + \frac{g'_2(x)}{g_2(x)} + \frac{g'_3(x)}{g_3(x)}$$
$$\implies f'(x) = f(x)\left[\frac{g'_1(x)}{g_1(x)} + \frac{g'_2(x)}{g_2(x)} + \frac{g'_3(x)}{g_3(x)}\right]$$
$$\implies f'(x) = g_1(x)g_2(x)g_3(x)\left[\frac{g'_1(x)}{g_1(x)} + \frac{g'_2(x)}{g_2(x)} + \frac{g'_3(x)}{g_3(x)}\right]$$
$$\implies f'(x) = g'_1(x)g_2(x)g_3(x) + g_1(x)g'_2(x)g_3(x) + g_1(x)g_2(x)g'_3(x),$$

q.e.d. How this generalizes to longer products, as in (3.125), should be apparent.

---

[84] See notes on the roles of the various factors in the Product Rule, page 298. Comments there generalize to more general products such as in Theorem 3.10.1 above.

Example 3.10.3

Thus we can perform the following quickly, this time in primed notation.

$$\frac{d}{dx}(x^3 e^x \sin x) = (x^3)' e^x \sin x + x^3 (e^x)' \sin x + x^3 e^x (\sin x)'$$
$$= 3x^2 e^x \sin x + x^3 e^x \sin x + x^3 e^x \cos x.$$
$$= x^2 e^x (3 \sin x + x \sin x + x \cos x)$$

If the derivatives called for in such a product are more complicated, we should revert to Leibniz notation, and a bit of rearranging, to aid in bookkeeping:

$$\frac{d}{dx}(f(x)g(x)h(x)) = f(x)g(x) \cdot \frac{dh(x)}{dx} + f(x)h(x) \cdot \frac{dg(x)}{dx} + g(x)h(x) \cdot \frac{df(x)}{dx} \tag{3.126}$$

With more complicated f, g, or h, Equation 3.126 has the advantage that the called derivative computations are the rightmost factors in each term, keeping them separate from the other terms, and so expanding those "sub-derivatives" is more convenient if they require further differentiation rules. This order is different from that given in Theorem 3.10.1, page 381.

Example 3.10.4

Later in the text we will often be computing derivatives with respect to time t, though that variable might not explicitly appear in the problem. Still it will make sense to apply $\frac{d}{dt}$ to quantities that do, in fact, depend on time t. So for instance there is the Ideal Gas Law formula from chemistry and physics that $PV = k \cdot nT$, where k is a constant (and so $\frac{dk}{dt} = 0$), P, V, n, and T are pressure, volume, the number of gas particles, and absolute temperature, respectively. This is the equation for a so-called *ideal gas*. In a mathematically sophisticated advanced chemistry text or article, it is not unusual to see a computation like

$$PV = k \cdot nT \quad \Longrightarrow \quad \frac{1}{P} \cdot \frac{dP}{dt} + \frac{1}{V} \cdot \frac{dV}{dt} = \frac{1}{n} \cdot \frac{dn}{dt} + \frac{1}{T} \cdot \frac{dT}{dt}.$$

Without logarithmic differentiation, this might seem rather mysterious and difficult to prove easily. However, one well versed in the technique would likely see the truth of this implication quickly, being practiced enough in the middle steps to anticipate the outcome. *Since all quantities are positive,* we can apply ln( ) to both sides, expand the logarithms, and apply $\frac{d}{dt}$:

$$PV = k \cdot T \Longrightarrow \quad \ln(PV) = \ln(k \cdot nT)$$
$$\Longrightarrow \quad \ln P + \ln V = \ln k + \ln n + \ln T$$
$$\Longrightarrow \quad \frac{d}{dt}[\ln P + \ln V] = \frac{d}{dt}[\ln k + \ln n + \ln T]$$
$$\Longrightarrow \quad \frac{1}{P} \cdot \frac{dP}{dt} + \frac{1}{V} \cdot \frac{dV}{dt} = 0 + \frac{1}{n} \cdot \frac{dn}{dt} + \frac{1}{T} \cdot \frac{dT}{dt}, \qquad \text{q.e.d.}$$

We would have a different equation involving these functions (of t) P, V, n, T if we simply applied $\frac{d}{dt}$ to both sides at the start, rather than first applying the natural logarithm function. If we chose that strategy we would require the Product Rule on each side. In fact, the equations we get with either method are equivalent under the original assumption that $PV = k \cdot nT$. To show the two equations involving the derivatives in fact say the same thing is an exercise in algebra.

3.10. THE NATURAL LOGARITHM II: FURTHER RESULTS

Recalling that simple powers inside of a logarithm become multiplying factors outside, for another quick example, this time from basic electricity, we can have:

$$P = \frac{E^2}{R} \implies \ln P = 2\ln E - \ln R \xrightarrow{d/dt} \frac{1}{P} \cdot \frac{dP}{dt} = \frac{2}{E} \cdot \frac{dE}{dt} - \frac{1}{R} \cdot \frac{dR}{dt}.$$

With practice one might learn to visualize, and therefore omit writing, the middle step.

3.10.2 Bases Other Than e: Logarithms

In this subsection, we look at derivatives of functions $\log_a x$ for more general a. This is a simple application of our change of base formula (3.104), page 368 (note that $\frac{1}{\ln a}$ is a constant):

$$\frac{d}{dx} \log_a x = \frac{d}{dx}\left[\frac{\ln x}{\ln a}\right] = \frac{d}{dx}\left[\frac{1}{\ln a} \cdot \ln x\right] = \frac{1}{\ln a} \cdot \frac{1}{x} = \frac{1}{x\ln a}.$$

We can also perform this computation with $|x|$ replacing x, and the same argument that showed that $\frac{d}{dx}\ln|x| = \frac{1}{x}$ will work here. In fact, since this function $\log_a x$ is just a constant multiple of $\ln x$, all the rules we had for $\ln x$ work here, with the constant carrying through. Thus the absolute value and Chain Rule versions are just

$$\boxed{\frac{d}{dx}\log_a |x| = \frac{d}{dx}\left[\frac{1}{\ln a}\ln|x|\right] = \frac{1}{x\ln a},} \tag{3.127}$$

$$\boxed{\frac{d}{dx}\log_a |u| = \frac{d}{dx}\left[\frac{1}{\ln a}\ln|u|\right] = \frac{1}{u\ln a} \cdot \frac{du}{dx}.} \tag{3.128}$$

Example 3.10.5

Consider the following derivative computations:

- $\dfrac{d}{dx}\log_{10} x = \dfrac{1}{x\ln 10}.$

- $\dfrac{d}{dx}\log_2\left|x\sqrt{x^2+1}\right| = \dfrac{d}{dx}\left[\log_2|x| + \dfrac{1}{2}\log_2(x^2+1)\right] = \dfrac{1}{x\ln 2} + \dfrac{1}{2} \cdot \dfrac{1}{(x^2+1)\ln 2} \cdot \dfrac{d}{dx}(x^2+1)$
  $= \dfrac{1}{x\ln 2} + \dfrac{1}{2} \cdot \dfrac{2x}{(x^2+1)\ln 2} = \dfrac{1}{x\ln 2} + \dfrac{x}{(x^2+1)\ln 2}.$

- $\dfrac{d}{dx}\log_3\left|\dfrac{\sin x}{2x+5}\right| = \dfrac{d}{dx}\left[\log_3|\sin x| - \log_3|2x+5|\right]$
  $= \dfrac{1}{\sin x \ln 3} \cdot \dfrac{d\sin x}{dx} - \dfrac{1}{(2x+5)\ln 3} \cdot \dfrac{d(2x+5)}{dx} = \dfrac{\cos x}{\sin x \ln 3} - \dfrac{2}{(2x+5)\ln 3}$
  $= \dfrac{1}{\ln 3}\cot x - \dfrac{2}{(2x+5)\ln 3}.$

- $\dfrac{d}{dx}\log_3(\log_5 x) = \dfrac{1}{\log_5 x \ln 3} \cdot \dfrac{d\log_5 x}{dx} = \dfrac{1}{\log_5 x \ln 3} \cdot \dfrac{1}{x\ln 5} = \dfrac{1}{\frac{\ln x}{\ln 5}\ln 3 \cdot x\ln 5} = \dfrac{1}{x\ln x \ln 3}.$

For this last example, we can cut short the calculus steps using algebraic properties of logarithms first:

$$\frac{d}{dx}\log_3(\log_5 x) = \frac{d}{dx}\log_3\left[\frac{\ln x}{\ln 5}\right] = \frac{d}{dx}\left[\log_3(\ln x) - \log_3(\ln 5)\right] = \frac{1}{\ln x \ln 3} \cdot \frac{d\ln x}{dx} - 0 = \frac{1}{x\ln x \ln 3}.$$

3.10.3 Bases Other Than e: Exponential Functions

Here we look at derivatives of exponential functions a^x ($a > 0$) and functions involving these. Two computations of the derivative of a^x are offered. Both illustrate useful techniques which are worth remembering. The first technique is logarithmic differentiation. We find $\frac{dy}{dx}$ under the assumption $y = a^x$. Note that $y > 0$ so no absolute values are needed.

$$\begin{aligned} y = a^x &\Longrightarrow \quad \ln y = \ln a^x \\ &\Longleftrightarrow \quad \ln y = x \ln a \\ &\Longrightarrow \quad \frac{d \ln y}{dx} = \frac{d[(\ln a)x]}{dx} \\ &\Longrightarrow \quad \frac{1}{y} \cdot \frac{dy}{dx} = \ln a \\ &\Longrightarrow \quad \frac{dy}{dx} = y \ln a \quad \Longleftrightarrow \quad \frac{dy}{dx} = a^x \ln a. \end{aligned}$$

Alternatively, we can use the fact that $a = e^{\ln a}$, and compute $\frac{d}{dx}[a^x]$ using the Chain Rule. Note that $a^x = \left[e^{\ln a}\right]^x = e^{(\ln a)x}$.

$$\frac{d}{dx} a^x = \frac{d}{dx}\left[e^{\ln a}\right]^x = \frac{d}{dx} e^{(\ln a)x} = e^{(\ln a)x} \cdot \frac{d[(\ln a)x]}{dx} = \left[e^{\ln a}\right]^x \cdot \ln a = a^x \ln a.$$

From either method, we get the derivative of a^x and its Chain Rule version:

$$\boxed{\frac{d a^x}{dx} = a^x \ln a,} \qquad \boxed{\frac{d a^u}{dx} = a^u \ln a \cdot \frac{du}{dx}.} \qquad (3.129)$$

Note that if $a = e$, we have $\ln a = \ln e = 1$, giving us $\frac{d}{dx} e^x = e^x \ln e = e^x \cdot 1 = e^x$ as before.

We proved the non-Chain Rule version of (3.129), from which the Chain Rule version follows in the usual manner. While the first technique of proof was the (now) familiar logarithmic differentiation, the second was to change the base, rewriting $a^x = (e^{\ln a})^x = e^{(\ln a)x}$. This algebraic trick requires some cleverness, and is not without its complications, but can be exploited in other venues, and allows us to use the simpler and ubiquitous derivative formula for e^x via an *algebraic* rewriting, rather than relying on the more obscure (but not unimportant) formula (3.129).

Example 3.10.6

We can now compute the following derivatives using (3.128), page 383, and (3.129):

- $\dfrac{d\, 2^x}{dx} = 2^x \ln 2.$

- $\dfrac{d}{dx} 3^{5x} = 3^{5x} \ln 3 \cdot \dfrac{d}{dx}[5x] = 3^{5x} \ln 3 \cdot 5 = 5(\ln 3) 3^{5x}.$

- $\dfrac{d}{dx}\left[\sin 10^x\right] = \cos 10^x \cdot \dfrac{d}{dx}\left[10^x\right] = \cos 10^x \cdot 10^x \ln 10 = 10^x \ln 10 \cdot \cos 10^x.$

- $\dfrac{d}{dx}\left[x^2 10^x\right] = x^2 \cdot \dfrac{d}{dx}\left[10^x\right] + 10^x \cdot \dfrac{d}{dx}\left[x^2\right] = x^2 \cdot 10^x \ln 10 + 10^x \cdot 2x = x \cdot 10^x [x \ln 10 + 2].$

3.10. THE NATURAL LOGARITHM II: FURTHER RESULTS

- $\frac{d}{dx}\left[2^{3^x}\right] = 2^{3^x} \ln 2 \cdot \frac{d}{dx}\left[3^x\right] = 2^{3^x} \ln 2 \cdot 3^x \ln 3 = 2^{3^x} 3^x \ln 2 \ln 3.$

- $\frac{d}{dx}\left[\tan^{-1} 2^x\right] = \frac{1}{(2^x)^2 + 1} \cdot \frac{d}{dx}\left[2^x\right] = \frac{1}{4^x + 1} \cdot 2^x \ln 2 = \frac{2^x \ln 2}{4^x + 1}.$

- $\frac{d}{dx}\left[2^{\tan^{-1} x}\right] = 2^{\tan^{-1} x} \ln 2 \cdot \frac{d}{dx}\left[\tan^{-1} x\right] = 2^{\tan^{-1} x} \ln 2 \cdot \frac{1}{x^2 + 1} = \frac{2^{\tan^{-1} x} \ln 2}{x^2 + 1}.$

- $\frac{d}{dx}\left[\ln 10^x\right] = \frac{1}{10^x} \cdot \frac{d}{dx}\left[10^x\right] = \frac{1}{10^x} \cdot 10^x \ln 10 = \ln 10.$

  Note that an alternative, arguably superior strategy would be to first rewrite the function:

  $\frac{d}{dx}\left[\ln 10^x\right] = \frac{d}{dx}\left[x \ln 10\right] = \ln 10,$ following from the usual Power Rule with a constant multiple $\ln 10$ carrying through the computation.

While the last derivative allowed us to first use the rules of logarithms and exponents, it should be pointed out that there are not useful rewritings for every possible case. In fact, that last derivative computation was the only one in Example 3.10.6 for which there was a useful way to algebraically rewrite the problem.

Now it happens that most of those would be fine candidates for logarithmic differentiation if we wanted to avoid the formulas for derivatives of exponential functions in bases other than e, namely (3.129), page 384. Indeed, only $\frac{d}{dx}[\sin 10^x]$ and $\frac{d}{dx}[\tan^{-1} 2^x]$ from Example 3.10.6 could not be computed directly with logarithmic differentiation. However, using our formulas for $\frac{da^x}{dx}$ and $\frac{da^u}{dx}$ will get us our results for the others more expeditiously than would logarithmic differentiation.

Before continuing, it should also be noted that the logarithm and exponential functions in bases $a \in (0,1) \cup (1, \infty)$ all have similar derivatives to those in base e, except for the extra factor of either $\ln a$ or $\frac{1}{\ln a}$, and so it is simply a matter of dividing or multiplying by $\ln a$:

$$\frac{d}{dx} \ln |x| = \frac{1}{x}, \qquad \frac{d}{dx} e^x = e^x,$$
$$\frac{d}{dx} \log_a x = \frac{1}{x \ln a}, \qquad \frac{d}{dx} a^x = a^x \ln a.$$

For the case $a = e$, we have $\ln a = \ln e = 1$, giving simple cases of the more general formulas.

3.10.4 Derivative of $(f(x))^{g(x)}$

So far, we have two derivative rules for functions that can loosely be defined as powers:

$$\frac{d}{dx}[x^n] = n \cdot x^{n-1}, \qquad n \in \mathbb{R} \qquad \text{(variable base, fixed exponent)},$$

$$\frac{d}{dx}[a^x] = a^x \ln a, \qquad a > 0 \qquad \text{(fixed base, variable exponent)}.$$

Crucially, in each of those cases either the base or the exponent is fixed, i.e., constant. However, we should expect very different derivative formulas for these since the functions behave very differently. For instance, polynomials do not have (linear) asymptotes while exponential functions do. Furthermore, if $n \in \{1, 2, 3, \ldots\}$, the function x^n has *polynomial growth*, while if $a > 1$ the function a^x has *exponential growth*, which is eventually much faster. There is also the exponential decay case ($a \in (0, 1)$), but for the moment let us consider the growths. For instance, x^2 grows without bound ($x^2 \to \infty$ as $x \to \infty$), but 2^x grows even faster, though we will have to wait for the sequel to actually prove this fact. Consider then a function like x^x, where neither base nor exponent is constant, in which the base grows without bound and so does the exponent. Here these two growths conspire in such a way that this new function will grow much faster than either x^2 or 2^x. This is not difficult to believe, for suppose $x = 100$. Then

$$\underbrace{100^2}_{=10^4} << \underbrace{2^{100}}_{\approx 1.26 \times 10^{30}} << \underbrace{100^{100}}_{=10^{200}}.$$

Here the notation "<<" means "is much less than," and as such is usually used subjectively, in much the same way that "$\approx$" is also subjective. We include it here for emphasis. The numerical results shown give just a small glimpse of the relative growth rates of these three functions, x^2, 2^x, and x^x as x grows larger (and positive).[85]

In what follows we are interested in computing derivatives of functions in which there is a base and an exponent, but both are allowed to vary. The usual method is logarithmic differentiation, but a formula is possible, and it carries some interesting intuition. In fact, this type of problem is often presented only as a logarithmic differentiation problem, so the reader should be both aware of that and able to compute these through logarithmic differentiation, but the formula we will derive is also worth knowing.[86]

Example 3.10.7

Find $\frac{d}{dx}\left[x^{\sin x}\right]$.

---

[85] Note that x^x is only continuous for $x > 0$. That is because, while it is defined for many negative numbers, it is undefined for many more. For instance, $(-3)^{-3}$ and $(-1)^{-1}$ make sense, but $(-1/2)^{-1/2}$ and $(-\pi)^{-\pi}$ do not. Nor does 0^0, unless we care to *define* it to be some number. In fact some algebra books do define it to be 1, but we will see in the sequel that there are other choices that make equal sense, so we will decline to define 0^0.

One way to define x^x and see that it is continuous for $x > 0$ is to rewrite this function as $x^x = \left(e^{\ln x}\right)^x = e^{(\ln x)x}$. In this last form we see that the only thing that can "break" the continuity is for x to be nonpositive. We will use that kind of technique for rewriting such a function on occasion in what follows, and indeed we used it before in one computation of $\frac{da^x}{dx}$.

[86] One could in fact use logarithmic differentiation to derive the Power Rule, Product Rule, or Quotient Rule, as we will see in the exercises. However, we do not abandon these rules since they are convenient and efficient, and can be easily implemented when called by other rules, while logarithmic differentiation requires whole sides of equations to be products, quotients, or powers, not for instance sums, differences, or other combinations that cannot have their logarithms expanded. Consider attempting, for instance, logarithmic differentiation for the problem of finding $\frac{dy}{dx}$ if

$$y = \sin\left(x^2 + 1\right) + \tan^{-1} x - \cos e^x + \sec x \tan x.$$

This would be a somewhat long but fairly simple problem using the older rules, but logarithmic differentiation would be useless here, except possibly for computing the derivative of the last, $\sec x \tan x$ term as a separate problem. Even for that term, while we *could* use logarithmic differentiation, a more efficient method would be to simply use the Product Rule.

3.10. THE NATURAL LOGARITHM II: FURTHER RESULTS

Solution: For convenience, we (equivalently) find $\frac{dy}{dx}$ where $y = x^{\sin x}$.

$$\begin{aligned} y = x^{\sin x} &\iff \ln y = \ln\left[x^{\sin x}\right] \\ &\iff \ln y = \sin x \cdot \ln x \\ &\implies \frac{d \ln y}{dx} = \frac{d}{dx}[\sin x \cdot \ln x] \\ &\implies \frac{1}{y} \cdot \frac{dy}{dx} = \sin x \cdot \frac{d \ln x}{dx} + \ln x \cdot \frac{d \sin x}{dx} \\ &\implies \frac{1}{y} \cdot \frac{dy}{dx} = \frac{\sin x}{x} + \ln x \cos x \\ &\iff \frac{dy}{dx} = y\left[\frac{\sin x}{x} + \ln x \cos x\right] = x^{\sin x}\left[\frac{\sin x}{x} + \ln x \cos x\right]. \end{aligned}$$

It is very interesting to note here that the derivative simplifies to

$$\begin{aligned} \frac{dy}{dx} &= \frac{\sin x \cdot x^{\sin x}}{x} + x^{\sin x} \ln x \cdot \cos x \\ &= (\sin x) x^{\sin x - 1} + x^{\sin x} \ln x \cdot \frac{d \sin x}{dx}. \end{aligned}$$

In this sum, the first term is what we would have if we assumed naively that the $\sin x$ exponent were a constant and we used the Power Rule, $\frac{d}{dx}[x^n] = n \cdot x^{n-1}$, with $n = \sin x$. The second term is what we would have if we instead assumed naively that the base x were a constant and we used the formula for the derivative of an exponential function, $\frac{d}{dx}[a^u] = a^u \ln a \cdot \frac{du}{dx}$. So it appears that the change we measure by applying $\frac{d}{dx}$ has two components, the first assuming that the exponent is constant and measuring that part of the change we get from the base changing, and the second assuming the base is constant and measuring the change we get from the exponent changing. This is very much akin to our earlier interpretation of the Product Rule. (See again the notes on the roles of the various factors in the Product Rule, page 298.) This gives us a general formula, which is slightly more complicated than one might anticipate from this example because of a Chain Rule involved in computing the component from the change in the base:

$$\boxed{\frac{d}{dx}\left[(f(x))^{g(x)}\right] = g(x) \cdot [f(x)]^{g(x)-1} \cdot \frac{d}{dx}[f(x)] + (f(x))^{g(x)} \ln f(x) \cdot \frac{d}{dx}[g(x)].} \qquad (3.130)$$

Again, the first term is computed as though the exponent were constant and the Power Rule employed, while the second term is computed as though the base were constant and the exponential rule were used. In both cases, Chain Rule versions were necessary to be most general.

Two possible proofs come to mind. One is to rewrite the original function with a constant base, so that

$$(f(x))^{g(x)} = \left[e^{\ln f(x)}\right]^{g(x)} = e^{(\ln f(x))(g(x))}, \qquad (3.131)$$

and use the Chain Rule, with a Product Rule inside, finally rewriting the result with the original base.

More in the spirit of how these problems are usually presented is a proof using logarithmic differentiation, which is what we used in the previous example. In the proof that follows we consolidate some of the steps. Note that we assume $f(x) > 0$ for continuity's sake.

$$y = (f(x))^{g(x)} \implies \ln y = g(x) \ln f(x)$$
$$\implies \frac{1}{y} \cdot \frac{dy}{dx} = g(x) \cdot \frac{d \ln f(x)}{dx} + \ln f(x) \cdot \frac{d g(x)}{dx}$$
$$\implies \frac{1}{y} \cdot \frac{dy}{dx} = g(x) \cdot \frac{1}{f(x)} \cdot \frac{d f(x)}{dx} + \ln f(x) \cdot \frac{d g(x)}{dx}$$
$$\implies \frac{dy}{dx} = y \left[g(x) \cdot \frac{1}{f(x)} \cdot \frac{d f(x)}{dx} + \ln f(x) \cdot \frac{d g(x)}{dx} \right]$$
$$\implies \frac{dy}{dx} = (f(x))^{g(x)} \left[g(x) \cdot \frac{1}{f(x)} \cdot \frac{d f(x)}{dx} + \ln f(x) \cdot \frac{d g(x)}{dx} \right]$$
$$\implies \frac{dy}{dx} = g(x) \cdot [f(x)]^{g(x)-1} \cdot \frac{d}{dx}[f(x)] + (f(x))^{g(x)} \ln f(x) \cdot \frac{d}{dx}[g(x)],$$

q.e.d.

While the resulting formula is rather long, it is easy enough to memorize if it is remembered in spirit: again, the first term treats $g(x)$ as a constant, and the second treats $f(x)$ as a constant. Still, it is predictable that many students would feel more comfortable either using logarithmic differentiation, or the change of base (to e) strategy of (3.131), page 387.

It is interesting to note that this new derivative formula (3.130), page 387, in fact generalizes the power and exponential rules.

1. If $g(x)$ is constant, so $\frac{d}{dx}[g(x)] = 0$, then we get the Power Rule, in its Chain Rule version.

2. If $f(x)$ is constant, so $\frac{d}{dx}[f(x)] = 0$, then we get the exponential rule, in its Chain Rule version.

Example 3.10.8

Find $\frac{d}{dx}\left[(\ln x)^x\right]$.

Solution: While we can use logarithmic differentiation here, we will use our general formula, first treating the exponent x as constant, and then the base $\ln x$ as constant.[87]

$$\frac{d}{dx}\left[(\ln x)^x\right] = x(\ln x)^{x-1} \frac{d}{dx} \ln x + (\ln x)^x (\ln(\ln x)) \cdot \frac{d}{dx}[x]$$
$$= x(\ln x)^{x-1} \cdot \frac{1}{x} + (\ln x)^x (\ln(\ln x)) \cdot 1$$
$$= (\ln x)^{x-1} + (\ln x)^x \ln(\ln x).$$

---

[87] In most mathematical literature, the short-hand for $\ln(\ln x)$ is simply $\ln \ln x$. Occasionally one sees such things as $\ln \ln \ln x \ln \ln x$, meaning $[\ln(\ln(\ln x))][\ln(\ln x)]$, for instance. For $\ln(\ln(\ln x))$ to be defined requires $x > e$, so $\ln x > 1$, so $\ln(\ln x) > 0$, allowing $\ln(\ln(\ln x))$ to be defined.

3.10. THE NATURAL LOGARITHM II: FURTHER RESULTS

Example 3.10.9

Compute $\frac{d}{dx}\left[x^{e^x}\right]$.

Solution:

$$\frac{d}{dx}\left[x^{e^x}\right] = e^x[x]^{e^x-1}\frac{dx}{dx} + x^{e^x}\ln x\frac{de^x}{dx}$$
$$= e^x[x]^{e^x-1}\cdot 1 + x^{e^x}\ln x \cdot e^x$$
$$= e^x[x]^{e^x-1} + e^x x^{e^x}\ln x.$$

Admittedly, it is rare to find the formula (3.130), page 387 for $\frac{d}{dx}(f(x))^{g(x)}$ in a standard calculus textbook. One could argue that the formula is sufficiently complicated that it is better to always use the logarithmic differentiation technique, since it is more flexible and, after all, has the built-in means to prove the formula. But (3.130) is included here because it contains some nice intuition and gives a shorter route to the desired derivatives. Indeed, while these types of derivatives are relatively rare in most practices, for those with the need to compute them routinely, the formula could have good utility.

Of course logarithmic differentiation had further utility than simply computing derivatives of functions of form $(f(x))^{g(x)}$, since the logarithm took products to sums, quotients to differences, and powers became multiplying factors, greatly simplifying the differentiation of complicated combinations of products, quotients, and powers in many circumstances, with a modest price to pay in complexity of the overall logarithmic differentiation process, as the examples in Subsection 3.10.1 (starting on page 378) demonstrated.

Exercises

For Exercises 1–8, compute the given derivative.

1. $\frac{d}{dx}\left[4^{\sin x}\right]$

2. $\frac{d}{dx}\left[\sin 4^x\right]$

3. $\frac{d}{dx}\log_2|\tan x|$

4. $\frac{d}{dx}\tan(\log_2 x)$

5. $\frac{d}{dx}\log_4 3^x$

6. $\frac{d}{dx}\log_2\left|\frac{x}{x^2+1}\right|$

7. $\frac{d}{dx}\log_2\left[\sin^2 x \cos^4 x\right]$

8. $\frac{d}{dx}\left[3^{2^x}\right]$

For Exercises 9–14, use logarithmic differentiation to prove the given formula. For example, to show $\frac{dx^n}{dx} = nx^{n-1}$, first set $y = x^n$ and then use logarithmic differentiation to find $\frac{dy}{dx}$.

9. $\frac{da^x}{dx} = a^x \ln a$, assuming $a > 0$.

10. $\dfrac{d[uv]}{dx} = u \cdot \dfrac{dv}{dx} + v \cdot \dfrac{du}{dx}$.

11. $\dfrac{d}{dx}\left[\dfrac{u}{v}\right] = \dfrac{v \cdot \frac{du}{dx} - u \cdot \frac{dv}{dx}}{v^2}$.

12. $\dfrac{d\,x^n}{dx} = n \cdot x^{n-1}$.

13. $\dfrac{d\sec x}{dx} = \sec x \tan x$, using the derivative formula for $\cos x$ and the fact that $\sec x = 1/\cos x$. (Recall also that $|a/b| = |a|/|b|$.)

14. $\dfrac{d\,e^x}{dx} = e^x$.

For Exercises 15–20, use logarithmic differentiation to compute the following.

15. $\dfrac{d}{dx}\sqrt{\dfrac{1+x}{1-x}}$. Note that on the domain of this function, we have $\sqrt{(1+x)/(1-x)} = \sqrt{|1+x|/|1-x|}$.

16. $\dfrac{d}{dx}\left[(10x^2 + 9)\right]$

17. $\dfrac{d}{dx}\left[(x^2+4)^3 (x^3+5)^6\right]$

18. $\dfrac{d}{dx}\left[\dfrac{x^5}{(x^4+8x-7)^3}\right]$

19. $\dfrac{d}{dx}\left[e^{2x}\sin 5x \cos 9x\right]$

20. $\dfrac{d}{dx}\left[\dfrac{\sin^5 x \cdot \sqrt[3]{x^2-2}}{2(6x-7)^4(2x+5)^3}\right]$

For Exercises 21–24, compute $\dfrac{dy}{dx}$ for each. You may use either logarithmic differentiation, or (3.130), page 387. (It would be useful to use both and examine how the results are in fact the same.)

21. $y = x^x$

22. $y = x^{1/x}$

23. $y = x^{x^x} = (x)^{(x^x)}$

24. $y = \left(\dfrac{\sin x}{x}\right)^x$

25. Prove (3.130), page 387, using (3.131), i.e., without logarithmic differentiation.

26. Show how (3.130) still returns the correct derivative by using it, and (separately) previous methods, to compute $\dfrac{d\,x^3}{dx}$.

27. Repeat for $\dfrac{d\,10^x}{dx}$.

28. Repeat for $\dfrac{d\,e^{x^2}}{dx}$.

29. Repeat for $\dfrac{d\sin^2 x}{dx}$.

30. Suppose $y = \dfrac{x^3\cos^7 x}{\sqrt{x^4+1}} + x^9 e^{3x} - \sec^{-1}x^2$. Writing $y = y_1 + y_2 - y_3$, find $\dfrac{dy}{dx}$. (Hint: Compute $\dfrac{dy_i}{dx}$ separately for $i = 1,2,3$. Note also that $|x^2| = x^2$.)

31. (Proof of Change of Base Formula (3.104), page 368) Here we prove that $\log_a x = \dfrac{\log_b x}{\log_b a}$ regardless of $a, b \in (0,1) \cup (1,\infty)$.

    (a) Let $y = \log_a x$.
    (b) Then $x = a^y$.
    (c) Apply $\log_b(\ )$ to both sides.
    (d) Use properties of logarithms to rewrite the new equation.
    (e) Solve for y.
    (f) Substitute for y using the original equation, q.e.d.

32. Repeat using $\ln(\ )$ in place of $\log_b(\ )$.

3.11 Further Interpretations of the Derivative

If one asks a calculus student what is the derivative, the most common answer will likely be "the slope of the tangent line," or perhaps "the slope of the curve," expressing the geometric significance of the derivative in the context of the graph of a function such as $y = f(x)$.

On the other hand, if one asks an engineer, scientist, or economist, one is more likely to hear "the rate of change." There are many quantities that are discussed, in even casual conversation, which are rates of change, even if they are not called that explicitly. Included in those are speed (or velocity) and acceleration, population (or any other) growth or decay, power, current, miles per gallon, and for that matter just about any concept that includes the word "per."

Describing a rate of change of a quantity requires answering also the question of "with respect to what?" We already discussed how velocity v is a rate of change of position s with respect to time t, so $v = ds/dt$. This is an *instantaneous* rate of change of position with respect to time, and as such is the result of computing the limit of $(\Delta s)/(\Delta t)$ of average velocities as $\Delta t \to 0$.

In general, a derivative dQ/dx can always be viewed as this instantaneous rate of change of Q with respect to x, assuming Q is ultimately a function of x. Now dQ/dx is again a limit of difference quotients $(\Delta Q)/(\Delta x)$ (see the discussion beginning with (3.4), page 212), so it carries with it the units of Q divided by the units of x and the intuition of being a limit of difference quotients. However, the best visualization of the meaning of dQ/dx depends upon the natures of the variables Q and x. For instance, with position s and time t being our "Q and x," respectively, a graph of s versus t, while interesting and occasionally useful, is perhaps not as illuminating as would be a visual demonstration of an object with position s moving as time t progresses.

In this short section we continue the discussion of Section 3.3, where we first explored various interpretations of the derivative, and the underlying intuitions of each. Because we have more derivative rules to work with, we can look at more diverse applications of the derivative.

3.11.1 Velocity and Acceleration

Recall that if $s(t)$ is the position along a numbered axis at time t, then velocity is given by

$$v(t) = \lim_{\Delta t \to 0} \frac{s(t + \Delta t) - s(t)}{\Delta t} = \frac{ds(t)}{dt},$$

or $\frac{ds}{dt}$ for short, and by the nature of the difference quotient this will carry units of length/time: if s is in meters (m) and t is in seconds (sec), then $v = ds/dt$ will be in m/sec.

Furthermore with acceleration, meaning a rate of change in velocity, we have[88]

$$a(t) = \lim_{\Delta t \to 0} \frac{v(t + \Delta t) - v(t)}{\Delta t} = \frac{dv(t)}{dt},$$

---

[88] There is a term for the rate of change in acceleration, variously called the *jerk, jolt, surge,* or *lurch*: $j = \frac{da}{dt}$. It is naturally a bit more difficult to visualize than velocity or acceleration, but one can certainly make an attempt. Consider, for instance, being a passenger in a car that has constant acceleration from a stop. If the acceleration were to suddenly change, it is not unreasonable to describe that change as a "jerk," or one of the other terms for the same concept.

or simply dv/dt, which will have units of (m/sec)/sec, or m/sec^2. While those units may seem a bit odd, mathematically they are correct and can perhaps make more sense when stated as "meters per second per second." For instance, for an object dropped into free fall near the surface of the earth, if not significantly susceptible to air resistance, it will have constant acceleration -9.8 m/sec^2, whereby initial velocity will be $v(0) = 0$, and one second later will be $v(1\text{ sec}) = -9.8$ m/sec, and then $v(2\text{ sec}) = -19.6$ m/sec, and so on, and so the velocity changes by -9.8 m/sec *per second*, i.e., $a = -9.8$ m/sec/sec $= -9.8$ m/sec^2. If we were to use English units, we would report $a = -32$ ft/sec^2.

Example 3.11.1

Find the vertical velocity and acceleration of an object whose height is given by the function $y(t) = -16t^2 + 64t + 80$. Here y is in feet and t is in seconds.

Solution: We simply compute the following, v having units of ft/sec and a having units of ft/sec^2:

$$v = \frac{dy}{dt} = -32t + 64,$$
$$a = \frac{dv}{dt} = -32.$$

Note some values for $v(t)$ before it hits the ground (which solving $y = 0$ would show occurs shortly after $t = 3$ sec):

$t =$	0	1	2	3
$y(t) =$	80	128	144	128
$v(t) =$	64	32	0	-32
$a(t) =$	-32	-32	-32	-32

A student of physics will encounter many notations for position, velocity, and acceleration in the course of studies, and so a robust understanding—independent of notation used—is one benefit of studying that field. In the next example, we look at uniform circular motion, and the velocity and acceleration in that context, and use a somewhat different notation due to the complexity of having more than one velocity and acceleration.

Example 3.11.2

Suppose a particle's position in the xy-plane is given by $\begin{cases} x = 5\cos 2t, \\ y = 5\sin 2t. \end{cases}$

1. Show that these points all lie on a circle, and find its equation.
2. Plot the positions of this particle for $t = 0, \frac{\pi}{4}, \frac{\pi}{2}, \frac{3\pi}{4}, \pi$.
3. Find the horizontal and vertical velocities and accelerations.

Solution: (1) The "trick" to show this particle lies on a circle is to square both sides of each equation, and notice something about their sum:

$$\begin{aligned} x^2 &= 25\cos^2 2t \\ y^2 &= 25\sin^2 2t \end{aligned} \quad \Longrightarrow \quad x^2 + y^2 = 25(\cos^2 2t + \sin^2 2t) \quad \Longrightarrow \quad x^2 + y^2 = 25,$$

and so the particle lies on the curve $x^2 + y^2 = 5^2$, i.e., the circle centered at the origin $(0,0)$ with radius 5. (In fact, as t varies the particle will traverse the circle counterclockwise repeatedly, as the next part hints.)

(2) First we make a table, and then we plot the points:

3.11. FURTHER INTERPRETATIONS OF THE DERIVATIVE

t	(x,y)	simplified
0	$(5\cos 0, 5\sin 0)$	$(5,0)$
$\pi/4$	$(5\cos\frac{\pi}{2}, 5\sin\frac{\pi}{2})$	$(0,5)$
$\pi/2$	$(5\cos\pi, 5\sin\pi)$	$(-5,0)$
$3\pi/4$	$(5\cos\frac{3\pi}{2}, 5\sin\frac{3\pi}{2})$	$(0,-5)$
π	$(5\cos 2\pi, 5\sin 2\pi)$	$(5,0)$

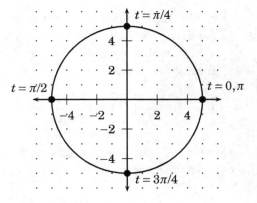

Next we plot these points to the right, labeling each. Note how the particle travels counterclockwise.

(3) Next we introduce some notation while we compute velocities and accelerations. For brevity, we omit explicitly stating the Chain Rule as we apply it.[89]

$$v_x = \frac{dx}{dt} = \frac{d}{dt}[5\cos 2t] = -10\sin 2t \quad \Longrightarrow \quad a_x = \frac{dv_x}{dt} = \frac{d}{dt}[-10\sin 2t] = -20\cos 2t,$$

$$v_y = \frac{dy}{dt} = \frac{d}{dt}[5\sin 2t] = 10\cos 2t \quad \Longrightarrow \quad a_y = \frac{dv_y}{dt} = \frac{d}{dt}[10\cos 2t] = -20\sin 2t.$$

3.11.2 First-Order and Second-Order Chemical Reactions

We discussed zero-order chemical reactions in Subsection 3.3.3, page 261. Recall that [A] represents the concentration of some reactant A. As before, we will let $X = [A]$.

- In a first-order reaction, we have $X = X_0 e^{-kt}$.

  Here $k > 0$ is some rate constant, and X_0 is the concentration of A at time $t = 0$. Then

  $$\frac{dX}{dt} = \frac{d}{dt}\left[X_0 e^{-kt}\right] = X_0 e^{-kt}\frac{d}{dt}[-kt] = -kX_0 e^{-kt}.$$

  If we recall that $X = X_0 e^{-kt}$, then this equation can become

  $$\frac{dX}{dt} = -kX.$$

- On the other hand, in a second-order reaction, we have $\dfrac{1}{X} = \dfrac{1}{X_0} + kt$.

---

[89]If one considers the *vector* quantities of position $\langle x(t), y(t)\rangle$, velocity $\langle v_x(t), v_y(t)\rangle$, and acceleration $\langle a_x(t), a_y(t)\rangle$, this computation becomes more interesting. For instance, the Pythagorean Theorem gives us the *magnitude* of the vectors $\langle x,y\rangle$, $\langle v_x,v_y\rangle$, and $\langle a_x,a_y\rangle$ as being $\sqrt{x^2+y^2} = \sqrt{25\cos^2 2t + 25\sin^2 2t} = \sqrt{25(\cos^2 2t + \sin^2 2t)} = \sqrt{25(1)} = 5$, and similarly $\sqrt{v_x^2+v_y^2} = \sqrt{10^2} = 10$, and $\sqrt{a_x^2+a_y^2} = \sqrt{20^2} = 20$, all three being constant. Also, the velocity vector is always tangential to the path, and for uniform circular motion, the acceleration vector $\langle a_x, a_y\rangle$ is in the direction opposite the position vector $\langle x,y\rangle$. (In this case, $\langle a_x, a_y\rangle = -20\langle x,y\rangle$.)

While this might not at first seem to justify the label "second-order," once we compute the derivative it will be more apparent. At least we can note that $1/X$ is increasing with time t, which means X is shrinking (note X, as a concentration, is never negative). Taking derivatives we see

$$\frac{d}{dt}\left[\frac{1}{X}\right] = \frac{d}{dt}\left[\frac{1}{X_0} + kt\right] \implies \frac{-1}{X^2} \cdot \frac{dX}{dt} = k \implies \frac{dX}{dt} = -kX^2.$$

Combining these derivations with what we derived previously for zero-order reactions, we can produce a chart, this time using the chemistry notation $X = [A]$:

order	concentration	rate
zero	$[A] = [A_0] - kt$	$\frac{d[A]}{dt} = -k$
first	$[A] = [A_0]e^{-kt}$	$\frac{d[A]}{dt} = -k[A]$
second	$\frac{1}{[A]} = \frac{1}{[A_0]} + kt$	$\frac{d[A]}{dt} = -k[A]^2$

So the choice of order corresponds to the exponent of $[A]$ in the rate equation. Beginning in Chapter 5 we will explore how the rate equations imply the concentration equations. So while the concentration equations may seem to show no pattern or justification for the orders, the rate equations show that the labels are justified. Generally, a reaction is considered to be of nth order if the rate equation is given by $\frac{d}{dt}[A] = -k[A]^n$, where n can be any exponent, though an arbitrary choice might not correspond to any actual chemical reactions.

It is not uncommon for theoretical equations for various phenomena to be simpler and more easily understood from the "derivative side," rather than an algebraic or otherwise non-calculus perspective.

3.11.3 Power

Power is a measure of how quickly energy is available, and as such is in some sense a measure of the rate of "flow" of energy. For instance, consider a 1-ton vehicle accelerating at a constant rate from 0 to 60 miles/hour. Whether that change in speed occurs in 10 seconds or 5 seconds, the energy required is the same. However, the energy is expended twice as fast to accelerate the vehicle to 60 miles/hour in 5 seconds, compared to doing the same work in 10 seconds. In other words, it takes twice the power to accelerate the vehicle to 60 miles/hour in half the time.[90]

Science textbooks define energy as *the ability to do work,* and work as *a force pushing or pulling an object through a distance.* Numerically, the energy required to do a unit of work will be the same as that unit measurement of the work done. For example, one joule of energy is required to do one joule of work.

---

[90] In both scenarios—accelerating the vehicle to 60 miles/hour in 10 seconds or 5 seconds, the car is subject to work, i.e., a push or pull over a distance. The car absorbs this work and gains *kinetic energy,* which is the energy an object has due to its motion (also proportional to mass m, by the formula $KE = \frac{1}{2}mv^2$), as a result of the work done to it. Since in both cases the vehicle emerges with the same kinetic energy, the same amount of work was performed on it. *Power* refers to how quickly that work could occur, i.e., how quickly the energy could be absorbed from the source of the force.

Example 3.11.3

Suppose a Ferris Wheel with a 40-ft radius is empty except for a 200-lb adult, whose height as measured from the ground changes with time t according to the function $h(t) = 45 + 40\sin(2\pi t)$, where t is in minutes. Thus the Ferris Wheel rotates once per minute. We can consider the power required to lift and lower this person in this manner (beyond what is required to move the Ferris Wheel when empty) by finding the change in the adult's potential energy with respect to time.

The potential energy is given by $E_p = mgh = Fh$, where F is the force of gravity on the adult, in this case $mg = F = 200\,\text{lb}$, and $h(t)$ is given. Thus the extra power P required to sustain the motion of this adult, as a function of time t, emerges from the computation

$$E_p = 200 h(t) = 200(45 + 40\sin(2\pi t))$$

$$\implies P = \frac{dE_p}{dt} = 200 \cdot 40\cos(2\pi t) \cdot \frac{d\, 2\pi t}{dt} = 16{,}000\pi \cos(2\pi t),$$

with units of ft-lb/min (or we could multiply by (1 min)/(60 sec) to have power in an arguably more intuitive ft-lb/sec, but we leave that to the interested reader). Some values of t, $h(t)$ and $P(t)$ are given next (exactly or to three significant digits) for one complete cycle of the ride:

t	0	1/8	1/4	3/8	1/2	5/8	3/4	7/8	1
$h(t)$	45	73.3	85	73.3	45	16.7	5	16.7	45
$P(t)$	50,300	35,500	0	−35,500	−50,300	−35,500	0	35,500	50,300

We see that when the adult is rising and "half-way up" in the ride ($t = 0, 1$), the most extra power is required (50,300 ft-lb/sec), while no extra power is needed the moment the adult is at the top of the ride ($t = 1/4$), or for that matter the bottom of the ride ($t = 3/4$). Negative "extra" power from the Ferris Wheel's motor is needed as the adult is lowering, meaning that the adult's weight is performing positive work on the Ferris Wheel (the adult's force from gravity is in the direction of his or her motion), until the adult reaches the bottom ($t = 3/4$), after which for a time the Ferris Wheel must perform positive work (requiring positive power P) on the adult again, and so on.[91]

3.11.4 Population Growth and Decay

Exponential growth and decay are applications of calculus in which one often uses a continuum model for a phenomenon that is actually discrete. While there are slightly differing opinions on what constitutes a discrete variable, it can be contrasted to a continuous one that can, for instance, take on a continuum of values within some interval of positive length. For instance length, (time) duration, and temperature are such that if they can take on values a and b, then they can take on any value within $[a, b]$. Discrete variables, by contrast, cannot. They will have either a finite number of values they can take on, or some infinite set of values that can be listed (or, as they say in mathematics, are "countably infinite"). Often, discrete variables are allowed to take on only integer values or some subset thereof. Examples include the possible answers on a true-or-false question, hotel ratings (zero to five "stars"), a government identity number, and so on. (For further discussion see Footnote 56, page 61.)

Another example is the number of individuals in a population. If one has all the data available to analyze an actual population, there is little need for calculus, except perhaps for

---

[91]Of course ideally a Ferris Wheel operator would attempt to distribute the riders in such a manner that at any given time those riders who require positive work (and therefore positive power) to be lifted are balanced with those requiring negative work because they are descending.

predictive analysis. However, if one wished to model a hypothetical population, the calculus can be useful.

Example 3.11.4

Suppose a population of 100 individuals is expected to double every 25 years. What is the growth rate, in persons per year, at the start, 10 years later, 25 years later, and 50 years later?

<u>Solution</u>: A reasonable model for the size of the population at year t after the 100 individuals form the population would be

$$P = 100 \cdot 2^{t/25}.$$

Note how $P(0) = 100$, $P(25) = 200$, $P(50) = 400$, and so on. This is a continuous approximation of a discrete problem. We can then compute

$$P'(t) = \frac{dP}{dt} = 100 \cdot 2^{t/25} \ln 2 \cdot \frac{d}{dt}\left[\frac{1}{25}t\right]$$
$$= 4\ln 2 \cdot 2^{t/25}.$$

From this we can compute

$$P'(0) = 4\ln 2 \qquad \approx 2.772$$
$$P'(10) = 4\ln 2 \cdot 2^{10/25} \approx 3.658$$
$$P'(25) = 4\ln 2 \cdot 2 \qquad \approx 5.545$$
$$P'(50) = 4\ln 2 \cdot 2^2 \qquad \approx 11.090$$

Note that $P'(50) = \left.\frac{dP}{dt}\right|_{t=50}$, and so it will carry units of persons/year.

This example is theoretically correct but conceptually awkward. First, it is inconceivable to produce 11.090 persons in a year. Second, this is actually an instantaneous rate of change, so it seems more intuitive in some ways to instead discuss the actual number of persons added during the year, so instead of $P'(50)$ it might be more interesting to compute

$$P(50) - P(49) = 100 \cdot 2^2 - 100 \cdot 2^{49/25} \approx 10.938,$$

though that is still not quite consistent with reality because persons only come in whole numbers. At this point, one might remember that this is likely to be an approximation already, and so a researcher might instead report from either observation that there will be an annual rate of approximately 11 persons/year at time $t = 50$ years. However, note that the computation that uses the calculus is a shorter computation than the one that computes the actual value $(\Delta P)/(\Delta t)$.[92]

A more reasonable continuous approximation with what is technically a discrete variable is radioactive decay. It is impractical to count the actual number of radioactive atoms in a visible sample, so we choose a scale of unit where the actual numbers we deal with are not large, as for instance we model the *mass* of the radioactive element, which more closely resembles a continuum.

Example 3.11.5

The somewhat unstable carbon-14 atom is produced when cosmic rays interact with nitrogen, and so many accept the theory that the proportion of ^{14}C in the atmosphere remains relatively constant,

---

[92]The interval to use to compute $(\Delta P)/(\Delta t)$ for this example is somewhat open to opinion. We used $(P(50) - P(49))/(1)$, but it is reasonable to opt for $(P(51) - P(50))/(1)$, or perhaps $(P(50.5) - P(49.5))/(1)$, this last one centering at $t = 50$. Either way, the calculus-utlizing computation of $P'(50)$ is a reasonable approximation.

3.11. FURTHER INTERPRETATIONS OF THE DERIVATIVE

while a plant absorbing it will therefore have the same proportion of ^{14}C when it dies regardless of the era in which it lived. Carbon-14 has a half-life of approximately 5,730 years. This means that an otherwise undisturbed sample with initially a total of N_0 of these atoms of ^{14}C will have approximately $\frac{1}{2}N_0$ atoms of ^{14}C after 5730 years, and $\frac{1}{2} \cdot \frac{1}{2} N_0$ atoms of ^{14}C after 11,460 years, and so on. If $N = N(t)$ is the number of atoms of ^{14}C after t years, then

$$N = N_0 \left(\frac{1}{2}\right)^{t/5730},$$

since $t/5730$ is the number of half-lives after t years. How many years before 5 percent of the ^{14}C has decayed? At what rate is it decaying at that time?

Solution: The first question is algebraic: We must find t so that $N = 0.95N_0$, i.e., so that

$$N_0 \left(\frac{1}{2}\right)^{t/5730} = 0.95 N_0$$

$$\Longrightarrow \quad \left(\frac{1}{2}\right)^{t/5730} = 0.95$$

$$\Longrightarrow \quad \ln\left[\left(\frac{1}{2}\right)^{t/5730}\right] = \ln 0.95$$

$$\Longrightarrow \quad \frac{t}{5730} \ln\left(\frac{1}{2}\right) = \ln(0.95)$$

$$\Longrightarrow \quad t = 5730 \ln(0.95)/\ln(0.5)$$

$$\Longrightarrow \quad t \approx 397 \text{ years}.$$

For the second question, we find $\frac{dN}{dt}$ and evaluate it at that time found previously:

$$N = N_0 \left(\frac{1}{2}\right)^{t/5730}$$

$$\Longrightarrow \quad \frac{dN}{dt} = N_0 \left(\frac{1}{2}\right)^{t/5730} \ln\left(\frac{1}{2}\right) \cdot \frac{d}{dt}\left(\frac{t}{5730}\right)$$

$$= -\frac{N_0 \ln 2}{5730} \left(\frac{1}{2}\right)^{t/5730} = \frac{-\ln 2}{5730} N.$$

Thus when $N = 0.95 N_0$ we have

$$\left.\frac{dN}{dt}\right|_{N=0.95N_0} = \frac{-\ln 2}{5730} 0.95 N_0$$

$$\approx -0.00011492 N_0,$$

or we could say that the sample is losing approximately 0.0115% of the original amount per year at that time.

Exercises

1. (Here we revisit uniform circular motion as in Example 3.11.2, page 392.) Consider the two-dimensional motion

   $$\begin{cases} x = 3\sin t, \\ y = 3\cos t, \end{cases}$$

   (a) Show that this motion lies along a circle of radius 3.

   (b) Plot and label the position for $t = 0, \pi/4, \pi/2, \pi/3, \pi/2,$ and 2π. In which direction is the particle moving around the circle?

   (c) Find v_x, v_y, a_x, a_y for $t = 0, \pi/2, \pi/3, \pi/2,$ and 2π.

2. Consider the two-dimensional motion of a particle given by

   $$\begin{cases} x = 2\sin 2t \\ y = 3\cos 2t. \end{cases}$$

   An ellipse centered at $(0,0)$ can be represented by the equation

   $$\frac{x^2}{a^2} + \frac{y^2}{b^2} = 1,$$

where $a, b > 0$.

(a) Show that the motion described above lies along an ellipse.

(b) Plot and label the position $(x(t), y(t))$ for times $t = 0$, $\pi/8$, $\pi/4$, $\pi/2$, $3\pi/4$, and π. In which direction (clockwise or counterclockwise) is the particle moving around the ellipse?

(c) Find v_x, v_y, a_x, a_y.

3. An object moves along a numbered axis with its position given according to the formula $s(t) = 4t^3 - 2t^2 - 8t - 4$.

   (a) Calculate the velocity function $v(t) = \frac{ds(t)}{dt}$.

   (b) Calculate the acceleration function $a(t) = \frac{dv(t)}{dt}$.

   (c) For what value(s) of t is the velocity equal to zero?

   (d) For what value(s) of t is the acceleration equal to zero?

4. The momentum p of an object is defined as being the product of its mass and its velocity. That is, $p = mv$, where p is measured in kg-m/s. If the mass of the object is constant, show that the rate of change of the kinetic energy of the object with respect to velocity is the object's momentum. (Recall kinetic energy is given by $\frac{1}{2}mv^2$.)

5. The force experienced by an object is the rate of change of its momentum (mv) with respect to time.

   (a) Assuming that an object's mass is constant, show that the force is given by the equation $F = ma$.

   (b) If the object's mass is not constant (e.g., as occurs with rockets as they expel burned fuel), find an equation for force F.

6. A population P is increasing according to the equation
$$P(t) = 23{,}000 e^{0.02t},$$
where t is measured in years.

   (a) Calculate $P(10)$, $P(11)$, and $P(12)$.

   (b) Find $P'(11) = \left.\frac{dP}{dt}\right|_{t=11}$.

   (c) Find the average change of the population per year from $t = 10$ to $t = 12$.

   (d) Compare your answers in parts (b) and (c). Are they similar?

7. The half-life of iodine-131 is 8.0197 days.

   (a) Assuming N_0 is the initial amount of ^{131}I, and N is the amount after t days, write an equation for N as a function of t.

   (b) What percentage of ^{131}I is left after one day?

   (c) What percentage is left after 10 days?

   (d) How old is a sample that has only 5 percent of the original left?

   (e) At what rate (as a percentage of the original quantity N_0) is this isotope decaying when $t = 0$?

   (f) Repeat the rate computation for $t = 1$.

   (g) Repeat for $t = 10$.

3.12. HYPERBOLIC FUNCTIONS (OPTIONAL)

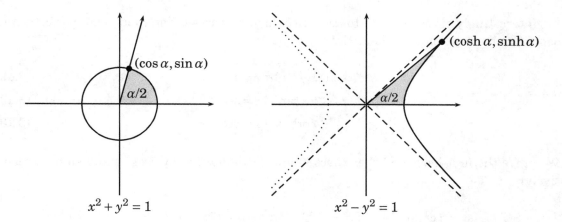

Figure 3.42: Illustration of analogous geometric definitions of hyperbolic functions versus trigonometric functions, both with areas $\alpha/2$, determining points on the curves. (Recall that a circular wedge with radius r, and angle θ, will have area $A = \frac{1}{2}r^2\theta$, so when $r = 1$ we have $A = \theta/2$.) In the circular case, $\alpha < 0$ indicates a clockwise rotation, and if the angle "sweeps" an area more than once, that area is added each time the angle sweeps through it. In the hyperbolic case on the right, we only consider Quadrants I and IV, and $\alpha < 0$ represents an area below the x-axis.

3.12 Hyperbolic Functions (Optional)

Here we look at some specialized functions that are at times convenient in applications. While the somewhat intriguing geometric definition of the hyperbolic sine and hyperbolic cosine are given in Figure 3.42, the more useful functional definitions are given next, which are equivalent to the geometric definitions, but we will not prove that here.

3.12.1 Definitions of Hyperbolic Functions

Definition 3.12.1 *We define the hyperbolic sine and cosine of a real number x by the equations*[93]

$$\sinh x = \frac{e^x - e^{-x}}{2}, \qquad \cosh x = \frac{e^x + e^{-x}}{2}. \qquad (3.132)$$

The other hyperbolic functions are defined in the usual ways as ratios of these (note some simplifications):

$$\tanh x = \frac{\sinh x}{\cosh x} = \frac{e^x - e^{-x}}{e^x + e^{-x}}, \qquad \coth x = \frac{\cosh x}{\sinh x} = \frac{e^x + e^{-x}}{e^x - e^{-x}}, \qquad (3.133)$$

$$\operatorname{sech} x = \frac{1}{\cosh x} = \frac{2}{e^x + e^{-x}}, \qquad \operatorname{csch} x = \frac{1}{\sinh x} = \frac{2}{e^x - e^{-x}}. \qquad (3.134)$$

---

[93] One interesting fact about these is that an ideal, uniformly dense cable fastened to two points of equal height and left to hang, will sag in the shape of the hyperbolic cosine's graph. Such a curve is quite commonly called a *catenary*, but has other, more obscure names such as *alysoid*, *chainette*, or *funicular*, giving further evidence of the impossibility of most mere mortals to keep up with all possible names for mathematical concepts.

Interestingly, these give rise to identities that are very similar, though not perfectly analogous to, the trigonometric identities:

$$\cosh^2 x - \sinh^2 x = 1, \tag{3.135}$$
$$1 - \tanh^2 x = \text{sech}^2 x, \tag{3.136}$$
$$\coth^2 x - 1 = \text{csch}^2 x. \tag{3.137}$$

We prove the first here, and the others follow (from division by $\cosh^2 x$ and $\sinh^2 x$, respectively):

$$\cosh^2 x - \sinh^2 x = \left(\frac{e^x + e^{-x}}{2}\right)^2 - \left(\frac{e^x - e^{-x}}{2}\right)^2 = \frac{e^{2x} + 2 + e^{-2x}}{4} - \frac{e^{2x} - 2 + e^{-2x}}{4} = \frac{2+2}{4} = 1, \text{ q.e.d.}$$

3.12.2 Derivatives of Hyperbolic Functions

Another interesting aspect of hyperbolic functions is how similar their derivative patterns are to those of the classical trigonometric functions. For instance,

$$\frac{d \sinh x}{dx} = \frac{1}{2} \cdot \frac{d}{dx}[e^x - e^{-x}] = \frac{1}{2}\left[e^x - e^{-x}\frac{d(-x)}{dx}\right] = \frac{e^x - e^{-x}(-1)}{2} = \frac{e^x + e^{-x}}{2} = \cosh x, \quad \text{while}$$

$$\frac{d \cosh x}{dx} = \frac{1}{2} \cdot \frac{d}{dx}[e^x + e^{-x}] = \frac{1}{2}\left[e^x + e^{-x}\frac{d(-x)}{dx}\right] = \frac{e^x + e^{-x}(-1)}{2} = \frac{e^x - e^{-x}}{2} = \sinh x.$$

These are again similar to the derivatives of the first basic trigonometric functions: $\frac{d \sin x}{dx} = \cos x$; $\frac{d \cos x}{dx} = -\sin x$.

For further exploration, we compute $\frac{d}{dx}\tanh x$. Note that twice we use $\frac{d}{dx}e^{-x} = -e^{-x}$, and at one point we use the fact that $A^2 - B^2 = (A+B)(A-B)$, with $A = e^x + e^{-x}$ and $B = e^x - e^{-x}$:

$$\frac{d}{dx}\tanh x = \frac{d}{dx}\left[\frac{e^x - e^{-x}}{e^x + e^{-x}}\right] = \frac{(e^x + e^{-x})\frac{d}{dx}(e^x - e^{-x}) - (e^x - e^{-x})\frac{d}{dx}(e^x + e^{-x})}{(e^x + e^{-x})^2}$$

$$= \frac{(e^x + e^{-x})(e^x + e^{-x}) - (e^x - e^{-x})(e^x - e^{-x})}{(e^x + e^{-x})^2}$$

$$= \frac{(e^x + e^{-x})^2 - (e^x - e^{-x})^2}{(e^x + e^{-x})^2}$$

$$= \frac{[(e^x + e^{-x}) + (e^x - e^{-x})][(e^x + e^{-x}) - (e^x - e^{-x})]}{(e^x + e^{-x})^2}$$

$$= \frac{2e^x \cdot 2e^{-x}}{(e^x + e^{-x})^2} = \frac{4}{(e^x + e^{-x})^2} = \left[\frac{2}{e^x + e^{-x}}\right]^2 = \text{sech}^2 x.$$

One could also have simply multiplied the numerator directly, but it is worth exploring the technique of writing any difference of squares $A^2 - B^2$ in its factored form $(A+B)(A-B)$.

The other derivatives of hyperbolic functions follow similarly, resembling those of the

3.12. HYPERBOLIC FUNCTIONS (OPTIONAL)

trigonometric functions except for different occurrences of the -1 factor:

$$\frac{d\sinh x}{dx} = \cosh x \qquad \frac{d\sinh u}{dx} = \cosh u \cdot \frac{du}{dx} \tag{3.138}$$

$$\frac{d\cosh x}{dx} = \sinh x \qquad \frac{d\cosh u}{dx} = \sinh u \cdot \frac{du}{dx} \tag{3.139}$$

$$\frac{d\tanh x}{dx} = \text{sech}^2 x \qquad \frac{d\tanh u}{dx} = \text{sech}^2 u \cdot \frac{du}{dx} \tag{3.140}$$

$$\frac{d\coth x}{dx} = -\text{csch}^2 x \qquad \frac{d\coth u}{dx} = -\text{csch}^2 u \cdot \frac{du}{dx} \tag{3.141}$$

$$\frac{d\,\text{sech}\,x}{dx} = -\text{sech}\,x\tanh x \qquad \frac{d\,\text{sech}\,u}{dx} = -\text{sech}\,u\tanh u \cdot \frac{du}{dx} \tag{3.142}$$

$$\frac{d\,\text{csch}\,x}{dx} = -\text{csch}\,x\coth x \qquad \frac{d\,\text{cscch}\,u}{dx} = -\text{csch}\,u\coth u \cdot \frac{du}{dx} \tag{3.143}$$

Computing derivatives that involve hyperbolic functions is just a matter of applying these new rules when relevant, along with the old rules as needed. However, we will not have a large number of examples or exercises applying these rules. Since the purpose of this section is to make the reader aware of rather than fluent in hyperbolic functions that arise in interesting but more specialized contexts (as compared to the trigonometric functions), we opt to have more than the usual proportion of results cited here derived or verified by the reader in the exercises at the end of the section. Unlike the more ubiquitous trigonometric functions, it is arguably more appropriate to derive these as needed than to recall them from rote memory.

3.12.3 Derivatives of Inverse Hyperbolic Functions

In fact, since all of these functions are based on combinations of exponential functions, in some sense they are redundant but worth defining for their rich algebraic and calculus structure. One can even find formulas for their inverses, using clever quadratic arguments and logarithms. We defer that for the moment, and instead list derivatives of the inverse hyperbolic functions, since they do resemble those of the inverse trigonometric functions.

So for instance we have the following, in non-Chain Rule versions. (By now the reader should be familiar with how to derive the Chain Rule versions of these from their non-Chain Rule versions.) The domains of these functions are given later on page 403.

$$\frac{d\sinh^{-1}x}{dx} = \frac{1}{\sqrt{x^2+1}} \qquad \frac{d\cosh^{-1}x}{dx} = \frac{1}{\sqrt{x^2-1}} \tag{3.144}$$

$$\frac{d\tanh^{-1}x}{dx} = \frac{1}{1-x^2} \qquad \frac{d\coth^{-1}x}{dx} = \frac{1}{1-x^2} \tag{3.145}$$

$$\frac{d\,\text{sech}^{-1}x}{dx} = \frac{-1}{x\sqrt{1-x^2}} \qquad \frac{d\,\text{csch}^{-1}x}{dx} = \frac{-1}{|x|\sqrt{1+x^2}} \tag{3.146}$$

Note how the difference between the derivative of an inverse hyperbolic function, and that of its inverse trigonometric counterpart, is often just a sign difference somewhere in the formula. We will explore why this is not surprising when we explore Taylor Series in the sequel, though some hint can be found in the analogous identities and derivatives.

Derivations of derivatives of inverse hyperbolic functions are similar to those of inverse trigonometric functions. For instance, to prove the first in (3.144), one proceeds as follows: Let $y = \sinh^{-1} x$, so that $\sinh y = x$. Applying d/dx to both sides of this equation, and eventually applying $\cosh^2 x - \sinh^2 x = 1$, yields

$$y = \sinh^{-1} x \implies \sinh y = x \implies \cosh y \cdot \frac{dy}{dx} = 1$$

$$\implies \frac{dy}{dx} = \frac{1}{\cosh y} = \frac{1}{\sqrt{\sinh^2 y + 1}} = \frac{1}{\sqrt{x^2 + 1}}, \quad \text{q.e.d.}$$

It was important that $\cosh y = \frac{1}{2}(e^y + e^{-y}) > 0$, so that we could use the positive square root case in the proof. Some more care must be taken with some of the other derivations, the details of which are left to the interested reader.[94]

3.12.4 Formulas for Inverse Hyperbolic Functions

It should be pointed out that the inverse hyperbolic functions can be constructed algebraically using special cases of simple algebraic techniques. For instance, we can solve for x in the equation $y = \sinh x$ as follows (note the use of the quadratic formula):

$$\begin{aligned}
y = \sinh x &\iff y = \frac{1}{2}\left(e^x - e^{-x}\right) \\
&\iff 2y = e^x - e^{-x} \\
&\iff 2y e^x = e^{2x} - 1 \\
&\iff 0 = (e^x)^2 - 2y(e^x) - 1 \\
&\iff e^x = \frac{-(-2y) \pm \sqrt{(-2y)^2 - 4(1)(-1)}}{2(1)} = \frac{2y \pm \sqrt{4y^2 + 4}}{2} = y \pm \sqrt{y^2 + 1} \\
&\iff e^x = y + \sqrt{y^2 + 1} \\
&\iff x = \ln\left(y + \sqrt{y^2 + 1}\right).
\end{aligned}$$

---

[94] Hyperbolic functions are mostly of interest in certain subfields of engineering and physics. Practitioners in those subfields may become very familiar with the quirks of these functions and their derivatives. For most others, it is simply important to know that they exist and to have some experience working with them so one can sense when a situation, and particularly one whose structure is "almost trigonometric," may instead be hyperbolic in nature, in which case an awareness of that fact can greatly expedite one's analysis, compared to resorting to the more fundamental functions which are the building blocks of these hyperbolic functions and their inverses.

That stated, most professional mathematicians use these rarely and can in good conscience resort to references when confronted by these. However, it is useful to have the experience of working with these at some point, as they do appear occasionally.

3.12. HYPERBOLIC FUNCTIONS (OPTIONAL)

Note how we cannot have the "$-$" case in the "$\pm$" because $e^x > 0$, while $y - \sqrt{y^2+1} < 0$ since $\sqrt{y^2+1} > \sqrt{y^2} = |y| \geq y$. For that reason, we can have only the "+" case. From this we get

$$\sinh^{-1} x = \ln\left(x + \sqrt{x^2+1}\right).$$

Similar techniques (where we find quadratic equations in e^x to solve), and the requirement that we choose segments of the original functions on which they are one-to-one (as we did with trigonometric functions), lead us to each of the following:

$$\sinh^{-1} x = \ln\left(x + \sqrt{x^2+1}\right), \qquad x \in \mathbb{R}, \qquad (3.147)$$

$$\cosh^{-1} x = \ln\left(x + \sqrt{x^2-1}\right), \qquad x \in [1, \infty), \qquad (3.148)$$

$$\tanh^{-1} x = \frac{1}{2} \ln\left(\frac{1+x}{1-x}\right), \qquad x \in (-1, 1), \qquad (3.149)$$

$$\coth^{-1} x = \frac{1}{2} \ln\left(\frac{x+1}{x-1}\right), \qquad x \in (-\infty, -1) \cup (1, \infty), \qquad (3.150)$$

$$\operatorname{sech}^{-1} x = \ln\left(\frac{1 + \sqrt{1-x^2}}{x}\right), \qquad x \in (0, 1], \qquad (3.151)$$

$$\operatorname{csch}^{-1} x = \ln\left(\frac{1}{x} + \frac{\sqrt{x^2+1}}{|x|}\right), \qquad x \in (-\infty, 0) \cup (0, \infty). \qquad (3.152)$$

It is an interesting exercise to use any one of these equations for an inverse hyperbolic function to derive its derivatives listed earlier. Such complications are somewhat involved, but have the satisfaction that they can be done using simply Power, Chain and logarithm-based differentiation rules, while we have no such options with the arctrigonometric functions.

3.12.5 Graphs of Hyperbolic Functions and Inverses

A few words should be included regarding the graphs of these functions. We mention here the graphs of the hyperbolic cosine and sine functions, and leave the others as exercises.

Example 3.12.1

Construct the graph of $y = \sinh x$.

<u>Solution</u>: It is useful to recall both forms for y:

$$y = \sinh x = \frac{1}{2}\left(e^x - e^{-x}\right).$$

If we let $f(x) = y$, then it is also useful to note that

$$f(-x) = \frac{1}{2}\left(e^{-x} - e^{-(-x)}\right) = \frac{1}{2}\left(e^{-x} - e^x\right) = -\frac{1}{2}\left(e^x - e^{-x}\right) = -f(x),$$

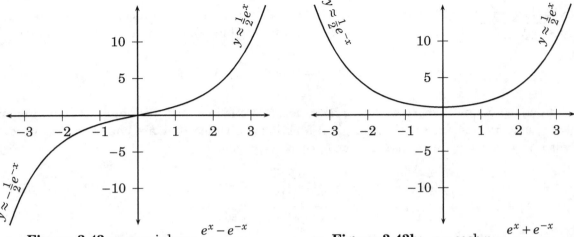

Figure 3.43a: $y = \sinh x = \dfrac{e^x - e^{-x}}{2}$

Figure 3.43b: $y = \cosh x = \dfrac{e^x + e^{-x}}{2}$

Figure 3.43: Graphs of $y = \sinh x$ and $y = \cosh x$, indicating also the asymptotic behaviors as $x \to \infty$ and as $x \to -\infty$.

and so the graph will be symmetric with respect to the origin. Furthermore,

$$\lim_{x \to \infty} y = \lim_{x \to \infty} \frac{e^x - e^{-x}}{2} \quad \frac{\frac{e^\infty - e^{-\infty}}{2}}{\frac{\infty - 0}{2}} \quad \infty$$

$$\lim_{x \to -\infty} y = \lim_{x \to -\infty} \frac{e^x - e^{-x}}{2} \quad \frac{\frac{e^{-\infty} - e^\infty}{2}}{\frac{0 - \infty}{2}} \quad -\infty$$

The second limit also follows from the first limit and the symmetry of the function. Also important is that because $x \to \infty \implies e^{-x} \to 0$, we get that for large $x > 0$, we have $y \approx \frac{1}{2}e^x$. Similarly, $x \to -\infty \implies e^x \to 0$, and so for $x < 0$ and large we have $y \approx -\frac{1}{2}e^{-x}$. Taken together, these facts describe a kind of asymptotic behavior of y, thus giving us some idea of the shape of the graph for large $|x|$.

Next we notice that $y' = \cosh x = \frac{1}{2}(e^x + e^{-x}) > 0$ for all x, and so this function is always increasing, and therefore is one-to-one.

Finally, $(0,0)$ is the only x- or y-intercept:

$$y = 0 \iff \frac{1}{2}(e^x - e^{-x}) = 0 \iff e^x = e^{-x} \iff e^{2x} = 1 \iff 2x = 0 \iff x = 0.$$

We put these facts together to sketch a graph of $y = \sinh x$, shown in Figure 3.43a.

Of course, the graph of $y = \sinh x$ little resembles that of $y = \sin x$, except for the common y-intercept and that both are symmetric with respect to the origin, i.e., $f(-x) = -f(x)$.

Example 3.12.2

The graph of $y = \cosh x = \frac{1}{2}(e^x + e^{-x})$ succumbs to similar analysis. The details are left to the exercises, and the graph is given in Figure 3.43b.

3.12. HYPERBOLIC FUNCTIONS (OPTIONAL)

Exercises[95]

1. Compute $\dfrac{d \sinh \sqrt{x}}{dx}$ and $\dfrac{d \sqrt{\sinh x}}{dx}$.

2. Compute $\dfrac{d \tanh(x^2)}{dx}$ and $\dfrac{d(\tanh x)^2}{dx}$.

3. Prove the first part of (3.142), that $\frac{d}{dx}\operatorname{sech} x = -\operatorname{sech} x \tanh x$ by using the derivatives of $\sinh x$ and $\cosh x$.

4. Repeat the previous problem but by first rewriting $\operatorname{sech} x$ using exponential functions.

5. Use the quotient rule and identities to show that $\frac{d}{dx}\coth x = -\operatorname{csch}^2 x$.

6. Show that
$$\frac{d \sinh^{-1}\left(\frac{1}{x}\right)}{dx} = \frac{d \operatorname{csch}^{-1} x}{dx}$$
by computing both directly and simplifying where appropriate.

7. Suppose that $\sinh x = -\sqrt{3}$. Using definitions and identities of the hyperbolic functions, namely (3.133)–(3.137), starting on page 399, to compute each of the following. (It will be helpful to notice that $x < 0$ and $\cosh x > 0$.)

    (a) $\cosh x =$
    (b) $\tanh x =$
    (c) $\coth x =$
    (d) $\operatorname{sech} x =$
    (e) $\operatorname{csch} x =$

8. Suppose $x = 5$. Using (3.147)–(3.152), compute each of the following exactly, and give the answer also to within 0.0001:

    (a) $\sinh^{-1} x =$
    (b) $\cosh^{-1} x =$
    (c) $\tanh^{-1} x =$
    (d) $\coth^{-1} x =$
    (e) $\operatorname{sech}^{-1} x =$
    (f) $\operatorname{csch}^{-1} x =$

9. Repeat the previous problem for $x = -5$, computing (where defined):

    (a) $\sinh^{-1} x =$
    (b) $\cosh^{-1} x =$
    (c) $\tanh^{-1} x =$
    (d) $\coth^{-1} x =$
    (e) $\operatorname{sech}^{-1} x =$
    (f) $\operatorname{csch}^{-1} x =$

10. Construct the graph of $f(x) = \cosh x$ by following the analysis of Example 3.12.1, page 403:

    (a) Show that $f(-x) = f(x)$. What kind of symmetry does this imply?
    (b) Find the limits of $f(x)$ as $x \to \infty$ and $x \to -\infty$.
    (c) Explain the asymptotic behavior $f(x) \approx \underline{}$ as $x \to \infty$ and (separately) as $x \to -\infty$.
    (d) Show that $f(x) > 0$ for all $x \in \mathbb{R}$.
    (e) Construct a sign chart for $f'(x)$ and discuss its implications (where $f(x)$ is increasing and decreasing, and any maximum or minimum points).

---

[95]The following exercises are somewhat more exploratory in nature, because this section covers the analogs of several sections of material for trigonometric and inverse trigonometric functions, and because these functions are not as ubiquitous as the previous functions we encountered in the textbook. It is important for the reader to be aware of these functions and that they have analogous—but not mirror-image—structures of their trigonometric counterparts, but to develop them with the same care that would similarly commit them to memory would be a distractingly lengthy, and arguably unnecessary process. For pedagogical reasons, it is therefore appropriate that the reader encounter some of the quirks of these functions through discovery, as the discovery process in this context is arguably more important than a rote memory of the results.

(f) Use these things to graph $f(x) = \cosh x$.

11. Using the graph of $y = \sinh x$, construct a graph of $y = \operatorname{csch} x$.

12. Using the graph of $y = \cosh x$, construct a graph of $y = \operatorname{sech} x$.

13. Find $\lim\limits_{x \to \infty} \tanh x$. (Hint: Write $\tanh x = \frac{e^x - e^{-x}}{e^x + e^{-x}}$, and then factor and cancel e^x from both the numerator and the denominator.)

14. Find $\lim\limits_{x \to -\infty} \tanh x$. (Hint: There are two ways to do this with current techniques, namely using a factoring similar to that in the previous exercise, or instead using the previous exercise and a symmetry argument.)

15. Using the previous exercise and from the form of its derivative, graph $f(x) = \tanh x$.

16. Graph $f(x) = \tanh^{-1} x$. (Use the previous exercise.)

17. Graph $f(x) = \coth x$.

18. Use a quadratic argument as we did in deriving (3.147), page 403, to derive (3.148), i.e.,
$$\cosh^{-1} x = \ln\left(x + \sqrt{x^2 - 1}\right).$$

19. Use the algebraic formula above for $\cosh^{-1} x$ to derive the second formula in (3.144), page 401, namely
$$\frac{d \cosh^{-1} x}{dx} = \frac{1}{\sqrt{x^2 - 1}}.$$

20. Use quadratic techniques as before to derive the formula for
$$\tanh^{-1} x = \frac{1}{2} \ln\left(\frac{1+x}{1-x}\right).$$
(Hint: Eliminate fractions through multiplication.)

21. Using only algebra, show that if we already know that $\sinh^{-1} x = \ln\left(x + \sqrt{x^2 + 1}\right)$, then it follows that $\operatorname{csch}^{-1} x = \ln\left(\frac{1 + \sqrt{x^2+1}}{x}\right)$. (Hint: Note that $\sinh^{-1} y = \operatorname{csch}^{-1} \frac{1}{y}$, because $\sinh \theta = y \iff \operatorname{csch} \theta = 1/y$.)

22. Show that
$$\cosh(A+B) = \cosh A \cosh B + \sinh A \sinh B.$$
(3.153)

23. Show that
$$\sinh(A+B) = \sinh A \cosh B + \cosh A \sinh B.$$
(3.154)

24. Use the previous exercises to show that
$$\cosh 2x = \sinh^2 x + \cosh^2 x.$$

25. Find two other formulas for $\cosh 2x$ using the previous exercise and the identity $\cosh^2 x - \sinh^2 x = 1$.

26. From the graph of $y = \sinh x$, construct the graph of $y = \sinh^{-1} x$.

27. Restricting $x \geq 0$ in the graph of $y = \cosh x$, construct a graph of $y = \cosh^{-1} x$.

28. Using the first derivative, show that $\sinh x$ is increasing on $\mathbb{R}$.

29. Similarly, show that $\sinh^{-1} x$ is increasing on $\mathbb{R}$.

30. Which of the inverse hyperbolic functions are

    (a) increasing on their domains?
    (b) decreasing on their domains?

Chapter 4

Using Derivatives to Analyze Functions; Applications

In this chapter, we develop methods for analyzing functions extensively by mining the information contained in their derivatives, and even their derivatives' derivatives, i.e., the functions' *second derivatives*, and consider even their third derivatives, and so on. We will have many new applications, both theoretical and real-world, for the insights developed here.

Of course, a very natural and general approach to analyzing functions is to consider their graphs in the Cartesian Plane, and much of this chapter is devoted to describing their graphs by analyzing their derivatives. That analysis will be important for applied problems as well.

Fortunately, we will be able to prove all of the results presented in this chapter, referring only to previously proved results and already mentioned facts we borrow from more advanced studies. Granted, we have already used some of the intuition without proof, for example when we noted that it seemed reasonable to believe

$$f'(x) > 0 \text{ on } (a,b) \implies f(x) \text{ increasing on } (a,b),$$
$$f'(x) < 0 \text{ on } (a,b) \implies f(x) \text{ decreasing on } (a,b)$$

and stated this as Theorem 3.2.6, page 238. In this chapter, we will finally prove this theorem, using the theoretically important Mean Value Theorem (MVT). While the MVT may at first seem to be a mere curiosity, it is astonishingly useful for proving many first derivative implications which are perhaps intuitive enough, but can be otherwise surprisingly difficult to prove. Furthermore, the MVT lends itself to many practical problems, particularly those involving average rates of change and inequalities.

The graphical significance of a function's first derivative as slope was discussed in the previous chapter and should seem straightforward. The significance of the second derivative is nearly as straightforward, though it is often necessary to break that analysis into more cases that depend on the value (or sign) of the first derivative. Further higher-order derivatives' effects on graphs are much less easily intuited, as their graphical implications somewhat rely on the values (or signs) of all previous derivatives, so we will not pursue these very far in this chapter. By contrast, the second derivative is somewhat intuitive, and is important enough for its own analysis, albeit in the context of being the derivative of the more easily understood first derivative.

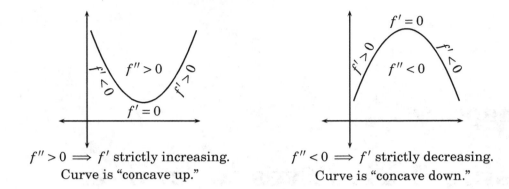

$f'' > 0 \implies f'$ strictly increasing.
Curve is "concave up."

$f'' < 0 \implies f'$ strictly decreasing.
Curve is "concave down."

Figure 4.1: Illustrations of the effects of the sign of f'' on the shape of the graph of f.

4.1 Higher-Order Derivatives and Graphing

In this section, we introduce notation for higher-order derivatives, develop some intuition regarding the first of these—which is the second derivative of a function—and use it to aid in graphing functions. We also use the opportunity to combine several analytical aspects of a function in order to sketch graphs which illustrate more and more aspects of the function's behavior, some of which we might not detect with such precision (or at all!) from a computer-generated graph, compared to the analysis we do here on paper.

4.1.1 Higher-Order Derivatives (Derivatives of Derivatives)

From physics, we learn that it is useful to take the derivative of velocity to find acceleration. Velocity being a derivative itself—of position—acceleration is thus the derivative of the derivative of position. There are a few ways of writing this:

$$a(t) = v'(t) = (s'(t))' = s''(t),$$

$$a = \frac{dv}{dt} = \frac{d}{dt}\left(\frac{ds}{dt}\right) = \frac{d^2 s}{(dt)^2} = \frac{d^2 s}{dt^2}.$$

The final form shown is a bit of an abuse of the notation, in that in other contexts $dt^2 = 2t\,dt$ (since $dt^2/dt = 2t$) is not the same as $(dt)^2$, but as often happens, it has become acceptable to shorten the notation in this context.[1] Note that $s''(t)$ is usually read "s double-prime of t." If we wished for a third derivative of a function $f(x)$, we can write it as $(f''(x))' = f'''(x)$,

---

[1] It should be pointed out that one of the Leibniz notation's strengths for first derivatives—the way its fractional form allows Chain Rule derivatives to be consistent with multiplicative rules (particularly cancellation) of arithmetic—does not extend to higher-order derivatives. It is usually not a problem, but the reader should be aware of this limitation. So for instance, $\frac{d^2 \sin u}{dx^2} \neq \frac{d^2 \sin u}{d^2 u} \cdot \frac{d^2 u}{dx^2}$, as the first term in that product is in fact meaningless. Instead, one would have to eventually use the Product Rule in computing

$$\frac{d^2 \sin u}{dx^2} = \frac{d}{dx}\left[\frac{d \sin u}{dx}\right] = \frac{d}{dx}\left[\cos u \cdot \frac{du}{dx}\right] = \cos u \cdot \frac{d^2 u}{dx^2} + \left(\frac{du}{dx}\right)\left(-\sin u \cdot \frac{du}{dx}\right) = \cos u \cdot \frac{d^2 u}{dx^2} - \sin u \left(\frac{du}{dx}\right)^2.$$

Similarly, a general Chain Rule for higher derivatives would be quite complicated, and one is better off computing those derivatives directly, using previous rules as they arise.

4.1. HIGHER-ORDER DERIVATIVES AND GRAPHING

or as $\frac{d}{dx}\left[\frac{d^2 f(x)}{dx^2}\right] = \frac{d^3 f(x)}{dx^3}$. Beyond the third derivative, we have further notations, chosen for convenience. For instance, the fourth derivative can be written with lower-case Roman numeral-type notation as $f^{\text{iv}}(x)$, or using a regular numeral but in parentheses, as in $f^{(4)}(x)$, to distinguish the fourth-order derivative of $f(x)$ from a simple power such as $(f(x))^4$. For another example,

$$f^{\text{vi}}(x) = f^{(6)}(x) = \frac{d^6 f(x)}{dx^6}.$$

To be clear, the last expression above is short for $\frac{d}{dx}\left[\frac{d}{dx}\left[\frac{d}{dx}\left[\frac{d}{dx}\left[\frac{d}{dx}\left[\frac{df(x)}{dx}\right]\right]\right]\right]\right] = \frac{d^6 f(x)}{(dx)^6}$, and the first expression is short for $((((((f(x))')')')')')'$.

For polynomials, each time we differentiate we get a lower-degree polynomial; for exponential functions e^x and the like, and the simplest trigonometric functions $\sin x$ and $\cos x$, each differentiation yields a function roughly as complicated as the original function; for functions that require the Chain Rule, Product Rule, or Quotient Rule, each differentiation step can yield a much more complicated function than before that step. A simple case of this phenomenon follows.

Example 4.1.1

Find the first two derivatives of $f(x) = \sin x^2$.

Solution: Here the first derivative is a simple Chain Rule, while the second derivative will require a Product Rule followed by a Chain Rule. We skip writing out separately some of the evaluations in the Chain Rule steps:

$$f(x) = \sin x^2 \implies f'(x) = \cos x^2 \cdot \frac{dx^2}{dx} = 2x \cos x^2$$

$$\implies f''(x) = 2x \frac{d \cos x^2}{dx} + \cos x^2 \cdot \frac{d\, 2x}{dx} = 2x(-\sin x^2)\frac{dx^2}{dx} + 2\cos x^2 = -4x^2 \sin x^2 + 2 \cos x^2.$$

Still-higher-order derivatives of $f(x) = \sin x^2$ would be even more complicated. However, for polynomials, when written in expanded (not factored) form it is clear that each differentiation step lowers the degree by one, until eventually all subsequent derivatives are zero.

Example 4.1.2

Compute all orders of derivatives for $f(x) = x^5 + 3x^4 - 8x^3 + 9x - 12$.

Solution: We take these in turn.

$f(x) = x^5 + 3x^4 - 8x^3 + 9x - 12 \implies f'(x) = 5x^4 + 12x^3 - 24x^2 + 9 \qquad \implies f^{(4)}(x) = 120x + 72$
$\implies f''(x) = 20x^3 + 36x^2 - 48x \qquad \implies f^{(5)}(x) = 120$
$\implies f'''(x) = 60x^2 + 72x - 48 \qquad \implies f^{(n)}(x) = 0$ for all $n \geq 6$.

Clearly, for an nth degree polynomial $f(x)$, we will have $m > n \implies f^{(m)}(x) = 0$, which is a useful observation in the sequel.

We will have further opportunities to compute higher-order derivatives throughout the text. We will pay particularly close attention to $f''(x)$ and its significance in this section.

4.1.2 Graphical Significance of the Second Derivative

For an open interval (a,b), and a function $f(x)$ defined for all $x \in (a,b)$, we have[2]

$$f' > 0 \text{ on } (a,b) \implies f \text{ is strictly increasing on } (a,b),$$
$$f' < 0 \text{ on } (a,b) \implies f \text{ is strictly decreasing on } (a,b).$$

Similarly, if the derivative of f' is positive on (a,b) then f' must be strictly increasing on (a,b), while if the derivative of f' is negative on (a,b) then f' must be strictly decreasing on (a,b). More formally,

$$f'' > 0 \text{ on } (a,b) \implies f' \text{ is strictly increasing on } (a,b),$$
$$f'' < 0 \text{ on } (a,b) \implies f' \text{ is strictly decreasing on } (a,b).$$

What this means graphically is illustrated in Figure 4.1, page 408. The shape of the curve is called *concave up* (or "u-shaped") when $f'' > 0$, and *concave down* (or "n-shaped") when $f'' < 0$. The presence of the letters "u" in "up" and "n" in "down" help some to remember these.

When both f' and f'' exist, are nonzero and do not change sign on an interval, there are four possible combinations of signs for f' and f'' on that interval. These cases, and their implications for the graphs, are illustrated below:

| $f' > 0$ | $f' > 0$ | $f' < 0$ | $f' < 0$ |
| $f'' > 0$ | $f'' < 0$ | $f'' > 0$ | $f'' < 0$ |

These four cases above can also be described as, respectively, increasing and concave up; increasing and concave down; decreasing and concave up; and decreasing and concave down. Figure 4.1, page 408 (at the start of this section), also illustrates how the signs of the first two derivatives of a function can often be used to find local (also called *relative*) maximum and minimum points. The tests are intuitive and easily remembered in substance, but since they are also well known by name we will mention them here, along with small graphics to immediately picture why they are true.

Theorem 4.1.1 (First Derivative Test): *If $f(x)$ is continuous on (a,b), and $x_0 \in (a,b)$ is such that*[3]

- $f' < 0$ *on* (a, x_0) *and* $f' > 0$ *on* (x_0, b), *then* $(x_0, f(x_0))$ *is a local minimum:* \/

---

[2] To be more precise, by $f' > 0$ on $(a,b) \implies f$ is strictly increasing on (a,b) we mean that

$$(\forall x \in (a,b))[f'(x) > 0] \implies (\forall x_1, x_2 \in (a,b))[x_1 < x_2 \longrightarrow f(x_1) < f(x_2)].$$

A similar implication is valid in the case we might abbreviate as $f' < 0 \implies f$ strictly decreasing.

[3] Recall by *local maximum* at $(x_0, f(x_0))$ we mean that there is some interval $(x_0 - \delta, x_0 + \delta)$ for some $\delta > 0$ such that $(\forall x \in (x_0 - \delta, x_0 + \delta))[f(x) \leq f(x_0)]$. It is similar for a local minimum but with $\leq$ replaced by $\geq$.

4.1. HIGHER-ORDER DERIVATIVES AND GRAPHING

- $f' > 0$ on (a, x_0) and $f' < 0$ on (x_0, b), then $(x_0, f(x_0))$ is a local maximum: ⟋⟍

We used this principle for constructing graphs in the previous chapter. With our consideration of the second derivative, we can now also consider the following:[4]

Theorem 4.1.2 (Second Derivative Test) *Suppose $f'(x_0) = 0$. Then the following hold.*

- *If $f''(x_0) > 0$, then $(x_0, f(x_0))$ is a local minimum:* ⌣

- *If $f''(x_0) < 0$, then $(x_0, f(x_0))$ is a local maximum:* ⌢

Example 4.1.3

Consider the function $f(x) = x^3 - 3x$, also the subject of Example 3.2.8 page 238 (graphed in Figure 3.8, page 240). We can easily use the first derivative test to find any local maximum and local minimum points by constructing a sign chart for its derivative, $f'(x) = 3x^2 - 3 = 3(x^2 - 1) = 3(x+1)(x-1)$.

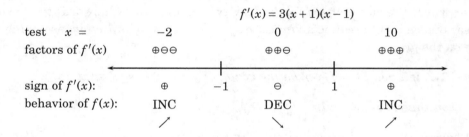

Clearly there is a local maximum at $(-1, f(-1)) = (-1, 2)$, and a local minimum at $(1, f(1)) = (1, -1)$. Neither are "global extrema" (also called *absolute extrema*), since the function is not bounded from above or below, as the limits as $x \to \infty$ and $x \to -\infty$ show: $x \to \infty \implies f(x) = x^3 - 3x \to \infty$, while $x \to -\infty \implies f(x) = x^3 - 3x \to -\infty$.

While the sign chart for the first derivative is powerful and interesting, for this particular function the Second Derivative Test would have been simple enough: $f''(x) = \frac{d}{dx}(3x^2 - 3) = 6x$, and so $f''(-1) = 6(-1) < 0$, while $f''(1) = 6(1) = 6 > 0$. So $f'(-1) = 0$ and $f(x)$ is concave down at $x = -1$, and we conclude $(-1, f(-1))$ is a local maximum point; similarly $f'(1) = 0$ and $f(x)$ is concave up at $x = 1$, and so that point must be a local minimum. Or, in the language of the Second Derivative Test,

$$f'(-1) = 0, \quad f''(-1) = -6 < 0 \implies (-1, f(-1)) \quad \text{is a local maximum:} \quad ⌢$$
$$f'(1) = 0, \quad f''(1) = 6 > 0 \implies (1, f(1)) \quad \text{is a local minimum:} \quad ⌣$$

Note that if $f''(x_0) = 0$, the Second Derivative Test is silent (vacuously true in fact). If both $f'(x_0) = 0$ and $f''(x_0) = 0$, a graph tends to be more "flat" at x_0 than if we have only $f'(x_0) = 0$. Consider the following examples where this occurs.

---

[4] See again Figure 4.1, page 408. Also note that $f''(x_0)$ existing requires that there exists $\delta > 0$ such that $f'(x)$ exists for all $x \in (x_0 - \delta, x_0 + \delta)$, or else the difference quotient whose limit would define $f''(x_0)$ would be undefined for some values Δx with $0 < |\Delta x| < \delta$, and then $\lim_{\Delta x \to 0}[(f'(x_0 + \Delta x) - f'(x_0))/\Delta x] = f''(x_0)$ would not exist.

1. Consider $f(x) = x^4 \implies f'(x) = 4x^3 \implies f''(x) = 12x^2$. Then $f'(0) = 0$, and $f''(0) = 0$. A sign chart for $f'(x) = 4x^3$ shows $(0, f(0))$ is a local (actually global) minimum, as $f' < 0$ on $(-\infty, 0)$, but $f' > 0$ on $(0, \infty)$. (That $(0, f(0))$ is necessarily a minimum follows because $f(0) = 0$ while $x \neq 0 \implies f(x) > 0$.)

2. On the other hand, for the case $f(x) = -x^4$, we have $f'(x) = -4x^3$, $f''(x) = -12x^2$, and so again we have $f'(0) = 0$ and $f''(0) = 0$, but $f(x) < 0$ for all $x \neq 0$, so $(0, f(0)) = (0, 0)$ is a local (again, actually global) maximum point. (Of course, $y = -x^4$ is just the vertical reflection of $y = x^4$ through the x-axis.) A First Derivative Test also shows this quickly.

3. $f(x) = x^3 \implies f'(x) = 3x^2 \implies f''(x) = 6x$. In this case, $f'(0) = 0$ and $f''(0) = 0$, but that point $(0, f(0)) = (0, 0)$ is neither a local maximum nor a local minimum, since $x < 0 \implies f(x) < 0$, while $x > 0 \implies f(x) > 0$. "From the left," $(0, 0)$ appears to be a maximum point, while "from the right," $(0, 0)$ appears to be a minimum point, so it is neither. (See the graph of $y = x^3$ on page 413.

It is interesting to look at points on the graph of a function $f(x)$ where the sign of f'' changes, i.e., where the curve's bend (or "concavity") changes from "upward" to "downward" or vice versa. Such a point is called an *inflection point*, or *point of inflection*, assuming it does actually lie on the graph:

Definition 4.1.1 *If a point $(x_0, f(x_0))$ on the graph of $y = f(x)$ is such that*

1. *$f(x)$ is continuous at x_0, and*

2. *$f''(x)$ has one sign (positive or negative) in some interval $(x_0 - \delta_1, x_0)$, and the opposite sign in $(x_0, x_0 + \delta_2)$, for some $\delta_1, \delta_2 > 0$,*

then $(x_0, f(x_0))$ is called an **inflection point** *of the graph of $y = f(x)$.*

If we were "driving a car" along such a curve from left to right, an inflection point would be where our steering wheel turns from one side of center (or the straight-ahead direction on the road), to the other. Some of our common functions illustrate these phenomena relating to the sign of the second derivative. For instance, for $f(x) = \sin x$, we have $f'(x) = \cos x$ and $f''(x) = -\sin x$. Thus for this case $f''(x) = -f(x)$, and so where $f(x)$ is positive, $f''(x)$ is negative and vice versa. Since $\sin x$ is defined (and even continuous) for all $x \in \mathbb{R}$, the points where $f''(x)$ changes signs are inflection points, namely at $x = \pm\pi, \pm 2\pi, \pm 3\pi, \cdots$, as we can see in the graph:

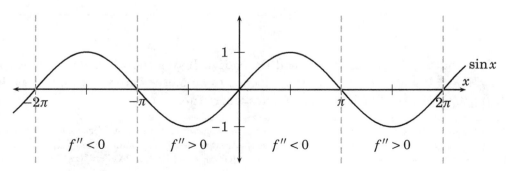

4.1. HIGHER-ORDER DERIVATIVES AND GRAPHING

By contrast, the function $f(x) = \ln x$ is concave down wherever defined, since $f(x) = \ln x \implies f'(x) = \frac{1}{x} \implies f''(x) = -1/x^2 < 0$ where $\ln x$ is defined. For that matter, this is also true of $f(x) = \ln|x|$, as graphed next:

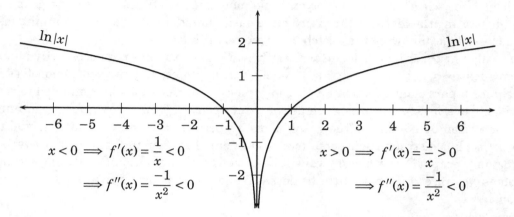

Even the functions x^2 and x^3, graphed next, are interesting for their first and second derivative properties. Note that $y = x^2$ is always concave up, while $y = x^3$ is always increasing but has an inflection point at $(0,0)$ where it changes from concave down to concave up.

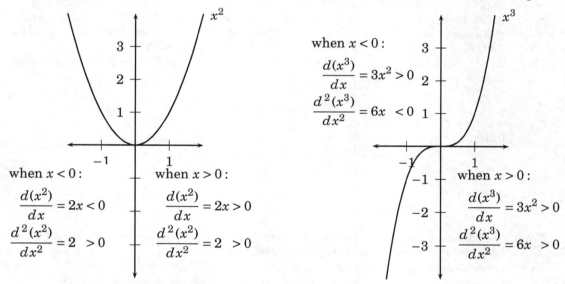

4.1.3 Detailed Graphs Showing All Discussed Behaviors

Functions that are more complicated can require a more systematic search for features such as intervals on which they increase/decrease, where they are concave up/down, and other interesting aspects, such as asymptotes, x-intercepts, and y-intercepts. Where possible, it is useful to have sign charts for

- $f'(x)$, to find where $f(x)$ is increasing/decreasing, and
- $f''(x)$, to find where $f(x)$ is concave up/down.

414 CHAPTER 4. USING DERIVATIVES TO ANALYZE FUNCTIONS; APPLICATIONS

One could argue a sign chart for $f(x)$ is also useful, but if we know the x-intercepts, then the sign charts for f' and f'' will imply a sign chart for f as well (see the next example).

Furthermore, if some of the graph's features are extraordinarily difficult to derive, as often happens with x-intercepts for example, one can sometimes sketch the graph of $y = f(x)$ illustrating the features that were more easily found and leave the remaining features unaddressed, i.e., allowing the sketch be vague regarding those.

Finally, the order in which one derives the various features is not necessarily important, as some features, such as the sign of f'' or some asymptotes, may present themselves immediately for a particular function but not for others. Later one may be reminded to consider previously overlooked features, such as the y-intercept $(0, f(0))$ where applicable. Such features are often noted only as the graph is drawn because they may be fairly trivial to find in practice, but distractions if noted too early in our "bookkeeping" of the characteristics of a function's graph. Nonetheless, we will be somewhat systematic in analyzing a function's graph as we proceed, at least with the derivative computations and their sign charts' ramifications.

Example 4.1.4 _____

Graph $f(x) = 2x^3 - 3x^2$, illustrating the signs of $f(x)$, $f'(x)$, $f''(x)$, all x- and y-intercepts, and the behavior as $x \to \pm\infty$. In doing so, label local extrema and inflection points.

Solution: We begin with the intercepts, noting that $f(x) = x^2(2x - 3) = 0$ at $x = 0, 3/2$. While we could construct a sign chart for $f(x)$, in fact once we note the locations of these intercepts, the addition of the sign chart for $f'(x)$ will be enough to determine the signs of $f(x)$ elsewhere.

Now $f'(x) = \frac{d}{dx}\left[2x^3 - 3x^2\right] = 6x^2 - 6x = 6x(x-1) = 0$ at $x = 0, 1$. From this we construct a sign chart for $f'(x)$ (test points omitted below):

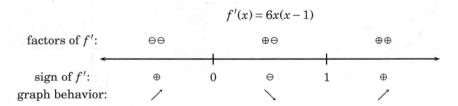

From the sign chart, clearly we have $(0, f(0)) = (0, 0)$ is a local maximum of $f(x)$, while $(1, f(1)) = (1, -1)$ is a local minimum. There are no further local extrema.

Next we consider the sign of $f''(x)$, by noting that $f''(x) = \frac{d}{dx}(f'(x)) = \frac{d}{dx}\left[6x^2 - 6x\right] = 12x - 6 = 0$ for $x = 1/2$, and is otherwise continuous, so we can construct the following sign chart for $f''(x)$:

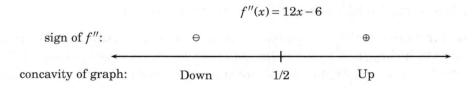

From the sign chart of f'', we see that $y = f(x) = 2x^3 - 3x^2$ has an inflection point at $(1/2, f(1/2)) = \left(\frac{1}{2}, \frac{1}{4} - \frac{3}{4}\right) = \left(\frac{1}{2}, -\frac{1}{2}\right)$.

4.1. HIGHER-ORDER DERIVATIVES AND GRAPHING

It is often useful to construct a combined sign chart—or function behavior chart—using all of the points used on the sign charts for $f'(x)$ and $f''(x)$. *(As mentioned before, it is often easier to defer the inclusion of signs of $f(x)$ until we are ready to construct the final graph.)* There are various ways to organize a combined sign chart for f' and f''. We offer the following:

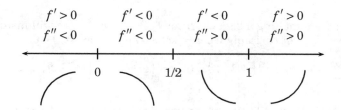

shape of graph:

From the shapes in our chart, and the fact that $f(x)$ is a polynomial and therefore defined and continuous on all of $\mathbb{R}$, we can deduce its behavior $x \to \pm\infty$, so it is not so necessary to compute those limits, but we do so here to verify these facts (noting the dominance of the x^3 terms):

$$\lim_{x \to \infty} f(x) = \lim_{x \to \infty} (2x^3 - 3x^2) = \lim_{x \to \infty}\left[x^3\left(2 - \frac{3}{x}\right)\right] \xlongequal{\infty \cdot 2} \infty,$$

$$\lim_{x \to -\infty} f(x) = \lim_{x \to -\infty} (2x^3 - 3x^2) = \lim_{x \to -\infty}\left[x^3\left(2 - \frac{3}{x}\right)\right] \xlongequal{-\infty \cdot 2} -\infty.$$

Again, these limits are implied by the combined sign chart above as well: Given the domain of f, and its graph's shapes and directions for large $|x|$, these limits could not be otherwise. Putting all these facts together we get the actual graph, plotted here, with the local maximum $(0,0)$, local minimum $(1,-1)$ and inflection point $(1/2, -1/2)$ all highlighted. Also recall the x-intercepts at $x = 0, 3/2$.

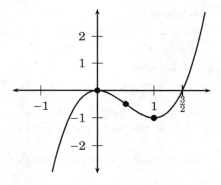

While the graph here is computer-generated, a reasonable "sketch" can be made to illustrate important features relevant to the calculus: where the graph is increasing or decreasing (and thus local extrema), and where it is concave up or down (and inflection points). One often makes a point to graph the x-intercepts (where $f(x) = 0$), and the y-intercept $((0, f(0)))$ as well. In the previous case, the latter is also an x-intercept, as happens when $f(0) = 0$.

Recall that a function $f(x)$ can change signs as x increases by either (1) passing through the value 0 in its output, or (2) having x pass through a value where $f(x)$ is discontinuous. It is similar with derivatives of any order for which we would like to produce a sign chart: We are interested in where the derivative's output (1) is passing through the value 0, or (2) is discontinuous, which for our cases will be those places where the derivative in question does not exist (as a real number). These are the points where a derivative can change signs.

Example 4.1.5

Sketch the graph of $f(x) = x - 3x^{2/3}$, indicating where the graph is increasing and decreasing, relative extrema, where the graph is concave up or down, inflection points, and any asymptotic behavior.

Solution: Rather than making a sign chart for $f(x) = x - 3x^{2/3}$, we will instead note all intercepts and let the derivative sign charts imply where $f(x) > 0$ and where $f(x) < 0$.

$$\underbrace{x - 3x^{2/3}}_{f(x)} = 0 \iff x^{2/3}\left(x^{1/3} - 3\right) = 0 \iff \left(x^{2/3} = 0\right) \vee \left(x^{1/3} = 3\right) \iff x \in \{0, 27\}.$$

Thus x-intercepts are $(0,0)$ and $(27,0)$, the former also being the y-intercept. Next we create a sign chart for $f'(x)$, which is easiest if $f'(x)$ is written as a fraction.

$$f'(x) = \frac{d}{dx}\left(x - 3x^{2/3}\right) = 1 - 2x^{-1/3} = \frac{x^{1/3} - 2}{x^{1/3}}.$$

Thus $f'(x) = 0 \iff x^{1/3} = 2 \iff x = 8$, while $f'(x)$ does not exist $\iff x = 0$. We need both points for our sign chart for f'. (Note $f(x)$ is continuous on all of $\mathbb{R}$.)

$$f'(x) = \frac{x^{1/3} - 2}{x^{1/3}}$$

factors of f':	$\ominus/\ominus$		$\ominus/\oplus$		$\oplus/\oplus$
sign of f':	$\oplus$	0	$\ominus$	8	$\oplus$
graph behavior:	↗		↘		↗

Since $f(x) = x - 3x^{2/3}$ is continuous (so, for instance, it has no vertical asymptotes or jump discontinuities), from the sign chart of $f'(x)$ we conclude that $f(x) = x - 3x^{2/3}$ has

- a local maximum at $(0, f(0)) = \left(0, 0 - 3 \cdot 0^{2/3}\right) = (0,0)$, and

- a local minimum at $(8, f(8)) = \left(8, 8 - 3 \cdot 8^{2/3}\right) = (8, 8 - 3 \cdot 4) = (8, -4)$.

We might wish to keep in mind when we graph this that, while $f'(8) = 0$ so the graph's slope is horizontal there, $f'(0)$ does not exist, so the graph is not smooth at $x = 0$. In fact, the graph "approaches vertical" as $x \to 0$, as we can observe from the following:

$$\lim_{x \to 0^-} f'(x) = \lim_{x \to 0^-} \frac{x^{1/3} - 2}{x^{1/3}} \xrightarrow{-2/0^-} \infty, \qquad \lim_{x \to 0^+} f'(x) = \lim_{x \to 0^+} \frac{x^{1/3} - 2}{x^{1/3}} \xrightarrow{-2/0^+} -\infty.$$

Next we consider concavity by computing $f''(x)$ and constructing its sign chart.

$$f''(x) = \frac{d}{dx}\left(1 - 2x^{-1/3}\right) = \frac{2}{3}x^{-4/3} = \frac{2}{3x^{4/3}} = \frac{2}{3\left(\sqrt[3]{x}\right)^4}.$$

This does not exist at $x = 0$ (which we actually should have known because $f'(0)$ does not exist), but $x \neq 0 \implies f''(x) > 0$, so this curve is concave up on any interval that does not contain $x = 0$. We can make a combined sign chart for f' and f'' as before:

4.1. HIGHER-ORDER DERIVATIVES AND GRAPHING

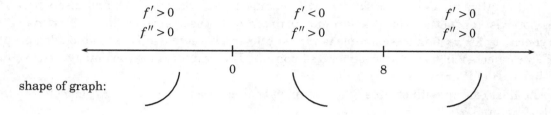

shape of graph:

Recalling again that $f(x)$ is a continuous function on $\mathbb{R}$, we can finally sketch it, though we might also wish to include the other x-intercept at $x = 27$, which makes for a somewhat compacted scale for a reasonable graph. Note the minimum point $(8, -4)$, and the x-intercepts at $x = 0, 27$ are emphasized.[5]

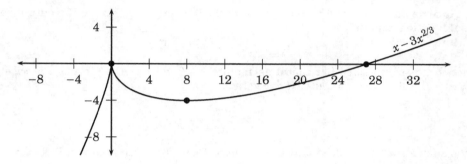

This function has no vertical or horizontal asymptotes, but we can see by the concavity and slopes that $x \to \infty \implies f(x) \to \infty$, while $x \to -\infty \implies f(x) \to -\infty$. We can also use limits to compute those behaviors:

$$\lim_{x \to \infty} f(x) = \lim_{x \to \infty} \left(x^{2/3} \left(x^{1/3} - 3 \right) \right) \xlongequal{\infty \cdot \infty} \infty,$$

$$\lim_{x \to -\infty} f(x) = \lim_{x \to -\infty} \left(x^{2/3} \left(x^{1/3} - 3 \right) \right) \xlongequal{\infty \cdot (-\infty)} -\infty.$$

Also noteworthy is how $f''(x) \to 0$ as $x \to \pm\infty$. All of these are somewhat reflected in the graph above as well. (If we were to extend the horizontal frame we might also notice $f'(x) \longrightarrow 1$ as $x \to \pm\infty$.)

The graph above illustrates that indeed we have to consider points where f' does not exist (or is otherwise discontinuous), as well as points where $f' = 0$, when we construct a sign chart for f'. But this is the case with any such real-number-inputting function's sign chart: we divide the real line $\mathbb{R}$ using points where the function in question (be it $f(x)$, $f'(x)$, $f''(x)$ or whatever) outputs zero or is discontinuous, i.e., the only two possibilities for the function to change sign as x-values increase along the real-line continuum.

In the example above, there was indeed a change of sign of $f'(x)$ at $x = 0$ where f' does not exist, eventually yielding a local maximum at $(0, f(0))$. Along with the local minimum at $(8, f(8))$ (where $f' = 0$), we have all local extrema accounted for.

---

[5]The point $(0, f(0))$ on the graph is often descriptively called a *cusp*. It is beyond the scope of this text to give the precise mathematical definition of cusp, but here its more colloquial meaning seems to be a good fit. At such a point f' does not exist, though there are other reasons for the nonexistence of a function's derivative, such as the discontinuity of the function, or its graph's slope approaching the vertical.

418 CHAPTER 4. USING DERIVATIVES TO ANALYZE FUNCTIONS; APPLICATIONS

Similarly, we have to check for concavity changes both where $f'' = 0$, and also where f'' does not exist (or is otherwise discontinuous), to complete the sign chart for f''. Now visually the concavity for $x < 0$ in the example is very subtle and barely visible on the included graph (which connects 2,000 computer-generated points). The graph is more obviously concave up for that part of it drawn for $x > 0$.

The next example illustrates other common complexities of computing $f''(x)$ and the inflection points.

Example 4.1.6

Graph $f(x) = e^{-x^2}$, illustrating signs of f' and f'', as well as asymptotic behavior.

Solution: At some point, we might wish to note $f(x) > 0$ for all $x \in \mathbb{R}$, due to the output range of the exponential function, and that
$$\lim_{x \to \pm\infty} f(x) \xrightarrow{e^{-\infty}} 0.$$

Thus the function has a two-sided horizontal asymptote $y = 0$, as $x \to \pm\infty$ (or, more precisely, $x \to \pm\infty \implies y = f(x) \to 0^+$). Next we compute its derivative:
$$f'(x) = \frac{d}{dx} e^{-x^2} = e^{-x^2} \cdot \frac{dx^2}{dx} = -2xe^{-x^2}.$$

Since $f'(x)$ exists for all $x \in \mathbb{R}$, we need only check where $f'(x) = 0$, which occurs only at $x = 0$. From this and the fact that the exponential part of $f'(x)$, namely e^{-x^2}, will be positive for all $x \in \mathbb{R}$, we can easily make a sign chart for $f'(x) = -2xe^{-x^2}$:

$$f'(x) = -2xe^{-x^2} = (-2x)\left(e^{-x^2}\right)$$

signs of factors: ⊕⊕   ⊖⊕

sign of f':   ⊕   0   ⊖
graph behavior: ↗   ↘

We can see clearly that there is a local (in fact global) maximum at $(0, f(0)) = (0, e^0) = (0, 1)$. This also happens to be the y-intercept.

Next we interest ourselves in the concavity, so we compute
$$f''(x) = \frac{d}{dx}\left[-2xe^{-x^2}\right] = -2x\frac{d}{dx}\left(e^{-x^2}\right) + e^{-x^2}\frac{d}{dx}(-2x)$$
$$= -2xe^{-x^2} \cdot (-2x) - 2e^{-x^2} = 4x^2 e^{-x^2} - 2e^{-x^2}$$
$$= 2e^{-x^2}\left(2x^2 - 1\right).$$

From this we get $f''(x) = 0 \iff 2x^2 = 1 \iff x^2 = \frac{1}{2} \iff x = \pm 1/\sqrt{2}$. Note that $\pm 1/\sqrt{2} \approx \pm 0.7071$.

$$f''(x) = 2e^{-x^2}(2x^2 - 1)$$

signs of factors: ⊕⊕   ⊕⊖   ⊕⊕

sign of f'':   ⊕ $-1/\sqrt{2}$ ⊖ $1/\sqrt{2}$ ⊕
concavity: up   down   up

We detected two inflection points, namely

4.1. HIGHER-ORDER DERIVATIVES AND GRAPHING

- $(-1/\sqrt{2}, f(-1/\sqrt{2})) = \left(-1/\sqrt{2}, e^{-1/2}\right) \approx (-0.7071, 0.6065)$, and
- $(1/\sqrt{2}, f(1/\sqrt{2})) = \left(1/\sqrt{2}, e^{-1/2}\right) \approx (0.7071, 0.6065)$.

As before, we synthesize a combined sign chart for f' and f'' as follows:

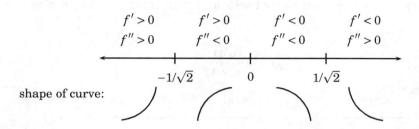

We then put all of this together, with the limiting behavior, also noticing the symmetry: If (x,y) is on the graph, then so is $(-x,y)$ because $f(x) = e^{-x^2}$ is an *even function*, i.e., $f(-x) = f(x)$. Also useful to note is that the y-intercept $(0, e^0) = (0, 1)$. From all of this we can construct a sketch of the graph:

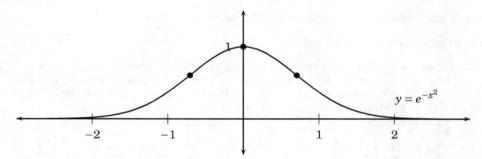

Highlighted are the maximum point at $(0, f(0)) = (0, 1)$, and the two inflection points located at $\left(\pm 1/\sqrt{2}, e^{-1/2}\right) \approx (\pm 0.7071, 0.6065)$.[6]

Rational functions also exhibit some interesting asymptotic behaviors, and can have complicated second derivatives. The next example has no vertical asymptotes, but does exhibit horizontal asymptotes and some interesting slope and concavity behavior.

Example 4.1.7

Follow the directions for the previous example for the function $f(x) = \dfrac{x}{x^2 + 1}$.

Solution: First we notice that $x^2 + 1 > 0$ for all $x \in \mathbb{R}$, so this function is defined, and for that matter continuous, on all of $\mathbb{R}$. Furthermore, $f(x) < 0$ for $x < 0$, while $f(x) > 0$ for $x > 0$. We might also notice that
$$\lim_{x \to \pm\infty} f(x) = \lim_{x \to \pm\infty} \frac{x}{x^2\left(1 + \frac{1}{x^2}\right)} = \lim_{x \to \pm\infty} \frac{1}{x\left(1 + \frac{1}{x^2}\right)} \xrightarrow{\frac{1/(\pm\infty(1+0))}{1/(\pm\infty)}} 0,$$

---

[6] The curve above is related to the bell-shaped *normal probability distributions* one encounters in studies of probability and statistics, the "standard normal curve" being given by the function $f(z) = \frac{1}{\sqrt{2\pi}} e^{-z^2/2}$, and the more general normal by $f(x) = \frac{1}{\sqrt{2\pi}\sigma} e^{-[(x-\mu)/\sigma]^2/2}$, where μ and σ are the mean and standard deviation.

that is, $f(x)$ has a two-sided horizontal asymptote $y = 0$. Next we compute

$$f'(x) = \frac{(x^2+1)\frac{dx}{dx} - x\frac{d(x^2+1)}{dx}}{(x^2+1)^2} = \frac{x^2+1-x\cdot(2x)}{(x^2+1)^2} = \frac{1-x^2}{(x^2+1)^2}.$$

Rewriting this somewhat we can more easily construct a sign chart for $f'(x)$, noting $f'(x) = 0$ for $x = \pm 1$:

$$f'(x) = \frac{(1+x)(1-x)}{(x^2+1)^2}$$

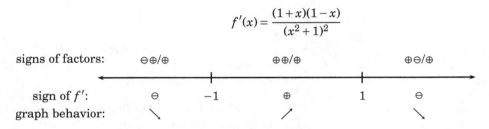

Since $f(x)$ is continuous on all of $\mathbb{R}$, from this chart we can clearly detect

- a local minimum at $(-1, f(-1)) = (-1, -1/2)$, and
- a local maximum at $(1, f(1)) = (1, 1/2)$.

At this point, we could have a reasonably good sketch, but from the sign chart for f', and the limiting behavior of $f(x)$ as $x \to \pm\infty$, we should expect inflection points as well. To find those, we must compute $f''(x)$, which will be more involved. For that computation we will use one of the intermediate forms for $f'(x)$ (note the factoring step in the third line):

$$f''(x) = \frac{d}{dx}\left[\frac{1-x^2}{(x^2+1)^2}\right] = \frac{(x^2+1)^2\frac{d}{dx}(1-x^2) - (1-x^2)\frac{d}{dx}(1+x^2)^2}{[(1+x^2)]^2}$$

$$= \frac{(x^2+1)^2(-2x) - (1-x^2)\cdot 2(1+x^2)(2x)}{(1+x^2)^4} = \frac{(-2x)(x^2+1)\left[x^2+1+2(1-x^2)\right]}{(x^2+1)^4}$$

$$= \frac{(-2x)(x^2+1)(3-x^2)}{(x^2+1)^4} = \frac{(-2x)(3-x^2)}{(x^2+1)^3}$$

$$\implies f''(x) = \frac{2x(x^2-3)}{(x^2+1)^3}.$$

The sign chart for $f''(x)$ should reflect possible sign changes of $f''(x)$ at $x = 0, \pm\sqrt{3}$:

$$f''(x) = \frac{2x(x^2-3)}{(x^2+1)^3}$$

signs of factors:	⊖⊕/⊕		⊖⊖/⊕		⊕⊖/⊕		⊕⊕/⊕
sign of f'':	⊖	$-\sqrt{3}$	⊕	0	⊖	$\sqrt{3}$	⊕
concavity:	down		up		down		up

Clearly we have inflection points at $(0, f(0)) = (0,0)$, at $\left(\sqrt{3}, f\left(\sqrt{3}\right)\right) = \left(\sqrt{3}, \frac{1}{4}\sqrt{3}\right)$, and at $\left(-\sqrt{3}, f\left(-\sqrt{3}\right)\right) = \left(-\sqrt{3}, -\frac{1}{4}\sqrt{3}\right)$, i.e., inflection at approximately $(0,0)$, $(1.732, 0.433)$ and $(-1.732, -0.433)$. Putting these two sign charts for f' and f'' together, we get the following chart showing general shapes of pieces of the graph:

4.1. HIGHER-ORDER DERIVATIVES AND GRAPHING

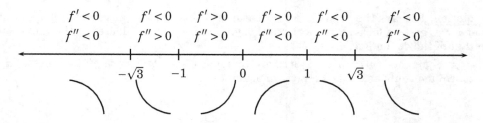

From this chart, the asymptotic behavior and the locations of local extrema and inflection points, we can sketch the graph:

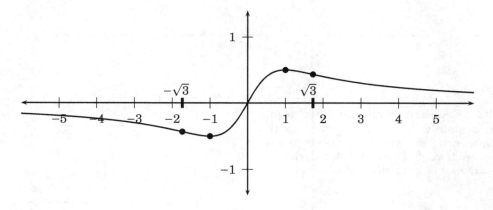

The function $f(x) = x/(x^2+1)$ is an *odd function*, meaning that $f(-x) = -f(x)$ for all x in the domain (which here is $\mathbb{R}$). Such a function's graph therefore shows what is called "symmetry with respect to the origin," meaning that if (x, y) is on the graph, so is $(-x, -y)$. Geometrically this means, for instance, that the part of the graph that resides in Quadrant I is accompanied by its "inverted image" in Quadrant III as if an inverting lens were placed at the origin, reproducing an "upside-down and backwards" image of the graph from Quadrant I through the origin and into Quadrant III. (Similarly, any part of the graph residing in Quadrant IV would have its inverted image also present in Quadrant II.) When present, such symmetry can be of help in producing the graph of $y = f(x)$.

The computations do show that if computing f' involves a Quotient Rule, then computing f'' will likely involve a more complicated Quotient Rule. Also note that the factored forms of $f(x)$ or $f'(x)$, which are often most convenient for creating their respective sign charts, might not be the simplest forms from which to compute their derivatives. The next example illustrates this also, and involves both a vertical asymptote and another linear asymptote that is neither vertical nor horizontal. Such linear but nonvertical and nonhorizontal asymptotes are called by various names such as **oblique asymptotes** or **slant asymptotes**.

Example 4.1.8

Graph $f(x) = \dfrac{x^2}{x-4}$, illustrating the signs of $f(x)$, $f'(x)$, and $f''(x)$, together with all asymptotic behavior.

Solution: Perhaps the most apparent features of this function's graph will be its vertical asymptote at $x = 4$, and its lack of any horizontal asymptotes since the degree of the numerator is greater than that of the denominator.

A sign chart for $f(x) = x^2/(x-4)$ is easily enough constructed, since $f(x) = 0$ only at $x = 0$, and $f(x)$ is discontinuous only at $x = 4$ (the vertical asymptote):

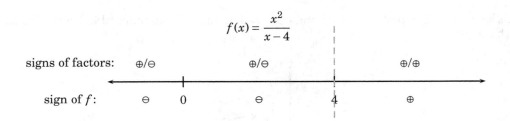

In fact, because the numerator has degree at least (actually greater than) the degree of the denominator, we can perform polynomial long division, and we will see that our derivative computations will be made easier:

For differentiation and some other purposes (explained next), it is clearly easier to use the form
$$f(x) = x + 4 + \frac{16}{x-4},$$
than to use the original form $f(x) = x^2/(x-4)$. Before we compute derivatives, note that
$$|x| \text{ large} \implies f(x) = x + 4 + \frac{16}{x-4} \approx x + 4,$$
and so the vertical distance between the graphs of $y = f(x)$ and $y = x + 4$ shrinks towards zero as $x \to \pm\infty$. Thus we have an **oblique asymptote**, which is the line $y = x + 4$, as an asymptote as $x \to \pm\infty$. This will be an important feature to include in our graph. Now we compute
$$f'(x) = \frac{d}{dx}\left[x + 4 + \frac{16}{x-4}\right] = 1 - \frac{16}{(x-4)^2} \cdot \frac{d(x-4)}{dx}$$
$$\implies f'(x) = 1 - \frac{16}{(x-4)^2}.$$

For the sake of building a sign chart, we can either find where $f' = 0$ and where f' does not exist from this expression, or we can recombine terms:
$$f'(x) = \frac{(x-4)^2 - 16}{(x-4)^2} = \frac{x^2 - 8x + 16 - 16}{(x-4)^2} = \frac{x(x-8)}{(x-4)^2}.$$

Thus we need $x = 0, 4, 8$ as our partitioning points on our sign chart for $f'(x)$. Also useful is using $x = 4$ to mark the location of the vertical asymptote, as we build our sign chart for f':

4.1. HIGHER-ORDER DERIVATIVES AND GRAPHING

$$f'(x) = \frac{x(x-8)}{(x-4)^2}$$

signs of factors: $\ominus\ominus/\oplus$ $\oplus\ominus/\oplus$ $\oplus\ominus/\oplus$ $\oplus\oplus/\oplus$

sign of f': $\oplus$ 0 $\ominus$ 4 $\ominus$ 8 $\oplus$

graph behavior: ↗ ↘ ↘ ↗

Noting that $x = 4$ is a vertical asymptote of $f(x)$ and the only discontinuity of $f(x)$, we can see clearly that we have two local extrema (since $x = 0, 8$ are not discontinuities):

- local maximum at $(0, f(0)) = (0, 0)$, and a
- local minimum at the point $(8, f(8)) = (8, 64/(8-4)) = (8, 64/4) = (8, 16)$.

We can use a simpler form of f' to compute

$$f''(x) = \frac{d}{dx}\left[1 - 16(x-4)^{-2}\right] = 32(x-4)^{-3}.$$

The sign chart for f'' follows readily, with $x = 4$ being the only dividing point.

$$f''(x) = \frac{32}{(x-4)^3}$$

sign of f'': $\ominus$ | $\oplus$

concavity: down 4 up

Combining the sign charts for f' and f'' would be fairly straightforward. Graphing the function to illustrate signs of f' and f'', along with the asymptotes, would yield the following:

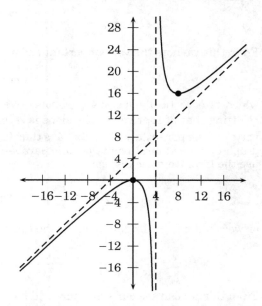

To review, we were given $f(x) = x^2/(x-4)$, which as it stands would require a Quotient Rule to find $f'(x)$, but polynomial long division yielded $f(x) = x + 4 + 16/(x-4)$, which is not only easier to differentiate, but also makes more obvious the presence and nature of the oblique asymptote since $|x|$ large implies $f(x) \approx x + 4$. However, the original form of $f(x)$ is better for constructing a sign chart of $f(x)$, should we care to.

When presented with a rational function $f(x) = p(x)/q(x)$, where p and q are polynomials, if the degree of p is at least as large as that of q, so that polynomial long division is possible, it is usually worth considering the form of $f(x)$ that arises from that long division, for purposes of differentiation and possible asymptotic behavior. The same can be said about $f'(x)$, though the effects of its asymptotics usually can be found in observations for $f(x)$ and $f''(x)$.

Nonlinear asymptotes are also a possibility. Consider the following example.

Example 4.1.9

Graph $f(x) = \dfrac{x^3}{x+2}$, illustrating signs of $f(x)$, $f'(x)$, and $f''(x)$, as well as all asymptotic behavior.

<u>Solution</u>: First we note the vertical asymptote at $x = -2$, and then construct a sign chart for $f(x)$:

424   CHAPTER 4. USING DERIVATIVES TO ANALYZE FUNCTIONS; APPLICATIONS

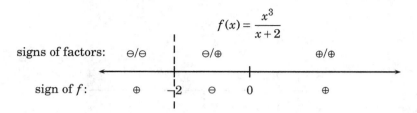

Anticipating the derivative computation, we again see that a viable alternative to the Quotient Rule is to perform division first. While long division can certainly accomplish this, here we instead take a slighly sophisticated shortcut, using $a^3 + b^3 = (a+b)(a^2 - ab + b^2)$ by adding and subtracting what would be the "b^3" term in the numerator:

$$f(x) = \frac{x^3}{x+2} = \frac{x^3 + 8 - 8}{x+2} = \frac{(x+2)(x^2 - 2x + 4) - 8}{x+2} = x^2 - 2x + 4 - \frac{8}{x+2}.$$

From this we note the very important asymptotic feature, namely that for large $|x|$ we have

$$f(x) \approx x^2 - 2x + 4,$$

which is a vertically-opening parabola. We can find its vertex using algebra techniques, or from realizing that at that vertex the slope is zero, so $0 = \frac{d}{dx}(x^2 - 2x + 4) = 2x - 2 = 2(x-1) \implies x = 1$. The vertex of the parabola $y = x^2 - 2x + 4$ is thus $(1, 1^2 - 2(1) + 4) = (1, 3)$. (If extra accuracy is desired for the graph of $f(x)$, several points on this parabolic asymptote can be drawn.) To compute $f'(x)$ we could use the form from the division, i.e.,

$$f'(x) = \frac{d}{dx}\left[x^2 - 2x + 4 - \frac{8}{x+2}\right] = 2x - 2 + \frac{8}{(x+2)^2},$$

or we could use the original form of the function to arrive at an already-combined fraction:

$$f'(x) = \frac{d}{dx}\left[\frac{x^3}{x+2}\right] = \frac{(x+2)(3x^2) - x^3(1)}{(x+2)^2} = \frac{3x^3 + 6x^2 - x^3}{(x+2)^2} = \frac{2x^3 + 6x^2}{(x+2)^2} = \frac{2x^2(x+3)}{(x+2)^2}.$$

From this we can see our sign chart will have possible sign-changing borders at $x = -3, -2, 0$.

$$f'(x) = \frac{2x^2(x+3)}{(x+2)^2}$$

signs of factors: ⊕⊖/⊕   ⊕⊕/⊖   ⊕⊕/⊖   ⊕⊕/⊕

sign of f':   ⊖   −3   ⊕   −2   ⊕   0   ⊕
graph behavior: ↘   ↗   ↗   ↗

Since $x = -3$ is a point of continuity of $f(x)$, from the sign chart above this must be a local minimum, specifically at $(-3, f(-3)) = (-3, -27/(-1)) = (-3, 27)$. Next we compute $f''(x)$:

$$f''(x) = \frac{d}{dx}\left[2x - 2 + 8(x+2)^{-2}\right] = 2 - 16(x+2)^{-3} \cdot 1 = \frac{2(x+2)^3 - 16}{(x+2)^3}$$

$$\implies f''(x) = \frac{2[(x+2)^3 - 8]}{(x+2)^3}.$$

While we could further factor the numerator, or simply solve $(x+2)^3 = 8$, it is reasonably clear that the numerator's value is zero when $x + 2 = 2$, i.e., when $x = 0$. The denominator is zero when $x = -2$.

4.1. HIGHER-ORDER DERIVATIVES AND GRAPHING

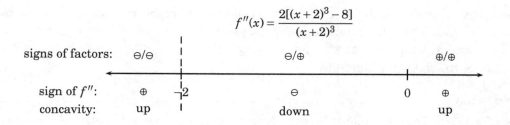

Clearly the concavity changes at the vertical asymptote $x = -2$, and also at the point $(0, f(0)) = (0,0)$ on the graph. Combining these two charts into one, we get the following:[7]

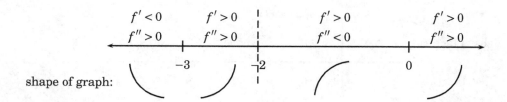

Recalling the vertical asymptote at $x = -2$, the nonlinear asymptote $y = x^2 - 2x + 4$, the local minimum at $(-3, 27)$, and the inflection point (also the x-intercept and y-intercept) at $(0, 0)$, we can reasonably sketch the graph, given next (and computer-generated):

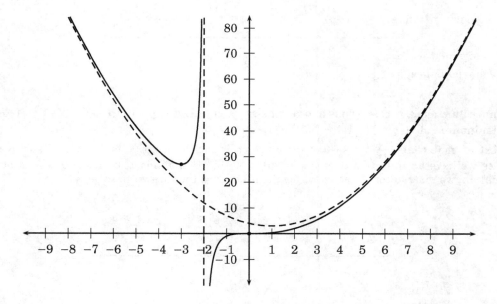

Note that we did not compute the limits as $x \to \pm\infty$, which are simple enough to compute, but which are also consequences of the other analyses, such as sign charts of f' and f''.

---

[7]Technically, $(0,0)$ is the only actual "inflection point," since it is an actual point on the curve, while there is no point on the curve where $x = -2$. But it is important to note that the concavity changes at the discontinuity at $x = -2$, as well as at the inflection point $(0,0)$.

When given a piecewise-defined function, a general approach is to consider each piece separately. The absolute value function is defined piecewise (one way for nonnegative inputs, and another way for negative inputs), but its properties can be at times seen in a simpler light. Consider the following example.

Example 4.1.10

Graph $f(x) = |x^3 - 3x|$, illustrating where the function is increasing and decreasing, concavity, and asymptotic behaviors.

Solution: We will do so by first analyzing and graphing the function inside the absolute values, temporarily calling it $g(x) = x^3 - 3x$, and the graph of $f(x) = |g(x)|$ will follow quickly from that of $g(x)$. Note that $g(x) = x(x^2 - 3)$, and so $g(x) = 0 \iff x \in \{0, -\sqrt{3}, \sqrt{3}\}$.

$$g(x) = x(x^2 - 3)$$

signs of factors: ⊖⊕   ⊖⊖   ⊕⊖   ⊕⊕

sign of g: ⊖ $-\sqrt{3}$ ⊕   0   ⊖ $\sqrt{3}$ ⊕

Next we consider $dg(x)/dx = \frac{d}{dx}(x^3 - 3x) = 3x^2 - 3 = 3(x^2 - 1)$, which is clearly zero at $x = \pm 1$. The sign chart for $dg(x)/dx$ then follows:

$$\frac{dg(x)}{dx} = 3(x+1)(x-1)$$

signs of factors: ⊕⊖⊖   ⊕⊕⊖   ⊕⊕⊕

sign of $dg(x)/dx$: ⊕ -1 ⊖   1   ⊕

behavior of graph of g: ↗   ↘   ↗

This inner function $g(x)$ clearly has a local maximum at $(-1, g(-1)) = (-1, (-1)^3 - 3(-1)) = (-1, 2)$, and a local minimum at $(1, g(1)) = (1, 1^3 - 3(1)) = (1, -2)$.

Next we note that $\frac{d^2 g(x)}{dx^2} = \frac{d}{dx}(3x^2) = 6x$, which is clearly negative for $x < 0$ and positive for $x > 0$, hence we have an inflection point at $(0, g(0)) = (0, 0)$. Combining our first and second derivative information, we make a combined chart for the function g (not drawn to scale):

| $g' > 0$ | $g' < 0$ | $g' < 0$ | $g' > 0$ |
| $g'' < 0$ | $g'' < 0$ | $g'' > 0$ | $g'' > 0$ |

        -1 0   1

Finally, we use this chart and the locations of extrema and the inflection point to graph the function $g(x)$ in gray, and then use the definition of the absolute value function to give us the graph of

$$f(x) = |g(x)| = \begin{cases} g(x) & \text{if } g(x) \geq 0, \\ -g(x) & \text{if } g(x) < 0, \end{cases}$$

as seen next:

4.1. HIGHER-ORDER DERIVATIVES AND GRAPHING

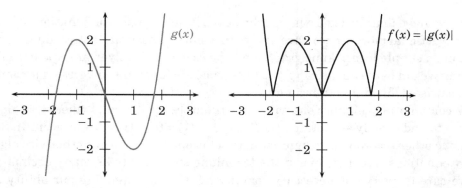

Points of interest on the graph of $f(x) = |x^3 - 3x|$ include the two maxima, namely $(\pm 1, 2)$, the three minima at $(0, 0)$ and $(\pm\sqrt{3}, 0)$—those also being the x-intercepts—and the inflection points $(\pm\sqrt{3}, 0)$.[8]

In this example, we strayed somewhat from the spirit of the previous examples, where we found directly the intervals where f is increasing, decreasing, concave up, and concave down by first finding points at which f' or f'' could change signs. The previous methods can be employed directly with this function $f(x) = |x^3 - 3x|$, though it requires extra work, such as first finding a piecewise-defined version of the function, in this case

$$f(x) = \begin{cases} -(x^3 - 3x) & \text{if} \quad x < -\sqrt{3}, \\ x^3 - 3x & \text{if} \quad -\sqrt{3} \leq x \leq 0, \\ -(x^3 - 3x) & \text{if} \quad 0 < x < \sqrt{3}, \\ x^3 - 3x & \text{if} \quad x \geq \sqrt{3}. \end{cases}$$

Then to look for possible points where, say, f' may change sign, we may first find ourselves considering, depending upon the "piece," $\frac{d}{dx}\left[\pm(x^3 - 3x)\right] = \pm(3x^2 - 3) = 0 \iff x = \pm 1$. We can then make a sign chart for each piece based in part on these things, but the boundaries of the pieces are also possible places for f' to change signs, so we must also consider the three points 0 and $\pm\sqrt{3}$ to partition our sign chart. The same applies to a sign chart for f''. This is unavoidable with some piecewise-defined functions, but those in which the outermost function is the absolute value function can be more easily graphed by considering the "inside" function's graph first, and reflecting any parts with negative heights across the x-axis and into the upper quadrants. In our chonsen analysis used in the example (analyzing the "inner" function's graph first), we see how the maxima, minima, and inflection points can emerge from examining first the graph of the "inner" function, and then transforming it to the graph representing its absolute value, rather than attempting direct analysis of the final function.

Since there are infinitely many functions we can consider, there are many different functional behaviors we can encounter. Even relatively simple combinations of well-understood functions can yield new functions with "emergent properties," which are not always clear at first glance from the new function's formula. Our derivative tests and simple observations regarding asymptotics can be employed to detect many of these emergent properties of both simple and complicated functions.

---

[8]Note that the inflection point of $g(x)$ is not an inflection point of $f(x) = |g(x)|$, but inflection points of $f(x)$ include the nonzero x-intercepts of $g(x)$. As a general rule, the concavity of $-g(x)$ will be opposite to that of $g(x)$, and this is reflected in those intervals where $f(x) = -g(x)$.

While we have mainly dealt with algebraic functions (simple combinations of powers of x, including fractional powers), we have also seen a simple observation regarding a trigonometric function, a simple logarithmic graph, and one case that involved an exponential function. Everything we did here can also apply to trigonometric functions, inverse trigonometric functions, piecewise-defined functions, and so on.

Some combinations of functions will, unfortunately, yield results where it is algebraically impossible to find exactly all solutions of $f(x) = 0$, $f'(x) = 0$, $f''(x) = 0$ and so on. In Section 4.5, we will have other techniques to approximate solutions to many such otherwise algebraically unwieldy equations, accurate (when the technique succeeds) to as many decimal places for which we care to carry out necessary computations, thus extending our ability to use the techniques of this section.

Before and after we develop those techniques we will have several other uses for the derivative.

Exercises

1. Show that $(fg)'' = f''g + 2f'g' + fg''$. In other words, derive the formula
$$\frac{d^2}{dx^2}(f(x)g(x)) =$$
$$f''(x)g(x) + 2f'(x)g'(x) + f(x)g''(x).$$
What do you suppose is the formula for $(fg)'''$?

For Exercises 2–21, graph each function in a manner that illustrates, where possible, signs of f, f', f'' (using sign charts for f' and f''), labeling all points where these signs change. Also indicate any asymptotic behaviors.

2. $f(x) = x^4 - x^2$

3. $f(x) = 3x^4 + 4x^3$

4. $f(x) = \dfrac{x^2 - 1}{x^2 + 1}$

5. $f(x) = \dfrac{x^2 + 1}{x^2 - 1}$

6. $f(x) = x(\sqrt{x} - 3)$

7. $f(x) = \dfrac{x^3}{x^2 - 1}$

8. $f(x) = \dfrac{x^4}{x^2 - 1}$

9. $f(x) = \dfrac{x^2 - x + 2}{x - 3}$

10. $f(x) = (x^2 - 9)^{2/3}$

11. $f(x) = \dfrac{x^2 + 1}{x^2}$

12. $f(x) = \dfrac{x^2}{x^2 + 3}$

13. $f(x) = \dfrac{x^4 - 1}{x^2}$

14. $f(x) = (x^{1/3} + 1)(x^{1/3} - 2)$

15. $f(x) = |x^2 - 2x - 3|$

16. $f(x) = x\sqrt{9 - x^2}$

17. $f(x) = \dfrac{x^2 + 1}{x^2 - 9}$

18. $f(x) = \dfrac{x^2 - 4}{x^2 - 9}$

19. $f(x) = x \ln x$. You can assume that $\lim\limits_{x \to 0^+} f(x) = 0$.

20. $f(x) = xe^{-x^2}$. You can assume that $\lim\limits_{x \to \pm\infty} f(x) = 0$.

21. $f(x) = x + \ln|x|$

4.1. HIGHER-ORDER DERIVATIVES AND GRAPHING

For Exercises 22–25, sketch the graph of $f(x)$ on the given interval, illustrating the signs of $f(x)$, $f'(x)$ and $f''(x)$.

22. $f(x) = x\sqrt{16-x^2}$, on $[-3,3]$.

23. $f(x) = |x^2 + 4x + 3|$, on $\mathbb{R}$.

24. $f(x) = x - \sin x$, on $[0, 2\pi]$.

25. $f(x) = 2x - \tan x$, on $(-\pi/2, \pi/2)$.

26. The position of a moving object at time t is given by $s(t) = 2t^3 + 3t^2 - 6t$, where $t \in [0, 10]$.

    (a) Find the time interval(s) over which the object's velocity is positive.

    (b) Find the time interval(s) over which the object's acceleration is positive.

27. The position of a moving object at time t is given by $s(t) = t^5 - 5t^4$, where $t \in [0, 10]$.

    (a) Find the time interval(s) over which the object's velocity is positive.

    (b) Find the time interval(s) over which the object's acceleration is positive.

28. Determine the x-coordinate of the only inflection point of a cubic polynomial $f(x) = ax^3 + bx^2 + cx + d$, where $a \neq 0$.

29. Consider a function of the form $f(x) = (x+3)^n$. Determine any inflection point(s) of the graph of $f(x)$ when

    (a) $n = 1$
    (b) $n = 2$
    (c) $n = 3$
    (d) $n = 4$
    (e) $n = 5$
    (f) $n = 6$

    (g) Make a conjecture as to when $f(x) = (x+3)^n$ has an inflection point given that n is a positive integer.

30. What condition(s) must be placed on $a, b, c, d, e \in \mathbb{R}$ so that the graph of $f(x) = ax^4 + bx^3 + cx^2 + dx + e$ has two inflection points?

31. Make a sketch of a function $f(x)$ that satisfies all of the following conditions:

    - $f(0) = f(4) = 0$,
    - $f'(x) > 0$ if $x < 2$,
    - $f'(2) = 0$,
    - $f'(x) < 0$ if $x > 2$,
    - $f''(x) < 0$ for all $x \in \mathbb{R}$.

32. Make a sketch of a function $f(x)$ that satisfies all of the following conditions:

    - $f(0) = f(4) = 0$,
    - $f'(x) > 0$ if $x < 2$,
    - $f'(2)$ does not exist,
    - $f'(x) < 0$ if $x > 2$,
    - $f''(x) > 0$ if $x \neq 2$.

33. Consider a function $f(x)$ that has a two-sided vertical asymptote at $x = 2$. Answer each of the following. (Hint: Consider the graphs.)

    (a) If $f'(x) > 0$ for $x \in (0,2) \cup (2,4)$, compute
    $$\lim_{x \to 2^-} f(x) \quad \text{and} \quad \lim_{x \to 2^+} f(x).$$

    (b) If $f''(x) > 0$ for $x \in (0,2) \cup (2,4)$, compute those same limits,
    $$\lim_{x \to 2^-} f(x) \quad \text{and} \quad \lim_{x \to 2^+} f(x).$$

34. Sketch the graph for $f(x) = \dfrac{|x|}{1+|x|}$, illustrating the signs of $f(x)$, $f'(x)$, and $f''(x)$. Be sure to illustrate any asymptotes. (Hint: Find a piecewise definition of $f(x)$.)

35. The probability density function for a normal distribution with mean $\mu \in \mathbb{R}$ and standard deviation $\sigma > 0$ is given by
$$f(x) = \frac{1}{\sigma\sqrt{2\pi}} e^{-\frac{1}{2}\left(\frac{x-\mu}{\sigma}\right)^2}$$
$$= \frac{1}{\sigma\sqrt{2\pi}} \exp\left[-\frac{1}{2}\left(\frac{x-\mu}{\sigma}\right)^2\right].$$

Note that here μ, σ and (of course) π are positive constants. Show that $x = \mu$ is the global maximum of this function. (This is somewhat similar to Example 4.1.6, page 418.)

36. Show that the normal density function from Exercise 35 has inflection points at $x = \mu \pm \sigma$.

37. Graph $f(x)$ from Exercise 35, noting that $f(x) \to 0^+$ as $x \to \pm\infty$. Choose reasonable lengths to represent μ and σ.

4.2. EXTREMA ON CLOSED INTERVALS

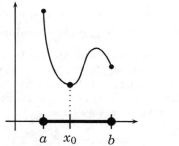
Interior minimum at x_0,
$f'(x_0) = 0$.
Maximum at endpoint a.

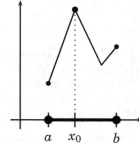
Interior maximum at x_0,
$f'(x_0)$ does not exist.
Minimum at endpoint a.

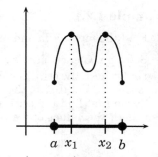
Interior maxima at x_1, x_2,
$f'(x_1), f'(x_2) = 0$.
Minimum at endpoints a, b.

Figure 4.2: Illustration that interior extrema can only occur at points where f' is zero or does not exist. Since a function continuous on a closed interval must have extrema for that interval (Extreme Value Theorem), any extremum must occur at an endpoint, or an interior point at which f' is zero or nonexistent, i.e., at an endpoint or a critical point.

4.2 Extrema on Closed Intervals

Recall the Extreme Value Theorem (page 107), stating that if $f(x)$ is continuous on a closed interval $[a, b]$, then there exist $x_{\min}, x_{\max} \in [a, b]$ such that

$$(\forall x \in [a,b])[f(x_{\min}) \le f(x) \le f(x_{\max})]. \tag{4.1}$$

In this section, we discuss how to locate these extrema. The main result should be intuitive, once all the possibilities are considered: that for such a function, continuous on $[a, b]$, the maximum value of its output must occur at an input that is an endpoint of the interval, or—if it is in the interior of the interval—at a point where $f' = 0$ or f' does not exist. It is similar for the minimum value. Points on a function's graph where $f' = 0$ or DNE are important enough to warrant their own definition.

Definition 4.2.1 *A point $(a, f(a))$ on the graph of $y = f(x)$ is called a* **critical point** *if either $f'(a) = 0$ or $f'(a)$ does not exist.*

Note that critical points are actual points $(a, f(a))$ on the graph $y = f(x)$ of the function f.

We will prove the following main theorem in the later subsections.

4.2.1 Main Theorem and Examples

Theorem 4.2.1 *For any function $f(x)$, continuous on a closed interval $[a, b]$, any such points $x_{\min}, x_{\max} \in [a, b]$ satisfying (4.1) of the EVT must be among the set that is the union of the critical points of f in (a, b) and the set of endpoints $\{a, b\}$ of the interval $[a, b]$.*

In other words, as long as the function is continuous on $[a, b]$, we need only scrutinize the critical points—where f' is zero or does not exist—in (a, b), and the endpoints a, b themselves. Checking the function values at these points exhausts all possibilities for the maximum and minimum values of $f(x)$ for those inputs $x \in [a, b]$ and the locations (inputs) of the extrema. See Figure 4.2, illustrating Theorem 4.2.1.

Example 4.2.1

Find the maximum and minimum values of $f(x) = x^3 - x$ for $x \in [0, 10]$, and their x-values.

Solution: Clearly $f(x)$ is continuous on $[0, 10]$ (it is a polynomial), so the theorem applies. Thus we need only check the critical points of $f(x)$, and the endpoints, of the interval $[0, 10]$. Since $f'(x) = 3x^2 - 1$ exists throughout $(0, 10)$, the only possible critical points will be those at which $f'(x) = 0$.

So we first find the critical points in $(0, 10)$, so we can check the values of f at those points. We then check the values of f at the endpoints $x = 0, 10$. For the critical points we have

$$f'(x) = 0 \iff 3x^2 - 1 = 0 \iff 3x^2 = 1 \iff x^2 = 1/3 \iff x = \pm 1/\sqrt{3}.$$

Now $-1/\sqrt{3} \notin (0, 10)$, so the only relevant critical point is $x = 1/\sqrt{3} \approx 0.577350269 \approx 0.577$.[9] Now we check the functional (output) values at the one critical point and the two endpoints.

critical point: $f\left(1/\sqrt{3}\right) = \left(1/\sqrt{3}\right)^3 - 1/\sqrt{3} = \dfrac{1}{3\sqrt{3}} - \dfrac{1}{\sqrt{3}} = \dfrac{1-3}{3\sqrt{3}} = \dfrac{-2}{3\sqrt{3}} \approx -0.385.$

endpoints: $f(0) = 0^3 - 0 = 0$, and $f(10) = 10^3 - 10 = 1000 - 10 = 990.$

From this we get that the minimum value of f on $[0, 10]$ is $-\dfrac{2}{3\sqrt{3}} \approx -0.385$, occurring at $x = 1/\sqrt{3} \approx 0.577$, and the maximum value is 990 occurring at $x = 10$.

To summarize geometrically, the highest and lowest points on the graph of the function $f(x) = x^3 - x$ for $x \in [0, 10]$ occur at the points $(10, 990)$ and $\left(1/\sqrt{3}, -2/(3\sqrt{3})\right) \approx (0.577, -0.385)$. Normally we will write instead

maximum: $f(10) = 990,$

minimum: $f\left(1/\sqrt{3}\right) = -2/3\sqrt{3} \approx -0.385.$

We could have graphed $y = f(x)$, and if we did so by hand (as in Section 4.1) we would have discovered possible locations of the extrema from the sign chart of f', and their values from the graph. (It would be difficult to choose a useful and consistent scale for the y-axis in this case.) An advantage of the method here is that it is not necessary to put in the same effort as we would to produce a graph since this method insures we need only check the function values at (in this case) three points to find the extreme values and their locations.

Example 4.2.2

Find the maximum and minimum values of $f(x) = \sin x + \cos x$ over the interval $x \in [0, 2\pi]$.

Solution: We note that this function is continuous, and so our theorem applies. Next we note that $f'(x) = \cos x - \sin x$ exists for all real x. Now we find those critical points within $(0, 2\pi)$, and examine f evaluated there and at the endpoints of our interval.

$$\begin{aligned} f'(x) = 0 &\iff \cos x - \sin x = 0 \\ &\iff \cos x = \sin x \\ &\iff 1 = \tan x \\ &\iff x \in \{\pi/4, 5\pi/4\}. \end{aligned}$$

---

[9] The approximation $x = 1/\sqrt{3} \approx 0.577350269$ is given so that some idea of the actual value is visible, and also to illustrate that, indeed, $1/\sqrt{3} \in [0, 10]$. For the sake of illustration, the approximate values of both inputs and outputs of the functions will be given routinely, though the analysis should use the exact values (such as $1/\sqrt{3}$) when possible before stating approximations.

4.2. EXTREMA ON CLOSED INTERVALS

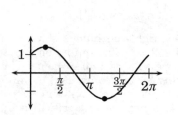

Figure 4.3a:
$f(x) = \cos x + \sin x, \quad x \in [0, 2\pi]$

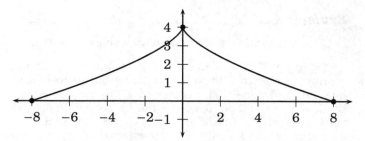

Figure 4.3b:
$f(x) = 4 - x^{2/3}, \quad x \in [-8, 8]$

Figure 4.3: Graphs of functions from Examples 4.2.2 (Figure 4.3a) and 4.2.3 (Figure 4.3b), showing locations of maximum and minimum values of their respective functions.

critical points:
$$f(\pi/4) = \sin\frac{\pi}{4} + \cos\frac{\pi}{4} = \frac{\sqrt{2}}{2} + \frac{\sqrt{2}}{2} = \sqrt{2} \approx 1.41421356,$$
$$f(5\pi/4) = \sin\frac{5\pi}{4} + \cos\frac{5\pi}{4} = -\frac{\sqrt{2}}{2} - \frac{\sqrt{2}}{2} = -\sqrt{2} \approx -1.41421356.$$

endpoints:
$$f(0) = \sin 0 + \cos 0 = 0 + 1 = 1,$$
$$f(2\pi) = \sin 2\pi + \cos 2\pi = 0 + 1 = 1.$$

We see that our maximum and minimum values occur at, repectively, $\left(\frac{\pi}{4}, \sqrt{2}\right) \approx (0.785, 1.414)$ and $\left(-\frac{\pi}{4}, -\sqrt{2}\right) \approx (-0.785, -1.414)$. This is illustrated in Figure 4.3.

The next example reminds us that critical points include those points on the graph at which f' does not exist.

Example 4.2.3

Find the maximum and minimum values of $f(x) = 4 - x^{2/3}$, for $x \in [-8, 8]$.

Solution: This function is defined and continuous on all of $\mathbb{R}$, so it is continuous on $[-8, 8]$. Now

$$f'(x) = -\frac{2}{3}x^{-1/3} = -\frac{2}{3\sqrt[3]{x}}.$$

While this is never zero, we note that $f'(x)$ does not exist at $x = 0$, which is thus a critical point in our interval. Now we compute f at the critical point and at the endpoints.

critical point: $f(0) = 4 - 0^{2/3} = 4$.

endpoints:
$$f(-8) = 4 - (-8)^{2/3} = 4 - \left[(-8)^{1/3}\right]^2 = 4 - (-2)^2 = 0,$$
$$f(8) = 4 - 8^{2/3} = 4 - \left[8^{1/3}\right]^2 = 4 - 2^2 = 0.$$

Thus the minimum value of $f(x)$ on $[-8, 8]$ is $f(\pm 8) = 0$, and the maximum value is $f(0) = 4$. This is illustrated in Figure 4.3.

Example 4.2.4

Find the maximum and minimum values of $f(x) = x - \sqrt{x}$ for the interval $x \in [0,4]$.

Solution: Here we have that f is continuous on $[0,4]$, since $f(x)$ is right-continuous at $x = 0$ and continuous at each of the other points in the interval (as per Definition 2.3.2 of continuity on a closed interval $[a,b]$, page 107). Furthermore, f is differentiable on $(0,4)$ as we will see, and so we need only check where $f' = 0$ on $(0,4)$, and the endpoints $x = 0,4$, when checking for maximum and minimum values of f on the full interval $[0,4]$. (While $f'(0)$ does not exist, that is not relevant since zero is an endpoint and we only consider interior points, i.e., those in $(0,4)$, when seeking critical points. Furthermore, 0 is an endpoint and so the value of $f(0)$ will already be considered when testing inputs to find the extrema.) First we compute

$$f'(x) = 1 - \frac{1}{2\sqrt{x}} = \frac{2\sqrt{x}-1}{2\sqrt{x}},$$

and see $f'(x) = 0$ when $\sqrt{x} = \frac{1}{2}$, i.e., when $x = \frac{1}{4} \in (0,4)$. Thus we check the relevant points—critical points and endpoints—and note the maximum and minimum values and their inputs' locations.

critical point: $f(1/4) = 1/4 - \sqrt{1/4} = 1/4 - 1/2 = -1/4$.

endpoints: $f(0) = 0 - \sqrt{0} = 0,\qquad f(4) = 4 - \sqrt{4} = 4 - 2 = 2$.

maximum: $f(4) = 2$.

minimum: $f(1/4) = -1/4$.

Example 4.2.5

Find the maximum and minimum values of $f(x) = \frac{x}{x^2+1}$ for the interval $x \in [0,2]$.

Solution: Note that the function is continuous over all of $\mathbb{R}$, since $x^2 + 1 \geq 1 > 0$, so that we have $(\forall x \in \mathbb{R})[x^2 + 1 \neq 0]$. Using the Quotient Rule to compute the derivative we get

$$f'(x) = \frac{(x^2+1)\frac{d}{dx}(x) - x\frac{d}{dx}(x^2+1)}{(x^2+1)^2} = \frac{x^2+1-x(2x)}{(x^2+1)^2} = \frac{1-x^2}{(x^2+1)^2},$$

which is defined everywhere, and $f'(x) = 0$ for $x = \pm 1$. The only critical point in $(0,2)$ is thus at $x = 1$.

critical point: $f(1) = \dfrac{1}{1^2+1} = \dfrac{1}{2} = 0.5$.

endpoints: $f(0) = 0/(0^2+1) = 0,\qquad f(2) = 2/(2^2+1) = 2/5 = 0.4$.

maximum: $f(1) = 1/2$.

minimum: $f(0) = 0$.

This function was graphed in Example 4.1.7, which starts on page 419.

Quite often we are only interested in either a maximum or minimum value of a function, and at times we do not need to have a closed, bounded interval. Consider the following example:

4.2. EXTREMA ON CLOSED INTERVALS

Example 4.2.6

Find the minimum value of $f(x) = 1 + x\ln x$.

Solution: Note that the domain is not specified, so we will look at the natural domain of $f(x)$, for which the natural logarithm present within the function requires $x \in (0, \infty)$.

Next we note that
$$f'(x) = x \cdot \frac{d\ln x}{dx} + (\ln x)\frac{dx}{dx} = x \cdot \frac{1}{x} + \ln x = 1 + \ln x.$$

Thus $f'(x) = 0 \iff \ln x = -1 \iff x = e^{-1}$.

We can then make a sign chart for $f'(x)$ if we like, though for this particular example we might instead note that
$$f''(x) = \frac{d}{dx}(1 + \ln x) = \frac{1}{x} > 0 \quad \text{for all } x > 0,$$

and so this function is concave up on its entire domain. From this we know that the one critical point, namely $(e^{-1}, f(e^{-1}))$, is a global minimum. The global minimum value of $f(x)$ is then
$$f(e^{-1}) = 1 + \frac{1}{e}\ln\left(\frac{1}{e}\right) = 1 - \frac{1}{e} = \frac{e-1}{e}.$$

If we would like an approximation of the actual point, we can use $\left(\frac{1}{e}, \frac{e-1}{e}\right) \approx (0.367879, 0.632121)$.

The graph of $f(x) = 1 + x\ln x$ is given on the right. In the sequel we will be able to show $x \to 0^+ \implies 1 + x\ln x \to 1$, but that observation is not necessary for this analysis. Indeed it was enough to know $f(x)$ is defined exactly for $x > 0$, is concave up ($f''(x) = 1/x > 0$) everywhere $f(x)$ is defined, and has zero derivative at $x = 1/e$, to conclude that point is a global minimum.

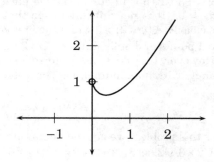

4.2.2 Applications (Max/Min Problems)

There are many applications where we wish to maximize or minimize some quantity that is dependent on another quantity in some functional relationship. In other words, there is some "output" $A(x)$ depending on an "input" x, where we wish to choose x to maximize or minimize $A(x)$. There is often some natural or practical domain of values x allowed by the context. Once we translate the application's parameters into the language of functions, we can find the desired extrema (or conclude one does not exist) based on the techniques already employed in this section. Some "applications" do come from a mathematical setting, as in our first example, while others are first set in a more "real world" context.

Example 4.2.7

Find the dimensions of the rectangle with the largest area with its base along the x-axis and that fits within the upper semicircle of $x^2 + y^2 = 9$.

Solution: It is useful to begin with an illustration:

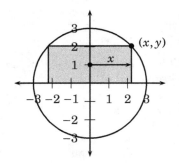

To have a function to maximize, we will take x to be the x-coordinate of the upper rightmost point of the rectangle. Then x represents half of the horizontal length of the rectangle. The total area is then given by $A = 2xy$. But we need A to be a function of one variable only, so we can use the previous techniques. From the equation of the circle, $y = \sqrt{9-x^2}$, and so

$$A = 2xy = 2x\sqrt{9-x^2}.$$

For this scenario, we must assume $x \in [0,3]$. Note that the endpoints $x \in \{0,3\}$ both yield zero area (consider the geometric implications of each), and so clearly neither will yield the global maximum. We thus look for critical points.

$$A = 2x\sqrt{9-x^2} \implies \frac{dA}{dx} = 2x\frac{d}{dx}\sqrt{9-x^2} + \sqrt{9-x^2} \cdot \frac{d\,2x}{dx} = 2x \cdot \frac{1}{2\sqrt{9-x^2}} \cdot (-2x) + 2\sqrt{9-x^2}$$
$$= \frac{-2x^2 + 2(9-x^2)}{\sqrt{9-x^2}} = \frac{2(9-2x^2)}{\sqrt{9-x^2}}.$$

From this we get $dA/dx = 0 \iff 9 - 2x^2 = 0 \iff 9 = 2x^2 \iff x^2 = 9/2$. Solutions of this equation are $x = \pm 3/\sqrt{2}$, but in our model we have $x \in [0,3]$ and so $x = 3/\sqrt{2} \approx 2.12$ is our relevant solution of $dA/dx = 0$. It is not difficult to see that $x = 3/\sqrt{2}$ must therefore yield the maximum value for A, since clearly $A\left(3/\sqrt{2}\right) > 0$ while the endpoints $x = 0, 3$ both yield $A = 0$. Since all we need to check are critical points and endpoints, $x = 3/\sqrt{2}$ must yield the maximum.

However, that does not quite solve our original problem, which asked for the dimensions of the rectangle with maximum area. The dimensions could be of the form $2x \times y$, where $x = 3/\sqrt{2}$ and

$$y = \sqrt{9-x^2} = \sqrt{9-(3/\sqrt{2})^2} = \sqrt{9-9/2} = \sqrt{9/2} = 3/\sqrt{2}.$$

So for this problem we get the maximum area occurs when $x = y = 3/\sqrt{2}$, but then the dimensions are $2 \cdot 3/\sqrt{2} \times 3/\sqrt{2}$, or $3\sqrt{2} \times 3/\sqrt{2} \approx 4.24 \times 2.12$.[10]

Another approach to finding the maximum value of A is to find the maximum value of A^2 and compute its square root. This works because $f(A) = A^2$ is increasing on $A \in [0,\infty)$, so that maximizing A is equivalent to maximizing A^2, and some of the calculus and arithmetic can be somewhat simpler, due to the absence of the radical in the expression for B.

To formalize this, let $B = A^2$. Then for $x \in [0,3]$, we have

$$B = \left[2x\sqrt{9-x^2}\right]^2 = 4x^2(9-x^2) = 36x^2 - 4x^4.$$

Setting $0 = \frac{dB}{dx} = 72x - 16x^3 = 8x(9-2x^2)$, we get $x = 0, \sqrt{9/2}$ as before, with $x = 3/\sqrt{2}$ maximizing $B = A^2$ and therefore maximizing A, as before.

Trigonometric functions also appear in many application problems, and so their derivatives are very important, as is the ability to solve (sometimes complicated) trigonometric

---

[10]Note that the rectangle we found was twice as wide (horizontally) as it was tall. This proportion will be the case regardless of the radius of the semicircle (see Exercise 24, page 446).

4.2. EXTREMA ON CLOSED INTERVALS

equations. In the following example, it is interesting to see how the constants 6 and 10 (or their ratio) appear in the final solution.

Example 4.2.8

Suppose we would like to span a 6-ft wall with a ladder to reach the side of a building 10 ft from the base of the wall. What is the shortest such ladder that can accomplish this?

Solution: We begin with a diagram of the situation and introduce some variables. In particular, we will let L be the length of the ladder and θ its angle of elevation from the horizontal.

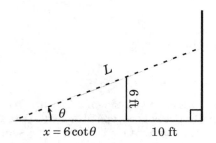

We require that L be a function of some variable so that we can use the techniques of this section to maximize L. Our first instinct may be to let x be the length of the smaller triangle, but finding L as a function of x is somewhat difficult. However, finding x as a function of θ is relatively simple, and then finding L as a function of θ is not much more difficult. We will suppress units until the end of our computations.

Because $6/x = \tan\theta$, we can easily write $x = 6\cot\theta$. From this,

$$\frac{6\cot\theta + 10}{L} = \cos\theta \implies L = (6\cot\theta + 10)\sec\theta$$

$$\implies L \left(6 \cdot \frac{\cos\theta}{\sin\theta} + 10 \right) \cdot \frac{1}{\cos\theta}$$

$$\implies L = 6\csc\theta + 10\sec\theta.$$

Clearly we have $\theta \in \left(0, \frac{\pi}{2}\right)$. While this is not a closed interval, it should be also clear from the geometry that $L \to \infty$ as $\theta \to 0^+$ and as $\theta \to \frac{\pi}{2}^-$, so we can expect an interior minimum. We next go looking for it by differentiating

$$\frac{dL}{d\theta} = -6\csc\theta\cot\theta + 10\sec\theta\tan\theta.$$

This exists for all $x \in \left(0, \frac{\pi}{2}\right)$, so we next search to find where $dL/d\theta = 0$:

$$\frac{dL}{d\theta} = 0 \implies 10\sec\theta\tan\theta = 6\csc\theta\cot\theta.$$

$$\implies \tan^3\theta = 3/5$$

$$\implies \tan\theta = \sqrt[3]{3/5}$$

$$\implies \theta = \tan^{-1}\sqrt[3]{3/5} \approx 40.15°.$$

Since it is L we are looking for, and it involves trigonometric functions of θ, we compute these by constructing a triangle with $\tan\theta = \frac{\sqrt[3]{3}}{\sqrt[3]{5}}$.

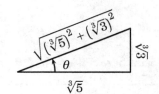

438   CHAPTER 4. USING DERIVATIVES TO ANALYZE FUNCTIONS; APPLICATIONS

The minimum value for L is therefore

$$L\left(\tan^{-1} \sqrt[3]{3/5}\right) = 6\csc\left(\tan^{-1} \sqrt[3]{3/5}\right) + 10\sec\left(\tan^{-1} \sqrt[3]{3/5}\right) = 6 \cdot \frac{\sqrt{\left(\sqrt[3]{5}\right)^2 + \left(\sqrt[3]{3}\right)^2}}{\sqrt[3]{3}} + 10 \cdot \frac{\sqrt{\left(\sqrt[3]{5}\right)^2 + \left(\sqrt[3]{3}\right)^2}}{\sqrt[3]{5}}$$

$$= 6 \cdot \sqrt{\left(\sqrt[3]{5/3}\right)^2 + 1} + 10 \cdot \sqrt{1 + \left(\sqrt[3]{3/5}\right)^2}$$

$$\approx 22.4,$$

and so the shortest ladder that can reach over the fence to touch the building is approximately 22.4 ft long.[11]

The next problem comes from physics and is solved in the abstract.

Example 4.2.9 _____

Consider a box weighing 100 lbs lying on a flat surface, where the coefficient of kinetic friction is $\mu_k \in (0,1)$. We wish to move the box across the surface by pulling it with a rope attached to one corner of the box, where the rope makes the angle $\theta \in [0, \pi/2]$ with the horizontal. At what angle should we pull the box so the net horizontal force on the box is maximized as it slides along the surface?

Solution:

The total horizontal force $\mathscr{F}$ on the box will be the difference between the horizontal force F_h from our pulling of the rope and the force caused by the friction: $\mathscr{F} = F_h - F_{\text{friction}}$. The force of friction is proportional to the net vertical force applied to the surface (which is the weight of the box minus the vertical pull of the rope), and is given by $F_{\text{friction}} = \mu_k \cdot (\text{weight} - F_v)$, where F_v is the vertical component of the force from the pull of the rope. Finally, if the rope makes an angle of θ from the horizontal and is pulled with a total force of F, then

$$F_v = F\sin\theta,$$
$$F_h = F\cos\theta,$$
$$F_{\text{friction}} = \mu_k(100 \text{ lbs} - F_v)$$
$$= \mu_k(100 \text{ lbs} - F\sin\theta),$$
$$\mathscr{F} = F_h - F_{\text{friction}}$$
$$= F\cos\theta - \mu_k(100 \text{ lbs} - F\sin\theta).$$

Since F and μ_k are fixed constants, we can write and differentiate the function $\mathscr{F}(\theta)$, for $\theta \in [0, \pi/2]$, as follows:

$$\mathscr{F}(\theta) = F\cos\theta - \mu_k(100 \text{ lbs}) + \mu_k F \sin\theta$$
$$\implies \mathscr{F}'(\theta) = -F\sin\theta + \mu_k F \cos\theta.$$

Note that here $\mathscr{F}'(\theta) = \frac{d}{d\theta}\mathscr{F}(\theta)$. Since $\mathscr{F}(\theta)$ is continuous on $[0, \pi/2]$ and differentiable on all of $(0, \pi/2)$,

---

[11]If one is only interested in the approximation, then finding $6\csc\theta + 10\sec\theta$ where $\theta = \tan^{-1}\sqrt[3]{3/5} \approx 40.14537714°$ (best left stored in calculator memory for the computation) can achieve this approximation somewhat faster.

4.2. EXTREMA ON CLOSED INTERVALS

we next look for critical points where $\mathscr{F}'(\theta) = 0$:

$$\mathscr{F}'(\theta) = 0 \iff -F\sin\theta + \mu_k F \cos\theta = 0$$
$$\iff -F\sin\theta = -\mu_k F \cos\theta$$
$$\iff \tan\theta = \mu_k.$$

Here we did assume $F \neq 0$ (or the problem is not terribly interesting). In fact, we should assume $F > 0$, so that the force of the friction is indeed opposite to the horizontal component of the force applied by the rope.

Next we check the endpoints and critical point of $\mathscr{F}(\theta)$ for $\theta \in [0, \pi/2]$ (using the diagram provided):

$$\mathscr{F}(0) = F\cos 0 - \mu_k(100\text{ lbs}) + \mu_k F \sin 0 = F - 100\mu_k \text{(lbs)},$$

$$\mathscr{F}(\pi/2) = F\cos\frac{\pi}{2} - \mu_k(100\text{ lbs}) + \mu_k F \sin\frac{\pi}{2} = \mu_k F - 100\mu_k \text{(lbs)},$$

$$\mathscr{F}(\tan^{-1}\mu_k) = F\cos(\tan^{-1}\mu_k) - \mu_k(100\text{ lbs}) + \mu_k F \sin(\tan^{-1}\mu_k) = \frac{F}{\sqrt{1+\mu_k^2}} - 100\mu_k\text{(lbs)} + \frac{\mu_k F \mu_k}{\sqrt{1+\mu_k^2}}$$

$$= \frac{F(1+\mu_k^2)}{\sqrt{1+\mu_k^2}} - 100\mu_k\text{(lbs)} = F\sqrt{1+\mu_k^2} - 100\mu_k\text{(lbs)}.$$

To justify our computations of $\cos(\tan^{-1}\mu_k)$ and $\sin(\tan^{-1}\mu_k)$, we can again employ the triangle-drawing technique from trigonometry, where we draw an appropriate right triangle where the angle, say ϕ, has the correct tangent, and we use the Pythagorean theorem to fill in the missing side (in this case the hypotenuse). Then we can read the other trigonometric angles of $\phi = \tan^{-1}\mu_k$ from the diagram:

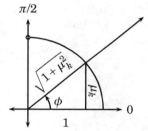

Since $\mathscr{F}$ at 0, $\pi/2$ and $\tan^{-1}\mu_k$ all have the term $-100\mu_k$(lbs), and the remaining terms each have the factor F, we need only check that $\sqrt{1+\mu_k^2}$ is greater than both 1 and μ_k. Since $\mu_k \in (0,1)$, we know $\sqrt{1+\mu_k^2} > \sqrt{1+0^2} = 1$, and similarly $\sqrt{1+\mu_k^2} > \sqrt{\mu_k^2} = \mu_k$, and so $\sqrt{1+\mu_k^2}$ is indeed greater than both 1 and μ_k.

We can conclude that the optimal angle we should pull on the rope to maximize net horizontal force acting on the box is thus

$$\theta = \tan^{-1}\mu_k.$$

Note the following from our computation:

- The weight of the box (100 lbs) was ultimately not relevant in the computation of the optimal θ. The value of F was also not relevant, as long as $F > 0$.

- A larger coefficient of kinetic friction μ_k will result in a steeper optimal angle.

- Not stated but clear from the physics is that if $\mathscr{F}(\theta) < 0$ then the box will not move from the force acting through the rope.

For any given problem, it is usually clear what the *dependent variable* of some function to optimize should be, but it may require some experimentation to find the *independent variable* which most easily allows us to find the form of the function, and to use derivative techniques for optimization. If a problem appears somewhat trigonometric in nature, one

440 CHAPTER 4. USING DERIVATIVES TO ANALYZE FUNCTIONS; APPLICATIONS

should always consider an independent variable which is an angle θ or similar, as we did in both Example 4.2.9 and Example 4.2.8. However, there are problems that may appear trigonometric in form, but can be solved more easily by using a length as an independent variable, as we do in the next example.

Example 4.2.10

Suppose a racer can swim 3 miles/hour and run 15 miles/hour. For an event the racer is to swim across a lake one mile wide and run to a point five miles from the point directly across the lake from his starting point (see illustration). How far from the point directly across the lake should be the point at which he emerges from the water to begin his run if he wishes to finish the race in minimum time?

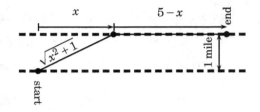

Solution: We will take x to be the distance "down-race" on the far shore at the point he emerges from the water. Then the lengths in miles for which he was, respectively, swimming and running are $\sqrt{x^2+1^2}$ (from the Pythagorean Theorem) and $5-x$. Since the respective speeds are 3 miles/hour and 15 miles/hour, his total time is the following, from which we compute the derivative dt/dx:

$$t = \frac{\sqrt{x^2+1}}{3} + \frac{5-x}{15} \implies \frac{dt}{dx} = \frac{1}{3} \cdot \frac{1}{2\sqrt{x^2+1}} \cdot \frac{d}{dx}\left(x^2+1\right) - \frac{1}{15}$$

$$\implies \frac{dt}{dx} = \frac{x}{3\sqrt{x^2+1}} - \frac{1}{15}.$$

Setting $dt/dx = 0$ gives us

$$\frac{x}{3\sqrt{x^2+1}} = \frac{1}{15} \implies 15x = 3\sqrt{x^2+1} \implies 5x = \sqrt{x^2+1} \implies 25x^2 = x^2+1 \implies 24x^2 = 1$$

$$\implies x = \pm\sqrt{1/24} = \pm\frac{1}{2\sqrt{6}} \approx \pm 0.2041.$$

Clearly, a negative value of x would have the racer backtracking, and so we discard the negative case and consider instead the critical point $x = 1/(2\sqrt{6})$. However, we should also consider some endpoints, which for this problem can be reasonably chosen to be $x = 0, 5$, so that we are finding $x \in [0,5]$ to minimize our racer's time:

$$t(0) = \frac{\sqrt{1}}{3} + \frac{5}{15} = \frac{1}{3} + \frac{1}{3} = \frac{2}{3} = 0.\overline{6},$$

$$t(5) = \frac{\sqrt{26}}{3} + 0 \approx 1.6997,$$

$$t\left(\frac{1}{\sqrt{24}}\right) = \frac{\sqrt{\frac{1}{24}+1}}{3} + \frac{5 - \frac{1}{\sqrt{24}}}{15} = \frac{\sqrt{\frac{25}{24}}}{3} + \frac{5\sqrt{24}-1}{15\sqrt{24}} = \frac{5 \cdot 5 + 5\sqrt{24} - 1}{15\sqrt{24}} = \frac{24 + 5\sqrt{24}}{15\sqrt{24}} = \frac{\sqrt{24}+5}{15} \approx 0.6599.$$

So the racer's best time will have him swimming slightly "downstream" (about 0.2 miles) from directly across the starting point, because $x = 1/\sqrt{24}$ (miles) minimizes the time function $t(x)$. The amount of time saved by doing so instead of swimming to the point $x = 0$ is thus

$$\left(\frac{2}{3} - \frac{\sqrt{24}+5}{15}\right) \text{ hour} \cdot \frac{3600 \text{ second}}{1 \text{ hour}} \approx 24\frac{1}{4} \text{ sec.}$$

4.2. EXTREMA ON CLOSED INTERVALS

(The distance he should swim downstream to minimize his race time would change if the speed of the lake's water were not assumed to be zero, or if the width of the lake or either of his race speeds—inside or outside of the water—were changed.)

Finally, we describe a short-cut when working with absolute values of functions, where $f(x) = |g(x)|$. We note first that there is never any harm in testing extra points for the maximum and minimum values of a function f, unless those points are outside of the interval $[a,b]$ on which we are trying to maximize or minimize f. It is better to test more than we need than to miss a point that might yield an extremum. We also note that

$$\frac{d|x|}{dx} = \begin{cases} 1 & \text{if } x > 0, \\ \text{DNE} & \text{if } x = 0, \\ -1 & \text{if } x < 0. \end{cases} \tag{4.2}$$

By the Chain Rule, we can expect that $\frac{d|g(x)|}{dx}$ will be $g'(x)$ on open intervals where $g(x) > 0$, might not exist if $g(x) = 0$, and will be $-g'(x)$ on open intervals where $g(x) < 0$. If we are looking for critical points of $f(x) = |g(x)|$, we can therefore look at critical points of $g(x)$, and those points where $g(x) = 0$, as possible critical points of $f(x) = |g(x)|$. This observation simplifies examples such as the following.

Example 4.2.11 _____

Find the maximum and minimum values of $f(x) = |x^2 + x - 2|$ on $[-3,3]$.

<u>Solution</u>: Clearly $f(x)$ is continuous on $[-3,3]$. We will check the values of $f(x)$ at the endpoints $x = \pm 3$, at the critical points of $x^2 + x - 2$, and where $x^2 + x - 2 = 0$. Now $\frac{d}{dx}(x^2 + x - 2) = 2x + 1 = 0$ at $x = -\frac{1}{2}$. Next, $x^2 + x - 2 = (x+2)(x-1) = 0$ at $x = -2, 1$. Combining these with the endpoints we get

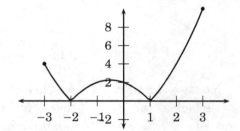

$$\begin{aligned}
f(-1/2) &= |1/4 - 1/2 - 2| = 9/4, \\
f(-2) &= |4 - 2 - 2| = 0, &\text{(minimum)} \\
f(1) &= |1 + 1 - 2| = 0, &\text{(minimum)} \\
f(-3) &= |9 - 3 - 2| = 4, \\
f(3) &= |9 + 3 - 2| = 10. &\text{(maximum)}
\end{aligned}$$

In this case, there is a local maximum at $x = -1/2$, but the maximum for the interval occurs at $x = 3$, that maximum value being $f(3) = 10$. There are two locations for the minimum, at $x = -2, 1$, where $f(x) = 0$. These can be seen in the graph.

4.2.3 Argument for Main Theorem (for Completeness)

The main theorem is Theorem 4.2.1, page 431, stating that a function $f(x)$, continuous on $[a,b]$, will achieve maximum and minimum values for $[a,b]$ somewhere in $[a,b]$, and that any such point where an extremum is achieved will either be an endpoint a or b, or an interior point $x \in (a,b)$ where $f'(x)$ is zero or does not exist. The theorem is ultimately quite intuitive.

442 CHAPTER 4. USING DERIVATIVES TO ANALYZE FUNCTIONS; APPLICATIONS

However, to prove the theorem requires a somewhat lengthy argument, which we give here, albeit in several steps.

We wish to find those points in $[a,b]$ at which a continuous function $f:[a,b] \longrightarrow \mathbb{R}$ achieves its maximum and minimum values possible for those $x \in [a,b]$. Recall from Section 2.3 that $f([a,b])$, the set of all possible outputs of $f(\ )$ resulting from inputs from $[a,b]$, will itself be a finite, closed interval, and thus will have a maximum and minimum value, i.e., there will indeed exist some $x_{\min}, x_{\max} \in [a,b]$ so that $f([a,b]) = [f(x_{\min}), f(x_{\max})]$, i.e.,

$$(\exists x_{\min}, x_{\max} \in [a,b])(\forall x \in [a,b])[f(x_{\min}) \leq f(x) \leq f(x_{\max})]. \tag{4.3}$$

This was the essence of the Extreme Value Theorem (EVT), which was Corollary 2.3.1, page 107.[12] The next theorem is stated in a manner reflective of its proof. However, its equivalent forms given in the corollaries are more useful in applications.

Theorem 4.2.2 *Suppose $f:[a,b] \longrightarrow \mathbb{R}$ is continuous, and that $f(x)$ achieves its maximum or minimum value for the interval $[a,b]$ at an interior point $x_0 \in (a,b)$. Under these assumptions, if $f'(x_0)$ exists, then $f'(x_0) = 0$, i.e.,*

$$f'(x_0) \text{ exists} \implies f'(x_0) = 0.$$

Proof: First we look at the case $f:[a,b] \longrightarrow \mathbb{R}$ continuous, $x_0 \in (a,b)$, and $f(x_0)$ is the maximum value of $f(x)$ on $[a,b]$. We need to show that if $f'(x_0)$ exists then $f'(x_0) = 0$. So we (further) suppose $f'(x_0)$ exists. This means

$$\lim_{\Delta x \to 0} \frac{f(x_0 + \Delta x) - f(x_0)}{\Delta x} = f'(x_0) \text{ exists.}$$

Thus the left and right side limits exist and must agree with this two-sided limit:

$$\lim_{\Delta x \to 0^-} \frac{f(x_0 + \Delta x) - f(x_0)}{\Delta x} = f'(x_0), \tag{4.4}$$

$$\lim_{\Delta x \to 0^+} \frac{f(x_0 + \Delta x) - f(x_0)}{\Delta x} = f'(x_0). \tag{4.5}$$

Now we carefully consider each of these two limits, (4.4) and (4.5). In particular we look at the signs of these. In both of these limits, since $f(x_0)$ is a maximum, we have $f(x_0 + \Delta x) \leq f(x_0)$, implying $f(x_0 + \Delta x) - f(x_0) \leq 0$. Thus the numerators in the limits are both nonpositive.

Next we look specifically at (4.4). Since $\Delta x \to 0^-$, we are looking at limits of fractions with denominators $\Delta x < 0$. Thus we have

$$f'(x_0) = \lim_{\Delta x \to 0^-} \frac{\overbrace{f(x_0 + \Delta x) - f(x_0)}^{\leq 0}}{\underbrace{\Delta x}_{<0}} \geq 0. \tag{4.6}$$

---

[12]It is very strongly suggested that the reader at least briefly review Section 2.3 at some point during the reading of this current section. It would likely be particularly helpful to reconsider the figures in that section to recall why it is necessary to have a continuous function *and* a closed interval for the conclusion of the Extreme Value Theorem to be guaranteed.

4.2. EXTREMA ON CLOSED INTERVALS

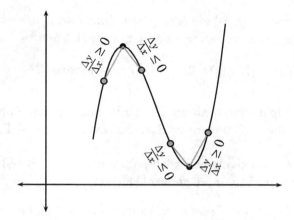

Figure 4.4: Illustration of behavior of *difference quotients* $(\Delta y)/(\Delta x)$, representing slopes of "secant" line segments (gray) joining points on the curve, where one endpoint is at a maximum or minimum. Near a maximum, these line segments have nonnegative slope immediately to the left, and nonpositive slope immediately to the right, of the maximum. The reverse is true at a minimum: nonpositive slope to the left, and nonnegative slope to the right. These observations are crucial in the proof that $f' = 0$ or does not exist at such interior maxima or minima.

On the other hand, when we instead look closely at the form of the limit in (4.5), we see

$$f'(x_0) = \lim_{\Delta x \to 0^+} \frac{\overbrace{f(x_0 + \Delta x) - f(x_0)}^{\leq 0}}{\underbrace{\Delta x}_{>0}} \leq 0. \qquad (4.7)$$

So, from (4.6) and (4.7) we have $(f'(x_0) \geq 0) \wedge (f'(x_0) \leq 0)$, i.e., $0 \leq f'(x_0) \leq 0$, so we have to conclude $f'(x_0) = 0$, q.e.d. for the case of an interior maximum.

The case where $f(x_0)$ is the minimum value of $f(x)$ for $x \in [a, b]$ is similar, except that the numerators in those same limits are all nonnegative, i.e., $f(x_0 + \Delta x) - f(x_0) \geq 0$ in the limits (4.6) and (4.7), giving $(f'(x_0) \leq 0) \wedge (f'(x_0) \geq 0)$, respectively, yielding again $f'(x_0) = 0$, q.e.d.[13]

It is useful to consider graphically why inequalities (4.6) and (4.7) should hold, as well as the conclusion of the theorem. This is illustrated in Figure 4.4 and the caption therein.

Now, recalling that $P \longrightarrow Q \iff (\sim P) \vee Q$ (see for instance (1.29), page 25) we can rewrite the conclusion of Theorem 4.2.2:

$$[f'(x_0) \text{ exists }] \longrightarrow [f'(x_0) = 0] \iff [f'(x_0) \text{ does not exist}] \vee [f'(x_0) = 0].$$

We can thus rewrite Theorem 4.2.2, in the equivalent form of a very important corollary:

---

[13] In the proof, it was necessary that $x_0 \in (a, b)$ so that there is room to the left and right of x_0 in the interval (a, b), and we can let $\Delta x \to 0^-$ and $\Delta x \to 0^+$ and eventually, for small enough $|\Delta x|$, have in both cases ($\Delta x > 0$ and $\Delta x < 0$) that $x_0 + \Delta x \in (a, b)$—so that $f(x_0 + \Delta x)$ is defined—as well. In contrast, if $x_0 \in \{a, b\}$, i.e., x_0 is an endpoint of $[a, b]$, we cannot say anything of its derivative without further information, because $x_0 + \Delta x$ is not guaranteed to be within the interval for both $\Delta x > 0$ and $\Delta x < 0$, no matter how small (but nonzero) Δx is.

Corollary 4.2.1 *Suppose $f : [a,b] \longrightarrow \mathbb{R}$ is continuous, and achieves its maximum or minimum output value for $[a,b]$ at an interior point $x_0 \in (a,b)$. Then*

$$[f'(x_0) = 0] \vee [f'(x_0) \text{ does not exist}].$$

So any *interior* (non-endpoint) extremum of a continuous function on a closed interval $[a,b]$ must occur at points where f' is zero or does not exist. This is illustrated in Figure 4.2, page 431.

As noted before, we collectively call such points where $f' = 0$ or DNE *critical points*. We redefine them here for a slightly more general context.

Definition 4.2.2 *For $f(x)$ defined and continuous on $(x_0 - \delta, x_0 + \delta)$, for some $\delta > 0$, the point $x = x_0$ is called a* **critical point** *of $f(x)$ if and only if $f'(x_0) = 0$ or $f'(x_0)$ does not exist.*

It helps later to establish now that, when analyzing a function $f(x)$ where we consider only those $x \in [a,b]$, by *critical points* we mean only those *interior* points $x_0 \in (a,b)$ where $f'(x_0)$ is zero or nonexistent, thus allowing x_0 to have some "wiggle room" both to the left and to the right within the interval in question (which allows for derivative computations). Using this definition, we can again rewrite Theorem 4.2.2, and Corollary 4.2.1:

For $f(x)$ continuous on $[a,b]$, any interior extremum must occur at a critical point.

This is still not quite the final and most useful version of this theorem. To arrive at the most complete statement, let us first recall (from Section 2.3) that if $f(x)$ is continuous on $[a,b]$, it will necessarily achieve its maximum and minimum values for that interval on that interval. Now we can import our corollary, which states that if one of these is achieved in the *interior* of $[a,b]$, i.e., on (a,b), then that point must be a critical point:

Assuming that x_0 is any point in $[a,b]$ at which f achieves either its maximum or minimum value for that interval $[a,b]$, then

$$x_0 \in [a,b] \iff (x_0 \in (a,b)) \vee (x_0 \in \{a,b\})$$
$$\implies \underbrace{(x_0 \text{ is a critical point of } f)}_{\text{Corollary 4.2.1, page 444}} \vee (x_0 \in \{a,b\}).$$

Thus x_0 must be a critical point of f in (a,b), or an endpoint of the interval $[a,b]$, q.e.d.

4.2. EXTREMA ON CLOSED INTERVALS

Exercises

For Exercises 1–14, find the maximum and minimum values of $f(x)$, and the x-values at which they occur, on the given closed interval $[a,b]$.

1. $f(x) = \frac{3}{2}x + 4$ on $[-2, 3]$.
2. $f(x) = 3x^2 - 12x + 11$ on $[0, 3]$.
3. $f(x) = \dfrac{x^2}{x^2 + 4}$ on $[-1, 1]$.
4. $f(x) = \sqrt[3]{x+2}$ on $[-2, 2]$.
5. $f(x) = \sin \pi x$ on $[0, 1]$.
6. $f(x) = \sin \pi x$ on $[0, 3]$.
7. $f(x) = 2\sin x + 8\cos x$ on $[-\pi/2, \pi/2]$.
8. $f(x) = (x+3)^{2/3}$ on $[-4, 1]$.
9. $f(x) = 3 - |x|$ on $[-1, 2]$.
10. $f(x) = \sqrt{9 - x^2}$ on $[-2, 1]$.
11. $f(x) = x^2\sqrt{9 - x^2}$ on its domain $[-3, 3]$.
12. $f(x) = x^3\sqrt{9 - x^2}$ on its domain $[-3, 3]$.
13. $f(x) = 4x - x^{3/2}$ on $[0, 16]$.
14. $f(x) = 2x^{2/3} - x^{5/3}$ on $[-1, 4]$.

15. Show that every quadratic polynomial function $f(x) = ax^2 + bx + c$, where $a \neq 0$, has exactly one critical point on $\mathbb{R} = (-\infty, \infty)$. Where is it located? When is it a minimum? When is it a maximum? (Consider f''.)

16. Consider the cubic polynomial function $f(x) = ax^3 + bx^2 + cx + d$, where $a \neq 0$, on the interval $\mathbb{R} = (-\infty, \infty)$.

    (a) Give an example of such an $f(x)$ that has no local maximum or minimum points.

    (b) Give an example of such a function that has exactly one maximum and one minimum point.

    (c) Is it possible for a cubic polynomial to have more than two local extrema? Why or why not?

    (d) Is it possible for a cubic polynomial to have exactly one local extremum? Why or why not?

17. Find two positive real numbers x and y such that their sum is 28 and their product is maximized.

18. The perimeter of a rectangle is 50 cm. Find the dimensions of the rectangle so that area is maximized.

19. A teacher gives a particularly difficult exam, and when the students' scores are very low, decides to use a "curve" to adjust their scores, whereby original scores will be inputted into the function $S(x) = 10\sqrt{x}$.

    (a) What are $S(0)$, $S(9)$, $S(16)$, $S(36)$, $S(49)$, $S(64)$, $S(81)$ and $S(100)$?

    (b) Which original score x would a student have to make before the curve in order that his or her score would make the largest increase in points from the teacher adopting this grading curve?

20. Suppose the rectangle in Exercise 18 has perimeter P. Find the length and width of the rectangle (in terms of P) so that area is maximized.

21. Suppose the sum of the lengths of three sides of a rectangle is 40 ft. What is the maximum area of the rectangle, and what are its dimensions?

22. The sum of two positive numbers is 16. What is the minimum value of the sum of their cubes?

23. Find the shortest possible distance from a point on the parabola $y = x^2$ to the point $(0, 2)$.

24. Show that when we replace the radius 3 in Example 4.2.7 (page 435) with a general radius R, then the rectangle with the largest area will still be twice as wide as it is tall.

25. The power output P of a battery, as measured in watts, is given by the equation $P = IV - I^2 r$, where I is the current running through the battery, V is the voltage of the battery, and r is the internal resistance of the battery. If $V = 12$ volts and $r = 2$ ohms, find the value of I that will maximize the value of P.

26. A goalie kicks a soccer ball, initially at rest, with an initial velocity of $v_0 = 30$ m/s, at an angle θ with respect to the ground. Find the value of θ that will maximize the horizontal distance, or "range" R, that the ball travels before hitting the ground if

$$R = \frac{v_0^2 \sin 2\theta}{g}, \quad 0 \le \theta \le \pi/2,$$

where $g > 0$ is the gravitational constant (acceleration due to gravity).

27. For a given selling price of p dollars, a company expects to sell $q = 58,500 - 6p^2$ of an item per year, for $p \in [0, 80]$. What price will yield the largest annual revenue?

28. In 1919, the German physicist Albert Betz published a computation regarding the maximum theoretical efficiency of a wind turbine, i.e., that theoretical maximum fraction of the kinetic energy in the wind that a turbine can capture. In his derivation, he assumed that the speed of the air at the turbine could be treated as the average of the entry speed v_1 and the exit speed v_2. With that, he concluded that the power P collected by the turbine satisfies

$$P = \frac{1}{4}\rho S v_1^3 \left(1 - R^2 + R - R^3\right),$$

where $R = v_2/v_1$ is the ratio of the wind velocity exiting the turbine to the wind velocity entering the turbine, and air density ρ and sweep area S of the turbine are constants. We can assume $R \in [0, 1]$.[14]

Assuming that ρ, S and v_1 are held constant, compute dP/dR and show that it is maximized when $R = 1/3$, i.e., when the wind exits the turbine at one-third of the velocity at which it enters the turbine.[15]

29. A beam with rectangular cross-section is to be cut from a cylindrical log with diameter 1 foot. The bearing

---

[14] Note that if $R = 1$, then the wind exits at the same speed in which it enters, and so no energy (and therefore no power) was extracted and $P = 0$. On the other hand, if $R = 0$ we have $P = \frac{1}{4}\rho S v_1^3$, but this also implies that the air stops when exiting the turbine, which is not ideal (or even possible) because it will cause "back-up" and therefore cause the kinetic energy of the air at the turbine to be lowered.

[15] Though not obvious, this ultimately implies that the maximum power we can extract with a wind generator is $16/27 \approx 59\%$ of the kinetic power available in the wind passing through the turbine. The interested student can find many sources for this discussion.

4.2. EXTREMA ON CLOSED INTERVALS

strength of the beam will be proportional to its width and the square of its height. What are the dimensions (width × height) that will give it the maximum strength?

30. A Norman window is one that consists of a lower, rectangular pane of some width w, together with an upper, semicircular pane with diameter w. Find the dimensions of a Norman window with maximum possible area if the perimeter of the window is 20 ft.

31. One side of a rectangular enclosure is constructed of brick at a cost of $45 per linear foot, while the other three sides consist of fencing at a cost of $15 per linear foot. Find the dimensions and total cost of such an enclosure if the area enclosed is 500 ft^2 and the cost is minimized.

32. Find the dimensions of the circular cylinder of maximum volume that can be inscribed in a sphere of radius r. Show that $h/r = \sqrt{2}$ and that $V_{\text{sphere}}/V_{\text{cylinder}} = \sqrt{3}$.

33. What are the dimensions (diameter and height) of the right circular cylindrical can of volume 1 Liter (1000 cm^3) with minimal surface area. (Hint: Make surface area a function of radius r.)

34. Find the right circular cylinder of maximum volume that can be inscribed inside a cone of radius R and height H. Show that its dimensions must satisfy $r = \frac{2}{3}R$ and $h = \frac{1}{3}H$.

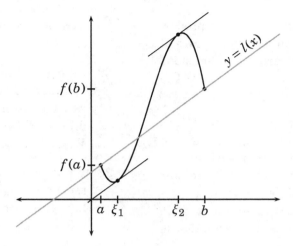

Figure 4.5: An illustration of the Mean Value Theorem (Theorem 4.3.1, page 449), with the function $f(x)$ being continuous on $[a,b]$ and differentiable on (a,b). The graph $y = l(x)$ is the line that interpolates $(a, f(a))$ and $(b, f(b))$, and necessarily has slope equal to the average rate of change of $f(x)$ over $a \leq x \leq b$. According to the Mean Value Theorem, there exists $\xi \in (a,b)$ so that $f'(\xi)$ is also equal to the slope of $l(x)$. For this particular function there are, in fact, two such values, labeled ξ_1 and ξ_2.

4.3 The Mean Value Theorem

The Mean Value Theorem (MVT) states the surprisingly useful fact that the average rate of change of a differentiable function over a closed, finite interval must occur as an instantaneous rate of change somewhere in that interval. It is akin to stating that if our average velocity on a (one-dimensional) trip were 60 miles per hour (mph), then our actual, instantaneous velocity had to be 60 mph at some time during the trip. It seems reasonable upon reflection, since if we were to (for instance) drop below the 60 mph average for a while, we would have to accelerate to a velocity greater than 60 mph to compensate, and in doing so the velocity's value would have to "pass through" the 60 mph value as an actual velocity at least once, assuming velocity is continuous (and therefore subject to the Intermediate Value Theorem, or IVT, Corollary 2.3.2, page 108). In fact, we will see that the theorem does not require a continuous derivative, but only that it exists.

In this section we recall what we mean by average rate of change of a function and determine the necessary hypotheses for which the intuition described occurs: that the average rate of change must actually occur as an instantaneous rate at some point. That is the essence of the Mean Value Theorem (MVT), which is illustrated in Figure 4.5, though its meaning might not be readily apparent before we delve into some preliminaries.

4.3.1 Linear Interpolation and Average Rate of Change

Suppose we are given a function $f : [a, b] \longrightarrow \mathbb{R}$, that is continuous on $[a, b]$. We can connect the two points $(a, f(a))$ and $(b, f(b))$ by a line, which is itself a function we call the *linear interpolation* of f between the two points. Graphically it is also called a *secant line* of the

4.3. THE MEAN VALUE THEOREM

curve $y = f(x)$, meaning literally a line connecting two points on a circle, but generalized to mean a line connecting any two points on any given graph.

The equation of the linear interpolation is readily calculated. It has slope $(f(b)-f(a))/(b-a)$, and so when we note that $(a, f(a))$ is one point on the graph, we can quickly write $y = l(x)$, where[16]

$$l(x) = f(a) + \left[\frac{f(b)-f(a)}{b-a}\right](x-a). \tag{4.8}$$

Compare, for instance, to (3.8), page 215. The linear interpolation is useful if we wish to approximate $f(x)$ for some x between or near a and b, but only have knowledge of f evaluated at a and b. In fact, we can also approximate a value for x given the desired output value for $f(x)$ by solving for x in $y = l(x)$. While the accuracy of the interpolation varies too much to be reliable without further information on the behavior of f on (a,b), often there are tables of pairs (x,y) for functions that behave approximately linearly between such pairs, and linear interpolation (using $y = l(x)$) is a fairly standard procedure.

At this point, we might again define the *average rate of change of $f(x)$ over the interval* $[a,b]$ as we did briefly in Chapter 3, specifically (3.4), page 212. It is the net change in f over $[a,b]$, divided by the length of the interval, to give the average change in $f(x)$ per unit length traversed in x, namely[17]

$$\text{"average change in } f(x) \text{ over } [a,b]\text{"} = \frac{f(b)-f(a)}{b-a}. \tag{4.9}$$

This is also the slope l' of the linear interpolation (4.8) of f on the same interval, i.e., the slope of the line through $(a,f(a))$ and $(b,f(b))$. (The reader should verify $l'(x) = (f(b)-f(a))/(b-a)$ by mentally differentiating (4.8), noting that a, b, $f(a)$ and $f(b)$ are constants.)

4.3.2 Mean Value Theorem

Here we give the statement of the Mean Value Theorem, and then look at an important Lemma known as Rolle's Theorem, which we then use in the proof of the Mean Value Theorem. Recall that f being differentiable means that $f'(x) = \lim_{\Delta x \to 0}[(f(x+\Delta x) - f(x))/(\Delta x)]$ exists and is finite.

Theorem 4.3.1 (Mean Value Theorem) *Suppose $f : [a,b] \longrightarrow \mathbb{R}$ is continuous on $[a,b]$ and differentiable on (a,b). Then there exists $\xi \in (a,b)$ such that*

$$f'(\xi) = \frac{f(b)-f(a)}{b-a}. \tag{4.10}$$

---

[16]Compare this, for instance, to (3.8), page 215. Indeed, anytime we have two points (x_1,y_1) and (x_2,y_2), we have the slope being $(y_2-y_1)/(x_2-x_1)$, and if we take this and one of the points, say (x_1,y_1), we get the equation of the line in the form

$$\frac{y-y_1}{x-x_1} = \frac{y_2-y_1}{x_2-x_1} \quad \Longrightarrow \quad y-y_1 = \left[\frac{y_2-y_1}{x_2-x_1}\right](x-x_1) \quad \Longrightarrow \quad y = y_1 + \left[\frac{y_2-y_1}{x_2-x_1}\right](x-x_1).$$

In (4.8) we have this, but with the two points being $(a,f(a))$ and $(b,f(b))$, instead of (x_1,y_1) and (x_2,y_2).

[17]Recall that the *average rate of change* of $f(x)$ over an interval $a \le x \le b$ is defined by the difference quotient $\frac{f(b)-f(a)}{b-a}$, as given in Definition 3.1.3, page 212 and elsewhere. The slope of the line segment joining $(a,f(a))$ and $(b,f(b))$ represents this average rate of change. It is a generalization of the more obvious definition of the average velocity $[s(t_f)-s(t_0)]/(t_f-t_0)$ over the interval $t \in [t_0,t_f]$, representing where the particle is at time t_f, minus where it was at time t_0, all divided by how much time it took to get there, i.e., divided by $t_f - t_0$.

This is illustrated in Figure 4.5 at the start of this section (page 448), in which there are two such values of ξ, namely ξ_1 and ξ_2.

Before we begin to prove this, note that what this says is that there is a point between a and b on which the slope of the curve matches that of the linear interpolation from $(a, f(a))$ to $(b, f(b))$, i.e., matches the average rate of change of $f(x)$ over $[a, b]$.

See again Figure 4.5 for an illustration of this theorem. Note the line $y = l(x)$, called the *interpolation line*, connecting $(a, f(a))$ and $(b, f(b))$. It is also called a *secant line* to the graph because it connects two points on the graph.

Now we turn to the proof, but in two stages. First we state and prove what is known as Rolle's Theorem, which is perhaps a bit more intuitive than the Mean Value Theorem.

Theorem 4.3.2 (Rolle's Theorem) *(Used to prove the Mean Value Theorem.) Suppose that $f : [a,b] \longrightarrow \mathbb{R}$ is continuous on $[a,b]$ and differentiable on (a,b). If $f(a) = f(b)$, then there exists $\xi \in (a,b)$ such that $f'(\xi) = 0$.*[18]

An illustration of Rolle's Theorem is left as an exercise. The reader could consider attempting one at this point in the discussion, or after a reading of the proof.

Proof: It is useful to consider two cases. First, that $f(x)$ is constant, with the common value $f(a) = f(b)$ on all of $[a,b]$, in which case $f'(\xi) = 0$ for all $\xi \in (a,b)$, and the conclusion of the theorem holds true.

For the other case, we suppose $f(x)$ is not a constant function on $[a,b]$, so it must achieve a maximum or minimum value for $x \in (a,b)$ (other than $f(a) = f(b)$), at some point $\xi \in (a,b)$. Thus ξ is a critical point of $f(x)$, and since $f'(\xi)$ exists, we must conclude $f'(\xi) = 0$, q.e.d.

We now finish the proof of the Mean Value Theorem, by applying Rolle's Theorem to the difference between the function $f(x)$ and its linear interpolation $l(x)$ between $(a, f(a))$ and $(b, f(b))$.

Proof: (MVT) Suppose $f(x)$ is continuous on $[a,b]$ and differentiable on (a,b). Define $l(x)$ to be the linear interpolation between $(a, f(a))$ and $(b, f(b))$ as in (4.8), page 449. Now consider the function

$$g(x) = f(x) - l(x).$$

We note that $g(x)$ is continuous on $[a,b]$, differentiable on (a,b). Furthermore, we note that $g(a), g(b) = 0$, so there exists some point in $\xi \in (a,b)$ such that $g'(\xi) = 0$, implying $f'(\xi) - l'(\xi) = 0$, or $f'(\xi) = l'(\xi)$. Now $l'(\xi) = (f(b) - f(a))/(b - a)$, so

$$f'(\xi) = l'(\xi) = \frac{f(b) - f(a)}{b - a}, \qquad \text{q.e.d.}$$

---

[18]Named for French Mathematician Michel Rolle (April 21, 1652–November 8, 1719), known best for this theorem for which he produced the first currently known published proof, though the result had apparently been observed earlier. He is also considered an early discoverer of what became known as Gaussian elimination (after the very important and influential German mathematician and physicist Johann Carl Friedrich Gauss, 1777–1855) for solving systems of linear equations, and an early critic of some of the accepted calculus due to perceived gaps between what was provable (at the time) and observed, though he later came to accept its truth.

4.3. THE MEAN VALUE THEOREM

4.3.3 Applications of MVT to Graphing Theorems

Recall that we "observed" in Chapter 3 that $f' > 0$ on an interval apparently implied f was increasing on that interval. Similarly, $f' < 0$ on the interval showed f was decreasing on that interval. With the Mean Value Theorem at our disposal, we are now in a position to prove these.

Theorem 4.3.3 *Suppose $a < b$, $f(x)$ is continuous on $[a,b]$ and $f'(x)$ exists on (a,b).*

(1) If $f'(x) > 0$ on (a,b), then $f(x)$ is increasing on $[a,b]$.

(2) If $f'(x) < 0$ on (a,b), then $f(x)$ is decreasing on $[a,b]$.

Proof: *First we prove (1). Let $a \le x_1 < x_2 \le b$. Then there exists $\xi \in (x_1, x_2)$ such that*

$$\frac{f(x_2) - f(x_1)}{x_2 - x_1} = f'(\xi) > 0 \implies f(x_2) - f(x_1) = \underbrace{f'(\xi)}_{>0} \cdot \underbrace{(x_2 - x_1)}_{>0} > 0$$

$$\implies f(x_2) > f(x_1).$$

This being true for all $x_1, x_2 \in [a,b]$ where $x_1 < x_2$ shows $f(x)$ is increasing on $[x_1, x_2]$. Case (2) is an easy modification of the argument used to prove case (1):

$$\frac{f(x_2) - f(x_1)}{x_2 - x_1} = f'(\xi) < 0 \implies f(x_2) - f(x_1) = \underbrace{f'(\xi)}_{<0} \cdot \underbrace{(x_2 - x_1)}_{>0} < 0$$

$$\implies f(x_2) < f(x_1).$$

Another useful and intuitive theorem is the following.

Theorem 4.3.4 *Suppose $f(x)$ is continuous on $[a,b]$ and differentiable on (a,b). Moreover, suppose $f'(x) = 0$ for all $x \in (a,b)$. Then $f(x)$ is constant on $[a,b]$.*

Proof: *The proof follows a similar strategy. Let $a \le x_1 < x_2 \le b$. Then there exists $\xi \in (x_1, x_2)$ such that*

$$\frac{f(x_2) - f(x_1)}{x_2 - x_1} = f'(\xi) = 0 \implies f(x_2) - f(x_1) = 0 \implies f(x_2) = f(x_1).$$

While the theorem itself is interesting, it seems obvious enough to be almost worthless. But in fact it has a very nice corollary, which is that if two functions have the same derivative on an interval, they must differ by a constant.

Corollary 4.3.1 *Suppose $f(x)$ and $g(x)$ are continuous on $[a,b]$ and differentiable on (a,b), and that $f'(x) = g'(x)$ on (a,b). Then there exists $C \in \mathbb{R}$ such that $f(x) = g(x) + C$.*

Proof: *Consider the function $h(x) = f(x) - g(x)$, which is also continuous on $[a,b]$, differentiable on (a,b), and $h'(x) = f'(x) - g'(x) = 0$ on (a,b), so there exists $C \in \mathbb{R}$ such that, on $[a,b]$, we have*

$$h(x) = C \iff f(x) - g(x) = C \iff f(x) = g(x) + C.$$

Example 4.3.1 *Note how, on an interval such as $\left(-\frac{\pi}{2}, \frac{\pi}{2}\right)$, we have*

$$\frac{d}{dx}\tan^2 x = 2\tan x \cdot \frac{d \tan x}{dx} = 2\tan x \sec^2 x \quad = 2\sec^2 x \tan x,$$

$$\frac{d}{dx}\sec^2 x = 2\sec x \cdot \frac{d \sec x}{dx} = 2\sec x \sec x \tan x = 2\sec^2 x \tan x.$$

Thus $\frac{d}{dx}\tan^2 x = \frac{d}{dx}\sec^2 x$ on $\left(-\frac{\pi}{2}, \frac{\pi}{2}\right)$, leading us to conclude $\tan^2 x = \sec^2 x + C$. Indeed, with $C = -1$ we have one of our basic trigonometric identities, namely $\tan^2 x = \sec^2 x - 1$. (Of course, this identity actually holds true anywhere $\cos x \ne 0$.)

Graphically, two functions with the same derivatives on an interval will be the same shape. One will just be a vertical translation of the other.

4.3.4 Numerical Applications

The existence of the value ξ with the prescribed derivative (equal to the average rate of change) can imply many things. Knowing bounds for the derivative over an interval implies we know bounds for $f'(\xi)$, and thus the average rate of change. We apply this next.

Example 4.3.2 *Suppose $f(x)$ is continuous on $[2,10]$, differentiable on $(2,10)$, and $f'(x) > 3$ on $(2,10)$. If $f(2) = 5$, find a lower bound for $f(10)$.*

Solution: There exists $\xi \in (2,10)$ such that $f'(\xi) = (f(10) - f(2))(10 - 2)$. Now

$$\frac{f(10) - f(2)}{10 - 2} = f'(\xi) > 3 \implies f(10) - f(2) = f'(\xi)(10-2) > 3(8) = 24$$

$$\implies f(10) > f(2) + 24 = 5 + 24 = 29.$$

We conclude that $f(10) > 29$.

Note that it was important that we multiplied the first equation by the positive quantity $10 - 2$.

The example is akin to asking, "If $s = 5$ ft when $t = 2$ sec, and $v > 3$ ft/sec at all times $t \in (2\text{sec}, 10\text{sec})$, then where must $s(10\text{sec})$ be?" If our starting position is at 5 ft, and our velocity is greater than 3 ft/sec for a time interval of length 8 sec, then we must have traveled more than 24 ft in the "positive" direction in that time, leaving us to the right of the position marked 29 ft.

In fact, this method works for more general problems.

Example 4.3.3 *Supposing $f(x)$ is continuous on $[3,15]$, and differentiable on $(3,15)$ in such a way that $-2 \le f'(x) < 4$ on that interval, and $f(3) = 6$, find an interval containing $f(15)$.*

Solution: According to the Mean Value Theorem, there exists $\xi \in (3, 15)$ such that $f'(\xi) = (f(15) - f(3))/(15 - 3)$. We can use this to find bounds for $f(15)$ as follows:

4.3. THE MEAN VALUE THEOREM

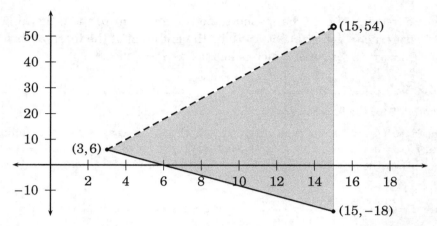

Figure 4.6: Illustration of the situation in Example 4.3.3. There, $f'(x) \in [-2, 4)$, $f(3) = 6$, and bounds for $f(15)$ are desired. Geometrically, the function is bounded by the two lines shown above, throughout all of $3 < x \leq 15$. The final bounds of $f(x)$ at $x = 15$ are labeled above. The upper line represents the "extreme" case $f'(x) = 4$ throughout $[3, 15]$, and the lower line the other "extreme" case where $f'(x) = -2$ throughout all of $[3, 15]$.

$$\frac{f(15) - f(3)}{15 - 3} = f'(\xi) \in [-2, 4)$$

$$\implies -2 \leq \frac{f(15) - f(3)}{15 - 3} < 4$$

$$\implies -2 \leq \frac{f(15) - 6}{12} < 4$$

$$\implies -24 \leq f(15) - 6 < 48$$

$$\implies -18 \leq f(15) < 54.$$

In fact, one can use a geometric interpretation of f' to construct lines of the slopes given by bounds on $f'(x)$ for $3 \leq x \leq 15$, and the values of $f(x)$ will fall between these lines. This is illustrated in Figure 4.6.

4.3.5 Verifying the MVT for Particular Cases

Most textbooks include exercises to verify the conclusion of the MVT, so we include a few such examples in that spirit here.

Example 4.3.4

Verify the hypotheses and conclusion of the Mean Value Theorem for the function $f(x) = x^2$ on the interval $[1, 9]$.

<u>Solution</u>: For $f(x) = x^2$, we see f is continuous on $[1, 9]$ and $f'(x) = 2x$ exists on $(1, 9)$. Thus the hypotheses of the MVT are met. Next we note that $f(1) = 1$ and $f(9) = 81$, and so

$$\frac{f(9) - f(1)}{9 - 1} = \frac{81 - 1}{9 - 1} = \frac{80}{8} = 10.$$

Next we solve the equation $f'(\xi) = 10 \iff 2\xi = 10 \iff \xi = 5 \in (1, 9)$. Thus $\xi = 5 \in (1, 9)$ satisfies the conclusion of the MVT.

Note that our value of $\xi = 5$ happened to be the midpoint of the interval $[1,9]$, i.e., $\xi = (1+9)/2$. It is not true in general that ξ will be the midpoint of the interval $[a,b]$, but this is the case for $f(x) = x^2$, the proof of which is left to Exercise 19.

Example 4.3.5

Verify the MVT for the function $f(x) = \tan x$ on the interval $[0, \pi/4]$.

Solution: Here $f(x) = \tan x$ is continuous on $[0, \pi/4]$ because $\cos x \neq 0$ on that interval. For the same reason, $f'(x) = \sec^2 x$ exists on $(0, \pi/4)$. Next we try to solve $f'(\xi) = (f(\pi/4) - f(0))/(\pi/4 - 0)$. In doing so, we will use the fact that $\cos \xi > 0$ for $\xi \in (0, \pi/4)$, and so $\sqrt{\cos^2 \xi} = \cos \xi$ in this context.

$$\sec^2 \xi = \frac{\tan(\pi/4) - \tan(0)}{\pi/4 - 0} = \frac{1 - 0}{\pi/4} \implies \sec^2 \xi = 4/\pi \implies \cos^2 \xi = \pi/4 \implies \cos \xi = \frac{1}{2}\sqrt{\pi}$$

$$\implies \xi = \cos^{-1}\left(\frac{1}{2}\sqrt{\pi}\right) \approx 0.4816604795.$$

Since $(0, \pi/4) \approx (0, 0.7853981634)$, we see that $\xi = \cos^{-1}\left(\frac{1}{2}\sqrt{\pi}\right) \in (0, \pi/4)$ satisfies the conclusion of the Mean Value Theorem.

Example 4.3.6

Consider the function $f(x) = 1/x$ on the interval $[-1, 1]$. We cannot use the MVT on this interval because f is not continuous on that whole interval. If we compute

$$\frac{f(1) - f(-1)}{1 - (-1)} = \frac{1 - (-1)}{2} = \frac{2}{2} = 1,$$

we will see that $f'(\xi) = -1/\xi^2 \neq 1$ regardless of ξ (since $f' < 0$ wherever defined). Since the hypotheses of the MVT do not hold (as in fact $f'(x)$ DNE at $x = 0 \in [-1, 1]$ as well), we should not be surprised that the conclusion does not hold either. A quick sketch of the function would further confirm this.

4.3.6 An Application of Linear Interpolation

There are occasions when scientists or engineers must refer to tables of values of either physical quantities or of "special functions," the values of which are very difficult to compute using ordinary methods. One uses such a table to look up a particular value of a variable, say y, which corresponds to a value of another variable, say x, on the table. However, there are many times when a y-value is needed for an x-value which is "in between" those listed on the table. For instance, if there are listed pairs (x_1, y_1) and (x_2, y_2) we can see on the table but we require a value of y for some value $x \in (x_1, x_2)$, we may need to make an educated guess for an approximate value of the y-value we need. Conversely, we may need to find an x-value corresponding to some y between y_1 and y_2. If the table is complete enough that the gaps in x- and y-values are relatively small, and we have some theoretical reason for believing there is no wild variation in the behavior of the values of the variables "in between" the values on the table, then a linear interpolation to approximate the in-between values seems reasonable.

4.3. THE MEAN VALUE THEOREM

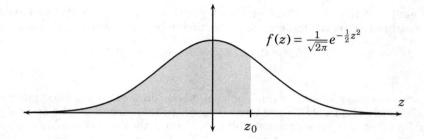

Figure 4.7: For a normal probability distribution, the probability that the variable z is less than or equal to z_0 is given by the shaded area under the function $f(z) = (2\pi)^{-1/2} e^{-z^2/2}$. Computing that area is a nontrivial task and so in practice researchers historically used tables of values rather than attempting to compute such quantities themselves. In more modern times, there are software packages and internet sources to compute such probabilities "in real time." However, more obscure quantities are still often found in published tables.

Example 4.3.7

What follows is an excerpt from a typical z-table of probabilities for a normally distributed random variable, as in Figure 4.7. In the left column are the values for z_0 before we add the hundredths place to the z_0 value. So for instance, the probability that z is less than or equal to 1.60 is 0.9452, i.e., $P(z \le 1.60) = 0.9452$, while $P(z \le 1.63) = 0.9484$, with these probabilities given to within ± 0.0001. A working chart of such probabilities will typically cover one or two textbook pages.[19]

z_0	.00	.01	.02	.03	.04	.05	.06	.07	.08	.09
⋮	⋮	⋮	⋮	⋮	⋮	⋮	⋮	⋮	⋮	⋮
1.60	.9452	.9463	.9474	.9484	.9495	.9505	.9515	.9525	.9535	.9545
⋮	⋮	⋮	⋮	⋮	⋮	⋮	⋮	⋮	⋮	⋮
2.3	.9893	.9896	.9898	.9901	.9904	.9906	.9909	.9911	.9913	.9916
⋮	⋮	⋮	⋮	⋮	⋮	⋮	⋮	⋮	⋮	⋮

(a) Approximate $P(z \le 1.638)$.

(b) Approximate z_0 such that $P(z \le z_0) = 0.95$.

(c) Approximate z_0 such that $P(z \le z_0) = 0.99$.

Solution:

(a) Here we see that table contains probabilities $P(z \le 1.63) = 0.9484$ and $P(z \le 1.64) = 0.9495$. We can consider these entries as ordered pairs $(1.63, 0.9484)$ and $(1.64, 0.9495)$. We wish to find the y-value on a line containing these two points such that the x-value is 1.638.

There are multiple ways of accomplishing this. Here we will use the formula for the linear interpolation (4.8), page 449, which for this case we derive and simplify:

$$l(x) = 0.9484 + \frac{0.9495 - 0.9484}{1.64 - 1.63}(x - 1.63) = 0.9484 + \frac{0.0011}{0.01}(x - 1.63)$$
$$\implies l(x) = 0.9484 + 0.11(x - 1.63),$$

---

[19] Note that a probability of 0.50 would mean a 50% probability of an event occurring. Probabilities are given as values in $[0, 1]$, with a probability of zero meaning an event is theoretically impossible by chance, and a probability of one meaning an event is theoretically certain to occur if left to chance.

456   CHAPTER 4. USING DERIVATIVES TO ANALYZE FUNCTIONS; APPLICATIONS

for which $l(1.638) = 0.9484 + 0.11(1.638 - 1.63) = 0.9484 + 0.11(0.008) = 0.94928 \approx 0.9493$.
This sandwiches the approximate point proportionately between the known points:[20]

$$(1.63, 0.9484), \quad (1.638, 0.94928), \quad (1.64, 0.9495).$$

(b) Here we are looking for an "x-value" for a point lying on the line connecting those with the nearest "y-values," namely $(1.64, 0.9495)$ and $(1.65, 0.9505)$. In this case, since the desired y-coordinate is the exact middle value of those for the two points, we expect the desired x-value to also be the middle value of the given points' x-values, namely $x = 1.645$. However, we can also work this out with the linear interpolation, or the slope argument given in Footnote 20 below. Note that our three points are $(1.64, 0.9495)$, $(x, 0.95)$, and $(1.65, 0.9505)$:

$$\frac{0.95 - 0.9495}{x - 1.64} = \frac{0.9505 - 0.9495}{1.65 - 1.64}$$
$$\implies \frac{x - 1.64}{0.0005} = \frac{0.01}{0.001}$$
$$\implies x - 1.64 = 0.005$$
$$\implies x = 1.645,$$

as expected.

(c) Here we are attempting to find an "x" value so that $(x, 0.99)$ is on the chart. The closest y-values on the chart are the points $(2.32, 0.9898)$ and $(2.33, 0.9901)$. We can either use a linear interpolation $y = l(x)$ and solve for x (which is useful if we need to find many points on the same interpolation line), but in this case it is again easier to use the slope equations as in (b).

However in this case the interpolated point is not the midpoint between the two known points, but is 2/3 of the distance between the left and right points. To be more computational, we will use the slope equalities as in (b), with the points being $(2.32, 0.9898)$, $(x, 0.99)$, and $(2.33, 0.9901)$:

$$\frac{0.99 - 0.9898}{x - 2.32} = \frac{0.9901 - 0.9898}{2.33 - 2.32}$$
$$\implies \frac{x - 2.32}{0.0002} = \frac{0.01}{0.0003}$$
$$\implies x - 2.32 = 0.00\overline{6}$$
$$\implies x = 2.32\overline{6},$$

For this case, we would likely report $z_0 = x \approx 2.327$.

In fact, one could switch the roles of variables so that the probabilities are "x-values" and the z_0-values are "y-values," whichever is more convenient. In (b) and (c) it would have been slightly easier to do so, so that the desired "x-values" were not in the denominators.

---

[20] If we only need to find the location of a single point on the interpolation line given two other points, it suffices to note that the slope between any pairs of the points must be equal. If we are looking for y so that $(1.63, 0.9484)$, $(1.638, y)$, and $(1.64, 0.9495)$ are on the same line, we can use, for instance,

$$\frac{0.9495 - 0.9484}{1.64 - 1.63} = \frac{y - 0.9484}{1.64 - 1.63}.$$

The advantage to finding the form of $l(x)$ is that it can be used with multiple values of x.

When the process outputs a number with more significant digits than was inputted, it may be preferable to report the output with the lower number of digits, but if it is to be used in a later computation, it is better to use all of the digits the process yields in the intermediate steps, and then round only the final answer.

Exercises

1. Draw an illustration for Rolle's Theorem similar to the figure given at the start of the section for the MVT (Figure 4.5).

2. Prove that Rolle's Theorem is equivalent to the Mean Value Theorem in this way:

    (a) In the proofs, we already show that Rolle's Theorem $\Longrightarrow$ MVT. This fact can therefore be assumed.

    (b) Show that Rolle's Theorem is a special case of MVT, and thus MVT $\Longrightarrow$ Rolle's Theorem.

For Exercises 3–11, verify that the MVT holds for the given f and given interval, by verifying the hypotheses and finding all relevant values of ξ for the conclusion.

3. $f(x) = x^2 - 5x + 6$ on the interval $[0, 4]$

4. $f(x) = x^2 - 5x + 6$ on the interval $[2, 3]$

5. $f(x) = \sin x$ on the interval $[0, \pi]$

6. $f(x) = \sin x$ on the interval $[0, \pi/2]$

7. $f(x) = \ln x$ on the interval $[1, e]$

8. $f(x) = x^3$ on the interval $[2, 4]$

9. $f(x) = \sqrt{x}$ on the interval $[4, 9]$

10. $f(x) = \frac{1}{x}$ on the interval $[-8, -7]$

11. $f(x) = e^x$ on the interval $[-2, 2]$

12. Consider $f(x) = |x|$. Note how $f'(x)$ is never equal to $[f(1) - f(-2)]/[1 - (-2)]$. Does this contradict the MVT? Why or why not?

13. Repeat the previous exercise for $f(x) = \tan x$ for $x \in [0, \pi]$.

14. Suppose $2 \le f'(x) < 5$ on $[2, 10]$, and that $f(2) = 4$. Find the range of possible values for $f(10)$.

15. Suppose $2 \le f'(x) < 5$ on $[2, 10]$ and $f(10) = 100$. Find the range of possible values for $f(2)$.

16. Suppose $2 \le f'(x) < 5$ on $[0, 10]$ and $f(2) = 4$. For any $x^* \in [0, 10]$, find the range of possible values for x^*. (Your answer should be some kind of interval where the left and right endpoints are both functions of x^*.)

17. Show that $x = 0$ is the only solution of the equation $\sin x = 2x$. Do so using the following steps:

    (a) Define $f(x) = 2x - \sin x$. Note $f(x) = 0 \iff \sin x = 2x$.

    (b) Show that $f(0) = 0$.

    (c) Compute $f'(x)$. For what values of x does $f'(x)$ exist?

    (d) Let $a = 0$. Suppose $b \ne 0$ is *another* solution of $f(x) = 0$.

    (e) Show that the Mean Value Theorem contradicts the existence of b. (Note $-1 \le \sin x \le 1$.)

18. A slightly more sophisticated argument than the one in the previous problem shows that $\sin x = x$ has exactly one solution. Construct such an argument thusly:

    (a) Let $f(x) = x - \sin x$. What is the significance of the statement $f(x) = 0$?

    (b) What is $f(0)$ and what is its significance?

    (c) What is the sign of f' on $(-\pi/4, \pi/4)$?

(d) What does this say about the possibility of another solution on $(-\pi/4, \pi/4)$?

(e) Is the function ever decreasing?

(f) What does this say about the possibility of another solution?

19. Show that if $f(x) = x^2$ and $a < b$, then ξ from the Mean Value Theorem conclusion will always be equal to $(a+b)/2$. (However, also see the next exercise.)

20. Show that for a general quadratic function $f(x) = Ax^2 + Bx + C$, ξ from the MVT for an interval $[a, b]$ will always be equal to the midpoint of that interval, i.e., $\xi = (a+b)/2$.

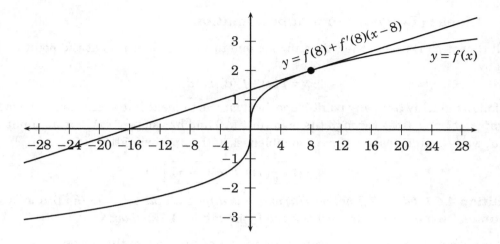

Figure 4.8: Partial graph of $f(x) = \sqrt[3]{x}$, along with the linear approximation (tangent line) at $x = 8$. The two graphs are very close to each other near $x = 8$ (and coincide at $x = 8$), but part company as we stray farther from $x = 8$. They may (and in fact do) eventually come together again for negative values of x, but that is just a quirk of this particular function, while the approximation is known to be useful near $x = 8$ by the nature of the tangent line.

4.4 Differentials and the Linear Approximation Method

The main idea of this section is how the tangent at $(a, f(a))$ can be used to approximate the function $f(x)$ near $x = a$.[21]

A related (indeed equivalent) notion is how differences in function values can be approximated by *differentials*, which we also define here.

One aspect of the theory is both a strength and liability: Different notations are used in application. We will consider three such presentations of essentially the same concept.[22]

---

[21]It should be pointed out that the terms *linear interpolation* and *linear approximation* have different but related meanings. In the former, we take two points on the graph of a function $f(x)$ and find the equation of the line $y = l(x)$ that connects these two points. While we may use that for approximating other values of the function (or values of x for desired values of the function), the term *linear approximation* is reserved for cases where a *tangent line at a single point* on the graph $y = f(x)$ is used to approximate other values of the function.

In other words, the *linear interpolation* uses a secant line for approximation purposes, while the *linear approximation* uses a tangent line for the same purposes. Therefore only the latter therefore requires some knowledge of calculus.

[22]One should consider why it would be useful to use approximations of functions. The simple answer is that when we do so, we usually approximate a complicated function with a simpler one. In doing so, we usually increase—sometimes greatly increase—computational speed, at a cost of accuracy. Sometimes the benefit is worth the cost, as for instance when we are attempting to intersect an intercontinental nuclear missile with another missile chasing it, the latter perhaps needing only to get consistently closer to the former until interception. In such a scenario, a fast algorithm repeatedly predicting the "next" position and trajectory of the nuclear missile might be more desirable than a slower but more accurate one predicting a much later position (and easily deceived by adding some randomness to the chased missile's flight path). Real-world applications being approximations anyway (at least to us mortals), another layer of approximation might not be much sacrifice in such a scenario.

There are also innumerable theoretical settings in which it is much simpler to approach a problem using approximating schemes, and then use some limit process—if available—to bring us to the exact answer ultimately.

4.4.1 Linear (Tangent Line) Approximation

Recall that if $f'(a)$ exists, then the tangent line to the graph of $f(x)$ at the point $(a, f(a))$ is given by

$$y = f(a) + f'(a)(x - a). \tag{4.11}$$

This follows quickly from any point-slope form for the tangent line there, using the facts that this tangent line passes through the point $(a, f(a))$ and has slope $f'(a)$. For an input value x near a, we can often observe the reasonableness of the approximation[23]

$$\boxed{f(x) \approx f(a) + f'(a)(x - a).} \tag{4.12}$$

Definition 4.4.1 *Given a function $f(x)$ which is differentiable at $x = a$, the **linear approximation of $f(x)$ centered at $x = a$** is given by Equation (4.12) above.*

One such approximation is given in Figure 4.8 at the start of this section.

Example 4.4.1 _____

Use a linear approximation of $f(x) = \sqrt[3]{x}$ to approximate $\sqrt[3]{8.5}$.

<u>Solution</u>: Here $f(x) = x^{1/3}$ and $f'(x) = \frac{1}{3}x^{-2/3}$. Using $a = 8$ (because $f(8)$ and $f'(8)$ are easily computed and 8 is near 8.5), we get

$$f(8) = 8^{1/3} = 2,$$
$$f'(8) = \frac{1}{3}(8)^{-2/3} = \frac{1}{3} \cdot \frac{1}{4} = \frac{1}{12}.$$

Thus, for x near 8, we have

$$f(x) \approx f(8) + f'(8)(x - 8), \text{ i.e.,}$$
$$f(x) \approx 2 + \frac{1}{12}(x - 8).$$

Using this we get

$$\sqrt[3]{8.5} = f(8.5) \approx 2 + \frac{1}{12}(8.5 - 8) = 2 + \frac{1}{12} \cdot \frac{1}{2} = 2 + \frac{1}{24} = 49/24 = 2.041666666\ldots.$$

Thus, by our linear approximation (of $f(x) = \sqrt[3]{x}$ centered at $x = 8$), we computed $\sqrt[3]{8.5} \approx 2.0417$.

_____

The actual value of $\sqrt[3]{8.5}$ is approximately 2.04082755, so our linear approximation is accurate to better than 0.001, which in this context means four significant digits when $x = 8.5$.

Any linear approximation is a statement regarding outputs $f(x)$ for x close to the point $x = a$ (at which the linear approximation is just the tangent line to the graph). Such an approximation is likely to worsen in accuracy as we leave the immediate vicinity of $x = a$, although the speed and degree to which this happens depend upon the function—in particular, how closely the graph follows the trend of the tangent line at $x = a$, as we move our input x away from the center a.

Following is a list of approximate values of $\sqrt[3]{x}$ as computed by this method centered at $x = 8$, and then as computed directly by computer (with the first 8 significant digits shown).

_____

[23]Note that (4.11) is *exact* at $x = a$.

4.4. DIFFERENTIALS AND THE LINEAR APPROXIMATION METHOD

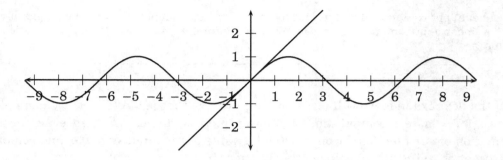

Figure 4.9: Partial graph of $f(x) = \sin x$, and the linear approximation at $x = 0$, which is $y = x$. Though not clear from the printed resolution here, the functions only coincide at $x = 0$. The proof of that fact is essentially outlined in Exercise 18 of Section 4.3.

Linear Approximation ($\approx$) versus Actual Value ($=$)

		$f(7) \approx 1.9166667$	$f(7) = 1.9129311\cdots$
$f(8) \approx 2$	$f(8) = 2$	$f(6) \approx 1.8333333$	$f(6) = 1.8171205\cdots$
$f(8.5) \approx 2.0416667$	$f(8.5) = 2.0408275\cdots$	$f(5) \approx 1.75$	$f(5) = 1.7099759\cdots$
$f(9) \approx 2.0833333$	$f(9) = 2.0800838\cdots$	$f(4) \approx 1.6666667$	$f(4) = 1.5874010\cdots$
$f(10) \approx 2.1666667$	$f(10) = 2.1544346\cdots$	$f(3) \approx 1.5833333$	$f(3) = 1.4422495\cdots$
$f(11) \approx 2.25$	$f(11) = 2.2239800\cdots$	$f(2) \approx 1.5$	$f(2) = 1.2599210\cdots$
$f(12) \approx 2.3333333$	$f(12) = 2.2894284\cdots$	$f(1) \approx 1.4166667$	$f(1) = 1$
$f(13) \approx 2.4166667$	$f(13) = 2.3513346\cdots$	$f(0) \approx 1.3333333$	$f(0) = 0$
$\vdots$	$\vdots$	$\vdots$	$\vdots$
$f(20) \approx 3$	$f(20) = 2.7144176\cdots$	$f(-8) \approx 0.6666667$	$f(-8) = -2$
$f(30) \approx 3.8333333$	$f(30) = 3.1072325\cdots$	$f(-16) \approx 0$	$f(-16) = -2.5198451\cdots$
$f(64) \approx 6.6666667$	$f(64) = 4$	$f(-64) \approx -4$	$f(-64) = -4$
$f(1000) \approx 84.6666667$	$f(1000) = 10.$	$f(-1000) \approx -82$	$f(-1000) = -10.$

We see that the approximation based on the behavior at $x = 8$ (i.e., the linear approximation at $x = 8$) stays reasonably close to the actual values of $f(x)$ until we stray far from $x = 8$ (except that in this case, the graphs come back together briefly at $x = -64$ and again part ways). The actual graph of $f(x) = \sqrt[3]{x}$, together with the tangent line emanating from $(8, f(8))$, are graphed in Figure 4.8, page 459.[24]

One of the most useful linear approximations in physics is used to approximate $\sin x$ for small $|x|$, i.e., when $|x - 0|$ is small, i.e., when x (measured in radians) is near zero:

Example 4.4.2

Find the linear approximation for $f(x) = \sin x$ at $x = 0$. (See Figure 4.9 above.)

Solution: We will use the formula $f(x) \approx f(a) + f'(a)(x - a)$ with $a = 0$ and $f(x) = \sin x$. Thus

$$\left.\begin{array}{l} f(0) = \sin 0 = 0, \\ f'(0) = \cos 0 = 1. \end{array}\right\} \quad \Longrightarrow \quad f(x) \approx 0 + (1)(x - 0) \quad \Longrightarrow \quad f(x) \approx x.$$

---
[24]Of course "close" and "far" are subjective notions of proximity, and acceptable tolerances differ from context to context.

462 CHAPTER 4. USING DERIVATIVES TO ANALYZE FUNCTIONS; APPLICATIONS

So the linear approximation of $f(x) = \sin x$, centered at $x = 0$, is simply $f(x) \approx x$. The graphs of $y = \sin x$, and this linear approximation $y = x$, are given in Figure 4.9, page 461. The approximation is very good for $|x| < 1$.[25]

In the next example we will use a similar strategy that again lets us use a simple function to approximate a computationally more complicated one. We then evaluate this approximation of our function for one particular value of the input variable, and compare the approximation to the actual value of the function there.

Example 4.4.3

Suppose a laser at ground level points to the base of a building 300 feet away. If the laser beam is then turned so that it still points to the building, but with an angle of elevation of 5°, then approximately how high on the building is the point illuminated by the beam?

Solution: The height h on the building is a function of θ, the angle of elevation of the beam, that is, the angle formed by the beam and the horizontal. Since $h(\theta)/(300 \text{ feet}) = \tan\theta$, we can write

$$h(\theta) = 300 \text{ feet} \cdot \tan\theta. \qquad (4.13)$$

If we use radian measure for θ, then

$$h'(\theta) = 300 \text{ feet} \cdot \sec^2\theta.$$

For $\theta = 0$, we have $h'(0) = 300$ feet. It is instructive to note that the units of $dh(\theta)/d\theta$ are formally feet/radian (though we often omit the ultimately dimensionless unit of radian). With this in mind, we can note that $5° = 5° \cdot \frac{\pi(\text{rad})}{180°} = \frac{\pi}{36}$ (radian), and so

$$h\left(\frac{\pi}{36}\right) \approx h(0) + h'(0)\frac{\pi}{36} \text{ (radian)} = 0 + 300 \frac{\text{feet}}{\text{radian}} \cdot \frac{\pi}{36} \text{ (radian)} = \frac{300\pi}{36} \text{ feet}$$
$$\implies h\left(\frac{\pi}{36}\right) \approx 26.18 \text{ feet}.$$

The actual height of the laser beam is $h(5°) = 300 \text{ feet} \cdot \tan 5° \approx 26.25$ feet.

If we look at this example closely, we see that replacing 5° with any angle θ in *radian* measure, then we can claim the approximation

$$h(\theta) \approx 300 \text{ feet} \cdot \theta. \qquad (4.14)$$

From the example we see that this approximation of the actual height is very good for $\theta = 5° = \pi/36$ (radian). In fact it compares well even for larger θ, but certainly not for all θ. Clearly from (4.13), $h(\theta) = 300 \text{ ft} \cdot \tan\theta \longrightarrow \infty$ as $\theta \to \frac{\pi}{2}^-$, which does not happen with our approximation (4.14). What happened is that the rate of change $dh/d\theta = 300\sec^2\theta$ did not stay at all constant, and in fact blew up also as $\theta \to \frac{\pi}{2}^-$.

---

[25] Note that $|x| < 1$ in radians (as we use for the calculus) corresponds approximately to $|x| < 57°$, but to use $\sin x \approx x$ to approximate $\sin 48°$, for instance, we need to convert back to radians: $\sin 48° = \sin\frac{48° \cdot \pi}{180°} = \sin\frac{48\pi}{180} \approx \frac{48\pi}{180} \approx 0.837758041$. The actual value of $\sin 48°$, rounded to seven places, is 0.7431448, so our approximation's error is still under 13%.

4.4. DIFFERENTIALS AND THE LINEAR APPROXIMATION METHOD

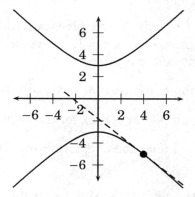

Figure 4.10: Graph for Example 4.4.4, showing a partial graph of the curve $y^2 - x^2 = 9$ and the (dashed) tangent line at $(4, -5)$, which is also the linear approximation to the local (implicit) function defined near there.

4.4.2 Linear Approximations and Implicit Functions

Because we can find $\frac{dy}{dx}$ on implicit curves written as equations, we can find tangent lines and therefore linear approximations to such curves. In such cases, perhaps y is not globally a function of x, but instead may be only a local function of x. Still, the overall method is valid: using a tangent line to approximate a curve locally. We offer two examples here.

Example 4.4.4

Consider the curve $y^2 - x^2 = 9$. Approximate y as a function of x near $(4, -5)$ using a linear approximation.

Solution: We use the usual implicit differentiation technique as in Section 3.6:

$$y^2 - x^2 = 9 \implies \frac{d}{dx}\left[y^2 - x^2\right] = \frac{d}{dx}[9]$$
$$\implies 2y \cdot \frac{dy}{dx} - 2x = 0$$
$$\implies 2y \cdot \frac{dy}{dx} = 2x \implies \frac{dy}{dx} = \frac{2x}{2y} = \frac{x}{y}.$$

For the point $(4, -5)$—which a quick check shows *is* on the curve—we have slope

$$\left.\frac{dy}{dx}\right|_{(4,-5)} = \left.\frac{x}{y}\right|_{(4,-5)} = \frac{4}{-5} = -\frac{4}{5}.$$

The tangent line is given by

$$y = -5 - \frac{4}{5}(x - 4).$$

Thus, on the curve near $(4, -5)$, we have $y \approx -5 - \frac{4}{5}(x - 4)$.

This graph and the linear approximation at $(4, -5)$ is given in Figure 4.10.

In the previous example, we could have solved for y without much difficulty, though we would need to be careful: $y^2 - x^2 = 9 \iff y = \pm\sqrt{9 + x^2}$, and in this case we need that part of the curve where $y = -\sqrt{x^2+9}$, from which a simple Chain Rule computation gives us $dy/dx = -x/\sqrt{x^2+9}$, whose value is again $-4/5$ at $x = 4$. However, there was some computational advantage in using the implicit differentiation:

(1) The differentiation was simpler (for instance, no radicals).

(2) So too was the evaluation simpler at $(4,-5)$, for the same reasons.

(3) We avoided the algebraic complication of determining if we needed the positive or negative case of the radical.

On balance, we usually have the choice of paying for complications earlier or later in a problem, depending on our approach. For instance, by solving for y algebraically (and carefully) we then have dy/dx as a function of x only, which has some appeal. It is a matter of personal choice for such cases. However, in the next example we would be unlikely to be able to solve for y algebraically, and must work with the equation which only implicitly (and locally) defines y as a function of x.

Example 4.4.5

Recall that in Example 3.6.7, page 324, we had the implicit curve given by $5x + x^2 + y^2 + xy = \tan y$, and we found the following formula for dy/dx, which we then evaluate at $(0,0)$:

$$\frac{dy}{dx} = \frac{5+2x+y}{\sec^2 y - 2y - x} \implies \frac{5+2x+y}{\sec^2 y - 2y - x}\bigg|_{(0,0)} = \frac{5}{\sec^2 0} = \frac{5}{1} = 5,$$

and so the tangent line to the curve at $(0,0)$ has slope 5. Thus near $(0,0)$, we can say that the local function is given approximately by $y \approx 0 + 5(x-0)$, or $y \approx 5x$, which is a huge advantage over trying to find an actual formula for y near $x = 0$.

4.4.3 Differentials

Next we define differentials. Eventually we will give numerical and geometric meaning to all of the terms in the definition, but first we define them only formally:

Definition 4.4.2 *Given a function $f(x)$, the **differential** of $f(x)$, namely $df(x)$, is given by*

$$df(x) = f'(x)dx, \qquad (4.15)$$

dx *being the **differential of** x, and where the prime "$'$" represents that the derivative is taken with respect to the underlying variable, which here is x.*

This is consistent with our previous use of Leibniz notation:

- $dx = d(x) = (x)'dx = 1 \cdot dx = dx$, as we would hope for consistency's sake, and

- $\frac{df(x)}{dx} = f'(x) \iff df(x) = f'(x)dx$, if we allow that we can, at least formally, multiply both sides by dx.

Now we look at some quick computations that follow from this definition:

4.4. DIFFERENTIALS AND THE LINEAR APPROXIMATION METHOD

- $d\sin x = \cos x\, dx,$
- $dx^2 = 2x\, dx,$
- $d\sqrt{x} = \dfrac{1}{2\sqrt{x}}\, dx = \dfrac{dx}{2\sqrt{x}},$

- $d\csc\theta = -\csc\theta\cot\theta\, d\theta,$
- $d\left(\dfrac{1}{x}\right) = -\dfrac{1}{x^2}\, dx = \dfrac{-dx}{x^2},$
- $d\ln u = \dfrac{1}{u}\, du = \dfrac{du}{u},$

- $d\left[\dfrac{x}{x+1}\right] = \dfrac{d}{dx}\left[\dfrac{x}{x+1}\right]dx = \left[\dfrac{(x+1)\frac{d}{dx}(x) - (x)\frac{d}{dx}(x+1)}{(x+1)^2}\right]dx = \dfrac{(x+1)(1) - (x)(1)}{(x+1)^2}\, dx$
$= \dfrac{1}{(x+1)^2}\, dx = \dfrac{dx}{(x+1)^2},$

- $d(x\tan x) = \dfrac{d}{dx}[x\tan x]\, dx = \left[x\cdot\dfrac{d\tan x}{dx} + \tan x\cdot\dfrac{d(x)}{dx}\right]dx = (x\sec^2 x + \tan x)\, dx.$

All of these become familiar derivative exercises if we divide these equations by dx, du, $d\theta$, and so on. We can also fashion differential versions of all of our rules by taking the derivative rules and multiplying both sides by the appropriate differential, say, dx. In particular, there are Product, Quotient, and Chain Rules:

$$d(uv) = u\, dv + v\, du, \tag{4.16}$$

$$d\left[\dfrac{u}{v}\right] = \dfrac{v\, du - u\, dv}{v^2}, \tag{4.17}$$

$$df(u(x)) = f'(u(x))u'(x)\, dx. \tag{4.18}$$

With these, we can compute many more differentials directly:

- $d(x\tan x) = x\, d\tan x + \tan x\cdot dx = x\sec^2 x\, dx + \tan x\, dx,$ which is exactly the result of our previous efforts to compute this differential: $d(x\tan x) = (x\sec^2 x + \tan x)\, dx.$

- $d\sin x^2 = \cos x^2 \cdot 2x\, dx = 2x\cos x^2\, dx.$

Note that (4.18) above is completely consistent with the idea that $df(u) = f'(u)\, du$:

- By definition: $df(u) = f'(u)\, du.$

- If it happens that u is a function of x, i.e., $u = u(x)$, then

$$df(u) = df(u(x)) = \dfrac{df(u(x))}{dx}\, dx = \dfrac{df(u(x))}{du(x)}\cdot\dfrac{du(x)}{dx}\, dx = \underbrace{f'(u(x))}_{f'(u)}\underbrace{u'(x)\, dx}_{du} = f'(u)\, du.$$

- We now have two methods for computing quantities such as $\dfrac{d\sin x^2}{dx^2}$:

  1. Considering x^2 as a variable in its own right:

$$\dfrac{d\sin x^2}{dx^2} = \cos x^2.$$

466 CHAPTER 4. USING DERIVATIVES TO ANALYZE FUNCTIONS; APPLICATIONS

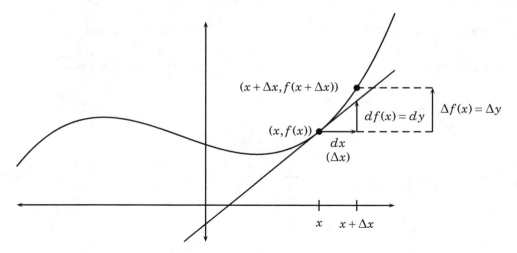

Figure 4.11: Illustration of the geometric meaning of $\frac{df(x)}{dx}$, giving further geometric meaning to both dx and $df(x)$. For any value x in the domain, if $f'(x)$ is defined, we have a tangent line of slope $f'(x) = \frac{df(x)}{dx}$. Algebraically, we could then write $df(x) = f'(x)dx$, where dx can represent a "run" and $df(x)$ the resultant "rise" along the tangent line. Note that $df(x)$ is an approximation of the actual "rise" $\Delta f(x)$ (not necessarily positive) of the function, as the input changes from x to $x + dx = x + \Delta x$. Moreover, that tangent line's rise $df(x)$ is the same as what is given by the linear approximation, as in (4.20).

2. Using the definition of differentials (Definition 4.4.2, page 464):

$$\frac{d\sin x^2}{dx^2} = \frac{(\sin x^2)'\,dx}{(x^2)'\,dx} = \frac{\cos x^2 \cdot 2x\,dx}{2x\,dx} = \cos x^2.$$

This again shows the robustness of the Leibniz notation (at least for first derivatives), with the apparent validity of algebraic manipulations of differentials as well as derivatives. Computations of differentials become ubiquitous as we develop integration techniques in later sections.

So far, we have only looked at these differentials formally. Now we will emphasize that these differentials can in fact be interpreted *numerically*.

Recall that $\frac{df(x)}{dx}$ measures how $f(x)$ changes as x changes. More precisely, the fraction $\frac{df(x)}{dx}$ gives us the instantaneous rate of change in $f(x)$ as x changes, at a particular value of x. (This is akin to $\frac{ds(t)}{dt}$ measuring velocity—that is, how position $s(t)$ is changing as t is changing—at a particular value of t.) Note also that $\frac{df(x)}{dx}$ can be interpreted as the slope of the graph of $f(x)$ at the particular value x. Hence, $\frac{df(x)}{dx}$ represents an instantaneous "rise/run."

Now we will let dx represent an actual, though variable, "run," i.e., a change in the input from a fixed value of x, while $df(x)$ will be the resultant "rise," *at the rate* $\frac{df(x)}{dx}$, i.e., $df(x)$ will represent the resulting "rise" along the tangent line which passes through $(x, f(x))$. This is illustrated in Figure 4.11.

The same justification for using linear approximations for functions allows us to use $df(x)$ to approximate an actual change in the function $f(x)$, as we perturb the x-value by a

4.4. DIFFERENTIALS AND THE LINEAR APPROXIMATION METHOD

small quantity dx. In fact, the actual change, the linear approximation, and the differentials are all related. If we call the perturbation in the input x by both names Δx and dx, we get

$$f(x+\Delta x) = f(x) + \frac{f(x+\Delta x) - f(x)}{\Delta x} \cdot \Delta x \approx f(x) + f'(x)\Delta x = f(x) + \frac{df(x)}{dx} \cdot dx.$$

$$\implies f(x+\Delta x) \approx f(x) + df(x). \tag{4.19}$$

This reflects exactly what occurs with the linear approximation:

$$f(x) \approx f(a) + \underbrace{f'(a)(x-a)}_{\frac{df(x)}{dx}\big|_{x=a} \cdot dx}, \tag{4.20}$$

where the part of $x + \Delta x$ in (4.19) is played by x in (4.20), the part of x is played by a, and Δx is played by $(x - a)$. Furthermore, $f'(a)(x - a)$ represents $df(x)$ evaluated at $x = a$ with $dx = (x - a)$, so the right-hand side of (4.20) is $f(a)$ plus the approximate perturbation of $f(x)$ from $f(a)$ as x strays from a.

4.4.4 Applications of Differentials

The overall theme of the previous subsection can be summarized by the statement "the differential $df(x)$ approximates the change $\Delta f(x)$ for a small change $dx = \Delta x$ in x," i.e.,

$$\Delta f(x) \approx df(x) = f'(x)dx. \tag{4.21}$$

This is sometimes lost among the varying notations. For instance, we can rewrite (4.21) even further, recalling the definition of $\Delta f(x)$, and that of the derivative:[26]

$$f(x+\Delta x) - f(x) \approx f'(x)\Delta x,$$

which of course follows from the definition of the derivative (when we divide by Δx), and the nature of limits:

$$\Delta x \to 0 \implies \frac{f(x+\Delta x) - f(x)}{\Delta x} \longrightarrow f'(x).$$

Since $f'(x) = df(x)/dx$, that essentially proves (4.21).[27]

In this section we will concentrate on the approximate change in the function's output, and how one illustrates the reasonableness of that approximation of the change in several examples.

Example 4.4.6

Consider a square of side length x. Its area is then given by $A(x) = x^2$. Now suppose we have a square of side length 4 cm, and wish to increase that side length by 0.2 cm. Use differentials to approximate the increase in area, and explain geometrically why it is a reasonable approximation.

Solution: Here we will illustrate both the calculus and the geometric intuition of this approach.

---

[26] Our linear approximation could be rewritten to be centered at x, as some textbooks write

$$f(x+\Delta x) = f(x+dx) \approx f(x) + f'(x)dx = f(x) + f'(x)\Delta x,$$

but in this subsection we focus on approximating the *change* in the function, rather than its new value.

[27] While the notational back and forth between Δ-notations and differential notations can be dizzying, it is important that the reader become comfortable with both. Comfort and fluency come with exposure and practice.

Here we will use $x = 4\,\text{cm}$ and $dx = 0.2\,\text{cm}$. The change in area ΔA as x increases from 4 to 4.2 is thus approximated

$$\Delta A \approx dA = 2x\,dx$$
$$= 2(4\,\text{cm})(0.2\,\text{cm}) = 1.6\,\text{cm}^2.$$

Thus the area added by lengthening the sides of the square from $4.0\,\text{cm}$ to $4.2\,\text{cm}$ is approximately $dA = 1.6\,\text{cm}^2$.

From the illustration on the right, we can see that the (light gray and white) area added to the original (gray) area is well approximated by the two narrow (light gray) rectangles of area $x\,dx + x\,dx = 2x\,dx$, the one small (white) square in the upper-right corner being the only part missing in our approximation. In fact, the new area is $x^2 + x\,dx + x\,dx + (dx)^2$, but since dx is small, $(dx)^2$ is much smaller, and might safely be disregarded when it is understood we are simply approximating $\Delta A = (x+dx)^2 - x^2 = 4.2^2 - 4^2 = 17.64 - 16 = 1.64$.

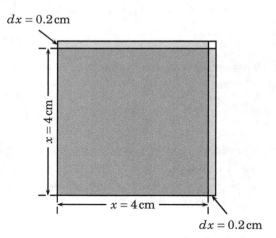

With the simplicity of this example, and with the ubiquity of compact computational devices, perhaps the practical value of using differential approximations for change seems minimal. Indeed, some of the appeal seems more for on-sight, mental approximations.

To this criticism we offer two responses, albeit not very well motivated at this point in the development. The first response is that, in fact, the approach is equivalent to that of linear approximations, which we have already argued can be very useful in fast algorithms in dynamic situations, such as destroying fast-moving projectile weapons like ballistic missiles. Second, the approach sheds much light on the later integral calculus, the theory of which seems much more transparent if one is already familiar with the *idea* of differential approximations to a cumulative, changing quantity such as area.

For now, note in particular that if dx is small, then the *percentage error* in using dA to approximate ΔA is quite small.[28]

Indeed, that missing square (in the upper right corner) in Example 4.4.6, which is the difference between ΔA and dA, is clearly a very small percent of the quantity ΔA, and that percentage error approaches zero as $\Delta x = dx \to 0^+$.

It should be pointed out that dx can also be negative, and the previous paragraph's conclusion still holds true, though the shading would be somewhat different, as $dx \to 0^-$.

Example 4.4.7

Consider a circle of radius $3.0\,\text{ft}$, which we then increase to $3.1\,\text{ft}$. Use differentials to approximate the increase in area, and explain geometrically why it is reasonable.

Solution: We proceed in a manner similar to that of the earlier example.

---

[28] Except in those cases where $\Delta A = 0$, for which any nonzero error in using dA to approximate ΔA might be called an "infinite" percent error, though to be precise, it should be called "undefined."

4.4. DIFFERENTIALS AND THE LINEAR APPROXIMATION METHOD

Here we will use $r = 3.0$ ft and $dr = 0.1$ ft. For a circle we have $A = \pi r^2$. The change in area ΔA as r increases from 3.0 ft to 3.1 ft is thus approximated by

$$\Delta A \approx dA = 2\pi r\, dr$$
$$= 2\pi(3.0\,\text{ft})(0.1\,\text{ft})$$
$$= 0.60\pi\,\text{ft}^2$$
$$\approx 1.88496\,\text{ft}^2$$

Thus the area added by increasing the radius from 3.0 ft to 3.1 ft is approximately 1.88 ft^2.

From the illustration on the right, we can see that the (light gray) area ΔA, added to the original (gray) area by adding dr to the radius, is well approximated by multiplying the inner circumference by the "thickness" dr.[29]

So our differential approximation dA is itself approximately 1.88 ft^2, while the exact value of ΔA is $\pi(3.1\,\text{ft})^2 - \pi(3.0,\text{ft})^2 = \pi(9.61 - 9)\text{ft}^2 \approx 1.916\,\text{ft}^2$.

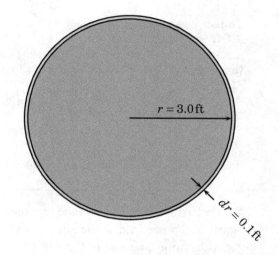

Example 4.4.8

A classical physics problem involves how far an object, launched from ground level at a speed v and angle θ from the horizontal, and assumed to be only negligibly affected by wind resistance, will travel horizontally before it impacts the ground. In a derivation which is not difficult to follow (but does involve a trigonometric identity), one finds that the distance D is given by the formula

$$D = \frac{v^2 \sin 2\theta}{g}.$$

Here $g \approx 9.8\,\text{m/sec}^2$ or $g \approx 32\,\text{ft/sec}^2$, the acceleration due to gravity. From the formula above, it is not difficult to see D is maximized when $2\theta = 90°$, i.e., $\theta = 45°$. (Note also what the formula says for $\theta \in \{0, \pi/2\}$, and why this makes sense physically.)

(a) Suppose $v = 400$ ft/sec and $\theta = 30°$. Use differentials to approximate the change in the distance D if the angle θ is decreased by $5°$.

(b) Suppose again that $v = 400$ ft/sec but that $\theta = 45°$ to give maximum distance. Use differentials to approximate how much farther the object would travel if we increase v by 10 ft/sec.

Solution: We take these in turn.

(a) Here v is taken to be a constant 400 ft/sec. Thus D is a function of θ alone, for which we can then compute the differential (noting we need the Chain Rule fact that $d\sin 2\theta = \cos 2\theta \cdot \frac{d2\theta}{d\theta}\,d\theta$):

$$D = \frac{(400\,\text{ft/sec})^2}{32\,\text{ft/sec}^2}\sin 2\theta \quad\Longrightarrow\quad dD = \frac{(400\,\text{ft/sec})^2}{32\,\text{ft/sec}^2}\cos 2\theta \cdot 2\,d\theta.$$

Recalling that to use the derivative formulas for the trigonometric functions, the angles must be in units of radians, and so $\theta = 30° = \pi/6$ and here $d\theta = -5° = -5° \cdot (\pi/180°)$. Rather than computing

---

[29] Imagine if we could "unroll" the added area. It would be very nearly rectangular, with the long sides of lengths approximately equal to $2\pi r$, the circumference of the original (inner) circle, and height dr.

an approximation for $d\theta$, we will use this exact expression as we compute[30]

$$dD = \frac{(400\,\text{ft/sec})^2}{32\,\text{ft/sec}^2} \cos 60° \cdot 2\left[-5° \cdot (\pi/180°)\right] \approx -436\,\text{ft}.$$

Thus a decrease in angle from 30° to 25° will decrease the distance traveled by approximately 436 ft.

(b) Here the angle $\theta = 45°$ is held constant, making D a function of v alone, from which we then compute the differentials, and evaluate with $v = 400$ ft/sec, and $dv = 10$ ft/sec:

$$D = \frac{1}{32\,\text{ft/sec}^2} v^2 \sin 90° = \frac{1}{32\,\text{ft/sec}^2} v^2$$
$$\implies dD = \frac{1}{32\,\text{ft/sec}^2} \cdot 2v\,dv = \frac{1}{32\,\text{ft/sec}^2} \cdot 2(400\,\text{ft/sec}) \cdot (10\,\text{ft/sec}) = 250\,\text{ft}.$$

So according to our differential approximation, the distance will increase approximately 250 ft.

The actual decline in distance traveled in (a) is closer to 387 ft ($\Delta D = 386.90869\ldots$ ft), and the actual increase in distance in (b) is closer to 253 ft ($\Delta D = 253.125$ ft).

Another approach would have been to use the Product Rule:

$$dD = d\left[\frac{v^2 \sin 2\theta}{g}\right] = \frac{1}{g}\left[v^2 \cos 2\theta \cdot 2\,d\theta + \sin 2\theta \cdot 2v\,dv\right],$$

and then in part (a) use $v = 400$, $dv = 0$, $\theta = 30°$, $d\theta = -5° = -5° \times \frac{\pi}{180°} = -\pi/36$ radians, while in part (b) we can use $\theta = 45°$, $d\theta = 0$, $v = 400$, and $dv = 10$. That approach would also allow us to use dD to approximate ΔD in case both θ and v vary only slightly.

4.4.5 Some Further Approximation Problems

The following examples should be intuitive, and eventually we will see that the underlying idea in each is the same. We will discuss this principle after we introduce the examples.

Example 4.4.9

Suppose a car travels along a highway, and a passenger notices the speedometer reads 72 miles per hour. About how far, in feet, will the car travel in the next second?

Solution: We have very little information here, but a reasonable assumption is that the car will not change speed very much in that one second (barring catastrophe). To the extent that that is true, we can say the distance traveled is given approximately by the following:

$$\text{Distance} \approx \frac{72\,\text{mile}}{\text{hour}} \cdot 1\,\text{second} = \frac{72\,\text{mile}}{1\,\text{hour}} \cdot \frac{5280\,\text{feet}}{1\,\text{mile}} \cdot \frac{1\,\text{hour}}{3600\,\text{seconds}} \cdot 1\,\text{second} = 105.6\,\text{feet}.$$

---

[30]It is not incorrect, and is often more intuitive, to use the degree measure of an angle when computing one of the trigonometric functions of the angle; the trigonometric functions can input an angle regardless of the units, as when we compute $\cos 60° = \cos(\pi/3) = 1/2$. When the angle appears outside of the trigonometric function or we wish to compute derivatives of trigonometric functions, we must be careful that the units are appropriate for that context. The calculus formulas live naturally in a context of radian (dimensionless) measure, but the stand-alone trigonometric functions are processors of angles and therefore can, in principle, so process them regardless of the units used to express them.

4.4. DIFFERENTIALS AND THE LINEAR APPROXIMATION METHOD

The idea in this example is akin to the grade school formula (distance) = (rate) · (time), but we made an assumption here that the velocity would be approximately constant over that one second. If we were to couch it in differential symbols, we could set $s(t)$ to be the position along some axis at time t, and take for convenience $t = 0$ to be the moment at which the passenger notes the velocity is $s'(0) = 72\,\text{mi/hr}$, and so we can approximate Δs, the net displacement in that one second, i.e., $\Delta s = s(1\sec) - s(0)$. Using a differential to approximate this, we could write the following (where $t = 0$ and $dt = 1\sec$ and conversion factors are introduced at the end):

$$ds = \frac{ds}{dt} \cdot dt = \left(\frac{72\,\text{mi}}{\text{hr}}\right) \cdot (1\sec) \cdot \frac{5280\,\text{ft}}{1\,\text{mi}} \cdot \frac{1\,\text{hr}}{3600\sec} \approx 105.6\,\text{ft}.$$

Example 4.4.10

Suppose a manufacturer's research shows that the profit from making x items of a particular type should be
$$P(x) = -0.004x^3 + 10x^2 - 1000.$$
Suppose further that the manufacturer is initially planning on a production run of 100 items. How much more profit would he make if he produced 101 items instead?

Solution 1 (Exact): If the model is correct, the actual extra profit from making that 101st item would be the difference in profit from making 101 items and the profit from making 100 items:

$$P(101) - P(100) = [-0.004(101)^3 + 10(101)^2 - 1000] - [-0.004(100)^3 + 10(100)^2 - 1000]$$
$$= 96888.796 - 95000$$
$$= 1888.796.$$

Of course, this should be rounded to hundredths of a dollar (that is, to the nearest cent), so according to this model the manufacturer would make an extra $1,888.80 from that 101st item.

Of course, the model itself is likely an approximation based on research. The apparent precision of the expected extra profit is open to further scrutiny (as with any economic model). In any event, another approach, this time definitely an approximation, is given next:

Solution 2 (Approximate): Note that

$$P'(x) = \frac{d}{dx}[-0.004x^3 + 10x^2 - 1000] = -0.012x^2 + 20x.$$

Furthermore, $P'(100) = -0.012(100^2) + 20(100) = 1880$ (dollars/item). In other words, assuming x can take on all values on the relevant continuum, the profit is changing at a theoretical instantaneous rate of $1,880/item when the number of items is 100. We can use this to approximate that the next (101st) item will cause a growth in the profit of approximately $1880. If we wish to use differential notation, where $x = 100$ and $dx = 1$ (with units of items), we could write $dP = P'(x)dx = \$1,880/\text{item} \cdot (1\,\text{item}) = \$1,880$.

Clearly the approximate solution was simpler computationally, and we did not give up much accuracy in using it instead. Recall that the first solution actually reflected the exact definition of marginal profit ($P(101) - P(100)$), while the second reflected the definition of what we called proxy marginal profit ($P'(100) \cdot 1 = P'(100)$), the latter being what many textbooks define to be marginal profit. Since the model for $P(x)$ is usually approximate itself, either method is usually considered legitimate.

Example 4.4.11

The period of a pendulum, with very small angular displacement, and near Earth's surface is given by the formula

$$T = 2\pi\sqrt{\frac{L}{g}},$$

where T is measured in seconds, L is the length of the pendulum in meters, and $g = 9.8\,\text{m/sec}^2$ is the acceleration due to gravity at Earth's surface.

Suppose the length of a pendulum initially is 1.4 m. It is then lengthened by 1%. Compute the differential dT.

Solution: Here we assume T is a function of L alone, i.e., $T = T(L)$, and then $dT = T'(L)dL$:

$$dT = \frac{d}{dL}\left[2\pi\sqrt{\frac{L}{g}}\right]dL = 2\pi \cdot \frac{1}{2}\left(\frac{L}{g}\right)^{-1/2} \cdot \frac{1}{g}\,dL = \frac{\pi}{g}\sqrt{\frac{g}{L}}\,dL = \pi\sqrt{\frac{g}{g^2 L}}\,dL$$

$$\implies \quad dT = \pi\sqrt{\frac{1}{Lg}}\,dL.$$

Using $dL = \Delta L = 0.01L$, and eventually $L = 1.4\,\text{m}$ and $g = 9.8\,\text{m/sec}^2$, we have

$$\Delta T \approx dT = \pi\sqrt{\frac{1}{Lg}} \cdot (0.01L) = 0.01\pi\sqrt{\frac{L}{g}} = 0.01\pi\sqrt{\frac{1.4}{9.8}} \approx 0.0118741041.$$

Normally we would not use so many significant digits, but when we compute the actual change ΔT of the period as L increases from 1.4 m to $(1.01)(1.4\,\text{m}) = 1.414\,\text{m}$ we have

$$T(1.4) = 2\pi\sqrt{1.4/9.8} \qquad \approx 2.374820823,$$
$$T(1.414) = 2\pi\sqrt{1.414/9.8} \qquad \approx 2.38666539,$$
$$T(1.414) - T(1.4) = 2\pi\left[\sqrt{1.414/9.8} - \sqrt{1.4/9.8}\right] \approx 0.0118445664,$$

and so, to a useful number of significant digits, $dT \approx 0.01187$ and $\Delta T \approx 0.01184$. Both could be rounded to 0.012 and have as many significant digits as most of the data given in the problem.

Alternatively, we could have computed that $dT \approx 0.00378\pi$, and $\Delta T \approx 0.00377\pi$.

Some texts put such things in the context of *error*, whereby the "error in $f(x)$" is given by $\Delta f(x) = f(x + \Delta x) - f(x)$, and the "error in x" is given by Δx. Then the *approximate error* in $f(x)$ is given by $df(x) = f'(x)dx$, again with $dx = \Delta x$. (This is somewhat useful when computing approximate tolerances, for instance in approximating a tolerance $\Delta f(x) \approx df(x)$ in $f(x)$ given a tolerance dx in x.)

Such texts then often look at *relative error* being $(\Delta f(x))/f(x)$, measuring what fraction of the value of $f(x)$ is the error $\Delta f(x)$. This is then approximated by $\frac{df(x)}{f(x)}$.

Finally, such texts often define the *percent error* in $f(x)$, where the relative error is multiplied by 100%, which is another expression for 1. Hence,[31]

$$\text{Percent Error} = \frac{\Delta f(x)}{\Delta x} \cdot 100\% \approx \frac{df(x)}{f(x)} \cdot 100\% = \frac{f'(x)dx}{f(x)} \cdot 100\%.$$

---

[31] One interesting thing about the definition of the approximate relative and percent errors is that they can also be written as $d\ln|f(x)| = d\ln|f(x)| \cdot 100\%$.

Exercises

For Exercises 1–10, find a formula for the linear approximation for the given function $f(x)$ centered at the given point $x = a$.

1. $f(x) = \sqrt{x}$, $a = 100$.
2. $f(x) = \ln x$, $a = 1$.
3. $f(x) = \ln|x|$, $a = -1$.
4. $f(x) = \ln x$, $a = e$.
5. $f(x) = \sin^{-1} x$, $a = 1/2$.
6. $f(x) = \tan^{-1} x$, $a = 1$.
7. $f(x) = x\cos x$, $a = 0$.
8. $f(x) = e^x$, $a = 0$.
9. $f(x) = \tan x$ centered at
   (a) $a = 0$,
   (b) $a = \pi/4$,
   (c) $a = -\pi/4$.
10. $f(x) = \cos x$ centered at
    (a) $a = 0$,
    (b) $a = \pi/2$,
    (c) $a = -\pi/3$.

11. Find the linear approximation centered at $(-3, 4)$ for the implicit function given by $x^2 + y^2 = 25$. Also, draw the curve and the linear approximation.

12. Find the linear approximation for the implicit function $\sin x = \cos y$ at $(\pi/4, \pi/4)$.

For Exercises 13–22, compute the differential using the appropriate variables.

13. dx^4
14. $d\sqrt{x}$
15. $d\sin\theta$
16. $d\csc\phi$
17. $d\sec\xi\tan\xi$
18. $d\left(\frac{1}{u}\right)$
19. $d\sin 2x$
20. $d\left(x^2 + 1\right)^8$
21. $d\left(\frac{x}{1+\cos x}\right)$
22. $d\left(\frac{e^x}{\ln x}\right)$

23. Recall that $df(x) = f'(x)dx$, that is, $df(x) = \frac{df(x)}{dx} \cdot dx$. Use this type of logic to prove (4.16), page 465, namely $d(uv) = u\,dv + v\,du$.

24. Do the same for (4.17), page 465, showing that $d\left(\frac{u}{v}\right) = \frac{v\,du - u\,dv}{v^2}$.

25. Do the same for (4.18), page 465, that $df(u(x)) = f'(u(x))u'(x)dx$.

26. The formula for the volume of a cube of side length x is $V = x^3$. Suppose a cube has side length 2 ft, and we wish to increase the side lengths to 2.1 ft.

    (a) Identify Δx and find the actual change in volume ΔV that results.
    (b) Use differentials to approximate ΔV.
    (c) Explain what dV could represent geometrically, in reference to the original cube, and why that is a reasonable approximation of ΔV. See Example 4.4.6, page 467.

27. The formula for the interior volume of a sphere is $V = \frac{4}{3}\pi r^3$. The formula for the surface area of a sphere is $S = 4\pi r^2$. Suppose a sphere of ice 2 m in diameter is suspended in a liquid and begins to melt, while retaining its shape. At one point in time, it has lost 10 cm of its diameter.

(a) Compute exactly how much of its volume was lost.

(b) Instead, use differentials to approximate how much volume was lost.

(c) Explain why, geometrically, it is reasonable that the formula for the surface area of the sphere appears. (See Footnote 29, page 469.)

28. A measuring device measures a spherical object's radius to be 0.8 cm. The measurement is correct to the nearest 0.01 cm.

    (a) Give an estimate of the error in the volume measurement of the object by using a differential.

    (b) If the radius is actually 0.005 cm less than the measured radius, what is the actual error in the volume of the object?

29. Using differentials, estimate the change in the volume of a cylinder where the radius is one-fourth of the height and the radius changes from 10 cm to 10.2 cm (and the radius/height ratio stays the same). Also find the actual change in volume and compare it to the differential approximation.

30. A light bulb has a resistance of $R = 20\,\Omega$. The wattage W is given by the formula $W = I^2 R$, where I is the current in amperes flowing through the bulb. Use differentials to estimate the change in wattage if the current changes from 1.5 amperes to 1.6 amperes. Also find the actual change in wattage and compare these values.

31. The speed of a wave v is equal to the product of its frequency f (measured in Hertz, Hz, sometimes written 1/sec) and its wavelength λ (measured in meters). Use differentials to estimate the change in wavelength $\Delta\lambda$ if $\lambda = 580 \times 10^{-9}$ m and the frequency changes from 2.40×10^{14} Hz to 2.45×10^{14} Hz, but the speed v remains constant. Also find the actual change $\Delta\lambda$ and compare.

32. A calculus professor knows that the next exam will be very difficult, and decides to grade it on a "curve," whereby if s is the "raw score," meaning the percentage of questions a student answers correctly, then the recorded "curved" score will be $R(s) = 10\sqrt{s}$. Calculators are not allowed for the exam.

    (a) What would be the recorded scores for raw scores of $s = 0$, $s = 100$ and $s = 25$?

    (b) While sitting for the examination, a student is fairly confident of being able to answer 49% of the questions correctly, thus achieving a recorded score of $R(49) = 10\sqrt{49} = 70$. The student would like a higher grade, and would like to estimate how much higher the recorded would be score if the student could achieve a raw score that was 5 percentage points higher. Since no calculators are allowed, the student decides to use differentials or (equivalently) a linear approximation to approximate the new score. What is the student's approximation for how many extra points higher the recorded score would be if the raw score could be raised five percentage points, i.e., from 49% to 54%?

    (c) How many actual points (rounded to the nearest tenth) would that student's adjusted score rise by if the raw score increased from 49% to 54%?

4.5. NEWTON'S APPROXIMATION METHOD

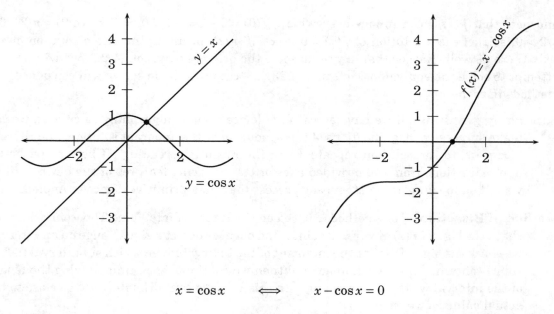

$$x = \cos x \quad \Longleftrightarrow \quad x - \cos x = 0$$

Figure 4.12: Figure showing the existence of a solution of $x = \cos x$, which occurs where the graphs of $y = x$ and $y = \cos x$ intersect, or equivalently a solution of $f(x) = 0$ where $f(x) = x - \cos x$.

4.5 Newton's Approximation Method

While there are many types of equations we can solve algebraically—and students typically spend much time in their young years learning to solve them—there are numerous equations we cannot solve algebraically. We can often demonstrate that there exist solutions, and perhaps even discover the number of solutions, without actually knowing the exact solutions. Within industry, approximate solutions often suffice, though the level of precision needed varies from problem to problem. When the problem is very complex, or an approximate solution is needed quickly, efficiency becomes an important consideration. Efficiency, accuracy, and prospects for success are all factors when choosing a method, and what works well for one problem might not work well—or even at all—for another.

In this section we will look at an approximating scheme attributed to Newton, though the modern interpretation has been ascribed to Thomas Simpson (of Simpson's Rule, Section 5.7). We will also examine other methods for comparison.

4.5.1 Approximation Methods for Solving Equations

Consider the equation
$$x = \cos x. \tag{4.22}$$

We can tell from the first figure of Figure 4.12 that there is a solution of this equation somewhere in $x \in [0, 2]$. In fact, by the Intermediate Value Theorem (page 108), we can look at

$$f(x) = x - \cos x \tag{4.23}$$

and note that $f(x)$ is continuous everywhere, $f(0) = 0 - 1 = -1 < 0$, while $f(\pi/2) = \pi/2 - 0 = \pi/2 > 0$, so there is a solution of $f(x) = 0$ in $x \in [0, \pi/2]$, meaning there is a solution of $x = \cos x$ there as well. Rather than trying to solve the original equation (4.22), we will instead attempt to solve the equivalent equation (4.23), $x - \cos x = 0$. To do so, we will consider three methods in turn.

Solving by graphing: If we have access to a device such as a graphing calculator, we can make successive magnifications of the window near the apparent solution, and thereby "zoom in" to the solution repeatedly, in the meantime reading off better and better approximations from the provided axes or curve tracing features of the device. While visually appealing, it is not an easy process to train a primitive device to implement.

Method of Bisection: This method is based on the Intermediate Value Theorem, where we check the sign of $f(x)$ at various points, and once we detect a sign change on an interval, we check the sign of $f(x)$ at the midpoint of the interval to see which subinterval (left or right) contained the sign change, and then we continue, each time halving the length of the interval with the sign change. In this way, we can ultimately get as close to the actual value as we wish.

Example 4.5.1

For instance, if we wish to attempt this method with our current problem, we would compute

$$\left. \begin{array}{rcl} f(0) & = & -1 < 0 \\ f(2) & = & 2 - \cos 2 \approx 1.090702573 > 0 \end{array} \right\} \text{sign change in } [0, 2].$$

Next we consider the value of $f(x)$ at the midpoint of $[0,2]$, and include the previous values for clarity:

$$\left. \begin{array}{rcl} f(0) & = & -1 < 0 \\ f(1) & = & 1 - \cos 1 \approx 0.4596796941 > 0 \\ f(2) & = & 2 - \cos 2 \approx 1.090702573 > 0 \end{array} \right\} \text{sign change in } [0, 1].$$

Next we compute $f(x)$ at the midpoint of $[0,1]$, again including the two endpoints for clarity:

$$\left. \begin{array}{rcl} f(0) & = & -1 < 0 \\ f(0.5) & = & 0.5 - \cos 0.5 \approx -0.3775825619 < 0 \\ f(1) & = & 1 - \cos 1 \approx 0.4596796941 > 0 \end{array} \right\} \text{sign change in } [0.5, 1].$$

Next we compute $f(x)$ and the midpoint of $[0.5, 1]$ and compare it to the endpoints:

$$\left. \begin{array}{rcl} f(0.5) & = & 0.5 - \cos 0.5 \approx -0.3775825619 < 0 \\ f(0.75) & = & 0.75 - \cos 0.75 \approx 0.0183111311 > 0, \\ f(1) & = & 1 - \cos 1 \approx 0.4596796941 > 0 \end{array} \right\} \text{sign change in } [0.5, 0.75].$$

Next we compute $f(x)$ and the midpoint of $[0.5, 0.75]$, namely $\frac{1}{2}(0.5 + 0.75) = 0.625$:

$$\left. \begin{array}{rcl} f(0.5) & = & 0.5 - \cos 0.5 \approx -0.3775825619 < 0 \\ f(0.625) & = & 0.625 - \cos 0.625 \approx -0.1859631195 < 0 \\ f(0.75) & = & 0.75 - \cos 0.75 \approx 0.0183111311 > 0 \end{array} \right\} \text{sign change in } [0.625, 0.75].$$

This tells us the solution is somewhere in $[0.625, 0.75]$, the midpoint of which is $\frac{1}{2}(0.625+0.75) = 0.6875$, and so on.

If we take $x_0 = 0$, $x_1 = 2$, and then consider the midpoints of the relevant intervals generated by the sign changes $x_2 = 1$, $x_3 = 0.5$, $x_4 = 0.75$, $x_5 = 0.625$, and so on, we see a sequence of x-values,

4.5. NEWTON'S APPROXIMATION METHOD

which get progressively closer to an (the) actual solution. A sign chart for the $f(x_n)$ values helps to illustrate this phenomenon:

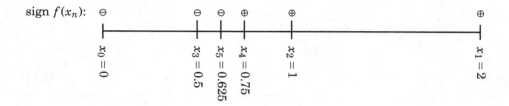

While this method can be easily programmed into a computer or similar device, it is not usually the most efficient computationally, meaning it takes more steps to get the same accuracy as the next method, which takes some advantage of the actual slope of the function, i.e., the function's derivative.

Newton's Method: With Newton's method, we begin with an initial "guess," which we call x_1. We then look at the tangent line to $f(x)$ at $(x_1, f(x_1))$, the slope of this line being $f'(x_1)$. We then take x_2 to be the x-intercept of this tangent line (which would be a solution if the curve's slope were constant!), in essence "following" the tangent line to the x-axis to find x_2. We then take the tangent line at $(x_2, f(x_2))$ and follow it to the x-axis to find x_3, and follow the tangent line at $(x_3, f(x_3))$ to the x-axis to find x_4, and so on, to produce a sequence $x_1, x_2, x_3, \ldots$, which will often converge very quickly towards a point x which solves $f(x) = 0$. This is illustrated for our example, namely solving $x = \cos x$, i.,e., $f(x) = x - \cos x = 0$, using $x_1 = 0$, and the x_n following for $n = 2, 3, 4, \ldots$. This is illustrated in Figure 4.13, page 478. Note that this requires that $f'(x_n) \neq 0$ for $n = 1, 2, 3, \ldots$.

Algebraically, for Newton's Method we let x_n be our "nth approximation," and then its tangent line to $y = f(x)$ at $x = x_n$ is given by

$$y = f(x_n) + f'(x_n)(x - x_n),$$

which when we set equal to zero—to find the x-intercept—gives us (assuming $f'(x_n) \neq 0$)

$$f(x_n) + f'(x_n)(x - x_n) = 0 \iff f'(x_n)(x - x_n) = -f(x_n)$$
$$\iff x - x_n = -\frac{f(x_n)}{f'(x_n)}$$
$$\iff x = x_n - \frac{f(x_n)}{f'(x_n)}.$$

Taking x_{n+1} to be this x-value at which the tangent line through $(x_n, f(x_n))$ intersects the x-axis, we get the following recursive formula:

$$x_{n+1} = x_n - \frac{f(x_n)}{f'(x_n)}. \tag{4.24}$$

478 CHAPTER 4. USING DERIVATIVES TO ANALYZE FUNCTIONS; APPLICATIONS

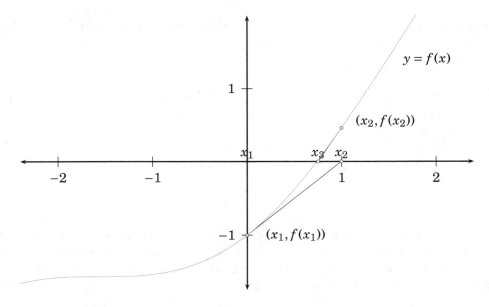

Figure 4.13: Figure showing two iterations of Newton's Method for finding approximate solutions of $f(x) = x - \cos x = 0$, i.e., solutions of $x = \cos x$, with an initial guess of $x_1 = 0$.

Example 4.5.2

For our present example in which $f(x) = x - \cos x$ and $f'(x) = 1 + \sin x$, our recursion relationship (4.24) becomes

$$x_{n+1} = x_n - \frac{x_n - \cos x_n}{1 + \sin x_n}.$$

So far, we have considered $x_1 = 0$ to be our "first guess." In what follows, we list values of x_n where we begin first with $x_1 = 0$, and then also consider $x_1 = 3$ and $x_1 = 4$, this last one being apparently unsuccessful. Note that non-integer approximations are given to ten significant digits.

n	x_n	n	x_n	n	x_n
1	0	1	3	1	4
2	1	2	-0.4965581783	2	-15.13524412
3	0.7503638678	3	2.131003844	3	16.07064125
4	0.7391128909	4	0.6896627208	4	-10.28559751
5	0.7390851334	5	0.7396529975	5	-4.80680582
6	0.7390851332	6	0.7390852044	6	-2.350795113
7	0.7390851332	7	0.7390851332	7	3.348266027
8	0.7390851332	8	0.7390851332	8	-2.095887668
9	0.7390851332	9	0.7390851332	9	9.740319348

For this particular example, it is clear that a good "first guess" x_1 can cause the algorithm to converge quickly to a stable value. However, using $x_1 = 4$ here leads to unstable oscillation in the sequence $\{x_n\}$ and no apparent convergence. The behavior is not always predictable at first glance. (The reader is invited to note that $x_1 = 2$ actually converges faster to the stable value of x_n than does $x_1 = 0$, though the solution is closer to 0 than 2.) To be sure we have an approximate solution, one can check $\cos 0.7390851332 \approx 0.7390851332$, as desired.

4.5. NEWTON'S APPROXIMATION METHOD

Note that when the method does converge, it often does so quite quickly: We managed to get recurring approximations agreeing to 10 significant digits after 5–6 iterations. To use our method of bisection and have accuracy of $\pm 10^{-10}$ we would need to bisect the interval $[0,2]$ some n times, giving a subinterval of length $2/2^n$ where $2/2^n \leq 10^{-10}$, i.e., $2^{1-n} \leq 10^{-10}$, which means $(1-n)\log 2 \leq -10$, or $(n-1)\log 2 \geq 10$, or $n \geq 1 + 10/\log 2 \approx 34.2$, so we would need 35 iterations of the Method of Bisection to guarantee the same accuracy for this problem. That is certainly not a problem for a computer program to accomplish quickly, but the 5–6 steps required here show Newton's Method to be more efficient for this particular problem.[32]

4.5.2 More on Newton's Method

In the typical application of Newton's Method, one attempts to solve an equation that has been put into the form $f(x) = 0$, makes a "guess" for x_1, runs the algorithm and inspects for convergence. If there will be convergence, it is usually readily apparent. If not, that is usually clear as well and another attempt is made to choose an x_1 that will cause convergence. There are some functions that will not allow for convergence, and a method such as the Method of Bisection can be attempted. (We will consider such situations later.) The electronic ability to graph the function in question can be very helpful as well, even if only to determine where to begin whichever method is chosen.

---

[32]Three common ways to run the Newton's Method algorithm are through the use of a calculator, a computer programming language or a spreadsheet. With most graphing calculators today, we can use the $\boxed{\text{ANS}}$ and $\boxed{\text{ENTER}}$ keys to run the recursive steps easily. For instance, if we type our x_1 value $\boxed{0}$ and $\boxed{\text{ENTER}}$, the display shows 0 as our "answer." We then can type

$$\text{ANS-(ANS-COS(ANS))/(1+SIN(ANS))} \boxed{\text{ENTER}}$$

and the calculator will return the output from entering our first "answer" ($x_1 = 0$) after it is run through the right-hand side of (4.24), yielding x_2. At that point, x_2 is our new "answer," so if we type $\boxed{\text{ENTER}}$ again it will repeat the line with ANS= x_2, outputting our new "answer," namely x_3 and so on. Repeatedly pressing $\boxed{\text{ENTER}}$ then outputs x_4, x_5, and so on.

On a spreadsheet, one can enter 0 into cell $A1$, and then in $A2$ enter

$$= A1 - (A1 - \cos(A1))/(1 + \sin(A1)),$$

causing the contents of $A2$ to be given by this formula. One can then "copy" cell $A2$ and "paste" its formulaic contents simultaneously into $A3$, $A4$, and so on, and the spreadsheet will likely understand this action to mean each cell should reference the one above it using a similar formula, i.e., so that

$$A2 = A1 - (A1 - \cos(A1))/(1 + \sin(A1)),$$
$$A3 = A2 - (A2 - \cos(A2))/(1 + \sin(A2)),$$
$$A4 = A3 - (A3 - \cos(A3))/(1 + \sin(A3)),$$

and so on. This will in fact make it easier to try new first guesses for $A1$ (think "x_1"), since editing $A1$ to be a different number will automatically "update" $A2$, $A3$, and so on. Such methods are discussed further in Subsection 5.7.2 where calculator-only methods become unwieldy.

Using either the programmable features of a graphing calculator, or an actual computer programming language, are alternative strategies, as is simply using calculator memory (or re-typing) and processing the numbers by brute force.

Of course, regardless of the chosen method, we have to be sure that the calculating device is reading angles in radians.

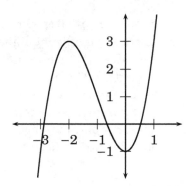

Figure 4.14: Graph of $f(x) = x^3 + 3x^2 - 1$ for Example 4.5.3. Note the intervals on which there is a sign change, and the approximate placements of the solutions of $f(x) = 0$, i.e., where $x^3 = 1 - 3x^2$.

There can also be several solutions of $f(x) = 0$, and perhaps some graphical analysis can help us to see how many solutions exist and approximately where they are located.

Example 4.5.3

Find all real solutions of $x^3 = 1 - 3x^2$.

Solution: We begin as before by making this a question about "zeros" of a relevant function $f(x) = x^3 + 3x^2 - 1$. Assuming we have no graph to work with, we can nonetheless seek out intervals on which the function changes signs through some experimentation:

x	-3	-2	-1	0	1	2	3
$f(x)$	-1	3	1	-1	3	19	53

We see sign changes in $[-3,-2]$, $[-1,0]$, and $[0,1]$. We know from algebra that there can be at most three solutions of $f(x) = 0$ since $f(x)$ is a third-degree polynomial, so we need only look for approximations of the solutions in these three intervals. We will choose x_1 to be the midpoint in each case. We will use these in our recursive formula, which for this problem becomes

$$x_{n+1} = x_n - \frac{f(x_n)}{f'(x_n)} \quad \Longrightarrow \quad x_{n+1} = x_n - \frac{x_n^3 + 3x_n^2 - 1}{3x_n^2 + 6x_n}.$$

Using $x_1 = -2.5, -0.5, 0.5$ respectively, we generate the following tables:

n	x_n
1	-2.5
2	-3.066666667
3	-2.900875604
4	-2.879719904
5	-2.879385325
6	-2.879385242
7	-2.879385242

n	x_n
1	-0.5
2	-0.6666666667
3	-0.6527777778
4	-0.6527036468
5	-0.6527036447
6	-0.6527036447
7	-0.6527036447

n	x_n
1	0.5
2	0.5333333333
3	0.5320906433
4	0.5320888862
5	0.5320888862
6	0.5320888862
7	0.5320888862

We conclude that approximate solutions of $x^3 = 1 - 3x^2$ are $x \approx -2.879385242, -0.6527036447$, and 0.5320888862. The graph of $f(x) = x^3 + 3x^2 - 1$ is given in Figure 4.14.

4.5. NEWTON'S APPROXIMATION METHOD

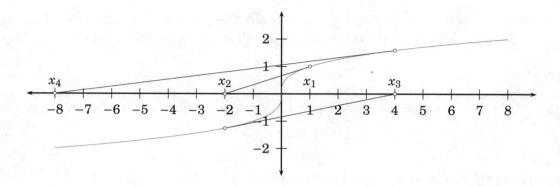

Figure 4.15: Figure showing the tangent lines producing the succession of the $x_1, x_2, x_3, \ldots$ generated by the Newton's Method recursion formula for $f(x) = \sqrt[3]{x} = 0$ if $x_1 = 1$, as in Example 4.5.4. As the tangent lines become more horizontal, the x_n values diverge from each other.

From both the geometric interpretation (of following the tangent lines back to the x-axis) and the recursion formula, we can see that a flatter curve, i.e., with smaller $f'(x_n)$—assuming the same size $f(x_n)$—will cause x_{n+1} to be more distant from x_n. In the previous example, at $x = 0$ we would have $f'(x) = 0$ and therefore the tangent line would never intercept the x-axis. We would also find ourselves with a zero denominator in the recursion formula. A common example showing graphically how the x_n can diverge is given next.

Example 4.5.4

Suppose we wish to use Newton's Method to solve $\sqrt[3]{x} = 0$. While we can see the solution is $x = 0$, instead we will suppose we do not know this, and instead make our first guess of $x_1 = 1$.

Our recursion relation becomes:

$$x_{n+1} = x_n - \frac{f(x_n)}{f'(x_n)}$$
$$= x_n - \frac{x_n^{1/3}}{\frac{1}{3}x_n^{-2/3}}$$
$$= x_n - 3x_n.$$

Running our algorithm, we get:

n	x_n
1	1
2	-2
3	4
4	-8
5	16

and so on. Indeed, we can see a simple pattern where $x_{n+1} = -2x_n$, and the x_n diverge. This coincides with the slopes of the tangent lines shrinking in size, the effect of which is shown graphically in Figure 4.15.

As has been pointed out, when the slope of the curve decreases in size, the method is likely to produce points x_n which are more spread apart, and therefore less likely convergent. However, the method can be quite useful when algebraic methods are not up to the task. It is helpful to have a general idea of the behavior of the function's graph, such as the number of solutions.

482 CHAPTER 4. USING DERIVATIVES TO ANALYZE FUNCTIONS; APPLICATIONS

As seen in previous examples, at times Newton's Method can potentially well-approximate a solution but may be sensitive to the input choice. Consider the following applied problem.

Example 4.5.5

Suppose there are two computer sort routines available for a given type of data, and if n is the number of data to be sorted, the times in nanoseconds required for the two sorts are, respectively,

$$T_1(n) = 20n^2,$$
$$T_2(n) = 1000n \ln n.$$

Find the size of the data after which the second algorithm is always faster than the first.

Solution: In order to use the variable n as we do in the formula for Newton's Method, we will instead solve where $T_1(x) = T_2(x)$. However, we can make our work somewhat simpler by looking more closely at this equation:

$$T_1(x) = T_2(x) \iff 20x^2 = 1000x \ln x \iff 20x(x - 50 \ln x) = 0.$$

Clearly $x = 0$ is not interesting (and is in fact not in the domain of $T_2(x)$), so we can instead define $f(x) = x - 50 \ln x$ and solve $f(x) = 0$. Since $f'(x) = 1 - \frac{50}{x}$ we have our recursion relation

$$x_{n+1} = x_n - \frac{f(x_n)}{f'(x_n)} = x_n - \frac{x_n - 50 \ln x_n}{1 - \frac{50}{x_n}}.$$

(For computational purposes we will not multiply by x_n/x_n to clear the fraction in the denominator, because it will complicate the numerator undesirably.) Using this recursion relation, we then consider various values for x_1.

n	x_n
1	10
2	-16.282313662
3	Undefined
4	Undefined
5	Undefined
6	Undefined
7	Undefined

n	x_n
1	70
2	568.486667359
3	292.911261655
4	282.157881324
5	282.115899422
6	282.115898749
7	282.115898749

Our first attempt ended in an error because after the first tangent line intersected the x-axis at a negative value x_2, our algorithm would have us then computing natural logarithms of negative numbers. However, our second attempt succeeded.

We might note also that $f(282.115898749) \approx 0$, and $f'(x) = 1 - \frac{50}{x} > 0$ for $x > 50$, and so $f(x) > 0$ past this x-intercept (approximated by our Newton's Method algorithm). Rounding to the next natural number of data we can say

$$x \geq 283 \implies f(x) > 0 \iff x > 50 \ln x \implies 20x^2 > 1000x \ln x.$$

In terms of the original question, $n \geq 283 \implies T_1(n) > T_2(n)$.

The previous example shows that sometimes we can solve a simpler equation than the one that naturally presents itself. It also shows that the method can be of some service in solving inequalities, given other relevant analysis (such as we used in the last paragraph above).

Exercises

1. In Chapter 3 we encountered the function $f(x) = x^4 - 6x^2 + 8x$, which had one obvious x-intercept at $(0,0)$. Use Newton's Method to approximate the other (negative) x-intercept accurate to $\pm 10^{-8} = \pm 0.00000001$. (The function is graphed in Figure 3.9, page 242.)

2. Newton's Method produces a method to compute square roots, which was used for many years by students who had no access to calculators. For instance, to compute $\sqrt{3}$, one would look at the function $f(x) = x^2 - 3$ and use the method to approximate where $f(x) = 0$. In this way, use Newton's Method to approximate $\sqrt{3}$ accurate to $\pm 10^{-9}$ or better.

3. Repeat the previous exercise but approximate instead $\sqrt{24}$ accurate to $\pm 10^{-9}$ or better.

4. Use a similar approach with Newton's Method to compute $\sqrt[3]{9}$ accurate to $\pm 10^{-9}$ or better.

5. (For this exercise a spreadsheet would be more convenient than a calculator. Be sure all angles are measured in radians.) Consider the equation $\tan x = x$.

   (a) Sketch rough graphs of $y = \tan x$ and $y = x$. Note that $x = 0$ is clearly one solution.
   (b) We will use Newton's Method to find a positive solution. In doing so, note what happens if we attempt the method with various initial values $x_1 = 4.1, 4.2, 4.3, 4.4, 4.5, 4.6, 4.7, 4.8,$ and 4.9. (For which of these values of x_1 does Newton's Method converge, and to what?)
   (c) Repeat with x_1 ranging from 7.3 to 8.0 in increments of 0.1.

6. Using only a scientific (non-graphing) calculator or similar device, graph $f(x) = x^3 + 2x^2 - 5x - 1$ as follows:

   (a) Find the critical points (using the Quadratic Formula) and produce a sign chart for $f'(x)$. Use this to identify local extrema.
   (b) Find any inflection points and produce a sign chart for $f''(x)$.
   (c) Use Newton's Method to find all (three) x-intercepts.
   (d) Sketch the graph, reflecting all of this information, along with the y-intercept.

7. Two populations grow at different rates, given by the functions
$$P_1(t) = 10{,}000e^t,$$
$$P_2(t) = 100{,}000\sqrt{t}.$$
   Here t is in years. In how many years will the first population "catch up" to the second? Use Newton's Method to compute your answer.

8. Two computer algorithms working on n data require different times to process the data. The times needed by each to process the data are given by
$$T_1(n) = 1000n^3,$$
$$T_2(n) = 20{,}000n^2 \ln n.$$
   Besides $n = 1$, after how many data is the second algorithm consistently faster than the first? That is, find N so that $n \geq N \implies T_2(n) < T_1(n)$. (Hint: See Example 4.5.5, page 482.)

9. Suppose $f(x) = x^2 - 1$.

    (a) Sketch the graph of $y = f(x)$.

    (b) Derive the recursive formula given by Newton's Method for this function.

    (c) Find x_2, x_3, x_4, and so on accurate to $\pm 10^{-9}$ if $x_1 = 2$.

    (d) Describe how the algorithm to generate $x_2, x_3, x_4, \ldots$ would behave if $x_1 = 1$.

    (e) Repeat, but for $x_1 = 0$.

10. Suppose $f(x) = 2x^{1/3}$.

    (a) Derive and simplify the recursive relationship given by Newton's Method for this function.

    (b) Show that the values of x_n will diverge for any initial value $x_1 \neq 0$. What happens if $x_1 = 0$?

11. Suppose $f(x) = x^3$.

    (a) Derive the recursive relationship given by Newton's Method for this function.

    (b) Let $x_1 = 1$. Compute x_2, x_3, and x_4.

    (c) Find a formula for x_n in terms of n. (Your formula should not explicitly include x_{n-1}.)

    (d) What is the limiting value of x_n?

12. Suppose $f(x) = x^3 - 6x^2 + 7x + 2$.

    (a) Derive the recursive relationship given by Newton's Method for this function.

    (b) Let $x_1 = 1$. Compute x_2, x_3 and x_4.

    (c) Show that the values of x_n alternate between two values.

    (d) Make a sign chart for $f'(x)$ and use it to sketch $y = f(x)$. (You will need to use the quadratic formula or a similar technique.)

    (e) Interpret the results of (b) and (c) in light of (d).

13. Suppose we wish to produce a graph of $f(x) = x^2 + 4\cos x$ on $[-\pi, \pi]$. Note the function's symmetry.

    (a) Make a sign chart for $f'(x)$, $x \in [-\pi, \pi]$. Clearly $f'(0) = 0$, but Newton's Method will be required to find two other critical points. (There is no need to make a sign chart for $f(x)$; it will follow from the other charts.)

    (b) Make a sign chart for $f''(x)$ on $x \in [-\pi, \pi]$. (This should not require Newton's Method.)

    (c) Use this information to plot a graph of $f(x)$ on this interval.

14. Suppose we wish to solve $\tan^{-1} x = x$, with x in degrees. Of course $x = 0$ is one solution, but we wish to find another.

    (a) Let $g(x)$ represent the arctangent function of x, but with output in degrees. Then we have $g(x) = k \cdot \tan^{-1} x$. Find k.

    (b) What is the derivative of $g(x)$? Should $g(x)$ be a faster-changing function than $\tan^{-1} x$, or more slowly changing?

    (c) Write the Newton's Method recursion formula for the equation $g(x) = x$.

    (d) Run the Newton's Method algorithm with $x_1 = 1$, $x_1 = 10$, and $x_1 = -10$. Do the approximate solutions make sense?

4.6. CHAIN RULE III: RELATED RATES

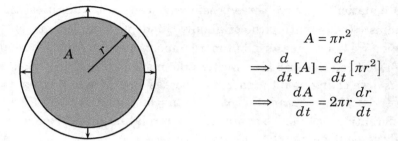

Figure 4.16: Illustration of how a relationship between area A and radius r implies a relationship that includes their rates of change with respect to time.

4.6 Chain Rule III: Related Rates

For some students of mathematics, related rates is the topic in which the calculus "comes alive" because the settings are dynamic: Given how one variable changes with time, we measure how this implies that related variables are then changing with time. Related rates are ultimately founded upon a few simple ideas:

1. That variables—particularly those relating to physical quantities—are often
   (a) related to each other, and perhaps more importantly,
   (b) ultimately are functions of time t, and
2. the Chain Rule.

If changing variables are related to each other, so too are their rates of change related, hence the usual name given to the topic of this section: *related rates*. Figure 4.16 gives one such illustration, discussed next, but the possibilities are endless.

4.6.1 Simple Examples of the Principle

For instance, if we have a circle whose radius r expands (or contracts) with time t, then that radius is a function of t. But then so is the area A of the circle, as r and A are related by $A = \pi r^2$. Furthermore, as a function of time, the equation $A = \pi r^2$ is actually an equality of two *functions of time* t; without even knowing their exact relationships with t, we nonetheless know A and πr^2 are the same functions of t, and so their derivatives should be the same as well:[33]

$$A = \pi r^2 \quad \implies \quad \frac{d}{dt}[A] = \frac{d}{dt}\left[\pi r^2\right] \quad \implies \quad \frac{dA}{dt} = 2\pi r \cdot \frac{dr}{dt}.$$

---

[33]We omitted the algebraically intuitive middle step, which we included more when first introducing the Chain Rule, because by now the reader should understand the need to include the derivative $\frac{dr}{dt}$ in the computation

$$\frac{d}{dt}\left[\pi r^2\right] = \frac{d}{dr}\left[\pi r^2\right] \cdot \frac{dr}{dt} = 2\pi r \cdot \frac{dr}{dt}.$$

In the last equation, we have three possibly nonconstant quantities: the two rates $\frac{dA}{dt}$, $\frac{dr}{dt}$ and the radius r (not counting the originally "hidden" variable time, t). If we know any two of these, then we know the third, but more importantly the equation *relates the rates* $\frac{dA}{dt}$ and $\frac{dr}{dt}$, though that relationship also involves the non-rate variable r. So if we were to ask, "At what rate is area changing if the radius is increasing at 2 m/min when $r = 5$ m?" then we can compute $\frac{dA}{dt} = 2\pi(5\text{m}) \cdot (2\text{m/min}) = 20\pi \text{ m}^2/\text{min} \approx 62.8 \text{ m}^2/\text{min}$.[34]

We can differentiate both sides of any equation relating time-dependent variables in order to get a relationship among their time derivatives, though such relationships might also involve the variables themselves in undifferentiated form:

volume of a sphere	$V = \frac{4}{3}\pi r^3$	$\Longrightarrow$	$\frac{dV}{dt} = 4\pi r^2 \frac{dr}{dt}$
momentum	$p = mv$	$\Longrightarrow$	$\frac{dp}{dt} = m\frac{dv}{dt} + v\frac{dm}{dt}$
ideal gas law (k is a constant)	$PV = knT$	$\Longrightarrow$	$P\frac{dV}{dt} + V\frac{dP}{dt} = k\left[n\frac{dT}{dt} + T\frac{dn}{dt}\right]$

These equations from differentiation all involve dynamic changes (changes with time) of the variables involved. Some texts explain that those variables are ultimately functions of a "hidden variable" t. The calculus leading to those equations involves the Chain Rule, but also the other differentiation rules as they apply (note the Product Rule). For a complicated situation, there is usually some geometric or algebraic work to set the stage for the calculus, but the calculus itself is straightforward if the rules are followed exactly as they are stated.

Example 4.6.1

Suppose a rocket's height y in miles at time t in seconds after the rocket lifts off is given by $y = \frac{1}{320}t^2$. Technically, to make the units work out correctly, we actually should give this equation as

$$y = \frac{1}{320 \text{ sec}^2/\text{mi}} t^2. \qquad (4.25)$$

This way, for t given in seconds we see that y will indeed be given in miles. From this equation we can compute the rocket's vertical velocity (by applying $\frac{d}{dt}$ to both sides) as follows:

$$\frac{d}{dt}y = \frac{d}{dt}\left[\frac{1}{320 \text{ sec}^2/\text{mi}} \cdot t^2\right] \Longrightarrow \frac{dy}{dt} = \frac{1}{320 \text{ sec}^2/\text{mi}} \cdot 2t$$

$$\Longrightarrow \frac{dy}{dt} = \frac{1}{160} \cdot \frac{\text{mi}}{\text{sec}^2} \cdot t. \qquad (4.26)$$

Not that if t is in units of seconds, then dy/dt will be in mi/sec. This makes for simple enough computations of the rocket's velocity along a continuum of values of $t > 0$, as long as the rocket's height is still given by (4.25). A short table of such values of y and dy/dt is given below:

t	0	1	2	3	4	5	$\cdots$	10	20	30	40	50
y	0	$\frac{1}{320}$	$\frac{4}{320}$	$\frac{9}{320}$	$\frac{16}{320}$	$\frac{25}{320}$	$\cdots$	0.3125	1.25	2.8125	5	7.8
dy/dt	0	$\frac{1}{160}$	$\frac{2}{160}$	$\frac{3}{160}$	$\frac{4}{160}$	$\frac{5}{160}$	$\cdots$	0.0625	0.125	0.1875	0.250	0.3125

---

[34]While such a computation might seem only theoretically interesting, in fact it becomes relevant when one encounters such things as chemical spills.

4.6. CHAIN RULE III: RELATED RATES

Note that the final column's entries including units are, respectively, 50 seconds, 7.8 miles, and 0.3125 miles/second. If we prefer to read that last ($t = 50$ sec) entry in miles/hour, we simply convert

$$\left.\frac{dy}{dt}\right|_{t=7.8 \text{ sec}} = 0.3125 \, \frac{\text{mi}}{\text{sec}} \cdot \frac{3600 \text{ sec}}{\text{hour}} = 1125 \text{ miles/hour}.$$

Recall that the vertical line in the expression $\left.\frac{dy}{dt}\right|_{t=7.8 \text{ sec}}$ is read, "evaluated when" or "evaluated at." Also note that in practice, for brevity one often does not write units at each step, resting assured that the calculus is completely consistent with units.[35]

Example 4.6.2

Suppose that two cars start at the same point, and that one drives north at 60 mi/hr while the other drives east at 80 mi/hr. Two hours later, how quickly is the distance between them increasing?

<u>Solution</u>: Here we can let y be the distance north that the first car traveled at time t, and x be the distance the second car traveled east at time t. The distance D between them is related by the Pythagorean Theorem, which we can then differentiate:

$$D^2 = x^2 + y^2 \implies \frac{d}{dt}\left[D^2\right] = \frac{d}{dt}\left[x^2 + y^2\right]$$

$$\implies 2D\frac{dD}{dt} = 2x\frac{dx}{dt} + 2y\frac{dy}{dt}$$

$$\implies D\frac{dD}{dt} = x\frac{dx}{dt} + y\frac{dy}{dt}.$$

When $t = 2$ hr, we have $x = 120$ mi, $y = 160$ mi, and $D = \sqrt{x^2 + y^2} = \sqrt{14{,}400 \text{ mi}^2 + 25{,}600 \text{ mi}^2} = \sqrt{40{,}000 \text{ mi}^2} = 200$ mi. With this, and what we were told of the rates $dx/dt = 60$ mi/hr and $dy/dt = 80$ mi/hr, the rate equation becomes (after dividing by miles)

$$200\frac{dD}{dt} = (120)(60 \text{ mi/hr}) + (160)(80 \text{ mi/hr}) \implies \frac{dD}{dt} = \frac{(120)(60 \text{ mi/hr}) + (160)(80 \text{ mi/hr})}{200}$$

$$= \frac{20{,}000 \text{ mi/hr}}{200} = 100 \, \frac{\text{mi}}{\text{hr}}.$$

---

[35] For many equations, units are contained in the variables and there is no need for a constant multiplier to account for units. If the units are introduced correctly from the beginning, as we did when we rewrote the motion as Equation (4.25), page 486, to more carefully include units, we can follow through with them as when we then derived the resulting equation (4.26). The student interested in studying how units follow through in the computations is encouraged to do so, on occasion. In doing so, it is important to notice how they carry over into derivatives from the difference quotients. For instance,

$$\left.\frac{dy}{dt}\right|_{t=t_0} = \lim_{\Delta t \to 0} \frac{y(t_0 + \Delta t) - y(t)}{\Delta t} = \lim_{\Delta t \to 0} \frac{\Delta y}{\Delta t},$$

and so dy/dt will have units that are those of y divided by those of t. Most of the time we will not include units in our computations, but there will be times when we may either verify they are valid or include them if they are, on balance, instructive.

488 CHAPTER 4. USING DERIVATIVES TO ANALYZE FUNCTIONS; APPLICATIONS

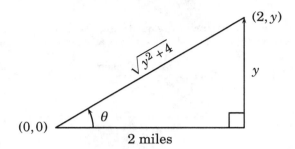

Figure 4.17: Figure for first related rates discussion in Subsection 4.6.2.

4.6.2 Choosing the Form of the Relationship

We often have choices for equivalent equations relating variables. In the previous example we could have chosen $D = \sqrt{x^2 + y^2}$, instead of $D^2 = x^2 + y^2$. Usually one will be simpler for differentiation but require more work to apply. Differentiating the first is more complicated, but the variable D does not appear except in the rate dD/dt, and so we do not need to know the value of D in that case. Sometimes further variables (e.g., some angle θ in the triangle in question) can be introduced to simplify a later step, but it may complicate some other part.

Example 4.6.3 _____

For a slightly more complicated example than Example 4.6.1, page 486, involving a launching rocket, now suppose we have a camera at ground level 2 miles from the site of the rocket's liftoff, as seen in Figure 4.17. If we would like to find the speed in degrees/sec that the camera's angle of elevation has to change to keep the rocket centered in the viewing area, we have to relate the angle of elevation θ to the variable y, which is simple enough:

$$\tan\theta = \frac{y}{2\text{ mi}}.$$

So that units work easily with the calculus, we will temporarily measure θ in radians, and we can differentiate this equation with respect to t as follows:

$$\frac{d}{dt}\tan\theta = \frac{d}{dt}\left[\frac{y}{2\text{ mi}}\right] \implies \sec^2\theta \cdot \frac{d\theta}{dt} = \frac{1}{2\text{ mi}} \cdot \frac{dy}{dt}$$
$$\implies \frac{d\theta}{dt} = \frac{1}{2\text{ mi}}\cos^2\theta \cdot \frac{dy}{dt}. \tag{4.27}$$

If we were asked for the rate of change in the camera's angle after the rocket had flown for 1 second, we would be looking for $d\theta/dt$ for that instant, but note that we would need to find both y and θ for that instant to use them in (4.27). Using a little bit of cleverness, we could instead note the hypotenuse in Figure 4.17 is $\sqrt{y^2+4}$, making $\cos\theta = 2/\sqrt{y^2+4}$, and so (4.27) can also be written

$$\frac{d\theta}{dt} = \frac{1}{2\text{ mi}} \cdot \frac{4}{y^2+4} \cdot \frac{dy}{dt} \implies \frac{d\theta}{dt} = \frac{2}{y^2+4} \cdot \frac{dy}{dt}. \tag{4.28}$$

This gives us a relationship among $d\theta/dt$, dy/dt and y, in which we do not need to know the value of θ. Alternatively, we could have used inverse trigonometric functions from the beginning:

$$\theta = \tan^{-1}\frac{y}{2} \implies \frac{d\theta}{dt} = \frac{1}{\frac{y^2}{4}+1} \cdot \frac{1}{2} \cdot \frac{dy}{dt} = \frac{2}{y^2+4} \cdot \frac{dy}{dt}. \tag{4.29}$$

Note that this is the same as (4.28). If we use the situation from Example 4.6.1, recall from (4.25) and

4.6. CHAIN RULE III: RELATED RATES

(4.26), page 486, that $y = \frac{1}{320}t^2$ and $dy/dt = \frac{t}{160}$. From these we would get:

$$\frac{d\theta}{dt} = \frac{2}{\left(\frac{1}{320}t^2\right)^2 + 4} \cdot \frac{1}{160} \cdot t = \frac{t}{\left(\frac{1}{320}t^2\right)^2 + 4} \cdot \frac{1}{80} \cdot \frac{320^2}{320^2} = \frac{1280t}{t^4 + 409{,}600}. \tag{4.30}$$

Now we could have instead found θ as a function of t using (4.25):

$$\theta = \tan^{-1}\frac{y}{2} = \tan^{-1}\frac{\frac{1}{320}t^2}{2} = \tan^{-1}\frac{t^2}{640}$$

$$\implies \frac{d\theta}{dt} = \frac{1}{\frac{t^4}{640^2} + 1} \cdot \frac{2t}{640} = \frac{640 \cdot 2t}{t^4 + 640^2} = \frac{1280t}{t^4 + 409{,}600}. \tag{4.31}$$

We see that each of our methods yield equivalent results for $d\theta/dt$, but there can be some variation in the complexity of the computation. Usually, when the relationship among variables is manipulated to be less complicated in one respect, the work to actually use the relationship among rates (i.e., "related rates") is made more complicated; the harder work can be done early or late.

For instance, suppose we wish to know how fast the camera's angle is rising at $t = 1$ second. We can use (4.31) with $t = 1$ immediately to get

$$\left.\frac{d\theta}{dt}\right|_{t=1} = \frac{1280(1)}{(1)^4 + 409{,}600} = \frac{1280}{409{,}601},$$

with the units in radians/second. To convert this to degrees/second, we multiply

$$\frac{1280}{409{,}601}\frac{\text{rad}}{\text{sec}} \cdot \frac{180°}{\pi \text{ rad}} \approx 0.179°/\text{sec}.$$

If we were instead going to use (4.28), we would compute

$$\left.\frac{d\theta}{dt}\right|_{t=1} = \frac{2}{y^2 + 4} \cdot \frac{dy}{dt},$$

but we would need to find y and dy/dt at $t = 1$. This is not terribly difficult because we have (4.25) and (4.26), i.e., that $y = \frac{1}{320}t^2$ and $dy/dt = \frac{t}{160}$, which give us

$$y|_{t=1} = \frac{1}{320}(1)^2 = \frac{1}{320}, \qquad \left.\frac{dy}{dt}\right|_{t=1} = \frac{1}{160}(1) = \frac{1}{160},$$

and so we have

$$\left.\frac{d\theta}{dt}\right|_{t=1} = \frac{2}{y^2+4} \cdot \left.\frac{dy}{dt}\right|_{t=1} = \frac{2}{\left(\frac{1}{320}\right)^2 + 4} \cdot \frac{1}{160} = \frac{1}{\frac{1}{102400} + 4} \cdot \frac{1}{80} = \frac{102{,}400}{1 + 409600} \cdot \frac{1}{80} = \frac{1280}{409{,}601},$$

as before in radians/second, i.e., approximately $0.179°/\text{sec}$. This indeed does illustrate that simpler forms for the relationships among rates often require extensive *ad hoc* computations to use for actual numerical purposes. If one were to produce a table of values of $d\theta/dt$ for various values of t, it is best to use the explicit formula (4.31). We do so next:

$\frac{d\theta}{dt}$ \ t	0	5	10	15	20	25	30	35	40	50	100
rad/sec	0	.016	.031	.042	.045	.040	.031	.023	.017	.010	.001
°/sec	0	.894	1.748	2.390	2.575	2.291	1.804	1.344	.988	.551	.073

Note how the camera at first has zero angular velocity, how that angular velocity increases for a time as the rocket's speed increases, and how the angular velocity then decreases until it almost

490   CHAPTER 4. USING DERIVATIVES TO ANALYZE FUNCTIONS; APPLICATIONS

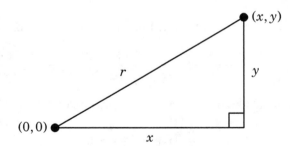

Figure 4.18: If x, y, and r are the legs and hypotenuse of a right triangle, then $x^2 + y^2 = r^2$ according to the Pythagorean Theorem. This implies that $2x\frac{dx}{dt} + 2y\frac{dy}{dt} = 2r\frac{dr}{dt}$, if x, y and r are functions of t. In fact, if we allow x and y to be respectively horizontal and vertical *displacements* from the origin $(0,0)$, with r the distance from $(0,0)$, the Pythagorean Theorem still holds and so does the equation relating the rates.

---

stops, as the rocket is flying very fast but the camera will be pointing almost vertically. This should seem somewhat intuitive.

The previous example showed that there can be different equations relating the rates of variables, because there can be different equations relating the variables. All such equations should be logically consistent given the context. The next example illustrates this in a somewhat less complicated setting.

Example 4.6.4

Figure 4.18 shows another example in which the form of the relation among variables determines the resultant relationship involving rates. Consider the following relations—equivalent for the situation in the figure—and the results of applying $\frac{d}{dt}$ to both sides.

$$r^2 = x^2 + y^2 \quad \stackrel{d/dt}{\Longrightarrow} \quad 2r\frac{dr}{dt} = 2x\frac{dx}{dt} + 2y\frac{dy}{dt},$$

$$r = \sqrt{x^2 + y^2} \quad \stackrel{d/dt}{\Longrightarrow} \quad \frac{dr}{dt} = \frac{1}{2\sqrt{x^2+y^2}}\left(2x\frac{dx}{dt} + 2y\frac{dy}{dt}\right),$$

$$y = \sqrt{r^2 - x^2} \quad \stackrel{d/dt}{\Longrightarrow} \quad \frac{dy}{dt} = \frac{1}{2\sqrt{r^2-x^2}}\left(2r\frac{dr}{dt} - 2x\frac{dx}{dt}\right),$$

$$x = \sqrt{r^2 - y^2} \quad \stackrel{d/dt}{\Longrightarrow} \quad \frac{dx}{dt} = \frac{1}{2\sqrt{r^2-y^2}}\left(2r\frac{dr}{dt} - 2y\frac{dy}{dt}\right).$$

The first of these four is the simplest with which to compute derivatives with respect to time. The next has the advantage that the *variable r* itself does not appear in the derivative equation (even though its *rate dr/dt* does). Therefore, if computing r for a particular problem would be onerous, the second algebraic equation (before applying d/dt) involving r, x, and y may be preferred to work with. Similarly, the third has no y in the differentiated form (though dy/dt appears), and the last equation has no x (except within dx/dt).

This example gives a reasonably stark illustration that choosing the form of the relationship to differentiate can have implications for how difficult it is to compute related rates.

4.6. CHAIN RULE III: RELATED RATES

However, the previous example where we found $d\theta/dt$ also illustrates how the price for ease of differentiation is sometimes paid in difficulty of actually using the consequent rates relation. Fortunately, they are consistent and thus it is often the case that any correct relation can be differentiated into a useful relation among rates.

To be clear, in the previous example, each of the differentiations can be broken into smaller steps. For instance the second one could read

$$\frac{d}{dt}[r] = \frac{d}{dt}\sqrt{x^2+y^2}$$
$$= \frac{1}{2\sqrt{x^2+y^2}} \cdot \frac{d}{dt}[x^2+y^2]$$
$$= \frac{1}{2\sqrt{x^2+y^2}}\left[\frac{d(x^2)}{dt} + \frac{d(y^2)}{dt}\right]$$
$$= \frac{1}{2\sqrt{x^2+y^2}}\left[\frac{d(x^2)}{dx}\cdot\frac{dx}{dt} + \frac{d(y^2)}{dy}\cdot\frac{dy}{dt}\right]$$
$$= \frac{1}{2\sqrt{x^2+y^2}}\left[2x\frac{dx}{dt} + 2y\frac{dy}{dt}\right],$$

as before. If desired this can be somewhat simplified:

$$\frac{dr}{dt} = \frac{1}{\sqrt{x^2+y^2}}\left[x\frac{dx}{dt} + y\frac{dy}{dt}\right] = \frac{x\frac{dx}{dt}+y\frac{dy}{dt}}{\sqrt{x^2+y^2}}.$$

Many important steps followed from the Chain Rule, such as the first step, and that step where we computed $\frac{dx^2}{dt} = 2x\frac{dx}{dt}$. Thus, an equation relating variables holding true forces another equation relating the rates of change of these variables to hold true, hence the title of this section.[36]

Example 4.6.5

Using the illustration in Figure 4.18, page 490, suppose that at some particular time t_0 we have $x = 30$ m and $y = 40$ m, but x is decreasing at 20 m/sec, while y is increasing at 10 m/sec. How is r changing at that time?

Solution: First we collect the numerical data we are given:

$$x = 30\,\text{m}, \quad dx/dt = -20\,\text{m/sec} \quad (\text{since } x \text{ is decreasing}),$$
$$y = 40\,\text{m}, \quad dy/dt = 10\,\text{m/sec} \quad (\text{since } y \text{ is increasing}).$$

We do not have r immediately (though we can find it algebraically from $x^2 + y^2 = r^2$), so we will use the previous computation for dr/dt:

$$\frac{dr}{dt} = \frac{x\frac{dx}{dt}+y\frac{dy}{dt}}{\sqrt{x^2+y^2}} = \frac{30\cdot(-20)+40\cdot(10)}{\sqrt{(30)^2+(40)^2}} = \frac{-200}{50} = -4.$$

We see that the distance r is decreasing at a rate of 4 m/sec at that instant. (The reader should verify that, if units are included in this computation, then this result will be in units of m/sec.)

4.6.3 Finding a Relationship among Variables

Before one answers the question of which form of the relationship among variables to differentiate for a particular application, one must first find at least one such relationship. The

---

[36]Some texts also call this process *implicit differentiation*, mainly because it resembles the work we did in Section 3.6. Related Rates and implicit differentiation are both really just specific applications of the Chain Rule.

492 CHAPTER 4. USING DERIVATIVES TO ANALYZE FUNCTIONS; APPLICATIONS

clues can often come from the geometry or other given constraints. Some of the more classical examples are given here.

Example 4.6.6

A 170-cm-tall pedestrian walks at a speed of 1 m/sec directly away from a street light whose bulb is 6 m high from the ground. When the pedestrian is 20 m from the base of the light pole,

(a) how quickly is their shadow from the bulb lengthening, and

(b) how quickly is its leading tip moving along the ground?

Solution: We begin with an illustration of the geometry of the setting, introducing some variables for relevent lengths.

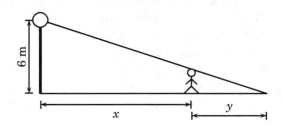

From our given information we have $\frac{dx}{dt} = 1$ m/sec. For (a), we wish to find $\frac{dy}{dt}$ when $x = 20$.

From similar triangles, we have

$$\frac{y}{1.7} = \frac{x+y}{6} \implies 6y = 1.7x + 1.7y$$
$$\implies 4.3y = 1.7x$$
$$\implies 4.3 \cdot \frac{dy}{dt} = 1.7 \cdot \frac{dx}{dt}$$
$$\implies \frac{dy}{dt} = \frac{1.7}{4.3} \cdot \frac{dx}{dt}.$$

Since we were given that $\frac{dx}{dt} = 1$ m/sec, we conclude $\frac{dy}{dt} = \frac{1.7}{4.3} \cdot 1$ m/sec ≈ 0.40 m/sec, which answers (a).

To answer (b), we note that the distance from the base of the light to the leading tip of the shadow will be $s = x + y$, and so $\frac{ds}{dt} = \frac{dx}{dt} + \frac{dy}{dt} \approx 1$ m/sec + 0.40 m/sec = 1.4 m/sec.

This example is interesting for a few reasons. For one, the distance x from the pole to the pedestrian does not appear in the rate equation (though its derivative dx/dt does), and so the information that the walker was 20 m from the base of the pole had no importance in either parts (a) or (b). The example also reminds us that a desired speed might not be the most easily found rate, as here dy/dt was easily found but represented the growth of the shadow's length, and not the speed of its leading tip. Finally, it was important in the differentiation step to treat the given lengths 6 m and 170 cm = 0.70 m as *constants,* while the lengths x and y (and s) were *variables,* and $x = 20$ m was one sample value of one of the variables.

The example also is somewhat noteworthy in its use of similar triangles to set up equations from which we could take time derivatives to relate rates. The next example also employs similar triangles towards that same end.

Example 4.6.7

Given an inverted (downward-pointing) right circular conical tank with base diameter 10 meters and height 15 meters, suppose it is being filled with water at a rate of 2 meter^3 per minute. Find how quickly the water level is rising when the depth of the water is 1 meter, 2 meters, 5 meters, 9 meters, approaching 10 meters, and approaching 0 meters.

4.6. CHAIN RULE III: RELATED RATES

Solution: We will let h be the height of the water in the tank, and note that we are seeking dh/dt for certain instances of h. We are also given a rate of volume change, which we can write as $dV/dt = 2$ (volume increases at 2 m³/min), so a natural thing to do here is to relate V to h and differentiate. In this and many other cases, it is best to begin with a picture:[37]

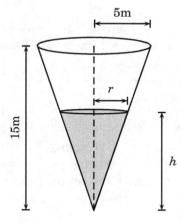

The next step is to find a relationship between V and h. This can perhaps most easily be done by introducing a third variable r, which will be the radius of the water cone when the depth is h. We can then write V in terms of both r and h from a formula in geometry:

$$V = \frac{1}{3}\pi r^2 h.$$

In fact, we can differentiate this with respect to t, but we would then have a dr/dt term, which would be somewhat problematic (since we only wish to have terms involving V, h, or their time derivatives). The next observation is key.

From trigonometry or the geometry of similar triangles we can "eliminate" r: $\frac{r}{5} = \frac{h}{15} \implies r = \frac{h}{3}$. Putting this into our volume formula, we now have $V = \frac{1}{3}\pi r^2 h = \frac{1}{3}\pi \left(\frac{h}{3}\right)^2 h$, i.e.,

$$V = \frac{\pi}{27}h^3.$$

Now we can take time derivatives:

$$\frac{d}{dt}V = \frac{d}{dt}\left[\frac{\pi}{27}h^3\right] = \frac{\pi}{27} \cdot 3h^2 \cdot \frac{dh}{dt} = \frac{\pi}{9}h^2 \cdot \frac{dh}{dt}.$$

We are given $\frac{dV}{dt}$ and wish to solve for $\frac{dh}{dt}$ for several different values of h. Solving for $\frac{dh}{dt}$ we get

$$\frac{dh}{dt} = \frac{9}{\pi h^2} \cdot \frac{dV}{dt}.$$

Now we use our given information that $\frac{dV}{dt} = 2$, giving us the following for $h \in (0, 10)$:

$$\frac{dh}{dt} = \frac{9}{\pi h^2} \cdot 2 = \frac{18}{\pi h^2}.$$

From this we can compute

$$\left.\frac{dh}{dt}\right|_{h=1} = \frac{18}{\pi} \approx 5.730 \text{ m/min,} \qquad \left.\frac{dh}{dt}\right|_{h=5} = \frac{18}{25\pi} \approx 0.229 \text{ m/min,}$$

$$\left.\frac{dh}{dt}\right|_{h=2} = \frac{18}{4\pi} \approx 1.423 \text{ m/min,} \qquad \left.\frac{dh}{dt}\right|_{h=9} = \frac{18}{81\pi} \approx 0.071 \text{ m/min.}$$

---

[37]In practice, the picture and the equations are usually constructed in parallel, as the equations demand more information from the picture and vice versa.

Furthermore, it is clear that

$$x \to 10^- \implies \frac{dh}{dt} = \frac{18}{\pi h^2} \longrightarrow \frac{18}{100\pi} \approx 0.057 \text{ m/min},$$
$$x \to 0^+ \implies \frac{dh}{dt} = \frac{18}{\pi h^2} \longrightarrow \infty.$$

We see a rapid rising of the water's height h, indeed theoretically "infinitely" fast at the beginning, and then a significant slowing of the rate of the rising in h as it takes more water to fill higher levels of the tank.

In this example, we could have left r in the equation for V and later used $\frac{1}{5}r = \frac{1}{15}h \implies \frac{1}{5}\frac{dr}{dt} = \frac{1}{15}\frac{dh}{dt}$ when solving for dh/dt, but it was arguably easier to have only V and h being related algebraically, and so they and their derivatives could be more simply related after the differentiation step. As mentioned before, there can be many correct paths to arrive at an answer to a related-rates question, but it is often the case that one path can be much more efficient, or at least more straightforward, than another.

For most novices, the most difficult part of solving a related-rates problem lies in the steps taken before derivatives are computed, i.e., in finding an actual equation relating the variables in question. If looking for an actual value of a rate for a variable, say p, while given the rate for another related variable q—that is if we want $\frac{dp}{dt}$ but know $\frac{dq}{dt}$—we must find some relationship between q and p, which can be differentiated with respect to t, in such a way that that the information given suffices for us to somehow find $\frac{dp}{dt}$. In many cases geometric considerations are often keys: the Pythagorean Theorem, area and volume formulas, and similar triangles and other trignometric considerations commonly arise in these derivations. In other cases, such as the next example, some nongeometric relationship exists among the variables, and we differentiate that relationship with respect to t.

4.6.4 Further Examples

There are occasions when we know some of the rates at a particular instant but do not necessarily know the entire scope of the relationship among the variables. The Leibniz notation can be particularly helpful here, though one has to be sure that, say, when writing dQ/dt that Q can be in theory ultimately a function of t.

Example 4.6.8

Suppose we have an ideal gas, which is governed by the equation

$$PV = nRT,$$

where P is pressure, V is volume, n is the number of particles, and T is the absolute temperature (for instance, in Kelvins), and R is a physical constant that also includes within it the necessary units to make the equation valid from a dimensional standpoint. Note that each of these quantities must be nonnegative. If the units of P are atmospheres, of V are liters, of n are moles, and of T are Kelvins, then the research shows that we can take

$$R \approx 0.08206 \frac{\text{L}\cdot\text{atm}}{\text{mol}\cdot\text{K}}.$$

4.6. CHAIN RULE III: RELATED RATES

While we could take time derivatives $\frac{d}{dt}$ of both sides, i.e., $\frac{d}{dt}(PV) = \frac{d}{dt}(nRT)$, using the Product Rule and noting that R is a constant, it is a bit faster to use logarithmic differentiation and take the natural logarithm of both sides first (noting that all variables are positive):

$$\ln(PV) = \ln(nRT)$$
$$\implies \ln P + \ln V = \ln n + \ln R + \ln T$$
$$\overset{d/dt}{\implies} \frac{1}{P} \cdot \frac{dP}{dt} + \frac{1}{V} \cdot \frac{dV}{dt} = \frac{1}{n} \cdot \frac{dn}{dt} + \frac{1}{T} \cdot \frac{dT}{dt}.$$

Technically, n is a "discrete" variable since it can only take on integer values, and so its derivative does not make sense. However, in practice we use units such as mols (1 mol$\approx 6.02214179 \times 10^{23}$) of particles, and so in practice n takes on decimal values with theoretically tens of significant digits possible, and as a result it is quite acceptable to act as if n were a "continuous" variable.[38]

There are eight quantities in this derivative formula, so if we know any seven we can use algebra to find the eighth. In fact, since we also have the original relationship $PV = nRT$, we can eliminate any of the variables P, V, n, T as well. Similarly, if we know one of these is held constant, then its derivative is zero and so we would have one less to be concerned about.

For instance, suppose we keep constant temperature and number of particles. Then we have

$$\frac{1}{P} \cdot \frac{dP}{dt} = -\frac{1}{V} \cdot \frac{dV}{dt}.$$

Using then $V = nRT/P$ we would have

$$\frac{1}{P} \cdot \frac{dP}{dt} = -\frac{P}{nRT} \cdot \frac{dV}{dt}$$
$$\implies \frac{dP}{dt} = -\frac{P^2}{nRT} \cdot \frac{dV}{dt}.$$

One interesting aspect of this is that if volume increases ($dV/dt > 0$), then pressure decreases ($dP/dt < 0$). While intuitive, it not only emerges from the equations, but also we can compute the actual (instantaneous) rate at which that occurs.

If at a particular instant $P = 2.0$ atm and is decreasing at 0.52 atm/min, and $V = 5.0$ L, we can find the instantaneous rate of change in V:

$$\frac{dV}{dt} = -\frac{V}{P} \cdot \frac{dP}{dt} = -\frac{5.0 \text{ L}}{2.0 \text{ atm}} \cdot \frac{-0.52 \text{ atm}}{\text{min}} \approx 1.3 \text{ L/min}.$$

We see that the volume at that instant is increasing at 1.3 L/min.

This example illustrates how differentiation with respect to time can elicit relationships among variables and their rates of change with respect to time even though we do not have explicit knowledge of the exact nature of their dependence on time: We do not have any formulas for how P, V, n, and T actually depend on time, but we can still say something about the relationship among their time derivatives. We can do that because we know how the variables themselves relate to each other, and because it is clear that these variables are,

---

[38] In theory, the derivative of an integer-valued or discrete-valued function will be either zero (on any open time interval on which the function is constant), or will not exist (at any point where the function has a discontinuity). Consider, for instance, the "greatest integer function" $f(x) = [\![x]\!] = \max\{z \in \mathbb{Z} \mid z \leq x\}$, whose derivative is zero between integer values of x but does not exist at those integer values because at integer input values, $f(x)$ is discontinuous and therefore its derivative $f'(x)$ does not exist.

496 CHAPTER 4. USING DERIVATIVES TO ANALYZE FUNCTIONS; APPLICATIONS

ultimately, functions of time: $P = P(t)$, $V = V(t)$, $n = n(t)$, and $T = T(t)$. Thus their derivatives dP/dt, dV/dt, dn/dt, and dT/dt all make sense, even if we do not know formulas for them.

Example 4.6.9

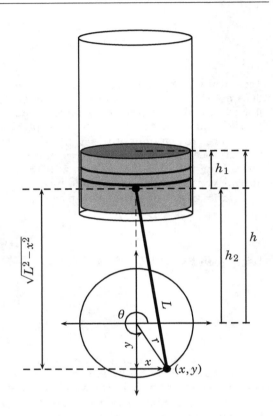

We wish to find the velocity dh/dt of the piston in the diagram on the right, from knowledge of its position and the angular velocity $\omega = d\theta/dt$ of the circular crank which is centered at the origin in the xy-plane. The piston moves up and down a cylinder while attached to a circular crank by a piston rod of length L.

The lower point of the piston rod attaches to the crank at the point (x,y), where the definitions of $\cos\theta$ and $\sin\theta$ give us

$$x = r\cos\theta,$$
$$y = r\sin\theta.$$

Taking h_2 to be the y-coordinate at the top of the piston rod, we first note the vertical length of the right triangle whose vertices lie at (x,y), $(0,y)$, and the top of the piston rod is given by the Pythagorean Theorem to be $\sqrt{L^2 - x^2}$. From this we have

$$h_2 = y + \sqrt{L^2 - x^2}.$$

(In the illustrated piston position, $y < 0$.) Since $h = h_1 + h_2$, this becomes $h = h_1 + y + \sqrt{L^2 - x^2}$, which when we reintroduce θ becomes

$$h = h_1 + r\sin\theta + \sqrt{L^2 - r^2\cos^2\theta}.$$

Noting that h_1, r, and L are constants, we now find dh/dt. Here we were able to write $h = h(\theta)$, and so one interpretation of our differentiation is that we are computing $\frac{dh}{dt} = \frac{dh}{d\theta} \cdot \frac{d\theta}{dt}$.[39]

$$\frac{dh}{dt} = \frac{d}{dt}\left[h_1 + r\sin\theta + \sqrt{L^2 - r^2\cos^2\theta}\right] = r\cos\theta \cdot \frac{d\theta}{dt} + \frac{1}{2\sqrt{L^2 - r^2\cos^2\theta}} \cdot \frac{d}{dt}[L^2 - r^2\cos^2\theta]$$

$$= r\cos\theta \cdot \frac{d\theta}{dt} + \frac{1}{2\sqrt{L^2 - r^2\cos^2\theta}} \cdot \left[-r^2 \cdot 2\cos\theta \cdot \frac{d}{dt}\cos\theta\right]$$

$$= r\cos\theta \cdot \frac{d\theta}{dt} + \frac{1}{2\sqrt{L^2 - r^2\cos^2\theta}} \cdot \left[-r^2 \cdot 2\cos\theta \cdot (-\sin\theta)\frac{d\theta}{dt}\right]$$

$$\implies \frac{dh}{dt} = \left[r\cos\theta + \frac{r^2\sin\theta\cos\theta}{\sqrt{L^2 - r^2\cos^2\theta}}\right]\frac{d\theta}{dt}.$$

Not crucial, but worth noting is that when we recall $x = r\cos\theta$ and $y = r\sin\theta$, we can also write

$$\frac{dh}{dt} = \left[r\cos\theta + \frac{r^2\sin\theta\cos\theta}{\sqrt{L^2 - r^2\cos^2\theta}}\right]\frac{d\theta}{dt} = \left[x + \frac{xy}{\sqrt{L^2 - x^2}}\right]\frac{d\theta}{dt}.$$

---

[39]We could have also used $h = h_1 + y + \sqrt{L^2 - x^2}$ and had an expression for dh/dt which would involve dy/dt and dx/dt, but computing those emergent derivatives would eventually have us computing $\frac{dy}{dt} = \frac{dy}{d\theta} \cdot \frac{d\theta}{dt}$, and similarly for x.

4.6. CHAIN RULE III: RELATED RATES

However, we will use the first form for dh/dt. Now let us consider an example with some numbers. We will assume that $r = 5\,\text{cm}$, $L = 15\,\text{cm}$, and $\omega = \frac{d\theta}{dt} = 6{,}000\,\text{rpm}$. ($h_1$ is irrelevant, which is a fact worth pondering.) We will report our final velocities in units of meters/second. To accomplish this, we must convert all lengths to meters, and the angular velocity ω to radians/second.

$$\frac{d\theta}{dt} = \omega = 6000\,\frac{\text{rev}}{\text{min}} \times \frac{2\pi\,\text{rad}}{\text{rev}} \times \frac{\text{min}}{60\,\text{sec}} = 200\pi\,\frac{\text{rad}}{\text{sec}}.$$

And so for instance, at $\theta = 30° = \pi/6$ we have

$$\frac{dh}{dt} = \left[(0.05\,\text{m})\cos\frac{\pi}{6} + \frac{(0.05\,\text{m})^2 \sin\frac{\pi}{6}\cos\frac{\pi}{6}}{\sqrt{(0.15\,\text{m})^2 - (0.05\,\text{m})^2 \cos^2\frac{\pi}{6}}}\right](200\pi\,\text{rad/sec}) \approx 31.9\,\text{m/sec}.$$

More values are given in the table:

θ	0	$\frac{\pi}{6}$	$\frac{\pi}{4}$	$\frac{\pi}{3}$	$\frac{\pi}{2}$	$\frac{2\pi}{3}$	$\frac{3\pi}{4}$	$\frac{5\pi}{6}$	π
$\frac{dh}{dt}$	31.4	31.9	27.6	20.3	0	−20.3	−27.6	−31.9	−31.4
θ	π	$\frac{7\pi}{6}$	$\frac{5\pi}{4}$	$\frac{4\pi}{3}$	$\frac{3\pi}{2}$	$\frac{5\pi}{3}$	$\frac{7\pi}{4}$	$\frac{11\pi}{6}$	2π
$\frac{dh}{dt}$	−31.4	−22.5	−16.8	−11.1	0	11.1	16.8	22.5	31.4

The values for dh/dt are anti-symmetric with respect to "top dead center" ($\theta = \pi/2$) on the upper semicircle of crank position (x,y), and again with respect to "bottom dead center" ($\theta = 3\pi/2$) on the lower semicircle, though there is no obvious full symmetry between the two semicircles. When one considers the action of the piston rod during a full cycle of θ, both of these observations are reasonable (and even predictable), as well as being apparent from the alternative representation $dh/dt = \left[x + \frac{xy}{\sqrt{L^2-x^2}}\right]\frac{d\theta}{dt}$.

One useful lesson from this Example 4.6.9 is that it is often simpler to analyze the problem using representations of constants, in this case several lengths (r, L, h_1) than to insert the actual values, which can cause unnecessary clutter in the equations. Of course it also has the advantage of allowing us to change those lengths without requiring us to repeat the same analysis with different numbers.

It should be pointed out that in some applications, particularly using a single-cylinder engine, it is quite likely that $d\theta/dt$ will not remain constant (as the crank is loaded by several varying sources), though our equations did not assume that it would. (Only our example application assumed it.)

An interesting—but not easy—problem would be to compute at what angle θ is the piston speed maximized. One could assume $d\theta/dt$ is constant, compute $\frac{d}{d\theta}\left[\frac{dh}{dt}\right]$, and find where it is zero, which may require a Newton's Method approximation, depending on the relationship between L and r.

4.6.5 An Approximate Modeling Example

While some of the beauty of calculus is that it gives exact answers for idealized problems, a reality of engineering and science is that real-world problems are usually not ideal, and

498   CHAPTER 4. USING DERIVATIVES TO ANALYZE FUNCTIONS; APPLICATIONS

measurements are usually not exact. While on one hand this can be somewhat unsatisfying, on the other hand it sometimes allows us to relax the standards of exactitude to find a simpler—even if somewhat more approximate—mathematical model. We take our model for what it is worth, but it may be worth a great deal in information and intuition regarding the actual, real-world problem.

Example 4.6.10

Suppose we have a coil of sheet metal of thickness 1 millimeter, rolled on a cylinder with "height" (between bases) of 10 centimeters, mounted on a horizontal spindle. Assuming that the coil begins with 2-meter diameter and is unrolled at 1.5 meters/second, find an equation relating the speed with which the coil spins as it unwinds as a function of the length that has been removed.

Solution: We are looking for a formula for $d\theta/dt$ in terms of the length s of metal that has been pulled from the coil. There are many approaches to this problem, though each approach is likely to consider the same relations among the variables.

One problem with this situation is that in the ideal case, there is no "nice" formula for the relation between θ and s. If we consider the problem of how the metal was coiled onto the cylinder, we realize that while the inner-most winding was mounted as a near-perfect circular cylinder, the next winding starts on the "bump" where the first coil ends, and the metal runs into itself. Theoretically, the metal must bend around this bump, requiring a small extra amount of metal, but experience indicates that the "bump" will be cushioned and diffused by the slightly bending metal until each layer is rolled on what is very nearly a perfect cylinder. We will ignore this small amount of extra metal and assume that the relationship between θ and s is ultimately very nearly quadratic. An initial rationale is that the change in length should be proportional to the radius ($s = r\theta$ when r is approximately constant), but r is increasing with θ as well, making s increase with $\theta \cdot \theta$. While not a precise argument, we will see that this is reasonable as we look at the first few cases of $\theta = 2n\pi$.

For the first layer, $s = 10$ cm $\cdot \theta$, $0 \leq \theta < 2\pi$. After that first layer, the new radius is 10.1 cm, and so we can take $s = 10$ cm $\cdot \theta + 10.1$cm $\cdot (\theta - 2\pi)$, $2\pi \leq \theta < 4\pi$. This continues, and so we see a pattern (in cm):

$$0 \leq \theta < 2\pi \implies s = 10\theta,$$
$$2\pi \leq \theta < 4\pi \implies s = 10 \cdot 2\pi + 10.2(\theta - 2\pi),$$
$$4\pi \leq \theta < 6\pi \implies s = 10 \cdot 2\pi + 10.2 \cdot 2\pi + 10.4(\theta - 4\pi),$$
$$6\pi \leq \theta < 8\pi \implies s = 10 \cdot 2\pi + 10.2 \cdot 2\pi + 10.4 \cdot 2\pi + 10.6(\theta - 6\pi).$$

As mentioned before, this is already not exact, since it does not take into account the small length of metal needed to bend over the "bump" that should occur at each $\theta = 2n\pi$. However, if we consider some data points (θ, s), we see these can contain $(0,0)$, $(2\pi, 20\pi)$, $(4\pi, 40.4\pi)$, $(6\pi, 61.2\pi)$.

We now attempt to connect these points with a quadratic function

$$s = a\theta^2 + b\theta + c,$$

and use these data to find the coefficients a, b, c, and further test them on subsequent data. Using the point $(0,0)$ we see that $0 = c$. Next we use $(2\pi, 20\pi)$ and $(4\pi, 40.4\pi)$. These give a system

$$\begin{array}{rcl} 4\pi^2 a + 2\pi b &=& 20\pi, \\ 16\pi^2 a + 4\pi b &=& 40.4\pi. \end{array}$$

4.6. CHAIN RULE III: RELATED RATES

From Cramer's Rule, using 2×2 determinants,[40] we have

$$a = \frac{\begin{vmatrix} 20\pi & 2\pi \\ 40.4\pi & 4\pi \end{vmatrix}}{\begin{vmatrix} 4\pi^2 & 2\pi \\ 16\pi^2 & 4\pi \end{vmatrix}} = \frac{80\pi^2 - 80.8\pi^2}{16\pi^3 - 32\pi^3} = \frac{-0.8\pi^2}{-16\pi^3} = \frac{1}{20\pi},$$

$$b = \frac{\begin{vmatrix} 4\pi^2 & 20\pi \\ 16\pi^2 & 40.4\pi \end{vmatrix}}{\begin{vmatrix} 4\pi^2 & 2\pi \\ 16\pi^2 & 4\pi \end{vmatrix}} = \frac{161.6\pi^3 - 320\pi^3}{16\pi^3 - 32\pi^3} = \frac{-158.4\pi^3}{-16\pi^3} = 9.9.$$

This gives us the quadratic model

$$s = \frac{1}{20\pi}\theta^2 + 9.9\theta. \tag{4.32}$$

Note that for $\theta = 6\pi$, we would have $s = \frac{36\pi^2}{20\pi} + 9.9 \cdot 6\pi = 1.8\pi + 59.4\pi = 61.2\pi$ as we required, i.e., $(6\pi, 61.2\pi)$ is one such point in the graph of the function.

Equation (4.32) "smooths out" our model for the length of metal coming off of the roll as θ increases despite the presence of "bumps" where the depth of the metal on the roll jumps when given in units of layers of metal. It also lets us relate the time derivatives:

$$\frac{ds}{dt} = \left(\frac{\theta}{10\pi} + 9.9\right)\frac{d\theta}{dt}. \tag{4.33}$$

From this we get

$$\frac{d\theta}{dt} = \frac{\frac{ds}{dt}}{\frac{1}{10\pi}\theta + 9.9}. \tag{4.34}$$

Note that here s is given in cm, θ in radians, and time t in seconds. Also note that if $\frac{ds}{dt} > 0$ is constant, then $\frac{d\theta}{dt}$ decreases with time, as experience would also suggest.

---

[40]Named for Gabriel Cramer (July 31, 1704–January 4, 1752), mathematician of Geneve (in modern-day Switzerland). Cramer's Rule as written becomes inefficient for larger systems of equations, but is useful when the "coefficients" become either very complicated, or themselves variable parameters.

Exercises

For Exercises 1–19, unless otherwise stated, assume all variables are functions of t and differentiate the formula with respect to t (apply $\frac{d}{dt}$ to both sides), finding an equation relating the variables' time derivatives, perhaps including the (non-differentiated) variables in your final equation. (Here π is always a constant.)

1. $x^3 + y^3 = 10$
2. $3x - 5y = 15$
3. $x = \sqrt{A}$
4. $A = x^2$ (compare to the previous exercise)
5. $A = \pi r^2$
6. $V = \frac{4}{3}\pi r^3$
7. $S = 4\pi r^2$
8. $S = 2\pi r$
9. $S = 6x^2$
10. $x = r\cos\theta$
11. $y = x\tan\theta$
12. $\theta = \sin^{-1}(y/4)$
13. $P = k/V$, k constant
14. $PV = k$, k constant (compare to the previous exercise)
15. $p = 2x + 2y$
16. $\ln V = \ln(\pi/3) + 2\ln r + \ln h$, (compare to the next exercise)
17. $V = \frac{1}{3}\pi r^2 h$ (compare to the previous exercise)
18. $p = mv$
19. $m = \dfrac{m_0}{\sqrt{1 - \frac{v^2}{c^2}}}$, m_0, c constants

The following are applied related-rates exercises, divided by general field.

Geometry

20. Water is filling a spherical balloon at a rate of $1.2\,\mathrm{m}^3/\mathrm{min}$. Recall that $V = \frac{4}{3}\pi r^3$.

    (a) What is the volume of the balloon when the radius is 20 cm?

    (b) How fast is the radius of the balloon changing when the radius is 20 cm?

21. A snowball is melting at a rate of $20\,\mathrm{cm}^3/\mathrm{hr}$. Assuming that it remains spherical and that at one time the radius is decreasing at 5 cm/hr, find the value of the radius at that time.

22. How fast is the radius of a melting snowball changing if the volume is changing at $10\,\mathrm{cm}^3/\mathrm{hr}$ and the radius is 5 cm? Assume that the snowball remains spherical.

23. A hot-air balloon is released 200 m from an observer on the ground, and travels upward (vertically) at a rate of 0.3 m/sec. When the balloon is 300 m above the ground, at what rate is the angle of inclination from the observer to the balloon increasing? (Give your answer in radians/sec and degrees/sec.)

24. The base of an isosceles triangle is 3 m. Find the rate of change of the altitude of the triangle when the altitude is 2 m and the base angles are each increasing at 0.2 rad/sec.

4.6. CHAIN RULE III: RELATED RATES

25. At what rate is the area of an equilateral triangle changing at a moment when each side is 30 cm and decreasing at a rate of 2 cm/sec?

26. Consider a baseball diamond where the distance from one base to the next is 90 ft. (The four bases actually form a square.) How fast is a player running from first base to second base when he is 20 ft from first base if at that moment his rate of change of his distance to home plate (opposite second base) is 5.4 ft/sec?

27. A rectangle is inscribed in a circle of radius 10 inches. The length of the rectangle is increasing at a rate of 0.5 in/min. What is the rate of change of the width of the rectangle when the length is 3 in? (Hint: The diagonal of the rectangle will also be a diameter of the circle.)

28. Continuing from the previous exercise, what is the rate of change of the area of the rectangle at that moment? (Hint: The Product Rule is involved.)

29. A 1.7-m-tall person walks with a speed of 0.4 m/sec away from a lamp post whose bulb is 5 m high. When the person is 2 m from the base of the post, how fast is the tip of the shadow moving along the ground?

30. A ladder 10 ft long is leaning against a wall. Suppse that the bottom of the ladder is sliding away from the wall at 2 ft/sec.

    (a) How quickly is the ladder moving down the wall when the top of the ladder is 8 ft from the ground?

    (b) If h is the height of the ladder against the wall, what is the trend in dh/dt as $h \to 0^+$?

31. Sand is dropped from a conveyor belt onto a conical pile where the height of the pile is always twice the radius. How fast is the volume changing with respect to time if the radius is changing at the rate of 1.2 inches/sec and the height of the pile is 12 inches? Recall $V = \frac{1}{3}\pi r^2 h$.

32. The width of a rectangle is always three times its length. At what rate is its area increasing if the length is 6 cm and the length is increasing at 2 cm/sec?

33. The length of a rectangle is increasing at 5 cm/sec while the width is decreasing at 3 cm/sec. What is the rate of change in the area at a moment when the length is 40 cm and the width is 20 cm?

34. You are flying a kite with a taut string (nearly straight) let out from a spool held 5 ft above the ground, and let out at 3 ft/sec. The kite moves horizontally (neither rising nor falling) 70 ft above the ground level. What is the horizontal speed of the kite at the moment you have let out 150 ft of string?

35. The radius of a cone is increasing at 3 cm/sec but the height is decreasing at 4 cm/sec. When the radius is 10 cm and the height is 18 cm, what is the rate of change of the volume?

36. A cube with faces meeting at only right angles and each edge length the same is decreasing in volume by 2 ft³/min when its volume is 1000 ft³. For that moment in time, answer the following:

    (a) How quickly is each side length decreasing?

    (b) How quickly is each face's area shrinking?

Physics (Astronomy, Electronics, Mechanics, Optics)

37. An object moves in a circular orbit according to the equation $x^2 + y^2 = 25$, where x and y are in feet (and so the actual equation can be written $x^2 + y^2 = 25$ ft^2). When the object is at the point $(-3, -4)$, the horizontal velocity is 1 ft/sec. At that moment, what is the vertical velocity?

38. The distance d_i of an image produced by a double convex lens with focal length 20 cm from the lens producing the image is given by $d_i = \frac{(20\text{ cm})(d_0)}{d_0 - 20\text{ cm}}$, where d_0 is the distance of the object from the lens on the other side of the lens. (See Example 2.6.8, page 155. We can omit the units for the calculus computations, since they are consistent here: $d_i = \frac{20 d_0}{d_0 - 20}$.) Suppose that the distance of the object from the lens is increasing at a rate of 0.5 cm/sec.

    (a) Is the image moving towards or away from the lens, and at what rate, when $d_0 = 30$ cm?

    (b) As $d_0 \to \infty$, what is the trend in d_i?

39. The voltage of a thermocouple is given by the equation
$$V = 2.5T + 0.005T^2,$$
where T is the temperature in °C and V is measured in volts. If the temperature is increasing at the rate of 1.5°C/min, how fast is the voltage changing when $T = 50$°C?

40. The resistance (measured in ohms) of a resistor is given by the equation $R = 24 + 0.004T^2$, where T is the temperature in °C. If the rate of change of the resistance is 2 ohms/min and the temperature is increasing at 5°C/min, what is the temperature at that moment?

41. The resistance R of two resistors R_1 and R_2 connected in parallel is given by the equation
$$\frac{1}{R} = \frac{1}{R_1} + \frac{1}{R_2},$$
where R, R_1, and R_2 are measured in ohms. If R_1 is increasing at 2 ohms/sec and R_2 is decreasing at 1.5 ohms/sec, find the rate of change of R when $R_1 = 30$ ohms and $R_2 = 25$ ohms.

42. The relationship between voltage V measured in volts, current I measured in amps, and resistance R measured in ohms is given by Ohm's Law $V = IR$. Suppose at one moment in time we have $I = 40$ amps, $R = 10$ ohms, the resistance is decreasing at 2 ohms/min and the voltage is rising at 10 volts/min. What is the rate of change of the current I at that moment?

43. The kinetic energy E_k of an object of mass m and velocity v is given by $E_k = \frac{1}{2}mv^2$. If m is in units of kg and v is in m/sec, then E_k is in units of joules J. The change in energy is given in watts, or joules/second (1 W=1 J/sec). Suppose a 50 kg object's acceleration is 4 m/sec^2 when $v = 30$ m/sec. Then what is the power needed to accomplish this at that moment?

44. The force of gravity between two objects of mass m_1 and m_2, with r as the distances between their centers of mass, is given by
$$F = G\frac{m_1 m_2}{r^2},$$
which will be given in newtons if m_1, m_2 are in kilograms and r is in meters, with G (not to be confused with $g \approx 9.8$ m/sec^2) given by
$$G = 6.67408 \times 10^{-11} \text{ m}^3 \text{ kg}^{-1} \text{ sec}^{-2}.$$

4.6. CHAIN RULE III: RELATED RATES

The mass of the Earth is 5.97237×10^{24} kg. The radius of the Earth averages 6.3710×10^6 m, and so this would be the average value of r at the surface of the Earth. Suppose a space shuttle is lifting vertically and eventually reaches a vertical velocity of 5,000 m/sec at 120 km above the earth's surface. What is the rate of change in the force of gravity that a 100 kg astronaut will undergo at that moment?[41]

45. The Earth's center is approximately 93,000,000 miles from the sun's center. Suppose we place the sun at the origin of an inertial (nonaccelerating, including nonrotating) plane of the Earth's revolution around the sun, and define θ to be the angle that the ray from the origin to the Earth's center makes with one axis of this plane (e.g., the postive x-axis), then we know that $d\theta/dt = 1$ revolution/year.

    (a) Convert $d\theta/dt$ into radians/second.
    (b) Using $s = r\theta$, where s is the arc length traveled as the Earth passes from the postive x-axis through an angle θ, find the Earth's speed as it travels around the sun.

46. The mass of a particle moving at velocity v is given by the equation
$$m = \frac{m_0}{\sqrt{1-\frac{v^2}{c^2}}} = m_0\left(1-\frac{v^2}{c^2}\right)^{-1/2},$$
where m_0 is the mass of the particle at rest and $c \approx 3 \times 10^8$ m/sec is the speed of light. Find the rate of change of the mass per time of a 2-gram particle if the velocity if $0.6c$ and the object is accelerating at $0.01c$/sec. (Hint: You will not need the actual value of c.)

Biology

47. The area of a circular-shaped skin infection is growing with a circular pattern at a rate of 5 mm^2/hr. Once the radius is 1 cm, how fast is the edge of the infection moving?

48. A population undergoing exponential growth such that its doubling time is d and its initial population is P_0 will be governed by $P(t) = P_0 2^{t/d}$. An invading bacterium enters its host and begins to multiply at such a rate that its population doubles every half-hour until checked by antibiotics or its population becomes unsustainable. The initial infection contains 150 bacteria cells, and so the population of bacteria can be given by
$$P(t) = 150 \cdot 2^{t/(1/2)} = 150 \cdot 2^{2t},$$
where t is in hours. How long after the initial infection is the bacteria reproducing at one million per hour?

49. Perhaps a more sophisticated model of population size is *logistic growth,* which takes into account that a population can grow large enough that its individual organisms compete with each other for limited resources. The usual model is the following, for $t \geq 0$:
$$P(t) = \frac{KP_0 e^{rt}}{K + P_0\left(e^{rt}-1\right)}. \qquad (4.35)$$

Here P_0 is the initial population, $r > 0$ scales the rate with time, K is the "carrying capacity" of the environment, t is time, and $P(t)$ is the population at time t. (Only P and t are variables.)

---

[41] In practice, the total velocity is in fact 8,000 m/sec, and there are forces from the vehicle's acceleration that will dominate the astronaut's perception of net force compared to the change in the force of gravity.

(a) Show that $P(0) = P_0$.

(b) Show that $\lim_{t \to \infty} P(t) = K$, which is the maximum population size the environment can sustain indefinitely (hence the term *carrying capacity*).

(c) Compute $\frac{dP}{dt}$.

(d) Show that $\lim_{t \to \infty} \left[\frac{dP}{dt} \right] = 0$.

50. Suppose we have a population undergoing logistic growth (4.35), with an initial population in thousands of $P_0 = 1$, a carrying capacity in thousands of $K = 5$, and a rate $r = 0.2$. A graph of this population would then be given by the following:

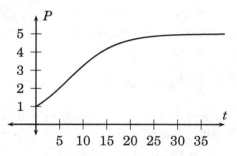

(a) Write the function $P(t)$ for this case (see (4.35), page 503).

(b) Find $\frac{dP}{dt}$ for $t = 5$.

(c) Repeat for $t = 15$.

(d) Repeat for $t = 25$.

(e) Find $\lim_{t \to 0^+} \left[\frac{dP}{dt} \right]$.

51. The wind chill temperature W (in degrees Fahrenheit) is related to the air temperature T and the wind velocity v (measured in miles/hour) by the formula

$$W = 35.74 + 0.6215T - 35.75v^{0.16} + 0.4275Tv^{0.16}.$$

Find the rate of change of the wind chill temperature if the air temperature remains constant at 50°F and the wind speed is 10 mi/hour and increasing at 2 mi/hour per hour.

52. A group of aquatic biologists is studying the effects of the number N of fish introduced into a recently dug, man-made freshwater lake, in particular studying the total mass growth T and the growth G in the average mass growth per fish. Note that $T = GN$. For numerical convenience, they use units of grams per week for the growth rates T and G, and tens of fish for N. In sampling the fish in the lake they note that a useful model is $G = 500 - 50N$, for $N > 0$.

(a) What does this this say about the growth rate of fish when $N = 5$ versus $N = 10$ or $N = 12$? What might be a biological explanation for this trend?

(b) Write T as a function of N alone.

(c) Compute $\frac{dT}{dt}$ if $N = 4$ and $\frac{dN}{dt} = 3$.

(d) Repeat for $N = 12$ and $\frac{dN}{dt} = 4$.

(e) Repeat for $N = 12$ and $\frac{dN}{dt} = -2$.

Multipart Biology

Exercises 53–58 concern predator-prey dynamics, in which we get relationships between the number x of the population of some type of prey, and the number y of the population of its predator, in which the growth of the number of prey can cause an increase in the number of predators, until the predator population is large enough to cause a decrease in the population of prey, until that trend causes a decrease in the population of predators again (through starvation), and so on. The Lotka-Volterra theory of this phenomenon includes models given in the literature by the form

$$\delta x - \gamma \ln x + \beta y - \alpha \ln y = c, \quad (4.36)$$

where $\alpha, \beta, \gamma, \delta > 0$, and c is some constant. Suppose in one such system we have $\alpha = \frac{2}{3}$,

$\beta = \frac{4}{3}$, and $\gamma = \delta = 1$. For simplicity, we will multiply (4.36) with these parameters by 3:

$$3x - 3\ln x + 4y - 2\ln y = C. \qquad (4.37)$$

Assume for some ecosystem that $(1,1)$ is on the predator-versus-prey graph of (4.37). This helps us to determine C, and gives us the following approximate graph, which is traversed counter-clockwise with increasing time t:

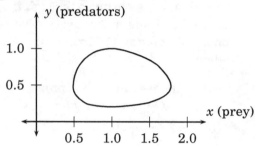

Answer, in turn, the following questions regarding this system.

53. Find the form of (4.36) for this predator-prey system (4.37). We will assume x, y are in units of thousands of organisms, and t is in units of weeks. (Hint: Use the given fact that $(1,1)$ is on the graph to find C.)

54. Differentiate the equation from Exercise 53 with respect to t in order to find an equation relating the rates $\frac{dx}{dt}$ and $\frac{dy}{dt}$. (It will be easier to do so after first multiplying both sides of the equation by 3.)

55. One approximate point on the graph is $(1.5, 0.2649)$. If $\frac{dy}{dt} = 0.05$ at that moment, what is $\frac{dx}{dt}$? What might be the biological explanation? (It may be interesting to consider which is the cause, and which is the effect.)

56. Another such point on the graph is $(1.5, 0.8447)$. Suppose at that point we have $\frac{dy}{dt} = 0.05$. Again, what is $\frac{dx}{dt}$? What might be the biological explanation?

57. Yet another approximate point on the graph is $(0.558, 0.75)$. If $\frac{dx}{dt} = -0.05$, what is $\frac{dy}{dt}$ and a possible biological explanation?

58. A final point to consider is $(0.504, 0.4)$. If $\frac{dx}{dt} = .05$, then what is $\frac{dy}{dt}$ and how could we explain biologically the relationship between dx/dt and dy/dt at that point?

Chemistry

59. The adiabatic law (no heat entering a system) for the expansion of air is $PV^{1.4} = k$, where P is pressure and V is the volume of the confined air. Suppose at some time the pressure is 50 lb/in^2 and the volume is 105 in^3.

    (a) Find k, including units.

    (b) At what rate is the volume changing at a time when pressure is 50 lb/in^2 and decreasing at 4 lb/in^2/sec? Is the volume increasing or decreasing?

60. Boyle's Law states that the pressure of a gas varies inversely as the volume of the gas if the temperature remains constant. If a gas occupies a volume of 1.4 m^3 at a pressure of 150 kPa (kilopascals) and the volume is increasing at a rate of 0.004 m^3/sec, how fast is the pressure changing when the volume is 1.1 m^3? (Hint: $P = \frac{k}{V}$ for some constant k, i.e., $PV = k$.)

61. The pH of a solution is given by pH $= -\log[H^+]$, where $[H^+]$ is the concentration of the hydrogen ion in mol/liter. Suppose at one particular instant we have $\frac{d(\text{pH})}{dt} = -0.2$/min, and $[H^+] = 5$ mol/L. Find the rate of change in $[H^+]$ at that instant.

Multipart Physics

For Exercises 62–72, consider the diagram that follows, representing an exercise machine with which an athlete lifts the left end of a pivoting bar of length 1 m through an arc of directed length s. A weight of mass $m = 100$ kg slides up and down a vertical track (dashed line) 0.5 m from the bar's pivot, and slides along the bar as it is lifted and lowered. The height of the weight is h. Note that from the illustration, clearly $\theta \in [0, \pi/2]$ (and in fact θ is more restricted than that, as will emerge in the exercises). We will ignore the mass of the bar and any friction in the apparatus.

Let E_p be the potential energy of the mass m, and E_k be the kinetic energy of the mass m. We will take the total energy to be $E = E_p + E_k$. The exercises that follow should be answered in turn.

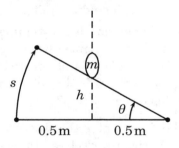

62. Find θ in radians as a function of s.

63. Find h as a function of s.

64. What is the maximum practical value of s in this diagram? (That is, at what value of s is the weight on the non-pivot end of the bar?)

65. Suppose the athlete pushes the bar through an arc of length 0.6 m. How high was the weight lifted?

66. Referring to the previous problem, how much work was required? (Hint: To find this, find the increase in potential energy E_P of the weight. See the next exercise.)

67. The work required to lift the weight a height h will be $E = E_P = hw$, where E is the work and w is the weight. Here $w = 100$ kg $\times$ 9.8 m/sec^2 = 980 N. Find E_p as a function of s.

68. Assuming the athlete can lift the weight in such a way as to move through the arc at a rate of one meter every three seconds, i.e., $ds/dt = (1/3)$ m/sec, find dh/dt when

    (a) $s = 0$ (assume the formula for ds/dt and dh/dt are valid for $s = 0$);

    (b) $s = 0.5$ m;

    (c) $s = 1$ m;

    (d) as $s \to \left(\frac{\pi}{2} \text{ m}\right)^-$, if we assume we can extend the bar as far as we like from the pivot, but the athlete still pushes the bar through the arc of radius 1 m.

69. For the values of s in the previous exercise, find the kinetic energy E_k of the weight. Note that $E_k = \frac{1}{2}mv^2$, where in this case $v = dh/dt$. (You can use v from the previous problem for each value of s.)

70. Assuming still that $ds/dt = (1/3)$ m/sec, find an expression for kinetic energy E_k of the weight as a function of s.

71. Find the total energy $E = E_p + E_k$ for the values of s as in the previous exercise.

72. Find an expression for the power the athlete must provide as a function of s for the previous exercise. Note that power is given by $P = dE/dt = dE/ds \cdot ds/dt$. Evaluate at $s = 0, 0.5$ m, 1 m as before.

Chapter 5

Basic Integration and Applications

In this chapter we consider the problem of recovering a function from knowledge of its derivative; or, equivalently, for a given function we will try to find another function whose derivative is that given function. The general process is sometimes called *antidifferentiation*, for obvious reasons. It is more commonly called *integration*, for reasons we will see later, especially starting in Section 5.3.

The main purpose of this chapter is to develop the first basic techniques for computing antiderivatives $F(x)$ for a given function $f(x)$, i.e., given $f(x)$ we look for $F(x)$ so that

$$F'(x) = f(x). \tag{5.1}$$

As we will see early in this chapter and throughout the next, *antidifferentiation* (finding some such $F(x)$), or *integration*, is often less straightforward computationally than is differentiation (finding $f'(x)$). However, there are easily as many applications of antidifferentiation as there are of differentiation so it is a worthwhile process. In the first section we will limit ourselves to two applications:

1. Given the slope $f'(x)$, find the function $f(x)$ by antidifferentiation. Moreover, given $f''(x)$, find $f'(x)$, and then $f(x)$.[1]

2. Given velocity v, find position s. Moreover, given acceleration a, find v and then s. (Recall $v = s'$ and $a = v' = s''$.)

Later in the chapter we will look at the geometric significance of antiderivatives $F(x)$ of a function $f(x)$. Just as the geometric meaning of slope gave us a useful perspective for arriving at derivative theorems (Mean Value Theorem, First Derivative Test, etc.), so too will the antiderivatives benefit from geometric analysis. To perform that analysis will require us to consider another major theoretical device, namely *Riemann sums*, which—together with the Fundamental Theorem of Calculus—will open the topic of integration to innumerable applications. To illustrate the reasonableness of the Fundamental Theorem of Calculus, we will again look closely at the velocity-position connection as well.

In the sequel we will develop more advanced techniques of antidifferentiation.

---

[1] In this application, the part of $f(x)$ in (5.1) is played by $f'(x)$, while the part of $F(x)$ is played by $f(x)$.

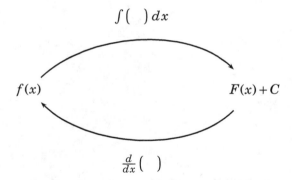

Figure 5.1: Inverse-style relationship between actions of the differential operator $\frac{d}{dx}(\ )$ and the integral operator $\int (\ )\,dx$, where $F'(x) = f(x)$, i.e., $F(x)$ is an *antiderivative* of $f(x)$.

For now, we will concentrate on the actual computation of antiderivatives of the more basic types. In the first section we will limit ourselves to those that arise from our known derivative formulas. In subsequent sections we will then explore the substitution technique, which is the antidifferentiation analog to the Chain Rule in differentiation, and is thus arguably the most important of the integration techniques. It will be developed at length.

5.1 Indefinite Integrals (Antiderivatives): Introduction

In this section we introduce *antiderivatives*, which are exactly what the name implies. These are also called *indefinite integrals* for reasons which will become clear in a later section.

5.1.1 Indefinite Integrals and Constants of Integration

Definition 5.1.1 *First consider a function $f(x)$ which is defined on an open interval (a,b). Another function $F(x)$, also defined on (a,b) is called an* **antiderivative** *of $f(x)$ on the same interval if and only if $F'(x) = f(x)$ on (a,b), i.e., $(\forall x \in (a,b))[F'(x) = f(x)]$.*

If instead $f(x)$ is defined on a closed interval $[a,b]$, we call $F : [a,b] \longrightarrow \mathbb{R}$ an **antiderivative** *of $f(x)$ on $[a,b]$ if and only if*

$$F'(x) = f(x), \qquad \text{for } x \in (a,b),$$
$$\lim_{\Delta x \to 0^+} \frac{F(a+\Delta x) - F(a)}{\Delta x} = f(a), \qquad \text{and}$$
$$\lim_{\Delta x \to 0^-} \frac{F(b+\Delta x) - F(b)}{\Delta x} = f(b).$$

In other words, on an open interval we require $F'(x) = f(x)$, while on the closed interval we also require the right derivative of $F(x)$ to be $f(a)$ at $x = a$, and the left derivative of $F(x)$ to be $f(b)$ at $x = b$.

Notice that (by Theorem 3.2.1, page 229) the definition implies that $F(x)$ is continuous on the interval in question (since where $F' = f$ exists, F must be continuous). For a simple

5.1. INDEFINITE INTEGRALS (ANTIDERIVATIVES): INTRODUCTION

example, consider $f(x) = 2x + 3$ on any open (or closed) interval of positive length. An antiderivative of $f(x)$ can be $F(x) = x^2 + 3x$, since then $F'(x) = 2x + 3 = f(x)$. However, another perfectly good antiderivative can be $F(x) = x^2 + 3x + 5$, or $F(x) = x^2 + 3x - 100{,}000$, since the derivative of the trailing constant term will always be zero. In logical terms we can write

$$F(x) = x^2 + 3x \implies F'(x) = 2x + 3,$$
$$F(x) = x^2 + 3x + 5 \implies F'(x) = 2x + 3,$$
$$F(x) = x^2 + 3x - 100{,}000 \implies F'(x) = 2x + 3, \text{ but}$$
$$F(x) = x^2 + 3x + C, \text{ for some } C \in \mathbb{R} \iff F'(x) = 2x + 3. \tag{5.2}$$

That (5.2) is an equivalence we will prove shortly. To signify that equivalence, we write

$$\int (2x + 3)\, dx = x^2 + 3x + C, \qquad C \in \mathbb{R}. \tag{5.3}$$

We call the right-hand side of (5.3) *the most general antiderivative*, or just *the antiderivative*, of $2x+3$ (with respect to x). It is also called the *indefinite integral* of $2x+3$ (again with respect to x), and we will eventually migrate to using that term as our default.[2] The constant C is called the *constant of integration*, since it must be included to achieve all solutions to the question of what is an antiderivative of $f(x) = 2x + 3$.

It is useful to note that there is an analogy to the operation of differentiation contained in the symbols (see Figure 5.1, page 508):

- $\dfrac{d}{dx}\bigl(\ \ \bigr)$ symbolizes computing the *derivative* of "( )" with respect to x;

- $\displaystyle\int \bigl(\ \ \bigr) dx$ symbolizes computing the *antiderivative* of "( )" with respect to x.

Just as $\frac{d}{dx}$ was considered a "differential operator," $\int (\ )\,dx$ is considered an "integral operator," inputting a function of x and outputting its general antiderivative. When we write (5.3), that is, $\int (2x+3)\,dx = x^2 + 3x + C$, the expression on the left can be broken into:

1. "$\int$," the *integral symbol*, introduced by Leibniz from an old style, German "long s" (so-called for pronunciation, not length) for reasons we will discuss later;

2. "$2x+3$," the *integrand*, whose antiderivatives we seek; and

3. "dx," the differential of x, signifying we are computing the antiderivative respect to x.

The integral symbol $\int$ and the differential dx together form the integral operator $\int (\ )\,dx$. When there is no ambiguity, the parentheses are omitted. Also, it is common to treat the differential dx as a multiplier of the integrand, for reasons that will become more clear after Section 5.3, and so it is common to see notation such as

---

[2]The indefinite integral has a strong connection to the very important *definite integral*, which is a measure of accumulated change as computed from the instantaneous rate of change. This computation is our eventual goal and—to restate the introduction to this chapter—the connection between antiderivatives (indefinite integrals) and accumulated change (definite integrals) is precisely the subject of the Fundamental Theorem of Calculus.

$$\int x\,dx = \frac{1}{2}x^2 + C, \qquad \int dx = x + C, \text{ and} \qquad \int \frac{dx}{x^2} = -\frac{1}{x} + C.$$

These are shorthand for $\int(x)\,dx = \frac{1}{2}x^2 + C$, $\int 1\,dx = x + C$, and $\int\left(\frac{1}{x^2}\right)dx = -\frac{1}{x} + C$.

We now note that, for example, all antiderivatives of $2x + 3$ are necessarily of the form $x^2 + 3x + C$. To prove this, on some interval define $F(x) = x^2 + 3x$, which we can easily see is an antiderivative of $2x + 3$ on that interval (since $F'(x) = 2x + 3$). Next suppose $G(x)$ is another such antiderivative, i.e., that $G'(x) = 2x + 3$, on that same interval. Then on that interval, F and G must differ by a constant (since $(F - G)'$ is the zero function):

$$\frac{d}{dx}[G(x) - F(x)] = G'(x) - F'(x) = (2x + 3) - (2x + 3) = 0 \implies G(x) - F(x) = C,$$

for some $C \in \mathbb{R}$. Thus any antiderivative $G(x)$ must be of the form $G(x) = F(x) + C$, q.e.d.[3]

Conversely, any function of the form $x^2 + 3x + C$ will clearly have derivative $2x + 3$, so the set of *all* antiderivatives of $2x + 3$ is the set of functions of form $x^2 + 3x + C$, where $C \in \mathbb{R}$ is arbitrary.

To be clear on the notation, we now insert the following definition.

Definition 5.1.2 *If $F(x)$ is an antiderivative of $f(x)$, with respect to x, on the interval I, then on that interval we write*

$$\int f(x)\,dx = F(x) + C, \tag{5.4}$$

where C is an arbitrary constant of integration. The process of computing an antiderivative is called **integration** *(while the process of computing a derivative is called* differentiation*).*

Example 5.1.1

Consider $f(x) = 2\sin x \cos x$. One antiderivative is $F(x) = \sin^2 x$, since

$$F'(x) = \frac{d\sin^2 x}{dx} = \frac{d(\sin x)^2}{dx} = 2\sin x \cdot \frac{d\sin x}{dx} = 2\sin x \cos x.$$

However, another antiderivative is $G(x) = -\cos^2 x$, since

$$G'(x) = \frac{d(-\cos^2 x)}{dx} = -\frac{d(\cos x)^2}{dx} = -\left[2\cos x \cdot \frac{d\cos x}{dx}\right] = -[2\cos x(-\sin x)] = 2\sin x \cos x.$$

Note that

$$F(x) - G(x) = \sin^2 x - \left(-\cos^2 x\right) = \sin^2 x + \cos^2 x = 1,$$

so we see that F and G do actually differ by a constant. To report the most general antiderivative of $f(x) = 2\sin x \cos x$, either of the following are valid (but understood to have different "C's"):

$$\int 2\sin x \cos x\,dx = \sin^2 x + C, \qquad \text{and} \qquad \int 2\sin x \cos x\,dx = -\cos^2 x + C.$$

---

[3]Recall that a function with the zero function for its derivative on an interval I must be constant on that interval, i.e., $h' = 0$ in (a, b) implies $h(x)$ is constant in (a, b). Applying this to $h = F - G$ we can argue

$$(\forall x \in I)[(F(x) - G(x))' = 0] \implies (\exists C \in \mathbb{R})(\forall x \in I)[F(x) - G(x) = C].$$

5.1. INDEFINITE INTEGRALS (ANTIDERIVATIVES): INTRODUCTION

Especially when dealing with trigonometric functions—with all their interconnections through various identities—it is common to find very different-looking forms of the general antiderivative, all of which differ by constants from each other. It is occasionally important to be alert for apparent discrepancies, which are explained by this nature of the general antiderivative.

Example 5.1.2

Suppose $f(x) = x+1$. Then both forms that follow are general antiderivatives:

$$\int (x+1)\,dx = \frac{1}{2}x^2 + x + C,$$

$$\int (x+1)\,dx = \frac{1}{2}(x+1)^2 + C.$$

We can see this by taking derivatives of each. We can also see this if we label $F(x) = \frac{1}{2}x^2 + x$ and $G(x) = \frac{1}{2}(x+1)^2$, so the antiderivatives given are just $F(x)$ and $G(x)$ plus constants, respectively, and then compute

$$F(x) - G(x) = \left[\frac{1}{2}x^2 + x\right] - \left[\frac{1}{2}\left(x^2 + 2x + 1\right)\right] = -\frac{1}{2},$$

so these do differ by a constant, as expected.

5.1.2 Power Rule for Integrals

Where the rules for computing derivatives were straightforward (which is not to say immediately "easy"), those for computing antiderivatives are not so algorithmic. Indeed, the methods are varied. Nonetheless, they are necessary to learn for a reasonably complete understanding of standard calculus, and we begin with the *Power Rule for integrals*:

$$\int x^n \, dx = \frac{1}{n+1} x^{n+1} + C, \qquad n \neq -1, \tag{5.5}$$

$$\int \frac{1}{x} \, dx = \ln|x| + C. \tag{5.6}$$

Here the intervals in question are those upon which x^n is defined. To check (5.6), we simply notice $\frac{d}{dx} \ln|x| = \frac{1}{x}$. For (5.5) we compute:

$$\frac{d}{dx}\left[\frac{1}{n+1} x^{n+1}\right] = \frac{1}{n+1} \cdot (n+1) x^{[(n+1)-1]} = x^n, \qquad \text{q.e.d.}$$

In performing this computation, we must notice that $n+1$ and $1/(n+1)$ are multiplicative constants, and are thus preserved throughout the computation (and do not, for instance, require Product/Quotient/Chain Rules since they do not vary). We also note that the right-hand side of (5.5) is meaningless for the case $n = -1$, so we need (5.6) for that case.

When checking any antiderivative by differentiation, it is customary not to include the arbitrary additive constant, since its derivative is zero. However, it is certainly correct to include it, as in $\frac{d}{dx}[\ln|x| + C] = \frac{1}{x} + 0 = \frac{1}{x}$.

We can apply the Power Rule immediately as in the following:

$$\int x^2\,dx = \frac{1}{3}x^3 + C,$$

$$\int x^3\,dx = \frac{1}{4}x^4 + C,$$

$$\int \frac{1}{x^2}\,dx = \int x^{-2}\,dx = \frac{1}{-1}x^{-1} + C = -x^{-1} + C = \frac{-1}{x} + C,$$

$$\int \frac{x}{x^2}\,dx = \int \frac{1}{x}\,dx = \ln|x| + C.$$

We can check all of these by taking derivatives of our answers, with respect to x (i.e., by applying d/dx).[4] As with derivatives, many functions that are powers of the variable are not explicitly written as such. Furthermore, as with derivatives, the variable name in antiderivative formulas does not matter as long as it is matched in the differential:

$$\int t^9\,dt = \frac{1}{10}t^{10} + C,$$

$$\int \sqrt{u}\,du = \int u^{1/2}\,du = \frac{1}{\frac{1}{2}+1}u^{[\frac{1}{2}+1]} + C = \frac{2}{3}u^{3/2} + C,$$

$$\int \frac{1}{z\sqrt[4]{z}}\,dz = \int z^{-5/4}\,dz = \frac{1}{\frac{-5}{4}+1}z^{[\frac{-5}{4}+1]} + C = \frac{z^{-1/4}}{-\frac{1}{4}} + C = \frac{-4}{\sqrt[4]{z}} + C.$$

To check these, we can apply respectively d/dt, d/du, and d/dz. Before we go on, we note that (5.5), page 511, interpreted formally, applies to the case $n = 0$ as well: $\int x^0\,dx = \frac{1}{0+1}x^{0+1} + C$, i.e.,

$$\int 1\,dx = x + C. \tag{5.7}$$

Taking the derivative of the right-hand side of (5.7) quickly shows it is in fact true. This integral is often, perhaps at first confusingly, abbreviated without the factor "1" included:

$$\int dx = x + C. \tag{5.8}$$

As with the Chain Rule computations, the differential dx is formally treated as a factor as in the following:

$$\int \frac{dx}{x^3} = \int x^{-3}\,dx = \frac{1}{-2}x^{-2} + C = \frac{-1}{2x^2} + C.$$

Later we will see this treatment of dx justified and exploited in several contexts.

---

[4]It is a very useful exercise to check these antiderivatives by quick mental derivative calculations. Since the derivative formulas have been used extensively to this point, the processes of computing antiderivatives can be well-informed by their connections to known derivative techniques. In particular, mistakes in computing antiderivatives can often be immediately detected and corrected. Perhaps even more importantly, the approximate form of an antiderivative can often be anticipated. For instance, we know that the derivative of a fourth-degree polynomial is necessarily a third-degree polynomial. It should be clear (though some argument is necessary to prove) that the antiderivative of a third-degree polynomial is necessarily a fourth-degree polynomial. Anticipating the final form is also very useful in substitution problems, which are introduced in Section 5.8 and ubiquitous thereafter.

5.1. INDEFINITE INTEGRALS (ANTIDERIVATIVES): INTRODUCTION

Now we state two very general results, which may seem obvious but are worth exploring with some care because of a technical consideration regarding the constant of integration.[5] These are also useful to us immediately because they allow us to use the Power Rule multiple times to compute the derivatives of polynomials.

Theorem 5.1.1 *Suppose that $F'(x) = f(x)$ and $G'(x) = g(x)$ on some interval in consideration, and that $k \in \mathbb{R}$ is a fixed constant. Then on that same interval we have*

$$\int [f(x) + g(x)]\, dx = F(x) + G(x) + C, \tag{5.9}$$

$$\int [k \cdot f(x)]\, dx = kF(x) + C. \tag{5.10}$$

where $C \in \mathbb{R}$ is a constant of integration.

These can be proved by taking derivatives. For instance, $\frac{d}{dx}[kF(x)] = k \cdot \frac{d}{dx} F(x) = kF'(x) = kf(x)$. Note that this theorem can be rewritten

$$\int [f(x) + g(x)]\, dx = \int f(x)\, dx + \int g(x)\, dx, \tag{5.11}$$

$$\int kf(x)\, dx = k \int f(x)\, dx, \quad k \ne 0, \tag{5.12}$$

$$\int 0\, dx = C. \tag{5.13}$$

A moment's reflection shows that we do need both (5.12) and (5.13) to cover both cases summarized in (5.10), else we lose the constant of integration if we let $k = 0$ in (5.12). (Note that, indeed, $\frac{d}{dx} C = 0$, which verifies (5.13).) More importantly, these new forms (5.11)–(5.13) are not inconsistent with those of Theorem 5.1.1 when we consider the arbitrary constants. For instance, if we assume $F'(x) = f(x)$ and $G'(x) = g(x)$ as in the theorem, then we can write (5.11) as follows:

$$\int f(x)\, dx + \int g(x)\, dx = (F(x) + C_1) + (G(x) + C_2)$$
$$= F(x) + G(x) + \underbrace{(C_1 + C_2)}_{\text{``}C\text{''}} = F(x) + G(x) + C,$$

where C_1 and C_2 are arbitrary constants, and their sum will also be an arbitrary constant which we can name C.

With this theorem and the Power Rule, we can now compute the indefinite integrals of polynomials and other linear combinations of powers (that is, sums of constant multiples of powers of x).

Example 5.1.3

Consider the following integrals:

---

[5] Many calculus students become careless about this constant of integration and just "tack it on the end" when computing antiderivatives. However, its correct placement is crucial in several contexts, so it is useful to be vigilant from the beginning of integration study. Carelessness in its placement can cause trouble already in this section, but will be particularly troublesome in subsequent third-semester calculus and in differential equations studies.

- $\int (4x^2 - 9x + 7)\, dx = 4 \cdot \dfrac{x^3}{3} - 9 \dfrac{x^2}{2} + 7 \cdot x + C = \dfrac{4}{3}x^3 - \dfrac{9}{2}x^2 + 7x + C,$

- $\int \left(\dfrac{3x-9}{x^3}\right) dx = \int (3x^{-2} - 9x^{-3})\, dx = 3 \cdot \dfrac{x^{-1}}{-1} - 9 \cdot \dfrac{x^{-2}}{-2} + C = -\dfrac{3}{x} + \dfrac{9}{2x^2} + C,$

- $\int (5x^2)^3\, dx = \int 125 x^6\, dx = 125 \cdot \dfrac{x^7}{7} + C = \dfrac{125}{7} x^7 + C,$

- $\int (x^2 + 7)^2\, dx = \int (x^4 + 14x^2 + 49)\, dx = \dfrac{x^5}{5} + 14 \cdot \dfrac{x^3}{3} + 49x + C,$

- $\int \sqrt{2x}\, dx = \int \sqrt{2}\sqrt{x}\, dx = \int 2^{1/2} x^{1/2}\, dx = 2^{1/2} \cdot \dfrac{x^{3/2}}{3/2} + C = \dfrac{2\sqrt{2}}{3} x^{3/2} + C.$

When we use an integration rule such as the Power Rule for integrals, as with derivatives it is important that the variable with respect to which the antiderivative is computed matches the variable in the function. For instance,

$$\int x^3\, dx = \frac{1}{4} x^4 + C, \qquad \int w^3\, dw = \frac{1}{4} w^4 + C, \qquad \text{but} \qquad \int (5x-11)^3\, dx \neq \frac{1}{4}(5x-11)^4 + C.$$

What "goes wrong" in the last integral is the antidifferentiation analog of what goes wrong to produce the inequality that follows (which is that we need the Chain Rule to make the variables of differentiation match):

$$\frac{d}{dx}\left[(5x-11)^4\right] \neq 4(5x-11)^3;$$

$$\frac{d}{dx}\left[(5x-11)^4\right] = 4(5x-11)^3 \cdot \frac{d(5x-11)}{dx} = 4(5x-11)^3 \cdot 5 \neq 4(5x-11)^3,$$

The problem is that the differential, dx, is that of x and not $(5x-11)$. In the next section, we will address a kind of integral version of the Chain Rule (commonly known as *integration by substitution* for reasons that will be clear later), which would make reasonably short work of this integral $\int (5x-11)^3\, dx = \frac{1}{20}(5x-11)^4 + C$. Without that later substitution method, we may need to expand $(5x-11)^3$ as a polynomial, or even guess the form of the solution and check that it works, and possibly make adjustments. Either way, it should suffice to point out that we need to take care in using integral formulas such as the Integral Power Rule, page 511.[6]

---

[6]Without fully developing the substitution technique here, we will note that we can rewrite the integral $\int (5x-11)^3\, dx$ so the differential matches the term $(5x-11)$. The argument would be the analog of our early Chain Rule expansions, such as found in Example 3.4.1, page 274. The idea is that $d(5x-11) = 5\, dx$ (recall the meaning of $d(5x-11)/dx$), and so $dx = \dfrac{d(5x-11)}{5}$, which allows us to rewrite the integral as follows:

$$\int (5x-11)^3\, dx = \int (5x-11)^3\, \frac{d(5x-11)}{5} = \frac{1}{5} \cdot \frac{1}{4} \cdot (5x-11)^4 + C = \frac{1}{20}(5x-11)^4 + C.$$

Except for the factor of $\frac{1}{5}$ in the second interval, that rewriting had an integral Power Rule form.

Our integration by substitution method will be more systematic than this computation. Consequently, it will read better and be less error-prone. That method will be called on extensively from then on.

5.1. INDEFINITE INTEGRALS (ANTIDERIVATIVES): INTRODUCTION

5.1.3 Finding C (Where Possible)

Many times we are interested in a particular antiderivative. This is then a question of finding the particular "C" we need. Recall that all antiderivatives of a function $f(x)$ (on a particular interval on which $f(x)$ is continuous) differ by a constant, so here we use some other information (where available) to "fix," i.e., determine, the constant.

Example 5.1.4

Find $f(x)$ so that $f'(x) = 2x$ and $f(3) = 7$.

Solution: For a problem such as this, it is common to write

$$f(x) = \int f'(x)\,dx,$$

where it is understood that we will eventually find the exact antiderivative so that the function is well-defined. For our particular problem, one might continue to write

$$f(x) = \int 2x\,dx = 2 \cdot \frac{x^2}{2} + C = x^2 + C.$$

Now we find the particular C for this case, and we do this by inputting the "datum" (sometimes called "data point") $f(3) = 7$. Graphically this means that the point $(3, 7)$ is on the curve. Since $f(x) = x^2 + C$, we can find C using this datum:

$$\begin{aligned} f(3) = 7 &\iff & 3^2 + C &= 7 \\ &\iff & 9 + C &= 7 \\ &\iff & C &= -2. \end{aligned}$$

Thus $f(x) = x^2 - 2$.

A graphical way of interpreting this example is to realize that all the curves $y = x^2 + C$ are parabolas, and in fact are just vertical shifts of the curve $y = x^2$. Our task in Example 5.1.4 was then to find which shift satisfies both $f'(x) = 2x$ and $f(3) = 7$. In Figure 5.2, page 516, the functions $y = x^2 + C$ are graphed for various values of C. Note that none of the graphs intersect each other. Once we require the graph to pass through a particular point—in this case the point $(3, 7)$—we "pin down" the one curve passing through that point, meaning that we determine exactly one curve from the family of curves that satisfy $f'(x) = 2x$, as graphed in Figure 5.2, and also satisfying the condition that $f(3) = 7$.

Finding a particular antiderivative is also very useful in kinematics. For instance, if we know the velocity function $s'(t) = v(t)$, we can find the position function s if we are also given one position datum to "fix" the constant. With the understanding that the constant is to be determined, it is often written:

$$s(t) = \int s'(t)\,dt = \int v(t)\,dt. \tag{5.14}$$

A common datum to prescribe is that $s(0) = s_0$ (where s_0 is some fixed number), but any data that "pins down" the function will suffice.

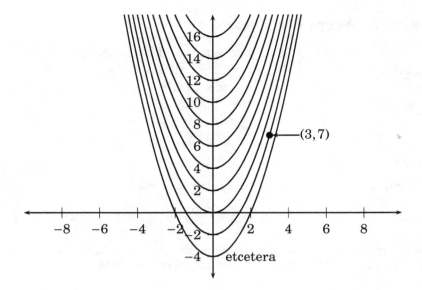

Figure 5.2: Partial view of the family of curves $y = x^2 + C$ satisfying $dy/dx = 2x$. For Example 5.1.4, page 515, we needed to find the value of C satisfying $f'(x) = 2x$, i.e., $f(x) = x^2 + C$, so that $(3,7)$ was on the curve. Note that the slopes of all curves given above are the same for a given x-value, but only one passes through $(3,7)$, namely $f(x) = x^2 - 2$.

Example 5.1.5

Suppose $v = t^2 + 11t - 25$, and $s(1) = 4$. Find $s(t)$.

Solution:

$$s(t) = \int v(t)\,dt = \int \left(t^2 + 11t - 25\right) dt = \frac{t^3}{3} + \frac{11t^2}{2} - 25t + C.$$

Using $s(1) = 4$ we get

$$\frac{1^3}{3} + \frac{11 \cdot 1^2}{2} - 25(1) + C = 4 \iff \frac{1}{3} + \frac{11}{2} - 25 + C = 4,$$

and so $C = 4 + 25 - \frac{11}{2} - \frac{1}{3} = 29 - \frac{35}{6} = \frac{174}{6} - \frac{35}{6} = \frac{139}{6}$. Finally, this gives us

$$s(t) = \frac{t^3}{3} + \frac{11t^2}{2} - 25t + \frac{139}{6}.$$

Now we will derive a well-known formula of physics, namely one-dimensional motion assuming constant acceleration.

Example 5.1.6

Suppose that acceleration is given by a constant, say $s''(t) = a$ (where a is fixed, i.e., $a(t) = a$ is constant). Suppose further that $s(0) = s_0$ and $v(0) = v_0$. Now we work "backwards" from the acceleration

5.1. INDEFINITE INTEGRALS (ANTIDERIVATIVES): INTRODUCTION

towards the position function (via the velocity function) as follows:

$$v(t) = s'(t) = \int s''(t)\,dt = \int a\,dt = at + C_1.$$

Note that the last computation required that acceleration, a, be constant. So far, we have $v(t) = at + C_1$ for some $C_1 \in \mathbb{R}$. Using the datum $v(0) = v_0$, we then have

$$a \cdot 0 + C_1 = v_0 \iff C_1 = v_0.$$

This gives us the following equation—well known to physics students—for the constant acceleration case:

$$v(t) = at + v_0. \tag{5.15}$$

Now we integrate (5.15), again taking care to treat constants a, v_0 and variable t correctly:

$$s(t) = \int s'(t)\,dt = \underbrace{\int v(t)\,dt = \int (at + v_0)\,dt}_{(5.15)} = a \cdot \frac{t^2}{2} + v_0 t + C_2.$$

Finally, using $s(0) = s_0$, we get

$$a \cdot \frac{0^2}{2} + v_0(0) + C_2 = s_0 \implies C_2 = s_0.$$

Thus

$$s = \frac{1}{2}at^2 + v_0 t + s_0. \tag{5.16}$$

It is important to note that (5.16) followed under the special condition that acceleration is constant (such as occurs when an object is in free-fall in a constant gravitational field, with no other resistance). Nonconstant acceleration will not give (5.15) or (5.16). However, the method for computing v and s, given a, is the same when a is not constant, and is given next.

1. Find $v(t) = \int a(t)\,dt$, using one datum regarding velocity at a particular time, to fix the constant of integration.

2. Find $s(t) = \int v(t)\,dt$, using another datum regarding position at a particular time, to fix the second constant of integration.

Actually, two position data can fix the constants as well, since we can just carry the first constant into the second calculation, and then we will have two equations with two unknowns (the constants of integration), and then solve for both constants.

Example 5.1.7

Suppose $a(t) = 3t^2$, $s(0) = 3$ and $s(1) = 5$. Find $v(t)$ and $s(t)$.

Solution: First we will find $v(t)$, to the extent that we can:

$$v(t) = \int a(t)\,dt = \int 3t^2\,dt = 3 \cdot \frac{t^3}{3} + C_1 = t^3 + C_1.$$

Our data $s(0) = 3$ and $s(1) = 5$ cannot be used (yet) to determine C_1, so next we find the form of $s(t)$ to the extent we can so far:

$$s(t) = \int v(t)\,dt = \int \left(t^3 + C_1\right) dt = \frac{t^4}{4} + C_1 t + C_2.$$

So we know that $s(t) = \frac{1}{4}t^4 + C_1 t + C_2$, for some C_1, C_2. Using the facts that $s(0) = 3$ and $s(1) = 5$, we get the following system of two equations in two unknowns:

$$\begin{cases} \frac{0^4}{4} + C_1(0) + C_2 = 3 \\ \frac{1^4}{4} + C_1(1) + C_2 = 5 \end{cases} \iff \begin{cases} C_2 = 3 \\ C_1 + C_2 = \frac{19}{4} \end{cases}$$

From the second form of the system, we see $C_2 = 3$, and so $C_1 = \frac{19}{4} - C_1 = \frac{19}{4} - 3 = \frac{7}{4}$. Putting all this together we first get

$$s(t) = \frac{t^4}{4} + \frac{7t}{4} + 3,$$

from which we can calculate $v(t) = s'(t)$ (or just read $v(t)$ off of our first integral calculation, inserting $C_1 = \frac{7}{4}$) to get

$$v(t) = t^3 + \frac{7}{4}.$$

5.1.4 First Trigonometric Rules

With every derivative formula for functions comes an analogous antiderivative formula, which is more or less the derivative formula in reverse. Sometimes the reverse is more obvious than other times. For instance, the Power Rule formula for derivatives is sometimes seen algorithmically as "*multiply* by the exponent ('bring the power down') and *decrease* the exponent by one," as in

$$\frac{d}{dx}\left[x^n\right] = n \cdot x^{n-1}.$$

If we are careful to reverse the process, we need to do the inverse steps in reverse order: *increase* the exponent by one and then *divide* by the exponent. That is the essence of (5.5), page 511, for those cases where $n \neq -1$:

$$\int x^n\,dx = \frac{x^{n+1}}{n+1} + C.$$

(Recall for $n = -1$ our formula involves a logarithm.)

By other sophisticated arguments, in the next section we will see a kind of reverse Chain Rule. In the sequel we will also come across what can be loosely called a reverse Product Rule called integration by parts, although there really is no good analog of the Product Rule with integrals *per se*.[7]

---

[7] Integration by parts is really an integration technique that takes advantage of the Product Rule for derivatives—or more precisely a permutation (rearrangement) of the Product Rule for derivatives—but is not itself a Product Rule for integrals; it does not by itself give a formula for $\int f(x)g(x)\,dx$. Instead, it gives a formula

5.1. INDEFINITE INTEGRALS (ANTIDERIVATIVES): INTRODUCTION

The formulas presented in this subsection are immediate consequences of our trigonometric derivative formulas. For instance, we have the following pairs of formulas:

$$\frac{d\sin x}{dx} = \cos x \iff \int \cos x\,dx = \sin x + C,$$

$$\frac{d\cos x}{dx} = -\sin x \iff \int (-\sin x)\,dx = \cos x + C.$$

This second integration formula is more awkward than necessary, since it is more likely we would like an antiderivative for $\sin x$ directly. We could multiply both sides by -1, and rename the new constant C, or just notice that $\frac{d}{dx}(-\cos x) = \sin x$, to come to the formula

$$\int \sin x\,dx = -\cos x + C.$$

As before, we can always check these by taking the derivative of the right-hand side. Recalling our six basic trigonometric derivative formulas, and making adjustments for negative sign placements as before, we have the following pairs of derivative/integral formulas:

$$\frac{d\sin x}{dx} = \cos x \iff \int \cos x\,dx = \sin x + C, \tag{5.17}$$

$$\frac{d\cos x}{dx} = -\sin x \iff \int \sin x\,dx = -\cos x + C, \tag{5.18}$$

$$\frac{d\tan x}{dx} = \sec^2 x \iff \int \sec^2 x\,dx = \tan x + C, \tag{5.19}$$

$$\frac{d\cot x}{dx} = -\csc^2 x \iff \int \csc^2 x\,dx = -\cot x + C, \tag{5.20}$$

$$\frac{d\sec x}{dx} = \sec x \tan x \iff \int \sec x \tan x\,dx = \sec x + C, \tag{5.21}$$

$$\frac{d\csc x}{dx} = -\csc x \cot x \iff \int \csc x \cot x\,dx = -\csc x + C. \tag{5.22}$$

With these and our previous rules, we have some limited ability to compute integrals involving trigonometric functions.

Example 5.1.8

Consider the following integrals:

that can be summarized by

$$\int f(x)g'(x)\,dx = f(x)g(x) - \int g(x)f'(x)\,dx,$$

which follows from integrating—applying $\int (\ )\,dx$ to both sides of—the following rearrangement of the Product Rule,

$$f(x)g'(x) = [f(x)g(x)]' - g(x)f'(x).$$

Because the Product Rule for derivatives is what makes the technique of integration by parts valid, many authors describe it as a kind of analog of the Product Rule, though again, it is not a direct formula for the integral of a product like we had for the derivative of a product.

Still, it is a very useful technique that we will spend some time developing in the sequel, when we have other methods to draw upon for the inevitable intermediate computations.

In the sequel, there will be still several other techniques that are not at all simple reverses of derivative rules, and for which checking by differentiating (computing the derivative of) the answer can be as difficult as, or more difficult than, the integration technique itself.

- $\int \left[x^2 + \sin x - \dfrac{1}{x} \right] dx = \dfrac{x^3}{3} - \cos x - \ln|x| + C$

- $\int \cos w\, dw = \sin w + C$

- $\int \dfrac{\sin x}{\cos^2 x}\, dx = \int \left[\dfrac{1}{\cos x} \cdot \dfrac{\sin x}{\cos x} \right] dx = \int \sec x \tan x\, dx = \sec x + C$

- $\int \tan^2 \theta\, d\theta = \int \left(\sec^2 \theta - 1 \right) d\theta = \tan \theta - \theta + C.$

We are fortunate if a trigonometric integral has a form which is just the derivative of one of the six basic trigonometric functions. Even when that is the case, it often requires some rewriting, as the latter pair of integrals above demonstrate.

5.1.5 Integrals Yielding Inverse Trigonometric Functions

These follow from derivative formulas, though the third requires some eventual explanation:

$$\int \frac{1}{\sqrt{1-x^2}}\, dx = \sin^{-1} x + C, \tag{5.23}$$

$$\int \frac{1}{x^2+1}\, dx = \tan^{-1} x + C, \tag{5.24}$$

$$\int \frac{1}{x\sqrt{x^2-1}}\, dx = \sec^{-1} |x| + C. \tag{5.25}$$

Note how we employ only three of the six arctrigonometric functions in (5.23)–(5.25). In fact these are sufficient. Recall for instance that the arccosine and arcsine have derivatives which differ by the factor -1. For simplicity, it is more commonly written $\int \frac{-1}{\sqrt{1-x^2}}\, dx = -\sin^{-1} x + C$, rather than using the arccosine function as the antiderivative, i.e., $\int \frac{-1}{\sqrt{1-x^2}}\, dx = \cos^{-1} x + C$, though the latter is certainly legitimate. Indeed, since $\sin^{-1} x + \cos^{-1} x = \frac{\pi}{2}$, we see $\cos^{-1} x$ and $-\sin^{-1} x$ differ by a constant. In fact one could rewrite (5.23) as $\int \frac{1}{\sqrt{1-x^2}}\, dx = -\cos^{-1} x + C$. The choice can sometimes depend upon which range of angles we wish the antiderivative function to output, though for this case we can adjust that with the constant C.

Similarly one usually writes $\int \frac{-1}{x^2+1}\, dx = -\tan^{-1} x + C$, though $\cot^{-1} x + C$ (for a "different" C) is also legitimate. It is similar for arcsecant and arccosecant: We usually avoid the arccosecant function as an antiderivative.

But for the arcsecant/arccosecant pair there is another small complication. Note how Equation (5.25) has the absolute value on the antiderivative rather than the x-term of the denominator in the integrand, so it does not appear to be just a restatement of the derivative rule for the arcsecant: $\frac{d}{dx} \sec^{-1} x = \frac{1}{|x|\sqrt{x^2-1}}$. To see that (5.25) is still correct, note that $|x| = x$ if $x > 0$, and $|x| = -x$ if $x < 0$. Taking the derivative of $\sec^{-1} |x|$ for those two cases, as we did in the computation of $\frac{d}{dx} \ln|x|$ (see page 371) we can see that we do get $\frac{1}{x\sqrt{x^2-1}}$ both times. But

5.1. INDEFINITE INTEGRALS (ANTIDERIVATIVES): INTRODUCTION

it should also be noted that, while not often seen, it would be legitimate to have the absolute value inside, rather than outside, the integral, as in $\int \frac{1}{|x|\sqrt{x^2-1}} dx = \sec^{-1} x + C$.[8]

These formulas (5.23)–(5.25) will become much more important in future sections. For now it is important to realize that these particular function forms do have (relatively) simple antiderivatives. At this point in the development, we are not prepared to make full use of these forms, but we need to be aware of them. Sometimes a simple manipulation produces a function containing one of these forms.

Example 5.1.9

Compute $\int \frac{x^2}{x^2+1} dx$.

<u>Solution</u>: With the aid of long division, we can see that[9]

$$\frac{x^2}{x^2+1} = 1 - \frac{1}{x^2+1},$$

and so

$$\int \frac{x^2}{x^2+1} dx = \int \left[1 - \frac{1}{x^2+1}\right] dx = x - \tan^{-1} x + C.$$

5.1.6 Integrals Yielding Exponential Functions

Next in our list of integrals that arise from differentiation formulas are those that yield exponential functions. (Here, as usual, $a \in (0,1) \cup (1,\infty)$.)

$$\int e^x dx = e^x + C, \tag{5.26}$$

$$\int a^x dx = \frac{a^x}{\ln a} + C. \tag{5.27}$$

The first of these, (5.26), is the more obvious. Both can be verified through differentiation, as we often do with these simpler antiderivative computations. Recalling that $\frac{da^x}{dx} = a^x \ln a$, and that $\ln a$ is a constant, we compute:

$$\frac{d}{dx}\left[\frac{a^x}{\ln a}\right] = \frac{1}{\ln a} \cdot \frac{da^x}{dx} = \frac{1}{\ln a} \cdot a^x \ln a = a^x, \quad \text{q.e.d.}$$

---

[8] To further complicate things, we could notice that $\sec^{-1} x = \cos^{-1} \frac{1}{x}$, so it can occur that computational software will output an arccosine function, as in $\int \frac{1}{x\sqrt{x^2-1}} dx = \cos^{-1} \frac{1}{|x|} + C$, and then this can be rewritten (with a "different C") $\int \frac{1}{x\sqrt{x^2-1}} dx = -\sin^{-1} \frac{1}{|x|} + C$. The software might also omit the absolute values, theoretically assuming $x > 0$. Still, the standard written computation would produce the expected form $\sec^{-1} |x| + C$.

[9] A popular alternative technique for a fraction like that in our integrand is to strategically add and subtract a term in the numerator, which produces a term in the numerator identical to (or a multiple of) the denominator, and the extra term, from which we can make two fractions:

$$\frac{x^2}{x^2+1} = \frac{x^2+1-1}{x^2+1} = \frac{x^2+1}{x^2+1} - \frac{1}{x^2+1} = 1 - \frac{1}{x^2+1}.$$

While this is a very useful technique for such a simple case, it does not easily extend to more complicated cases. Long division—when the degree of the numerator is at least that of the denominator—can always be employed.

Example 5.1.10

We compute some antiderivatives involving these and other rules. Some "simplifications" are matters of preference. Note e^x is an exponential function, while x^e is a power.

- $\int [1 + x + e^x]\, dx = x + \dfrac{x^2}{2} + e^x + C.$

- $\int 2^x\, dx = \dfrac{2^x}{\ln 2} + C.$

- $\int \left(3 \cdot 2^x + x^2 + e^x + x^e\right) dx = 3 \cdot \dfrac{2^x}{\ln 2} + \dfrac{x^3}{3} + e^x + \dfrac{x^{e+1}}{e+1} + C.$

- $\int \dfrac{2^x - 3^x}{5^x}\, dx = \int \left[\dfrac{2^x}{5^x} - \dfrac{3^x}{5^x}\right] dx = \int \left[\left(\dfrac{2}{5}\right)^x - \left(\dfrac{3}{5}\right)^x\right] dx = \dfrac{\left(\frac{2}{5}\right)^x}{\ln \frac{2}{5}} - \dfrac{\left(\frac{3}{5}\right)^x}{\ln \frac{3}{5}} + C.$

- $\int 5^{2x+1}\, dx = \int 5 \cdot (5^2)^x\, dx = 5 \int 25^x\, dx = \dfrac{5 \cdot 25^x}{\ln 25} + C = \dfrac{5^{2x+1}}{2 \ln 5} + C.$

- $\int (1 + 3^x)^2\, dx = \int [1 + 2 \cdot 3^x + 3^{2x}]\, dx = \int [1 + 2 \cdot 3^x + 9^x]\, dx = x + \dfrac{2 \cdot 3^x}{\ln 3} + \dfrac{9^x}{\ln 9} + C.$

In the last integral, we used the fact that $(3^x)^2 = 3^{x \cdot 2} = 3^{2 \cdot x} = \left(3^2\right)^x = 9^x$.

In later sections we develop the initial techniques that allow us to find antiderivatives of functions which are not known to be derivatives of common functions, but are related to them. In particular we look for reverse Chain Rules but the technique to do so has many more applications, and is used extensively in the rest of the text.

5.1.7 Integrals Yielding Inverse Hyperbolic Functions

Here we just note some interesting integral formulas which arise from the derivative formulas for the hyperbolic functions, from Subsection 3.12.3, starting on page 401.

$$\int \frac{1}{\sqrt{x^2 + 1}}\, dx = \sinh^{-1} x + C. \tag{5.28}$$

$$\int \frac{1}{\sqrt{x^2 - 1}}\, dx = \cosh^{-1} x + C. \tag{5.29}$$

$$\int \frac{1}{1 - x^2}\, dx = \tanh^{-1} x + C. \tag{5.30}$$

$$\int \frac{1}{x\sqrt{1 - x^2}}\, dx = -\operatorname{sech}^{-1} x + C. \tag{5.31}$$

$$\int \frac{1}{x\sqrt{1 + x^2}}\, dx = -\operatorname{csch}^{-1} |x| + C. \tag{5.32}$$

These formulas (5.28)–(5.32) are useful now and then, though in the sequel we will find that there are other techniques for finding antiderivatives of these integrands. Those techniques can be rather involved, and so some nontrivial effort can be spared if one can retrieve these formulas when such integrands are encountered.

5.1. INDEFINITE INTEGRALS (ANTIDERIVATIVES): INTRODUCTION

Exercises

Verify the following antiderivative formulas by differentiation (applying $\frac{d}{dx}$ or $\frac{d}{dt}$, whichever is appropriate, to the right-hand sides).

1. $\int e^{2t} dt = \frac{1}{2} e^{2t} + C$

2. $\int \frac{e^{\sqrt{x}}}{\sqrt{x}} dx = 2 e^{\sqrt{x}} + C$

3. $\int \cos 2t \, dt = \frac{1}{2} \sin 2t + C$

4. $\int \ln x \, dx = x \ln x - x + C$

5. $\int \sec x \, dx = \ln|\sec x + \tan x| + C$

6. $\int \tan^{-1} x \, dx = x \tan^{-1} x - \frac{\ln(x^2+1)}{2} + C$

7. $\int \sin^2 x \, dx = \frac{1}{2}(x - \sin x \cos x) + C$

8. $\int \sec 7x \tan 7x \, dx = \frac{1}{7} \sec 7x + C$

9. $\int \tan x \, dx = \ln|\sec x| + C$

10. $\int x e^x \, dx = x e^x - e^x + C$

11. $\int \frac{1}{x(1+(\ln x)^2)} dx = \tan^{-1}(\ln x) + C$

12. $\int \frac{2}{x^2-1} dx = \ln\left|\frac{x-1}{x+1}\right| + C$ (Hint: expand the logarithms first.)

Calculate the following indefinite integrals, in most cases by rewriting the integrand to achieve forms found earlier in the section.

13. $\int \frac{1}{x^{2/3}} dx$

14. $\int \frac{\cos x}{\sin^2 x} dx$

15. $\int \frac{1}{\cos^2 x} dx$

16. $\int (x^2 + 3x + 9) dx$

17. $\int (x^2 + 1)^3 dx$

18. $\int \sqrt{9w} \, dw$

19. $\int \frac{1}{\sqrt{9w}} dw$

20. $\int \frac{(x+1)^2}{x^2} dx$

21. $\int (-5 \sin x) dx$

22. $\int 10^x dx$

23. $\int \cot^2 t \, dt$

24. Suppose $f'(x) = -2 \sin x$. Draw several possible functions $f(x)$ with this derivative, and find and highlight the particular function $f(x)$ such that $f(0) = 1$.

25. Suppose $a(t) = -2 \cos t$, $v(0) = 3$ and $s(0) = 7$. Find $s(t)$.

26. Suppose $f''(x) = 4x^3 - 3x^2 + 2x - 9$, $f'(-1) = 3$ and $f(0) = 4$. Find $f(x)$.

For 27–35, find the function satisfying the given criteria.

27. $f'(x) = 3x^2 + 2x + 5$, $f(1) = 2$.

28. $g'(z) = \sqrt{z}$, $g(4) = \frac{19}{3}$.

29. $s'(t) = \frac{1}{t^2+1}$, $s(1) = \pi/2$.

30. $f'(x) = 1 + \cos x$, $f(\pi/2) = 6$.

31. $f'(x) = \sec x \tan x$, $f(0) = 4$.

32. $f'(x) = 5e^x$, $f(\ln 3) = 11$.

33. $f'(x) = 4$, $f(2) = 5$.

34. $f'(x) = 0$, $f(2) = 5$.

35. $f'(x) = \sin x + \cos x$, $f(\pi/4) = 3$.

5.2 Summation (Sigma) Notation

This short section looks at the Σ-notation common for signifying summations. For example,

$$\sum_{i=1}^{n} a_i = \underbrace{a_1}_{i=1} + \underbrace{a_2}_{i=2} + \underbrace{a_3}_{i=3} + \cdots + \underbrace{a_{n-1}}_{i=n-1} + \underbrace{a_n}_{i=n}. \tag{5.33}$$

It is no accident that the "Greek S" is used to signify a sum. In (5.33), i is the *index* of the summation, ranging along the integers from 1 to n, i.e., $i = 1, 2, 3, \ldots, n$. If $a_1 = 3$, $a_2 = 5$, $a_3 = 1$, $a_4 = 7$, and $a_5 = -2$, we can write

$$\sum_{i=1}^{5} a_i = a_1 + a_2 + a_3 + a_4 + a_5$$
$$= 3 + 5 + 1 + 7 - 2 = 14.$$

Often the summation desired is of the form $\sum_{i=1}^{n} f(i)$, where $f : \{1, 2, \ldots, n\} \longrightarrow \mathbb{R}$, though the domain of f need not begin at $i = 1$. For instance,

$$\sum_{i=1}^{5} i^2 = \underbrace{1^2}_{i=1} + \underbrace{2^2}_{i=2} + \underbrace{3^2}_{i=3} + \underbrace{4^2}_{i=4} + \underbrace{5^2}_{i=5} = 1 + 4 + 9 + 16 + 25 = 55,$$

$$\sum_{i=0}^{3} (2i+1) = \underbrace{[2(0)+1]}_{i=0} + \underbrace{[2(1)+1]}_{i=1} + \underbrace{[2(2)+1]}_{i=2} + \underbrace{[2(3)+1]}_{i=3} = 1 + 3 + 5 + 7 = 16,$$

$$\sum_{i=0}^{4} 2 = \underbrace{2}_{i=0} + \underbrace{2}_{i=1} + \underbrace{2}_{i=2} + \underbrace{2}_{i=3} + \underbrace{2}_{i=4} = 5(2) = 10.$$

As these make clear, some care must be taken regarding the range of values for the index.

So far we have used i for the index in each example. Other indices are also common, usually lower-case Latin or Greek letters. It should also be pointed out that these summations are, in many ways, *operators* $\sum(\ )$, notationally similar to the derivative and integral operators $\frac{d}{dx}[\ ]$ and $\int (\ )dx$, so some of the same notation conventions apply as, for instance, parentheses are included or omitted for clarity:

$$\sum_{i=1}^{5}(i+9) = (1+9) + (2+9) + (3+9) + (4+9) + (5+9) = 10 + 11 + 12 + 13 + 14 = 60,$$

$$\sum_{i=1}^{5} i + 9 = \left(\sum_{i=1}^{5} i\right) + 9 = (1 + 2 + 3 + 4 + 5) + 9 = 24,$$

$$\sum_{i=1}^{3} 2i = \sum_{i=1}^{3}(2i) = 2 + 4 + 6 = 12,$$

$$\sum_{i=1}^{5}(i+1)^2 = 2^2 + 3^2 + 4^2 + 5^2 + 6^2 = 4 + 9 + 16 + 25 + 36 = 90,$$

$$\left(\sum_{i=1}^{5}(i+1)\right)^2 = (2 + 3 + 4 + 5 + 6)^2 = 20^2 = 400.$$

The notation has many uses. It can be found nearly anywhere a large number of similar quantities are routinely added, such as in accounting or on spreadsheets. In statistics, if we have data $x_1, x_2, \ldots, x_n$, then the *sample mean* (or *average*) $\overline{x}$ and *sample standard deviation* s are, respectively,

$$\overline{x} = \frac{x_1 + x_2 + \cdots + x_n}{n} = \frac{\sum_{i=1}^{n} x_i}{n} = \frac{1}{n} \sum_{i=1}^{n} x_i, \qquad s = \sqrt{\frac{\sum_{i=1}^{n} (x_i - \overline{x})^2}{n-1}}.$$

Example 5.2.1

Suppose we are given data from the table below:

i	1	2	3	4	5	6	7	8
x_i	97	54	89	96	96	87	59	68

From this data we can compute

$$\overline{x} = \frac{1}{8} \sum_{i=1}^{8} x_i = \frac{1}{8}(x_1 + x_2 + x_3 + x_4 + x_4 + x_6 + x_7 + x_8)$$

$$= \frac{1}{8}(97 + 54 + 89 + 96 + 96 + 87 + 59 + 68) = \frac{1}{8}(646) = 80.75.$$

Next we can make a table for the values needed in the sum for the standard deviation s:[10]

i	1	2	3	4	5	6	7	8
x_i	97	54	89	96	96	87	59	68
$x_i - \overline{x}$	16.25	-26.75	8.25	15.25	15.25	6.25	-21.75	-12.75

From these values we can compute

$$\sum_{i=1}^{8} (x_i - \overline{x})^2 = 16.25^2 + (-26.75)^2 + 8.25^2 + 15.25^2 + 15.25^2 + 6.25^2$$

$$+ (-21.75)^2 + (-12.75)^2 = 2187.5,$$

$$\implies \sqrt{\frac{1}{n-1} \sum_{n=1}^{8} (x - \overline{x})^2} = \sqrt{\frac{1}{7}(2187.5)} \approx 17.68.$$

Some arithmetic properties of summations can be seen with little or no difficulty, such as

$$\sum_{i=1}^{n} (a_i + b_i) = \sum_{i=1}^{n} a_i + \sum_{i=1}^{n} b_i, \tag{5.34}$$

$$\sum_{i=1}^{n} k a_i = k \cdot \sum_{i=1}^{n} a_i \tag{5.35}$$

$$\sum_{i=1}^{n} k = nk. \tag{5.36}$$

---

[10]The standard deviation s is a measure of how much, on average, a datum will stray from the "center" $\overline{x}$. There are theoretical reasons why it is a square root of a sum of squares, and we divide by $n-1$ instead of n so it is not a true "average." The details are left to a course in probability theory.

5.2. SUMMATION (SIGMA) NOTATION

These quickly become clear enough that they are rarely cited, though it is worth formally proving these as an exercise in careful proof writing, so we do so next. The first of these amounts to a simple regrouping of the sum, the second is the distributive property, and the third is a simple counting of terms being added:

$$\sum_{i=1}^{n}(a_i+b_i) = (a_1+b_1)+(a_2+b_2)+\cdots+(a_n+b_n)$$

$$= (a_1+a_2+\cdots+a_n)+(b_1+b_2+\cdots+b_n) = \left(\sum_{i=1}^{n}a_i\right)+\left(\sum_{i=1}^{n}b_i\right),$$

$$\sum_{i=1}^{n}(ka_i) = ka_1+ka_2+\cdots+ka_n = k(a_1+a_2+\cdots+a_n) = k\cdot\left(\sum_{i=1}^{n}a_i\right),$$

$$\sum_{i=1}^{n}k = \underbrace{k+k+\cdots+k}_{n\text{ copies}} = n\cdot k, \qquad \text{q.e.d.}$$

Note that in (5.36), the number of terms is important, so that $\sum_{i=0}^{n}k = (n+1)k$, while $\sum_{i=2}^{5}k = 4k$. In general, $\sum_{i=m}^{n}k = (n-m+1)k$, since we have to count the "endpoint" terms $i=m,n$ as well as those in between.[11] In all cases, the summation is a sum of several copies of the same constant.

It is interesting to note that there are formulas for sums of positive powers of the index. The first few are as follows:

$$\sum_{i=1}^{n}i = 1+2+3+\cdots+n = \frac{n(n+1)}{2}, \tag{5.37}$$

$$\sum_{i=1}^{n}i^2 = 1^2+2^2+3^2+\cdots+n^2 = \frac{n(n+1)(2n+1)}{6}, \tag{5.38}$$

$$\sum_{i=1}^{n}i^3 = 1^3+2^3+3^3+\cdots+n^3 = \frac{n^2(n+1)^2}{4}. \tag{5.39}$$

We will prove (5.39) after some examples and leave the proofs of the others to the exercises. With these, we can compute for example

$$\sum_{i=1}^{5}i^2 = \frac{5(5+1)(2(5)+1)}{6} = \frac{5\cdot 6\cdot 11}{6} = 55,$$

which we can also compute directly: $1^2+2^2+3^2+4^2+5^2 = 55$. With these formulas (5.37)–(5.39), together with the more obvious formulas (5.34)–(5.36), page 526, we can more readily compute some commonly occurring summations without resorting to adding each individual term.

Example 5.2.2

Consider the following computations (see (5.37)–(5.39) and (5.34)–(5.36)):

---

[11] This is akin to the fact that if an event occurs on days 1 through 12 in a month, that event occurs for 12 days, though one is tempted to assume it happened for $12-1=11$ days. Similarly, an event occurring on days 20–30 will occur on $30-20+1=11$ days. It is a matter of counting both "endpoints." Using instead the difference $30-20$ effectively and erroneously removes the 20th day of the month from our count.

- $\sum_{i=1}^{100} i = \frac{100(100+1)}{2} = 5050$

- $\sum_{i=1}^{10}(3+i^2) = \sum_{i=1}^{10} 3 + \sum_{i=1}^{10} i^2 = 10 \cdot 3 + \frac{10(10+1)(2(10)+1)}{6}$
  $= 30 + \frac{10 \cdot 11 \cdot 21}{6} = 30 + 385 = 415,$

- $\sum_{i=1}^{40}[(2i+3)^2] = \sum_{i=1}^{40}[4i^2+12i+9] = 4\sum_{i=1}^{40} i^2 + 12\sum_{i=1}^{40} i + \sum_{i=1}^{40} 9$
  $= 4 \cdot \frac{(40)(41)(81)}{6} + 12 \cdot \frac{(40)(41)}{2} + 40 \cdot 9 = 88,560 + 9,840 + 360 = 98,760,$

- $\sum_{i=50}^{100} i^2 = \sum_{i=1}^{100} i^2 - \sum_{i=1}^{49} i^2 = \frac{(100)(101)(201)}{6} - \frac{(49)(50)(99)}{6}$
  $= 338,350 - 40,425 = 297,925.$

This last summation we were able to rewrite as a difference, canceling the terms we do not need in our original sum.

Sometimes there is value in "adjusting the indices." While this is not well motivated in this section, we consider it here for purposes of early introduction. Note that the following summations are the same, though the subscripts are, respectively, i, $i+1$, and $i-2$:

$$\sum_{i=1}^{5} a_i = \underbrace{a_1}_{i=1} + \underbrace{a_2}_{i=2} + \underbrace{a_3}_{i=3} + \underbrace{a_4}_{i=4} + \underbrace{a_5}_{i=5},$$

$$\sum_{i=0}^{4} a_{i+1} = \underbrace{a_1}_{i=0} + \underbrace{a_2}_{i=1} + \underbrace{a_3}_{i=2} + \underbrace{a_4}_{i=3} + \underbrace{a_5}_{i=4},$$

$$\sum_{i=3}^{7} a_{i-2} = \underbrace{a_1}_{i=3} + \underbrace{a_2}_{i=4} + \underbrace{a_3}_{i=5} + \underbrace{a_4}_{i=6} + \underbrace{a_5}_{i=7}.$$

This can simplify computations such as the following:

$$\sum_{i=1}^{10}(i+3)^2 = \sum_{i=4}^{13} i^2 \quad \text{(check the actual numbers being summed)}$$
$$= \sum_{i=1}^{13} i^2 - \sum_{i=1}^{3} i^2$$
$$= \frac{(13)(14)(27)}{6} - \frac{(3)(4)(7)}{6} = \frac{1}{6}(4914 - 84) = \frac{4830}{6} = 805.$$

While this example may not be all that compelling, with larger numbers its worth is likely to be more convincing. Furthermore, this technique of re-indexing the numbers in the sum has its usefulness in many other contexts and will be refined and expanded as needed later.

Finally, for completeness we now look at the typical method of proving (5.39), page 527. This type of formula is usually proven using *mathematical induction*. What this entails is

5.2. SUMMATION (SIGMA) NOTATION

(1) directly proving one or more of the "first" cases, and (2) showing that anytime we have, say, the nth case we automatically get the $(n+1)$st case. So we define statements $P_1, P_2, P_3, \ldots$ so that the statement P_n is defined as follows:

$$P_n: \quad \sum_{i=1}^{n} i^3 = \frac{n^2(n+1)^2}{4}.$$

Now on its face, P_n could be true or false for a particular n, but our strategy will be to show that

(1) P_1 is true, and that

(2) $P_n \implies P_{n+1}$.

This second part is called the *induction step*.

Before continuing, an explanation of mathematical induction is in order. As a general rule, if we prove P_1 is true, **and** that for all $n \in \mathbb{N}$ we have $P_n \implies P_{n+1}$ (i.e., P_n being true would force P_{n+1} to be true also), then by this implication (2) since P_1 is true we must have that P_2 is true, and by the implication (2) we further have P_3 is true, and from that we next have P_4 is true, and so on. This eventually proves, say, $P_{1,000,000}$ is true because there is this "chain of truth" from P_1 to P_2 to P_3, and so on, so after 999,999 such steps, we would reach the conclusion that $P_{1,000,000}$ is also true. We are forced to conclude then that P_n is true for each $n \in \mathbb{N}$, because the truth of P_n would be reached after $n-1$ invocations of the induction step (2). That is the essence of the strategy known as "proof by induction" in mathematics, though there are variations of it.

Getting back to this particular proof, the statement P_1 would be that $\sum_{i=1}^{1} i^3 = \frac{1^1(1+1)^2}{4}$, which is clearly true because it is equivalent to $1^3 = \frac{1^2(2)^2}{4}$, i.e., $1 = 1$, which is true (obviously).

The induction step (2) has a simple, yet sophisticated little proof. We want to show $P_n \implies P_{n+1}$, so to do that we suppose (hypothetically) P_n, i.e., that the "nth case" is true, and then show that this would imply the "$(n+1)$st" case follows.

So if P_n is true, i.e., $\sum_{i=1}^{n} i^3 = \frac{n^2(n+1)^2}{4}$, then

$$\sum_{i=1}^{n+1} i^3 = \left(\sum_{i=1}^{n} i^3\right) + (n+1)^3 \stackrel{\text{by } P_n}{=\!=\!=} \frac{n^2(n+1)^2}{4} + (n+1)^3 = (n+1)^2 \left[\frac{n^2 + 4(n+1)}{4}\right]$$

$$= \frac{(n+1)^2(n+2)^2}{4} = \frac{(n+1)^2((n+1)+1)^2}{4},$$

which is the statement P_{n+1} when examined closely:

$$P_{n+1}: \quad \sum_{i=1}^{n+1} i^3 = \frac{(n+1)^2((n+1)+1)^2}{4}.$$

Thus we showed $P_n \implies P_{n+1}$, proving the induction step (2), and so with P_1 being true we get P_n is true for all $n \in \mathbb{N}$, q.e.d.

Since this proof served the purpose to introduce induction as a proof style, as well as to use the method for a particular example (proving (5.39)), it is more verbose than it would be

Exercises

For Exercises 1–9, compute the sum by writing out every term and simplifying the result.

1. $\sum_{i=1}^{6} i$

2. $\sum_{i=0}^{5} 2^i$

3. $\sum_{i=0}^{5} 2^{-i}$

4. $\sum_{i=4}^{10} i^2$

5. $\sum_{i=1}^{11} (-1)^i$

6. $\sum_{i=1}^{12} \cos\frac{i\pi+1}{2}$

7. $\sum_{i=1}^{10} 2+3$

8. $\sum_{i=5}^{10} 3$

9. $\sum_{i=2}^{6} \left[\frac{1}{i+1} - \frac{1}{i-1} \right]$ (notice what cancels when you write out the terms in long-hand)

For Exercises 10–15, use techniques such as those used in Example 5.2.2, page 527, to compute the sums:

10. $\sum_{i=1}^{200} i$

11. $\sum_{i=100}^{200} i$

12. $\sum_{i=1}^{10} i^3$

13. $\sum_{i=1}^{20} (2i+1)$

14. $\sum_{i=1}^{10} [i(i-1)]$

15. $\sum_{i=1}^{20} \left[i(2i-3)^2 \right]$

16. By writing out all of the terms, prove the following and compute the sum (using the second form of the sum):

$$\sum_{i=0}^{9} (i+1)^3 = \sum_{i=1}^{10} i^3.$$

17. By writing out several terms, show that the following computation is

---

[12] A more streamlined proof by induction would look more like the following.

Theorem: For any $n \in \mathbb{N}$, $\sum_{i=1}^{n} i^3 = \dfrac{n^2(n+1)^2}{4}$

Proof: First we note that it is true for $n=1$: $1^3 = \frac{1 \cdot 2^2}{4}$. Next we assume it is true for the nth case. Then

$$\sum_{i=1}^{n+1} i^3 = \underbrace{\sum_{i=1}^{n} i^3}_{n\text{th case}} + (n+1)^3 = \frac{n^2(n+1)^2}{4} + (n+1)^3 = \frac{(n+1)^2(n^2 + 4(n+1))}{4} = \frac{(n+1)^2((n+1)+1)^2}{4}.$$

Thus by induction the formula holds for all $n \in \mathbb{N}$, q.e.d.

Another variation of induction occurs where one begins by proving the first few cases, say P_1, P_2, P_3, and then uses a "weaker" induction step, where the implication might use more than the information from the statement immediately prior, i.e., where instead of $(\forall n \in \mathbb{N})[P_n \longrightarrow P_{n+1}]$, one uses

$$[(\forall k \in \mathbb{N})(k \leq M \longrightarrow P_k)] \implies P_{M+1}.$$

In other words, the induction step is to show that the next statement P_{M+1} is true under the assumption that all previous statements in the list $P_1, P_2, \cdots, P_M$ are true, rather than just P_M.

The basic spirit is that the truth of the later statements is "bootstrapped" off of the truths of the previous statements, and that for any statement P_n, no matter how large is $n \in \mathbb{N}$, its truth will be established in finitely many valid implication steps from the truths of the previous statements. Thus there are many possible variations of structure for induction proofs. (Some proofs require even and odd cases of n to be proved separately, for instance.)

5.2. SUMMATION (SIGMA) NOTATION

valid and compute the sum:

$$\sum_{i=14}^{30}(i-1)^3 = \sum_{i=13}^{29} i^3$$
$$= \sum_{i=1}^{29} i^3 - \sum_{i=1}^{12} i^3.$$

18. Using a strategy similar to the problems above, compute the following:

    (a) $\sum_{i=3}^{20}(i-2)^2$

    (b) $\sum_{i=5}^{20}(i+2)^2$ (for this one, you may wish to write a summation as a difference of two other summations)

19. Statistics texts offer an alternative computation of the sample standard deviation s, which does not require first computing the sample mean $\bar{x}$. The formula is given often in terms of the *sample variance* s^2, which allows us to compute the sample standard deviation $s = \sqrt{s^2}$. A formula for variance is

$$s^2 = \frac{n \cdot \sum_{i=1}^{n} x_i^2 - \left(\sum_{i=1}^{n} x_i\right)^2}{n(n-1)}.$$

Use this formula to compute s^2 and then s from Example 5.2.1, page 526.

20. Prove (5.37), page 527, by induction.

21. Prove (5.38), page 527, by induction.

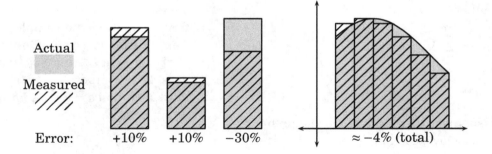

Figure 5.3: Visual illustration of four examples of percent errors, the first two being equal as percentages, the third being three times as large as those as a percent error but also an underestimate rather than an overestimate, and the last being also an underestimate, where we use the sum of areas of rectangles to approximate the area directly under the curve and above the horizontal axis. These themes recur often in this section.

5.3 Riemann Sums and the Second Fundamental Theorem of Calculus (FTC)

Anytime a theorem is called "fundamental" in its field, we expect it to be somewhat deep, ultimately intuitive, very important, and not trivial to prove. These all apply to the Fundamental Theorem of Calculus (FTC), which we will use extensively, though an actual proof of the theorem is beyond the scope of this text, and will not be found here.[13]

In fact, the theorem presented in this section is often called the Second Fundamental Theorem of Calculus, or FTC Part 2, because in a more rigorous setting its proof usually comes after that of the less ubiquitous First FTC, introduced in Section 5.4. The FTC Part 2 is used more than Part 1, is equally intuitive, and will be what we usually mean by "The Fundamental Theorem of Calculus," or FTC. However we will use FTC to refer to both.

Instead of attempting an actual proof of this part of the FTC, we present a case where it is, more or less, obvious (or at least very believable), and then generalize somewhat to less obvious cases that follow nonetheless. Along the way there are several concepts to define and explore. Each justifies careful study and occasional revisits, even for professionals.

For our explanation, we begin with the background concepts of relative and percent error, show how they stay controlled within summations, show how the study of approximate displacements from known velocities leads us to Riemann Sums and a case for the FTC, and then generalize that observation for the full conclusion of the FTC.

For basic skills, Subsections 5.3.4 and 5.3.6 are the most useful, but the theory found in the other subsections yields extremely important insight and should be revisited repeatedly.

---

[13]Most, if not all, science and engineering calculus textbook authors attempt an argument for why the FTC is true but do not prove it. Some give partial proofs with gaps explained away by intuitive arguments that fall short of proofs, while others give arguments that are more technical and closer to an actual proof but perhaps less intuitive. Typically, all such arguments leave gaps that require a junior or senior level of university real analysis or topology to fill with a proper proof.

Our argument here leans towards the intuitive type, with its claim of truth based upon the relationship between position and velocity, also not constituting a rigorous proof but (hopefully) offering a strongly intuitive argument. We leave the rigorous proofs for junior- or senior-level classes.

5.3.1 Error, Absolute Error, Relative Error, and Percent Error

For a simple example of these types of errors, consider a man who weighs 200 lb, weighed on a scale that indicates his weight to be 210 lb. In such a case, we say the *error* is 10 lb. If the scale indicated his weight to be 190 lb., we would say the error is −10 lb, though in both cases we would say the *absolute error*, i.e., the absolute value of the error, is 10 lb. For a 200-lb man, perhaps this seems "relatively" small, but if we are weighing a newborn child, 10 lb is clearly an unacceptable error. Thus it is also important to note what fraction of his weight the error represents, so we compute the *relative absolute error*, namely (10 lb)/(200 lb), or 0.05. Now as a percentage (or "parts per hundred"), we multiply by 100% (which is just another expression for 1) and find the absolute percent error to be 0.05 × 100% = 5%. Note that the relative and percent error are unitless, in the sense that the "lbs" cancel. Some of these measurements of error are defined as follows:[14]

$$\text{Absolute Error} = |(\text{Measured Quantity}) - (\text{Actual Quantity})| \qquad (5.40)$$

$$\text{Error} = (\text{Measured Quantity}) - (\text{Actual Quantity}) \qquad (5.41)$$

$$\text{Relative Error} = (\text{Error})/(\text{Actual Quantity}) \qquad (5.42)$$

$$\text{Percent Error} = (\text{Relative Error}) \cdot 100\%. \qquad (5.43)$$

One can also define "relative absolute error" and "percent absolute error" in the obvious ways. Since it is informative to have a sign (+/−) associated with the error, we will usually use the term "error" rather than absolute error (in the sense of absolute values), when there is no confusion.

Relative and percent errors are easily visualized, and in fact judging when a relative or percentage error is "small" or "large" is fairly easy given an accurate illustration of the quantities involved. For some examples see Figure 5.3, page 532.

Now we consider what would be the cumulative effect on a summation if the measured amount were consistently a given percentage higher than the actual.

Example 5.3.1

Suppose the actual quantity we desire to know is $\sum_{i=1}^{n} a_i$, where we attempt to measure each of the a_i, and each is measured to be b_i, where b_i is a 5% overestimation of a_i in each case. Then the measured summation will be

$$\sum_{i=1}^{n} b_i = \sum_{i=1}^{n} (1.05 a_i) = 1.05 \sum_{i=1}^{n} a_i.$$

In other words, if the b_i all overestimate the respective a_i by exactly 5%, then the summation $\sum b_i$ overestimates the summation $\sum a_i$ by exactly 5%.

This example simply illustrates the distributive property of multiplication over summations. We can conclude similarly that a consistent underestimation of a_i by 5% ($b_i = 0.95 a_i$) would result in exactly a 5% underestimation of the summation ($\sum b_i = 0.95 \sum a_i$). More generally, since these would be the "extreme" cases, we can further state that if, in the sense

---

[14] Some texts do not use absolute values in defining "absolute error," but use the term "absolute" to distinguish the error from relative or percent error. Some authors only report the absolute value of any error, in which case information on whether an approximation is an overestimation or underestimation must be found elsewhere. As always, it is important to know what is meant by these common terms in context.

of absolute values, the percent absolute error in a_i is less than 5% (+/−), then the percent absolute error in the sum must also be less than 5% (+/−). This leads us to the more general conclusion:

Theorem 5.3.1 *If each a_i is estimated by a respective b_i within p% (absolute) error, then it follows that $\sum a_i$ is also estimated within p% (absolute) error by $\sum b_i$.*

That fact will be important as we argue for the reasonableness of the FTC (Part 2).

5.3.2 A Physics Example to Motivate Riemann Sums and the FTC

Here we consider an abstract motion problem. We wish to find the net displacement of an object in one-dimensional motion, over the time interval $[t_0, t_f]$. For a classical problem, we expect the velocity $v(t)$ over this time interval to be continuous. For technical reasons (explained in later footnotes), we also assume for now that velocity is always positive, i.e., $v(t) > 0$ on $[t_0, t_f]$. Note that we would then have $v(t) \in [v_{\min}, v_{\max}]$, where $v_{\min}, v_{\max} > 0$ are the minimum and maximum values of $v(t)$, respectively, for $t \in [t_0, t_f]$.

The actual net displacement over the time interval $[t_0, t_f]$ is $s(t_f) - s(t_0)$. In our case this will be positive, since the velocity $v(t) = s'(t)$ is assumed to be positive, implying $s(t)$ is increasing. To construct a scheme approximating this net displacement based upon the velocity function, we partition the time interval $[t_0, t_f]$ into subintervals $[t_{i-1}, t_i]$, where

$$t_0 < t_1 < t_2 < t_3 < \cdots < t_{n-1} < t_n = t_f,$$

and approximate the displacement $s(t_i) - s(t_{i-1})$ for each subinterval. The width of the ith subinterval $[t_{i-1}, t_i]$ will be Δt_i (meaning the change in t over the ith subinterval):

$$\Delta t_i = t_i - t_{i-1}. \tag{5.44}$$

Suppose we wish this scheme to approximate the total displacement $s(t_f) - s(t_0)$ to within $p\%$. If we can choose our time subintervals to be each small enough that the velocity varies less than $p\%$ within each interval, we can make such an approximation of $s(t_f) - s(t_0)$ as follows.

First consider for such an interval $[t_{i-1}, t_i]$ a point $t_i^* \in [t_{i-1}, t_i]$ at which we *sample* the velocity, that sample velocity being $v(t_i^*)$. If that interval is short enough, then the velocity change over the interval will be small (by the definition of continuity). We can even choose the interval to be small enough that all velocities on that interval are within $p\%$ of the *average velocity* $[s(t_i) - s(t_{i-1})]/(\Delta t_i)$ for that subinterval (which exists at some time within the interval by the Mean Value Theorem). Then we will have a small percentage error resulting from assuming average velocity for the interval is approximately $v(t_i^*)$, regardless of our choices of the sample times t_i^* from their respective intervals. This assures that the displacement change over that interval can be approximated by

$$s(t_i) - s(t_{i-1}) \approx v(t_i^*) \Delta t_i,$$

to within that same percentage error $p\%$.

5.3. RIEMANN SUMS AND THE SECOND FUNDAMENTAL THEOREM (FTC)

Our scheme is thus to approximate the net displacement on each subinterval $[t_{i-1}, t_i]$, and then to sum these.[15]

Subinterval	Sample	Width	Approximate Displacement		Actual Displacement
$[t_0, t_1]$	$\ni \quad t_1^*$	Δt_1	$v(t_1^*)\Delta t_1$	$\approx$	$s(t_1) - s(t_0)$
$[t_1, t_2]$	$\ni \quad t_2^*$	Δt_2	$v(t_2^*)\Delta t_2$	$\approx$	$s(t_2) - s(t_1)$
$[t_2, t_3]$	$\ni \quad t_3^*$	Δt_3	$v(t_3^*)\Delta t_3$	$\approx$	$s(t_3) - s(t_2)$
$\vdots$	$\vdots$	$\vdots$	$\vdots$	$\vdots$	$\vdots$
$[t_{n-2}, t_{n-1}]$	$\ni \quad t_{n-1}^*$	Δt_{n-1}	$v(t_{n-1}^*)\Delta t_{n-1}$	$\approx$	$s(t_{n-1}) - s(t_{n-2})$
$[t_{n-1}, t_n]$	$\ni \quad t_n^*$	Δt_n	$v(t_n^*)\Delta t_n$	$\approx$	$s(t_n) - s(t_{n-1})$
		Sum:	$\sum_{i=1}^{n} v(t_i^*)\Delta t_i$	$\approx$	$s(t_n) - s(t_0)$

When the entries in the last two columns are summed, respectively, we get much cancellation in the last column, resulting in the approximation:

$$\sum_{i=1}^{n} v(t_i^*)\Delta t_i \approx \sum_{i=1}^{n} s(t_i) - \sum_{i=0}^{n-1} s(t_i) = s(t_n) - s(t_0),$$

which, since $t_n = t_f$, we can rewrite as

$$\boxed{\sum_{i=1}^{n} v(t_i^*)\Delta t_i \approx s(t_f) - s(t_0).} \tag{5.45}$$

If in all subintervals $[t_{i-1}, t_i]$ we have $v(t)$ varying by no more than $p\%$ from the average velocity $[s(t_i) - s(t_{i-1})]/(\Delta t_i)$ for the subinterval, then the error in replacing each $s(t_i) - s(t_{i-1})$ with $v(t_i^*)\Delta t_i$ is no more than $p\%$ (+/−). This is accomplished by choosing each Δt_i so small that $v(t)$ varies by less than $p\%$ over the interval $[t_{i-1}, t_i]$. (This should seem reasonable, though it is beyond the scope of this text to prove this.) Then we can be assured that the approximation (5.45), rewritten here as

$$s(t_f) - s(t_0) = \sum_{i=1}^{n}[s(t_i) - s(t_{i-1})] \approx \sum_{i=1}^{n} v(t_i^*)\Delta t_i, \tag{5.46}$$

will be accurate within $p\%$ of the actual value of $s(t_f) - s(t_0)$, by Theorem 5.3.1, page 534. As we will see later, the right-hand side of (5.46) is an example of a *Riemann Sum*.

Next we argue (without proof) that the percent error for each subinterval $[t_{i-1}, t_i]$ in (5.46) should shrink to zero as the subinterval lengths shrink to zero—and thus $v(t)$ varies

---

[15]Some theorems in topology regarding the concepts of *uniform continuity* and *compactness* assure that given any positive number p, we can construct a partition $t_0 < t_1 < \cdots t_{n-1} < t_n = t_f$ which guarantees both that (1) each $v(t_i^*)$ approximates the average velocity for the ith subinterval $[t_{i-1}, t_i]$ to within p percent, and (2) that the number of subintervals required whereby each is small enough for (1) to occur will nonetheless be finite.

Assuming that $v(t)$ is continuous and positive on $[t_0, t_f]$ is necessary for (1), but the continuity is reasonable for a velocity (absent, say, infinite forces) and the criterion of positivity can be dealt with later by considering a new velocity $V(t) = v(t) + k$ where $k > |\min\{v(t) | t \in [t_0, t_f]\}|$, proving a result for that function $V(t)$ and then following its implications for the case $v = V - k$.

less and less over each subinterval—and our choices of the $v(t_i^*)$ matter less and less. So if we force each Δt_i to shrink toward zero (forcing the number n of subintervals to "approach infinity"), we get the limit[16]

$$\boxed{\lim_{\substack{\max\{\Delta t_i\} \to 0^+ \\ (n \to \infty)}} \sum_{i=1}^{n} v(t_i^*)\Delta t_i = s(t_f) - s(t_0).} \qquad (5.47)$$

At this point, we recall that since $s'(t) = v(t)$, we know that $s(t)$ is an antiderivative of $v(t)$, which we will see is quite relevant to (5.47). (Later we will see how this generalizes as in (5.52) below for any continuous function $f : [a, b] \longrightarrow \mathbb{R}$.)

Notationally, we would write the limit on the left-hand side of (5.47) as $\int_{t_0}^{t_f} v(t)\,dt$, and so we can append this fact to (5.47) and write

$$\boxed{\int_{t_0}^{t_f} v(t)\,dt \overset{\text{definition}}{=\!=\!=\!=} \lim_{\substack{\max\{\Delta t_i\} \to 0^+ \\ (n \to \infty)}} \sum_{i=1}^{n} v(t_i^*)\Delta t_i = s(t_f) - s(t_0).} \qquad (5.48)$$

The notation on the left is called *the definite integral, from t_0 to t_f, of $v(t)$ with respect to t*, or variations of this. Because of the form of the right-hand side of (5.48), it should not be surprising that these two uses for the term *integral*—one having to do with limits of certain sums and the other having to do with antiderivatives—are related in an equation such as (5.48). We will shortly introduce this new integral notation more formally. Before getting into the generalities of these things, we also point out a useful observation regarding choices of antiderivatives in problems such as these (namely that any antiderivative will do).

Example 5.3.2

Suppose $v(t) = 2t + 1$ for $t \in [0, 10]$ and we wish to find the net displacement of the object over the time interval $[0, 10]$. Note that $F(t) = t^2 + t$ is **an antiderivative** of $v(t)$. Thus $s(t) = F(t) + C$ for some $C \in \mathbb{R}$ (by Corollary 4.3.1, page 451), and so we can compute (noting how the constants "C" cancel):

$$s(10) - s(0) = (F(10) + C) - (F(0) + C) = F(10) - F(0) = (10^2 + 10) - (0^2 + 0) = 110.$$

In particular, the value of C did not matter in the computation above.

Thus we can find $s(10) - s(0)$ by finding **any** antiderivative (not necessarily $s(t)$ itself) of $v(t)$, and taking the difference when it is evaluated at $t = 10$ and at $t = 0$, because any two antiderivative choices for $F(t)$ will differ by a constant, and so when we take the difference $F(t_f) - F(t_0)$, that additive constant cancels. This will be a recurring theme throughout the text: The extra constant C will cancel in such computations of definite integrals, because they require a difference of two values of the same antiderivative.

---

[16] The reasoning here is sound because we assumed v was continuous and positive on the closed interval $[t_0, t_f]$. There are two technicalities in generalizing this reasoning for a general continuous v. First, there may be a case where velocity is momentarily zero, and if that occurs at time t_i^*, our choice of $v(t_i^*)\Delta t_i$ would be off by 100%. Second, if the actual displacement were zero and we chose some t_i^* such that $v(t_i^*) \neq 0$, then our error is—in a sense—of infinite percent. For those situations, we rely on the fact that we can then choose the interval small enough that the absolute error (as opposed to percent error) in the approximation $s(t_i) - s(t_{i-1}) \approx v(t_i^*)\Delta t_i$ is as small as we like, while keeping the percent error small in the other intervals. Later, some geometric illustrations will show this is reasonable, though a rigorous proof is beyond the scope of this text.

5.3. RIEMANN SUMS AND THE SECOND FUNDAMENTAL THEOREM (FTC)

5.3.3 General Riemann Sums and the Fundamental Theorem of Calculus

The sum on the right-hand side of (5.46), page 535, is one example of what is known as a *Riemann Sum*. More generally, for **any** $f(x)$ defined on $[a,b]$, we define *Riemann sums* to include any sum of the form[17]

$$\sum_{i=1}^{n} f(x_i^*)\Delta x_i, \tag{5.49}$$

where we *partition* $[a,b]$ into subintervals $[x_{i-1}, x_i]$, with

$$a = x_0 < x_1 < x_2 < x_3 < \cdots < x_{n-1} < x_n = b, \tag{5.50}$$

and where the width Δx_i and *sample point* x_i^* of the ith subinterval $[x_{i-1}, x_i]$ are given by

$$\Delta x_i = x_i - x_{i-1}, \qquad x_i^* \in [x_{i-1}, x_i]. \tag{5.51}$$

What we have argued in the context of velocities, where it seemed reasonable, is for historical reasons actually known as the Second Fundamental Theorem of Calculus, or FTC Part 2:

Theorem 5.3.2 (Second FTC): *If $f(x)$ is continuous on $[a,b]$, and $F(x)$ is an antiderivative of $f(x)$ on $[a,b]$, then we have the following finite limit, existing regardless of the choices of $x_i^* \in [x_{i-1}, x_i]$, for Riemann sums as defined in (5.49)–(5.51):*

$$\lim_{\substack{\max\{\Delta x_i\} \to 0^+ \\ (n \to \infty)}} \left(\sum_{i=1}^{n} f(x_i^*)\Delta x_i \right) = F(b) - F(a). \tag{5.52}$$

Also note that if $a \neq b$ then $\max\{|\Delta x_i|\} \to 0^+ \implies n \to \infty$, since when we shrink the subintervals' maximum length, we must increase their number. At this point, we introduce some very important notation (which the reader should memorize):

$$\boxed{\int_a^b f(x)\,dx \xrightarrow{\text{definition}} \lim_{\max\{\Delta x_i\} \to 0^+} \left(\sum_{i=1}^{n} f(x_i^*)\Delta x_i \right),} \tag{5.53}$$

$$\boxed{F(x) \bigg|_a^b \xrightarrow{\text{definition}} F(b) - F(a).} \tag{5.54}$$

With these we can rewrite the Fundamental Theorem of Calculus (5.52) in the new notation:

$$\boxed{\int_a^b f(x)\,dx = F(x)\bigg|_a^b = F(b) - F(a).} \tag{5.55}$$

---

[17]Named for Georg Friedrich Bernhard Riemann (usually shortened to Bernhard Riemann), 1826–1866, a German mathematician with very important contributions to calculus and differential geometry, the latter of which laid important groundwork for later physicists, such as Albert Einstein in his derivation of the equations of general relativity. Riemann's work is therefore one example of how the work of curious mathematicians can produce mathematical results that long predate many real-world physical problems that give the mathematics its deeper relevance. There are numerous such cases of "life imitating math" as well as "math imitating life." The importance of his Riemann Sums was felt more immediately.

Partitions for Riemann Sums are usually read left to right, as in (5.50), though there are uses for having partitions that read right to left for an interval $[b,a]$, where $a = x_0 > x_1 > x_2 > \cdots > x_{n-1} > x_n = b$, in which case each $\Delta x_i < 0$, and then the FTC still holds except we take the limit as $\max\{|\Delta x_i|\} \to 0^+$. There is also the trivial case where $a = b$ and each $\Delta x_i = 0$. Regardless, (5.52) remains unchanged. For most illustrations, we will assume $a < b$ and each $\Delta x_i > 0$, but the reader should consider visualizations of these other cases $a > b$ and $\Delta x_i < 0$ (where we "march left" down the interval as we sum the terms).

The process by which one computes the definite integral is called **integration**, the idea being that we are "integrating together" all of the terms $f(x_i^*)\Delta x_i$, but in a special way whereby we in fact look at the limit of such sums as $\max|\Delta x_i| \longrightarrow 0$, and so one can imagine one is adding more and more terms $f(x_i^*)\Delta x_i$ but the widths $|\Delta x_i|$ of the intervals are shrinking. So we shrink the intervals but still sum the (growing number of) terms $f(x_i^*)\Delta x_i$.[18]

To distinguish the integral symbol $\int$ in this context from its use in Section 5.1, the quantity on the left-hand sides of (5.53) and (5.55) is called *the definite integral of $f(x)$ with respect to x, from $x = a$ to $x = b$.*[19]

Definition 5.3.1 *For a function $f(x)$, continuous on $[a,b]$ or $[b,a]$, whichever applies, define the **definite integral** of f over the interval from a to b by the following notation and its numerical definition given by the equation*

$$\boxed{\int_a^b f(x)\,dx \stackrel{\text{definition}}{=\!=\!=\!=} \lim_{\max\{|\Delta x_i|\}\to 0^+} \left(\sum_{i=1}^n f(x_i^*)\Delta x_i\right).}$$

By the Fundamental Theorem of Calculus, this can be computed using (5.55).

To summarize (with a more common, computational statement left in darker type), if we have $f(x)$ is continuous on $[a,b]$ and $F(x)$ is an antiderivative of $f(x)$ on that interval, then

$$\boxed{\int_a^b f(x)\,dx \stackrel{\text{definition}}{=\!=\!=\!=} \lim_{\max\{|\Delta x_i|\}\to 0^+} \left(\sum_{i=1}^n f(x_i^*)\Delta x_i\right) \stackrel{\text{FTC}}{=\!=\!=\!=} F(x)\Big|_a^b \stackrel{\text{definition}}{=\!=\!=\!=} F(b) - F(a).}$$

Thus the four quantities in this box are equal by theorem or definition, assuming $f(x)$ is continuous on $[a,b]$ and $F(x)$ is an antiderivative of $f(x)$ on that interval.

Of course, the FTC gives a very strong connection between the two uses of the integration symbol $\int$: in antiderivatives (indefinite integrals) when no endpoints are given, and in limits

---

[18]A common, and very enticing, error (of a venial and not mortal kind) is to think of integration as adding an infinite number of signed rectangular areas $f(x_i^*)\Delta x_i$ with zero widths $|\Delta x_i|$. In other words, let the widths shrink to zero and then add (integrate) those signed area terms $f(x_i^*)\Delta x_i$. The error in this thinking is that it is the reverse of what is occurring conceptually. The correct perspective is to note how we add those (finitely many) terms of Riemann Sums $\sum_{i=1}^n f(x_i^*)\Delta x_i$, and *then* compute the limit as their number increases to infinity, *due to their widths shrinking to zero*, i.e., as $\max\{|\Delta x_i|\}\to 0$. (We compute limits of Riemann Sums, not Riemann Sums of limits.) To the casual reader we may appear guilty of this error later, when we look at *infinitesimal rectangles* of "width" dx and signed area $f(x)dx$ at each x, but the analysis that validates that powerful analytical approach should always be understood as originating from the limits of Riemann Sums, and the invocation of the Fundamental Theorem of Calculus.

[19]This is often verbalized somewhat lazily as "the integral, from a to b, of $f(x)\,dx$," though "dx" should be verbalized "with respect to x." Better still would be, "the (definite) integral, as x ranges from a to b, of $f(x)$ with respect to x," though that may strike some as verbose.

5.3. RIEMANN SUMS AND THE SECOND FUNDAMENTAL THEOREM (FTC)

of Riemann Sums (definite integrals), which by FTC is the same as the difference of any antiderivative's values at the endpoints a, b.[20]

5.3.4 Computing Basic Definite Integrals

Here we use the Fundamental Theorem of Calculus (FTC) to compute definite integrals $\int_a^b f(x)\,dx$. Recall that we can only use FTC if $f(x)$ is continuous on $[a, b]$, and we know the form of the antiderivatives $F(x)$ on the given interval.

- $\int_0^2 x^3\,dx = \left.\dfrac{x^4}{4}\right|_0^2 = \dfrac{2^4}{4} - \dfrac{0^4}{4} = \dfrac{16}{4} - 0 = 4.$

- $\int_{-2}^2 3x^2\,dx = 3 \cdot \left.\dfrac{x^3}{3}\right|_{-2}^2 = \left.x^3\right|_{-2}^2 = 8 - (-8) = 16.$

- $\int_{-1}^0 (2x+1)^2\,dx = \int_{-1}^0 \left(4x^2 + 4x + 1\right)dx = \left.\left[\dfrac{4}{3}x^3 + 2x^2 + x\right]\right|_{-1}^0 = [0] - \left[-\dfrac{4}{3} + 2 - 1\right] = \dfrac{1}{3}.$

Note how $\int (2x+1)^2\,dx$ requires us (for now) to expand the integrand to use the Power Rule because $(2x+1)$ does not "match" the differential dx, i.e., we are not computing an antiderivative with respect to $(2x+1)$, but with respect to x. This is the integration analog to the same lesson one learns in discussions up to and including the Chain Rule for differentiation, where care must be taken to observe the exact variable of differentiation or, in this case, antidifferentiation (integration). Further examples with a variety of variables follow.

- $\int_0^\pi \sin\theta\,d\theta = \left.[-\cos\theta]\right|_0^\pi = [-\cos\pi] - [-\cos 0] = [-(-1)] - [-1] = 1 + 1 = 2.$

- $\int_1^e \dfrac{1}{t}\,dt = \left.\ln t\right|_1^e = \ln e - \ln 1 = 1 - 0 = 1.$

- $\int_0^1 \dfrac{1}{u^2+1}\,du = \left.\tan^{-1} u\right|_0^1 = \tan^{-1} 1 - \tan^{-1} 0 = \dfrac{\pi}{4} - 0 = \dfrac{\pi}{4}.$

If f is not continuous on $[a, b]$, we cannot use the FTC to compute $\int_a^b f(x)\,dx$:

- $\int_{-1}^1 \dfrac{1}{x}\,dx$ has an integrand discontinuous at $x = 0 \in [-1, 1]$.

- $\int_0^{2\pi} \tan\theta\,d\theta$ has discontinuities at $\theta = \pi/2, 3\pi/2 \in [0, 2\pi]$.

There are techniques for handling such cases, but those are deferred for now.

---

[20]The endpoints a and b in the definite integral are often referred to as the lower and upper *limits of integration*, arguably an unfortunate term since "limit" usually refers to very different concepts. Perhaps better words in this context would be *boundary points*, or (as we use here) *endpoints* of integration.

5.3.5 Geometric Interpretation of Definite Integrals

Using the same notation as before, the very important geometric interpretation of this very important limit,

$$\lim_{\substack{\max\{|\Delta x_i|\}\to 0^+ \\ (n\to\infty)}} \left(\sum_{i=1}^n f(x_i^*)\Delta x_i\right) = \int_a^b f(x)\,dx,$$

is that it describes the "net signed area" between the graph of a function $f(x)$ and the x-axis, along the interval $x \in [a,b]$. See the first graph in Figure 5.4, page 541.

Recall that a function $f(x)$ gives the *height* of the curve $y = f(x)$ at a specific value of x in the domain of f. This "height" can be positive, negative, or zero at any such given value of x. For the moment, we consider only nonnegative functions, with therefore nonnegative heights, which yield nonnegative areas (if $a < b$) bounded on one side by the graph of the given function and on the other side by the x-axis over the given interval $[a,b]$.

Since the heights of a function often vary within any interval, we cannot simply use a rectangular "base times height" formula for computing any one such area in question. However, we can approximate the area using rectangles whose heights are derived from the function, and whose bases lie along the x-axis. As before, we break the interval $[a,b]$ in question into a partition of n subintervals with $n+1$ endpoints $x_0, x_1, \ldots, x_n$ so that

$$a = x_0 < x_1 < x_2 < \cdots < x_{n-1} < x_n = b,$$

and sample the height of the function on each subinterval to approximate the area between the curve and the x-axis for that subinterval. We do this by choosing a value $x_i^* \in [x_{i-1}, x_i]$, whose height is then given by $f(x_i^*)$. The signed area of this ith approximating rectangle will then be $f(x_i^*)\Delta x_i$, where (again) we have $\Delta x_i = x_i - x_{i-1}$.

Adding the areas of all such approximating rectangles gives us a Riemann Sum approximation of the net area between the curve and the interval $[a,b]$ on the x-axis:

$$\text{Net Shaded Area} \approx \sum_{i=0}^n f(x_i^*)\Delta x_i.$$

One such approximation scheme is illustrated in Figure 5.4, page 541. That scheme uses x_i^* to be the midpoint of the ith interval $[x_{i-1}, x_i]$. It also uses a constant width Δx_i for each interval.

The next figure, namely Figure 5.5, page 541, shows how the approximating rectangles improve in accuracy as their number grows. The second approximation uses twice as many rectangles as the first, and clearly the percentage error has decreased as the number of rectangles increased. Both used a *right-endpoint approximation* scheme, i.e., where $x_i^* = x_i$. This is the most common Riemann Sum scheme because of its simplicity, though of course it is unlikely that the right endpoint of an interval is where we would find the "average" height for that interval. Nonetheless, it will give us the correct area as $n \to \infty$ and $\max\{|\Delta x_i|\} \to 0^+$.

Indeed, Figure 5.5 shows how much error can be reduced when the number of rectangles increases. In that particular case, since the function is increasing, using the right-endpoint method whereby $x_i^* = x_i$ we get that the Riemann Sums overestimate the actual areas. However, we decrease the actual and percent error dramatically when we, say, increase the number of rectangles by a factor of 2. According to the Fundamental Theorem of Calculus, when

5.3. RIEMANN SUMS AND THE SECOND FUNDAMENTAL THEOREM (FTC)

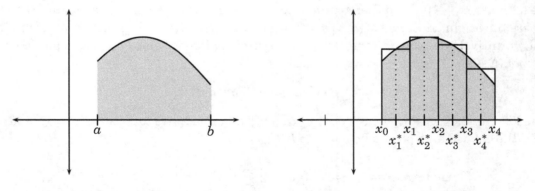

$$\text{Area} \approx f(x_1^*)(x_1 - x_0) + f(x_2^*)(x_2 - x_1) + f(x_3^*)(x_3 - x_2) + f(x_4^*)(x_4 - x_3)$$
$$= f(x_1^*)\Delta x_1 + f(x_2^*)\Delta x_2 + f(x_3^*)\Delta x_3 + f(x_4)\Delta x_4$$
$$= \sum_{i=1}^{4} f(x_i^*)\Delta x_i, \qquad \text{where } \Delta x_i = x_i - x_{i-1}.$$

Figure 5.4: Figure for general Riemann Sum, in the case of a positive function f. The actual area between the curve and the x-axis on some interval $[a,b]$ is approximated by a sum of areas of rectangles, where for each subinterval $[x_{i-1}, x_i]$ height is approximated by sampling one height $f(x_i^*)$ of the function in the interval, with $x_i^* \in [x_{i-1}, x_i]$ (the ith subinterval). The area of the ith rectangle will be $f(x_i^*)(x_i - x_{i-1}) = f(x_i^*)\Delta x_i$. When we add these together we get a *Riemann Sum*, approximating the actual net area (shaded).

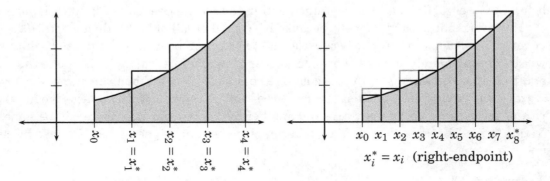

Figure 5.5: Illustration of typically improving area approximations of Riemann Sums when the subinterval lengths shrink and the number of subintervals increases. Both approximations use a *right-endpoint* scheme, where $x_i^* = x_i$ is used for sampling the height of f in the ith subinterval $[x_{i-1}, x_i]$. In the illustrations, the number of subintervals was doubled from 4 to 8, and clearly the percent error (represented here by the non-shaded areas within the rectangles as a fraction of the shaded areas) shrinks. The shaded area is the area to be approximated. Here both sets of rectangles' areas overestimate the desired area under the curve but clearly the second scheme—with twice as many rectangles—has significantly less overestimation than the first. (Gray lines in the graph on the right show where the larger rectangles on the left would cover.) As the number of rectangles is allowed to grow towards infinity, and their widths shrink to zero, the error will shrink towards zero.

we let $\max\{\Delta x_i\} \to 0^+$, and therefore $n \to \infty$, we will get a value which is equal to $F(b) - F(a)$, where the original interval is $[a,b]$ and F is an antiderivative of the function f on that interval. Intuitively (looking graphically at our approximation schemes), it seems also true that as $\max\{\Delta x_i\} \to 0^+$ and $n \to \infty$ we also get

$$\sum f(x_i^*) \Delta x_i \longrightarrow \text{Shaded Area}.$$

The FTC applies for any choices of $x_i^* \in [x_{i-1}, x_i]$, and so applies to the case $x_i^* = x_i$, and so by FTC the shaded area should be equal to $F(b) - F(a)$:

$$\text{Shaded Area} \xequal{\text{Intuition}} \lim_{\substack{\max\{\Delta x_i\} \to 0^+ \\ (n \to \infty)}} \left(\sum_{i=1}^n f(x_i^*) \Delta x_i \right)$$

$$\xequal{\text{Definition}} \int_a^b f(x)\,dx \xequal{\text{FTC}} F(x)\bigg|_a^b \xequal{\text{Definition}} F(b) - F(a).$$

In our case the star above can be removed and x_i inserted for x_i^*, since any choices of $x_i^* \in [x_{i-1}, x_i]$ will still, in the limit, have the Riemann Sums approach the same value (as stated by the FTC), namely $\int_a^b f(x)\,dx$.[21]

5.3.6 Computing Areas Using FTC or Riemann Sums

The previous argument is long and subtle, and the reader is encouraged to ponder it frequently. The upshot is that, geometrically, $\int_a^b f(x)\,dx$ represents the *net signed area* between the curve $y = f(x)$, $a \le x \le b$, and the x-axis (or that interval on the x-axis); the function $f(x)$ gives the height at each $x \in [a,b]$, with the "base" (of the region whose area we are computing) being the interval $[a,b]$ as it is contained within the x-axis. See again Figure 5.4, page 541.

However, that area is "signed" because if $f(x) < 0$ on all of $[a,b]$, then $\int_a^b f(x)\,dx$ will be negative as well, as we can see because each $f(x_i^*)\Delta x$ will be negative but its absolute value will be approximately the area between $f(x)$ and the ith interval $[x_{i-1}, x_i]$ along the x-axis, and this approximation will improve as $n \to \infty$ and $\Delta x \to 0^+$. When the curve is below the x-axis (thus having negative height), the "area" will be represented by a negative number. If part of the curve is above the x-axis and another part below the axis, there will be some area "cancellation." In this subsection, we will compute these net signed areas bounded by curves and the x-axis.

Example 5.3.3

Find the area bounded by the parabola $y = x^2$ and the x-axis along the interval $0 \le x \le 2$. (This is the same as the "net signed area" since $f(x) \ge 0$.)

Solution: While it helps to draw this to visualize the situation, it is not actually necessary. The function $f(x) = x^2$ is nonnegative, so any Riemann Sum approximation of the area will not contain

---

[21] In classical geometry we begin with areas of rectangles and from there derive areas of polygons. Then a limit argument that dates back to Archimedes (though the term "limit" came much later) gives us areas of circles.

For the areas considered here—namely those bounded by the horizontal axis and the graph of a continuous, nonnegative function defined on an interval $[a,b]$—it is technically simpler to just *define* the area in question as the integral $\int_a^b f(x)\,dx$, i.e., the limit of Riemann Sums. It seems consistent with our intuition of what "area" should mean, but given that area is not classically *defined* in a way that includes such cases, the integral definition seems as good as any.

negative "heights" of the rectangles. Once we are sufficiently convinced that shrinking widths and growing numbers of such rectangles will, in the limit, approach the actual area, we can invoke the FTC to compute the area, as is illustrated:

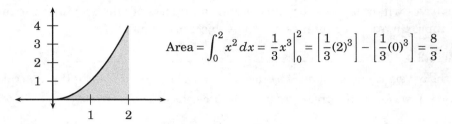

$$\text{Area} = \int_0^2 x^2\, dx = \frac{1}{3}x^3 \bigg|_0^2 = \left[\frac{1}{3}(2)^3\right] - \left[\frac{1}{3}(0)^3\right] = \frac{8}{3}.$$

If instead we wish to compute the area through limits of Riemann Sums directly, we would divide $[0,2]$ into n subintervals with endpoints $0 = x_0 < x_1 < x_2 < \cdots < x_{n-1} < x_n = 2$, and let $n \to \infty$. Before taking the limit, we note that the width of each subinterval would be $\Delta x_i = \frac{2-0}{n} = \frac{2}{n}$. Since this is the same for each subinterval, we will call that common width $\Delta x = \frac{2}{n}$. Furthermore, we can take any $x_i^* \in [x_{i-1}, x_i]$, so for simplicity we will take $x_i^* = x_i$ (the right endpoint) for each interval, which we further compute to be $x_i = 0 + i\Delta x = i \cdot \frac{2}{n} = \frac{2i}{n}$, for $i = 1, 2, \ldots, n$. Thus

$$\text{Area} = \lim_{\substack{\max\{|\Delta x_i|\} \to 0^+ \\ (n \to \infty)}} \sum_{i=1}^n f(x_i^*)\Delta x_i = \lim_{\substack{\{|\Delta x|\} \to 0^+ \\ (n \to \infty)}} \sum_{i=1}^n f(x_i)\Delta x = \lim_{n \to \infty} \sum_{i=1}^n \left[\left(\frac{2i}{n}\right)^2 \frac{2}{n}\right]$$

$$= \lim_{n \to \infty} \sum_{i=1}^n \frac{8i^2}{n^3} = \lim_{n \to \infty} \frac{8}{n^3} \sum_{i=1}^n i^2 = \lim_{n \to \infty} \left[\frac{8}{n^3} \cdot \frac{n(n+1)(2n+1)}{6}\right]$$

$$= \lim_{n \to \infty} \frac{8(2n^3 + 3n^2 + n)}{6n^3} = \frac{16}{6} = \frac{8}{3}.$$

We used (5.38), page 527, namely that $\sum_{i=1}^n i^2 = n(n+1)(2n+1)/6$. Note that when we write $\sum_{i=1}^n f(x_i)\Delta x$, in that expression n is a constant, and so if it appears as a factor (multiplier) inside of the summation, then it can be brought out (factored). However, no term involving i can be factored outside of the summation, because i is not constant within the summation, but changes values in the range $i = 1, 2, \ldots, n$.

For right-endpoint Riemann Sums (referring to the choice $x_i^* = x_i \in [x_{i-1}, x_i]$), where we wish to have a partition of the interval $[a,b]$ into n subintervals of equal length, with endpoints labeled $a = x_0 < x_1 < x_2 < \cdots < x_{n-1} < x_n = b$, we will always have the formulas

$$\Delta x = \frac{b-a}{n}, \tag{5.56}$$

$$x_i = a + i \cdot \Delta x. \tag{5.57}$$

As we can easily see, computing that area represented by $\int_0^2 x^2\, dx$ is much easier with the FTC than with the definition of the integral, that being the limit of Riemann Sums with $\max\{\Delta x_i\} \to 0^+$ and therefore $n \to \infty$, though it is perhaps satisfying to see both are viable options for this example, and their results agree. However, there are many opportunities for mistakes in setting up the Riemann Sums and in computing the limits as $n \to \infty$.

544　CHAPTER 5. BASIC INTEGRATION AND APPLICATIONS

Yet there are cases where finding an antiderivative is difficult or impossible with the functions we are used to, and so we have use for *approximating* the definite integral using Riemann Sums or other approximation methods. The more advanced techniques for finding antiderivatives will occupy us for one chapter (and parts of another) in the sequel while the integral approximation techniques are the subject of Section 5.7.

Example 5.3.4

Find the net signed area bounded by the curve $y = x^3$ and the x-axis for the interval $-2 \leq x \leq 2$.

Solution: If we follow the precedent from the previous example, we get

$$\text{Area} = \int_{-2}^{2} x^3 \, dx = \left.\frac{1}{4}x^4\right|_{-2}^{2} = \left[\frac{1}{4}(2)^4\right] - \left[\frac{1}{4}(-2)^4\right] = \frac{16}{4} - \frac{16}{4} = 0.$$

This may seem a bit odd at first, so we can also consider the right-endpoint Riemann Sums and compute their limits to verify this. Here $\Delta x_i = \Delta x = (b-a)/n = (2-(-2))/n = 4/n$, and $x_i = a + i \cdot \Delta x = -2 + i \cdot 4/n$. Thus, when we compute the Riemann Sums and take $n \to \infty$, we get

$$\sum_{i=1}^{n} f(x_i)\Delta x_i = \sum_{i=1}^{n} \left[\left(-2 + i \cdot \frac{4}{n}\right)^3 \cdot \frac{4}{n}\right] = \frac{4}{n}\sum_{i=1}^{n}\left((-2)^3 + 3(-2)^2\frac{4i}{n} + 3(-2)\frac{16i^2}{n^2} + \frac{64i^3}{n^3}\right)$$

$$= \frac{4}{n}\sum_{i=1}^{n}\left(-8 + \frac{48}{n}i - \frac{96}{n^2}i^2 + \frac{64}{n^3}i^3\right)$$

$$= \frac{-32}{n}(n) + \frac{192}{n^2} \cdot \frac{n(n+1)}{2} - \frac{384}{n^3} \cdot \frac{n(n+1)(2n+1)}{6} + \frac{256}{n^4} \cdot \frac{n^2(n+1)^2}{4}$$

$$\longrightarrow -32 + \frac{192}{2} - \frac{384 \cdot 2}{6} + \frac{256}{4}$$

$$= -32 + 96 - 128 + 64 = 0.$$

So the limit of the Riemann Sums agrees with the difference of values of some antiderivative when evaluated at the endpoints -2 and 2, namely that $\int_{-2}^{2} x^3 \, dx = 0$.

This still seems odd—that the area would be zero—until we note that there should be a cancellation of two "areas" that are identical, except that their signs are opposites. We can calculate the individual areas separately:

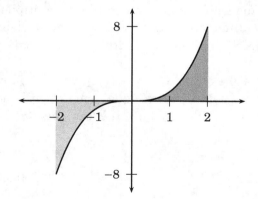

$$\int_{-2}^{0} x^3 \, dx = \left.\frac{1}{4}x^4\right|_{-2}^{0} = \frac{0^4}{4} - \frac{(-2)^4}{4} = -4,$$

$$\int_{0}^{2} x^3 \, dx = \left.\frac{1}{4}x^4\right|_{0}^{2} = \frac{2^4}{4} - \frac{0^4}{4} = 4,$$

Net area $= -4 + 4 = 0.$

We see again that using the *definition* of the definite integral is often quite difficult computationally, whereas the Fundamental Theorem of Calculus makes for short work if the

5.3. RIEMANN SUMS AND THE SECOND FUNDAMENTAL THEOREM (FTC)

antiderivative is simple. However, the definition is important conceptually and can help conceptually in other circumstances.

It is also important to note how areas of different sign can "cancel" in the right circumstances. If we are to believe that we can extend the general geometric notion of area so that it can have a positive or negative sign, and that the area of a region is the sum of its partial areas, then we should accept the first equality given next, and therefore the final result:

$$\int_{-2}^{2} x^3 \, dx = \int_{-2}^{0} x^3 \, dx + \int_{0}^{2} x^3 \, dx = -4 + 4 = 0.$$

When we have an antiderivative formula for the entire interval (such as $[-2,2]$ in the previous example) there is no need. However, sometimes we have antiderivative formulas for individual subintervals (as in Example 5.3.5 which follows) and other times there are geometric considerations which make a result simpler to obtain. For instance, in the previous example we could have noted the symmetry (with respect to the origin) of the *odd function* (i.e., one for which $f(-x) = -f(x)$ for all x in the domain), $f(x) = x^3$, and the symmetry of the interval, and noted that there was exactly as much "positive area" as there was "negative area," and therefore the net area would be zero.[22]

We used the following intuitive theorem, which we state without proof:

Theorem 5.3.3 *If $f(x)$ is continuous on $[a,c]$ and $b \in (a,c)$, then*

$$\int_{a}^{c} f(x) \, dx = \int_{a}^{b} f(x) \, dx + \int_{b}^{c} f(x) \, dx. \tag{5.58}$$

This theorem is useful in many contexts, but particularly useful when the function we are integrating is defined piece-wise, as in the following example.

Example 5.3.5

Compute the area under the curve of the function

$$f(x) = \begin{cases} x^2 & \text{if } x \leq 1, \\ \sqrt{x} & \text{if } x \geq 1 \end{cases}$$

over the interval $[0,2]$.

Solution: Here we have a function that is given by one formula for one interval of x-values, and another formula for another interval, and the area we wish to compute lies along an interval overlapping these. In a case such as this, we break the area's interval into two pieces, where each has a simple formula—with a useful antiderivative—for the the bounding function on that subinterval. Here we will use the following:

---

[22] In the sequel there are situations in which it is important not to use the argument about "canceling areas" if there is a chance one of the areas is infinite, as can happen near vertical asymptotes, for instance. We want to be careful not to be tempted to compute $\infty - \infty$ as being zero, for instance. (See for instance Example 2.8.1, page 177, and the related discussions.)

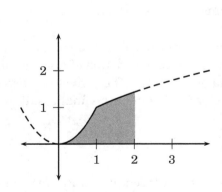

$$\int_0^2 f(x)\,dx = \int_0^1 f(x)\,dx + \int_1^2 f(x)\,dx$$
$$= \int_0^1 x^2\,dx + \int_1^2 x^{1/2}\,dx$$
$$= \left[\frac{1}{3}x^3\Big|_0^1\right] + \left[\frac{2}{3}x^{3/2}\Big|_1^2\right]$$
$$= \left[\frac{(1)^3}{3} - \frac{0^3}{3}\right] + \left[\frac{2}{3}(2)^{3/2} - \frac{2}{3}(1)^{3/2}\right]$$
$$= \frac{1}{3} + \frac{4\sqrt{2}-2}{3}$$
$$= \frac{4\sqrt{2}-1}{3} \approx 1.55228475.$$

Note that in this example, either formula was valid for computing $f(1)$ in the sense that $f(1) = 1 = (1)^2 = \sqrt{1}$. Indeed, the function $f(x)$ was continuous (so the FTC applies), as are both functions $x \mapsto x^2$ and $x \mapsto \sqrt{x}$ at $x = 1$. Thus there was no difficulty in using the formula $f(x) = x^2$ for $[0, 1]$ and $f(x) = \sqrt{x}$ for $[1, 2]$, even though $x = 1$ is shared by them.

In fact, for a single point such as $x = 1$, the "area" under the curve will be zero, so we are allowed some flexibility in using whatever formula for $f(x)$ matches everywhere in the interval, except perhaps at a finite number of points (themselves determining zero area between the curve and the x-axis). It is especially useful if we use a formula for $f(x)$ that represents a continuous function on the interval, so we can employ the FTC and go searching for an antiderivative.

Example 5.3.6

Suppose $f(x) = \begin{cases} x^2, & \text{if } x \ne 1, \\ 5, & \text{if } x = 1. \end{cases}$ Find $\int_0^3 x^2\,dx$.

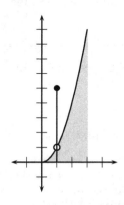

Solution Here we have a single point at which the function is discontinuous, namely $x = 1$. However, we should be able to convince ourselves that the area under that single point is zero, and so it can be ignored, and we can compute the area as if $f(1) = (1)^2$, in which case we could use the FTC:

$$\text{Area} = \int_0^3 x^2\,dx = \frac{x^3}{3}\Big|_0^3 = \frac{27}{3} - 0 = 9.$$

In fact if we go back to our Riemann Sum definition of $\int_0^3 f(x)\,dx$, we would see that even if we chose $x_i^* = 1$ for one of our intervals, the term $f(x_i^*)\Delta x_i$ would have its influence shrink

5.3. RIEMANN SUMS AND THE SECOND FUNDAMENTAL THEOREM (FTC)

to zero in the limit as $n \to \infty$, i.e., as $\max\{\Delta x_i\} \to 0^+$. We will use that same idea in the next example.

Example 5.3.7

Suppose $f(x) = \frac{|x|}{x}$, and we wish to find $\int_{-1}^{1} f(x)dx$. The function is undefined at $x = 0$, but intuitively the "area under the curve at $x = 0$" is itself zero, because the width of that one point is zero.[23]

So we can let $f(0)$ be redefined to be any finite value and compute the integral as in the previous example, ignoring the possible presence of $f(0)\Delta x_i$ in the Riemann sums whose limits we are ultimately computing.

However, we will have different expressions for $f(x)$ for the cases $x < 0$ and $x > 0$, at least if we want expression forms for which we can use our antiderivative formulas. So for this example we look at $\int_{-1}^{0} f(x)dx$ and $\int_{0}^{1} f(x)dx$ separately. Except at $x = 0$ the expressions for $f(x)$ have well-known antiderivatives, and so we "fill in" $f(0)$ for each one separately, with the values that would make $f(x)$ continuous at $x = 0$ on the respective intervals:

$$\int_{-1}^{1} f(x)dx = \int_{-1}^{0} f(x)dx + \int_{0}^{1} f(x)dx$$
$$= \int_{-1}^{0} (-1)dx + \int_{0}^{1} 1\, dx$$
$$= (-x)\Big|_{-1}^{0} + (x)\Big|_{0}^{1}$$
$$= -0 - [-(-1)] + [1 - 0]$$
$$= -1 + 1 = 0.$$

That the areas would "cancel," and indeed what their values are such that they would cancel, is clear when this function is graphed.

5.3.7 Total Area

Here we demonstrate how to compute "total area" between the graph of a function $f(x)$ defined on $[a,b]$ and the interval $[a,b]$ along the x-axis without regard to its sign: To be sure all area is counted as positive, we simply insert absolute values around the integrand:

$$\text{Total area} = \int_{a}^{b} |f(x)|\,dx. \tag{5.59}$$

This keeps the graph of the new integrand from dipping below the x-axis, so there will be no "negative" area counted within the integral, and therefore no area cancellation.

In evaluating the integral in (5.59), we have to realize that we can have different expressions for $|f(x)|$ for different subintervals, depending on the signs of $f(x)$ on the intervals. As in the previous examples, we may need to divide the interval in question into subintervals where $f(x)$ does not change sign—hopefully subintervals in which we can use antidifferentiation formulas applied to either $f(x)$ if $f(x) \geq 0$ in the subinterval, or $-f(x)$ if $f(x) \leq 0$ there.

---

[23]This is a subtle point that can easily be over-generalized, i.e., one can draw too many conclusions from this observation that $\int_{0}^{0} f(x)dx = 0$ regardless of $f(x)$. In fact, the integral makes no sense if $f(0)$ is undefined, but we expect the area to be zero if $f(0)$ is *any* real number, so it seems not unreasonable to disregard the behavior of $f(x)$ at a single point. Most advanced texts will allow for the integral in Example 5.3.7 even though $f(0)$ is undefined.

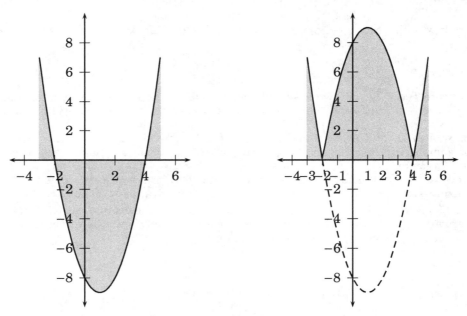

Figure 5.6: Graphs of $y = f(x) = x^2 + 2x - 8$ and the total (unsigned) area between the curve and the x-axis for $x \in [-3,5]$, along with the graph $y = |f(x)| = |x^2 + 2x - 8|$, with an equal (unsigned) area between that graph and the x-axis for $x \in [-3,5]$. See Example 5.3.8.

Example 5.3.8

Find the total area bound by the x-axis, the graph of the function $f(x) = x^2 - 2x - 8$, and the lines $x = -3$ and $x = 5$.

Solution: We note that $f(x) = (x-4)(x+2) = 0$ at $x = -2, 4$, and it is not difficult to see $f(x)$ changes sign at those two values. In fact $f(x) < 0$ on $(-2,4)$, $f(x) = 0$ at $x = -2, 4$, and $f(x)$ is positive elsewhere (see Figure 5.6). Thus the desired total area is given by

$$\int_{-3}^{5} |f(x)|\,dx = \int_{-3}^{-2} |f(x)|\,dx + \int_{-2}^{4} |f(x)|\,dx + \int_{4}^{5} |f(x)|\,dx$$
$$= \int_{-3}^{-2} f(x)\,dx + \int_{-2}^{4} (-f(x))\,dx + \int_{4}^{5} f(x)\,dx.$$

Taking these in turn,

$$\int_{-3}^{-2} (x^2 - 2x - 8)\,dx = \left[\frac{x^3}{3} - x^2 - 8x\right]\Big|_{-3}^{-2} = \frac{-8}{3} - 4 + 16 - \frac{-27}{3} + 9 - 24 = -3 + \frac{19}{3} = \frac{10}{3},$$

$$\int_{-2}^{4} [-(x^2 - 2x - 8)]\,dx = \left[\frac{-x^3}{3} + x^2 + 8x\right]\Big|_{-2}^{4} = \frac{-64}{3} + 16 + 32 - \frac{8}{3} - 4 + 16 = 36,$$

$$\int_{4}^{5} (x^2 - 2x - 8)\,dx = \left[\frac{x^3}{3} - x^2 - 8x\right]\Big|_{4}^{5} = \frac{125}{3} - 25 - 40 - \frac{64}{3} + 16 + 32 = \frac{61}{3} - 17 = \frac{10}{3}.$$

Summing these we get $\int_{-3}^{5} |f(x)|\,dx = \frac{10}{3} + 36 + \frac{10}{3} = 36 + \frac{20}{3} = 42 + \frac{2}{3} = \frac{128}{3} = 42.\overline{6}$.

A useful fact that helps to catch errors is that each definite integral above should have a

5.3. RIEMANN SUMS AND THE SECOND FUNDAMENTAL THEOREM (FTC)

nonnegative value. For a more complicated problem it can be useful to produce a sign chart for the function. (See also Footnote 24, page 550.)

5.3.8 Physics Application: Net Displacement versus Distance Traveled

Related to the previous section is the concept of computing an object's total distance traveled, given its velocity function. Recall that, over an interval $t_0 \leq t \leq t_f$, the *net displacement* of an object with position function $s(t)$ is given by $s(t_f) - s(t_0)$. Note that the fundamental theorem of calculus, and the definition of velocity given by $v(t) = s'(t)$, gives us

$$s(t_f) - s(t_0) = \int_{t_0}^{t_f} s'(t)\,dt = \int_{t_0}^{t_f} v(t)\,dt. \tag{5.60}$$

Example 5.3.9

We can compute the net displacements for the following velocity functions and their time intervals:

- $v(t) = \dfrac{1}{t}$, $1 \leq t \leq 10 \implies s(10) - s(1) = \int_1^{10} \dfrac{1}{t}\,dt = \ln t \Big|_1^{10} = \ln 10 - \ln 1 = \ln 10 \approx 2.302585.$

- $v(t) = \sqrt{t}$, $0 \leq t \leq 100 \implies s(100) - s(0) = \int_0^{100} t^{1/2}\,dt = \dfrac{2}{3} t^{3/2} \Big|_0^{100} = \dfrac{2}{3}\left[100^{3/2} - 1^{3/2}\right] = \dfrac{2}{3}(1000 - 1)$
$= \dfrac{2 \cdot 999}{3} = 666.$

If we wished to measure instead the *total distance* traveled by the object, we have to be sure that a negative-only velocity would still result in a positive distance traveled (even though the displacement would be negative).

$$\text{Distance traveled} = \int_{t_0}^{t_f} |v(t)|\,dt. \tag{5.61}$$

Example 5.3.10

Suppose a particle's velocity is given by $v(t) = 3\sin t$, for $t \in [0, 2\pi]$. Find the total distance travelled during this time interval.

Solution: First we note a sign change in $\sin t$ occurs at $t = \pi$, and so

$$|v(t)| = |3\sin t| = \begin{cases} 3\sin t, & \text{for } t \in [0, \pi) \\ -3\sin t, & \text{for } t \in [\pi, 2\pi). \end{cases}$$

When computing the total distance traveled, we must break the integral into two subintegrals with nonnegative integrands:[24]

$$\int_0^{2\pi} |v(t)|\,dt = \int_0^{2\pi} |3\sin t|\,dt = \int_0^{\pi} |3\sin t|\,dt + \int_{\pi}^{2\pi} |3\sin t|\,dt$$

$$= \int_0^{\pi} 3\sin t\,dt + \int_{\pi}^{2\pi} (-3\sin t)\,dt$$

$$= -3\cos t\Big|_0^{\pi} + 3\cos t\Big|_{\pi}^{2\pi}$$

$$= [-3\cos(\pi) - (-3\cos 0)] + [3\cos(2\pi) - 3\cos\pi] = 3 + 3 + 3 + 3 = 12.$$

If we had neglected the absolute values, we would instead have computed the *net (or total) displacement,* which would be

$$\int_0^{2\pi} 3\sin t\,dt = (-3\cos t)\Big|_0^{2\pi} = -3\cos 2\pi + 3\cos 0 = -3 + 3 = 0.$$

This is zero because the object's position was the same at $t = 0$ and $t = 2\pi$, though it had moved a total distance of 12 units (6 units right, followed by 6 units left) during $[0, 2\pi]$.

5.3.9 A Word on Units

We saw in previous chapters how the appropriate units and dimensions emerge from real-world derivative computations. They do so as well with integral computations, properly interpreted. Both derivatives and integrals are limits that naturally inherit the correct units from the variables involved, as long as those were assigned correct units themselves. For instance, if the units of $v(t)$ are meter/sec, and the time unit for t is sec, then the integral $\int_{t_0}^{t_f} v(t)\,dt$, equal to $s(t_f) - s(t_o)$ by the FTC (assuming $v(t)$ is continuous on $[t_0, t_f]$), will be in meters. This is clear from the Riemann Sums, whose limits give us that definite integral

$$\int_{t_0}^{t_f} v(t)\,dt = \lim_{\substack{\max\{\Delta t_i\}\to 0^+ \\ (n\to\infty)}} \sum_{i=1}^{n} \underbrace{v(t_i^*)}_{\text{meter/sec}} \underbrace{\Delta t_i}_{\text{sec}}.$$

Just as in the derivatives where $\frac{\Delta s}{\Delta t}$ takes on dimensions of length/time, which survive the limit process as this quotient approaches $\frac{ds}{dt}$, so too does this occur with the definite integral, and so it is acceptable to consider, say, dt to have the same units as Δt and therefore the same units as t, and $v(t)$ to have kinds of units it had before, such as in the following:

$$\int_{t_0}^{t_f} \underbrace{v(t)}_{\text{meter/sec}} \underbrace{dt}_{\text{sec}} = \underbrace{s(t_f) - s(t_0)}_{\text{meter}}.$$

---

[24] A shortcut method for computing $\int_a^b |f(x)|\,dx$ is to find those numbers $c_1, c_2, \ldots, c_m \in (a,b)$ where $f(c_i) = 0$, i.e., the only points where $f(x)$ could possibly change signs in $[a,b]$, and then compute each subintegral $\int_a^{c_1} f(x)\,dx$, $\int_{c_1}^{c_2} f(x)\,dx$, ..., $\int_{c_m}^{b} f(x)\,dx$, and then sum their absolute values. This is especially useful if there is a "nice" antiderivative formula that works for $f(x)$ throughout $[a,b]$. Then each signed area represented by one of the subintegrals has no area cancellation within it. By summing their absolute values we can be sure there is no cancellation among them, and the result would be the total geometric (unsigned) area.

5.3. RIEMANN SUMS AND THE SECOND FUNDAMENTAL THEOREM (FTC)

The final units at each stage of the computation are clearly meters, as we should expect.

Similarly, if we are considering an area computation as before, where we have $f(x)$ measuring a height (or directed length), and x a horizontal position, i.e., directed length from the y-axis, then $\int_a^b f(x)\,dx$ would be of dimension length × length = length^2.

As we see in subsequent sections, the definite integral has a very wide range of uses in analysis of physical quantities, and in those cases the units and dimensions do work out correctly, though sometimes a physical constant is necessary to account for all of the units. Such constants are already built into the underlying theory before the integrals are introduced, so the calculus is consistent.

Exercises

For Exercises 1–14, use the FTC to compute the value of the definite integral.

1. $\int_0^1 x\,dx$

2. $\int_2^6 5\,dx$

3. $\int_{-1}^1 (x^2 - 2x + 7)\,dx$

4. $\int_0^2 (x^3 - 8x^2 + 5x - 9)\,dx$

5. $\int_0^1 (2x-5)^2\,dx$

6. $\int_{-2}^2 (9-8x)\,dx$

7. $\int_0^9 \sqrt{x}\,dx$

8. $\int_{-8}^8 \sqrt[3]{x}\,dx$

9. $\int_{-1}^1 \frac{1}{x^2+1}\,dx$

10. $\int_1^{e^2} \frac{dx}{x}$

11. $\int_{1/3}^{1/2} \frac{1}{x^2}\,dx$

12. $\int_{\pi/6}^{\pi/4} \csc\theta \cot\theta\,d\theta$

13. $\int_{-\pi}^{\pi} \sin\theta\,d\theta$

14. $\int_0^{\pi/4} \sec^2\phi\,d\phi$

For Exercises 15–16, compute the given definite integral of the piecewise-defined function.

15. $\int_{-2}^5 |x|\,dx$

16. $\int_{-1}^1 f(x)\,dx$ where

$$f(x) = \begin{cases} 2x+1 & \text{if } x < 0 \\ 1-x^2 & \text{if } x \geq 0 \end{cases}$$

For Exercises 17–24,

(a) Graph the region whose net signed area is represented.

(b) Use the geometry to compute the net signed area.

(c) Show that the FTC gives the same area found in (b) above.

17. $\int_1^6 4\,dx$

18. $\int_0^4 5t\,dt$

19. $\int_2^5 (5-x)\,dx$

20. $\int_3^5 (2x-4)\,dx$

21. $\int_{-2}^{2} x^3 \, dx$

22. $\int_{0}^{2\pi} \sin x \, dx$

23. $\int_{-\pi/4}^{\pi/4} \tan x \, dx$. Later we will note that $\int \tan x \, dx = \ln|\sec x| + C$.

24. $\int_{-3}^{3} \sqrt{9 - x^2} \, dx$. In the sequel we have techniques to show that, for $a > 0$, we have $\int \sqrt{a^2 - x^2} \, dx = \frac{a^2}{2} \sin^{-1}\left(\frac{x}{a}\right) + \frac{x}{2}\sqrt{a^2 - x^2} + C$.[25]

25. $f(x)$ is continuous on $[a,b]$ and $F(x)$ is an antiderivative of $f(x)$ on the interval $[a,c]$, and $b \in (a,c)$. Using the FTC, prove Theorem 5.3.3, that
$$\int_a^c f(x)\,dx = \int_a^b f(x)\,dx + \int_b^c f(x)\,dx.$$
Also, draw and label an illustration to explain why this is reasonable.

26. Suppose $v(t) = t$ for $t \in [0,10]$, and we wish to compute $s(10) - s(0)$.

    (a) If we divide $[0,10]$ into $n = 5$ equal subintervals $[t_0, t_1]$, $[t_1, t_2]$, $\cdots$, $[t_4, t_5]$ and take $t_i^* = t_i$, show that $\sum_{i=1}^{n} v(t_i^*) \Delta t_i = 60$.

    (b) Do the same for $n = 10$, showing the sum is 55.

    (c) Do the same for $n = 20$, showing the sum is 52.5.

    (d) Compute $\int_0^{10} t \, dt$ using the Fundamental Theorem of Calculus.

    (e) Show that we get the same answer as in 26d if we let $n \to \infty$ in our Riemann Sums. (See for instance the second half of Example 5.3.3, page 542. You may also need a formula from (5.37)–(5.39), page 527.)

27. Consider the rectangle whose vertices are given in the Cartesian coordinate system as $(0,0)$, $(L,0)$, $(0,W)$, and (L,W), where $L, W > 0$.

    (a) Graph the region.

    (b) Set up the integral that represents the area as determined in part (a).

    (c) Evaluate the integral in part (b).

    (d) Verify that (c) corresponds to the formula for the area of a rectangle of length L and width W.

28. Consider a triangle whose vertices are given in the Cartesian coordinate system as $(0,0)$, $(0,b)$, and $(h,0)$ where $h, b > 0$.

    (a) Graph the region.

    (b) Set up the integral that represents the area. Recall that the hypotenuse of the triangle is part of the line connecting $(0,b)$ to $(h,0)$. Use the point-slope form to find the equation of this line.

    (c) Evaluate the integral of part (b).

    (d) Verify that this integral corresponds to the formula for the area of a triangle of base b and height h.

---

[25]While we do not yet have the ability to derive this as yet (the techniques required being available in the sequel), for this problem the geometry is much simpler. To see the region, one can take $y = \sqrt{9-x^2}$, square both sides, and get the equation of a well-known geometric object, though we must be aware that we need $y \geq 0$.

5.3. RIEMANN SUMS AND THE SECOND FUNDAMENTAL THEOREM (FTC)

29. Consider the rectangular region from Exercise 27.

    (a) Revolve the region about the x-axis and determine the geometric shape of the volume formed.

    (b) Evaluate the volume (in terms of L and W) by using a known geometric volume formula.

    (c) It will be shown later that this volume is given by the integral
    $$V = \pi \int_0^L [W^2]\, dx.$$
    Evaluate this integral.

    (d) Show that (b) and (c) give the same result.

30. Consider the triangular region from Exercise 28.

    (a) Revolve the region about the x-axis and determine the geometric shape of the volume formed.

    (b) Evaluate the volume (in terms of h and b) by using a known geometric volume formula.

    (c) It will be shown later that this volume is given by the integral
    $$V = \pi \int_0^h \left(-\frac{b}{h}x + b\right)^2 dx.$$
    Evaluate this integral.

    (d) Show that (b) and (c) give the same result.

5.4 First Fundamental Theorem of Calculus; Further Results

In this section we consider several theorems on definite integrals. These are intuitive enough on their faces, but it is also worth considering what their proofs might look like.

5.4.1 First FTC: Explanations

The First Fundamental Theorem of Calculus asserts the existence of an antiderivative, as needed in the Second FTC. Because the Second FTC is more commonly used, we introduced it earlier. For completeness and the introduction of other classes of computations, we include here a discussion of the First FTC. Since the two theorems are so closely related, they are called collectively The Fundamental Theorem of Calculus, or FTC, Parts 1 and 2.

Theorem 5.4.1 (Fundamental Theorem of Calculus, Part 1) *Suppose $f(x)$ is a continuous function on the closed interval $[a,b]$. If we define*

$$F(x) = \int_a^x f(t)\,dt, \tag{5.62}$$

then $F(x)$ exists and is continuous on $[a,b]$. Furthermore, for all $x \in (a,b)$ we have $F'(x) = f(x)$:

$$x \in (a,b) \implies F'(x) = \frac{d}{dx}\left[\int_a^x f(t)\,dt\right] = f(x). \tag{5.63}$$

Before we look at two arguments for the reasonableness of the theorem, a note on interpretation is in order. In the integrals $\int_a^x f(t)\,dt$ appearing in (5.62) and (5.63), since x is an endpoint (or "limit") of the integral, it cannot also be the variable of integration; we must use another variable that ranges from a to x within the integral (or underlying Riemann Sums). Indeed, it would make no sense to have an integral of the form $\int_a^x f(x)\,dx$ because the variable in the function and differential is supposed to *range* from a to x, and we cannot "let x range from a to x." So another variable, such as t, is used as the "variable of integration," as if we have a horizontal "t-axis" with a variable interval $[a,x]$ on that axis.[26]

The simplest way to show the reasonableness—but not a rigorous proof—of the First FTC is to use the Second FTC, itself shown reasonable from a velocity-position standpoint in the previous section. Similarly, one could argue for the reasonableness of the Second FTC as a result of that of the First FTC (hence the names). Still, it is useful to point out that if we were to *define* $F(x) = \int_a^x f(t)\,dt$, then we could *argue* that $F' = f$ on (a,b) is reasonable as follows:

---

[26]There are two standard (but hardly mandatory) practices for choosing the variable of integration in these situations: using t, unless the variable found in the endpoints (or "limits") of integration is itself t; or using the Greek analog of the variable which is an endpoint.

For instance, if we wished to have the integral be ultimately a function of t, the variable of integration is often the Greek letter τ (tau). If we want the integral to be a function of x, we usually will not use the Greek letter χ (chi) because (at least when handwritten) it too closely resembles x, so instead t or the Greek letter ξ (xi) is often used. Hence, in technical work it is common to see instances such as

$$s(t) = s(0) + \int_0^t v(\tau)\,d\tau \quad \text{and} \quad f(x) = f(a) + \int_a^x f'(\xi)\,d\xi.$$

5.4. FIRST FUNDAMENTAL THEOREM OF CALCULUS; FURTHER RESULTS

Explanation 1. We will leave the existence and continuity of $F(x)$ to the second explanation. Let $\Phi(x)$ be an antiderivative of $f(x)$ on $[a,b]$, so $\Phi'(x) = f(x)$ on (a,b) and $\Phi(x)$ has right-derivative $f(a)$ at $x = a$, and left-derivative $f(b)$ at $x = b$. Then for $x \in (a,b)$ we have

$$\frac{d}{dx}\left[\int_a^x f(t)dt\right] \xrightarrow{\text{Second FTC}} \frac{d}{dx}[\Phi(x) - \Phi(a)] = \Phi'(x) - 0 = f(x) - 0 = f(x).$$

This is basically (5.63) following from the Second FTC (from the last section). Moreover, this is the same as saying $F(x)$ from (5.62) satisfies $F(x) = \Phi(x) - \Phi(a)$ and has derivative $F'(x) = f(x)$ for $x \in (a,b)$, demonstrating that so does $F(x)$ since it differs from $\Phi(x)$ by a constant: $F'(x) = \frac{d}{dx}[F(x)] = \frac{d}{dx}[\Phi(x) - \Phi(a)] = \Phi'(x) - 0 = f(x)$ on (a,b), q.e.d.

Explanation 2. Here we try to explain the reasonableness of this First FTC in its own right. To do so, we begin with the notion that $\int_a^x f(t)dt$, as a limit of Riemann sums, represents the signed area between the curve $y = f(t)$ and the horizontal (t-) axis along the interval $[a,x]$. In our illustration, for simplicity we assume $f > 0$ and $\Delta x > 0$.

Suppose f is continuous on $[a,b]$ and $x \in (a,b)$. We then take $F(x) = \int_a^x f(t)dt$, i.e., the area under the graph of $y = f(t)$ for $a \le t \le x$, as shaded dark gray. The combined gray and light gray areas comprise $F(x + \Delta x)$.

The light gray area under the curve would then be equal to $F(x+\Delta x)-F(x)$, and if $\Delta x > 0$ is small enough, this would be very well approximated (in the sense of percent error) by the area in the thin rectangle within the light gray area, with height $f(x)$ and width Δx. Thus $F(x + \Delta x) - F(x) \approx f(x)\Delta x$, and from the illustration it is reasonable that the percent error in this approximation will approach 0 as $\Delta x \to 0$. *This percent error will remain the same if we divide by Δx*, i.e., for the approximation

$(F(x+\Delta x)-F(x))/\Delta x \approx f(x)$ with percent error for this approximation also approaching zero as $\Delta x \to 0$, and so $f(x) = F'(x)$, q.e.d.:

$$F'(x) = \lim_{\Delta x \to 0} \frac{F(x+\Delta x) - F(x)}{\Delta x} = f(x).$$

This demonstrates (albeit nonrigorously) that $F'(x)$ exists for each $x \in (a,b)$. It also follows that $F(x)$ is continuous on (a,b) since differentiability implies continuity (Theorem 3.2.1, page 229). However, it is very easy to find reasonable the continuity of F, even on all of $[a,b]$ (since the area it represents should change continuously, especially since the continuity of f on $[a,b]$ implies it is bounded there), and that F also has appropriate one-sided derivatives at a and b, as needed for the Second FTC. With the illustration, these facts are as easily believable as our previous (velocity-position) argument for the Second FTC. Furthermore, upon reflection it is clear from our illustration that the area represented by $\int_a^b f(t)dt$ is the same as that represented by the differences of areas $F(b) - F(a)$ (the latter being zero), since we took $F(x)$ to be an area function. This completes Explanation 2.[27]

---

[27]Explanation 1 demonstrates how the First FTC is compatible with the Second FTC, and follows from it, but

These explanations are complex enough that a full understanding is not likely to occur with one reading, and so these should be revisited on occasion to help the reader grasp the rich contexts in which these facts and insights interconnect.

5.4.2 Derivative Computations Using the First (and Second) FTC

Here we use a somewhat different, but more robust, explanation of how to compute "derivatives of integrals." Those for which we can use the First FTC directly are plentiful and discussed first, but a slightly more complicated technique, which uses more of the conclusions of the (earlier discussed) Second FTC, generalizes them most robustly, though for the simplest examples like the following it may seem excessive.

Example 5.4.1

Compute $\dfrac{d}{dx}\left[\displaystyle\int_0^x \cos t\, dt\right]$.

Solution: We will discuss three methods for computing this. For more complicated cases, the third is the most robust.

(a) Clearly the integrand, being continuous on all of $\mathbb{R}$, will be continuous on any interval with endpoints 0 and x, and in an open interval containing x (as in the First FTC). Because of the form of the derivative we are to compute, we can use the First FTC (5.63), page 554, directly (with $a = 0$):

$$\frac{d}{dx}\left[\int_0^x \cos t\, dt\right] = \cos x.$$

(b) Since we know the antiderivative of $\cos t$ with respect to t, we can instead compute this as follows:

$$\frac{d}{dx}\left[\int_0^x \cos t\, dt\right] = \frac{d}{dx}\left[\sin t\Big|_0^x\right] = \frac{d}{dx}[\sin x - 0] = \cos x.$$

(c) In fact, we did not need to know the antiderivative of the cosine function; we only needed to know such an antiderivative exists (which it does, also by the First FTC), say $F(t)$, such that $F'(t) = \cos t$, and then we have

$$\frac{d}{dx}\left[\int_0^x \cos t\, dt\right] = \frac{d}{dx}[F(x) - F(0)] = F'(x) - 0 = \cos x.$$

Each of the techniques (a)–(c) deserves further explanation.

To use the First FTC directly in (a) required not only a continuous integrand, but that the left endpoint (or "lower limit of integration") was constant, and the upper endpoint—which actually could be less than or equal to the lower endpoint and this method would still apply—was our variable of differentiation (x in this case).[28]

---

as was mentioned, in fact one usually uses the First FTC to prove the Second FTC. None of the explanations offered here are rigorous proofs, because a true proof of either FTC would require tools of analysis and topology which are found in upper-level undergraduate or introductory graduate mathematics courses. So here we task ourselves with simply showing the reasonableness and internal consistency of these concepts.

[28] It can be easily shown that the First FTC actually generalizes to cases where $a > b$, and $x \in (b,a)$, in which case the "lower" endpoint is greater then the "upper" endpoint. The Riemann sums in the limit defining the integral $\int_a^x f(t)dt$ would have negative values of Δt_i, and the partition would read $a = t_1 > t_2 > t_3 > \cdots > t_{n-1} > t_n = x$. All of the arguments about percentage error and such would transfer to this kind of case, with only a sign change, which would be consistent across the argument.

5.4. FIRST FUNDAMENTAL THEOREM OF CALCULUS; FURTHER RESULTS

To use the method in (b), we needed to know the antiderivative of $\cos t$ (and that $\cos t$ was continuous over the interval of integration, which it is). We also used that $\sin 0$ is a constant, though it was even simpler than that because $\sin 0 = 0$.

To use the method of (c), we only needed to know that an antiderivative for $\cos t$ exists on the interval with endpoints 0 and x, which is guaranteed by the continuity of the cosine function and the First FTC. We also used that $F(0)$ is a constant, and therefore has derivative zero.

Example 5.4.2

Here we compute the following derivatives, using the method in (c).

- $\dfrac{d}{dx}\left[\displaystyle\int_0^x \cos^2 \xi\, d\xi\right]$. Here we can use the First FTC directly, but we can also suppose $F(\xi)$ is an antiderivative of $\cos^2 \xi$ (with respect to the input, in this case ξ), so that $F'(\xi) = \cos^2 \xi$. Then

$$\frac{d}{dx}\left[\int_0^x \cos^2 \xi\, d\xi\right] \xsquigarrow{F'(\xi)=\cos^2\xi} \frac{d}{dx}\left[F(\xi)\Big|_0^x\right] = \frac{d}{dx}[F(x)-F(0)] = F'(x) - 0 = \cos^2 x.$$

- $\dfrac{d}{dx}\left[\displaystyle\int_x^{2\pi} \sin t\, dt\right]$. Here we cannot use the exact statement of the First FTC directly because the endpoints of integration have been switched. However, we can again take some function $F(t)$ so that $F'(t) = \sin t$ and compute

$$\frac{d}{dx}\left[\int_x^{2\pi} \sin t\, dt\right] \xsquigarrow{F'(t)=\sin t} \frac{d}{dx}\left[F(t)\Big|_x^{2\pi}\right] = \frac{d}{dx}[F(2\pi)-F(x)] = 0 - F'(x) = -\sin x.$$

We could also use the fact that $\int_a^b f(t)\,dt = -\int_b^a f(t)\,dt$, which we will explore later (but is a straightforward observation regarding the underlying Riemann Sums), and then invoke the First FTC, computing $\frac{d}{dx}\left[\int_x^{2\pi} \sin t\, dt\right] = \frac{d}{dx}\left[-\int_{2\pi}^x \sin t\, dt\right] = -\sin x$ as before.

- $\dfrac{d}{dx}\left[\displaystyle\int_{\sin x}^{x^3} e^{z^2}\, dz\right]$. Assume $F'(z) = e^{z^2}$, i.e., $F(z)$ is an antiderivative of e^{z^2} with respect to z:

$$\frac{d}{dx}\left[\int_{\sin x}^{x^3} e^{z^2}\, dz\right] \xsquigarrow{F'(z)=e^{z^2}} \frac{d}{dx}\left[F(x^3)-F(\sin x)\right] = F'(x^3)\cdot \frac{dx^3}{dx} - F'(\sin x)\cdot \frac{d\sin x}{dx}$$

$$= e^{(x^3)^2}\cdot 3x^2 - e^{(\sin x)^2}\cdot \cos x = 3x^2 e^{x^6} - e^{\sin^2 x}\cos x.$$

If the integrand is continuous over the interval of integration (as for instance $f(z) = e^{z^2}$ is over an interval with endpoints $\sin x$ and x^3 in our last computation), we know such an antiderivative F exists and we can simply make note of its derivative, therefore omitting the paragraph-style explanations we had. The technique is illustrated again next:

Example 5.4.3

$$\frac{d}{dt}\left[\int_{\cos t}^{\sin t} \sin^2 x\, dx\right] \xsquigarrow{F'(x)=\sin^2 x} \frac{d}{dt}[F(\sin t)-F(\cos t)] = F'(\sin t)\cdot \frac{d\sin t}{dt} - F'(\cos t)\cdot \frac{d\cos t}{dt}$$

$$= \sin^2(\sin t)\cos t - \sin^2(\cos t)(-\sin t) = \sin^2(\sin t)\cos t + \sin^2(\cos t)\sin t.$$

The first equality is crucial, as is the correct application of the Chain Rule in the second equality. The logic of the procedures is what is crucial. However, if one were performing such computations routinely, arguably worth noting is a general rule that follows from that logic:[29]

Theorem 5.4.2 *Suppose f is continuous on an open interval containing the endpoints $g(x)$ and $h(x)$, and that $g'(x)$ and $h'(x)$ exist. Then*

$$\frac{d}{dx}\left[\int_{g(x)}^{h(x)} f(\xi)d\xi\right] = f(h(x))\cdot h'(x) - f(g(x))\cdot g'(x)$$
$$= f(h(x))\cdot \frac{d\,h(x)}{dx} - f(g(x))\cdot \frac{d\,g(x)}{dx}. \qquad (5.64)$$

With the assumptions of the differentiabilities of $g(x)$ and $h(x)$, and the bit of "wiggle room" afforded by having their values within an open interval on which f is continuous, we are assured of the differentiability of the integral.

This theorem would have made slightly quicker work of the computations in the previous examples, as for instance we could have computed (noting that all involved functions are continuous and differentiable on all of $\mathbb{R}$):

- $\dfrac{d}{dx}\left[\displaystyle\int_0^x \cos^2 \xi\, d\xi\right] = \cos^2 x \cdot \dfrac{dx}{dx} - \cos^2 0 \cdot \dfrac{d0}{dx} = \cos^2 x.$

- $\dfrac{d}{dx}\left[\displaystyle\int_x^{2\pi} \sin t\, dt\right] = \sin(2\pi)\cdot \dfrac{d\,2\pi}{dx} - \sin x \cdot \dfrac{dx}{dx} = -\sin x.$

- $\dfrac{d}{dx}\left[\displaystyle\int_{\sin x}^{x^3} e^{z^2} dz\right] = e^{(x^3)^2}\cdot \dfrac{d\,x^3}{dx} - e^{(\sin x)^2}\cdot \dfrac{d\sin x}{dx} = 3x^2 e^{x^6} - e^{\sin^2 x}\cos x.$

Of course, even slightly complicated formulas are easily misremembered, whereas the logic of writing the definite integral as a difference of antiderivatives, and the application of the Chain Rule, should become basic knowledge, from which these computations can be made without reference to Equation (5.64). (Granted, if one knows the logic of the FTC and Chain Rule, then upon moderate reflection (5.64) in some sense explains itself.)

On the other hand, we cannot use this logic to compute the following integral:

$$\frac{d}{dx}\left[\int_{-e^x}^{e^x} \frac{1}{t^2}dt\right] \;\neq\; \frac{1}{(e^x)^2}\cdot \frac{d\,e^x}{dx} - \frac{1}{(-e^x)^2}\cdot \frac{d(-e^x)}{dx} = \frac{1}{e^x} + \frac{1}{e^x} = \frac{2}{e^x},$$

the *inequality* being correct because the integrand $1/t^2$ is not even defined at the midpoint of the interval $[-e^x, e^x]$ regardless of x. Indeed, later study would show that the definite integral within the brackets is infinite regardless of $x \in \mathbb{R}$, and so that derivative cannot exist, and we are reminded that the conclusion of a theorem is only guaranteed if its hypotheses are true!

---

[29]Here we used F as *an* antiderivative of the integrand function, not necessarily that specific antiderivative $F(x) = \int_a^x f(t)dt$ from (5.62), page 554. Recall that it does not matter here, since any two antiderivatives differ by a constant, which cancels in our computations with definite integrals.

5.4. FIRST FUNDAMENTAL THEOREM OF CALCULUS; FURTHER RESULTS

5.4.3 Other General Theorems of Definite Integrals

The following theorems can follow from our two Fundamental Theorems of Calculus, or from observations regarding the underlying Riemann Sums. We will discuss both approaches to demonstrating these things.

Theorem 5.4.3 *Assuming $f(x)$ is continuous on the closed interval with endpoints a and b, then*

$$\int_b^a f(x)\,dx = -\int_a^b f(x)\,dx. \tag{5.65}$$

This follows from the second FTC and the remarks therein. If one believes the statement of the second FTC, then a proof is relatively quick. We simply let $F(x)$ be an antiderivative of $f(x)$ on the given interval, and then

$$\int_b^a f(x)\,dx = F(a) - F(b) = -(F(b) - F(a)) = -\int_a^b f(x)\,dx, \quad \text{q.e.d.}$$

However, while somewhat involved in its mechanics, it is worthwhile to consider a Riemann Sum-style argument. For that, we consider how we would make partitions for the two integrals. In fact, we can use the same partitions, but with their subinterval endpoints being labeled slightly differently and their respective signed "widths" being opposites. (In other words, we sum the terms in the opposite order because we travel opposite directions as we traverse the partitions.)

For simplicity, we will assume $a < b$ (because our argument will have a symmetry by which it would also work for $b < a$). Under this assumption, it makes sense to consider the interval $[a, b]$. Using different variables to be sure all of our symbols are well defined, we could have the partitions

$$a = x_0 < x_1 < x_2 < \cdots < x_{n-1} < x_n = b \text{ for } \int_a^b f(x)\,dx,$$

$$b = \xi_0 > \xi_1 > \xi_2 > \cdots > \xi_{n-1} > \xi_n = a \text{ for } \int_b^a f(\xi)\,d\xi = \int_b^a f(x)\,dx.$$

where we can take $\xi_i = x_{n-i+1}$ and $\Delta \xi_i = -\Delta x_{n-i+1}$, and $\xi_i^* = x_{n-i+1}^*$, where $i = 1, 2, \ldots, n$. In other words, as we add terms $f(x_i^*)\Delta x_i$ as we move rightward through the list of partition intervals for the integral $\int_a^b f(x)\,dx$, we can simultaneously move leftward through the list for the integral $\int_b^a f(\xi)\,d\xi$ (which is the same as $\int_b^a f(x)\,dx$). In terms of Riemann Sums, we have

$$\underbrace{\sum_{i=1}^n f(\xi_i^*)\Delta\xi_i}_{b=\xi_0 > \xi_1 > \cdots > \xi_n = a} = \sum_{i=1}^n \left[f(x_{n-i+1}^*)(-\Delta x_{n-i+1})\right] = -\sum_{i=1}^n f(x_{n-i+1}^*)\Delta x_{n-i+1} = -\underbrace{\sum_{j=1}^n f(x_j^*)\Delta x_j}_{a = x_0 < x_1 < \cdots < x_n = b},$$

where in the final summation we sum the same terms as the one before it, but in the opposite order. When we take the limits as $\max\{|\Delta x_i|\} = \max\{|\Delta \xi_i|\} \to 0$ (and therefore $n \to \infty$), the left-hand side approaches $\int_b^a f(\xi)\,d\xi$ and the right-hand side approaches $-\int_a^b f(x)\,dx$, so these must be equal. Since the variable of integration is irrelevant (be it ξ or x), we again demonstrated that $\int_b^a f(x)dx = -\int_a^b f(x)dx$, q.e.d.

Next we further develop Theorem 5.3.3, page 545, namely that if the following integrals make sense (exist) we have

$$\int_a^c f(x)dx = \int_a^b f(x)dx + \int_b^c f(x)dx. \tag{5.66}$$

The theorem has added appeal in that it somewhat resembles the "cancellation" one gets with certain kinds of fraction multiplication, as in $\frac{c}{a} = \frac{b}{a} \cdot \frac{c}{b}$, though one must be sure that the conditions for that cancellation are met. If we briefly suppress the integrand and differential, we could write

$$\int_a^c = \int_a^b + \int_b^c,$$

which has its appeal. Either way, the intuition should be clear: that to achieve the area for the range of x between a and c, we can instead look at the area from a to b, and the area from b to c separately, and then sum them. In fact, by the Second FTC this is rather straightforward, for if $F(x)$ is an antiderivative of $f(x)$, (5.66) simply becomes

$$F(c) - F(a) = F(b) - F(a) + F(c) - F(b),$$

which is pretty obviously true once one notices what cancels. Furthermore, by our computation it is clearly not required that b be between a and c, but only that F be an antiderivative which is valid for an interval containing all three values a, b, c. In the case of definite integrals, the theorem relies on the existence of the underlying integrals (while in the case of the fractions, it is necessary that denominators be nonzero).

Example 5.4.4

Suppose $\int_1^2 f(x)dx = 2$, $\int_2^3 f(x)dx = 5$ and $\int_1^{10} f(x)dx = 13$. Find $\int_1^3 f(x)dx$ and $\int_3^{10} f(x)dx$.

Solution: We must compute these using the given definite integrals.

- For the first desired integral, we compute $\int_1^3 f(x)dx = \int_1^2 f(x)dx + \int_2^3 f(x)dx = 2 + 5 = 7$.

- For this second integral, we can use $\int_1^3 f(x)dx + \int_3^{10} f(x)dx = \int_1^{10} f(x)dx$, for which we know the values of first and third integrals and wish to know the second, and so we can compute

$$\int_3^{10} f(x)dx = \int_1^{10} f(x)dx - \int_1^3 f(x)dx = 13 - 7 = 6.$$

It is worth noting the intuition behind that last integral equation stands on its own, for if each of the following integrals are defined, we have (where we suppress the terms $f(x)dx$):

$$\int_a^c - \int_a^b = \int_b^c, \quad \text{and} \quad \int_a^c - \int_b^c = \int_a^b. \tag{5.67}$$

These can be pictured as (net, signed) area subtraction phenomena, as for instance the area for $x \in [a,b]$ is subtracted from the area for $x \in [a,c]$, leaving the area for $x \in [b,c]$.

5.4. FIRST FUNDAMENTAL THEOREM OF CALCULUS; FURTHER RESULTS

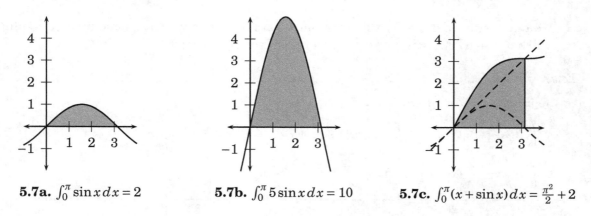

5.7a. $\int_0^\pi \sin x\, dx = 2$ **5.7b.** $\int_0^\pi 5\sin x\, dx = 10$ **5.7c.** $\int_0^\pi (x+\sin x)\, dx = \frac{\pi^2}{2}+2$

Figure 5.7: Graphic illustration of Theorem 5.4.4. In Figure 5.7b we see how a constant multiple of a function similarly scales any definite integral of that function, and in Figure 5.7c we see how the areas under different functions add when we take the area under the sum of those functions. Both computations follow perhaps more readily from the FTC or Riemann Sum considerations than from the geometry.

5.4.4 Linear Combinations of Integrands

Another useful theorem regarding definite integrals is the following, assuming all integrals in question exist, as occurs under the assumptions of the theorem.

Theorem 5.4.4 *Suppose f and g are continuous on $[a,b]$ and $k \in \mathbb{R}$. Then*

$$\int_a^b kf(x)\,dx = k \cdot \int_a^b f(x)\,dx, \tag{5.68}$$

$$\int_a^b [f(x)+g(x)]\,dx = \int_a^b f(x)\,dx + \int_a^b g(x)\,dx. \tag{5.69}$$

From the mechanics of the Second FTC and our previous antiderivative computations, these may seem obvious enough, though the latter is probably less obvious geometrically. Equations (5.68) and (5.69) can be summarized in the following equation, for any $\alpha, \beta \in \mathbb{R}$:[30]

$$\int_a^b [\alpha f(x)+\beta g(x)]\,dx = \alpha \int_a^b f(x)\,dx + \beta \int_a^b g(x)\,dx. \tag{5.70}$$

---

[30]Equivalences those like discussed here, namely (5.68) ∧ (5.69) ⟺ (5.70), appear in many subjects in which these two notions of "linearity" are often utilized: one that states that a linear operator (such as d/dx or $\int_a^b (\ )\,dx$) preserves constant multipliers and sums, and one that says the operator preserves any "linear combinations." (A linear combination of $f(x)$ and $g(x)$ is a function of the form $\alpha f(x)+\beta g(x)$, with $\alpha, \beta \in \mathbb{R}$.) To show (5.70) ⟹ (5.68) ∧ (5.69) would have us first suppose (5.70) is true and note that $\alpha = k, \beta = 0$ is just one case and yields (5.68), while $\alpha = \beta = 1$ is another case and yields (5.69).

Conversely, we can show (5.68) ∧ (5.69) ⟹ (5.70) by noting that if we assume both (5.68) and (5.69), then for any $\alpha, \beta \in \mathbb{R}$ we would have

$$\int_a^b [\alpha f(x)+\beta g(x)]\,dx \overset{(5.69)}{=\!=\!=} \int_a^b \alpha f(x)\,dx + \int_a^b \beta g(x)\,dx \overset{(5.68)}{=\!=\!=} \alpha \int_a^b f(x)\,dx + \beta \int_a^b g(x)\,dx, \quad \text{q.e.d.}$$

562 CHAPTER 5. BASIC INTEGRATION AND APPLICATIONS

We have occasional use for this theorem, especially when we have some further results for which to use it. For now we just offer a pair of simple examples.

Example 5.4.5

Suppose $\int_0^1 f(x)dx = 8$ and $\int_0^1 g(x)dx = 5$. Then

- $\int_0^1 [3f(x)+2g(x)]dx = 3\int_0^1 f(x)dx + 2\int_0^1 g(x)dx = 3(8)+2(5) = 34.$
- $\int_0^1 [f(x)-g(x)]dx = \int_0^1 f(x)dx - \int_0^1 g(x)dx = 8-5 = 3.$

The second result in the example follows from a special case of (5.69), page 561, in which $\alpha = 1$ and $\beta = -1$, giving us

$$\int_a^b [f(x)-g(x)]dx = \int_a^b f(x)dx - \int_a^b g(x)dx. \tag{5.71}$$

5.4.5 Integrating Odd and Even Functions over Symmetric Intervals

Here we note that there are some functions whose geometries allow us to take some shortcuts in computing their integrals, sometimes avoiding antiderivative computations entirely. Recall f is (respectively) *odd* if $f(-x) = -f(x)$ and *even* if $f(-x) = f(x)$, for all x in the domain.

Example 5.4.6

Compute $\int_{-8}^8 \sqrt[3]{x}\,dx$ and $\int_{-\pi}^{\pi} \sin^5\theta\,d\theta$.

Solution: The first we can compute directly, and the second we will see how to compute directly in the sequel. However, in either case the integrals have value zero, and in both cases some effort is saved by making that observation.

To see this, note that if $f(x) = \sqrt[3]{x}$ and $g(\theta) = \sin^5\theta$, both are *odd functions*, meaning that $f(-x) = -f(x)$ and $g(-\theta) = -g(\theta)$. This means that their graphs are "symmetric with respect to the origin," itself meaning that if there was an inverting lens at the origin, then the image on one side of the lens (origin) is the upside-down, backward image of the other: If (x,y) is on the graph of the function, then so is $(-x,-y)$.

In the case of $f(x) = \sqrt[3]{x}$, we have $f(-x) = \sqrt[3]{-x} = -\sqrt[3]{x} = -f(x)$, and so $y = f(x)$ has that symmetry with respect to the origin. Similarly, if $g(\theta) = \sin^5\theta$, then $g(-\theta) = \sin^5(-\theta) = (\sin(-\theta))^5 = (-\sin\theta)^5 = -\sin^5\theta = -g(\theta)$. Since both functions are continuous on all of $\mathbb{R}$, the integrals in question exist (as limits of Riemann Sums). Next we look at the geometries of their graphs:

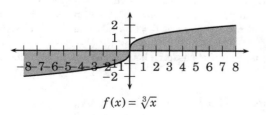

$f(x) = \sqrt[3]{x}$

It is not difficult to see that

$$\int_{-8}^0 \sqrt[3]{x}\,dx = -\int_0^8 \sqrt[3]{x}\,dx,$$

$$\Longrightarrow \int_{-8}^8 \sqrt[3]{x}\,dx = \int_{-8}^0 \sqrt[3]{x}\,dx + \int_0^8 \sqrt[3]{x}\,dx = 0.$$

5.4. FIRST FUNDAMENTAL THEOREM OF CALCULUS; FURTHER RESULTS

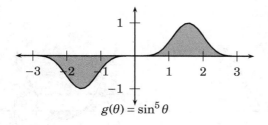

Similarly it is clear that
$$\int_{-\pi}^{0} \sin^5\theta\, d\theta = -\int_{0}^{\pi} \sin^5\theta\, d\theta,$$
$$\Longrightarrow \int_{-\pi}^{\pi} \sin^5\theta\, d\theta = \int_{-\pi}^{0} \sin^5\theta\, d\theta + \int_{0}^{\pi} \sin^5\theta\, d\theta = 0.$$

The graphs are instructive but not necessary; we only needed to verify that they were odd functions with this symmetry with respect to the origin, and that they were continuous and so their integrals existed. That the integrals made sense as signed net areas made our conclusion more apparent. This leads to a theorem:

Theorem 5.4.5 *Suppose $f(x)$ is continuous on $[-a,a]$ and an odd function on that interval. Then*
$$\int_{-a}^{a} f(x)\, dx = 0.$$

A proof from Riemann Sums can be achieved, as can a proof using later principles, but the geometry makes it more obvious. In all cases, one shows that the integral over $[-a,0]$ is the additive inverse of the integral over $[0,a]$, and so their sum is zero.

This is especially useful if we have an odd, continuous function whose antiderivative is difficult or impossible to write as a combination of our usual functions, and we have "symmetric limits (endpoints)" of integration. It can also be useful if any term of our integrand is odd and the limits symmetric:

Example 5.4.7

Compute $\int_{-10}^{10} \left(x^2 + e^{x^2} \sin x\right) dx$.

Solution: First we note that if $g(x) = e^{x^2} \sin x$, then
$$g(-x) = e^{(-x)^2} \sin(-x) = e^{x^2}(-\sin x) = -e^{x^2} \sin x = -g(x),$$

and so the second term in the integral is an odd function. Thus
$$\int_{-10}^{10} \left(x^2 + e^{x^2} \sin x\right) dx = \int_{-10}^{10} x^2\, dx + \int_{-10}^{10} e^{x^2} \sin x\, dx = \frac{1}{3}x^3\Big|_{-10}^{10} + 0 = \frac{1000}{3} - \frac{-1000}{3} = \frac{2000}{3}.$$

We are not likely to find a simple antiderivative for $e^{x^2} \sin x$, so it is fortunate that this function is odd, and that our interval has zero as its midpoint.

Also worth noting is that when we have an even function, i.e., one in which $f(-x) = f(x)$, then we have symmetry with respect to the y-axis, and so the graph to the right of the y-axis is the mirror-image of the graph to the left. Thus we have the following:

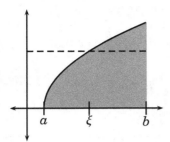 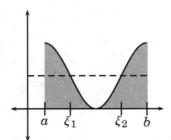 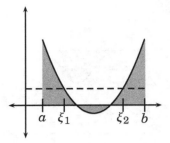

Figure 5.8: Three illustrations of function averages. The dashed horizontal lines indicate the average values (or heights) of the respective functions on the intervals, and the values ξ indicate where the functions achieve those respective values.

Theorem 5.4.6 *Suppose $f(x)$ is continuous on $[-a,a]$, and is an even function on that interval. Then*
$$\int_{-a}^{a} f(x)\,dx = 2\int_{0}^{a} f(x)\,dx.$$

This theorem could have made our computation at the end of the last example slightly quicker, since $f(x) = x^2$ is an even function:

$$\int_{-10}^{10} x^2\,dx = 2\int_{0}^{10} x^2\,dx = 2\left.\frac{x^3}{3}\right|_{0}^{10}$$
$$= 2\cdot\left[\frac{1000}{3} - 0\right] = \frac{2000}{3}.$$

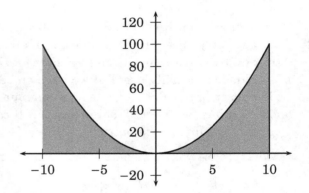

This is particularly useful when the obvious antiderivative's value is zero at $x = 0$.[31]

5.4.6 Average Function Value and MVT for Integrals

A very natural definition for the average of a function value over a closed interval is the following (see Figure 5.8):

Definition 5.4.1 *Suppose $f(x)$ is continuous on $[a,b]$. Then the **average value** (or **integral average**) of $f(x)$ on $[a,b]$ is defined to be*

$$\frac{1}{b-a}\int_{a}^{b} f(x)\,dx. \tag{5.72}$$

This is consistent, for example, with the average velocity over a time period $[t_0, t_f]$ when a particle's velocity is $v(t)$. Except for the division by $t_f - t_0$, the following is simply a statement

---

[31] In fact, an interesting corollary of the Chain Rule is that the derivative of any odd and differentiable function is an even function, and the derivative of any even and differentiable function is an odd function. Conversely, a continuous even function will have an odd antiderivative, and a continuous odd function will have an even antiderivative (infinitely many, in fact, since the additive constants are themselves even functions).

5.4. FIRST FUNDAMENTAL THEOREM OF CALCULUS; FURTHER RESULTS

of the Second Fundamental Theorem of Calculus, though it can also be read that the familiar average velocity from before is the same as the integral average velocity as in (5.72):

$$\frac{s(t_f)-s(t_0)}{t_f-t_0} = \frac{1}{t_f-t_0}\int_{t_0}^{t_f} v(t)\,dt.$$

The geometric intuition is slightly different (as it often is). We can instead look at the integral average as the height that would give the same (net, signed) area if the function were instead constant over that interval $[a,b]$. Or we could think of (5.72) as giving the *average height* of the graph of $y = f(x)$ over the interval $a \le x \le b$.

What is interesting is that if the function $f(x)$ is continuous on that closed interval $[a,b]$, then there exists at least one point $\xi \in (a,b)$ at which the function will equal its average for the interval $[a,b]$. This follows from the previous Mean Value Theorem (MVT) of Section 4.3 (specifically page 449). We state it next as a theorem:

Theorem 5.4.7 (Mean Value Theorem (MVT) for Integrals) *Suppose $f(x)$ is continuous on $[a,b]$. Then there exists $\xi \in (a,b)$ such that*

$$f(\xi) = \frac{1}{b-a}\int_a^b f(x)\,dx. \tag{5.73}$$

Proof: To prove this, we begin by letting $F(x)$ be an antiderivative of $f(x)$ on the interval $[a,b]$. Then F is continuous on $[a,b]$, and $F'(x) = f(x)$ on (a,b). From the earlier MVT, we have that there exists $\xi \in (a,b)$ such that

$$F'(\xi) = \frac{F(b)-F(a)}{b-a}, \qquad \text{i.e.,} \qquad f(\xi) = \frac{1}{b-a}\int_a^b f(x)\,dx, \qquad \text{q.e.d.}$$

Figure 5.8, page 564, shows three examples where the values (heights) of the integral averages are indicated by the dashed lines, along with values of ξ at which the functions achieve their average values for the intervals. We include some examples to demonstrate this phenomenon, though the main use of the theorem is theoretical.

Example 5.4.8

Find the average value of $f(x) = \frac{1}{x}$ on $[1,10]$, and any point(s) $\xi \in (1,10)$ at which it is achieved.

Solution: First we compute the average value of f on $[1,10]$, and then find $\xi \in (1,10)$ so that $f(\xi)$ is equal to this average. The relevant graph, including the average value (dashed line) and ξ at which it occurs, is illustrated on the right. (Vertical and horizontal scales are different.)

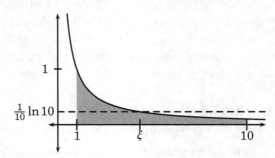

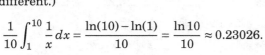

Next we find ξ, for which there is only one value in this case:

$$\frac{1}{\xi} = \frac{\ln 10}{10} \quad \Longrightarrow \quad \xi = \frac{10}{\ln 10} \approx 4.342944819.$$

Example 5.4.9

Consider $f(x) = \sqrt{x}$. Find the average value of $f(x)$ on the interval $[0,9]$, and find all points $\xi \in (0,9)$ so that $f(\xi)$ is equal to that average.

Solution: We proceed as before:

$$\frac{1}{9-0}\int_0^9 \sqrt{x}\,dx = \frac{1}{9}\int_0^9 x^{1/2}\,dx = \frac{1}{9}\left[\frac{2}{3}x^{3/2}\Big|_0^9\right]$$
$$= \frac{2}{27}\left(9^{3/2} - 0^{3/2}\right) = \frac{2 \cdot 27}{27} = 2.$$

Next we solve $f(\xi) = 2$, i.e., $\sqrt{\xi} = 2$, whose solution is simply $\xi = 4$. There are no other solutions.

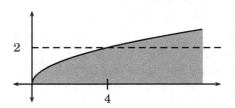

It is worth again noting that our latest definition of average is compatible with previous definitions. For instance, the previous example could easily be reframed using position and velocity. Suppose $v(t) = \sqrt{t}$. Then the average velocity for the time interval $[0,9]$ would be $(s(9) - s(0))/(9 - 0)$. While we do not know $s(t)$ exactly, we know that it is of the form $s(t) = \frac{2}{3}t^{3/2} + C$, and so

$$\frac{s(9) - s(0)}{9 - 0} = \frac{\left[\frac{2}{3} \cdot 9^{3/2} + C\right] - \left[\frac{2}{3} \cdot 0^{3/2} + C\right]}{9} = \frac{1}{9} \cdot \frac{2}{3} \cdot 9^{3/2} = \frac{2}{27} \cdot 27 = 2.$$

This is—except for the temporary presence of the additive constant C—exactly the same computation as in the example, i.e., $\frac{1}{9-0}\int_0^9 v(t)\,dt$, both yielding the average velocity over $0 \le t \le 9$. Note how in both computations, if s is in meters and t is in seconds (for instance), then the units for average velocity would be meters/sec. That is obvious in the difference quotient if we inserted the units into the computation $(s(9) - s(0))/(9 - 0)$, and in light of Subsection 5.3.9 (page 550), also clear from the integral $\frac{1}{9-0}\int_0^9 \sqrt{t}\,dt$, though the integrand should probably be written $\sqrt{t/\text{sec}} \cdot \frac{\text{meters}}{\text{sec}}$. The denominator $9-0$ would actually be $9\,\text{sec} - 0 = 9\,\text{sec}$, and dt would be in units of sec. Also, $\xi = 2\,\text{sec}$, as ξ is a particular value of t.

So we can frame one of our questions of function averages in terms of position and velocity, and use our previous definition of average velocity from Chapter 3, or use our integral average definition, as we do in the following example.

Example 5.4.10

Find the average velocity over the time interval $[0, \pi/2]$ if $v(t) = \cos t$.

Solution: We proceed as before. The average of the velocity $v(t) = \cos t$ over $[0, \pi/2]$ being

$$\frac{1}{\pi/2}\int_0^{\pi/2} \cos t\,dt = \frac{2}{\pi}\left[\sin t\right]\Big|_0^{\pi/2} = \frac{2}{\pi}\left[\sin \frac{\pi}{2} - \sin 0\right] = \frac{2}{\pi}[1 - 0] = \frac{2}{\pi}.$$

5.4. FIRST FUNDAMENTAL THEOREM OF CALCULUS; FURTHER RESULTS

In the previous example, when we look for $\xi \in [0, \pi/2]$ such that $v(\xi) = 2/\pi$, we can simply solve $\cos \xi = 2/\pi$, and since $[0, \pi/2]$ is in the range of the arccosine function, we have $\xi = \cos^{-1}(2/\pi) \approx 0.8806892354$

There can be some interesting pathologies in such examples related to integral averages. For instance, if the interval in the previous example were some positive integer multiple of the period of the cosine function, then the integral would be zero, and so thus would be the integral average, and that would be achieved at any ξ that is an odd multiple of $\pi/2$. It is similar for the sine function. However, this is not the case for every periodic function; but since the sine and cosine functions have similar areas above and below in any given period, definite integrals over those periods will be zero.

Another interesting case is that of Example 5.3.7, page 547, where the function is $f(x) = |x|/x$, undefined at $x = 0$ (and therefore discontinuous there), and with integral average $\frac{1}{1-(-1)} \int_{-1}^{1} f(x)\,dx = \frac{1}{2}(0) = 0$. However, there is no $\xi \in (-1, 1)$ so that $f(\xi) = 0$. This does not violate the Integral MVT because that theorem applies only to continuous integrands.

5.4.7 Integral Bounds

Here we have three theorems that are somewhat intuitive. We state the first two together. The first assumes that M is an upper bound of f on $[a, b]$, and the second assumes m is a lower bound of f on $[a, b]$. Such bounds exist due to the continuity of f, so we could in fact take $M = \max\{f(x) \mid x \in [a, b]\}$ and $m = \min\{f(x) \mid x \in [a, b]\}$ and the theorem would hold, but in the event we have upper or lower bounds but do not know the least upper bound or greatest lower bound, we still have this general theorem:

Theorem 5.4.8 *Suppose $f(x)$ is continuous on $[a, b]$. Then the following implications hold:*

$$(\forall x \in [a,b])[f(x) \leq M] \implies \int_a^b f(x)\,dx \leq M(b-a), \tag{5.74}$$

$$(\forall x \in [a,b])[f(x) \geq m] \implies \int_a^b f(x)\,dx \geq m(b-a). \tag{5.75}$$

A proof would be a straightforward affair with Riemann Sums. For instance, to prove (5.74) we can note that, as $\max\{\Delta x_i\} \to 0$ and thus $n \to \infty$, we get the following:

$$\underbrace{\sum_{i=1}^{n} f(x_i^*)\Delta x_i}_{\displaystyle\int_a^b f(x)\,dx} \quad \leq \quad \sum_{i=1}^{n} M\Delta x_i \quad = \quad M(b-a)$$

$$\therefore \int_a^b f(x)\,dx \leq M(b-a), \quad \text{q.e.d.}$$

Note that we used $f(x_i^*) \leq M$ for the inequality in our diagram. Interestingly, the equation $\sum_{i=1}^{n} M\Delta x_i = M(b-a)$ in that diagram does not rely on n or the partition of $[a, b]$. A similar proof will give us (5.75). From these we also immediately the following, for any function $f(x)$ continuous on $[a,b]$, since it will have a maximum and minimum value for the interval on the interval:

$$(b-a)\cdot\min\{f(x)\,|\,x\in[a,b]\} \le \int_a^b f(x)\,dx \le (b-a)\cdot\max\{f(x)\,|\,x\in[a,b]\}. \tag{5.76}$$

Example 5.4.11

Find upper and lower bounds for $\int_1^{10} \ln x\,dx$.

Solution: Since $\ln x$ is increasing on $[1,10]$, with minimum value $\ln 1 = 0$ and maximum value $\ln 10$, we have $0(10-1) \le \int_1^{10} \ln x\,dx \le (\ln 10)(10-1)$, i.e.,

$$0 \le \int_1^{10} \ln x\,dx \le 9\ln 10.$$

We can generalize the previous theorem to the case of variable bounds, for instance the case $f(x) \le g(x)$, in the following:

Theorem 5.4.9 *Suppose $f(x)$ and $g(x)$ are continuous on $[a,b]$. Then we have the following implication:*

$$(\forall x \in [a,b])[f(x) \le g(x)] \implies \int_a^b f(x)\,dx \le \int_a^b g(x)\,dx. \tag{5.77}$$

This is clear from Riemann Sums or a geometric illustration of the respective areas under the curves of $f(x)$ and $g(x)$. Note that it is implied that $a \le b$ in the theorem's statement.

Example 5.4.12

Consider the integral $\int_{-\pi}^{\pi} x^2 \cos x\,dx$. Since $x^2 \ge 0$ and $\cos x \in [-1,1]$, we can say that

$$-x^2 \le x^2 \cos x \le x^2 \implies \int_{-\pi}^{\pi}(-x^2)\,dx \le \int_{-\pi}^{\pi} x^2 \cos x\,dx \le \int_{-\pi}^{\pi} x^2\,dx$$
$$\implies -\frac{2}{3}\pi^3 \le \int_{-\pi}^{\pi} x^2 \cos x\,dx \le \frac{2}{3}\pi^3.$$

Here we twice used $\int_{-\pi}^{\pi} x^2\,dx = \frac{1}{3}x^3\big|_{-\pi}^{\pi} = \frac{1}{3}\pi^3 - \frac{1}{3}(-\pi)^3 = \frac{2}{3}\pi^3$.

We finish this section with one final theorem, which may best be described as illustrating how the various subareas within an integral can partially or fully cancel each other and shrink the absolute size of the net area:

Theorem 5.4.10 *If $f(x)$ is continuous on $[a,b]$, then*

$$\left|\int_a^b f(x)\,dx\right| \le \int_a^b |f(x)|\,dx. \tag{5.78}$$

5.4. FIRST FUNDAMENTAL THEOREM OF CALCULUS; FURTHER RESULTS

Proof: Recall $|x| \leq k \iff -k \leq x \leq k$, and so (5.78) is equivalent to

$$-\int_a^b |f(x)|\,dx \leq \int_a^b f(x)\,dx \leq \int_a^b |f(x)|\,dx. \tag{5.79}$$

The truth of the inequality (5.79) can be seen from the previous theorem when we realize that $-|f(x)| \leq f(x) \leq |f(x)|$, and so these three functions' integrals over $[a,b]$ will carry the same inequalities, i.e., $\int_a^b [-|f(x)|]\,dx \leq \int_a^b f(x)\,dx \leq \int_a^b |f(x)|\,dx$, which is the same as (5.79), q.e.d.

Another intuitive way of looking at the inequality (5.78) is to note that, while both quantities are clearly nonnegative, in the one on the left we have the possibility of some areas of different signs canceling within the integral, while on the right this is impossible, since the integrand is never negative. If the function never changes signs over the interval $[a,b]$, then the inequality (5.78) is actually an equality.

For a simple physics application, consider

$$|s(t_f) - s(t_0)| = \left|\int_{t_0}^{t_f} s'(t)\,dt\right| \leq \int_{t_0}^{t_f} |s'(t)|\,dt \implies |s(t_f) - s(t_0)| \leq \int_{t_0}^{t_f} |v(t)|\,dt, \tag{5.80}$$

i.e., the distance between the starting and ending points of the motion for the interval $[t_0, t_f]$ is no more than the total distance traveled.

5.4.8 Last Remarks on These Topics

While we presented several theorems in this section, on reflection each should seem self-evident. With a bit of geometric intuition regarding definite integrals as net signed areas, and the statement of the Second FTC, none of these results (including the First FTC) should seem surprising, and indeed some should be readily predictable. We mention them here so that the reader is more likely to consider these general insights when looking at future problems, rather than introduce them as they arise in other contexts.

We also state these "obvious" results with some care, to remind the reader that these conclusions require some hypotheses, such as the continuity of the integrands. For instance, $\int_{-1}^{1} \frac{1}{x}\,dx$ does not exist, though one may be tempted to apply—erroneously—the conclusion of the FTC even though the hypotheses are not met. On the other hand, $\int_{-1}^{1} \frac{|x|}{x}\,dx$ exists, but the Integral Mean Value Theorem does not apply to that integral. In both cases this is because the integrand is not continuous on $[-1,1]$. So indeed it is important to take care that our general intuition is tempered by the realization that there can be some rather pathological cases—which do not fit the hypotheses of the theorems—that can tempt us to apply the "obvious" results stated in this section to cases where they do not actually apply. But absent such pathologies, the results here are rather intuitive and easily remembered, or even simply observed when needed.

Exercises

1. Compute $\dfrac{d}{dx}\displaystyle\int_0^x t^2\,dt$ two ways:

   (a) Using the first FTC.

   (b) Using the second FTC (computing $\int_0^x t^2\,dt$ and differentiating).

2. Do the same for $\dfrac{d}{dx}\displaystyle\int_1^x \dfrac{1}{t}\,dt$. Note that this requires $x > 0$.

For Exercises 3–20, compute the given derivative, preferably without computing any actual antiderivatives. (Assume the given integral is continuous on the relevant interval of integration.)

3. $\dfrac{d}{dx}\left[\displaystyle\int_2^x e^{t^2}\,dt\right]$

4. $\dfrac{d}{dx}\left[\displaystyle\int_0^x t^2\sin t\,dt\right]$

5. $\dfrac{d}{dx}\left[\displaystyle\int_0^{x^2} t^2\sin t\,dt\right]$

6. $\dfrac{d}{dt}\left[\displaystyle\int_0^t \tau^3\sin^2\tau\,d\tau\right]$

7. $\dfrac{d}{dt}\left[\displaystyle\int_0^{t^3} \tau^3\sin^2\tau\,d\tau\right]$

8. $\dfrac{d}{dx}\left[\displaystyle\int_{\sin x}^{x^3} \tau^3\sin^2\tau\,d\tau\right]$

9. $\dfrac{d}{dx}\left[\displaystyle\int_1^{\sin x} (z^3-3)^5\,dz\right]$

10. $\dfrac{d}{dx}\left[\displaystyle\int_{x^4}^2 \sqrt{1-z^2}\,dz\right]$

11. $\dfrac{d}{dx}\left[\displaystyle\int_0^x \sin^2 z\,dz\right]$

12. $\dfrac{d}{dt}\left[\displaystyle\int_0^t \ln(\xi+1)\,d\xi\right]$

13. $\dfrac{d}{dx}\left[\displaystyle\int_x^{100} t^2\,dt\right]$

14. $\dfrac{d}{dx}\left[\displaystyle\int_x^{2\pi} \sin^2 z\,dz\right]$

15. $\dfrac{d}{dx}\left[\displaystyle\int_0^{x^2} \sqrt{t}\,dt\right]$

16. $\dfrac{d}{dx}\left[\displaystyle\int_0^{x^3} \cos^2 t\,dt\right]$

17. $\dfrac{d}{dx}\left[\displaystyle\int_0^{\tan x} \dfrac{1}{\xi^2+1}\,d\xi\right]$

18. $\dfrac{d}{dx}\left[\displaystyle\int_{e^x}^{e^{2x}} \ln t\,dt\right]$

19. $\dfrac{d}{dx}\displaystyle\int_{-x}^x [t^3\sin t^2\,dt]$

20. $\dfrac{d}{dx}\left[\displaystyle\int_{\ln x}^x e^t\,dt\right]$

21. Find $F'(x)$ if $F(x) = \int_x^{x+2}\sin t^2\,dt$.

22. Find $F'(x)$ if $F(x) = \int_x^{x+2}\ln t\,dt$.

23. Find $F'(x)$ if $F(x) = \int_{x^2}^{x^4}\ln t^3\,dt$.

For Exercises 24–28, suppose that $\int_1^{10} g(x)\,dx = 5$, $\int_{10}^{12} g(x)\,dx = 3$, and $\int_1^{20} g(x)\,dx = 2$. Compute the following.

24. $\int_1^{12} g(x)\,dx$

25. $\int_{12}^{20} g(x)\,dx$

26. $\int_{10}^{20} g(x)\,dx$

27. $\int_{12}^{10} [5g(x)]\,dx$

28. $\int_{10}^{10} [6g(x)]\,dx$

For Exercises 29–32, compute the given integral assuming that $\int_2^4 f(x)\,dx = 3$ and $\int_2^4 g(x)\,dx = 2$.

29. $\int_2^4 [f(x) - g(x)]\,dx$

30. $\int_2^4 [5f(x) - 2g(x)]\,dx$

5.4. FIRST FUNDAMENTAL THEOREM OF CALCULUS; FURTHER RESULTS

31. $\int_4^2 [2f(x) - 5g(x)]\, dx$

32. $\int_2^4 [g(x) + 3]\, dx$

For Exercises 33–45, consider the following curve, where the probability density function $f(z) = \frac{1}{\sqrt{2\pi}} e^{-z^2/2}$ for the standard normal curve is plotted (not to scale), and the following approximate integrals are known.

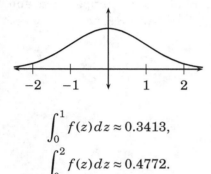

$\int_0^1 f(z)\, dz \approx 0.3413,$

$\int_0^2 f(z)\, dz \approx 0.4772.$

Answer each question or find the requested integral approximation.

33. Show that $f(z)$ is an even function.

34. $f(0) =$

35. $\int_1^2 f(z)\, dz \approx$

36. $\int_{-1}^1 f(z)\, dz \approx$

37. $\int_{-2}^0 f(z)\, dz \approx$

38. $\int_{-2}^1 f(z)\, dz \approx$

39. $\int_{-2}^2 f(z)\, dz \approx$

40. $\int_1^0 f(z)\, dz \approx$

41. $\int_{-1}^0 f(z)\, dz \approx$

42. $\int_2^{-2} f(z)\, dz \approx$

43. $\int_3^3 f(z)\, dz \approx$

44. $\int_0^1 e^{-z^2/2}\, dz \approx$

45. $\int_0^2 e^{-z^2/2}\, dz \approx$

For Exercises 46–54, compute the average value of $f(x)$ on the given interval, and find all ξ on that interval where $f(\xi)$ equals that average value.

46. $f(x) = \sin x,\ x \in [0, \pi/2]$

47. $f(x) = \sin x,\ x \in [0, \pi]$

48. $f(x) = x^2,\ x \in [-10, 10]$

49. $f(x) = x^2,\ x \in [0, 10]$

50. $f(x) = \frac{1}{x},\ x \in [1, e]$

51. $f(x) = \frac{1}{x^2+1},\ x \in [-1, 1]$

52. $f(x) = x^3,\ x \in [-1, 1]$

53. $f(x) = x^2 + 2x + 3,\ x \in [1, 5]$

54. $f(x) = \sqrt[3]{x}$ on $[1, 8]$

55. Answer the following:

    (a) Compute $\dfrac{d}{dx}\left[\int_2^x (t^2 + t + 3)\, dt\right]$.

    (b) Compute $\int_2^x \dfrac{d}{dt}\left[t^2 + t + 3\right]\, dt$.

    (c) Are they the same? Why?

56. Answer the following, assuming $f'(t)$ exists:

    (a) Compute $\dfrac{d}{dx}\left[\int_a^x f(t)\, dt\right]$.

    (b) Compute $\int_a^x \dfrac{d}{dt}[f(t)]\, dt$.

    (c) Are they the same? Why?

57. Assuming f', g', and h' exist everywhere, answer the following:

    (a) Compute $\dfrac{d}{dx}\left[\int_{g(x)}^{h(x)} f(t)\, dt\right]$.

    (b) Compute $\int_{g(x)}^{h(x)} \dfrac{d}{dt}[f(t)]\, dt$.

    (c) Are they the same?

58. Suppose $f(x)$ and $g(x)$ are continuous on $[a, b]$, with $a < b$. Prove that

    (a) the average value of $(f(x) + g(x))$ on $[a, b]$ is the sum of the average value of $f(x)$ on $[a, b]$ and the average value of $g(x)$ on $[a, b]$; and that

    (b) also on $[a, b]$, the average value of $5f(x)$ is 5 times the average value of $f(x)$.

For Exercises 59–62, suppose $f(x)$ is continuous on $[a,b]$, where $a < b$, and that $\int_a^b f(x)\,dx = 0$.

59. Is $f(x) = 0$ for all $x \in [a,b]$? Explain.

60. Is $f(x) = 0$ for at least one value of $x \in [a,b]$? Explain.

61. Is $\int_a^b |f(x)|\,dx = 0$? Explain.

62. Is $\left|\int_a^b f(x)\,dx\right| \leq \int_a^b |f(x)|\,dx$? Explain.

63. Suppose $f(x)$ is continuous on $[1,5]$, and that $-2 \leq f(x) \leq 3$ for all $x \in [1,5]$.

    (a) Find an upper bound for $\int_1^5 f(x)\,dx$.

    (b) Find a lower bound for $\int_1^5 f(x)\,dx$.

64. Find an upper bound for $\int_0^{10} x^2 |\sin x|\,dx$.

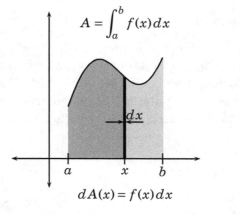
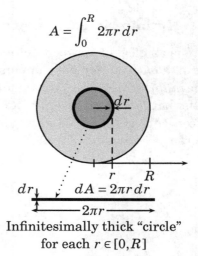

Infinitesimally thin rectangle at each $x \in [a,b]$

Infinitesimally thick "circle" for each $r \in [0,R]$

Figure 5.9: Two uses of infinitesimals to compute total areas. In the first, we can visualize a "continuum" of infinitesimal rectangles accumulating to fill in the shaded area as $x \to b^-$. In the second, we see a similar "continuum" of infinitesimal circular rings accumulating to fill in the area of the entire circle as $r \to R^-$. These rings, or technically *annuli* (singular *annulus*), which when cut and "unrolled" are nearly rectangular, *to within any small percentage error if dr is small enough*, have lengths $2\pi r$ and thicknesses dr and thus area $dA = 2\pi r\, dr$.

5.5 Infinitesimals and Definite Integrals in Geometry

In this section, we look more closely at the definite integral notation, and consider a very elegant interpretation of it, using the notion of integrating "infinitesimals." This allows us to find many applications of the integral calculus through very simple visualizations.

The idea of considering a quantity to be "infinitesimally small," or "so small there is no way to measure it," is a classical idea which is, in some sense, absurd, but the mechanics of the uses of the infinitesimal notation and visualization is worth considering, especially if we can put the mechanics on firm footing.

The techniques of infinitesimals were more popular in earlier, classical treatments of calculus. Because they gain their rigor ultimately from references to limits of Riemann Sums of continuous functions over closed intervals, and thus the Fundamental Theorem of Calculus (FTC), many modern treatments dispense entirely with infinitesimals in favor of Riemann Sum treatments. Indeed, we will refer often to the underlying Riemann Sum limit theory, but the technique of infinitesimals is powerfully illustrative and so we resurrect it here. We then apply it to numerous applied problems.

With most of the examples put forward here, where we set up a definite integral representing some quantity, there are three interpretations that reinforce each other's validity: an infinitesimals approach, a Riemann Sums approach, and an argument for the derivative of a function which represents the accumulation of some quantity. At first we will show arguments for all three, while demonstrating how they reinforce each other. Eventually we will settle on the infinitesimals approach, but with an eye on the others.

5.5.1 Infinitesimals and Areas under Curves

There is an elegant viewpoint often used in interpreting definite integrals $\int_a^b f(x)\,dx$, which calls on a once incompletely understood notion from the early days of calculus, namely that of the *infinitesimals*. For such an interpretation to be correct in a particular context, it is best to refer back to the viewpoint of Riemann Sums and their limits, at least when first applying the concept, which we do here and compare it to Riemann Sums and both Fundamental Theorems of Calculus (FTCs).[32]

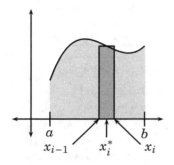 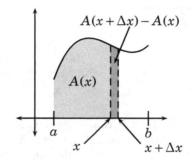 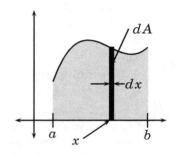

Riemann Sum Limits **First Fundamental Theorem Style** **Infinitesimals**

In the Riemann Sum limit paradigm of the definite integral, we choose a partition $a = x_0 < \cdots < x_n = b$, and use approximating rectangles of signed areas $f(x_i^*)\Delta x_i$ (with $\Delta x_i = x_i - x_{i-1}$ and $x_i^* \in [x_{i-1}, x_i]$) along the partition, and argue that as the maximum Δx_i approaches zero (and $n \to \infty$), we get

$$\sum_{i=1}^{n} f(x_i^*)\Delta x_i$$

approaching the signed area between the curve and the horizontal axis.

The First FTC proof considered the function $A(x)$ to be equal to the (signed) area between the curve $y = f(x)$ and the interval $[a, x]$ along the horizontal axis. In the argument for the First FTC, we noted how the difference quotient approximation

$$\frac{A(x + \Delta x) - A(x)}{\Delta x} \approx f(x)$$

is quite reasonable in the sense of percentage error (at least when $f(x) \neq 0$) and therefore this becomes an equality in the limit, which proved that the derivative of this area function $A(x)$ must be $f(x)$, i.e., that $A'(x) = f(x)$.

With the perspective of infinitesimals, at any $x \in [a, b]$ we consider an infinitesimal rectangular area element dA of width dx and height $f(x)$. On reflection this is clearly consistent with our previous definition of a differential, whereby $dA = A'(x)\,dx$, in a sense somewhat similar to the argument for the First FTC. Eventually we feel confident in "observing" that $dA = f(x)\,dx$ from the graph, knowing its significance as a differential, but interpreting it geometrically to be the (signed) infinitesimal area dA.

---

[32]It is proper that all discussions of these interpretations should ultimately refer back to limits of Riemann Sums, since the integral $\int_a^b f(x)\,dx$ is *defined* to be a limit of such Riemann Sums. The other two approaches are, in theory, just (very) useful visualizations, with infinitesimals sometimes the more useful visual device.

5.5. INFINITESIMALS AND DEFINITE INTEGRALS IN GEOMETRY

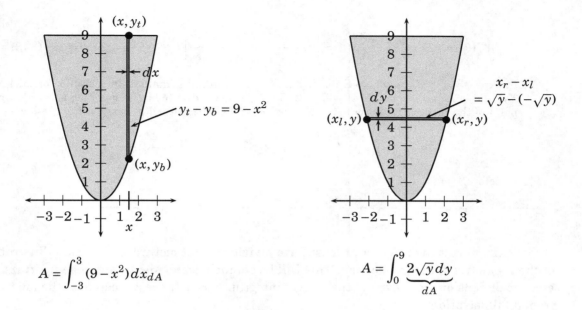

Figure 5.10: Two methods of using differentials to find the area bounded by the curves $y = 9$ and $y = x^2$. In the first we have an infinitesimal rectangle of width dx at any given $x \in [-3, 3]$, the other dimension of the rectangle being $y_t - y_b = 9 - x^2$. In the second method we have horizontal rectangles of thickness dy at any given $y \in [0, 9]$, the other dimension being $x_r - x_l = 2\sqrt{y}$. See Example 5.5.1.

5.5.2 Areas Bounded by Curves

In this subsection we set up, and usually evaluate, definite integrals that represent areas bounded by curves. We do this by finding infinitesimally thin rectangles, with the desired area collectively composed of such infinitesimally thin rectangles. The thickness of such rectangles will be either dx or dy for these cases.

Example 5.5.1 _____

Consider the area bounded by the graphs of $y = x^2$ and $y = 9$. To algebraically find where they intersect, one solves the system of equations $y = x^2$, $y = 9$, which is fairly straightforward: We simply set "$y = y$," i.e., $x^2 = 9$, whose solution is clearly $x = \pm 3$, $y = 9$, i.e., $\{(-3, 9), (3, 9)\}$. Since in the region $x \in [-3, 3]$ we have the parabola $y = x^2$ below the line $y = 9$ except at the endpoints, ours is a situation as in Figure 5.10.

There are two relatively straightforward approaches to using infinitesimals to set up an integral to compute this area. The first is to consider it as an integration of infinitesimally thin rectangles of thickness dx, and the second is consider the desired area as an integration of infinitesimally thin rectangles of thickness dy. Our choice will determine the differential (dx or dy) and therefore the form of the desired definite integral. Both are illustrated in Figure 5.10.

- thickness = dx: At each $x \in [-3, 3]$, consider an infinitesimally thin rectangle of thickness dx, whose length is the distance between the top curve $y = 9$ and the bottom curve $y = x^2$. We will call this length $y_t - y_b = 9 - x^2$, where y_t is the y-value at the "top" of the rectangle, and y_b is the y-value at the "bottom" of the rectangle. Thus,

$$A = \int_{-3}^{3}(9-x^2)dx = \left[9x - \frac{x^3}{3}\right]\Big|_{-3}^{3} = \left[27 - \frac{27}{3}\right] - \left[-27 - \frac{-27}{3}\right] = 27-9-(-27-9) = 18-(-18) = 36.$$

- thickness $= dy$: Here we have, for any $y \in [0,9]$, an infinitesimally thin rectangle of thickness dy, and length $x_r - x_l = \sqrt{y} - (-\sqrt{y}) = 2\sqrt{y}$. Here we used x_r to be the x-value at the rightmost point of the rectangle, and x_l to be the x-value at the leftmost point. Integrating these we have

$$A = \int_0^9 \left(\sqrt{y} - (-\sqrt{y})\right)dy = \int_0^9 2\sqrt{y}\,dy = \frac{4}{3}y^{3/2}\Big|_0^9 = \frac{4}{3}(27) - \frac{4}{3}(0) = 36,$$

as before.

It is important to draw—or at least have a vivid mental picture of—a visual illustration of the geometric figure in question. This will be of much use in identifying the important geometric dimensions to construct the correct integral. For instance, we can note the following from an illustration:

1. *The variable whose differential gives us the (infinitesimal) width, or "thickness" of the desired rectangles will be the variable of integration.* Since the entire definite integral—the integrand, differential and endpoints—must be in terms of that variable, identifying the width of the infinitesimal rectangles (as dx or dy) is quite valuable in guiding us in the next steps of constructing the appropriate definite integral.

2. Related to the thickness of a rectangle (as a guiding principle) is that *if the rectangles are vertical (meaning the longer dimension is vertical), then the thicknesses must be dx* and so the integral—including endpoints—must be entirely in terms of x. On the other hand, *horizontal rectangles will have thickness dy* and the entire integral must be in terms of y.

    (a) If the rectangles are vertical, their longer dimension will be $y_t - y_b$ where y_t is the greater y-value, or "top" value of y, at the top of the rectangle, and y_b the lower, or "bottom" value. Since the width will be dx, we must have $y_t - y_b$ as a function of x.

    (b) Similarly, any horizontally longer rectangle lengths should be of the form $x_r - x_l$, where $x_r > x_l$, and so x_r represents the rightmost x-coordinate along an infinitesimally thin, horizontal rectangle, and x_l the leftmost. Since the width will be dy, $x_r - x_l$ will be a function of y.

As in the example, we may have a choice between an integral (or integrals) in the variable y and an integral (or integrals) in the variable x. Sometimes the choice can make a large difference in the difficulty of the integral.

5.5. INFINITESIMALS AND DEFINITE INTEGRALS IN GEOMETRY

Example 5.5.2

Consider the area bounded by the graphs of $y = 0$, $y = \sin x$, $x = 0$, and $x = \pi/2$. This too can be broken up into vertical or horizontal rectangles.[33]

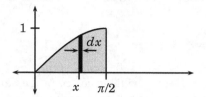 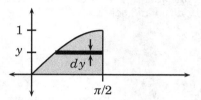

Vertical rectangles: Here, for each $x \in [0, \pi/2]$ we have a rectangle of thickness dx. The length is the difference between the y-value on the "upper curve" $y = \sin x$, and the y-value on the "lower curve" $y = 0$. Thus the length of the vertical rectangles is $\sin x - 0 = \sin x$, and so $dA = \sin x \, dx$, and our area is given by the integral

$$\int_0^{\pi/2} \sin x \, dx = (-\cos x)\Big|_0^{\pi/2} = (-\cos(\pi/2)) - (-\cos 0) = 0 - (-1) = 1.$$

Horizontal rectangles: Here, for each $y \in [0, 1]$ we have a rectangle whose thickness is dy. Its length is the difference between the x-value on the "right curve" $x = \pi/2$, and the x-value on the "left curve" $y = \sin x$. However, since the thickness is dy, we need the eventual integral to be entirely in terms of y. In this case, the "left curve" must be given by $x = \sin^{-1} y$. This allows the length to be given by $\pi/2 - \sin^{-1} y$, giving us $dA = [\pi/2 - \sin^{-1} y] \, dy$, and thus

$$A = \int_0^1 [\pi/2 - \sin^{-1} y] \, dy.$$

We do not currently have techniques available to compute the antiderivative $\int \sin^{-1} y \, dy$ (though such techniques are discussed in the sequel), and so we will for now be content to realize that the integral that uses vertical rectangles is easily computed, and so for this example we will favor that approach.[34]

Clearly there can be advantages in choosing vertical rectangles over horizontal rectangles, or vice versa. In the next example, it is somewhat simpler to use horizontal rectangles and integrate with respect to y, as we find the area *between two curves*.

---

[33] By "vertical" rectangle we mean one whose longest sides are vertical, and by "horizontal" rectangle we mean one whose longest sides are horizontal. In the context of infinitesimal rectangular areas, a vertical rectangle will have width dx and a horizontal rectangle will have width dy. Because some texts prefer "width" to be always vertical or always horizontal, we sometimes use the term "thickness." That term will be used for other, usually infinitesimal, dimensions later.

[34] In fact, with more advanced techniques we can compute $\int \sin^{-1} y \, dy = y \sin^{-1} y + \sqrt{1 - y^2} + C$, which can be checked by applying $\frac{d}{dy}$ to both sides, and so our area integral using horizontal rectangles becomes

$$\int_0^1 \left[\frac{\pi}{2} - \sin^{-1} y\right] dy = \left[\frac{\pi}{2} y - y \sin^{-1} y - \sqrt{1 - y^2}\right]\Big|_0^1 = \left[\frac{\pi}{2} - 1 \cdot \frac{\pi}{2} - 0\right] - [0 - 0 - 1] = 1,$$

as before. However, in this case the vertical rectangles produce a much simpler computation.

Example 5.5.3

Find the area bounded by $x = y^2$ and $y = x - 2$.

Solution: First we must find where these intersect. Since it is easy to isolate x in each equation, we will do so and set "$x = x$."

$$\left.\begin{array}{r}x = y^2 \\ y = x - 2\end{array}\right\} \implies \left.\begin{array}{r}x = y^2 \\ x = y + 2\end{array}\right\} \implies y^2 = y + 2.$$

Solving, we get $y^2 - y - 2 = 0 \iff (y-2)(y+1) = 0 \iff y \in \{2, -1\}$. Using either curve's equation to find the x-values for these y-values, we conclude that the graphs intersect at $(4, 2)$ and $(1, -1)$.

Now the first curve $x = y^2$ is a parabola and the other $y = x - 2$ is a line, both easily graphed, which we do here twice to accommodate our two approaches, for vertical or horizontal rectangles.

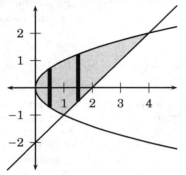 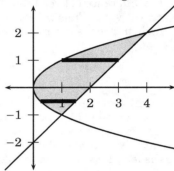

Vertical Rectangles: For vertical rectangles (above left), which would have width (or "thickness") dx, the formula for the height will differ depending on the value of x; for $x \in [0, 1]$ both the top and the bottom of the rectangle lie on the graph of $x = y^2$, while for $x \in [1, 4]$, the top lies on the graph of $x = y^2$ and the bottom lies on $y = x - 2$. The area can still be found with vertical rectangles, but we would need two integrals if we are to have a single formula for each integrand:

$$A = \int_0^1 \left[\sqrt{x} - (-\sqrt{x})\right] dx + \int_1^4 \left[\sqrt{x} - (x - 2)\right] dx = \int_0^1 2\sqrt{x}\, dx + \int_1^4 \left[\sqrt{x} - x + 2\right] dx$$

$$= \left[\frac{4}{3}x^{3/2}\right]\Big|_0^1 + \left[\frac{2}{3}x^{3/2} - \frac{1}{2}x^2 + 2x\right]\Big|_1^4 = \frac{4}{3} + \left[\frac{2}{3}(4)^{3/2} - \frac{1}{2}(4)^2 + 2(4)\right] - \left[\frac{2}{3} - \frac{1}{2} + 2\right]$$

$$= \frac{4}{3} + \frac{2}{3}(8) - \frac{1}{2}(16) + 8 - \frac{2}{3} + \frac{1}{2} - 2 = \frac{8 + 32 - 48 + 48 - 4 + 3 - 12}{6} = \frac{27}{6} = \frac{9}{2}.$$

Horizontal Rectangles: Here we have an infinitesimal rectangle at each $y \in [-1, 2]$, with thickness of dy. The horizontal length is given by "x" on the curve $y = x - 2$, minus "x" on the curve $x = y^2$. The rightmost curve can be written $x = y + 2$, and so our length is $(y + 2) - y^2$, and so the area of the infinitesimal rectangle will be $dA = (y + 2 - y^2)dy$. This formula holds for all of $y \in [-1, 2]$, i.e., for the whole area in question, and so we can write the area as a single definite integral

$$A = \int_{-1}^2 (y + 2 - y^2) dy = \left[\frac{1}{2}y^2 + 2y - \frac{1}{3}y^3\right]\Big|_{-1}^2 = \left[2 + 4 - \frac{8}{3}\right] - \left[\frac{1}{2} - 2 + \frac{1}{3}\right] = 6 - \frac{8}{3} - \frac{1}{2} + 2 - \frac{1}{3} = \frac{9}{2},$$

as before.

Clearly, in this case it is simpler to use horizontal rectangles, and compute a single definite integral in y.

5.5. INFINITESIMALS AND DEFINITE INTEGRALS IN GEOMETRY

It is always suggested to consider all available alternatives when computing a desired quantity, such as the areas in these examples. There will still be occasions in which the geometry does not allow for the area to be represented by a single integral of a function not defined piecewise, and then multiple integrals are required.

5.5.3 Area of a Circle

It is worth noting, and pondering, how the area of a circle can be easily computed using an argument by infinitesimals. The basic idea was illustrated in Figure 5.9, page 573, at the start of this section, and is reproduced here on the right. We assume that the circle has radius $R > 0$.

The basic argument is that we can break the circle into infinitesimal rings, or annuli. At each $r \in [0,R]$, we consider a circular ring of (say, inner) radius r and thickness dr. The actual area occupied by such a ring will be very well approximated by $2\pi r\, dr$, because if it were "unrolled" it would be very nearly rectangular, with one length being $2\pi r$ and the "height" (or "thickness") being dr. To a percentage error approaching zero as the thickness dr approaches zero, we have indeed $\Delta A \approx dA = 2\pi r\, dr$. (Put another way, $\lim_{\Delta r \to 0}(\Delta A/\Delta r) = dA/dr = 2\pi r$, the argument being similar to the argument for the First FTC.) Therefore we can integrate these to achieve the total area:

$$A = \int_0^R 2\pi r\, dr = \pi r^2 \Big|_0^R = \pi R^2, \qquad \text{q.e.d.}$$

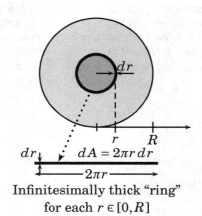

Infinitesimally thick "ring" for each $r \in [0,R]$

5.5.4 Volume and Cavalieri's Principle

Assume that the following three solids have the same cross-sectional area at each height y. According to Cavalieri's Principle, they must therefore have the same total volume:[35]

If the cross-sectional area at each height $y \in [a,b]$ is given by $A(y)$, then a simple argument with infinitesimals leads us to conclude

$$V = \int_a^b A(y)\,dy. \tag{5.81}$$

---

[35] Named for Italian mathematician Bonaventura Francesco Cavalieri (1598–1647), whose work foreshadowed some of the elementary (but nonetheless deep) notions of infinitesimals and therefore integral calculus. This principle was also discovered by Chinese mathematician Zu Gengzhi (circa 450–520).

This will represent the volume for each solid and is consistent with Cavalieri's Principle that they be equal (since $A(y)$ is the same function for each solid). The next example better illustrates the reasonableness of this integral representation of the volume.

Example 5.5.4

Use infinitesimals to set up an integral to compute the volume of a right circular cone with base radius R and height h.

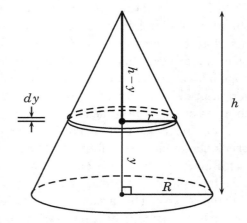

Solution: Our illustration is on the right. We note that at each height $y \in [0, h]$, the cross-sectional area is πr^2, where the radius r is a function of y. If we give such a circle a "thickness," or "height" dy, then it forms an infinitesimally thin cylinder of volume $dV = \pi r^2 \, dy$.[36]

What is left to do is to find the radius r as a function of y, and then set up and compute the resulting integral. From the diagram, we can see that by a similar triangles argument that $(h-y)/h = r/R$, or equivalently $r/(h-y) = R/h$. From either, we get $r = (h-y)\frac{R}{h}$, and so

$$V = \int_0^h \pi \left[(h-y)\frac{R}{h}\right]^2 dy = \pi \cdot \frac{R^2}{h^2} \int_0^h (h^2 - 2hy + y^2) dy = \pi \cdot \frac{R^2}{h^2} \left[h^2 y - hy^2 + \frac{1}{3}y^3\right]\Big|_0^h$$

$$= \pi \cdot \frac{R^2}{h^2} \left[h^3 - h^3 + \frac{1}{3}h^3\right] - 0 = \pi \cdot \frac{R^2}{h^2} \cdot \frac{1}{3} h^3 = \frac{1}{3}\pi R^2 h.$$

This is a well-known volume formula for a cone, one interesting observation being that the volume of a cone is one-third that of a cylinder with the same base and height.

5.5.5 Volumes of Revolution: Disc/Washer Method

One interesting class of solids is those that are generated (or "swept") by revolving an area around a central axis. A solid object occupying exactly that volume of space the area sweeps through is then called a "solid of revolution," and the space itself called a "volume of revolution." (The terms are often interchanged.)

In such a solid, a plane perpendicular to the axis of revolution and intersecting the solid would intersect it in cross-sectional areas which will be unions of circles centered along the axis of revolution.

While there are general formulas for computing these volumes, their number can get rather large in order to cover all common situations, and so here we recommend considering each case separately, though general approaches do emerge.

---

[36] Recall that the volume of a cylinder of radius r and height h is given by $\pi r^2 h$, i.e., the area πr^2 of the circular base is multiplied by the height h of the cylinder.

5.5. INFINITESIMALS AND DEFINITE INTEGRALS IN GEOMETRY

Example 5.5.5

Consider the volume generated by revolving the area bounded by $f(x) = \sqrt{x}$, the x-axis and the line $x = 4$ around the x-axis.

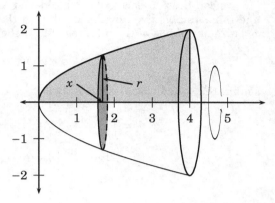

The region described is drawn in light gray on the right. When we revolve it around the x-axis, it sweeps through a volume of space, forming a solid with a parabolic profile.

To find its volume, consider at each $x \in [0,4]$ a disc of infinitesimal thickness dx, whose radius is the distance from the axis of revolution to the edge of the solid. We thus exhaust the solid by such discs (such as the one drawn in dark gray on the right). The area of such a disc will in this case be $A(x) = \pi r^2 = \pi (f(x))^2$, and the volume of the infinitesimal disc will $A(x)dx$. Therefore the total volume will be

$$V = \int_0^4 A(x)\,dx = \int_0^4 \pi(f(x))^2\,dx = \int_0^4 \pi\left(\sqrt{x}\right)^2\,dx = \int_0^4 \pi x\,dx = \left.\frac{\pi x^2}{2}\right|_0^4 = \pi(4)^2 - \pi(0)^2 = 16\pi \approx 50.27.$$

It is one of the more remarkable and aesthetically pleasing aspects of the calculus that we can find volumes of such a figure simply by finding the area of each cross section. In a later subsection we will demonstrate yet another method of finding this volume, but for now we will look at how to compute volumes with another variety of cross section—namely "washer" or *annulus*—which often arises with solids of revolution.

Example 5.5.6

Given the same plane area as in the previous example, find the volume of the solid found by revolving the area instead around the y-axis.

Solution: The region is graphed as shown. At each level $y \in [0,2]$, the cross-sectional area of the solid will be an annulus (i.e., the base of a "washer"). It will have an outer radius and an inner radius.

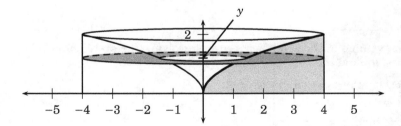

The area of an annulus is the area between its outer and inner circles, which is that of its outer circle, but with the inner circle's area deleted. If the outer and inner radii are, respectively, R and r, then this area in between is

$$A = \pi R^2 - \pi r^2 = \pi\left(R^2 - r^2\right).$$

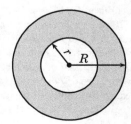

If we have a *washer* with height h and this annulus of area A as a base, then the *volume* of the washer is
$$V = Ah = \pi(R^2 - r^2)h.$$

We can thus, at each $y \in [0,2]$, consider a washer of infinitesimal height dy, with outer radius 4 and inner radius given by the x-value on the curve $y = \sqrt{x}$. As a function of y that inner radius is $x = y^2$. The volume of the infinitesimal washer (shown in dark gray on the illustration of the solid) is
$$dV = A\,dy = \pi \left[4^2 - \left(y^2\right)^2 \right] dy,$$
so that the volume of the solid is thus
$$V = \int_0^2 \pi \left[4^2 - \left(y^2\right)^2 \right] dy = \pi \int_0^2 \left[16 - y^4 \right] dy = \pi \left[16y - \frac{y^5}{5} \right]\Big|_0^2$$
$$= \pi \left[32 - \frac{32}{5} \right] - \pi[0] = \pi \cdot \frac{160 - 32}{5} = \frac{128\pi}{5} \approx 80.42.$$

Among other things, it is worth noting the difference in total volume between when we revolved the same area around the x-axis versus the y-axis. Around the x-axis, the volume was $16\pi \approx 50.27$, while when we revolved it around the y-axis our result was $128\pi/5 \approx 80.42$. That the second solid's volume was larger is not surprising when we look at how the radius of revolution for "most points" was longer in the second than in the first. Consequently, the same area swept through a larger volume in the second example.

We also mention here that in a way the washer method is just a generalization of the Disc Method (where applicable), in the sense that the Disc Method could be construed as simply a special case of the washer method, in which the inner radius is $r = 0$.

We next consider cases where we revolve this same area around some vertical or horizontal lines that are not coordinate axes. What changes is simply the expression for each radius. In all cases it is useful to consider cross-sectional areas. If they are circles or annuli, then the Disc/Washer method can in principle give us the volume. In all cases it is important to consider an illustration of area and the volume swept by that area.[37]

Example 5.5.7

Consider the volume of the solid obtained by revolving the area bounded by $y = \sqrt{x}$, the x-axis and the line $x = 4$, around the line $y = 3$.

Solution: As before we use an illustration:

---

[37] Recall that anytime we have a solid for which we have an expression for the cross-sectional areas, we can integrate that area to find the volume. For vertical cross-sectional areas $A(x)$, we have $dV = A(x)dx$ and $V = \int_{x_{\min}}^{x_{\max}} A(x)dx$. If we have horizontal cross-sectional areas $A(y)$, we have $dV = A(y)dy$ and thus $V = \int_{y_{\min}}^{y_{\max}} A(y)dy$. Sometimes $A(x)$ is a simpler expression to antidifferentiate than $A(y)$, or vice versa, so it is good practice to consider both before settling on a method of computing the volume.

The upcoming *shell method* gives another technique whereby we exhaust the volume by a means other than the cross-sectional area. This leads to some confusion among students. The key then is to consider all of the options very carefully, and in particular to consider the "thickness" dx or dy, which dictates what must be the variable of integration, as well as the expressions used for dV in the various overall methods.

5.5. INFINITESIMALS AND DEFINITE INTEGRALS IN GEOMETRY

Here, for each $x \in [0,4]$, we have a washer of thickness dx, and an outer and inner radius.

The outer radius is the difference in y-values between the line $y = 3$ and the line $y = 0$. In other words, the outer radius is $R = 3$ throughout the solid.

The inner radius, on the other hand, is the difference in y-values between the line $y = 3$ and the curve $y = \sqrt{x}$. Thus the inner radius is $r = 3 - \sqrt{x}$.

This gives a volume of the washer to be

$$dV = \pi \left[R^2 - r^2 \right] dx = \pi \left[3^2 - \left(3 - \sqrt{x} \right)^2 \right] dx$$
$$= \pi \left[9 - \left(9 - 6\sqrt{x} + x \right) \right] dx = \pi \left[6\sqrt{x} - x \right] dx.$$

The volume of the solid is thus given by

$$V = \pi \int_0^4 \left[6\sqrt{x} - x \right] dx = \pi \left[6 \cdot \frac{2}{3} x^{3/2} - \frac{1}{2} x^2 \right]\Big|_0^4 = \pi \left[4 \cdot 8 - \frac{1}{2}(16) \right] - \pi[0] = 24\pi.$$

It is interesting to note that this solid has a slightly smaller volume than the previous one. Indeed, at $24\pi = \frac{120}{5}\pi$, we see it is $\frac{8}{5}\pi$ less than the volume of the previous solid, which was $\frac{128}{5}\pi$. In decimal approximations, this solid at a volume of approximately 75.40 is indeed less than the previous solid at approximately 80.42.

For our last solid using this region, we will revolve it around the vertical axis $x = 5$.

Example 5.5.8 _____

Find the volume of the solid obtained by revolving the area bounded by $y = \sqrt{x}$, the x-axis and the line $x = 4$, around the line $x = 5$.

<u>Solution</u>: Again, we make an illustration of the solid, this time with a washer of thickness dy at each $y \in [0,2]$. The outer radius is the difference in x-values between $x = 5$ and $y = \sqrt{x}$, and so $R = 5 - y^2$. Clearly, $r = 5 - 4 = 1$. Thus $dV = \pi \left(R^2 - r^2 \right) dy = \pi \left((5 - y^2)^2 - 1^2 \right) dy$, for $y \in [0,2]$, giving us

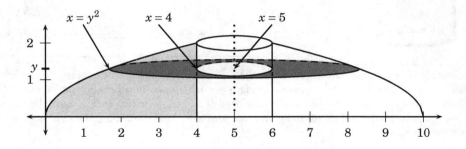

$$V = \int_0^2 \pi \left[(5-y^2)^2 - 1^2\right] dy = \pi \int_0^2 \left[25 - 10y^2 + y^4 - 1\right] dy = \pi \left[24y - \frac{10}{3}y^3 + \frac{y^5}{5}\right]\Big|_0^2$$
$$= \pi \left[48 - \frac{80}{3} + \frac{32}{5}\right] - 0 = \pi \frac{48 \cdot 15 - 80 \cdot 5 + 32 \cdot 3}{15} = \frac{416\pi}{15} \approx 87.13.$$

We will see in the next subsection that there is often another way to decompose a solid of revolution into infinitesimal shapes we can integrate to find the volume of the whole solid. One should always consider both techniques, because sometimes one method has a complication that is not present in the other. Sometimes a complication is geometric—which makes *setting up* the correct integral difficult with that technique—and sometimes it is in the difficulty of *computing* the resulting integral. Sometimes one technique has *both* advantages over the other. Thus it is always useful to at least consider both techniques and any other technique that presents itself. We will revisit this thought as we continue through the next subsection.

5.5.6 Volumes of Revolution: Cylindrical Shell Method

Consider a cylinder of radius r and height h. The simplest computation of its volume is to multiply the area of its base and its height: $V = \pi r^2 h$.

Now consider an open, *cylindrical shell*, similar to a soup can without the circular top or bottom. If its inner radius is r, height is h, and the side thickness is dr, then the approximate volume of the material needed to manufacture such a shell is given by the product of the circumference, height, and thickness. For reasons we will see later, we will write this volume as

$$dV = 2\pi r h \, dr. \tag{5.82}$$

This can be seen as an approximation whose percent error shrinks to zero as $dr \to 0^+$. The illustration shows how a cylindrical shell that is slit vertically and "unrolled" will be very nearly a rectangular parallelepiped with length $2\pi r$ (the circumference of the cylinder), height h and "thickness" (or "depth") dr:[38]

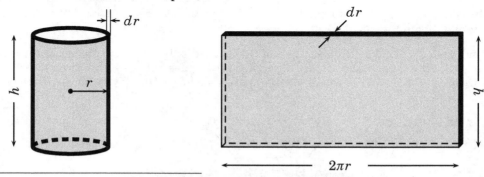

---

[38]A *parallelepiped* is a hexahedron (six-sided, three-dimensional object whose sides are all plane regions) whose sides (or *faces*) are all parallelograms. A *right parallelepiped* is a parallelepiped whose faces are all rectangles. In such cases, the volume is equal to the area of any one side (or "base"), multiplied by the distance ("height") between the plane containing the given side and the plane containing the opposite, parallel side. A cube is the simplest case.

5.5. INFINITESIMALS AND DEFINITE INTEGRALS IN GEOMETRY

The unrolled cylinder wall will not be exactly a right parallelepiped (or thin "cube") because the outer and inner radii will be different, and the sides of the unrolled wall, which were formed by the cut, cannot both meet the other sides at 90° angles, but the computation for the volume will have very small percentage error when approximated by $2\pi r h\, dr$ if dr is very small. Indeed, the percentage error will approach zero as dr approaches zero (from the right). We use this fact in the next example.

Example 5.5.9

Suppose the area between the graphs of $y = \sqrt{x}$, $y = 0$, $x = 0$ and $x = 4$ is revolved around the y-axis. Use an integral that utilizes the volumes of cylindrical shells to compute the volume of the generated solid.

Solution: This is the same volume we found in Example 5.5.6, page 581. However, here we will decompose it into a union of infinitesimal cylindrical shells, one for each $x \in [0,4]$. One such infinitesimal cylinder is drawn. The thickness of its side will be dx in this case, and so its volume will be, in the sense of infinitesimals, $dV = 2\pi(\text{radius})(\text{height})\,dx$.

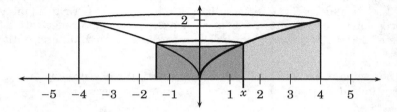

For this particular case, the radius is the difference between the x-coordinate at the rightmost edge of the cylindrical shell, and the x-coordinate at the axis of revolution. For a given $x \in [0,4]$, this radius will then be $x - 0 = x$.

The height at a given $x \in [0,4]$ will be given by the y-value on the graph $y = \sqrt{x}$ (minus the height along the x-axis, i.e., zero), and so the height of the cylindrical shell will be $\sqrt{x}$.

The thickness of the shell being dx, we have $dV = 2\pi(x)\left(\sqrt{x}\right)dx$, and so

$$V = \int_0^4 2\pi x^{3/2}\,dx = \left[2\pi \cdot \frac{2}{5}x^{5/2}\right]\bigg|_0^4 = \frac{4}{5}\pi \cdot 4^{5/2} = \frac{4}{5}\pi \cdot 32 = \frac{128}{5}\pi \approx 80.42,$$

as we computed with the washer method in Example 5.5.6, page 581.

It may take a little more imagination to visualize the solid in question as the union of hollow, cylindrical shells, than as a union of washers (as this solid was portrayed in Example 5.5.6, page 581). It also takes a bit more imagination to visualize how the "volume" of an infinitesimal cylindrical shell (i.e., the material used to produce it, not including its interior or bases) would be the product (circumference)(height)(thickness), which is then $2\pi(\text{radius})(\text{height})(\text{thickness})$. However, it does often make for a simpler integral. In this particular case, our integrals were respectively

$$V = \int_0^2 \pi\left[4^2 - \left(x^2\right)^2\right]dy = \int_0^2 \pi\left[16 - y^4\right]dy = \left[\pi\left(16y - \frac{1}{5}y^5\right)\right]\bigg|_0^2 = \cdots = \frac{128\pi}{5}, \quad \text{and}$$

$$V = \int_0^4 2\pi x\sqrt{x}\,dx \quad = \int_0^4 2\pi x^{3/2}\,dx \quad = \left[2\pi \cdot \frac{2}{5}x^{5/2}\right]\bigg|_0^4 \quad = \cdots = \frac{128\pi}{5}.$$

For this particular solid, one practiced in these methods would find the shell method somewhat simpler in both set-up and computation, though both methods have their appeal.

For some of the examples to follow, we have not developed the techniques for computing the antiderivative needed to evaluate the integrals which emerge. Indeed, Section 5.8 and subsequent sections, as well as the sequel to this text, will greatly expand the types of antiderivative computations we can accomplish. However, if we can discover a definite integral which represents a volume (or any other quantity), it is still useful even if we cannot find the antiderivative because there are methods for approximating a definite integral. The most obvious method is to use Riemann Sums, but there are other methods that are usually much more accurate with fewer computational steps. These are developed in Section 5.7. It is therefore worthwhile to set up the integrals even if they cannot be easily evaluated by the Second Fundamental Theorem of Calculus.

Example 5.5.10

Consider the solid of revolution generated by revolving around the y-axis the area bounded by $y = 2\sin x^2$, $y = 0$, $x = 0$, and $x = \sqrt{\pi}$. Use the cylindrical shell method to write an integral representing the volume of this solid. To evaluate the integral, use the following antiderivative formula (derived with techniques of Section 5.9), which can be easily verified through differentiation: $\int x \sin x^2 \, dx = -\frac{1}{2}\cos x^2 + C$.

Solution: The region is drawn below in light gray, partially obscured by one infinitesimal cylindrical shell illustrated (dark gray), at a value $x \in [0, \sqrt{\pi}]$ of thickness dx.

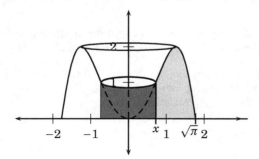

$$V = \int_0^{\sqrt{\pi}} 2\pi x \cdot 2\sin x^2 \, dx$$
$$= \left[-2\pi \cos x^2\right]\Big|_0^{\sqrt{\pi}} \quad \text{(explained previously)}$$
$$= (-2\pi \cos \pi) - (-2\pi \cos 0)$$
$$= 2\pi + 2\pi$$
$$= 4\pi.$$

One interesting observation regarding this latest solid is that one could also rotate the area bounded by $y = 2\sin x^2$, $y = 0$, but with $x \in [-\sqrt{\pi}, \sqrt{\pi}]$ instead of $x \in [0, \sqrt{\pi}]$, and naively integrate these cylindrical shells with one for each $x \in [-\sqrt{\pi}, \sqrt{\pi}]$. However, in doing so we would have twice the actual volume because each geometric cylinder would appear twice in the integral: once for each $x \in [0, \sqrt{\pi}]$, and once for each $x \in [-\sqrt{\pi}, 0]$. Furthermore, the radius would be x for $x \in [0, \sqrt{\pi}]$, but $-x$ for $x \in [-\sqrt{\pi}, 0]$, i.e., the radius would be $|x|$ in both cases. Then our integral, which counts a cylinder for each $x \in [-\sqrt{\pi}, \sqrt{\pi}]$, would be

$$\int_{-\sqrt{\pi}}^{\sqrt{\pi}} 2\pi |x| \cdot 2\sin x^2 \, dx = \int_{-\sqrt{\pi}}^{0} 2\pi(-x) \cdot 2\sin x^2 \, dx + \int_0^{\sqrt{\pi}} 2\pi x \cdot 2\sin x^2 \, dx$$
$$= \left[2\pi \cos x^2\right]\Big|_{-\sqrt{\pi}}^{0} + \left[-2\pi \cos x^2\right]\Big|_0^{\sqrt{\pi}} = 4\pi + 4\pi = 8\pi,$$

which is *not* the correct volume. Indeed, each cylinder appears twice in the computation, hence the integral does represent twice the correct volume of the solid. It is thus important that the cylinders do exhaust the entire solid, but without redundancy.[39]

Example 5.5.11

Compute the volume of the figure described in Example 5.5.7, page 582, but this time compute it using the Cylindrical Shell Method.

Solution: (Due to the limitations of illustrating a three-dimensional object on a two-dimensional paper, the reader is encouraged to peruse the illustration of this solid in the context of Example 5.5.7.)

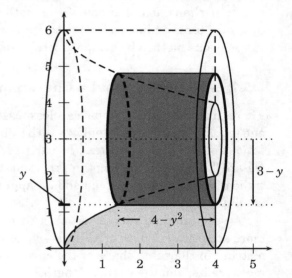

Here we must have horizontal shells, which therefore will have thicknesses dy, and so we will integrate over the variable y. We will have an infinitesimal cylinder for each $y \in [0,2]$.

We see the radii will be of the form $3 - y$, because that is the plane height difference between the center of a shell and its outer edge.

The "height" of the cylinder will be the difference in x-values between the line $x = 4$ and the curve $y = \sqrt{x}$. Since this must be in terms of y (for our eventual integral), we can write these two curves bookending the cylinder's "height" with $x = 4$ and $x = y^2$, so in terms of y, the cylinder's "height" is $4 - y^2$. Putting these together we get

$$dV = 2\pi(\text{rad})(\text{height})(\text{thickness})$$
$$= 2\pi(3-y)(4-y^2)dy, \qquad y \in [0,2].$$

Thus

$$V = \int_0^2 2\pi(3-y)(4-y^2)dy = 2\pi \int_0^2 \left[12 - 3y^2 - 4y + y^3\right] dy = 2\pi \left[12y - y^3 - 2y^2 + \frac{1}{4}y^4\right]\bigg|_0^2$$
$$= 2\pi[(24 - 8 - 8 + 4) - 0] = 2\pi(12) = 24\pi,$$

as we computed (using the Disc/Washer Method) in Example 5.5.7.

It is always advisable to consider both Disc/Washer and Cylindrical Shell methods when the task is to compute the volume of a solid of revolution. Sometimes both methods yield reasonable integrals, as with the previous example showing the Shell Method, along with Example 5.5.7, page 582, where the same volume is computed using the Disc/Washer Method. Note how if one method yields an integral in the variable x, for the same solid, the other

---

[39] Another way to integrate $\int_{-\sqrt{\pi}}^{\sqrt{\pi}} 2\pi|x| \cdot 2\sin x^2\, dx$ is to notice that the integrand is an *even function*, so that its graph is symmetric with respect to the y-axis, and so

$$\int_{-\sqrt{\pi}}^{\sqrt{\pi}} 2\pi|x| \cdot 2\sin x^2\, dx = 2\int_0^{\sqrt{\pi}} 2\pi x \cdot 2\sin x^2\, dx = 2 \cdot 4\pi = 8\pi,$$

the second integral in this line being the same as the computation performed in Example 5.5.10, page 586.

method's integral would necessarily be one in y, and vice versa. In either case, the thickness (dx or dy) dictates the variable of integration.

It should also be noted that finding an integral representing the volume is not a single-step process. It must be done with care, whereby discs, washers, or cylinders are considered, and tested to see if the solid can be "exhausted" by them. A reasonable graphical illustration of the solid should be produced and consulted often in the derivation; the infinitesimal thickness, and the other dimensions (in the same variable as the thickness) must be found from the illustration, again with great care as there are many opportunities for mistakes.

It is not unreasonable, in fact, to attempt to produce the final integrals for both methods, and to perhaps abandon one of the methods if it becomes clear that the integral would be unwieldy (or even impossible) to compute with our integration methods. It can also be useful to verify one method by computing the same volume with the other method, when possible.[40]

5.5.7 Arc Length (and a Cautionary Note Regarding Infinitesimals)

There are times when simple carelessness or a cavalier approach to infinitesimals can cause one to make an incorrect analysis with infinitesimals, leading to erroneous integral representations of physical quantities. In particular, we have to be sure that the infinitesimal *pieces* we are "integrating" are not approximated in such as way as to be off by a persistent multiplicative factor, but instead should approximate the quantity in question to an arbitrarily small percentage error.[41]

A very common such error often occurs in initial approaches to so-called *arc length*—or more accurately, length of a curve—computations. We seek the length of a curve as if it were a string in the same shape of the curve, and as such we could then straighten the string to measure its, and the curve's, length.

Consider, for instance, a curve of the form $y = f(x)$. Since a curve's direction is often not horizontal, dx is not a good approximation for a small piece of the curve's length. Indeed, the first graph in Figure 5.11, page 589, shows that the infinitesimal length ds of arc can be dramatically different from dx; in that particular case, where the slope is 5/4, we have $ds = \sqrt{(dx)^2 + (0.8 dx)^2} = \sqrt{1.64}\, dx$, or $dx = \sqrt{1/1.64}\, ds \approx 0.78\, ds$, so dx is consistently an underestimation of ds by around 22%. This would also occur if we used Riemann Sums: Each subinterval's approximation would be necessarily an approximately 22% underestimation, and so therefore would be the sum. However, in both graphs in Figure 5.11 we can see how the hypotenuse of the right triangle with shorter sides dx and dy closely "hugs" the curve when dx is small, in such a way that percent error in using that hypotenuse to represent ds seems to shrink to zero as dx shrinks to zero. Thus we conclude the following (the second equation following algebraically if $dx > 0$):

$$ds = \sqrt{(dx)^2 + (dy)^2} = \sqrt{1 + \left(\frac{dy}{dx}\right)^2}\, dx. \tag{5.83}$$

---

[40]Eventually we will show how to compute an approximation of a definite integral when we cannot find an antiderivative.

[41]This is also the case with an approach by limits of Riemann Sums, and indeed that is from where the analysis should ultimately spring forth, although there are times when the *derivative* connection is more strongly suggestive via infinitesimals, as with our discussion of the First Fundamental Theorem of Calculus.

5.5. INFINITESIMALS AND DEFINITE INTEGRALS IN GEOMETRY

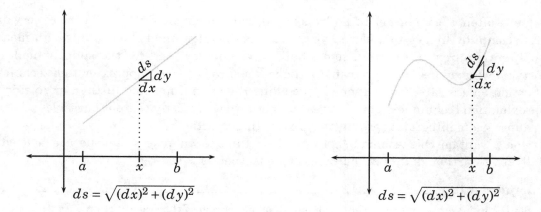

Figure 5.11: Illustrations of the infinitesimal arc lengths $ds = \sqrt{(dx)^2+(dy)^2}$. In the case of a straight line curve, the match-up is exact regardless of the length dx. Furthermore, trying to use $ds = dx$ (erroneously!), for instance, would be incorrect by a persistent factor, namely the slope of the line, and that percent error would not decrease as $dx \to 0^+$. In the curved case (exaggerated here slightly for purposes of illustration), as $dx \to 0^+$, it is not difficult to see that, at least for a differentiable function $y = f(x)$, the percent error in using $\sqrt{(dx)^2+(dy)^2}$ shrinks to zero as $dx \to 0^+$, and so it is safe to write $ds = \sqrt{(dx)^2+(dy)^2}$ as an equality in the infinitesimal sense.

From this we can get that anytime we have a continuous function $f : [a,b] \to \mathbb{R}$, where f' exists on $[a,b]$, then the *arc length* (or simply curve's length) for the graph of $y = f(x)$, $x \in [a,b]$ is given by

$$s = \int_{x=a}^{x=b} ds = \int_{x=a}^{x=b} \sqrt{(dx)^2+(dy)^2} = \int_a^b \sqrt{1 + \left(\frac{dy}{dx}\right)^2}\, dx, \tag{5.84}$$

the first form (with the radical) being somewhat easily remembered from the geometric interpretation (Pythagorean Theorem) and the second form following from simple algebra. However, many textbooks then list the following formula for computational purposes:

$$s = \int_a^b \sqrt{1+(f'(x))^2}\, dx. \tag{5.85}$$

Now this discussion parallels one in which we argue that

$$\Delta s_i \approx \sqrt{1+\left(f'\left(x_i^*\right)\right)^2}\, \Delta x_i, \tag{5.86}$$

and so

$$s = \lim_{\substack{\max\{\Delta x_i\} \to 0 \\ (n \to \infty)}} \sum_{i=1}^n \sqrt{1+\left(f'\left(x_i^*\right)\right)^2}\, \Delta x_i \stackrel{\text{definition}}{=\!=\!=\!=\!=} \int_a^b \sqrt{1+(f'(x))^2}\, dx, \tag{5.87}$$

as before, again based on the idea that our percentage error is small in (5.86) if Δx_i is kept small enough. This is no less—and perhaps more—convincing than the argument from infinitesimals, though the actual argument for the integrals must refer back to Riemann Sums since it is through them that the integral is ultimately defined.

Quite often, the integrals we get from (5.85), even for simple functions $f(x)$, are very difficult to compute. In fact, to find a curve that forces the integrand to have a "nice" antiderivative often requires $f(x)$ to be a somewhat "ugly" function, and for the simpler nonlinear functions we may need to approximate the value of (5.85) using approximation techniques of Section 5.7 (as already mentioned), including Riemann Sums and further, more efficient approximation techniques. For that reason, here we will mostly content ourselves to set up the appropriate integrals representing arc length quantities.

One advantage this technique has over the volume techniques is that there is no need for an illustration, though it may be interesting nonetheless.

Example 5.5.12

Set up an integral that represents the length of one period of the sine function.

Solution: Here $f(x) = \sin x$, so $f'(x) = \cos x$. One period being, for instance, $x \in [0, 2\pi]$ we can simply appeal to (5.84), or equivalently (5.85). These require us to compute $\frac{d}{dx} \sin x = \cos x$, from which we write

$$s = \int_0^{2\pi} \sqrt{1+((\sin x)')^2}\, dx = \int_0^{2\pi} \sqrt{1+\cos^2 x}\, dx.$$

We have no simple way to find an antiderivative. Using approximation methods we will develop later, one could approximate this integral

$$\int_0^{2\pi} \sqrt{1+\cos^2 x}\, dx \approx 7.64,$$

which should seem reasonable given the nature of the sine curve when drawn to scale:

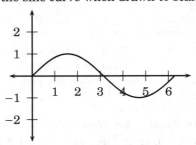

Note that the integrand is always at least 1, since $\sqrt{1+(f')^2} \geq \sqrt{1} = 1$, and so we will never have cases where different parts of the integral "cancel" each other. However, we will get the wrong sign for the arc length if we integrate from right to left ($dx < 0$), so it is understood we always integrate from left to right.[42]

While the integrand in the arc length formula is not often one we can easily antidifferentiate, the integral is at least straightforward to construct. For instance, we can quickly set up such integrals for most examples, only requiring that $f'(x)$ exists on the given interval.

- $\int_{-1}^{1} \sqrt{1+\left((x^2)'\right)^2}\, dx = \int_{-1}^{1} \sqrt{1+(2x)^2}\, dx = \int_{-1}^{1} \sqrt{1+4x^2}\, dx$ represents the arc length of the simple parabolic graph $y = x^2$ for the interval $-1 \leq x \leq 1$. We will be able to compute exactly this integral with techniques in the sequel.[43]

---

[42] Indeed, when we remove dx from under the radical, we are actually removing $\sqrt{(dx)^2} = |dx|$, which we take to be dx precisely because we are assuming $a \leq b$.

[43] More advanced integration techniques (verifiable here by careful differentiation) would allow us to compute:

$$\int_{-1}^{1} \sqrt{1+4x^2}\, dx = \left[\frac{1}{2}x\sqrt{1+4x^2} + \frac{1}{4}\ln\left|\sqrt{1+4x^2}+2x\right|\right]\Bigg|_{-1}^{1} = \sqrt{5} + \frac{1}{4}\ln\left(\sqrt{5}+2\right) - \frac{1}{4}\ln\left(\sqrt{5}-2\right) \approx 2.9578857.$$

5.5. INFINITESIMALS AND DEFINITE INTEGRALS IN GEOMETRY

- $\int_{-1/2}^{1/2} \sqrt{1+\left(\left(\sqrt{1-x^2}\right)'\right)^2}\,dx = \int_{-1/2}^{1/2}\sqrt{1+\left(\dfrac{-2x}{2\sqrt{1-x^2}}\right)^2}\,dx = \int_{-1/2}^{1/2}\sqrt{1+\dfrac{x^2}{1-x^2}}\,dx$

$= \int_{-1/2}^{1/2}\sqrt{\dfrac{1-x^2+x^2}{1-x^2}}\,dx = \int_{-1/2}^{1/2}\dfrac{1}{\sqrt{1-x^2}}\,dx = \left[\sin^{-1}x\right]\Big|_{-1/2}^{1/2} = \dfrac{\pi}{6}-\dfrac{-\pi}{6}=\dfrac{\pi}{3}$

represents arc length of that part of the upper unit circle for which $-\frac{1}{2}\le x\le \frac{1}{2}$. Recall that the circumference of the unit circle is 2π, so this represents one-sixth of that whole circle's circumference. (With a little effort one can see that if we draw the arc in question, it is a circular sector with included angle of $\sin^{-1}(1/2) - \sin^{-1}(-1/2) = \pi/6 - (-\pi/6) = \pi/3$, which is indeed one-sixth of the complete circle of total included angle 2π.)[44]

Just as with volumes of revolution, at times it might be desirable to have an integral in the y-variable instead of the x-variable. In such as case, we would extract a factor of dy (rather than dx as we did in (5.84)) and thus rewrite

$$ds = \sqrt{(dx)^2+(dy)^2} = \sqrt{\left(\dfrac{dx}{dy}\right)^2+1}\,dy = \sqrt{1+\left(\dfrac{dx}{dy}\right)^2}\,dy.$$

Example 5.5.13

Set up integrals in both x and y to find the arc length of the (parabolic) graph $y=x^2$, for $0\le x\le 4$.

Solution: To integrate in x, we write an integral as we did in our first "bullet point" on page 590, but with a different interval of integration. We use $dy/dx = 2x$ and compute

$$\int_0^4 \sqrt{1+(2x)^2}\,dx = \int_0^4 \sqrt{1+4x^2}\,dx.$$

To integrate in y, we note that on this curve we have $x=\sqrt{y}$, and $x\in[0,4] \iff y\in[0,16]$. Furthermore, $dx/dy = \dfrac{1}{2\sqrt{y}}$, and so our arc length is also represented by

$$\int_0^{16}\sqrt{1+\left(\dfrac{1}{2\sqrt{y}}\right)^2}\,dy = \int_0^{16}\sqrt{1+\dfrac{1}{4y}}\,dy = \int_0^{16}\sqrt{\dfrac{4y+1}{4y}}\,dy.$$

In either case, the integral will have value approximately 16.819.

Neither of the integrals in the above example is particularly simple to compute. In fact, the first requires slightly fewer steps, but both require techniques from the sequel. Despite this, it is useful to know that there is the option of considering an integral in y or x, if feasible.

---

[44]With our current tools we cannot yet derive the arc length of the entire upper unit semicircle with this integration method because the resulting integral, namely $\int_{-1}^{1}\dfrac{1}{\sqrt{1-x^2}}\,dx$, would be what in the sequel we would call *improper*. We see that the integrand approaches ∞ as $x\to -1^+$ or $x\to 1^-$. In the sequel we describe a reasonable technique of handling this, which is to consider integrals of the form $\int_{\alpha}^{0}\dfrac{1}{\sqrt{1-x^2}}\,dx$ and $\int_{0}^{\beta}\dfrac{1}{\sqrt{1-x^2}}\,dx$, where $\alpha\in(-1,0)$ and $\beta\in(0,1)$, and consider what occurs to these, respectively, as $\alpha\to -1^+$ and $\beta\to 1^-$, and if these limits exist as finite numbers, then our original integral is the sum (π) of these limits. The techniques for analyzing improper integrals is ultimately straightforward but has surprisingly many and varied technicalities to work around as they arise in practice.

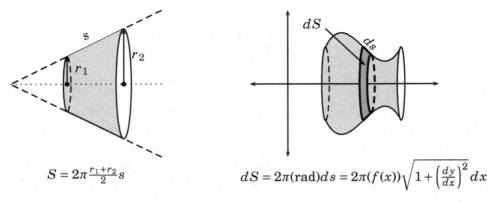

Figure 5.12a. Cone frustum.   **Figure 5.12b.** Surface area element.

Figure 5.12: The surface area of a "frustum" of a cone is often given by $\pi(r_1+r_2)s$, which is the same as the area of a rectangle with height s and base that is the average of the circumferences of the two circles at either end of the frustum, as reflected in the formula given in Figure 5.12a. The infinitesimal area element of a solid of revolution such as found in Figure 5.12b can be taken to be the infinitesimal arc length ds (for a given x in this case) multiplied by $2\pi(\text{rad})$, where "rad" is the radius of that element at the location x. If the axis of rotation is the x-axis and the curve is given by $y = f(x)$, then $\text{rad} = |f(x)|$. For these cases we assume $f(x) > 0$ and so $\text{rad} = f(x)$.

5.5.8 Surface Areas of Revolution

In this subsection we discuss the surface area of a solid of revolution, not counting the "top and bottom bases." The formula can be seen as a natural consequence of the arc length formula from the previous subsection.

Figure 5.12a shows the surface area of a "frustum" of a right circular cone. We will not derive the formula in this section, though it comes from knowledge of the formulas for areas of circular sectors.[45]

In the case of the infinitesimal surface area element dS (note the S is upper-case) as in Figure 5.12b, assuming dy/dx exists along the curve, then that surface represented by dS is very nearly the shape of a frustum of a cone, and the two radii are very nearly the same. Indeed, to a percentage error which will approach zero as dx shrinks to zero, we would have the slant height being ds (lower-case s), and the frustum-like area element dS would then have surface area

$$dS = 2\pi(\text{rad})ds = 2\pi|f(x)|\sqrt{1+\left(\frac{dy}{dx}\right)^2}dx = 2\pi(f(x))\sqrt{1+\left(\frac{dy}{dx}\right)^2}dx, \qquad (5.88)$$

in the case of that figure. Of course, one has to adjust depending upon the function, region, and general configuration of the particular situation to arrive at the correct formula for dS

---

[45] The area formula for frustum of a right circular cone has some quick intuitive appeal, in that it equals the area of a right circular cylinder's curved surface if s represents the height of the cylinder (instead of the "slant height" of the frustum), and $r = r_1 = r_2$, as is the case with a cylinder. In the case of the frustum, we can instead use the average radius for r: "r"$= (r_1 + r_2)/2$.

5.5. INFINITESIMALS AND DEFINITE INTEGRALS IN GEOMETRY

(as well as ds).[46]

Example 5.5.14

Suppose we have the parabolic region $y = x^2$, $x \in [-2, 2]$ revolved around the y-axis. This generates a "parabolic dish" shape such as used in many microphones and antennas (or "antennae").

In fact, the entire figure is generated if we revolve the curve $y = x^2$ over the range $[0, 2]$, and "generated twice" if we revolve the entire curve $y = x^2$, $x \in [-2, 2]$ around the y-axis.

In this case, for a given $x \in [0, 2]$, the infinitesimal length of arc will be

$$ds = \sqrt{1 + \left(\frac{dy}{dx}\right)^2}\, dx = \sqrt{1 + (2x)^2}\, dx = \sqrt{1 + 4x^2}\, dx.$$

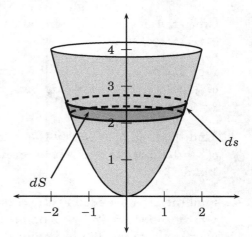

Now dS is the infinitesimal area we get from revolving that length ds around the x-axis. For that particular x-value, the radius will be rad = $|x| = x$, since $x \in [0, 2]$, and then

$$dS = 2\pi(\text{rad})ds = 2\pi x \sqrt{1 + 4x^2}\, dx.$$

Since we have one strip area dS for each $x \in [0, 2]$, we integrate these to compute the total surface area S. In doing so, we will note that $\frac{d}{dx}(1 + 4x^2)^{3/2} = \frac{3}{2}(1 + 4x^2)^{1/2} \cdot 8x = 12x(1 + 4x^2)^{1/2}$, which differs from our integrand by a constant factor $\pi/6$. Thus we could easily adjust, and verify if necessary, that

$$S = \int_0^2 2\pi x \sqrt{1 + 4x^2}\, dx = \left[\frac{\pi}{6}(1 + 4x^2)^{3/2}\right]\Big|_0^2 = \frac{\pi}{6}\left[17^{3/2} - 1\right] \approx 36.1769.$$

In this case, it was important that we used only the interval $0 \leq x \leq 2$ for integration: It is sufficient to generate the surface, and no part of the surface is generated more than once. If we used $-2 \leq x \leq 2$ for our interval of integration, we would have computed exactly twice the area, because we would have had an infinitesimal length of arc ds for both x and $-x$, for all $x \in [0, 2]$ in that case. However, for the radius we would have used $|x|$.

Indeed, we could have used the interval $[-2, 0]$ for this surface area, but the radii for each $x \in [-2, 0]$ would have been $|x| = -x$:

$$S = \int_{-2}^0 2\pi(-x)\sqrt{1 + 4x^2}\, dx = \left[\frac{-\pi}{6}(1 + 4x^2)^{3/2}\right]\Big|_{-2}^0 = \frac{-\pi}{6}\left[-1 + 17^{3/2}\right] \approx 36.1769,$$

as before.

---

[46]Perhaps worth noting is also that dS could be visualized as representing a near-rectangular region, if "snipped and unrolled," with "base" $2\pi(\text{rad})$ and height ds. Since the radius is assumed to have very small (infinitesimal!) percentage change along the infinitesimal area element dS, this explanation is perhaps also believable.

Exercises

1. Find the total (unsigned) area under the sine curve $y = \sin x$ for $x \in [0, \pi]$.

2. Find the total (unsigned) area between the sine curve $y = \sin x$ and the x-axis for $x \in [0, 2\pi]$.

3. Find the area bounded by the graph of $y = 1/x^2$ and the x-axis between $x = 1/2$ and $x = 2$.

4. Find the area bounded by the graph of $y = 1/x$ and the x-axis between $x = 1$ and $x = e$.

5. Find the area bounded by the curves $y = 3x^2$ and $y = 12$ by using an integral involving infinitesimally thin rectangles of thickness dx.

6. Find the area bounded by the graphs of $y = 5 - x^2$ and $y = x - 1$ by using infinitesimally thin rectangles of thickness dx.

7. Repeat Exercise 5 but with rectangles of thickness dy.

8. Repeat Exercise 6 but with rectangles of thickness dy.

9. Find the area bounded by the graphs of $y = x^2 - x$ and $y = x$.

10. Consider the area bounded by $y = \sqrt{x}$ and $y = x^2$.

    (a) Find this area using an integral in x.

    (b) Find this area using an integral in y. Note that the calculus (not just the geometry) would indicate that these are the same.

11. Find the area bounded by the graphs $x = 9 - y^2$ and $x = y^2 - 9$.

12. Find the positive area bounded by the graph of $y = x^3 + x^2 - 5x + 3$ and the x-axis. (Note $x = 1$ is one x-intercept.)

13. Consider the function $y = \sin^{-1} x$ for $x \in [0, 1]$, as drawn below.

    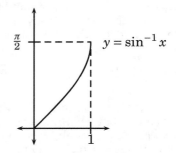

    (a) Find the area bounded by the curve and the y-axis, for $y \in [0, \pi/2]$. (Use horizontal rectangles.)

    (b) Use this result, and the area of the rectangle drawn in the graph, to find $\int_0^1 \sin^{-1} x \, dx$. (The area we desire, and the area found in (a), will sum to the area of the rectangle.)

14. Use the method of the previous problem to compute $\int_1^e \ln x \, dx$.

    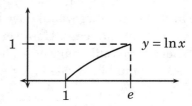

15. If the surface area of a sphere of radius r is given by $S = 4\pi r^2$, use this to find a formula for the volume of a sphere of radius R. (Consider a sphere as a union of "spherical shells," which can be slit and laid flat, each with "thickness" dr. See Subsection 5.5.3.)

5.5. INFINITESIMALS AND DEFINITE INTEGRALS IN GEOMETRY

16. If the area of the lateral surface (i.e., the curved surface, not including the flat circular "bases") of a cylinder of radius r is $A = 2\pi rh$, show how this implies that the volume of the cylinder of radius R and height h is is $V = \pi R^2 h$. (Hint: Break the cylinder of radius R into cylindrical shells of radius r for $r \in [0,R]$. Set up an integral in r and evaluate.)

17. For each of the following sentences,

    - rewrite the sentence using the correct choice from the parentheses, and
    - draw an illustration to demonstrate the correctness of your answer. (Sometimes an illustration might work for more than one sentence, and can be labeled as such.)

    (a) If the infinitesimal discs or washers are horizontal, then the axis of rotation is (vertical/horizontal).

    (b) If the axis of rotation is vertical, then the thickness of the discs or washers is (dx/dy).

    (c) If the infinitesimal discs or washers are vertical, then the axis of rotation is (vertical/horizontal).

    (d) If the axis of rotation is horizontal, then the thickness of the discs or washers is (dx/dy).

    (e) If the discs or washers are horizontal, then the variable of integration will be (x/y).

    (f) If the discs or washers are vertical, then the variable of integration will be (x/y).

18. Do the same for the cylindrical shell method (replacing "discs or washers" with "cylindrical shells").

    (a) If the axis of rotation is vertical, then the thickness of the cylindrical shell will be (dx/dy).

    (b) If the axis of rotation is horizontal, then the thickness of the cylindrical shell will be (dx/dy).

    (c) If the cylinders are vertical, then the variable of integration is (x/y).

    (d) If the cylinders are horizontal, then the variable of integration is (x/y).

19. Use the Disc/Washer Method to derive that the volume of a sphere of radius R is $V = \frac{4}{3}\pi R^3$. Do this by revolving the area bounded by $y = \sqrt{R^2 - x^2}$ and the x-axis around the x-axis.

20. Use the Disc/Washer Method to derive the volume of the solid obtained by rotating the ellipse $\frac{x^2}{a^2} + \frac{y^2}{b^2} = 1$ around the x-axis. (Hint: Solve for y and then it will read much as the previous problem.) Note that the formula for the volume of a sphere is a special case $a = b = R$.

21. Consider the area from Example 5.5.3, page 578. Suppose we wish to revolve that area around the y-axis.

    (a) Which method (Disc/Washer or Cylindrical Shell) more easily allows us to represent the volume as a single integral?

    (b) Set up the appropriate integral representing that volume.

22. Repeat the previous exercise but assume instead we are revolving around the x-axis. (Note that it is enough to assume $y \geq 0$ due to overlap that can occur as the region is revolved.)

23. Find the volume generated by revolving the area bounded by the graphs of $y = x^2$, $x = 3$, and the positive x-axis around the x-axis.

24. Repeat the previous problem but with the y-axis as the axis of rotation.

25. Use the Disc Method to find the volume generated if the area bounded by the graphs of $y = 4 - x^2$, $y = 0$, and the positive x-axis is revolved around the x-axis.

26. Use the Shell Method to find the volume generated if the area bounded by the graphs of $y = 4 - x^2$, $y = 0$, and the positive x-axis is revolved around the y-axis.

27. Find the volume generated if the area bounded by the graphs of $y = x^3$, $y = 0$, and $x = 2$ is revolved around the y-axis. Use the Shell Method.

28. Repeat the previous exercise using the Disc Method.

29. Set up and simplify (without computing) the integral that represents the arc length of $y = x^{4/3}$, $x \in [1,4]$. (Do not evaluate the integral.)

30. Set up and simplify the integral that gives the arc length of $y = \cos x$, $x \in [0, \pi]$. (Do not evaluate the integral.)

31. Consider the function $f(x) = \frac{3}{4}x + 1$. Find the arc length of the graph of $y = f(x)$ for $x \in [0,8]$ the following way:

    (a) Using the integral formula for the arc length.

    (b) Using the Pythagorean Theorem. (It may be useful to make a rough sketch of the function and the desired length.)

5.6 Physical Applications of Infinitesimals, Definite Integrals

Here we extend the previous section's discussion regarding analysis by infinitesimals, to examine how they can lead to further applications of integration. We do so without having the more advanced techniques for computing more complicated integrals—those discussed later in this chapter and in the sequel text—but we instead lay more groundwork for how to approach applied problems so we will have applications ready for those techniques.

5.6.1 Velocity and Position, Revisited

Because the relationship between position and velocity so well illustrates the intuition of calculus, we visit it again but in the context of infinitesimals. Consider the following integral equation, which is intuitive—and some would say obvious—from several perspectives, assuming $v(t)$ is continuous on $[t_0, t_f]$:[47]

$$\int_{t_0}^{t_f} \underbrace{v(t)\,dt}_{ds(t)} = s(t_f) - s(t_0). \tag{5.89}$$

Recall that in Chapter 3 (more specifically Definition 3.1.1, page 210) we showed how it seemed appropriate that $v(t) = \frac{ds(t)}{dt}$. Then s is an antiderivative of v and (5.89) becomes just a statement of the Fundamental Theorem of Calculus for this special case. We can also look at the differential form of the notion $\frac{ds(t)}{dt} = v(t)$, i.e.,

$$d\,s(t) = v(t)\,dt, \tag{5.90}$$

Now let us somewhat dissect this notation as it stands. From the perspective of infinitesimals we could say that "$ds(t)$ is an infinitesimal change of position at time t caused by an infinitesimal change dt in time, when the velocity was $v(t)$." Note that there is an assumption that velocity is, for these purposes, relatively constant as time changes by this infinitesimal amount dt, and so the change of position would be $v(t)\,dt$, to a percentage error that is as small as we like.[48]

This thinking allows one to (naively) look at $\int_{t_0}^{t_f} v(t)\,dt$ as an infinite sum of infinitesimal quantities $ds(t)$, one such infinitesimal for each $t \in [t_0, t_f]$, and these somehow accumulating to represent the actual quantity $s(t_f) - s(t_0)$.

Again, this also makes sense if we keep in mind that this integral represents a limit of Riemann Sums of the form $\sum_{i=1}^{n} v(t_i^*)\Delta t_i$, as $\max\{\Delta t_i\} \to 0^+$ and $n \to \infty$, and so the values of $v(t_i^*)\Delta t_i$ can be forced to have percent errors as small as we like in approximating $s(t_i) - s(t_{i-1})$ by taking the maximum $|\Delta t_i|$ to be small.

---

[47]It is a kind of running joke that mathematicians have their own idea of the meaning of "obvious." Much of mathematics is indeed "obvious" to the trained observer who has pondered it long enough that it seems transparent. A fact's being clearly true upon one's first encounter with it is not a necessary condition for that fact to be called "obvious" in mathematical conversation. (One story has a famous professor putting theorems and their proofs on a board for a class, writing from notes that a particular step is obvious, hesitating, thinking for a while, leaving the room for the office, coming back sometime later declaring, "Yes, it is obvious," and moving on.)

[48]Again, the percentage error argument works when the velocity is continuous and nonzero. A different argument, which we do not pursue here, takes care of the case where v is continuous and can assume the value zero. In such a case, the percentage error may be undefined, but the actual error can be held as small as we like.

As we will see eventually, this kind of analysis is quite powerful for discovery purposes in a multitude of circumstances beyond displacement and area problems, though to be sure of its validity for other cases, a Riemann Sum analysis should be included, where one sees if a percentage error argument is convincing. For instance, when looking at $\int_a^b f(x)\,dx$, one considers "infinitesimal rectangles of infinitesimal widths dx, and heights $f(x)$, these rectangles having signed areas $f(x)\,dx$, at each value of $x \in [a,b]$." These are illustrated in the first graph in Figure 5.9, page 573.

Countless other examples can be found, where we don't need the *exact* formula for a "piece" of the accumulated quantity we need, but **if we have an approximation that has percentage error that shrinks to zero** when we break our quantity (such as displacement or area) into pieces whose number approaches infinity but whose individual contributions shrink to zero, then our integral formula for that desired cumulative quantity is correct. This is more obvious when the definite integral in question is viewed as a limit of Riemann Sums, but the use of infinitesimals has its appeal.

(It is also worth recalling that $\int_{t_0}^{t_f} |v(t)|\,dt$ represents the total distance traveled over the time interval $[t_0, t_f]$, while $\int_a^b |f(x)|\,dx$ represents the total (nonsigned) area between the x-axis and the curve $y = f(x)$. These things were discussed in Section 5.3.)

5.6.2 Work and Energy

Nearly every physics textbook refers to the the following very useful definitions:

Energy: The ability to perform work.

Work: A force is said to do work if, while acting, there is a displacement of the point of application in the direction of the force.

In the simplest case of one-dimensional force and motion, this is quantified in the following way. Suppose the position s is along a coordinate axis, and force F is constant. Then the work done by the force over the displacement Δs is given by

$$W = F\Delta s. \tag{5.91}$$

For instance, if the axis is horizontal with the positive direction to the right, then the force will be considered positive if it is in the rightward direction, and negative if in the leftward direction. In terms of (nonzero) signs, we have four possibilities. In the first two, the force and displacement are in the same directions, and so F and Δs have the same signs, and the work W done by the force is positive, i.e., $F\Delta x > 0$:

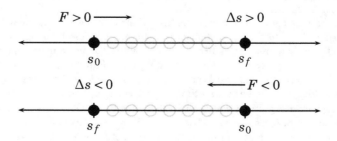

In the next cases, the force F and displacement Δs are of opposite sign, and so $W = F\Delta s < 0$, i.e., work is negative:

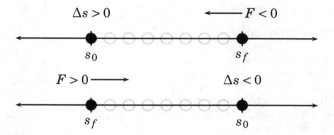

So if the force is in the direction of motion, the work will be positive; if the force is in the opposite direction of the motion, the work will be negative. For another perspective, if the force is in the direction of the motion, then it is reinforcing the motion and adding energy to the object, while if it is in the opposite direction of the motion, the force is opposing the motion and robbing the object of energy. (There may be other phenomena simultaneously adding or robbing energy from the object. We are only interested in the contribution by the particular force in question.)

For instance, a motor accelerating a car from a stop will drive the tires to cause the car to experience a force in the direction of the desired motion, and thus the motor must ultimately exert positive work. However, the brakes can then cause a force in the opposite direction of the car's motion to slow it, and thus will apply negative work to the car. Interestingly, because of the conservation of energy (described later), the car is simultaneously transferring its energy to the brakes, seen mostly in the form of thermal energy resulting in a temperature rise within the brakes.

Similarly, a person pushing a car forward will be performing work on the car, while the car is removing energy from the person (mostly chemical energy within the muscle cells). On the other hand, if the person is trying to stop a car from rolling, this requires negative work done to the car, and that energy is transferred to the person, though probably not in any useful form (except perhaps to warm them up a bit, and hopefully not run them over).[49]

For the most part (with some interesting nuclear events being the dramatic exception), energy is expected to be "conserved," i.e., neither created nor destroyed, although it may be transformed among different forms, such as mechanical energy (energy of movement), thermal energy, electromagnetic energy, chemical energy, nuclear energy, and others. The distinctions are sometimes blurred since, for instance, thermal energy may consist of molecules vibrating or moving quickly, which is arguably just mechanical energy on a smaller geometric scale, and indeed it is the thermal energy in a combustion motor of a vehicle that produces the mechanical energy of the vehicle's motion.

Example 5.6.1

Suppose we need to place a 100-lb bag of cement off of the floor of a warehouse and onto a shelf 4-ft high. How much work (energy) is required to do this? (Neglect the energy needed to move the cement bag horizontally onto the shelf.)

---

[49] Anyone who has pushed cars more than a few times will likely notice that it is fairly easy to push a car a small distance, but once Δs gets very large (e.g., a city block length) it is a very tiring process, as the pusher(s) must overcome considerable internal friction of the rolling mechanism, making $W = F\Delta s$ a quickly growing quantity.

Solution: The total work here will be simply the force required to lift the cement through the distance, multiplied by that distance (both being positive). Thus the total work will be

$$W = F\Delta s = 100 \text{ lb} \times 4 \text{ ft} = 400 \text{ ft-lb}.$$

If we wish to use our calculus to set up an integral representing this quantity, we can consider the diagram on the right, with the work required at level $s \in [0,4]$ to lift the cement by a vertical displacement ds to be $dW = F(s)ds$, where $F(s) = 100$ lb is the force applied to the cement bag. Thus the total work W is given by

$$W = \int_0^{4 \text{ ft}} (100 \text{ lb})\, ds = \left[(100 \text{ lb})s\right]\Big|_0^{4 \text{ ft}} = 400 \text{ ft-lb}.$$

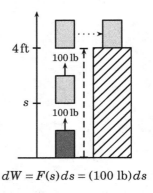

$dW = F(s)\, ds = (100 \text{ lb})\, ds$

Units are often omitted within integrals, and a note may be mentioned afterwards regarding what would be the final units, in a style such as the following:

$$\int_0^4 100\, ds = \left[100s\right]\Big|_0^4 = 400 - 0 = 400, \quad \text{with units of ft-lb.}$$

It is worth noting that one type of energy is the energy an object can have due to its motion, i.e., its *kinetic energy*. With a little effort, one can derive that the kinetic energy of an object with mass m and velocity v is given by[50]

$$E_k = \frac{1}{2}mv^2. \tag{5.92}$$

Note how if we double the mass m, we double the kinetic energy E_k, but if we double the velocity v, we quadruple the kinetic energy E_k. Also, E_k does not depend on the direction of the motion, but only the speed: speed $= |v| = \sqrt{v^2}$, and so we could write $E_k = \frac{1}{2}m(\text{speed})^2$.

Equation (5.92) can also be useful because in many cases the work done to an object will be the same as its change in kinetic energy. We will use this after the next example.[51]

---

[50] One method for deriving Equation (5.92) is to consider the kinetic energy to be equal to the work necessary to achieve that velocity $v = v_f$ from a standstill ($v_0 = 0$). For simplicity, we can further assume $s_0 = 0$ (and thus $s = \Delta s$) and that we accelerate the mass m with a constant acceleration a. Then $v = at \implies t = \frac{v_f}{a}$, and $s_f = \frac{1}{2}at^2 = \frac{1}{2}a\left(\frac{v}{a}\right)^2 = \frac{1}{2} \cdot \frac{v^2}{a}$. Thus the work required is

$$\int_0^{s_f} F(s)\, ds = \int_0^{s_f} ma\, ds = mas_f = ma \cdot \frac{1}{2} \cdot \frac{v^2}{a} = \frac{1}{2}mv^2.$$

Since kinetic energy is the energy due to the motion, it should not matter how we achieved the velocity $v_f = v$ (we just chose a simple method of acceleration), and thus we can conclude $E_k = \frac{1}{2}mv^2$, q.e.d.

[51]This assumes the absence of any other forces (e.g., friction) or energy conversions (e.g., burning to change chemical energy to heat), and so work can indeed translate directly to kinetic energy.

5.6. PHYSICAL APPLICATIONS OF INFINITESIMALS, DEFINITE INTEGRALS 601

Example 5.6.2

Consider a 14-lb bowling ball dropped off of a rooftop that is 30 ft high. How much kinetic energy does the ball have when it hits the ground? (Neglect the effects of wind resistance and the size of the ball.)

Solution: If we wish to consider the downward direction to be negative, and the force of the gravity to be in that same direction, because the force is constant, we can easily compute

$$W = (-14\,\text{lb}) \times (-30\,\text{ft}) = 420\,\text{ft-lb}.$$

To produce an integral, we would have s range from 30 ft down to 0, and $F(s) = -14$ lb be constant. We would still have $dW = F(s)\,ds = (-14\,\text{lb})\,ds$, and then

$$W = \int_{30}^{0} (-14)\,ds = \left[-14s\right]\Big|_{30}^{0} = 0 - (-420) = 420,$$

with units in ft-lb.

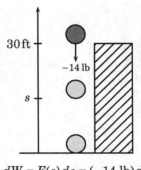

$dW = F(s)\,ds = (-14\,\text{lb})\,ds$

Interestingly, there are two ways of computing the velocity of the bowling ball at the bottom of its path: one from an acceleration computation, and the other from an energy argument. In the first, we find how much time the ball takes to fall and use that time to find out how much it would have accelerated, neglecting air resistance and other dissipative forces. To do so we would use Equation (5.16), page 517, namely $s = s_0 + v_0 t + \frac{1}{2}at^2$, with $s_0 = 30$ ft, $v_0 = 0$, and $a = -32$ ft/sec². When the bowling ball hits the ground, $s = 0$ and so

$$0 = 30\,\text{ft} + 0 + \frac{1}{2}(-32\,\text{ft/sec}^2)t^2 \quad\Longrightarrow\quad t^2 = 30 \cdot 2/32\,\text{sec}^2 = \frac{60}{32}\text{sec}^2 = \frac{15}{8}\text{sec}^2$$

$$\Longrightarrow\quad t = \sqrt{15/8}\,\text{sec} \approx 1.369\,\text{sec}.$$

From that, we can use Equation (5.15), page 517, with the relevant values: $v = v_0 + at = 0 - 32\,\text{ft/sec}^2 \cdot \sqrt{15/8} \approx -43.817\,\text{ft/sec}$.

But we can also find the velocity of the bowling ball at any relevant height by knowing its kinetic energy $E_k = \frac{1}{2}mv^2$ (Equation (5.92), page 600) there, which we know because we know how much work gravity did to the ball to bring it to that height (from its initial position at rest, i.e., zero kinetic energy, at position $s = 30$ ft). Thus we just need to solve for

v as follows:[52]

$$E_k = 420 \text{ ft-lb} \implies \frac{1}{2}mv^2 = 420 \text{ ft-lb}$$

$$\implies \frac{1}{2} \cdot \frac{14 \text{ lb}}{32 \text{ ft/sec}^2} v^2 = 420 \text{ ft-lb}$$

$$\implies v^2 = 2 \cdot \frac{32 \text{ ft/sec}^2}{14 \text{ lb}} \cdot 420 \text{ ft-lb} = 1920 \left(\frac{\text{ft}}{\text{sec}}\right)^2$$

$$\implies v = \pm\sqrt{1920 \left(\frac{\text{ft}}{\text{sec}}\right)^2} \approx \pm 43.8 \frac{\text{ft}}{\text{sec}}.$$

Since the ball is dropping and traveling downward, we have $v < 0$, and so $v \approx -43.8$ ft/sec.

In fact, a similar analysis shows that if we were to throw the bowling ball upward from the ground so that it would achieve the height 30 ft and no higher, to counter the negative work done by gravity it would need exactly the kinetic energy it would possess initially if it were thrown vertically with velocity $v \approx 43.8$ ft/sec.

The next obvious question is how to compute the total work done by a nonconstant force. For instance, consider a spring that satisfies Hooke's Law, in which $F = -ks$, where $s = 0$ is the resting (noncompressed, nonextended) state of the spring. For such a spring, if position is positive (to the right of zero), i.e., $s > 0$, then the direction of the force exerted by the spring is negative (i.e., to the left). Similarly, if position is negative (left of zero), i.e., $s < 0$ then the direction of the force exerted by the spring is positive (rightward). Such a force is sometimes referred to as *restorative*. The farther the spring is from its resting state, the stronger is the restorative force:

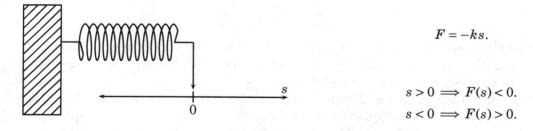

$$F = -ks.$$

$$s > 0 \implies F(s) < 0.$$
$$s < 0 \implies F(s) > 0.$$

Example 5.6.3

The two parallel springs in the front suspension ("fork") of a particular motorcycle are listed as having 4.5 inches of downward travel available from the unloaded position (with the motorcycle vertical, stationary, and without riders), and have a spring rate given as $k = 0.58$ kg/mm. How much work must be expended to completely compress the forks? Use standard metric units.

Solution: First we must convert the units of kg to newtons. Indeed, kg is a unit of mass and not force (or weight), and so we will assume that the manufacturer meant by 1 kg the amount of gravitational force encountered by 1 kg at the earth's surface, where 1 kg=9.8 N. We should also use units of meters (m) instead of millimeters. We next reason that since there are two identical springs,

---

[52]Recall that the weight of an object is equal to its mass multiplied by the acceleration due to gravity, which is usually taken to be 9.8 m/sec^2, or 32 ft/sec^2, neither of which is exact at ground level. Since gravity does vary with height, these values are usually accepted for purposes of textbook physics problems. The metric mass unit is the kilogram, while the English unit is the slug. If we take w to be the weight of an object, then $w = mg \implies m = w/g$.

the force to compress them both simultaneously will be the same as twice the force needed to compress one such spring. From these considerations, we compute the constant for the combined springs:

$$k = 2 \cdot \frac{0.58\,\text{kg}}{\text{mm}} \cdot \frac{9.8\,\text{N}}{\text{kg}} \cdot \frac{1000\,\text{mm}}{\text{m}} = 11{,}368\,\text{N/m}.$$

This will be in effect for the entire travel of the forks, which is much less than one meter. (While here we show five significant digits for k, our initial data had fewer and so we will eventually truncate our final answer to reflect that. It is always best to do as little truncation as possible until after the final computation.) We also need to convert the length 4.5 inches of the springs' compression into meters:

$$4.5\,\text{in} \cdot \frac{2.54\,\text{cm}}{\text{in}} \cdot \frac{\text{m}}{100\,\text{cm}} = 0.1143\,\text{m}.$$

Since we are considering the force needed to compress the springs, rather than the force exerted by the springs in resisting that compression, the sign of the external force exerted *onto the springs* to cause the compression will be the opposite of the sign of the force exerted *by the springs*.

If we let s be the signed length of the springs, where $s < 0$ means they are compressed, then the work needed to compress the springs from s to $s+$ (where here $ds < 0$) will be $dW = -F(s)\,ds = k(s)\,ds = 11{,}368s\,ds$, and so the total work exerted by these springs from $s = 0$ to $s = -4\,\text{in} = -0.1143\,\text{m}$ will be

$$W = \int_0^{-0.1143} 11{,}368 s\, ds = \left[5684 s^2\right]\Big|_0^{-0.1143} = 5684(-0.1143)^2 - 0 \approx 74\,\text{N-m} = 74\,\text{J}.$$

Note how the force—not counting the weight of the motorcycle—used to compress the fork springs varies from 0 (when the motorcycle is upright, stationary, and has no extra weight), $0.1143\,\text{m} \times 11{,}368\,\text{N/m} \approx 1300\,\text{N}$ of force, or approximately 300 lb of force, at full compression. Such large forces can easily occur if the front wheel encounters a sharp enough bump at a high speed. While that force seems too high to account for a total work of only 74 N-m, it must be recalled that the displacements Δs possible near full compression are quite small. A small force acting through a large displacement can require the same energy (i.e., perform the same work) as a large force over a shorter displacement. (Indeed, this is why levers operate as they do.) We require the integral to compute the total work performed by a variable force over any displacement.

Another type of work problem requiring an integral is one in which the total work is distributed unevenly across the object in question. Consider the following example.

Example 5.6.4

Suppose a cable of length 100 ft is coiled on the ground and is to be pulled up length-wise into the vertical position until the lower end is suspended 50 ft above the ground. If the weight density of the cable is 2 lb/ft, how much work is required to lift the rope into its final position?

Solution: Here we will consider how much work is required to lift an infinitesimal length dy of cable from the ground to its final position. It will be useful, therefore, to have a vertical axis, as on the right. Note y will be in feet.

At each height $y \in [50\,\text{ft}, 150\,\text{ft}]$, consider an infinitesimal length dy of cable. This will have weight $2\,\text{lb/ft}\,dy$, and will have been lifted to the height y, requiring work $dW = (2\,\text{lb/ft} \cdot dy) \cdot y$.

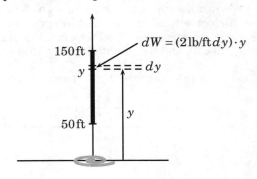

Note that the units of $dW = 2\,\text{lb/ft}\,dy \cdot y$ will be $(\text{lb/ft}) \cdot \text{ft} \cdot \text{ft} = \text{ft-lb}$. The total amount of work to lift the cable is therefore

$$\int_{50}^{150} 2y\,dy = \left[y^2\right]\Big|_{50}^{150} = 150^2 - 50^2$$
$$= 22{,}500 - 2500 = 20{,}000,$$

with units of ft-lb. So the total work required to lift the cable into that position was 20,000 ft-lb.

A similar problem, where we consider the work required on an infinitesimal piece of the material in question and integrate these, can be found in the next example.

Example 5.6.5

Suppose a water tower is constructed in the shape of an inverted right circular cone of height 15 meters and width 10 meters at the top. Furthermore, the vertex of the cone is 30 meters from the ground. How much energy (work) is required to fill the tank with water. Ignore the work needed to lift the water left in the pipeline leading to the tank. (Use $1\,\text{g/cm}^3$ for the volume density of water.)

<u>Solution</u>: As is often advisable, we begin with an illustration. In it we introduce variables r and h, for reasons we will note later, and we also note the radius at the top of the tank is half of the diameter given, and is thus 5 m. We also provide a vertical axis measuring the height (in meters) from the ground.

For each $y \in [30,45]$, consider a horizontal, infinitesimal disc of water within the tank, that disc having volume $dV = \pi(\text{radius})^2\,dy$. For simplicity, we will use the Greek letter $\rho = 1\,\text{g/cm}^3$ to represent the density of the water, and use g for the gravitational constant $9.8\,\text{m/sec}^2$, and so the weight of the water in that infinitesimal disc is $(\rho\,dV)g = \rho g\,dV$. The work needed to place the water there from the ground level is then its weight times the height it is raised, i.e.,

$$dW = y \cdot \rho g\,dV = y\rho g\pi(\text{radius})^2 dy.$$

Now we have to write the radius in terms of y, and integrate dW along the interval $30 \le y \le 45$.

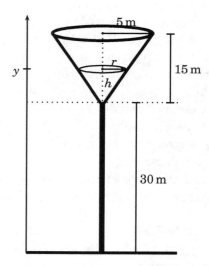

We will use similar triangles to find r in terms of y. In our diagram we can use, for instance, $r/h = (5\,\text{m})/(15\,\text{m})$, or $r/h = 1/3$, and so $r = \frac{1}{3}h$. Also from the diagram, $h = y - 30\,\text{m}$, and so $r = \frac{1}{3}(y - 30\,\text{m})$.

Before setting up the final integral, we should be sure our units are compatible. We will still assume y is measured in meters, in particular with $30\,\text{m} \le y \le 45\,\text{m}$, and then (noting there is the *unit* g (for gram), and the gravitational constant g):

$$dW = \pi\rho g y r^2\,dy = \pi(1\,\text{g/cm}^3)(9.8\,\text{m/sec}^2)y(y - 30\,\text{m})^2\,dy$$

$$= \pi \cdot \underbrace{\frac{1\,\text{g}}{\text{cm}^3} \cdot \frac{\text{kg}}{1000\,\text{g}} \cdot \left(\frac{100\,\text{cm}}{\text{m}}\right)^3}_{\text{units of kg/m}^3} \cdot \underbrace{\frac{9.8\,\text{m}}{\text{sec}^2} \cdot y \cdot (y - 30\,\text{m})^2\,dy}_{\text{units of m}^5/\text{sec}^2} = \underbrace{9800\pi y(y - 30)^2\,dy}_{\text{units of kg-m}^2/\text{sec}^2\text{ suppressed}}.$$

While it is perfectly proper to include the units in our final expression of dW, doing so will be somewhat unwieldy in our final integral computation, but with units of kg-m²/sec² = kg-m/sec² · m = N-m, or joules (J) of work. With this in mind, we can now integrate more compactly:

$$W = \int_{30}^{45} 9800\pi y(y-30)^2\, dy = 9800\pi \int_{30}^{45} y(y^2 - 60y + 900)\, dy = 9800\pi \int_{30}^{45} (y^3 - 60y^2 + 900y)\, dy$$

$$= 9800\pi \left[\frac{1}{4}y^4 - 20y^3 + 450y^2\right]\bigg|_{30}^{45} = 9800\pi\left[\left(\frac{45^4}{4} - 20(45)^3 + 450(45)^2\right) - \left(\frac{45^4}{4} - 20(45)^3 + 450(45)^2\right)\right]$$

$$= 9800\pi[46,406.25] \approx 14.3 \times 10^9 \text{ joules.}$$

For some perspective, recall that one watt is a joule/second, and so one kilowatt-hour is equal to 3,600,000 joules:

$$1\,\text{kW-hr} = 1000\,\frac{\text{J}}{\text{sec}} \times 1\,\text{hr} \times \frac{3600\,\text{sec}}{\text{hr}} = 3,600,000\,\text{J}.$$

And so the energy needed to fill the water tower's tank, as given in kW-hr, is approximately

$$14.3 \times 10^9\,\text{J} \div (3,600,000\,\text{J}/(\text{kW-hr})) = 14.3 \times 10^9\,\text{J} \times \frac{1}{3,600,000} \times \frac{\text{kW-hr}}{\text{J}} \approx 400\,\text{kw-hr}.$$

Therefore, a 10,000-watt (10 kW) water pump would require around 40 hours to fill the tank.[53]

5.6.3 Hydrostatic Pressure and Total Force

Pressure is given in units of force per unit area. A flat area of size A that is subject to a constant pressure of magnitude P, given in units of force/area, will experience a total force of magnitude $F = A \cdot P$, with units of force/area × area = force. *Hydrostatic* refers to water pressure at different depths, though many of the principles here work for other fluids. (Here we only consider liquids, though *fluids* can also refer to gases.)

The simplest and most common model for a contained liquid is that of a fluid but incompressible continuum of matter, in which pressure at a given point within the fluid exerts equal force in each direction, either against objects within the fluid, its container, or other fluid particles (here considered to be points in the continuum). In this way, any force that is applied to one region of the fluid is "spread out," though the magnitude of the force can vary

---

[53]There is a concept in physics called "potential" or "stored" energy, and water towers provide a very useful illustration of the idea. The energy needed to pump the water into the towers can be recovered, because the water can be made to perform work as it flows back downward. Energy (work) is required to move water through pipes to and within homes and other structures, and so rather than providing energy on an as-needed basis with pumps large enough to provide sufficient flow at peak demand times, it is more practical to store that energy by pumping the water to heights at which gravity will force it through the pipes sufficiently even during peak demand times. Smaller pumps can be used continuously, including during times of little or no demand, thus storing the potential energy for times of higher demand. This strategy also usefully allows water to be supplied during electrical power disruptions.

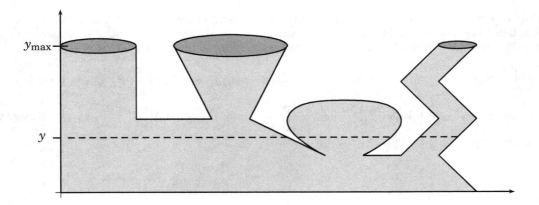

Figure 5.13: For a given liquid, the pressure at a level y in an open tank will only depend upon the depth $y_{max} - y$, regardless of that shape of the tank. While it is true that the pressure is due to the weight of the liquid (and atmospheric pressure, which is usually not relevant for reasons explained in the text), any nonvertical sides of the tank also bear some of the weight of the liquid and so ultimately it is only the depth from the top level of fluid that determines pressure. Because the curved "reservoir" part of the tank is open to the rest of the tank, its pressure will depend on how high the liquid is elsewhere in the tank as well, and will in fact be equal to pressure anywhere else in the tank that is at the same level y.

from point to point, as the "particles" press against *each other* and other objects. In particular, the force a particle pushes with will vary with the height of the "column" of particles above it: A deeper particle has more particles directly above it, adding more weight with which to provide pressure. If there is no net motion of the fluid, each of these forces would be in equilibrium, but again, clearly those particles nearer the top will have less force per unit area than the deeper particles. If a particle is in a "reservoir" below the surface of the fluid, but has a channel leading to the main part of the fluid, the pressure at each height in the reservoir will be the same as at those heights in the main part of the fluid, due to the way pressure is distributed.[54]

A fluid in a container with an open "top," as in Figure 5.13, will have pressure at each depth due to the weight of the fluid and the atmospheric pressure outside of the container, added to find total pressure. However, the atmospheric pressure is rarely an issue because the sides of the container also experience atmospheric pressure, and so there is an equalization on the sides of the container of that part of the force due to atmospheric pressure, and it is the *net force* caused by the *net pressure* that is of interest in most applications. For our discussion, we will therefore neglect atmospheric pressure.

At room temperature, water weighs approximately 62 lb/ft^3. From this we can compute

---

[54] While this model assumes a fluid is a continuum, of course a fluid at a given time is composed mostly of "empty space," with the matter usually being molecules in constant motion, which interact often and rapidly, for instance in exchanging momentum as they collide. Even temperature is really a measure of the *average* kinetic energy of these particles. However, these particles are so small, and the "empty spaces" have particles passing through them so regularly, that for purposes of studying pressures and resultant forces, we can assume a fluid forms a uniform continuum, though pressure can vary within the continuum.

that a 1 ft × 1 ft × 1 ft cubic container will have a total of 62 lb of force on its base. Because this force is exerted by the *fluid*, namely water, we can claim that this resulted from 62 lb/ft^2 of pressure at the base, all due to the weight of the water. A proportionately taller or shorter cube of water would have proportionately more or less weight at the bottom, and therefore proportionately more or less pressure there. It is therefore not difficult to see that the pressure at each depth will be proportional to the depth.

Indeed, at any point at a depth h (for "height" of the water above it), we can construct an infinitesimal column with cross-sectional area dA, and the force from the water will be given by weight of the water in the volume $h\,dA$.

In fact, this can be found from first principles. If the density of the fluid is ρ, in dimensions of mass/volume, then the mass of a volume V of the fluid is ρV. Since the weight (force of gravity on the fluid due to its mass) is given by mg, where g is the gravitational constant (roughly 32 ft/sec^2 at Earth's surface), then the weight of a volume of the fluid is $mg = \rho V g = \rho g V$. Thus a column of fluid of height h directly above an infinitesimal horizontal area dA will have weight $\rho h\, dA \cdot g = \rho g h\, dA$. In the case of water, the density varies slightly with temperature but we will ignore that effect. The weight of the column of water we take to be $(\rho g) h\, dA = (62\,\text{lb/ft}^3) \cdot h\, dA$. Since this represents the *force* due to gravity that the column of water produces on that area dA, we can divide by the area over which it would act as if this column were in a container of the same shape. Dividing by the area dA gives us[55]

$$P = \rho g h \qquad \text{for a general fluid;} \qquad (5.93)$$

$$P = (62\,\text{lb/ft}^3)\, h \qquad \text{for water.} \qquad (5.94)$$

Moreover, at a particular depth the force is the same in all directions, and we use that to determine the total force from this pressure on any surface bounding the fluid, such as a water tank wall or the boundary of a dam.

Example 5.6.6

Consider a water tank of height 5 ft and for which one rectangular wall measures 10 ft along the horizontal base. What is the total force of the water pushing against this wall? (See illustration.)

Solution: It is customary to coordinatize such a problem. Here we will have the y-axis run along one vertical boundary of the wall in question. At each height $y \in [0, 5\,\text{ft}]$, we will consider the force on an infinitesimal rectangular area $dA = (10\,\text{ft})\,dy$, as shown in the illustration. The depth of that area element is $5\,\text{ft} - y$, and the pressure there is $(62\,\text{lb/ft}^3)(\text{depth})$, given in lb/ft^2. Thus the force on that infinitesimal area element dA will be

$$dF = P(y)\,dA = (62\,\text{lb/ft}^3)(5\,\text{ft} - y)(10\,\text{ft})\,dy.$$

These are repeated next, to the right of the illustration, with the units suppressed:

---

[55] A somewhat obscure fact is that mass in the English system of measurement is given in units of *slugs* (where it is in kilograms in the metric system). Thus the density of water is roughly 1.93 slugs/ft^3, and so when we multiply by 32 ft/sec^2, we get for water 62 lb/ft^3. Note that slugs × ft/sec^2 = lb.

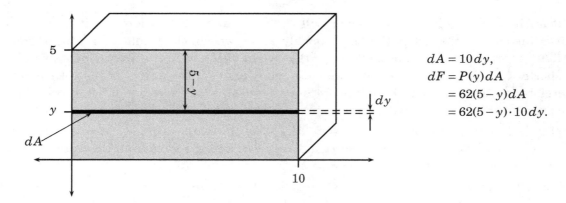

$$dA = 10\,dy,$$
$$dF = P(y)\,dA$$
$$= 62(5-y)\,dA$$
$$= 62(5-y)\cdot 10\,dy.$$

The total force will then be

$$F = \int_{y=0}^{y=5} dF = \int_0^5 62(5-y)\cdot 10\,dy = 620\int_0^5 (5-y)\,dy$$
$$= 620\left(5y - \frac{1}{2}y^2\right)\Big|_0^5 = 620\left[\left(5(5) - \frac{1}{2}(5)^2\right) - \left(5(0) - \frac{0^2}{2}\right)\right]$$
$$= 620\left(25 - \frac{25}{2}\right) = 620\cdot\frac{25}{2} = 7750.$$

Thus the total force from the fluid pressure on the wall is 7,750 lb.

From this example, we see that such a tank would indeed require walls that could withstand fairly large forces. When such a wall fails, the effects can be rather violent for people and objects in the immediate vicinity, and so such tanks require quite sturdy construction.

With this context, it is also interesting to note that if this is a free-standing tank, then that part of the force on the inside of the wall due to atmospheric pressure "pushing" down on the top of the fluid would be offset by the atmospheric pressure on the outside of the wall; the atmospheric forces would be the same and thus cancel each other's effects, since atmospheric pressure would be nearly identical throughout the height of the tank, including on the water's surface at the top. For a buried tank, the atmospheric pressure is an issue, though the ground fill would likely negate it. However, in either scenario, the wall would indeed need to be strong enough to withstand atmospheric pressure, a feat also accomplished by ubiquitous living things, and so it is not usually an issue, except when that pressure suddenly changes. These are issues for other studies, so we will not mention them again here.

Example 5.6.7 _____

Consider a trough with parabolic end plates, 4 ft wide and 2 ft high. If the trough is full, what is the total hydrostatic force on one of these end plates?

Solution: Here we again "coordinatize" the relevant region over which we wish to find the total force. As before, for each height y we look at an infinitesimal area dA constructed as constructed next, though for this example we have the width of the area dA varying with y.

For convenience, we will place the vertex at the origin $(0,0)$, and then we will have $y = ax^2$. We need the total width at the top to be 4 ft, and need this parabola to pass through the points $(\pm 2, 2)$, so $y = ax^2$ becomes $2 = a(2)^2$, giving $a = \frac{1}{2}$. Thus our parabola's equation is $y = \frac{1}{2}x^2$.

5.6. PHYSICAL APPLICATIONS OF INFINITESIMALS, DEFINITE INTEGRALS

From this we also get $x^2 = 2y$, or $x = \pm\sqrt{2y}$. This is important because the length of the area represented by dA is $x_R - x_L$ here; we need these in terms of y, because we need dA in terms of y and dy. With a little cleverness, this is not difficult because $x_R - x_L = \sqrt{2y} - (-\sqrt{2y}) = 2\sqrt{2y}$. For our infinitesimal piece dF of the total force, we also need to have the pressure at each height, multiplied by the infinitesimal area element at that height. We get the following from the illustration, where the thick line represents the area dA, at vertical position y, that area having width dy:

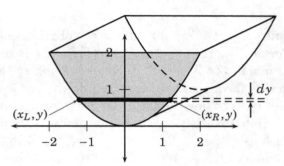

$$d(\text{area}) = (\text{length})(\text{width}),$$
$$\text{width} = dy,$$
$$\text{length} = x_R - x_L = 2\sqrt{2y},$$
$$\therefore dA = 2\sqrt{2y}\,dy.$$
$$\text{depth} = 2 - y,$$
$$\text{pressure} = 62(\text{depth}) = 62(2 - y).$$
$$\therefore dF = 62(2-y) \cdot 2\sqrt{2y}\,dy.$$

Since we have a piece dF of the total force for each $y \in [0,2]$, we then integrate these to get

$$F = \int_{y=0}^{y=2} dF = \int_0^2 62(2-y) \cdot 2\sqrt{2y}\,dy = 62 \int_0^2 (2-y)\left(2\sqrt{2}\right) y^{1/2}\,dy$$
$$= 124\sqrt{2} \int_0^2 \left(2y^{1/2} - y^{3/2}\right) dy = 124\sqrt{2} \left[2 \cdot \frac{2}{3} y^{3/2} - \frac{2}{5} y^{5/2}\right]\Big|_0^2$$
$$= \frac{496\sqrt{2}}{3} \cdot 2^{3/2} - \frac{248\sqrt{2}}{5} \cdot 2^{5/2} - 0 = \frac{496 \cdot 2^2}{3} - \frac{248 \cdot 2^3}{5}$$
$$= \frac{1984}{3} - \frac{1984}{5} = \frac{2}{15}(1984) = \frac{3968}{15} \approx 265.$$

The final answer will have units of pounds, and so the force on the trough will be approximately 265 lb.

Interestingly enough, if we were to seal the trough and turn it "upside-down," the force on the wall would be greater, because the longer area elements dA would be found at greater depths. We will see in Example 5.8.18, page 641, that if we could seal and turn this tank upside-down when it was full of water, we would get 434 lb of force on each parabolic side. However, the antiderivative we would be forced to compute in that scenario will require a slightly more advanced technique for computation.

5.6.4 Torque

Torque is a measure of the tendency of a force to rotate an object around a pivot. Consider a force of magnitude F pressing on an object in the direction the force could rotate the object, at a distance r (think "radius") from its pivot, as in the picture:

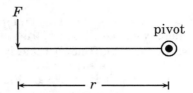

Then the *torque* produced by that force F on the pivot is represented by the lower-case Greek letter τ (tau), given by the equation

$$\tau = Fr. \qquad (5.95)$$

Units of torque resemble those of work (i.e., energy), but we will discuss later how they differ. For now we note that if F is in lbs and r is in feet, then it is often conventional to give the torque in terms of lb-ft (instead of ft-lb), so the casual but informed observer knows it is a unit of torque and not work or energy that is discussed. Metric units include N-m, but any unit of force times a unit of length is acceptable.

Note that there are infinitely many combinations of F and r that would yield, say, 100 ft-lb, such as $F = 50$ lb and $r = 2$ ft, or $F = 100$ lb and $r = 1$ ft. So a 50-lb force can produce the same torque as a 100-lb force if the pivot arm distance is twice as long for the 50-lb force as for the 100-lb force. (This is the essence of mechanical "leverage.")

Example 5.6.8

Suppose a rod of length 10 feet has a linear density of 1/32 slugs/ft (that is, 1.0 lb/ft). How much torque would the rod produce on a pivot at one end of the bar when it is suspended from that end horizontally? (This torque is due to gravity.)

Solution: We will assume that the bar is suspended from the origin of the x-axis and lies on the positive x-axis.

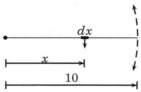

We can write the contribution to the torque of the infinitesimal length at the position x in either of the following ways:

$$d\tau = x(\lambda\,dx)g = x(1/32 \cdot dx)(32) = 1x\,dx,$$
$$d\tau = 1\text{ lb/ft} \cdot dx \cdot x = 1\text{ lb/ft} \cdot x\,dx.$$

We see that the units of $d\tau$ will be lb-ft. The total torque experienced by the pivot at the origin will then be (suppressing units)

$$\int_0^{10} x\,dx = \left.\frac{x^2}{2}\right|_0^{10} = \frac{10^2}{2} - 0 = 50$$

with units of lb-ft.

Notice how if all of the weight of the bar, namely $\int_0^{10} \lambda g\,dx = \int_0^{10} 1\,dx = 10$ (lb), were located at $x = 10$ (ft), then the torque would have been 10 lb × 10 ft = 100 lb-ft, but in the example most of the weight is by varying amounts closer to the pivot, eventually yielding exactly half as much torque as if it were all on the far end of the rod. A different, nonconstant density function $\lambda(x)$ could also change the total torque.

5.6.5 Torque and Work

When we consider the equation for work done by a force in the direction of its application, namely $W = Fs$, we can see that regardless of how large F is, if $s = 0$ then no work is done by the force. (For example, gravity does no work to a person standing on the ground if the ground is stationary.) Similarly, a force or torque will do no rotational work on an object if it is restrained from pivoting that object. However, if the torque ("turning force") acts through a circular path of angle θ, *and always points in the direction of that path* as the pivoting occurs, then it can be shown that the work done by that force is given by

$$W = \tau\theta. \tag{5.96}$$

We derive this on the right, noting that the most appropriate unit for θ is the radian because it is, in a sense, dimensionless, being defined as a directed arc length divided by its radius, both with length dimensions. Thus the units of an angle in radian measure will be of the form length/length, and so lengths "cancel." Thus W and τ in (5.96) appear to contain the same units (force times length), but τ includes radians (in its denominator, usually hidden) while the other does not.[56]

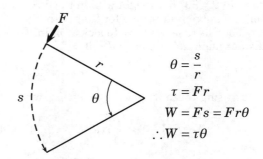

Note that in the diagram we assume that the force F will change direction throughout the path of rotation so that it is always perpendicular to the radial ray, and thus parallel to the path of rotation.

Example 5.6.9

Suppose the rotation is through an angle of $60° = \pi/3$ (rad). If $F = 100$ N and $r = 250$ cm, then we would have (including a conversion factor)

$$W = \tau\theta = Fr\theta = 100\,\text{N} \times 250\,\text{cm} \times \frac{\pi}{3} \times \frac{1\,\text{m}}{100\,\text{cm}} \approx 262\,\text{N-m}.^{57}$$

---

[56] Torque is, therefore, in fact a measure of energy (or work) per radian of rotation, and since radians are dimensionless, it appears that torque and energy (or work) have the same units. They do not if one realizes that torque must be *multiplied* by an angle to represent energy or work.

It is a very common misconception for novice physics or technical students to believe initially that torque and energy are ultimately the same, because their written units seem to carry the same dimensions. They do not, if we accept that torque has a hidden angular measure in its denominator.

This confusion helps to explain why torque is often given in units of force × length, while work or energy is often given in units of length × force, i.e., the distinction is made by reordering the units.

[57] It would arguably be more clear to write $\tau = 250$ N-m/rad, but it is customary *not* to write radians into the torque, even though this means that they appear in the work equation in θ and then disappear in the final answer. The disappearance in the final answer is then justified because a radian is, after all, dimensionless when considered as a length divided by a length.

In the next example, torque is not constant, but changes with the angle θ, i.e. $\tau = \tau(\theta)$.

Example 5.6.10

If a 200-lb man sits in a Ferris Wheel car at the highest position 75 feet from the center of the wheel, as his car lowers, the torque with which his weight acts on the wheel is given by

$$\tau = (200 \text{ lb})(75 \text{ ft})\sin\theta,$$

where θ is the angular displacement his car makes with the upward vertical direction, and $\theta \in [0, 180°]$. Find the total work his weight performs on the Ferris Wheel as he lowers from the top to the lowest position.

Solution: Note that at the highest ($\theta = 0$) or lowest ($\theta = 180°$), the force of his weight acts in a direction parallel to the "rod" on which his car is positioned, and therefore contributes zero torque. Also, at $\theta = 90°$ his weight acts in the direction perpendicular to that "rod" and therefore is giving the rod the maximum torque for that path.

For this case, we have to look at infinitesimal work preformed by the man's weight over an angular displacement $d\theta$, for each $\theta \in [0, \pi]$:

$$dW = \tau(\theta)d\theta = (200 \text{ lb})(75 \text{ ft})\sin\theta\, d\theta.$$

Integrating these we get the following, where $W(\theta)$ is the work done up to the moment the angle is θ:

$$W(\pi) - W(0) = \int_0^\pi \tau(\theta)d\theta = \int_0^\pi 15{,}000\text{(lb-ft)}\sin\theta\, d\theta$$

$$= -15{,}0000\text{(lb-ft)}\cos\theta\Big|_0^\pi = [15{,}000 - (-15{,}000)]\text{(ft-lb)} = 30{,}000 \text{ ft-lb}.$$

This result is less than if the maximum torque of 15,000 lb-ft could be maintained through the entire half-revolution, which would have yielded $15{,}000 \text{ lb-ft} \times \pi$ (rad) $\approx 47{,}124$ ft-lb of work. Since the torque was not constant, we had to look at the infinitesimal cases $dW = \tau(\theta)d\theta$ and integrate these.[58]

5.6.6 Power and Work

Recall that $P = \frac{dW}{dt}$, where power is the rate of work production and W is the work produced up to the time t. It follows that for an infinitesimal time interval of length dt at time t, we have

$$dW = P(t)dt \implies W(t_f) - W(t_0) = \int_{t_0}^{t_f} P(t)dt. \tag{5.97}$$

As usual we should note that $P(t)dt$ is in units such as watts × seconds, or more generally energy/time × time, and thus energy, as will be the integral (which is a limit of Riemann Sums of terms with these units). The change in work is thus in units of energy, as expected.

---

[58] A student of physics who is familiar with physical *conservation laws* could note that the 200-lb man descended 150 feet from his maximum height to the minimum, and in doing so lost 150 ft × 200 lb = 30,000 ft-lb of potential energy (or performed that much work) in so lowering, which again helps us to arrive at the figure of 30,000 ft-lb of work. In making such an argument, one has to be careful that the scenario is indeed one in which energy is conserved, and no part of the energy is transformed into some other form not mentioned, which is a discussion perhaps beyond the scope of this text.

5.6. PHYSICAL APPLICATIONS OF INFINITESIMALS, DEFINITE INTEGRALS

Example 5.6.11

Suppose the power of a particular nuclear device is given by $P(t) = 1.914 \times 10^{13}(e^t - 1)$ in watts, with t in seconds from the initial activation. How much energy is produced in the first two seconds of this reaction?

Solution: Here we simply integrate

$$\int_0^2 1.914 \times 10^{13}(e^t - 1)\,dt = 1.914 \times 10^{13} \left[(e^t - t) \Big|_0^2 \right]$$
$$= 1.914 \times 10^{13} \left[(e^2 - 2) - (1) \right] \approx 1.914 \times 10^{13}(4.389) \approx 8.40 \times 10^{13},$$

with units of joules. (Recall watt=joule/sec.)

5.6.7 Current and Charge

Suppose that the current flowing *into* one side (e.g., the cathode) of a battery or capacitor is I, given in amps, which are the same as coulombs/sec. If the charge on that side of the battery or capacitor is Q, then the change in Q over an infinitesimal length of time dt will be

$$dQ = I\,dt,$$

where $I = I(t)$ is a function of t.[59] This should make sense since I is in units of charge per unit time, and dt is in units of charge.

Example 5.6.12

Suppose a charger is connected to a large, discharged capacitor, and the current from the charger to the capacitor in amps (i.e., coulombs/second) is given by $I = 2|\sin t|$, where t is in seconds. Approximately how much charge will the capacitor absorb after one hour?

Solution: Here $dQ/dt = 2|\sin t|$, and so $dQ = 2|\sin t|\,dt$, and t ranges from 0 to 3600 sec. Thus we have the following wave form and integral:

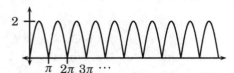

$$Q(1 \text{ hour}) - Q(0) = \int_0^{3600} 2|\sin t|\,dt.$$

Unfortunately, there is no "nice" antiderivative of $|\sin t|$. However, the period of the function within the integral is π, and so we can note that $3600/\pi \approx 1146$, so there are approximately 1,146 periods of the function for $[0, 3600]$. Furthermore, for the first period $t \in \sin[0, \pi]$ we have $\sin t \geq 0$, and therefore $|\sin t| = \sin t$ (which does have a nice antiderivative) on that interval. Thus

$$Q(1 \text{ hour}) - Q(0) \approx 1146 \int_0^\pi 2|\sin t|\,dt = 1146 \int_0^\pi 2\sin t\,dt = 2292\left[-\cos t \Big|_0^\pi \right] = 2292[1+1] = 4584 \approx 4600,$$

in units of coulombs.

---

[59]This differs slightly—by a sign—from (3.32), page 266, because here I is the current flowing into the battery or capacitor, while in that equation I was the current flowing out of it.

In the previous example it would be difficult to find the exact value of the integral (needing for instance to find the boundaries of the last partial period of the integrand in $[0, 3600]$), but since there is some vagueness regarding the number of significant digits available in the actual physical setting, we can content ourselves with the final approximation without embarrassment.

Exercises

1. The velocity (in m/s) of a body as a function of time is given by $v(t) = 0.6t^2 - 0.8t^3$.

   (a) To the extent possible, find the form of its position $s(t)$.

   (b) Calculate the displacement as the body moves from $t = 1$ sec to $t = 4$ sec.

2. A variable force acts on a particle in the horizontal direction and is given by $F(s) = 3s^2 + 2s + 5$, where s is the horizontal position measured in meters and F is measured in newtons. Calculate the work done on the particle as it moves from $s = 2$ m to $s = 10$ m.

3. Recall that Hooke's law for a spring is $F = -ks$, meaning that the force F by which the spring resists having its natural length changed is proportional to, and in the opposite direction from, that change s. Consequently, the force needed to *cause* that change is given by $F = ks$.

   (a) Suppose for a given spring we have $k = 600$ N/m. How much work is required to stretch this spring from $s = 0$ to $s = 20$ cm?

   (b) Derive the spring energy equation $W = \frac{1}{2}ks^2$, given that $W = \int_0^s F(\sigma)d\sigma$, $F(\sigma) = k\sigma$, and s is the maximum spring displacement from the natural length. Here $W(s)$ is then the work needed to achieve that displacement, and is also then considered the stored (potential) energy of the spring in that state.

   (c) Plot $F = ks$, with s on the horizontal axis and F on the vertical.

   (d) The area under the curve $F(s)$ versus s represents the work $W(s)$ described above. Show geometrically that $W = \frac{1}{2}ks^2$.

4. A horizontally acting force acts on a particle to produce a potential energy equal to the work done by the force, that potential energy in this case given by $U(x) = 3x^4 + 2x$. Since $U(x) = \int_0^x F(\xi)d\xi$, find the force function. (Note: The First Fundamental Theorem of Calculus can make short work of this.)

5. A force $F(x) = 3x^2 + 2x + k$ (given in newtons) acts on a body. Here k is a constant. This force produces a potential energy $U(x)$ (see the previous exercise).

   (a) To the extent possible, find the form of the potential function $U(x)$.

   (b) Find k if $U(5) = 220$ J.

6. The velocity function of an object moving in the horizontal direction is given by $v(t) = 20 - t^2 + t$, where v is measured in m/sec.

(a) Find the displacement of the body as it moves from $t = 3$ to $t = 6$ seconds.

(b) Find the total distance D traveled from $t = 3$ to $t = 6$ seconds. (Recall (5.61), page 549.)

7. Suppose a straight flagpole 25 feet long is lying on the ground and is to be lifted into a vertical position. The flagpole is of uniform weight density, and weighs a total of 100 lb.

    (a) What is the weight density, i.e., weight per unit length? (Give the answer in units of lb/ft.)

    (b) How much work will be needed to lift the pole into a vertical position (with one end on the ground)? Use an integral to compute this.

8. A cylindrical water tank with an internal diameter of 10 feet and an internal height of 20 feet is to be filled from water piped in at ground level. How much energy is required to fill this tank? (Assume the water weighs approximately 62 lb/ft^3.)

9. A cable weighing 10 lb/ft is to wrap around a semicircular drum of radius 5 ft, as in the illustration below.

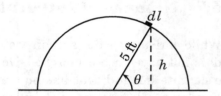

If the cable is originally delivered to the base of the drum at ground level, we wish to compute how much work is required to lift the cable to this new position along the curve of the drum. We will ignore the thickness of the cable and any extra work lost from friction in the process of lifting or sliding the cable into place. (In other words, we will compute the potential energy of the cable in its new position.)

(a) Recall $dl = r\,d\theta$. Show that the weight of the cable for an infinitesimal length dl along the cable in its final position is $50\,d\theta$ in pounds.

(b) Explain (by derivation) why the work required to lift that infinitesimal length dl of cable at angle θ is $250\sin\theta\,d\theta$.

(c) Compute the total work needed to lift the whole cable into this new position.

5.7 Numerical Integration: Approximating Definite Integrals

While derivative rules are fairly straightforward, by now it should be clear that integration does not have such comprehensive rules. In practice, it is not at all unusual to encounter a definite integral that has no simple formula for the antiderivative needed to apply the Fundamental Theorem of Calculus. Indeed, even a function as uncomplicated as $f(x) = e^{x^2}$ does not have an antiderivative we can express using a finite number of functions of the type we have already encountered.[60]

However, such functions do arise as integrands in both theory and application, and so it is necessary to find methods of approximating their definite integrals numerically (hence the title of this section). The methods we develop in this section are inspired by the geometric interpretation of the definite integral.

5.7.1 Previous Uses of Geometry for Integrals

While we often use integration to analyze geometry, at times it is useful to use the geometry to analyze an integral. Following are three instances where the geometric representation of the integral—as an expression of a net signed area—helps us to compute the integral.

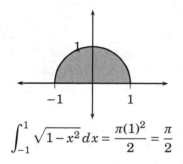

$$\int_{-1}^{1} \sqrt{1-x^2}\,dx = \frac{\pi(1)^2}{2} = \frac{\pi}{2}$$

The integral represents the area of a semi-circle of radius 1.

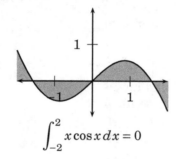

$$\int_{-2}^{2} x \cos x\,dx = 0$$

The function is continuous on $[-2, 2]$ and so the integral exists, but the integrand is an odd function and limits are symmetric. Positive and negative areas cancel.

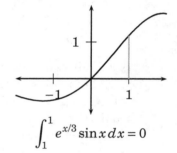

$$\int_{1}^{1} e^{x/3} \sin x\,dx = 0$$

The endpoints of the integral are the same, and so width of the given area is zero. Therefore the total area is zero.

In the sequel we explore how to find antiderivatives of these three integrals, though each antiderivative computation has several steps. An informed look at the geometry in each case makes the integral's value reasonably clear.

---

[60] It should be pointed out that it is usually desirable to have a formula for the antiderivative of the integrand, allowing for a theoretically "exact" value of the definite integral when the geometry is less helpful. In this section and in the sequel we find ways of approximating many cases where, even with the advanced integration techniques explored there, we cannot write the relevant antiderivative with an explicit formula. Here we approximate integrals geometrically, while in the sequel, with Taylor Series, we find other ways of expressing many functions, giving rise to further ways of approximating those functions, and the approximations are then easily integrated.

5.7. NUMERICAL INTEGRATION: APPROXIMATING DEFINITE INTEGRALS

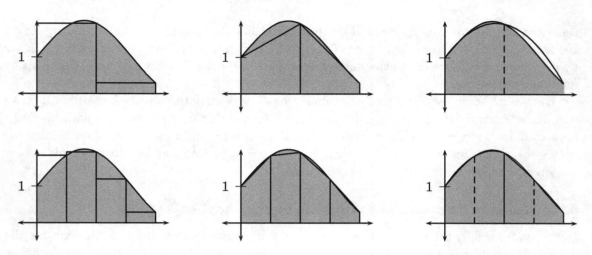

Figure 5.14a: Riemann Sums using 2 rectangles (top) and 4 rectangles (bottom) to approximate the shaded area which represents the integral.

Figure 5.14b: Approximations of the definite integral using areas of 2 trapezoids (top) and 4 trapezoids (bottom), connecting points on the curve.

Figure 5.14c: Approximations using the area under 1 parabola (top) connecting three points on the curve, and 2 parabolas (bottom), each connecting three points on the curve.

Figure 5.14: Three schemes for approximating $\int_0^\pi (1 + \sin 1.25x)\,dx$. In Figure 5.14a we approximate the area with an increasing number of rectangles, as in the definition of Riemann Sums; in Figure 5.14b we approximate the area with an increasing number of trapezoids, which better follow the flow of the graph; and finally in Figure 5.14c we use an increasing number of parabolas. Note how, for this particular graph, we get a very accurate approximation of the area representing the integral by using just two parabolas (or *parabolæ* in some older literature). In fact, it is difficult to see the gap between the parabolas and the graph. (The dashed lines show where the middle point of each parabola is located.)

In this section we will look at some geometric techniques for approximating definite integrals. Each technique can be motivated by a geometric approximation of the desired area using a simpler area. Each technique also shares the use of sampling points over regular intervals. In fact, the approximations are easily performed side by side using a tool such as a spreadsheet, because (as we will see) in practice the only differences are in how the sample heights are weighed.

We will not spend too much effort on approximating a definite integral using rectangles, though that method is so closely related to the definition of a definite integral as a limit of Riemann Sums (as the number of rectangles approaches ∞ and their widths approach zero). From Figure 5.14 it is clear that we can expect a trapezoidal approximation to more quickly give more accurate approximations in many cases, and expect an approximation by parabolas to be even more accurate, assuming the curve is approximately "piecewise quadratic," particularly on the "pieces" we choose for our partition of the interval of integration.

But we should also not be surprised if the formulas for the approximations become more complicated as we consider in turn the rectangular approximations, trapezoidal approximations, and parabolic approximations. Indeed this is the case, but if the data from the partitions is properly organized, the final computations do not increase very much in complexity. If one has access to a spreadsheet, it is fairly easy to compute all three approximations, as we will see. Historically, other mathematical packages computed approximations of integrals by using parabolic areas.

However, the approximation by parabolas has a famously tedious derivation, and so we will desist from delving into it as it would be a formidable distraction. We will, however, note that it has a name that is very well known among those who have had a rigorous first-semester calculus course. That third method of approximating definite integrals is best known as *Simpson's Rule*.[61]

5.7.2 Riemann Sums; Computations by Spreadsheets

Here we just briefly mention that one can always use any Riemann Sum to approximate the definite integral because, after all, the definite integral is just a limit of Riemann Sums, as their maximum interval width approaches zero. Eventually, one expects the trend of the Riemann Sum values, as the interval widths shrink (and their numbers grow), to approximate the definite integral with more and more accuracy.

In Figure 5.14a, page 617, we have two Riemann Sums approximating $\int_0^\pi (1+\sin 1.25x)\,dx$. Both use $x_i^* = x_i$ and are therefore called *right endpoint approximations*. Here the interval in question is $[0, \pi]$, and so we have two such approximations:

- $n = 2$, and so $\Delta x = \frac{\pi - 0}{2} = \pi/2$, and the Riemann Sum is then

$$\sum_{i=1}^{2} f\left(x_i^*\right) \Delta x = f(\pi/2)(\pi/2) + f(\pi)(\pi/2)$$

$$\approx (1.923879533)(1.570796327) + (0.292893219)(1.570796327)$$

$$= 3.482098495$$

$$\approx 3.48.$$

As we will see, there is not much use in giving too many significant digits in our final answer.

- $n = 4$, and so $\Delta x = \frac{\pi}{4}$. For this computation, we will make it more compact by factoring the constant $\Delta x = \pi/4$ from our sum in order to multiply by it once:

$$\sum_{i=1}^{4} f\left(x_i^*\right) \Delta x = f(\pi/4)(\pi/4) + f(\pi/2)(\pi/4) + f(3\pi/4)(\pi/4) + f(\pi)(\pi/4)$$

$$= (\pi/4)[f(\pi/4) + f(\pi/2) + f(3\pi/4) + f(\pi)]$$

$$\approx (0.785398163)[1.831469612 + 1.983879533 + 1.195090322 + 0.292893219]$$

$$= 4.118103861$$

$$\approx 4.12.$$

---

[61]Named for British mathematician and inventor Thomas Simpson (1710–1761), though used much earlier by many famous mathematicians and physicists. Perhaps Simpson's name was attached to it because of the popularity of his textbooks, which included the rule. One famous earlier user of the rule was the very important German mathematician and astronomer Johannes Kepler, and so sometimes the rule carries his name instead.

5.7. NUMERICAL INTEGRATION: APPROXIMATING DEFINITE INTEGRALS

In the next few sections, we will learn how to compute the antiderivative $\int(1+\sin 1.25x)\,dx = x - \frac{1}{1.25}\cos 1.25x + C$, which we can very easily verify by differentiation. For now, it is useful to know that the definite integral we are approximating can be computed exactly, if we assume 1.25 is exact, and so $1.25 = 5/4$ and $1.25\pi = 5\pi/4$:

$$\int_0^\pi (1+\sin 1.25x)\,dx = \left[x - \frac{4}{5}\cos\frac{5x}{4}\right]_0^\pi = \left[\pi - \frac{4}{5}\cos\frac{5\pi}{4}\right] - \left[0 - \frac{4}{5}\cos 0\right]$$
$$= \pi - \frac{4}{5}\cdot\left(-\frac{1}{\sqrt{2}}\right) + \frac{4}{5} \quad (5.98)$$
$$\approx 4.507278079,$$

accurate to 10 significant digits. It is not surprising that a Riemann Sum (≈ 4.12) with $n = 4$ will better approximate the true value of the definite integral, compared to a Riemann Sum (≈ 3.48) with $n = 2$.

It quickly gets obvious that such Riemann Sum computations can be very tedious, even if we record only the input values of the functions and the final result and allow an inexpensive scientific calculator to perform all of the computations at once. However, this still becomes less practical when we take Riemann Sums with larger values of n. A competent computer programming student could easily write code to compute such things, but perhaps the simplest tool for moderately large (or even smaller) computations is the ubiquitous spreadsheet software. Following is an illustration of how one could program a spreadsheet to compute the previous Riemann Sum with $n = 4$. First we have the spreadsheet showing each cell as entered, many with formulas:

	A	B	C	D	E	F
1	a,b,Dx	i	xi	f(xi)	wi	f(xi)wi
2						
3	0	0	=a3+b3*a5	=1+sin(1.25*c3)	0	=d3*e3
4	=pi()	=b3+1	=a3+b4*a5	=1+sin(1.25*c4)	1	=d4*e4
5	=pi()/4	=b4+1	=a3+b5*a5	=1+sin(1.25*c5)	1	=d5*e5
6		=b5+1	=a3+b6*a5	=1+sin(1.25*c6)	1	=d6*e6
7		=b6+1	=a3+b7*a5	=1+sin(1.25*c7)	1	=d7*e7
8						
9				Integral	Approx=	=sum(f3:f7)*a5

The spreadsheet with its actual values is shown next, after which we offer a short explanation of the entries as displayed here. Note that all cell entries (in white cells) were inserted (but not necessarily typed) by the spreadsheet's author, though the column and row labels (A–F and 1–9, respectively) are defaults designed into the spreadsheet program.[62]

---

[62]There are many stylistic variations of spreadsheet design and authorship. Within the spreadsheet cells displayed here, we used lower-case letters for reasons of space and ease of reading. The program will often switch these to upper-case letters. Within the cells, these are not case-sensitive.

	A	B	C	D	E	F
1	a,b,Dx	i	xi	f(xi)	wi	f(xi)wi
2						
3	0	0	0	1	0	0
4	3.1415926536	1	0.7853981634	1.8314696123	1	1.8314696123
5	0.7853981634	2	1.5707963268	1.9238795325	1	1.9238795325
6		3	2.3561944902	1.1950903220	1	1.1950903220
7		4	3.1415926536	0.2928932188	1	0.2928932188
8						
9				Integral	Approx=	4.1181038614

On this particular spreadsheet, horizontal rows are numbered 1–9, and vertical columns are labeled A–F. A cell's address is given by its column and row, such as B4, the cell in Column B, Row 4 (containing the formula b3+1, whose calculated value is 1). On this sheet, Row 1 contains labels describing the natures of the entries within their respective columns. For instance, A3 contains the left endpoint of the interval $[0, \pi]$ of integration, A4 shows the right endpoint, and A5 shows Δx for this particular Riemann Sum. When the inserted entry starts with "=," it is followed by a formula to compute the value in the cell. The number π is a built-in, constant function for this particular spreadsheet program, requiring no input and hence the empty parentheses in the first, "typed" form of the spreadsheet. We could have simply entered a numerical value, but most spreadsheets will have this constant built in.

A formula for the value of a cell can refer to the values in other cells as well, and so in Column B we have one method for increasing the value of the subscript i (for x_i) each time we move down that column. Once the entries in B3 and B4 are set, the contents of B4 (as a formula in this case) can be "copied" using "copy" from the edit menu, and then "pasted" into B5–B7 very quickly, and the spreadsheet editor will assume that the cell address is "relative," and so when B4 is "copied and pasted" into B5–B7, the formula entered into these cells will be adjusted (see original spreadsheet form). In this way, if we wished to have a larger regular partition of $[a, b] = [0, \pi]$, we could simply adjust A5 (i.e., Δx) and paste B4 into all entries beneath it until the endpoints x_i of the partition are exhausted. (We would then do the same for Columns C–F, with some adjustments for the final row.)

Note also that in Column C, where we compute the values of $x_i^* = x_i$, we use the formula $x_i = a + i\Delta x$. Since we wish to use A3 for (the endpoint) a and A5 for Δx each time, but may wish to write the formula for x_i for one occurrence and "cut and paste" it into the other cells, we need the addresses for a and Δx to be "static," i.e., kept the same for each cell we are "pasting" into. To achieve this, it is normal for a spreadsheet to allow the author to signify that a row or column is "static" for these purposes by using the symbol $ in front of the coordinate (row or column designation) that needs to remain fixed.[63]

---

[63] In fact the $ symbol in front of the column coordinate (in this case "a" or "A") does not require fixing, because we are pasting "down a column," which will not adjust the column label in formulas for the "pasted" cells (so we could have used a$3 and a$5 in this case.). The row numbers 3 and 5 do need to be fixed for such pasting.

5.7. NUMERICAL INTEGRATION: APPROXIMATING DEFINITE INTEGRALS

Column D should be self-explanatory, upon modest reflection. Column E is useful to decide which values $f(x_i)$ to include in the Riemann Sum. For instance, if we prefer a left-endpoint approximation we could let E3–E6 be 1 and E7 be zero. (Later we will change these weights for the Trapezoidal Rule and Simpson's Rule approximations.) Our Riemann Sum can then be written

$$\int_a^b f(x)\,dx \approx \sum_{i=1}^n f(x_i^*)\Delta x = \sum_{i=0}^n f(x_i)w_i \Delta x = (\Delta x)\sum_{i=0}^n f(x_i)w_i. \tag{5.99}$$

If we wish to change our definition of x_i^* from the right endpoint of the interval $[x_{i-1}, x_i]$ to the left, we can simply change the weights so that $w_0 = 1$ and $w_4 = 0$:[64]

	A	B	C	D	E	F
1	a,b,Dx	i	xi	f(xi)	wi	f(xi)wi
2						
3	0	0	0	1	1	1
4	3.1415926536	1	0.7853981634	1.8314696123	1	1.8314696123
5	0.7853981634	2	1.5707963268	1.9238795325	1	1.9238795325
6		3	2.3561944902	1.1950903220	1	1.1950903220
7		4	3.1415926536	0.2928932188	0	0
8						
9				Integral	Approx=	4.6734642287

If we wished to increase the number of intervals, we would adjust Δx (cell A5) and change the number of values of x_i, and therefore the number of rows. The edits required to do so would be fairly simple, considering the majority of the changes could be made through "cut and paste" operations.

5.7.3 Trapezoidal Rule

Geometrically, the Trapezoidal Rule approximates a definite integral with trapezoidal areas, as in Figure 5.14b, page 617. Beginning with the point $(x_0, f(x_0))$, if we draw the line segment to $(x_1, f(x_1))$, and from there to $(x_2, f(x_2))$, and so on, until we finally connect $(x_{n-1}, f(x_{n-1}))$ to $(x_n, f(x_n))$, we see how these can be "tops" of trapezoids with vertical sides that are perpendicular to a horizontal "foot" lying on the x-axis.

The approximating trapezoids encountered in this rule, as seen in Figure 5.14b, are a special case where two adjacent angles share a side that is contained in the x-axis, and both of these sides are perpendicular to the horizontal. In such a case, we can compute the area by multiplying the length of this horizontal "foot" contained in the x-axis by the average length of the two vertical sides, as illustrated next.

---

[64] Note that we do not list x_i^* in our spreadsheet, but only x_i, for $i = 0, 1, 2, \cdots, n$. When we change the weights w_i, we change the extent to which the values $f(x_i)$ are "counted" in the final sum (before multiplication by Δx). Those with weights $w_i = 0$ are, effectively, removed from the sum.

As we can see from the illustration on the right, the trapezoid has the same area as a rectangle with its "base" equal to the "foot" of the trapezoid, and its "height" equal to the average of the two heights y_1 and y_2. (The average of the heights y_1 and y_2 will be half-way between y_1 and y_2.) Indeed, if we superimpose the rectangular area of dimensions foot $\times \frac{1}{2}(y_1+y_2)$ on the trapezoidal area, that part of the trapezoidal area not "covered" by the rectangular area is the same as that part of the rectangular area that is not covered by the trapezoidal area.

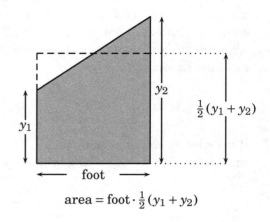

area = foot $\cdot \frac{1}{2}(y_1+y_2)$

We will see that a fairly simple formula emerges when we use such trapezoids to approximate an integral, as in Figure 5.14b. The pattern is a bit more obvious if we consider the case $n=4$ in that figure. In such a case, we use the partition $0 = x_0 < x_1 < x_2 < x_3 < x_4 = \pi$ (the actual values of the endpoints not crucial here), and when we add the areas of the trapezoids, due to some repetition we get

$$\int_a^b f(x)\,dx \approx \frac{1}{2}(f(x_0)+f(x_1))\Delta x + \frac{1}{2}(f(x_1)+f(x_2))\Delta x$$
$$+ \frac{1}{2}(f(x_2)+f(x_3))\Delta x + \frac{1}{2}(f(x_3)+f(x_4))\Delta x$$
$$\implies \int_a^b f(x)\,dx \approx \left[\frac{1}{2}f(x_0)+f(x_1)+f(x_2)+f(x_3)+\frac{1}{2}f(x_4)\right](\Delta x).$$

It is because $f(x_1)$, $f(x_2)$, and $f(x_3)$ each appear as vertical side lengths in adjacent trapezoids that they each appear twice in the first computation, and thus lack the multiplier 1/2 in the final line. One can interpret this final result in other ways besides that it follows directly from the trapezoidal approximations of the areas contributed by the individual subintervals (i.e., Figure 5.14b). For instance, we sample and fully weight the middle heights $f(x_1)$ through $f(x_3)$ but only give the leftmost and rightmost heights $f(x_0)$, $f(x_4)$ a half-weight each. Alternatively, we average the heights at the left and right endpoints of the interval of integration $[a,b] = [0,\pi]$ when choosing heights to sample. However one interprets this, it easily generalizes: Assuming $f(x)$ is continuous on $[a,b]$, $n \geq 2$ is an integer, $\Delta x = \frac{b-a}{n}$, $a = x_0 < x_1 < x_2 < \cdots < x_n = b$, and $x_i = a + i\Delta x$, then when we approximate the integral using trapezoidal areas as described, we have the approximation

$$\boxed{\int_a^b f(x)\,dx \approx (\Delta x)\left[\frac{1}{2}f(x_0)+f(x_1)+f(x_2)+\cdots+f(x_{n-1})+\frac{1}{2}f(x_n)\right].} \qquad (5.100)$$

While performing the computation in (5.100) would be tedious, it is doable. Moreover, it is easily performed with a computer program or spreadsheet. For instance, to approximate $\int_0^\pi (1+\sin 1.25x)\,dx$ as before but with a Trapezoidal Rule scheme, we can use our previous spreadsheet and simply change the weights w_i in Column E:

5.7. NUMERICAL INTEGRATION: APPROXIMATING DEFINITE INTEGRALS

	A	B	C	D	E	F
1	a,b,Dx	i	xi	f(xi)	wi	f(xi)wi
2						
3	0	0	0	1	0.5	0.5
4	3.1415926536	1	0.7853981634	1.8314696123	1	1.8314696123
5	0.7853981634	2	1.5707963268	1.9238795325	1	1.9238795325
6		3	2.3561944902	1.1950903220	1	1.1950903220
7		4	3.1415926536	0.2928932188	0.5	0.1464466094
8						
9				Integral	Approx=	4.395784045

Recall that $\int_0^\pi (1 + \sin 1.25x)\,dx$ has an exact answer, which to ten significant digits is equal to 4.507278079. Our right-endpoint Riemann Sum of $n = 4$ approximated this with an error of -0.3891742176, or approximately -8.6%, while the left-endpoint Riemann Sum with $n = 4$ approximated this with an error of $+0.1661861497$, or approximately $+3.7\%$ error. This Trapezoidal Rule approximation, on the other hand, yielded an error of -0.111494034 or approximately -2.5% error, so it is a noticeable improvement requiring very little extra work.

One should also expect significant improvement if the number of subintervals is increased. If we double them for this case, and use again the Trapezoidal Rule, our spreadsheet could look like the following:

	A	B	C	D	E	F
1	a,b,Dx	i	xi	f(xi)	wi	f(xi)wi
2						
3	0	0	0	1	0.5	0.5
4	3.1415926536	1	0.3926990817	1.4713967368	1	1.4713967368
5	0.3926990817	2	0.7853981634	1.8314696123	1	1.8314696123
6		3	1.1780972451	1.9951847267	1	1.9951847267
7		4	1.5707963268	1.9238795325	1	1.9238795325
8		5	1.9634954085	1.6343932842	1	1.6343932842
9		6	2.3561944902	1.1950903220	1	1.1950903220
10		7	2.7488935719	0.7097153227	1	0.7097153227
11		8	3.1415926536	0.2928932188	0.5	0.1464466094
12						
13				Integral	Approx=	4.4797446772

Cells B8–D11 can be simply produced through a "cutting and pasting" of B7–D7. Similarly, F8–F11 can be copied from F7. However, some cells do have to be redefined. For instance here A5=A4/8 (where it had been A4/4), and E3–E11 should be entered more or less manually. The cell for our final result would now be F13=A5*SUM(F3:F11).

The error in our final computation, using the Trapezoidal Rule with $n = 8$ for the integral $\int_0^\pi (1 + \sin 1.25x)\,dx$, was -0.0275334018, or approximately -0.6%, obviously a significant improvement over the already impressive -2.5% error when we used the Trapezoidal Rule with $n = 4$.[65]

5.7.4 Simpson's Rule

The idea behind Simpson's Rule is straightforward, namely using areas under parabolas that might better "hug" the curve than can the line segments used in the Trapezoidal Rule, and still have simple area formulas (by the FTC). The final formula is not complicated, and often gives surprisingly good approximations of definite integrals, and so we include it here. However, its derivation from first principles is exceptionally tedious, and omitted.[66]

Unlike the Trapezoidal Rule, in which there are n trapezoidal areas computed (where n is the number of equal-length subintervals in which we divide $[a,b]$), with Simpson's Rule there are $n/2$ parabolas used, and so the technique requires n to be an even number. Now we show the formulas for Simpson's Rule.

- First we look at the case of a single, (up to) quadratic function to approximate $\int_a^b f(x)\,dx$. In this case we have $n = 2$, $\Delta x = \frac{b-a}{2}$, $a = x_0 < x_1 < x_2 = b$, and $x_i = a + i\Delta x$, $i = 0, 1, 2$. For this case, we connect $(x_0, f(x_0))$, $(x_1, f(x_1))$, and $(x_2, f(x_2))$ by the quadratic interpolation of those three points, and after a tedious derivation we would get

$$\int_a^b f(x)\,dx \approx \frac{\Delta x}{3}[f(x_0) + 4f(x_1) + f(x_2)] = \frac{b-a}{6}\left[f(a) + 4f\left(\frac{a+b}{2}\right) + f(b)\right]. \quad (5.101)$$

- If instead we divide $[a, b]$ into n subintervals, where $n > 2$ is an even number, our pattern is

$$\boxed{\begin{aligned}\int_a^b f(x)\,dx \approx \frac{\Delta x}{3}[&f(x_0) + 4f(x_1) + 2f(x_2) + 4f(x_3) + 2f(x_4) + \cdots \\ &+ 4f(x_{n-3}) + 2f(x_{n-2}) + 4f(x_{n-1}) + f(x_n)].\end{aligned}} \quad (5.102)$$

---

[65]It should be pointed out that these computations which used spreadsheets were certainly doable without spreadsheets or other programmable devices. However, if one is confronted with several of these, or wishes to use a much finer partition with many more subintervals which would be tedious to work by hand, our examples here should demonstrate the utility of such programs.

One might also object to the display of so many significant digits for the intermediate steps when clearly the approximations will not yield so many. While such criticism is not without merit, particularly if intermediate computations are tabulated by hand, here it is not costly to include so many digits, and it makes it much less likely that round-off error will accumulate. The criticism may ring more true if the process were to be cataloged by hand, in which case it is still better practice to include more digits in intermediate steps than the expected accuracy of the final output. One should be especially careful about rounding the value of Δx, as it is involved in *multiplying* the sum of weighted values of the $f(x_i)$, *and* in the process of computing the x_i values. A rounding of Δx can therefore cause an accumulation of errors in rather important parts of the overall computation.

[66]Most engineering and science calculus texts include the formula but not the derivation. Simply computing the polynomial of degree 2 or less whose graph passes through three given points is complicated, and so to then integrate those and arrive at a general formula would be an exercise more in persistence and bookkeeping than enlightenment. (See Exercise 12.)

5.7. NUMERICAL INTEGRATION: APPROXIMATING DEFINITE INTEGRALS

For our integral $\int_0^\pi (1+\sin 1.25x)\,dx$, which we know to have value 4.507278079 to ten significant digits, we will compute the Simpson's Rule approximation for $n = 2$, as illustrated in the top of Figure 5.14c on page 617, and then compute the same for $n = 8$ after a simple editing of the previous spreadsheets.

For $n = 2$, and for simplicity using $f(x) = 1 + \sin 1.25x = 1 + \sin\frac{5x}{4}$ we have

$$\int_0^\pi (1+\sin 1.25x)\,dx \approx \frac{\pi}{6}[f(0)+4f(\pi/2)+f(\pi)] = \frac{\pi}{6}\left[1+4\left(1+\sin\frac{5\pi}{8}\right)+\left(1+\sin\frac{5\pi}{4}\right)\right]$$
$$\approx 4.706321177.$$

This is a remarkable degree of accuracy from using just one parabola's area to approximate the integral of the function. Indeed, the error is approximately +0.1990430978, or approximately +4.4%. With an edit of our previous spreadsheet, we can program a Simpson's Rule computation for $n = 8$ (four parabolas) where we had a Trapezoidal Rule (with eight trapezoids). For $n = 8$ we would have, in general,

$$\int_a^b f(x)\,dx \approx \frac{\Delta x}{3}[f(x_0)+4f(x_1)+2f(x_2)+4f(x_3)+2f(x_4)+4f(x_5)+2f(x_6)+4f(x_7)+f(x_8)].$$

For $\int_0^\pi (1+\sin 1.25x)\,dx$, and $n = 8$ we use this formula with $\Delta x = \pi/8$, $x_i = 0 + i\Delta x = i \cdot \frac{\pi}{8}$. For our spreadsheet, we thus have A5=pi()/8, the weights in Column E entered by hand according to the pattern presented, and, crucially, F13=A5/3*SUM(F3:F11). Other entries are the same as in the previous (Trapezoidal Rule) spreadsheet, or with code similar to the spreadsheet on page 619 but with the edits described above:

	A	B	C	D	E	F
1	a,b,Dx	i	xi	f(xi)	wi	f(xi)wi
2						
3	0	0	0	1	1	1
4	3.1415926536	1	0.3926990817	1.4713967368	4	5.8855869473
5	0.3926990817	2	0.7853981634	1.8314696123	2	3.6629392246
6		3	1.1780972451	1.9951847267	4	7.9807389067
7		4	1.5707963268	1.9238795325	2	3.8477590650
8		5	1.9634954085	1.6343932842	4	6.5375731367
9		6	2.3561944902	1.1950903220	2	2.3901806440
10		7	2.7488935719	0.7097153227	4	2.8388612910
11		8	3.1415926536	0.2928932188	1	0.2928932188
12						
13				Integral	Approx=	4.5077315546

Comparing the actual value, correct to 10 places, to this Simpson's Rule approximation, we get roughly four significant digits of accuracy with Simpson's Rule *for this case*:

Actual value $\int_0^\pi (1+\sin 1.25x)\,dx$:   4.507278079 (accurate to 10 significant digits)
Simpson's Rule, $n=8$:   4.507731555
Error: -0.000453476
Approximate percent error: -0.01%

Simpson's Rule can be very effective for computing definite integrals where the functions are well approximated by polynomials of degree at most two. Functions that grow very quickly, such as e^x for large x, or that have near-vertical tangents locally, can be somewhat problematic. As we will observe in the sequel, sine and cosine functions are quite close to quadratic in nature on small intervals, and so when Δx is small, Simpson's Rule is very effective for such functions.

5.7.5 Numerical Integration for Tabular Data

There are occasions when we are interested in some quantity, which is the definite integral of another quantity for which we do not have a formula and therefore cannot antidifferentiate, but for which we may have a table of data. For instance, if we have sensors that compute velocity periodically, we can use a Riemann Sum, Trapezoidal Rule, or Simpson's Rule to approximate the desired change in position.

Example 5.7.1

Suppose we have the following table of data for velocity of an object (for instance, a ping-pong ball during a game, as measured with a radar gun and recorded periodically):

t (sec)	0	1	2	3	4	5	6	7	8	9	10
v (m/sec)	13.4	14.1	6.4	-2.6	-4.6	2.2	11.6	14.9	9.1	-0.4	-5.0

Suppose we wish to approximate the total distance traveled in these ten seconds. In other words, suppose we wish to compute $\int_0^{10} |v(t)|\,dt$. Theoretically this would require a slightly different table of values:

t (sec)	0	1	2	3	4	5	6	7	8	9	10
$|v|$ (m/sec)	13.4	14.1	6.4	2.6	4.6	2.2	11.6	14.9	9.1	0.4	5.0

Not knowing the exact shape of the graph, but knowing that a Trapezoidal Rule or Simpson's Rule computation will tend to be more accurate than a simple Riemann Sum, we can approximate the integral one of those two ways:

Trapezoidal Rule: Here we can use

$$\int_0^{10} |v(t)|\,dt \approx \Delta x \left[\frac{1}{2}|v(0)| + |v(1)| + |v(2)| + \cdots + |v(8)| + |v(9)| + \frac{1}{2}|v(10)| \right]$$

$$= (1)\left[\frac{1}{2}(13.4) + 14.1 + 6.4 + 2.6 + 4.6 + 2.2 + 11.6 + 14.9 + 9.1 + 0.4 + \frac{1}{2}(5.0) \right] = 75.1.$$

Simpson's Rule: Here we instead compute

5.7. NUMERICAL INTEGRATION: APPROXIMATING DEFINITE INTEGRALS

$$\int_0^{10} |v(t)|\,dt \approx \frac{1}{3}\Big[|v(0)| + 4|v(1)| + 2|v(2)| + 4|v(3)| + 2|v(4)| + 4|v(5)|$$
$$+ 2|v(6)| + 4|v(7)| + 2|v(8)| + 4|v(9)| + |v(10)|\Big]$$
$$= \frac{1}{3}\Big[13.4 + 4(14.1) + 2(6.4) + 4(2.6) + 2(4.6) + 4(2.2)$$
$$+ 2(11.6) + 4(14.9) + 2(9.1) + 4(0.4) + 5.0\Big]$$
$$= 72.8\overline{6}.$$

To summarize, by the Trapezoidal Rule $\int_0^{10} |v(t)|\,dt \approx 75.1$, while with Simpson's Rule we get $\int_0^{10} |v(t)|\,dt \approx 72.9$.

It is often pointed out that Simpson's Rule will be exact if the function is itself of degree 2 or lower, or even if piecewise of degree 2 or lower if the "pieces" happen to coincide with those of the parabolas used in Simpson's Rule. If it is easily available, Simpson's Rule is usually preferred, though it is mildly more difficult to program than is the Trapezoidal Rule.

5.7.6 Error Bounds

While we will not prove these here, the following results regarding errors of these methods hold:

Trapezoidal Rule: If f'' exists and is continuous on $[a,b]$, and $|f''(x)| \leq M$ for all $x \in [a,b]$, then the error E in approximating the integral $\int_a^b f(x)\,dx$ using the Trapezoidal Rule will have the bound

$$|E| \leq \frac{(b-a)^3 M}{12n^2} \tag{5.103}$$

Simpson's Rule: If $f^{(4)}$ exists and is continuous on $[a,b]$, and $|f^{(4)}(x)| \leq M$ for all $x \in [a,b]$, then the error E in approximating the integral $\int_a^b f(x)\,dx$ using Simpson's Rule will have the bound

$$|E| \leq \frac{(b-a)^5 M}{180n^4}. \tag{5.104}$$

In either case, we have $n \to \infty \implies E \to 0$, assuming the derivatives in question exist. In the event that they do not, it is still possible for these schemes to converge to the correct value of the integral. Indeed, the integral being a limit of Riemann Sums using any sampling scheme for $x_i^* \in [x_{i-1}, x_i]$, clearly the Trapezoidal Rule must converge to the integral as $n \to \infty$, since that part of the sum where they differ must shrink to zero. It is not so obvious with Simpson's Rule, but it too will converge as $n \to \infty$. In either case, unless some further knowledge of the function's behavior is known, it is advisable to run several approximations with increasingly large values of n to observe if it appears the approximations are converging as n increases, as we did with Newton's Method approximation schemes in Section 4.5.

Exercises

1. Consider $\int_0^1 e^x \, dx$.

   (a) Use the Fundamental Theorem of Calculus to compute this exactly, and write the answer accurate to 8 places to the right of the decimal (accurate to $\pm 10^{-8}$).

   (b) Use the Trapezoidal Rule with $n = 4$ to approximate this integral.

2. Repeat Exercise 1 using $n = 8$.

3. Repeat Exercise 1 using Simpson's Rule with $n = 4$.

4. Repeat Exercise 1 using Simpson's Rule with $n = 8$.

5. Show that Simpson's Rule gives an exact answer for $\int_0^{10} x^2 \, dx$ with $n = 2$. Explain why this is expected.

6. Show that Simpson's Rule gives an exact answer for $\int_0^{10} x \, dx$ with $n = 2$. Explain why this is expected.

7. Approximate $\int_{-1}^{1} e^{-x^2} \, dx$ using Simpson's Rule and $n = 4$.

8. Repeat with $n = 8$.

9. Consider $\int_1^4 \frac{1}{x} \, dx$.

   (a) What is the exact value?

   (b) What is an approximation accurate to $\pm 10^{-8}$?

   (c) Use the Trapezoidal Rule with $n = 4$ to approximate this integral.

   (d) Use Simpson's Rule with $n = 4$ to approximate this integral.

10. A fire hose is used to empty a tall water tower. A flow-rate gauge is attached to the hose. Each minute the flow rate is recorded, in gallons per minute. Use a Trapezoidal Rule to approximate the total water which flowed through the hose over the total time interval. (We can let $V(t)$ be the volume of water that has flowed through the hose after time t, and so the table is technically listing dV/dt values, where we desire $V(10) - V(0) = \int_0^{10} (dV/dt) \, dt$.) Flow rate values for $t = 0, 1, 2, \cdots, 10$ are, respectively, 150, 130, 120, 140, 105, 130, 100, 90, 90, 85, 60.

11. Repeat the previous problem using Simpson's Rule.

12. Given points (x_1, y_1), (x_2, y_2) and (x_3, y_3), where the x_i values are all distinct, show that the graph of the following function passes through each point:

$$q(x) = y_1 \cdot \frac{(x - x_2)(x - x_3)}{(x_1 - x_2)(x_1 - x_3)} + y_2 \cdot \frac{(x - x_1)(x - x_3)}{(x_2 - x_1)(x_2 - x_3)} + y_3 \cdot \frac{(x - x_1)(x - x_2)}{(x_3 - x_1)(x_3 - x_2)}.$$

This is the quadratic interpolation used in the derivation of Simpson's Rule.

13. Using the formula for $q(x)$, find the quadratic function that passes through $(0, 3)$, $(2, -1)$, and $(3, 0)$. Simplify your answer.

14. Plot each group of three points, and find the simplified form of $q(x)$ for each. (There may be technical difficulties with the formula, which you should explain.)

    (a) $(0, 0)$, $(1, 2)$ and $(2, 4)$

    (b) $(0, 1)$, $(1, 2)$ and $(3, 1)$

    (c) $(1, 2)$, $(1, 3)$ and $(4, 3)$

5.8 Substitution with the Power Rule

Integration by *substitution*, also known as *change of variable*, is the most important of the general integrating techniques, finding its way into the other techniques as well. While we introduce it here, for now we limit the scope to Power Rules.

Before looking at this method formally, consider the following antiderivative statements, each of which refer to the same Power Rule (perhaps most familiar in the first case):

$$\int x^2 \, dx = \frac{x^3}{3} + C,$$

$$\int u^2 \, du = \frac{u^3}{3} + C,$$

$$\int (\sin x)^2 \, d(\sin x) = \frac{(\sin x)^3}{3} + C. \tag{5.105}$$

The last integral is simply asking for an antiderivative of $(\sin x)^2$ with respect to $\sin x$. Indeed, we can check the answer as before:

$$\frac{d}{d\sin x}\left[\frac{1}{3}(\sin x)^3\right] = \frac{1}{3} \cdot 3(\sin x)^2 = (\sin x)^2,$$

as we expect. Of course, we usually take derivatives and antiderivatives with respect to a variable, and not a function. However, the integral in (5.105) is not so unlikely to occur as one might think. Recall that $df(x) = f'(x)\,dx$ is the definition of the differential (see (4.15), page 464). Thus $d\sin x = \cos x\,dx$, and the integral in (5.105) can be written instead

$$\int (\sin x)^2 \cos x \, dx = \int (\sin x)^2 \frac{d\sin x}{dx}\, dx = \int (\sin x)^2 \, d\sin x = \frac{1}{3}(\sin x)^3 + C.$$

Indeed, it is not hard to see that the Chain Rule gives us $\frac{d}{dx}\left[\frac{1}{3}(\sin x)^3\right] = \frac{1}{3}\cdot 3(\sin x)^2 \cos x = \sin^2 x \cos x$.

In this section, we will concentrate on integrals of the form

$$\boxed{\int u^n \, du = \begin{cases} \frac{1}{n+1} \cdot u^{n+1} + C & \text{if } n \neq -1, \\ \ln|u| + C & \text{if } n = -1. \end{cases}} \tag{5.106}$$

As anticipated in the discussion, the content of the differential du may be more expansive than what we may expect from a single variable. The point of this section is to recognize when we have the form (5.106) and how to go about rewriting the integral into the proper form.

The reader should be forewarned: This method requires a fair amount of practice at first, and false starts are inevitable when learning the method. It is not a simple algorithm. **For each problem, the practitioner decides, or experiments to find, which substitution will produce an integral that can be computed with known rules.** In this section we limit ourselves to the Power Rule (5.106), but in subsequent sections we will delve into many other rules, and it is not always obvious which rule should be used for a given integral. With practice, one learns to look for clues, and anticipate what will occur several steps ahead, to see if there is indeed an integration rule that can apply.[67]

5.8.1 The Technique

Here we will look at some of the simpler problems of integration by substitution. As we proceed, several observations will be made regarding the method.

Example 5.8.1

Compute the indefinite integral $\int (x^2+1)^7 \cdot 2x\,dx$.

Solution: The technique is to introduce a new variable, u, with which we can write the original integral in a simpler form (which we can compute). We also have to take into account what will be the new differential, namely du:

$$u = x^2 + 1$$
$$\implies du = 2x\,dx.$$

(Recall that if u is a function of x, then $du = u'(x)dx$, consistent with $\frac{du}{dx} = u'(x)$.) Using this information, we can replace all the terms in the original integral: the $(x^2+1)^7$ becomes u^7, and the terms $2x\,dx$ collectively become du (see the previous implication arrow). Thus

$$\int (x^2+1)^7 \cdot 2x\,dx = \int u^7\,du = \frac{1}{8}u^8 + C.$$

This is all true, but **we** introduced u, while the original question asked for an antiderivative with respect to x. We only need to replace u in the final answer, using again $u = x^2+1$. Summarizing,

$$\int (x^2+1)^7 \cdot 2x\,dx = \int u^7\,du = \frac{1}{8}u^8 + C = \frac{1}{8}(x^2+1)^8 + C.$$

We can check our answer in this example by computing the derivative of our answer (using the Chain Rule), yielding $\frac{d}{dx}\left[\frac{1}{8}(x^2+1)^8\right] = \frac{1}{8} \cdot 8(x^2+1)^7 \cdot \frac{d}{dx}(x^2+1) = (x^2+1)^7 \cdot 2x$ as hoped. In fact, integration by substitution, at least in its simplest forms, is often called a type of reverse Chain Rule, as for instance we can rewrite the rule (5.106) as follows:

$$\int u^n\,du = \int u^n \cdot \left(\frac{du}{dx}\right) dx = \int \left[u^n \cdot \frac{du}{dx}\right] dx = \begin{cases} \frac{1}{n+1} \cdot u^{n+1} + C & \text{if } n \neq 1, \\ \ln|u| + C & \text{if } n = -1. \end{cases} \quad (5.107)$$

To see that this is correct, if we take the derivative of the claimed antiderivatives *with respect to x*, we see that we do indeed get $u^n \cdot \frac{du}{dx}$ in both cases from the Chain Rule.

In short, with integration by substitution we try to pick some function we call u, so that

---

[67] In fact, sometimes there is no rule that will produce an antiderivative formula using known functions (i.e., no "closed form" antiderivative), and then some approximation scheme will be necessary. When available, it is most desirable to have an exact antiderivative, and we can find one often enough that it is well worth studying these techniques.

5.8. SUBSTITUTION WITH THE POWER RULE

1. a main part of the integrand can be written as a simple function of u—one for which we know the antiderivative with respect to u—and, equally crucial, so that

2. the remaining variable terms of the integral can be safely absorbed into du (except for multiplicative constants, which we will see add only a slight complication).

If these are both satisfied, our substitution of u and du terms gives us a new, simple integral (entirely in terms of u and du!), which we can then compute, and then a simple algebraic substitution brings us back to a solution in x.

When working such a problem (as opposed to, say, *publishing* a problem and solution for professional consumption), a useful format is to (1) write the original integral; (2) write the substitution function u, together with its differential du, both separated from the original integral (e.g., to the side or below); (3) write the new form of the original integral, i.e., in u and du; (4) compute the antiderivative of this new integral in u as a continuation of the first step; and (5) resubstitute to arrive at the antiderivative in x. Hence a typical homework-style presentation of Example 5.8.1 might look like the following, with the choice of u and resulting du offset from (or below) the original integral and resulting integrals.[68]

$$\left[\begin{array}{l} u = x^2 + 1 \\ \Longrightarrow du = 2x\,dx \end{array}\right] \qquad \int (x^2+1)^7 \cdot \underline{2x\,dx} = \int u^7 \underline{du} = \frac{1}{8}u^8 + C = \frac{1}{8}(x^2+1)^8 + C.$$

That part of the integral that we hope to absorb into du is underscored in the original integral and the corresponding term in the new integral. The rest of the integral was just $(x^2+1)^7 = u^7$. In fact, when we choose u so that a major portion of the integral can be written u^n, then any other factors that are variable, along with the differential dx, must be absorbed into du or the substitution will fail (because the resulting integral will contain both x and u and no antiderivative rules will apply). We will continue to use this kind of spatial organization when we integrate by substitution in the examples that follow.

Example 5.8.2 _____

Compute the indefinite integral $\int \sin^2 x \cos x\,dx$.

<u>Solution</u>: Note that this integral can be written $\int (\sin x)^2 \cos x\,dx$. Now we proceed:

$$\left[\begin{array}{l} u = \sin x \\ \Longrightarrow du = \underline{\cos x\,dx} \end{array}\right] \qquad \int \sin^2 x \, \underline{\cos x\,dx} = \int u^2 \underline{du} = \frac{1}{3}u^3 + C = \frac{1}{3}\sin^3 x + C.$$

_____

This is in fact an example that began this section, using a different notation (see (5.105), page 629). Before continuing, we will make a very minor change to the integral in the first numbered example (Example 5.8.1, page 630), and show a simple way to extend our method to handle this.

---
[68] There is nothing special about the brackets or the placement of the work therein. We use brackets to the left for typesetting purposes, and to isolate the substitution variable identification (and its differential computation) from the actual integrals. The same holds for our underscoring of certain terms. The work in the brackets is often placed under the original integral in written homework or in lecture presentations.

Example 5.8.3

Compute $\int x(x^2+1)^7 \, dx$.

Solution: Here we will make the same substitution as before, but the du will have an extra factor of 2. Since constant factors are relatively easy to handle in derivative and antiderivative problems in general, we should not expect this extra factor of 2 to cause much difficulty. It will simply mean one extra step in the substitution computations.[69]

$$\begin{bmatrix} u = x^2 + 1 \\ \implies du = 2x\,dx \\ \implies \frac{1}{2}du = x\,dx \end{bmatrix} \qquad \int x(x^2+1)^7 \, dx = \int u^7 \cdot \frac{1}{2} du = \frac{1}{2} \cdot \frac{1}{8} u^8 + C = \frac{1}{16}(x^2+1)^8 + C.$$

This time, the extra nonconstant and differential terms of the original integral were, collectively, $x\,dx$. Though that product is not exactly du, it is a constant multiple of du. In our substitution we took an extra step and solved, again collectively, for $x\,dx = \frac{1}{2} du$.

The preceding example shows that we need to be flexible when looking for a possible Power Rule application. Not every integral where we can use the Power Rule will be of the strict form (5.107), page 630. Indeed, we need to be especially vigilant to notice that an integral may be of the form

$$\int k \cdot u^n \, du = \int k \cdot u^n \cdot \left(\frac{du}{dx}\right) dx. \qquad (5.108)$$

So when we make a substitution, we try not to be distracted by extra or missing multiplicative constants, as they will work themselves out in the substitution and final integration steps.

Example 5.8.4

Compute $\int x^3 \cos^5 x^4 \sin x^4 \, dx$.

---

[69]There is another method used by some texts to handle a problem such as this, which is to simply introduce the needed factor of 2 in the integral to complete the differential $du = 2x\,dx$, and compensate for the insertion of the new factor by simultaneously inserting a factor of $\frac{1}{2}$, which is simply carried through the rest of the calculation:

$$\int x(x^2+1)^7 \, dx = \int \frac{1}{2} \underbrace{(x^2+1)^7}_{u^7} \cdot \underbrace{2x\,dx}_{du} = \frac{1}{2} \int u^7 \, du = \frac{1}{2} \cdot \frac{1}{8} u^8 + C = \frac{1}{16}(x^2+1)^8 + C,$$

where again $u = x^2 + 1$, $du = 2x\,dx$.

This method is appealing because one rewrites the integrand into a form where it is, more or less, clearly a derivative of a Chain Rule function (perhaps multiplied by a constant, as with $\frac{1}{2}$ here).

We will avoid this method because, though it is not so challenging for simpler problems, it quickly becomes unreasonably difficult if an integral is complicated. Furthermore, the method presented in this text—in the authors' opinions—makes for much better preparation for more advanced methods, such as trigonometric substitution and integration by parts, both found in the sequel and other texts.

5.8. SUBSTITUTION WITH THE POWER RULE

<u>Solution</u>: It is perhaps more obvious how to proceed if we rewrite the integral in the form $\int x^3 (\cos x^4)^5 \sin x^4 \, dx$. Then we see that the u^n term will be $(\cos x^4)^5 = u^5$, where $u = \cos x^4$. Next we need to see if du can absorb the other nonconstant terms. Indeed it can:

$$\left[\begin{array}{c} u = \cos x^4 \\ \Longrightarrow \quad du = -\sin x^4 \cdot 4x^3 \, dx \\ \Longrightarrow \quad -\dfrac{1}{4} du = x^3 \sin x^4 \, dx \end{array}\right] \quad \begin{aligned} \int x^3 \cos^5 x^4 \sin x^4 \, dx &= \int u^5 \left(-\frac{1}{4}\right) du \\ &= -\frac{1}{4} \cdot \frac{1}{6} u^6 + C = -\frac{1}{24}(\cos x^4)^6 + C \\ &= \frac{-1}{24} \cos^6 x^4 + C. \end{aligned}$$

We should point out that this substitution would not have worked without both the x^3 and $\sin x^4$ factors accompanying the $\cos^5 x^4$ factor in the integral, lest the (calculus-generated) du term over-account or under-account for variable terms not absorbed into the (algebraic) u-substitution. (Recall that constant factors of the integrand can be moved into and out of the integral, but variables cannot.)

It is also worth noting that we could have used a two-step substitution. For instance, a student recognizing that x^4, and a multiple of its derivative in the form of x^3, both appear, might first make a substitution of the form $u = x^4$:

$$\left[\begin{array}{c} u = x^4 \\ \Longrightarrow \quad du = 4x^3 \, dx \\ \Longrightarrow \quad \dfrac{1}{4} du = x^3 \, dx \end{array}\right] \quad \begin{aligned} \int x^3 \cos^5 x^4 \sin x^4 \, dx &= \int \cos^5 u \sin u \, \frac{1}{4} du \\ &= \frac{1}{4} \int \cos^5 u \sin u \, du. \end{aligned}$$

This is already progress, since then we have a simpler integral, albeit requiring its own substitution:

$$\left[\begin{array}{c} w = \cos u \\ \Longrightarrow \quad dw = -\sin u \, du \\ \Longrightarrow \quad -dw = \sin u \, du \end{array}\right] \quad \begin{aligned} \frac{1}{4} \int \cos^5 u \sin u \, du &= \frac{1}{4} \int w^5 (-dw) \\ &= -\frac{1}{4} \cdot \frac{1}{6} w^6 + C \\ &= -\frac{1}{24} (\cos u)^6 + C. \end{aligned}$$

Of course, this gives the answer in terms of u, so we substitute back again, in terms of x. Summarizing,

$$\int x^3 \cos^5 x^4 \sin x^4 \, dx = \cdots = -\frac{1}{24} w^6 + C = -\frac{1}{24}(\cos u)^6 + C = -\frac{1}{24} \cos^6 x^4 + C.$$

The second approach is longer, but it has the advantage that we are not trying to rewrite the integral in one, all-encompassing (and thus more complicated) substitution step. Indeed, it is sometimes desirable to simplify an integral with substitution even if the resulting integral cannot be evaluated immediately. With most examples, we will use the first method, but the student working problems should be aware that the option of successive substitutions is perfectly valid: If the first substitution yields a simpler integral, it may be worth working with that integral even if the final form of the antiderivative is not obvious after one substitution.

Next we look at a few very common types of examples where the power of n is $1/2$, -1, and -2. These appear often enough that it is worth some effort to remember them specifically.

Example 5.8.5

Compute $\int \dfrac{x}{\sqrt{x^2-9}}\,dx$.

<u>Solution</u> Here we will take $u = x^2 - 9$, since the $du = 2x\,dx$ can absorb both the dx and the extra factor of x:

$$\left[\begin{array}{l} u = x^2 - 9 \\ \implies du = 2x\,dx \\ \implies \dfrac{1}{2}du = x\,dx \end{array}\right] \qquad \int \dfrac{x}{\sqrt{x^2-9}}\,dx = \int u^{-1/2} \cdot \dfrac{1}{2}du = \dfrac{1}{2} \cdot 2u^{1/2} + C = \sqrt{x^2-9} + C.$$

Example 5.8.6

Compute $\int \dfrac{\sin x}{\cos x}\,dx$.

<u>Solution</u>: Here we will take $u = \cos x$, since $du = -\sin x\,dx$ will absorb the other terms.

$$\left[\begin{array}{l} u = \cos x \\ \implies du = -\sin x\,dx \\ \implies -du = \sin x\,dx \end{array}\right] \qquad \int \dfrac{\sin x\,dx}{\cos x} = \int \dfrac{1}{u}(-du) = -\ln|u| + C = -\ln|\cos x| + C.$$

Note that if we instead took $u = \sin x$, then $du = \cos x\,dx$, but $\cos x$ is not a **multiplicative** factor in the original integral; the desired factor is $\frac{1}{\cos x}$, which is not contained in the du term if $u = \sin x$.

It should be remembered that checking these antiderivatives is as simple as computing the derivative of the answer. Here

$$\dfrac{d}{dx}[-\ln|\cos x|] = -\dfrac{1}{\cos x} \cdot \dfrac{d}{dx}\cos x = -\dfrac{1}{\cos x}(-\sin x) = \dfrac{\sin x}{\cos x},$$

as we hope. Of course our original integrand, and this computed derivative, can both be written as $\tan x$.

Note that we can write $-\ln|\cos x| = \ln|\cos x|^{-1} = \ln\left|(\cos x)^{-1}\right| = \ln|\sec x|$, so many calculus books (such as this one) contain the integration formula

$$\int \tan x\,dx = \ln|\sec x| + C. \tag{5.109}$$

(It is also interesting to "fill in the dots" for the computation $\frac{d}{dx}\ln|\sec x| = \cdots = \tan x$, verifying (5.109). See also Exercises 29 and 30, page 643.)

5.8. SUBSTITUTION WITH THE POWER RULE

Example 5.8.7

Compute $\int \dfrac{e^{3x}}{(e^{3x}+4)^2}\, dx$.

Solution: Here we note that the numerator e^{3x} of the integrand is the derivative of the term $e^{3x}+4$ inside of the power in the denominator, except for a multiplicative constant. Thus we will let $u = e^{3x}+4$, and du will absorb the function in the numerator:

$$\left[\begin{array}{r} u = e^{3x}+4 \\ \Longrightarrow \quad du = e^{3x}\cdot 3\,dx \\ \Longrightarrow \quad \tfrac{1}{3}\,du = e^{3x}\,dx \end{array}\right] \qquad \int \dfrac{e^{3x}}{(e^{3x}+4)^2}\,dx = \int \dfrac{1}{u^2}\cdot\dfrac{1}{3}\,du = \dfrac{1}{3}\int u^{-2}\,du = \dfrac{1}{3}\cdot(-1)u^{-1}+C$$
$$= -\dfrac{1}{3}(e^{3x}+4)^{-1}+C = \dfrac{-1}{3\left(e^{3x}+4\right)}+C.$$

At this point, we notice three common forms of integration by substitution:

$$\int \dfrac{u'(x)}{\sqrt{u(x)}}\,dx = 2\sqrt{u(x)}+C, \tag{5.110}$$

$$\int \dfrac{u'(x)}{u(x)}\,dx = \ln|u(x)|+C, \tag{5.111}$$

$$\int \dfrac{u'(x)}{[u(x)]^2}\,dx = \dfrac{-1}{u(x)}+C. \tag{5.112}$$

In all three cases, $u'(x)\,dx = du$, and we have simple Power Rules. In the first and third cases, there are multiplicative constants that occur. There is no real need to memorize these, but they occur often enough that their "mechanics" should become familiar. For that reason, these three results can become, if not memorized, at least easily recognized and cited.

The method also works for cases where the du term is just a constant multiple of dx:

Example 5.8.8

Compute $\int \dfrac{1}{(6-2x)^5}\,dx$.

Solution:

$$\left[\begin{array}{r} u = 6-2x \\ \Longrightarrow \quad du = -2\,dx \\ \Longrightarrow \quad \dfrac{-1}{2}\,du = dx \end{array}\right] \qquad \int \dfrac{1}{(6-2x)^5}\,dx = \int u^{-5}\cdot\dfrac{-1}{2}\,du = -\dfrac{1}{2}\cdot\dfrac{1}{-4}u^{-4}+C$$
$$= \dfrac{1}{8}(6-2x)^{-4}+C = \dfrac{1}{8(6-2x)^4}+C.$$

In the case that $du = dx$, this can often be anticipated and the experienced calculus student might omit the middle steps:

Example 5.8.9

Compute $\int (x+9)^4\,dx$.

Solution:

$$\left[\begin{array}{l} u = x+9 \\ \Longrightarrow du = dx \end{array}\right] \qquad \int (x+9)^4 \, dx = \int u^4 \, du = \frac{1}{5} u^5 + C = \frac{1}{5}(x+9)^5 + C.$$

Another way to look at this example is to realize that $d(x+9) = dx$, so we can write[70]

$$\int (x+9)^4 \, dx = \int (x+9)^4 \, d(x+9) = \frac{1}{5}(x+9)^5 + C.$$

In other words, dx is the same as $d(x+9)$, so we get the same if we interpret the original integral as an antiderivative with respect to $(x+9)$. Indeed, this is a shortcut one learns with practice—thinking but perhaps not writing the second step—but at first it is still best to write out the full substitution, as we did for this example, at least until one is proficient in the method as presented here. Of course, this is the analog to a Chain Rule where the "inner" derivative is 1:

$$\frac{d}{dx}\left[\frac{1}{5}(x+9)^5\right] = \frac{1}{5} \cdot 5(x+9)^4 \cdot \frac{d(x+9)}{dx} = \frac{1}{5} \cdot 5(x+9)^4 \cdot 1 = (x+9)^4, \quad \text{q.e.d.}$$

5.8.2 A Useful Twist on the Method

Recall our second example, namely Example 5.8.3 on page 632: $\int x(x^2+1)^7 \, dx$. We used a substitution $u = x^2 + 1$ because $du = 2x \, dx$ contained the extra factor of x in the integrand. The substitution eventually gave us $\int u^7 \cdot \frac{1}{2} \, du$, which was a simple Power Rule. Of course, we could have "simply" expanded the original function

$$x(x^2+1)^7 = x\left(x^{14} + 7x^{12} + 21x^{10} + 35x^8 + 35x^6 + 21x^4 + 7x^2 + 1\right)$$
$$= x^{15} + 7x^{13} + 21x^{11} + 35x^9 + 35x^7 + 21x^5 + 7x^3 + x$$

and integrated "term by term." However the substitution method was arguably easier, and the answer's simple form, $\frac{1}{16}(x^2+1)^8 + C$ would probably not be easily recognized from a strategy which expands the integrand first.

Now consider the integral $\int x(x-1)^{3/2} \, dx$. Here we cannot simply "expand" the integrand (even by brute force, as before), because of the fractional power term $(x-1)^{3/2}$, which is algebraically more difficult to deal with than positive integer powers. Furthermore, if we let $u = x - 1$, then $du = dx$, but this differential term cannot absorb the extra factor x. The key is to then notice that the original substitution offers a way out: While du cannot absorb the extra x factor, it can be written in terms of u, since $u = x - 1 \iff u + 1 = x$. Next we show how this can be utilized. Indeed, we will expand the new integrand, but what is interesting is how the algebraic difficulties of the $(x-1)^{3/2}$ term (namely that this is of the form $(a+b)^r$, $r \notin \mathbb{N}$) are transferred to the x term which, being a positive integer power, is then easier to handle. Here we write this out in the standard example format:

---

[70] Note that the *change* in $x + 9$ is the same as the change in x.

5.8. SUBSTITUTION WITH THE POWER RULE

Example 5.8.10

Compute $\int x(x-1)^{3/2}\,dx$.

Solution: Here we substitute writing some variable u in terms of x, **and** writing x in terms of u, of course in a mathematically consistent manner. Both substitutions are calculated in what follows, but separately. (This time we underscore the substitution for x, instead of the differential part.) Once the substitutions are completed, we can perform the multiplication to get two simple Power Rules:

$$\left[\begin{array}{c} u = x-1 \\ \Longrightarrow\ du = dx \\ \hline \text{Also,}\quad u = x-1 \\ \Longleftrightarrow u+1 = x \end{array}\right] \qquad \begin{aligned} \int \underline{x}(x-1)^{3/2}\,dx &= \int \underline{(u+1)}u^{3/2}\,du = \int \left(u^{5/2}+u^{3/2}\right)du \\ &= \frac{2}{7}u^{7/2} + \frac{2}{5}u^{5/2} + C \\ &= \frac{2}{7}(x-1)^{7/2} + \frac{2}{5}(x-1)^{5/2} + C. \end{aligned}$$

Though this answer is correct, one often factors what one can for the final answer:

$$= \frac{2}{35}(x-1)^{5/2}[5(x-1)+7] + C = \frac{2}{35}(x-1)^{5/2}(5x+2) + C.$$

A similar strategy allows us to compute the integral in the next example.

Example 5.8.11

Compute $\int \dfrac{x}{\sqrt{2x+1}}\,dx$.

Solution: We will work this problem twice using two different substitutions. The first is perhaps the more obvious, but the second has some appeal as well.

$$\left[\begin{array}{c} u = 2x+1 \\ \Longrightarrow\ du = 2\,dx \\ \Longrightarrow\ \frac{1}{2}du = dx \\ \hline \text{Also,}\quad u = 2x+1 \\ \Longrightarrow\ x = \frac{1}{2}(u-1) \end{array}\right] \qquad \begin{aligned} \int \frac{x}{\sqrt{2x+1}}\,dx &= \int \frac{\frac{1}{2}(u-1)}{u^{1/2}} \cdot \frac{1}{2}\,du \\ &= \frac{1}{4}\int \left(u^{1/2} - u^{-1/2}\right)du \\ &= \frac{1}{4}\left(\frac{2}{3}u^{3/2} - 2u^{1/2}\right) + C \\ &= \frac{1}{6}(2x+1)^{3/2} - \frac{1}{2}(2x+1)^{1/2} + C. \end{aligned}$$

Again one might factor, simplify, and rearrange the variable parts of the answer to arrive at

$$= \frac{1}{6}(2x+1)^{1/2}[(2x+1)-3] + C = \frac{1}{6}\sqrt{2x+1}(2x-2) + C = \frac{1}{3}(x-1)\sqrt{2x+1} + C.$$

For the alternative substitution, we let $u = \sqrt{2x+1}$. Note how much of the integrand is then absorbed into du (due to the relationship between the square root and its derivative).

$$\left[\begin{array}{c} u = \sqrt{2x+1} \\ \implies du = \dfrac{1}{2\sqrt{2x+1}} \cdot 2\,dx \\ \implies du = \dfrac{1}{\sqrt{2x+1}}\,dx \\ \hline u = \sqrt{2x+1} \\ \implies u^2 = 2x+1 \\ \implies \dfrac{1}{2}(u^2-1) = x \end{array}\right] \qquad \begin{aligned} \int \dfrac{x}{\sqrt{2x+1}}\,dx &= \int \underbrace{\dfrac{1}{2}(u^2-1)}_{x}\underbrace{du}_{\frac{dx}{\sqrt{2x+1}}} \\ &= \dfrac{1}{2}\left[\dfrac{1}{3}u^3 - u\right] + C \\ &= \dfrac{1}{6}u^3 - \dfrac{1}{2}u + C \\ &= \dfrac{1}{6}\left(\sqrt{2x+1}\right)^3 - \dfrac{1}{2}\sqrt{2x+1} + C \text{ (as before)}. \end{aligned}$$

It is important to notice that we used the equation for u to calculate du within a given substitution strategy. Also, the reader should begin to see that we can make some rather interesting substitutions, so long as we are consistent when replacing every term inside the integral. In doing so, it will become apparent (1) if it is even possible to use a given substitution to rewrite the integral, and (2) even if so, whether the new integral is one that we can actually compute.

5.8.3 Other Miscellaneous Power Rule Substitutions

So far, we have concentrated on algebraic (polynomial and rational-power), exponential, and trigonometric functions in our substitution problems. It is also worth examining how Power Rules can arise from integrals involving logarithmic and arctrigonometric functions, which we do in this subsection.

Example 5.8.12

Compute $\int \dfrac{(\ln x)^5}{x}\,dx$.

Solution: Here we see a factor $\frac{1}{x}$, which is the derivative of $\ln x$, so the latter will be u:

$$\left[\begin{array}{c} u = \ln x \\ du = \dfrac{1}{x}dx \end{array}\right] \qquad \int \dfrac{(\ln x)^5}{x}\,dx = \int u^5\,du = \dfrac{1}{6}u^6 + C = \dfrac{(\ln x)^6}{6} + C.$$

Note that, as a general rule, if we have a function $f(x)$ with antiderivative $F(x)$, then we have[71]

$$\left[\begin{array}{c} u = \ln x \\ \implies du = \dfrac{1}{x}dx \end{array}\right] \qquad \int \dfrac{f(\ln x)}{x}\,dx = \int f(u)\,du = F(u) + C = F(\ln x) + C. \qquad (5.113)$$

---

[71]In fact, we can replace $\ln x$ with $\ln|x|$ throughout in (5.113).

5.8. SUBSTITUTION WITH THE POWER RULE

Similar formulas apply to the arctrigonometric functions. Rather than list and commit to memorize them, it is better to look at the general idea that if, say, $\sin^{-1} x$ occurs in an integral, we would look immediately to see if its derivative, $\frac{1}{\sqrt{1-x^2}}$, also appears. In fact, that is a general guideline for all functions in this context (choose u so that du also appears in the integral, as a multiplicative factor).

Example 5.8.13

Compute $\int \dfrac{1}{\sqrt{1-x^2}\sin^{-1}x}\,dx$.

Solution: Note here that if $u = \sin^{-1} x$ then our du will account for $\frac{1}{\sqrt{1-x^2}}\,dx$:

$$\left[\begin{array}{l} u = \sin^{-1} x \\ \Longrightarrow du = \dfrac{1}{\sqrt{1-x^2}}\,dx \end{array}\right] \qquad \int \dfrac{1}{\sqrt{1-x^2}\sin^{-1}x}\,dx = \int \dfrac{1}{u}\,du = \ln|u| + C = \ln\left|\sin^{-1}x\right| + C.$$

Example 5.8.14

Compute $\int \dfrac{\sec^{-1} x}{x\sqrt{x^2-1}}\,dx$. Assume $x > 0$ (or more precisely, $x > 1$), so that $|x| = x$.

Solution:

$$\left[\begin{array}{l} u = \sec^{-1} x \\ du = \dfrac{1}{x\sqrt{x^2-1}}\,dx \end{array}\right] \qquad \int \dfrac{\sec^{-1} x}{x\sqrt{x^2-1}}\,dx = \int u\,du = \dfrac{1}{2}u^2 + C = \dfrac{\left(\sec^{-1} x\right)^2}{2} + C.$$

In this example, if instead $x < 0$ (actually $x < -1$), we would replace x by $-|x|$ in the denominator of the integrand, giving eventually $-\frac{1}{2}(\sec^{-1} x)^2 + C$ for the antiderivative.

5.8.4 Substitution in Definite Integrals

While thus far we have only discussed finding indefinite integrals using the substitution method, in fact we can extend the logic to definite integrals. All that remains is to change the values of the endpoints (or "limits") of integration to the corresponding values of the new variable.

Example 5.8.15

Compute the definite integral $\int_0^{\pi/4} \dfrac{\sec^2 x}{\sqrt{\tan x + 1}}\,dx$.

Solution: First we note that the integrand is continuous on all of $[0, \pi/4]$, because on that interval $\tan x \geq 0 \Longrightarrow \tan x + 1 \geq 1$.

Next we see that if we set $u = \tan x + 1$ we will have $du = \sec^2 x\, dx$, which absorbs the factor $\sec^2 x$ in the integrand. Finally, $x = 0 \implies u = \tan 0 + 1 = 1$, and $x = \pi/4 \implies u = \tan\frac{\pi}{4} + 1 = 2$. Thus

$$\int_0^{\pi/4} \frac{\sec^2 x}{\sqrt{\tan x + 1}}\, dx = \int_1^2 \frac{1}{\sqrt{u}}\, du = \int_1^2 u^{-1/2}\, du = 2u^{1/2}\Big|_1^2 = 2\sqrt{2} - 2\sqrt{1} = 2\sqrt{2} - 2.$$

Alternatively, but less efficiently, we can use the same substitution to compute the antiderivative in x first, and then evaluate at the endpoints:

$$\int \frac{\sec^2 x}{\sqrt{\tan x + 1}}\, dx = 2\sqrt{\tan x + 1} + C \implies \int_0^{\pi/4} \frac{\sec^2 x}{\sqrt{\tan x + 1}}\, dx = 2\sqrt{\tan x + 1}\Big|_0^{\pi/4} = 2\sqrt{2} - 2\sqrt{1} = 2\sqrt{2} - 2,$$

as before.

We see either method in this example ultimately leads to the same final computation, but the overall process is simpler and more efficient if finished in the variable u instead of detouring back to x. Confident and experienced calculus students eventually find the first method—where we completely rewrite the integral in terms of u, including the endpoints—to be simpler and more efficient, even if initially the second method appears more intuitive (because it combines two already-established skills). The reader is encouraged to use the first method, but also invited to keep in mind the second method, and to notice that the final computations are the same, though rearranged slightly. Since the antiderivative in the u-variable is usually simpler than the antiderivative in the x-variable, some of the "bookkeeping" of the final computations can be more easily spread out into smaller computations in the first method.[72]

In the next examples, we emphasize this "total substitution" method.

Example 5.8.16

Compute $\int_0^{\pi} \cos^4 5x \sin 5x\, dx$.

Solution: Technically, we should first notice that the integrand is continuous on $[0, \pi]$ (and indeed on all of $\mathbb{R}$), and so the FTC applies. (We will usually only note when this is *not* the case.) We next see that $\sin 5x$ is a constant multiple of the derivative of $\cos 5x$, and so we let $u = \cos 5x$ and the du will absorb the sine term:

$$\begin{array}{rcl} u &=& \cos 5x \\ \implies du &=& -5\sin 5x\, dx \\ \implies \frac{-1}{5} du &=& \sin 5x\, dx \\ \hline x = \pi \implies u &=& \cos 5\pi = -1 \\ x = 0 \implies u &=& \cos 0 = 1 \end{array} \qquad \begin{aligned} \int_0^{\pi} \cos^4 5x \sin 5x\, dx &= \int_1^{-1} u^4 \cdot \frac{-1}{5}\, du \\ &= \frac{-1}{5} \cdot \frac{1}{5} u^5 \Big|_1^{-1} \\ &= \frac{1}{25} - \left(\frac{-1}{25}\right) = \frac{2}{25}. \end{aligned}$$

---

[72]Some authors do use the second method in their texts because it combines previous skills without introducing the new concept of changing the endpoints into u-values. It does, however, mean a bit more back and forth in the flow of the logic, and slightly clumsier bookkeeping. For more complicated integrals, the method here is much more compact.

It is certainly valuable to the student to realize that there are often multiple ways of solving the same calculus problem, as we saw with derivatives as well (e.g., rearranging terms to use one differentiation rule instead of another).

5.8. SUBSTITUTION WITH THE POWER RULE

The integral in u, namely $\int_1^{-1} u^4 \cdot \frac{-1}{5}\, du$, is interesting in that the interval $[-1,1]$ is traversed backwards, but there is no real theoretical problem with that, as we discussed in Section 5.4 (see Theorem 5.4.3, page 559). The next example offers some further useful insights.

Example 5.8.17

Compute $\displaystyle\int_{-2}^{2} x(x^2+3)^3\, dx$.

Solution: Here we see that if $u = x^2 + 3$, then $du = 2x\, dx$ and this is a constant multiple of $x\, dx$:

$$\begin{array}{rcl} u & = & x^2+3 \\ \Longrightarrow du & = & 2x\, dx \\ \Longrightarrow \tfrac{1}{2} du & = & x\, dx \\ \hline x=2 \Longrightarrow u & = & 7 \\ x=-2 \Longrightarrow u & = & 7 \end{array} \qquad \int_{-2}^{2} x(x^2+3)\, dx = \frac{1}{2}\int_{7}^{7} u^3\, du = \frac{1}{8}u^4 \Big|_{7}^{7} = \frac{7^4}{8} - \frac{7^4}{8} = 0.$$

(See notes that follow.)

For the example above, there were several opportunities to note that the integral's value would be zero. We could have noted that the integrand is an odd function ($f(-x) = -f(x)$) and the endpoints are "symmetric" in the sense that the interval of integration is of the form $[-a, a]$. From this and the integrand's continuity, we can deduce that the integral's value will be zero. (This was discussed in the development of Theorem 5.4.5, page 563.)

It is also interesting to note that the integral in u has only a single point $u = 7$ in its "interval" of integration, and so it is of the form $\int_a^a g(u)\, du = 0$.

If we notice neither of these things, we will still arrive at zero for the integral's value, either as an integral in u or an integral in x, though as mentioned before it is a bit more awkward to find the antiderivative in x before using the FTC:

$$\int_{-2}^{2} x(x^2+3)\, dx = \frac{1}{8}(x^2+3)^4 \Big|_{-2}^{2} = \frac{1}{8}(7)^4 - \frac{1}{8}(7)^4 = 0.$$

Next we consider an applied problem that utilizes this integration method.

Example 5.8.18

Consider the trough in Example 5.6.7, page 608, but filled with water, sealed and turned upside-down. What is the total pressure on one of the parabolic walls?

Solution: As usual, we coordinatize, though this time we draw only the wall. We will put the "base" of this wall along the x-axis, centered at the origin $(0,0)$. In this case, we have a parabola of the form $y = ax^2 + 2$, and since the parabola passes through the point $(2,0)$, we have $0 = a(2)^2 + 2 = 4a + 2$, so $4a = -2$ and $a = -1/2$. The equation of the parabola is thus $y = -\frac{1}{2}x^2 + 2$.

The length of an area element at height y is given by $x_R - x_L$, where we find x_R, x_L by solving the equation of our parabola for x in terms of y:

$$y = -\frac{1}{2}x^2 + 2 \iff y - 2 = -\frac{1}{2}x^2 \iff x^2 = -2(y-2) \iff x = \pm\sqrt{4-2y}.$$

Clearly x_R will be the "+" case, and x_L the "−" case. We get $x_R - x_L = \sqrt{4-2y} - [-\sqrt{4-2y}] = 2\sqrt{4-2y}$. The depth of the area element will be $2 - y$. Summarizing this information, we have, for all $y \in [0,2]$, the following:

length $= x_R - x_L = 2\sqrt{4-2y}$,
width $= dy$,
$dA = (x_R - x_L)dy = 2\sqrt{4-2y}\,dy$,
depth $= 2 - y$,
$dF = $ (pressure)$dA = 62$(depth)(area)
$\therefore dF = 62(2-y) \cdot 2\sqrt{4-2y}\,dy$.

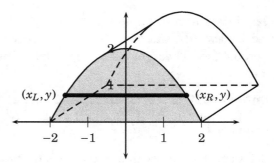

We next integrate dF for the whole length $0 \leq y \leq 2$ to find the total force the fluid exerts on the wall.

$$F = \int_{y=0}^{y=2} dF = \int_0^2 \left(62(2-y) \cdot 2\sqrt{4-2y}\right) dy.$$

This integral fits nicely into our technique here, where we will let $u = 4 - 2y$, and then the radical term is simply $\sqrt{u} = u^{1/2}$. We will have to adjust the other terms, but they will be sums of powers of u, except for dy, which will be a constant multiple of du.

$$\left[\begin{array}{l} \quad\quad u = 4 - 2y \\ \implies\quad du = -2\,dy \\ \iff \dfrac{-1}{2}du = dy. \\ \hline \quad\quad u = 4 - 2y \\ \implies\quad 2y = 4 - u \\ \implies\quad y = 2 - \dfrac{1}{2}u. \\ \hline y = 2 \implies u = 0, \\ y = 0 \implies u = 4. \end{array}\right]\quad \begin{aligned} F &= 62\int_0^2 2(2-y)\sqrt{4-2y}\,dy \\ &= 62\int_4^0 2\left(2 - \left(2 - \tfrac{1}{2}u\right)\right)\sqrt{u} \cdot \tfrac{-1}{2}du \\ &= 62\int_4^0 2(1/2)u \cdot u^{1/2} \cdot \tfrac{-1}{2}du \\ &= 62\int_4^0 (-1/2)u^{3/2}du = 62 \cdot \tfrac{-1}{2} \cdot \tfrac{2}{5}u^{5/2}\Big|_4^0 \\ &= -\dfrac{62 \cdot 0^{5/2}}{5} + \dfrac{62 \cdot 4^{5/2}}{5} = \dfrac{62 \cdot 32}{5} = 396.8, \end{aligned}$$

with units of pounds.

Compare this example with Example 5.6.7, page 608, in which the trough is turned vertically 180° from its position here, so that the parabolas open vertically, and the total force on that wall is approximately 265 lb. There the longer area elements dA are at shallower depths, while here the deeper elements are also the longer ones, resulting in a higher total force on the walls here.

Exercises

Compute the given indefinite integrals.

1. $\displaystyle\int \sin^5 x \cos x \, dx$

2. $\displaystyle\int \cos^4 x \sin x \, dx$

3. $\displaystyle\int_0^{\frac{1}{2}\sqrt{\pi}} x \sin^3 x^2 \cos x^2 \, dx$

4. $\displaystyle\int_0^{\pi/2} \cos^3 x \sin x \, dx$

5. $\displaystyle\int \cos^2 2x \sin 2x \, dx$

6. $\displaystyle\int (1-\cos 5x)^2 \sin 5x \, dx$

7. $\displaystyle\int (8-7x)^5 \, dx$

8. $\displaystyle\int_5^{13} \sqrt{3x-2} \, dx$

9. $\displaystyle\int_5^6 \frac{1}{9-2x} \, dx$

10. $\displaystyle\int_{-1}^1 \frac{3^x}{3^x+2} \, dx$

11. $\displaystyle\int_{-1}^1 x^3(x^4+10)^2 \, dx$

12. $\displaystyle\int [(2x+1)(x^2+x+9)^3] \, dx$

13. $\displaystyle\int \frac{\sec^2 x \, dx}{\sqrt{\tan x}}$

14. $\displaystyle\int_1^e \frac{\ln x}{x} \, dx$

15. $\displaystyle\int \frac{1}{x(\ln x)^2} \, dx$

16. $\displaystyle\int \frac{x}{x^2+4} \, dx$

17. $\displaystyle\int \frac{(\tan^{-1} x)^2}{1+x^2} \, dx$

18. $\displaystyle\int \cot x \, dx$

19. $\displaystyle\int_0^{\pi/4} \sec^2 x \tan^5 x \, dx$

20. $\displaystyle\int \csc^2 2x \cdot \sqrt{1-\cot 2x} \, dx$

21. $\displaystyle\int \frac{1}{\sqrt{x}} \cdot (5+\sqrt{x})^{10} \, dx$

22. $\displaystyle\int \frac{\cos^{-1} x}{\sqrt{1-x^2}} \, dx$

23. $\displaystyle\int_0^1 \frac{3x^2+5}{2x^3+10x+9} \, dx$

24. $\displaystyle\int e^x \tan e^x \sec^2 e^x \, dx$

25. $\displaystyle\int \frac{x^3}{\sqrt{x^4+1}} \, dx$

26. $\displaystyle\int \left(\frac{3-2x}{5}\right)^6 dx$

27. Revisiting Example 5.8.5, page 634, find the antiderivative by instead using $u = \sqrt{x^2-9}$. (Hint: du will itself require a Chain Rule, and contain more terms than usual.)

28. Perform the following computation without rearrangement for both cases.

    (a) $\displaystyle\int \sec x \cdot \sec x \tan x \, dx$

    (b) $\displaystyle\int \tan x \sec^2 x \, dx$

    (c) Explain why, though the answers "look" different, in fact they are the same.

29. Without rearrangement, compute $\displaystyle\int \frac{\sec x \tan x}{\sec x} \, dx$.

30. Perform the previous computation after first simplifying the fraction (perhaps into sines and cosines). Explain why the answers are the same.

For Exercises 31–37, use the methods of Subsection 5.8.2 to compute the integral.

31. $\int x\sqrt{1-x}\,dx$

32. $\int x(x+5)^5\,dx$

33. $\int \dfrac{x}{(x+1)^2}\,dx$

34. $\int_{-2}^{2} x\sqrt{2-x}\,dx$

35. $\int_{1}^{3} \dfrac{x}{\sqrt{3x+1}}\,dx$

36. $\int \dfrac{x}{\sqrt{2x+1}}\,dx$, with $u = 2x+1$

37. $\int \dfrac{x}{\sqrt{2x+1}}\,dx$, with $u = \sqrt{2x+1}$

38. Find the volume of the solid of revolution generated if the area bounded by $y = \sqrt{x^2+1}$ and the x-axis, for $x \in [0,1]$, is revolved around the y-axis.

39. Find the volume of a sphere of radius 5 by taking the semicircle $x^2 + y^2 = 25$, $x \geq 0$, and revolving it around the y-axis, using the Cylindrical Shell Method.

40. Find the volume of the solid of revolution generated if the area bounded by $y = (3x^2 + 2)^2$ and the x-axis, for $x \in [0,1]$, is revolved around the y-axis. Use the Shell Method and integration by substitution.

41. Find the arc length of the graph of $f(x) = -\ln(\cos x)$ for $x \in [0, \pi/3]$. (You will eventually need an integral such as that found in Section 5.1, Exercise 5, page 523.)

5.9 Second Trigonometric Rules

We first looked at the simplest trigonometric integration rules—those arising from the derivatives of the trigonometric functions—in Section 5.1 (Subsection 5.1.4, specifically page 519). Here we will complete the trigonometric rules in which one of the six basic trigonometric functions is the "outer" function. In fact, we have four of the six antiderivatives we need: Sine and cosine come quickly from the derivative formulas, and tangent and cotangent come from substitution arguments. As it turns out, secant and cosecant require a little more cleverness, and while we will not derive these from first principles, we will show that checking them is a quick and interesting derivative computation. Unfortunately (or fortunately, depending on one's perspective) there are variations of the antiderivatives of tangent, cotangent, secant, and cosecant. We will choose one form for each, but the well-informed student must be aware of the others to be prepared to discuss calculus topics among students and professionals with different backgrounds.[73]

5.9.1 Antiderivatives of the Six Trigonometric Functions

The antiderivatives of the six basic trigonometric functions are as follows:

$$\int \sin x \, dx = -\cos x + C, \tag{5.114}$$

$$\int \cos x \, dx = \sin x + C, \tag{5.115}$$

$$\int \tan x \, dx = \ln|\sec x| + C, \tag{5.116}$$

$$\int \cot x \, dx = -\ln|\csc x| + C, \tag{5.117}$$

$$\int \sec x \, dx = \ln|\sec x + \tan x| + C, \tag{5.118}$$

$$\int \csc x \, dx = -\ln|\csc x + \cot x| + C. \tag{5.119}$$

The first four of these can be verified mentally through quick derivative computations by anyone well enough versed in differentiation. The last two require more care, but are some-

---

[73]In fact, there are no strongly compelling reasons not to use

$$\int \tan x \, dx = -\ln|\cos x| + C, \tag{5.120}$$

$$\int \cot x \, dx = \ln|\sin x| + C, \tag{5.121}$$

which after all have slightly simpler verifications by differentiation than (5.116) and (5.117). Here we have opted to use the formulas (5.116) and (5.117) involving the secant and cosecant for a few reasons. First, they are themselves quite popular. Second, the reader used to (5.116) and (5.117) will be less likely to be confused when presented the simpler alternatives by a colleague (or future professor) with a different background, while the reader used to those simpler alternatives may have some initial difficulty if similarly presented our forms here. Finally, there is so much added structure, both calculus and algebraic, found in the context of the secant and cosecant functions that it is important to be familiar and comfortable with them.

Admittedly, however, if (5.116) and (5.117) were not so common, we would likely opt for the simpler forms (5.120) and (5.121).

what interesting to check. For instance, we can verify (5.118) with a standard derivative computation as follows:

$$\frac{d \ln|\sec x + \tan x|}{dx} = \frac{1}{\sec x + \tan x} \cdot \frac{d(\sec x + \tan x)}{dx} = \frac{1}{\sec x + \tan x} \cdot (\sec x \tan x + \sec^2 x)$$

$$= \frac{\sec^2 x + \sec x \tan x}{\sec x + \tan x} = \frac{\sec x(\sec x + \tan x)}{\sec x + \tan x} = \sec x, \text{ q.e.d.}$$

Unlike the antiderivatives of the tangent and cotangent functions, it is not at all clear how one would derive antiderivatives of the secant and cosecant functions, and so it is important to memorize those especially. Indeed, it is likely these were discovered through experimentation, and such results are often very time consuming to reproduce from first principles if one has to reinvent "the tricks," one of which will be explored in the exercises. In fact, we will later show a popular alternative antiderivative for the cosecant, and a not-so-popular alternative for the secant. The alternatives for the tangent and cotangent are similar in popularity to those we will use for our standards.

There is little we can do with just (5.114)–(5.119) as they stand, but we nonetheless explore a few examples quickly.

Example 5.9.1 _____

Next are two brief antiderivative computations involving our basic trigonometric integral formulas.

- $\displaystyle\int \frac{\sin^2 x + \cos x}{\sin x}\, dx = \int (\sin x + \cot x)\, dx = -\cos x - \ln|\csc x| + C.$

- $\displaystyle\int (x + \sec x)\, dx = \frac{x^2}{2} + \ln|\sec x + \tan x| + C.$

_____

Example 5.9.2 _____

Suppose $v(t) = 1 + \tan t$, and $s(\pi/3) = 7$. Find $s(t)$ and the range of t for which the solution is valid.

Solution: We know that $s(t)$ is an antiderivative of $v(t)$, so we write the following, realizing that we will use our one datum ($s(\pi/3) = 7$) to find the additive constant later.

$$s(t) = \int v(t)\, dt = \int (1 + \tan t)\, dt = t + \ln|\sec t| + C.$$

So far, we know that $s(t) = t + \ln|\sec t| + C$, and that $s(\pi/3) = 7$, so

$$7 = \frac{\pi}{3} + \ln\left|\sec \frac{\pi}{3}\right| + C \iff 7 = \frac{\pi}{3} + \ln 2 + C \iff 7 = \frac{\pi}{3} + \ln 2 + C,$$

and so $C = 7 - \frac{\pi}{3} - \ln 2$. With this, we can write

$$s(t) = t + \ln|\sec t| + 7 - \frac{\pi}{3} - \ln 2.$$

5.9. SECOND TRIGONOMETRIC RULES

Now $\sec t$ is continuous on the open intervals with endpoints which are odd multiples of $\pi/2$ (i.e., angles in between those which form a reference angle of $\pi/2$ from the horizontal axis). Since $\pi/3 \in (-\pi/2, \pi/2)$, over which $\sec t > 0$ and is continuous, we can in fact conclude

$$s(t) = t + \ln \sec t + 7 - \frac{\pi}{3} - \ln 2, \qquad t \in (-\pi/2, \pi/2).$$

5.9.2 Substitution and the Basic Trigonometric Functions, Part I

A student who uses calculus extensively is likely to eventually encounter an antiderivative problem where the form is particularly difficult or obscure, in which case it is common to refer to tables of integrals. These usually contain all the basic forms as well as those that would be difficult enough to warrant a search through such a reference.[74] It is interesting to note that most modern tables of integrals do not use the common variable x in the formulas, but instead use u, which is the most common variable for substitution-type problems. This is because substitution is so ubiquitous that it is assumed the reader might not need a form exactly as it is in the table, but rather needs one that becomes one of the forms (or a constant multiple of one of the forms) found in the table only after a substitution. In that spirit, the standard method of listing the antiderivatives of the basic six trigonometric functions is as follows:

$$\int \sin u \, du = -\cos u + C, \tag{5.122}$$

$$\int \cos u \, du = \sin u + C, \tag{5.123}$$

$$\int \tan u \, du = \ln|\sec u| + C, \tag{5.124}$$

$$\int \cot u \, du = -\ln|\csc u| + C, \tag{5.125}$$

$$\int \sec u \, du = \ln|\sec u + \tan u| + C, \tag{5.126}$$

$$\int \csc u \, du = -\ln|\csc u + \cot u| + C. \tag{5.127}$$

Of course, these are just our previous formulas (5.114)–(5.119) page 645, but with the entire integral written in the variable u instead of x. However, each of these properly interpreted

---

[74] In a typical calculus class, the professor usually has to answer the question of why students have to learn all of the difficult integration techniques when there are references available. The answer is several-fold, and we make just a few points addressing it here.

First, many problems are simple enough not to require reference, and using tables for every problem becomes akin to looking up every word in a foreign language dictionary as one reads a newspaper in that language, for instance, rather than learning the language first: Not only is it inefficient and time-consuming, but much of the spirit of the writing is probably lost. Second, it is not even possible to use tables for every integration problem simply because many problems that can be handled through the techniques will not closely match formulas in the tables. Third, on occasion there will be a technicality that the editors of the tables did not anticipate for a particular problem, or a mistake they did not catch, and so reliance on tables can be problematic. (On this last point, the same is true of some mathematical software packages that claim to compute integrals.)

contains a reverse Chain Rule, also known as a substitution-type form. So for instance, if $u = u(x)$, then $du = u'(x)dx$ and so we can read (5.124), i.e., $\int \tan u \, du = \ln|\sec u| + C$, as

$$\int \underbrace{\tan u(x)}_{\tan u} \underbrace{u'(x) dx}_{du} = \ln|\sec u(x)| + C,$$

verified by differentiation:

$$\frac{d}{dx} \ln|\sec u(x)| = \frac{1}{\sec u(x)} \cdot \frac{d \sec u(x)}{dx} = \frac{1}{\sec u(x)} \cdot \sec u(x) \tan u(x) \cdot \frac{d u(x)}{dx} = \tan u(x) \cdot u'(x),$$

q.e.d. So forms (5.122)–(5.127) are all forms in which a basic trigonometric function of some function $u(x)$, and the derivative $u'(x)$ and the differential dx are the nonconstant factors of the integral. We now look at several examples.

Example 5.9.3 _____

Compute $\int x \sin x^2 \, dx$.

Solution: As often occurs, the form is not exact but a constant multiple of one of our forms, this time (5.122), and furthermore the order of the factors is changed. Here we see that the factor x is a constant multiple of $u'(x)$ if $u(x) = x^2$, so the extra factor of x can be "absorbed" into the differential du after the substitution. This ultimately leaves us with the problem of finding the antiderivative of a sine function.

$$\left[\begin{array}{c} u = x^2 \\ \Longrightarrow \quad du = 2x\,dx \\ \Longleftrightarrow \quad \frac{1}{2} du = x\,dx \end{array} \right] \qquad \int x \sin x^2 \, dx = \int \sin u \cdot \frac{1}{2} du = -\frac{1}{2} \cos u + C = -\frac{1}{2} \cos x^2 + C.$$

This example—and most examples in this section—can be quickly checked by differentiation.

Example 5.9.4 _____

Compute $\int e^x \cot e^x \, dx$.

Solution: Here we see the derivative of the argument e^x of the cotangent function is also present as a multiplicative factor.

$$\left[\begin{array}{c} u = e^x \\ \Longrightarrow du = e^x dx \end{array} \right] \qquad \int e^x \cot e^x \, dx = \int \cot u \, du = -\ln|\csc u| + C = -\ln\left|\csc e^x\right| + C.$$

5.9. SECOND TRIGONOMETRIC RULES

Example 5.9.5

Compute $\int \dfrac{\sec\sqrt{x}}{\sqrt{x}}\,dx$.

<u>Solution</u>: Here the factor $\frac{1}{\sqrt{x}}$ is in fact a constant multiple of the derivative of $\sqrt{x}$, the argument of the trigonometric function. Thus we take $u = \sqrt{x}$, and then the resulting du will absorb the $\frac{1}{\sqrt{x}}$ term:

$$\left[\begin{array}{l} u = \sqrt{x} \\ \Longrightarrow\ du = \dfrac{1}{2\sqrt{x}}\,dx \\ \Longleftrightarrow\ 2\,du = \dfrac{1}{\sqrt{x}}\,dx \end{array}\right] \qquad \begin{aligned} \int \dfrac{\sec\sqrt{x}}{\sqrt{x}}\,dx &= \int \sec\sqrt{x}\cdot \dfrac{1}{\sqrt{x}}\,dx \\ &= \int \sec u \cdot 2\,du \\ &= 2\ln|\sec u + \tan u| + C \\ &= 2\ln|\sec\sqrt{x} + \tan\sqrt{x}| + C. \end{aligned}$$

Example 5.9.6

Compute $\int \dfrac{\cos(1+4\ln x)}{x}\,dx$.

<u>Solution</u>: Here we see the derivative of $(1+4\ln x)$ appearing as a factor as well, except for a constant factor.

$$\left[\begin{array}{l} u = 1 + 4\ln x \\ \Longrightarrow\ du = 4\cdot \dfrac{1}{x}\,dx \\ \Longleftrightarrow\ \dfrac{1}{4}\cdot du = \dfrac{1}{x}\,dx \end{array}\right] \qquad \begin{aligned} \int \dfrac{\cos(1+4\ln x)}{x}\,dx &= \int \cos(1+4\ln x)\cdot \dfrac{1}{x}\,dx \\ &= \int \cos u \cdot \dfrac{1}{4}\,du \\ &= \dfrac{1}{4}\sin u + C \\ &= \dfrac{1}{4}\sin(1+4\ln x) + C. \end{aligned}$$

These can become more complicated. One may find oneself simplifying an integral through a substitution, only to be presented with a new integral still requiring further substitution, as demonstrated next.

Example 5.9.7

Compute $\int x^2 \csc(\cos x^3)\sin x^3\,dx$.

<u>Solution</u>: To be clear, first we note that the integrand is the product of three factors:

$$x^2 \cdot \csc(\cos x^3)\cdot \sin x^3,$$

so the argument of the cosecant is $\cos x^3$. Now we will compute this two different ways. The first method requires two substitutions, which is an option that students must be aware is legitimate, assuming all computations are made carefully and consistently.

Method 1. Here we will first make a substitution $u = x^3$ to yield a simpler integral without any polynomial factors, though our new integral will still require some work.

$$\left[\begin{array}{r} u = x^3 \\ \Longrightarrow \quad du = 3x^2\,dx \\ \Longleftrightarrow \quad \frac{1}{3}du = x^2\,dx \end{array}\right] \qquad \int x^2 \csc\left(\cos x^3\right) \sin x^3\,dx = \int \csc(\cos u) \sin u \cdot \frac{1}{3}\,du$$

So at this point our problem reduces to computing $\int \csc(\cos u)\sin u \cdot \frac{1}{3}\,du$. To do so we use another substitution, noting that $\sin u$ is the derivative—up to a multiplicative constant—of $\cos u$ (with respect to u this time). To remain consistent, this second substitution must use a new variable (lest we give one letter two different meanings within the same problem, which would be contradictory!). So we call our new variable something other than u or x. A commonly used variable at this stage is w:

$$\left[\begin{array}{r} w = \cos u \\ \Longrightarrow \quad dw = -\sin u\,du \\ \Longleftrightarrow \quad (-1)\,dw = \sin u\,du \end{array}\right]$$

$$\begin{aligned}
\frac{1}{3}\int \csc(\cos u)\sin u\,du &= \frac{1}{3}\int \csc w\,(-1)\,dw \\
&= -\frac{1}{3}\left[-\ln|\csc w + \cot w|\right] + C \\
&= \frac{1}{3}\ln|\csc(\cos u) + \cot(\cos u)| + C \\
&= \frac{1}{3}\ln\left|\csc\left(\cos x^3\right) + \cot\left(\cos x^3\right)\right| + C.
\end{aligned}$$

Note how we computed the antiderivative in w, which we then replaced by its expression in u, and finally by the definition of u in terms of x.

Method 2. If we can see far enough ahead, we can combine both substitutions into one. For clarity we will use a different variable—namely z—here (though by convention one would usually use u). We choose $z = \cos x^3$, noting that its derivative, requiring the Chain Rule, will have a $\sin x^3$ and an x^2 term (ignoring multiplicative constants), which leaves us with a constant multiple of $\int \csc z\,dz$, for which we have a formula.

$$\left[\begin{array}{r} z = \cos x^3 \\ \Longrightarrow \quad dz = -\sin x^3 \cdot 3x^2\,dx \\ \Longleftrightarrow \quad -\frac{1}{3}dz = x^2 \sin x^3\,dx \end{array}\right] \qquad \begin{aligned} \int x^2 \csc\left(\cos x^3\right) \sin x^3\,dx &= \int \csc z \cdot \frac{-1}{3}\,dz \\ &= -\frac{1}{3}\cdot\left[-\ln|\csc z + \cot z|\right] + C \\ &= \frac{1}{3}\ln\left|\csc\left(\cos x^3\right) + \cot\left(\cos x^3\right)\right| + C. \end{aligned}$$

The second method in fact simply combines the two substitutions from the first method into one. Indeed, the formula for w in terms of x is the same as that of z:[75]

$$w = \csc(u) = \csc\left(\cos x^3\right) = z.$$

---

[75]Similarly, when we recall these variables are all functions of x, we also have $dw = dz$:

$$dw = \frac{dw}{du}\cdot\frac{du}{dx}\cdot dx = (-\sin u)\cdot\left(3x^2\right)dx = -\sin x^3 \cdot 3x^2\,dx = dz.$$

Again we see the power of the Leibniz notation in what is essentially a Chain Rule. Of course we should expect that $w = z \Longrightarrow dw = dz$. But also we see that while there are obvious algebraic consistencies in our substitution method, there are also consequent calculus consistencies which, while more subtle, are still correct when we perform all the computations correctly.

5.9. SECOND TRIGONOMETRIC RULES

When one is well practiced in substitution, the second method will likely be chosen. However, it is important also for the student to realize that even if a substitution does not achieve an integral that can be immediately computed, that does not mean that the particular substitution need be abandoned. If the new integral is simpler, then the first substitution can be worthwhile. In fact, in the sequel text (which concerns itself with second-semester calculus) we will have integration methods which will often *require* multiple substitutions. Of course, it is always important that all steps be carried out carefully, accurately, and consistently.

In this section we concentrated on those integrals that reduce to integrals of a single trigonometric function, perhaps with the aid of a substitution. In the sequel we will look at the many techniques for computing those integrals that contain several factors of trigonometric functions, and no other factors. The techniques of our present section will be called upon often, but these are only a small part of the needed knowledge for computing complicated "trigonometric integrals" in the sequel. But we have some other techniques already. For instance, there were the first trigonometric integral formulas we had in Section 5.1, Subsection 5.1.4, which arose from the derivative rules for the six basic trigonometric functions. We had another technique for dealing with some trigonometric integrals, which was substitution in the case where we could rewrite the trigonometric integral as a Power Rule-type integral.[76]

We would be remiss if we did not also remind the reader that there are substitution-type integrals that are deceptively simple, where we would let $u = kx$ for some constant $k \in \mathbb{R}$.

Example 5.9.8

Compute $\int \csc 5x$.

Solution: Here we let $u = 5x$, so that $du = 5\,dx$, and therefore $\frac{1}{5}du = dx$. One would thus work such a problem as follows:

$$\left[\begin{array}{r} u = 5x \\ \implies du = 5\,dx \\ \iff \frac{1}{5}du = dx \end{array}\right] \qquad \begin{aligned} \int \csc 5x\,dx &= \int \csc u \cdot \frac{1}{5}du \\ &= -\frac{1}{5}\ln|\csc u + \cot u| + C \\ &= -\frac{1}{5}\ln|\csc 5x + \cot 5x| + C. \end{aligned}$$

Such examples are deceptively simple because the derivative of kx does not explicitly appear, even accounting for extra multiplicative constants, though one should always entertain the notion of a "hidden factor of 1" in any expression.[77]

One must additionally always be aware that the Power Rule may be relevant to a given integral, as in the case that follows.

---

[76] In fact, there will be several other substitution-type arguments we will make for trigonometric integrals besides those that yield Power Rules.

[77] Many textbooks include a constant k in their formulas for antiderivatives of the trigonometric functions. For instance, such texts will write $\int \tan kx\,dx = \frac{1}{k}\ln|\sec kx| + C$.

Example 5.9.9

Compute $\int \dfrac{\sin 3x}{\cos^5 3x}\,dx$.

Solution: Here we see that the derivative of the cosine function is present as a factor, and we are left with a power of the cosine:

$$\begin{bmatrix} u = \cos 3x \\ \implies du = -\sin 3x \cdot 3\,dx \\ \iff \dfrac{-1}{3}du = \sin 3x\,dx \end{bmatrix}$$

$$\int \dfrac{\sin 3x}{\cos^5 3x}\,dx = \int (\cos 3x)^{-5} \sin 3x\,dx = \int u^{-5} \cdot \dfrac{-1}{3}\,du$$

$$= -\dfrac{1}{3} \cdot \dfrac{-1}{4} u^{-4} + C$$

$$= \dfrac{1}{12} \cos^{-4} 3x + C$$

$$= \dfrac{1}{12} \sec^4 3x + C.$$

The integration techniques encountered here, and especially in the sequel, are many and varied. We will see later how a slight change in an integrand can substantially change the antiderivative, its difficulty in computing it, or the technique used to achieve it. We have seen this phenomenon before. Consider for instance the following three groups of integrals:

$$\int \dfrac{x}{x^2+1}\,dx = \dfrac{1}{2}\ln(x^2+1) + C, \qquad \int \dfrac{x}{\sqrt{1-x^2}}\,dx = -\sqrt{1-x^2} + C,$$

$$\int \dfrac{1}{x^2+1}\,dx = \tan^{-1} x + C. \qquad \int \dfrac{1}{\sqrt{1-x^2}}\,dx = \sin^{-1} x + C.$$

$$\int \sec x\,dx = \ln|\sec x + \tan x| + C,$$

$$\int \sec^2 x\,dx = \tan x + C,$$

$$\int \sec^3 x\,dx = \dfrac{1}{2}(\sec x \tan x + \ln|\sec x + \tan x|) + C.$$

In fact, this last integral will have to wait until techniques from the sequel, and is quite long and technical. Even the verification by differentiation is nontrivial, and requires one to employ a trigonometric identity along the way. Suffice for now to simply note that the techniques and results, for even these first three powers of the secant, are all very different.

Computing antiderivatives in good time requires the ability to think ahead, to be able to recognize which technique or substitution will likely solve the particular problem. That in turn requires a fairly complete knowledge of the techniques, even to the extent that one can anticipate the outcomes of several later steps. Of course, practice is one key to gaining this understanding.

5.9. SECOND TRIGONOMETRIC RULES

Exercises

For Exercises 1–6, verify by differentiation each of our basic six trigonometric integrals in the forms we use, (5.114)–(5.119). For reference see the proof for the secant, page 646.

1. $\int \sin x \, dx = -\cos x + C$

2. $\int \cos x \, dx = \sin x + C$

3. $\int \tan x \, dx = \ln|\sec x| + C$

4. $\int \cot x \, dx = -\ln|\csc x| + C$

5. $\int \sec x \, dx = \ln|\sec x + \tan x| + C$

6. $\int \csc x \, dx = -\ln|\csc x + \cot x| + C$

For Exercises 7–33, compute the integral.

7. $\int x^2 \cos x^3 \, dx$

8. $\int x^3 \sin x^4 \, dx$

9. $\int x^4 \tan x^5 \, dx$

10. $\int x^6 \csc x^7 \, dx$

11. $\int_0^\pi \sin 4x \, dx$

12. $\int_0^\pi \cos 5x \, dx$

13. $\int \sec 3x \, dx$

14. $\int \tan 5x \, dx$

15. $\int \csc \frac{x}{2} \, dx$

16. $\int \sin \frac{x}{4} \, dx$

17. $\int \sin(\omega t) \, dt$ \quad (ω a constant)

18. $\int \cos[\omega(t - \phi)] \, dt$ \quad (ω, ϕ constants)

19. $\int x \sec(x^2 + 1) \, dx$

20. $\int_1^e \frac{\tan(\ln x)}{x} \, dx$

21. $\int_{2/\pi}^{3/\pi} \frac{\sin\left(\frac{1}{x}\right)}{x^2} \, dx$

22. $\int_0^{\frac{1}{2}\sqrt{\pi}} x \cos x^2 \, dx$

23. $\int_{\ln\sqrt{\pi/4}}^{\ln\sqrt{\pi/3}} e^{2x} \cot\left(e^{2x}\right) dx$

24. $\int \sqrt{x} \csc\left(x\sqrt{x}\right) dx$

25. $\int x^3 \cot\left(6e^{5x^4}\right) e^{5x^4} \, dx$

26. $\int x^2 \sec(\cos x^3) \sin x^3 \, dx$

27. $\int \frac{x}{\sin 3x^2} \, dx$

28. $\int 2^x \cos(3 \cdot 2^x) \, dx$

29. $\int (1 - \sec x)^2 \, dx$

30. $\int (\tan 7x - 2)^2 \, dx$

31. $\int \frac{1 - \sin^2 x}{\cos x} \, dx$

32. $\int \frac{\sin^2 x}{\cos x} \, dx$ (see previous problem)

33. $\int \dfrac{(1-\cos x)^2}{\sin x}\,dx$ (multiply, and see previous problems)

34. Recall that the period of the sine function $y = A\sin Bx$ is $2\pi/B$, where the amplitude is $|A|$. Find the total area for one period of a sine graph with amplitude 1 and period $2\pi/B$, i.e., compute
$$\int_0^{2\pi/B} |\sin Bx|\,dx.$$
Hint: The function is positive for the first half-period.

35. Using either the result above or its method, find the area of one period of a sine graph if the period is P.

36. Derive our formula for the integral of $\sec x$ by the following algebraic device, namely multiplying and dividing by $\sec x + \tan x$ within the integral, i.e.,
$$\int \sec x\,dx = \int \sec x \cdot \dfrac{\sec x + \tan x}{\sec x + \tan x}\,dx,$$
multiplying numerators, and then using an appropriate substitution argument.

37. An object is moving in uniform circular motion in the xy-plane along a path centered at $(0,0)$ with radius r. Suppose that the horizontal position is given by $x(t) = r\cos(2\pi f t)$ where f is the frequency (number of revolutions traveled per second) and t is measured in seconds.

    (a) Determine the horizontal velocity dx/dt of the object.

    (b) Determine the horizontal distance D traveled on the time interval $t \in [1/8, 1/4]$ (in seconds) when $f = 1$ rev/sec, $r = 5$ m.

    (c) Do the same with $f = 0.5$ rev/sec, $r = 6$ m, over the interval $0 \le t \le 1$ sec.

    (d) Explain why the previous answer should be the same as the diameter of the circle.

By differentiation, verify the following alternative integration formulas for Exercises 38–41.

38. $\int \tan x\,dx = -\ln|\cos x| + C.$

39. $\int \cot x\,dx = \ln|\sin x| + C.$

40. $\int \sec x\,dx = -\ln|\sec x - \tan x| + C.$

41. $\int \csc x\,dx = \ln|\csc x - \cot x| + C.$

5.10 Substitution with All Basic Forms

In this section, we will add to our forms for substitution and recall some rather general guidelines for substitution. Except for our four new trigonometric forms from Section 5.9, all forms in this chapter derive directly from derivative rules. These comprise what we call here the *basic* integration rules. Each is based on a single function specific to the rule. So for instance, we will have in our list the following:

$$\int \frac{1}{u^2+1} \, du = \tan^{-1} u + C.$$

As before, the usual variable of integration in the given problem will likely be x, but the *form* may be ultimately as before, except for multiplicative constants, where $u = u(x)$ and then $du = u'(x) \, dx$ contains another factor from the original integral. So for instance we might see

$$\begin{bmatrix} u = x^2 \\ \implies du = 2x \, dx \\ \iff \frac{1}{2} du = x \, dx \end{bmatrix} \quad \int \frac{x}{x^4+1} \, dx = \int \frac{1}{(x^2)^2 + 1} \cdot x \, dx = \int \frac{1}{u^2+1} \cdot \frac{1}{2} du$$

$$= \frac{1}{2} \tan^{-1} u + C = \frac{1}{2} \tan^{-1} x^2 + C.$$

One clue that we might try $u = x^2$ was that its derivative was a factor in the integrand in the form of the factor x, again excepting multiplicative constants, and so we wrote that factor separately next to the differential dx. As it turned out, the rest of the integrand could indeed be written as a function of $u = x^2$.

Reading the prevous problem backwards, the arctangent is the "outer function" of a Chain Rule differentiation problem, and x^2 was the "inner function." Put in terms of integration, the *form* was $\int \frac{1}{u^2+1} \, du$, excepting multiplicative constants, with the "inner function" $u = x^2$. The arctangent appeared because of the ultimate form of the integral, in terms of $u = x^2$.

But note that the arctangent can also appear as the "inner function," which we may wish to set equal to u. So for instance,

$$\begin{bmatrix} u = \tan^{-1} x \\ \implies du = \frac{1}{x^2+1} \, dx \end{bmatrix} \quad \int \frac{(\tan^{-1} x)^2}{x^2+1} \, dx = \int (\tan^{-1} x)^2 \cdot \frac{1}{x^2+1} \, dx$$

$$= \int u^2 \, du = \frac{u^3}{3} + C = \frac{(\tan^{-1} x)^3}{3} + C.$$

In this integral, we eventually invoked the Power Rule for integrals:

$$\int u^n \, du = \begin{cases} \frac{1}{n+1} u^{n+1} + C & \text{if } n \neq 1, \\ \ln|u| + C & \text{if } n = -1. \end{cases}$$

When desiring to compute an indefinite integral (i.e., an antiderivative), it is useful to consider many possible forms to which it may relate. For instance, the previous integral $\int \frac{x}{x^4+1} \, dx$ superficially resembles the integral computed next, but the result is very different:

$$\begin{bmatrix} u = x^4 + 1 \\ \implies du = 4x^3 \, dx \\ \iff \frac{1}{4} du = x^3 \, dx \end{bmatrix} \quad \int \frac{x^3}{x^4+1} \, dx = \int \frac{1}{u} \cdot \frac{1}{4} du = \frac{1}{4} \ln|u| + C = \frac{1}{4} \ln(x^4+1) + C.$$

(Note how we do not need absolute values in the final answer because $x^4 + 1 > 0$.)

It is interesting to juxtapose the three integrals with which we began this section:

$$\int \frac{x}{x^4+1} dx = \frac{1}{2} \tan^{-1} x^2 + C,$$

$$\int \frac{x^3}{x^4+1} dx = \frac{1}{4} \ln\left(x^4 + 1\right) + C,$$

$$\int \frac{\left(\tan^{-1} x\right)^2}{x^2+1} dx = \frac{\left(\tan^{-1} x\right)^3}{3} + C.$$

In all of these cases, we are looking for some $u = u(x)$ so that

- one major (nonconstant) factor of the integral can be simply written $f(u)$, i.e., u is an "inner function" of some composite function $f(u(x))$ which appears in the integrand,

- the remaining factors of the integrand will collectively be a constant multiple of $du = u'(x)dx$,

- and finally, so that $\int f(u)du$ is an integral we can compute, i.e., we know the antiderivative of f, say F:

$$\int f(u(x)) \cdot ku'(x)dx = k \int f(u)du = k \cdot F(u) + C = k \cdot F(u(x)) + C.$$

So of course identifying u is the key, and in doing so we have to be sure its derivative is also present, and finally that we are left with an integral—albeit in u—which we can handle.[78]

5.10.1 List of Basic Forms

Here we list all of the basic integral forms we use in this section and beyond. [79]

---

[78] This all assumes that there is a substitution that will make the integral into one of these simple forms. It is not always the case. An important integral that occurs in probability and other subjects is $\int e^{x^2} dx$, which cannot be changed by substitution into a useful form for simple integration. In fact it will not succumb to any of the usual integration methods in this text or the sequel. We will eventually find a way to deal with this integral, in the sequel, be rewriting the function as a Taylor Series, which can be integrated, but then the result is also a Taylor Series, and not a function we can write any other way from the functions we know.

In the meantime it is actually a good exercise to see why this integral cannot be forced into any of our methods. Indeed, seeing what goes wrong in such a case very well complements seeing what goes right in the cases where substitution, and later methods, do achieve an answer.

[79] We exclude in our main list the hyperbolic forms. The student interested in those can review their differentiation rules, which would then give rise to the hyperbolic integration forms. In the sequel we show how any integral giving rise to a purely inverse hyperbolic function of u can also be dealt with using other techniques, in particular "trigonometric substitution," though knowledge of the hyperbolic forms can allow for faster computations in some cases.

Even so, we have to be careful of the values for which we seek the antiderivative, because the inverse hyperbolic functions are sometimes defined by limiting the domains of the original hyperbolic functions, as we do with the trigonometric functions and their inverses. Sometimes that is more easily addressed with the inverse hyperbolic functions, and sometimes it is more easily addressed with the forms we arrive at through the techniques of the sequel.

5.10. SUBSTITUTION WITH ALL BASIC FORMS

$$\int u^n \, du = \frac{u^{n+1}}{n+1} + C, \quad n \neq -1 \quad (5.128)$$

$$\int \frac{1}{u} \, du = \ln|u| + C \quad (5.129)$$

$$\int \sin u \, du = -\cos u + C \quad (5.130)$$

$$\int \cos u \, du = \sin u + C \quad (5.131)$$

$$\int \tan u \, du = \ln|\sec u| + C \quad (5.132)$$

$$\int \cot u \, du = -\ln|\csc u| + C \quad (5.133)$$

$$\int \sec u \, du = \ln|\sec u + \tan u| + C \quad (5.134)$$

$$\int \csc u \, du = -\ln|\csc u + \cot u| + C \quad (5.135)$$

$$\int e^u \, du = e^u + C \quad (5.136)$$

$$\int a^u \, du = \frac{a^u}{\ln a} + C \quad (5.137)$$

$$\int \sec^2 u \, du = \tan u + C \quad (5.138)$$

$$\int \csc^2 u \, du = -\cot u + C \quad (5.139)$$

$$\int \sec u \tan u \, du = \sec u + C \quad (5.140)$$

$$\int \csc u \cot u \, du = -\csc u + C \quad (5.141)$$

$$\int \frac{1}{\sqrt{1-u^2}} \, du = \sin^{-1} u + C \quad (5.142)$$

$$\int \frac{1}{u^2+1} \, du = \tan^{-1} u + C \quad (5.143)$$

$$\int \frac{1}{u\sqrt{u^2-1}} \, du = \sec^{-1}|u| + C \quad (5.144)$$

It cannot be overly stressed that each form given assumes that a substitution may be required. So again, the following formulas say the same:

$$\int e^u \, du = e^u + C, \qquad \int e^{u(x)} u'(x) \, dx = \int e^{u(x)} \cdot \frac{du(x)}{dx} \, dx = e^{u(x)} + C.$$

This is verified using the Chain Rule:

$$\frac{d \, e^{u(x)}}{dx} = e^{u(x)} \cdot \frac{d \, u(x)}{dx} = e^{u(x)} u'(x).$$

Recognizing when we have a particular form is again key to using these formulas.

Example 5.10.1

Compute $\int x e^{x^2} \, dx$.

Solution: Here we see the derivative of x^2 appearing as the factor x, except for a constant multiple. Hence we let $u = x^2$, and the du will contain the other factor x, leaving an integral in one of our standard forms, namely (5.136), and nothing else except a multiplicative constant.

$$\begin{bmatrix} u = x^2 \\ \implies du = 2x \, dx \\ \iff \frac{1}{2} du = x \, dx \end{bmatrix} \qquad \int x e^{x^2} \, dx = \int e^u \cdot \frac{1}{2} \, du = \frac{1}{2} e^u + C = \frac{1}{2} e^{x^2} + C.$$

As usual, we seek a function u which gives a form $\int k \cdot f(u)\,du = \int k \cdot f(u) \cdot \frac{du}{dx}\,dx$, i.e., so that (1) the derivative of u is a factor in the integrand, and (2) the resulting integral is simple enough that we can (more) easily find the final form $k \cdot F(u) + C$.

Example 5.10.2

Compute $\int \frac{x^3}{\sqrt{1-x^8}}\,dx$.

Solution: At first it is tempting to let $u = 1 - x^8$ and hope this will become a Power Rule, except that such u implies $du = -8x^7$, which is very different from a constant multiple of the other factor here, namely x^3.

In fact, the other factor can be a good source of information about what to set equal to u. Indeed the factor x^3 will be part of the differential of $u = x^4$, and then we can recognize a form that will yield an arcsine ultimately, i.e., form (5.142) from our list.

$$\left[\begin{array}{c} u = x^4 \\ \implies du = 4x^3\,dx \\ \iff \frac{1}{4}du = x^3\,dx \end{array}\right] \quad \int \frac{x^3}{\sqrt{1-x^8}}\,dx = \int \frac{1}{\sqrt{1-(x^4)^2}} \cdot x^3\,dx = \int \frac{1}{\sqrt{1-u^2}} \cdot \frac{1}{4}\,du$$

$$= \frac{1}{4}\sin^{-1} u + C = \frac{1}{4}\sin^{-1}(x^4) + C.$$

As with the Power Rule, there are occasions where the derivative of our u is a nonzero constant, and thus a constant multiple of every other nonzero constant. While these integrals are arguably easier than the others we encounter here, and one might be able to guess their antiderivatives and check with a quick differentiation, their relative simplicity can be a source of confusion.

Example 5.10.3

Compute $\int e^{10x}\,dx$.

Solution: Here we simply let $u = 10x$.

$$\left[\begin{array}{c} u = 10x \\ \implies du = 10\,dx \\ \iff \frac{1}{10}du = dx \end{array}\right] \quad \int e^{10x}\,dx = \int e^u \cdot \frac{1}{10}\,du = \frac{1}{10}e^u + C = \frac{1}{10}e^{10x} + C.$$

Next we return to some examples where a "guess" at a final answer is possible, but it is still recommended that we perform the u-substitution to arrive at the antiderivative with more care.

5.10. SUBSTITUTION WITH ALL BASIC FORMS

Example 5.10.4

Compute $\int 2^{3^x} \cdot 3^x \, dx$.

Solution: Eventually we will need (5.137), page 657, but first we simply notice that the factor 3^x is a constant multiple of the derivative of the exponent in the first factor, so we let $u = 3^x$.

$$\left[\begin{array}{r} u = 3^x \\ \implies du = 3^x \ln 3 \, dx \\ \iff \frac{1}{\ln 3} du = 3^x \, dx \end{array} \right] \quad \int 2^{3^x} \cdot 3^x \, dx = \int 2^u \cdot \frac{1}{\ln 3} du = \frac{1}{\ln 3} \cdot \frac{1}{\ln 2} \cdot 2^u + C = \frac{2^{3^x}}{\ln 3 \cdot \ln 2} + C.$$

This example shows the importance of following the various constant factors through the integration. Students who rely upon guessing the answers, without performing the formal substitution steps, are much more likely to misplace one or more constant factors.

For further practice, we consider more basic examples.

Example 5.10.5

Compute $\int \frac{\tan \sqrt{x}}{\sqrt{x}} \, dx$.

Solution: The key here is that the derivative of the argument of the tangent is also present as a factor. Recall $\frac{d}{dx}\left(\sqrt{x}\right) = \frac{1}{2\sqrt{x}}$.

$$\left[\begin{array}{r} u = \sqrt{x} \\ \implies du = \frac{1}{2\sqrt{x}} dx \\ \iff 2 du = \frac{1}{\sqrt{x}} dx \end{array} \right] \quad \begin{array}{l} \int \frac{\tan \sqrt{x}}{\sqrt{x}} dx = \int \tan \sqrt{x} \cdot \frac{1}{\sqrt{x}} dx = \int \tan u \cdot 2 \, du = 2 \ln |\sec u| + C \\ = 2 \ln |\sec \sqrt{x}| + C. \end{array}$$

In the next section we will explore the arctrigonometric antiderivatives further by considering additional complications. For now we look at two such complications, first involving the arctangent (though the arcsine has a similar potential complication), and then the arcsecant, which has the same complications as the others and then one more.

Example 5.10.6

Compute $\int \frac{x^2}{1 + 25x^6} \, dx$.

Solution: There are two clues directing our choice of u. First, we see the factor x^2, which is a multiple of the derivative of x^3. Then we see that the denominator can be written as $1 + \left(5x^3\right)^2$, which

660   CHAPTER 5. BASIC INTEGRATION AND APPLICATIONS

we can put into the form yielding the arctangent, namely (5.143).

$$\left[\begin{array}{c} u = 5x^3 \\ \implies du = 15x^2\,dx \\ \iff \frac{1}{15}du = x^2\,dx \end{array}\right] \qquad \int \frac{x^2}{1+25x^6}\,dx = \int \frac{1}{1+(5x^3)^2}\cdot x^2\,dx = \int \frac{1}{1+u^2}\cdot \frac{1}{15}\,du$$

$$= \frac{1}{15}\tan^{-1} u + C = \frac{1}{15}\tan^{-1}(5x^3) + C.$$

The complication in this example is fairly benign: that the u term contains a multiplicative constant. Here we wanted $25x^6$ to be u^2 for the form (5.143), so we took $u = 5x^3$. Fortunately, this was consistent with du containing the x^2 term of the integrand, and that form (5.143) could actually be used. In the next section we will see how to deal with cases where the additive constant in the denominator of the integrand is not 1.[80]

Next we look at another complication that is somewhat specific to the arcsecant form.

Example 5.10.7 _____

Compute $\int \frac{1}{x\sqrt{9x^2-1}}\,dx$.

Solution: Because a new complication needs to be explained while the problem is solved, the organization will be slightly different from previous exercises, but every technique used next has appeared previously. Note that we are trying to fit this integral into form (5.144).

Here we want $u^2 = 9x^2$ to we have $\sqrt{u^2-1}$ as one factor in the denominator of our integrand. Thus will let $u = 3x \implies du = 3\,dx \iff \frac{1}{3}du = dx$. Our integral so far is then

$$\int \frac{1}{x\sqrt{u^2-1}}\cdot \frac{1}{3}\,du.$$

Now, all of our integral formulas require just one variable, though in fact this integral makes sense because of the relationship between x and u. But to use a formula we have to put it all into the new variable, namely u. So there is one term left, which is the factor x on the bottom, which has to be put into u-terms. For that, we go back to our original substitution and note that $u = 3x \iff x = \frac{1}{3}u$. Now we continue:

$$\int \frac{1}{x\sqrt{9x^2-1}}\,dx = \int \frac{1}{x\sqrt{u^2-1}}\cdot \frac{1}{3}\,du$$

$$= \int \frac{1}{\frac{1}{3}u\sqrt{u^2-1}}\cdot \frac{1}{3}\,du$$

$$= \int \frac{1}{u\sqrt{u^2-1}}\,du$$

$$= \sec^{-1}|u| + C$$

$$= \sec^{-1}|3x| + C.$$

---

[80]Note that we could also have used $u = -5x^3$, but then $du = -15x^2\,dx$, and so our answer would ultimately be (as the reader should verify) $-\frac{1}{15}\tan^{-1}(-5x^3) + C$. In fact, that is the same as the answer we got, since the arctangent is an "odd" function, that is, $\tan^{-1}(-z) = -\tan^{-1} z$.

5.10. SUBSTITUTION WITH ALL BASIC FORMS

With practice, our process of solving the previous problem could more resemble the following:

$$\left[\begin{array}{c} u = 3x \\ \implies du = 3\,dx \\ \implies \frac{1}{3}du = dx, \\ \text{Also,}\quad u = 3x \\ \iff \frac{1}{3}u = x \end{array}\right] \qquad \int \frac{1}{x\sqrt{9x^2-1}}dx = \int \frac{1}{x\sqrt{(3x)^2-1}}dx = \int \frac{1}{\frac{1}{3}u\sqrt{u^2-1}} \cdot \frac{1}{3}du$$
$$= \int \frac{1}{u\sqrt{u^2-1}}du = \sec^{-1}|u|+C = \sec^{-1}|3x|+C.$$

The next example illustrates why it is important to recall that $(a^n)^m = a^{m \cdot n}$, so that $e^{2x} = (e^x)^2$.

Example 5.10.8

Compute $\int \frac{e^x}{\sqrt{1-e^{2x}}}dx$.

<u>Solution:</u> Here we note a form $\int \frac{1}{\sqrt{1-u^2}} \cdot \frac{du}{dx} \cdot dx$ if we let $u = e^x$.

$$\left[\begin{array}{c} u = e^x \\ \implies du = e^x\,dx \end{array}\right] \qquad \int \frac{e^x}{\sqrt{1-e^{2x}}}dx = \int \frac{1}{\sqrt{1-e^{2x}}} \cdot e^x\,dx = \int \frac{1}{\sqrt{1-u^2}}du$$
$$= \sin^{-1}u + C = \sin^{-1}e^x + C.$$

On the other hand, we have to be careful that we do not try to "read" arctrigonometric or other forms into what may be simple Power Rules.

Example 5.10.9

Compute $\int \frac{e^{2x}}{\sqrt{1-e^{2x}}}dx$.

<u>Solution</u> Here we note that the derivative of the function under the radical is also a factor, leaving only constant multiple factors:

$$\left[\begin{array}{c} u = 1 - e^{2x} \\ \implies du = -e^{2x} \cdot 2\,dx \\ \implies \frac{-1}{2}du = e^{2x}\,dx \end{array}\right] \qquad \int \frac{e^{2x}}{\sqrt{1-e^{2x}}}dx = \int (1-e^{2x})^{-1/2} \cdot e^{2x}\,dx = \int u^{-1/2} \cdot \frac{-1}{2}du$$
$$= \frac{-1}{2} \cdot 2u^{1/2} + C = -\sqrt{1-e^{2x}} + C.$$

We see in these examples that there are many known forms $\int f(u)\,du = F(u) + C$, and it is helpful to consider the consequences of a particular choice of u to the integral forms which can (or cannot) arise from such a substitution. It is a skill which requires some time to cultivate, but it is a skill that the intersted student can master with practice, usually including many false starts. That is normal since the process is somewhat complicated to mentally absorb. It is also important to experience false starts to see what does not work, along the path to discovering what does. The notation is crucially important to get right, as it can be a very good guide to the solution (and away from errors) in the process.

Exercises

For Exercises 1–28, compute the integral.

1. $\int \dfrac{dx}{(x^2+1)\tan^{-1}x}$

2. $\int \dfrac{1+e^x}{x+e^x}\,dx$

3. $\int \dfrac{\sec(\ln x)}{x}\,dx$

4. $\int \dfrac{\sqrt[3]{\ln 9x}}{x}\,dx$

5. $\int_0^1 \dfrac{x^2}{x^3+1}\,dx$

6. $\int \dfrac{x^3+1}{x^2}\,dx$

7. $\int_0^1 \dfrac{x^2}{x^6+1}\,dx$

8. $\int \dfrac{2^x}{2^x+1}\,dx$

9. $\int \dfrac{dx}{x\ln x\sqrt{(\ln x)^2-1}}$

10. $\int \dfrac{e^{\sin^{-1}x}\,dx}{\sqrt{1-x^2}}$

11. $\int e^x e^{e^x}\,dx$

12. $\int \dfrac{\cot\left(\frac{1}{x}\right)\,dx}{x^2}$

13. $\int \dfrac{e^{1/x}}{x^2}\,dx$

14. $\int_1^{\sqrt{e}} \dfrac{dx}{x\sqrt{1-(\ln x)^2}}$

15. $\int \dfrac{\sec x\tan x}{\sqrt{1-\sec x}}\,dx$

16. $\int \dfrac{e^x}{(1+e^x)^2}\,dx$

17. $\int \csc^2 3x\cot^6 3x\,dx$

18. $\int \dfrac{1}{\sqrt{x}(x+1)}\,dx$

19. $\int_0^1 2^x 3^{2^x}\,dx$

20. $\int_0^1 \dfrac{2^x}{1+4^x}\,dx$

21. $\int 5^x \sec 5^x\,dx$

22. $\int \dfrac{1}{\sqrt{1-5x^2}}\,dx$

23. $\int \dfrac{e^{2x}\,dx}{e^{4x}+1}$

24. $\int \dfrac{e^{2x}\,dx}{e^{2x}+1}$

25. $\int \dfrac{e^{3x}\,dx}{\sqrt{1-e^{3x}}}$

26. $\int \dfrac{e^{3x}\,dx}{\sqrt{1-e^{6x}}}$

27. $\int_2^3 \dfrac{dx}{x\ln x}$

28. $\int \dfrac{dx}{x\ln x\ln(\ln x)}$

29. Compute $\int \dfrac{x-1}{\sqrt{1-4x^2}}\,dx$. (Hint: Break this into two integrals.)

30. Compute and compare:

    (a) $\int \dfrac{x}{x^2+1}\,dx$ \qquad (b) $\int \dfrac{1}{x^2+1}\,dx$.

31. Compute $\int xe^{x^2}\,dx$ two ways.

    (a) Let $u = x^2$ (as in Example 5.10.1).

    (b) Instead let $u = e^{x^2}$.

32. In the spirit of the previous exercise, compute $\int \dfrac{e^{1/x}}{x^2}\,dx$ two ways, i.e., using two different substitutions.

33. Consider $\int \sin x\cos x\,dx$.

    (a) Compute this using $u = \sin x$.

    (b) Compute this using $u = \cos x$.

    (c) Explain why the two answers are equivalent.

34. Though it cannot be rewritten into a basic form, it is interesting to attempt some substitutions for $\int e^{x^2}\,dx$. Attempt to rewrite $\int e^{x^2}\,dx$ using

    (a) $u = x^2$. Note why this fails.

    (b) $u = x$. Explain why this is never useful for a substitution attempt for any integral $\int f(x)\,dx$.

35. When a 30-megawatt back-up electrical power plant is activated from "standby" at time $t = 0$, its output is given approximately by the equation

    $$P(t) = \dfrac{30e^{0.6(t-10)}}{1+e^{0.6(t-10)}},$$

    with P in megawatts and t in hours.

    (a) Find the total energy produced in the first 20 hours after the plant is activated.

    (b) Show that $\lim_{t\to\infty} P(t) = 30$.

5.11 Further Arctrigonometric Forms

Here we will still use the same arctrigonometric forms we had before, namely

$$\int \frac{1}{\sqrt{1-u^2}}\,du = \sin^{-1} u + C, \qquad (5.145)$$

$$\int \frac{1}{u^2+1}\,du = \tan^{-1} u + C, \qquad (5.146)$$

$$\int \frac{1}{u\sqrt{u^2-1}}\,du = \sec^{-1}|u| + C. \qquad (5.147)$$

What is different in this section is that our integrals will need to be algebraically rewritten into these forms, and this will require more than the previous substitution.

Each of the integrals in (5.145), (5.146), and (5.147) need to be exactly as they are stated. For instance, replacing $1-u^2$ with $1+u^2$ or u^2-1 in (5.145) will give completely different antiderivatives. In fact, even the domain of the integrand would be completely different with any such changes! Similar changes would substantially alter the results in (5.146) and (5.147).

In this section, we will have integrands that we can algebraically rewrite so they conform to one of the forms (5.145), (5.146) or (5.147). In fact, there are only a couple of algebraic "tricks" that we introduce here. The first of these is to force the denominators to have the additive constant 1, where originally there may be another constant. This is accomplished through simple factoring techniques. The second technique is "completing the square," where appropriate, and then using the first technique to finish rewriting the integrand. With substitution there will often be further multiplicative constants to accommodate as well.

5.11.1 Factoring to Achieve "1"

Each of our integrals (5.145), (5.146), and (5.147) have the number 1 conspicuously appearing in the denominator, near the u^2 term. Any other nonzero *constant* there will have an effect on the vertical and horizontal scaling of the function in ways we cannot ignore in the formula. To compensate is fairly straightforward: Factor the constant, and see what should be called "u^2."

Example 5.11.1

Compute $\int \frac{1}{9+x^2}\,dx$.

Solution: Our first priority is to rewrite this so we have a form $1+u^2$ in the denominator.

$$\int \frac{1}{9+x^2}\,dx = \int \frac{1}{9\left(1+\frac{x^2}{9}\right)}\,dx = \int \frac{1}{9}\cdot \frac{1}{1+\frac{x^2}{9}}\,dx.$$

The factor $\frac{1}{9}$ can simply be carried along for the rest of the computation. The denominator of the other factor can be written $1+u^2$ (same as u^2+1 in our formula) if we take $u^2 = \frac{x^2}{9}$, which can be accomplished letting $u = \frac{x}{3}$.

$$\left[\begin{array}{c} u = \dfrac{x}{3} \\ \Longrightarrow\ du = \dfrac{1}{3} dx \\ \Longleftrightarrow\ 3\,du = dx \end{array}\right] \qquad \begin{aligned} \int \frac{1}{9+x^2}\,dx &= \frac{1}{9}\cdot\int \frac{1}{1+\frac{x^2}{9}}\,dx = \frac{1}{9}\int \frac{1}{1+u^2}\cdot 3\,du \\ &= \frac{3}{9}\tan^{-1} u + C = \frac{1}{3}\tan^{-1}\left(\frac{x}{3}\right) + C. \end{aligned}$$

This example illustrated much of the process: Algebraically manipulate by factoring to achieve "1" in the appropriate place, and then pick u so another strategic term is u^2. It is slightly more complicated with the forms yielding arcsine and arcsecant, due to the presence of the radical. In the next example we will show more detail than one might normally write.

Example 5.11.2

Compute $\int \dfrac{1}{\sqrt{25-4x^2}}\,dx$.

Solution: Here we must find a way to replace the constant 25 with 1 instead. We factor as before, but respect the operation of the radical as well.

$$\int \frac{1}{\sqrt{25-4x^2}}\,dx = \int \frac{1}{\sqrt{25\left(1-\frac{4x^2}{25}\right)}}\,dx = \int \frac{1}{\sqrt{25}\sqrt{1-\frac{4x^2}{25}}}\,dx = \int \frac{1}{5}\cdot\frac{1}{\sqrt{1-\frac{4x^2}{25}}}\,dx.$$

So the factor 25 under the radical becomes the factor 5 outside the radical. Otherwise it is the same process as the previous example. Now we continue, using a substitution which will result in $u^2 = \frac{4x^2}{25}$. For simplicity we take $u = \frac{2x}{5}$.

$$\left[\begin{array}{c} u = \dfrac{2x}{5} \\ \Longrightarrow\ du = \dfrac{2}{5}\cdot dx \\ \Longleftrightarrow\ \dfrac{5}{2}\,du = dx \end{array}\right] \qquad \begin{aligned} \int \frac{1}{\sqrt{25-4x^2}}\,dx &= \frac{1}{5}\int \frac{1}{\sqrt{1-\frac{4x^2}{25}}}\,dx = \frac{1}{5}\int \frac{1}{\sqrt{1-u^2}}\cdot\frac{5}{2}\,du \\ &= \frac{1}{2}\sin^{-1} u + C = \frac{1}{2}\sin^{-1}\left(\frac{2x}{5}\right) + C. \end{aligned}$$

This example again illustrates the role of the number 1 in the denominator, but also suggests a couple of new points that we make here. First, it is not obvious where the factors 2 and 5—being the square roots of the 4 and 25 appearing in the original—will be present in the final answer. There is a pattern for the arctangent form, and a different one for the arcsine form, but patterns can be forgotten if not used often enough, where the logic of manipulating the integral algebraically to get one of the three basic arctrigonometric forms should still be reproducible after the patterns—which we will explore at the end of this section—are forgotten. Second, we are approaching the boundary between integrals which are easily checked with differentiation, and those where the differentiation has at least as many algebraic difficulties as the integration. In such cases, it is usually better to have carefully written each

5.11. FURTHER ARCTRIGONOMETRIC FORMS

integration step so it can be audited for accuracy, rather than risk algebraic error in testing our answer. Consider a verification of the answer in this latest example (readers' steps may vary):

$$\frac{d}{dx}\left[\frac{1}{2}\sin^{-1}\left(\frac{2x}{5}\right)\right] = \frac{1}{2}\cdot\frac{1}{\sqrt{1-\left(\frac{2x}{5}\right)^2}}\cdot\frac{d}{dx}\left[\frac{2x}{5}\right] = \frac{1}{2}\cdot\frac{1}{\sqrt{1-\frac{4x^2}{25}}}\cdot\frac{2}{5}$$

$$= \frac{1}{5}\cdot\frac{1}{\sqrt{1-\frac{4x^2}{25}}} = \frac{1}{\sqrt{25}}\cdot\frac{1}{\sqrt{1-\frac{4x^2}{25}}} = \frac{1}{\sqrt{25-4x^2}}, \qquad \text{q.e.d.}$$

While such a verification is certainly possible, it is not likely one to be performed "mentally" with much confidence, as we may have been able to do with many previous computations. Indeed, there are enough constants to be accommodated that this verification should be done in careful writing. In the sequel we will see much more complicated rewritings of integrals, and verification will usually be much better accomplished by checking our individual steps in integration rather than by differentiating our answers.

Our next example takes this theme one step further. Recall that substitution in the arcsecant form had a slight complication, which was that the u-variable appeared both inside and outside the radical. This caused a minor complication in Example 5.10.7, page 660, for instance. A similar problem will occur in this next example.

Example 5.11.3

Compute $\int \frac{1}{x\sqrt{81x^2-16}}\,dx$.

<u>Solution</u>: As with the previous two examples, it is necessary to have a 1 in the place presently occupied by 16, so we will factor the 16 from the radical. The other algebraic difficulties will be taken care of by the substitution. Indeed, the remaining term under the radical must be u^2, and the rest of the form will follow, with residual multiplicative constants. (Note that, as before, u is chosen *after* the integral in x is manipulated to achieve the "1" in the correct position.)

$$\begin{bmatrix} u = \frac{9x}{4} \\ \implies du = \frac{9}{4}\cdot dx \\ \iff \frac{4}{9}du = dx \\ \text{also,} \quad u = \frac{9x}{4} \\ \iff x = \frac{4u}{9} \end{bmatrix} \qquad \int \frac{1}{x\sqrt{81x^2-16}}\,dx = \int \frac{1}{4x\sqrt{\frac{81x^2}{16}-1}}\,dx = \int \frac{1}{4\cdot\frac{4u}{9}\sqrt{u^2-1}}\cdot\frac{4}{9}du$$

$$= \int \frac{1}{4u\sqrt{u^2-1}}du = \frac{1}{4}\sec^{-1}|u|+C$$

$$= \frac{1}{4}\sec^{-1}\left|\frac{9x}{4}\right|+C.$$

Note how the term $\frac{81x^2}{16}$ under the radical became simply u^2, and then the term x outside the radical became $\frac{4u}{9}$, both consistent with $u = \frac{9x}{4}$. Also note that a factor of $\frac{4}{9}$ in the denominator canceled with the same factor multiplying the differential du.

This latest example again illustrates the points made before: that having the 1-term in the denominator is the key to the whole process, that the rest is taken care of by the u-substitution that follows, and finally, that checking by differentiation is nontrivial.

Another minor complication is that the numbers we must factor might not be perfect squares. The process is exactly the same, though perhaps some more care is required. (Note that the substitution is chosen *after* the integral is rewritten with the "1" in its proper position.)

Example 5.11.4

Compute $\int \dfrac{1}{\sqrt{5-2x^2}}\,dx$.

Solution: The process is exactly the same as before. The key is to factor the denominator to have a 1 in the place of the 5:

$$\left[\begin{array}{c} u = \sqrt{\dfrac{2}{5}}\cdot x \\ \Longrightarrow \quad du = \sqrt{\dfrac{2}{5}}\cdot dx \\ \Longleftrightarrow \sqrt{\dfrac{5}{2}}\cdot du = dx \end{array}\right]$$

$$\int \dfrac{1}{\sqrt{5-2x^2}}\,dx = \int \dfrac{1}{\sqrt{5}\cdot\sqrt{1-\dfrac{2x^2}{5}}}\,dx = \dfrac{1}{\sqrt{5}}\int \dfrac{1}{\sqrt{1-u^2}}\cdot\sqrt{\dfrac{5}{2}}\,du$$

$$= \dfrac{1}{\sqrt{2}}\sin^{-1}u + C = \dfrac{1}{\sqrt{2}}\sin^{-1}\left(\sqrt{\dfrac{2}{5}}\cdot x\right) + C.$$

5.11.2 Completing the Square

In the previous subsection, our first concern after identifying our target form was to rewrite the integrand to have the number 1 in the appropriate place in the denominator. In this subsection, our first concern is identifying, except for a multiplicative constant, what will be u^2. We do this by completing the square first, and then fixing the form to have the number 1 where we need it, and finally choosing u and working from there as before. As there are differing levels of difficulty in such problems, we will begin with one of the simplest and continue from there. It should be noted that the completing the square step is sometimes needed before determining that one of our three forms, (5.145), (5.146), or (5.147), can even be achieved. If not, and we are fortunate, another earlier method may work, though usually we should notice that before attempting the method here. If no earlier method will work, this method or perhaps one available in the sequel can solve the problem.

Recall that when completing the square, one adds and subtracts $(b/2)^2$, where the original polynomial is $x^2 + bx$, or more generally $x^2 + bx + c$:

$$x^2 + bx + c = x^2 + bx + \left(\dfrac{b}{2}\right)^2 - \left(\dfrac{b}{2}\right)^2 + c$$

$$= \left(x + \dfrac{b}{2}\right)^2 - \left(\dfrac{b}{2}\right)^2 + c.$$

As we will see, the fact that the coefficient of x^2 was 1 was key to the computation. If not, the leading coefficient will be factored from the x^2 and x terms. Our first examples will not require that initial factoring.

5.11. FURTHER ARCTRIGONOMETRIC FORMS

Example 5.11.5

Compute $\int \dfrac{1}{x^2+2x+2}\,dx$.

<u>Solution</u>: The hope is that we can somehow write the denominator as $1+u^2$, perhaps multiplied by some nonzero constant, without introducing any more variable factors. For this one, we are unusually fortunate. Note that here "b" is 2.

$$\left[\begin{array}{c} u = x+1 \\ \Longrightarrow du = dx \end{array}\right] \qquad \int \dfrac{1}{x^2+2x+2}\,dx = \int \dfrac{1}{x^2+2x+\left(\frac{2}{2}\right)^2 - \left(\frac{2}{2}\right)^2+2}\,dx$$

$$= \int \dfrac{1}{x^2+2x+1-1+2}\,dx$$

$$= \int \dfrac{1}{(x+1)^2+1}\,dx = \int \dfrac{1}{u^2+1}\,du$$

$$= \tan^{-1} u + C = \tan^{-1}(x+1)+C.$$

What made this last example particularly simple was that the additive constant outside of the perfect square was already 1, which is of course key to our arctrigonometric antiderivative forms. If not, we have to perform some division.

Example 5.11.6

Compute $\int \dfrac{1}{x^2+6x+17}\,dx$.

<u>Solution</u>: Here $b=6$, so we add and subtract $\left(\frac{b}{2}\right)^2 = 3^2 = 9$ to achieve the desired integral form.

$$\left[\begin{array}{c} u = \dfrac{x+3}{\sqrt{8}} \\ \Longrightarrow \quad du = \dfrac{1}{\sqrt{8}}\,dx \\ \Longleftrightarrow \sqrt{8}\,du = dx \end{array}\right] \qquad \begin{aligned} \int \dfrac{1}{x^2+6x+17}\,dx &= \int \dfrac{1}{x^2+6x+9-9+17}\,dx = \int \dfrac{1}{(x+3)^2+8}\,dx \\ &= \dfrac{1}{8}\int \dfrac{1}{\frac{(x+3)^2}{8}+1}\,dx = \dfrac{1}{8}\int \dfrac{1}{u^2+1}\cdot\sqrt{8}\,du \\ &= \dfrac{1}{\sqrt{8}}\tan^{-1} u + C = \dfrac{1}{\sqrt{8}}\tan^{-1}\left(\dfrac{x+3}{\sqrt{8}}\right) + C. \end{aligned}$$

As this last example illustrates, the final form of the antiderivative can be more complicated when completing the square is required. While it would be an interesting exercise to verify the answer by differentiation, perhaps verifying each individual step in the solution process would be a more efficient means of verifying the answer we derived.

For simplicity, we will continue with arctangent forms for the moment, as we look at the next complication, which is that the coefficient of x^2 is not equal to 1. In such a case, we factor that leading coefficient out of the entire polynomial, or at least out of the x^2 and x terms. It is then important to perform the addition and subtraction steps of completing the square within the factor with the x^2 and x terms; the addition and subtraction of the $(b/2)^2$ in the process must occur simultaneously and beside each other. Note that such a term has a different effect inside parentheses (or brackets) compared to outside, so we must have the addition and subtraction steps together in order that numerically they have no net effect.

Example 5.11.7

Compute $\int \dfrac{1}{5x^2 - 4x + 9}\, dx$.

<u>Solution</u>: Our first priority is to have the coefficient of x^2 to be 1, after which we complete the square and finish the problem.

$$\int \dfrac{1}{5x^2 - 4x + 9}\, dx = \int \dfrac{1}{5\left[x^2 - \tfrac{4}{5}x + \tfrac{9}{5}\right]}\, dx = \int \dfrac{1}{5\left[x^2 - \tfrac{4}{5}x + \left(\tfrac{2}{5}\right)^2 - \left(\tfrac{2}{5}\right)^2 + \tfrac{9}{5}\right]}\, dx$$

$$= \int \dfrac{1}{5\left[\left(x - \tfrac{2}{5}\right)^2 - \tfrac{4}{25} + \tfrac{45}{25}\right]}\, dx = \dfrac{1}{5}\int \dfrac{1}{\left[\left(x - \tfrac{2}{5}\right)^2 + \tfrac{41}{25}\right]}\, dx$$

(Note how we added and subtracted $(2/5)^2$ together, both within the brackets.) Now we need to manipulate this integral so there is a 1 in place of the fraction $\tfrac{41}{25}$, which we do as before, by factoring. Continuing,

$$\int \dfrac{1}{5x^2 - 4x + 9}\, dx = \cdots = \dfrac{1}{5}\int \dfrac{1}{\left[\left(x - \tfrac{2}{5}\right)^2 + \tfrac{41}{25}\right]}\, dx$$

$\left[\begin{array}{c} u = \dfrac{5}{\sqrt{41}}\left(x - \tfrac{2}{5}\right) \\[4pt] \Longrightarrow\ du = \dfrac{5}{\sqrt{41}}\, dx \\[4pt] \Longleftrightarrow\ \dfrac{\sqrt{41}}{5}\, du = dx \end{array}\right]$

$$= \dfrac{1}{5}\int \dfrac{1}{\tfrac{41}{25}\left[\underbrace{\tfrac{25}{41}\left(x - \tfrac{2}{5}\right)^2}_{\text{``}u^2\text{''}} + 1\right]}\, dx$$

$$= \dfrac{1}{5}\cdot\dfrac{25}{41}\int \dfrac{1}{u^2 + 1}\cdot\dfrac{\sqrt{41}}{5}\, du$$

$$= \dfrac{1}{\sqrt{41}}\tan^{-1} u + C$$

$$= \dfrac{1}{\sqrt{41}}\tan^{-1}\left[\dfrac{5}{\sqrt{41}}\left(x - \dfrac{2}{5}\right)\right] + C.$$

Such integrals require care, some planning, and an understanding of what constitutes progress towards a useful form. Clearly it would be laborious to verify our solution through differentiation. Indeed, it is much simpler to revisit the work that led to the final answer, if it is presented clearly. As with any calculus problem, clear and organized work helps tremendously in both the computation and in checking the validity of the work.

Also, the usefulness of proper use of the notation should not be underestimated. Imagine, if possible, how to guess at the solution in the last example without working through the notation correctly. It is unlikely anyone could "guess" the exact form of the answer (beyond that it might involve an arctangent), and if someone could, it would be very difficult to verify by computing its derivative.

Learning to use calculus notation precisely in a first-semester course will pay large dividends in subsequent calculus and related studies as well.

5.11. FURTHER ARCTRIGONOMETRIC FORMS

Exercises

For Exercises 1–10, evaluate the indefinite integral.

1. $\int \dfrac{dx}{\sqrt{5-x^2}}$

2. $\int \dfrac{1}{x^2+3}\,dx$

3. $\int \dfrac{1}{x\sqrt{36x^2-4}}\,dx$

4. $\int \dfrac{1}{\sqrt{9-4x^2}}\,dx$

5. $\int \dfrac{1}{\sqrt{6-3x^2}}\,dx$

6. $\int \dfrac{1}{x\sqrt{49x^2-4}}\,dx$

7. $\int \dfrac{e^{2x}}{5+e^{4x}}\,dx$

8. $\int \dfrac{1}{9+(x-2)^2}\,dx$

9. $\int \dfrac{\sec^2 x}{\sqrt{4-\tan^2 x}}\,dx$

10. $\int \dfrac{1}{\sqrt{e^{2x}-3}}\,dx$

For Exercises 11–15, evaluate the integral after completing the square.

11. $\int \dfrac{dx}{x^2-4x+9}$

12. $\int \dfrac{1}{\sqrt{4x-x^2}}\,dx$

13. $\int \dfrac{1}{(x+2)\sqrt{x^2+4x+3}}\,dx$

14. $\int \dfrac{1}{\sqrt{x}(2+x)}\,dx$ (Hint: let $u=\sqrt{x}$.)

15. $\int \dfrac{1}{\sqrt{6x-x^2}}\,dx$

After repeated substitution exercises for some integral forms that yield inverse trigonometric antiderivatives, some general patterns emerge. Indeed, some textbooks instruct their readers to memorize some more general formulas, such as those given in Exercises 16–18. Verify each of those by differentiating the right-hand sides (with respect to u). (For Exercise 18 it might be useful to recall the discussion surrounding Equation (3.92), page 349.) Assume $a > 0$ in each case.

16. $\int \dfrac{du}{\sqrt{a^2-u^2}} = \sin^{-1}\left(\dfrac{u}{a}\right) + C.$

17. $\int \dfrac{du}{a^2+u^2} = \dfrac{1}{a}\tan^{-1}\left(\dfrac{u}{a}\right) + C.$

18. $\int \dfrac{du}{u\sqrt{u^2-a^2}} = \dfrac{1}{a}\sec^{-1}\left|\dfrac{u}{a}\right| + C.$[81]

Use the formulas in Exercises 16–18 to compute the antiderivatives in Exercises 19–25 below.

19. Compute $\int \dfrac{dx}{\sqrt{5-x^2}}.$

Compare to the answer in Exercise 1.

20. Compute $\int \dfrac{1}{x^2+3}\,dx.$

Compare to the answer in Exercise 2.

21. Compute $\int \dfrac{1}{x\sqrt{36x^2-4}}\,dx.$

Compare to the answer in Exercise 3.

22. Compute $\int \dfrac{1}{\sqrt{6-3x^2}}\,dx.$

Compare to the answer in Exercise 5.

---

[81] Interestingly, and perhaps frustratingly, there are variations of the formula given in Exercise 18, each with its own way of dealing with the cases where u or a are negative. Some texts will list the antiderivative as $\frac{1}{a}\sec^{-1}\frac{|u|}{a}$, and others offer instead $\frac{1}{|a|}\sec^{-1}\frac{u}{a}$. It is an interesting exercise to note how these indeed have the same derivative.

23. Compute $\int \dfrac{e^{2x}}{5+e^{4x}}\,dx$.

Compare to the answer in Exercise 7.

24. Compute $\int \dfrac{\sec^2 x}{\sqrt{4-\tan^2 x}}\,dx$.

Compare to the answer in Exercise 9.

25. Compute $\int \dfrac{1}{\sqrt{e^{2x}-3}}\,dx$.

Compare to the answer in Exercise 10.

Answers to Odd-Numbered Problems

Chapter 1.

Exercises 1.1:

1. (a) P is true, (b) at least one of P, Q is false, (c) both P and Q are false, (d) P is true but Q is false, (e) one of P, Q is true and the other is false, (f) P and Q are both true.

3. (a) Q is false but P is true, (b) P is true but Q is false, (c) they do, (d)

P	Q	$\sim Q$	$\sim P$	$(\sim Q) \to (\sim P)$
T	T	F	F	T
T	F	T	F	F
F	T	F	T	T
F	F	T	T	T

5.

P	Q	$P \leftrightarrow Q$	$\sim(P \leftrightarrow Q)$
T	T	T	F
T	F	F	T
F	T	F	T
F	F	T	F

Same as P **XOR** Q.

7.

P	Q	$\sim Q$	$\sim P$	$(\sim Q) \leftrightarrow (\sim P)$
T	T	F	F	T
T	F	T	F	F
F	T	F	T	F
F	F	T	T	T

9.

P	Q	R	$Q \vee R$	$P \wedge (Q \vee R)$	$\sim$
T	T	T	T	T	F
T	T	F	T	T	F
T	F	T	T	T	F
T	F	F	F	F	T
F	T	T	T	F	T
F	T	F	T	F	T
F	F	T	T	F	T
F	F	F	F	F	T

11.

P	Q	R	P	$Q \wedge R$	$P \vee (Q \wedge R)$
T	T	T	T	T	T
T	T	F	T	F	T
T	F	T	T	F	T
T	F	F	T	F	T
F	T	T	F	T	T
F	T	F	F	F	F
F	F	T	F	F	F
F	F	F	F	F	F

13.

P	Q	R	$P \vee Q$	R	$(P \vee Q) \vee R$
T	T	T	T	T	T
T	T	F	T	F	T
T	F	T	T	T	T
T	F	F	T	F	T
F	T	T	T	T	T
F	T	F	T	F	T
F	F	T	F	T	T
F	F	F	F	F	F

They are the same.

15. Tautology.

P	P	$P \to P$
T	T	T
F	F	T

17. Tautology.

P	$\sim P$	$P \vee \sim P$
T	F	T
F	T	T

19. Contradiction.

P	$\sim P$	$P \leftrightarrow (\sim P)$
T	F	F
F	T	F

21. Tautology.

P	Q	$\sim P$	$P \wedge (\sim P)$	Q	$\to$
T	T	F	F	T	T
T	F	F	F	F	T
F	T	T	F	T	T
F	F	T	F	F	T

23. Tautology.

P	Q	$P \wedge Q$	P	$(P \wedge Q) \to P$
T	T	T	T	T
T	F	F	T	T
F	T	F	F	T
F	F	F	F	T

25. Neither.

P	Q	$P \wedge Q$	$P \to (P \wedge Q)$
T	T	T	T
T	F	F	F
F	T	F	T
F	F	F	T

27. (a)

P	Q	$P \to Q$	$Q \to P$	$(P \to Q) \vee (Q \to P)$
T	T	T	T	T
T	F	F	T	T
F	T	T	F	T
F	F	T	T	T

(b) "P causes Q or Q causes P," most understandings of the word "causes," could not possibly be a tautology, as any two statements P and Q which are obviously not related will demonstrate. (c) Yes.

Answers to Odd-Numbered Problems

Exercises 1.2: Truth values of equivalent statements are in bold face. Relevant entries for other types of problems may also be in bold face. Column headings are sometimes abbreviated for space considerations, e.g., by the given operation as acting on the statements of the two previous columns.

1.

P	$\sim P$	$\sim(\sim P)$
T	F	**T**
F	T	**F**

3.

P	Q	$P \to Q$	$\sim Q$	$\sim P$	$\to$
T	T	**T**	F	F	**T**
T	F	**F**	T	F	**F**
F	T	**T**	F	T	**T**
F	F	**T**	T	T	**T**

5.

P	Q	$P \to Q$	Q	P	$Q \to P$
T	T	**T**	T	T	**T**
T	F	**F**	F	T	**T**
F	T	**T**	T	F	**F**
F	F	**T**	F	F	**T**

$P \to Q$ and $Q \to P$ are not equivalent.

7.

7(a):

P	Q	R	$Q \vee R$	$P \wedge (Q \vee R)$	$P \wedge Q$	$P \wedge R$	$\vee$
T	T	T	T	**T**	T	T	**T**
T	T	F	T	**T**	T	F	**T**
T	F	T	T	**T**	F	T	**T**
T	F	F	F	**F**	F	F	**F**
F	T	T	T	**F**	F	F	**F**
F	T	F	T	**F**	F	F	**F**
F	F	T	T	**F**	F	F	**F**
F	F	F	F	**F**	F	F	**F**

7(b):

P	Q	R	$Q \wedge R$	$P \vee (Q \wedge R)$	$P \vee Q$	$P \vee R$	$\wedge$
T	T	T	T	**T**	T	T	**T**
T	T	F	F	**T**	T	T	**T**
T	F	T	F	**T**	T	T	**T**
T	F	F	F	**T**	T	T	**T**
F	T	T	T	**T**	T	T	**T**
F	T	F	F	**F**	T	F	**F**
F	F	T	F	**F**	F	T	**F**
F	F	F	F	**F**	F	F	**F**

9.

P	Q	R	$Q \wedge R$	$P \to (Q \wedge R)$	$P \to Q$	$P \to R$	$\wedge$
T	T	T	T	**T**	T	T	**T**
T	T	F	F	**F**	T	F	**F**
T	F	T	F	**F**	F	T	**F**
T	F	F	F	**F**	F	F	**F**
F	T	T	T	**T**	T	T	**T**
F	T	F	F	**T**	T	T	**T**
F	F	T	F	**T**	T	T	**T**
F	F	F	F	**T**	T	T	**T**

11.

P	Q	$P \to Q$	$\sim(P \to Q)$	P	$\sim Q$	$\wedge$
T	T	T	**F**	T	F	**F**
T	F	F	**T**	T	T	**T**
F	T	T	**F**	F	F	**F**
F	F	T	**F**	F	T	**F**

13.

P	Q	$P \to Q$	$\sim(P \to Q)$	$\sim Q$	$\sim P$	$P \wedge (\sim Q)$	$Q \wedge (\sim P)$	$\vee$
T	T	T	**F**	F	F	F	F	**F**
T	F	F	**T**	T	F	T	F	**T**
F	T	T	**F**	F	T	F	T	**T**
F	F	T	**F**	T	T	F	F	**F**

15. If P or Q being true implies P and Q are true, that is another way of saying you can't have one of P, Q true without the other being true, which is another way to interpret $P \leftrightarrow Q$.

P	Q	$P \vee Q$	$P \wedge Q$	$\rightarrow$	$P \leftrightarrow Q$
T	T	T	T	**T**	**T**
T	F	T	F	**F**	**F**
F	T	T	F	**F**	**F**
F	F	F	F	**T**	**T**

17.

(a)
P	$\mathcal{T}$	$P \vee \mathcal{T}$
T	T	**T**
F	T	**T**

(b)
P	$\mathcal{T}$	$P \wedge \mathcal{T}$
T	T	**T**
F	T	**F**

(c)
P	$\mathcal{F}$	$P \vee \mathcal{F}$
T	F	**T**
F	F	**F**

(d)
P	$\mathcal{F}$	$P \wedge \mathcal{F}$
T	F	**F**
F	F	**F**

19. (a) P (b) $\mathcal{T}$ (c) $\mathcal{T}$ (d) $\sim P$ (e) P (f) $\sim P$ (g) $\mathcal{F}$ (h) $\sim P$

21.

P	Q	R	$P \rightarrow Q$	$Q \rightarrow R$	$R \rightarrow P$	$\cdot \wedge \cdot \wedge \cdot$	$P \leftrightarrow Q$	$Q \leftrightarrow R$	$R \leftrightarrow P$	$\cdot \wedge \cdot \wedge \cdot$
T	T	T	T	T	T	**T**	T	T	T	**T**
T	T	F	T	F	T	**F**	T	F	F	**F**
T	F	T	F	T	T	**F**	F	F	T	**F**
T	F	F	F	T	T	**F**	F	T	F	**F**
F	T	T	T	T	F	**F**	F	T	F	**F**
F	T	F	T	F	T	**F**	F	F	T	**F**
F	F	T	T	T	F	**F**	T	F	F	**F**
F	F	F	T	T	T	**T**	T	T	T	**T**

23.

P	Q	R	$P \wedge Q$	R	$(P \wedge Q) \rightarrow R$	$P \rightarrow R$	$Q \rightarrow R$	$(P \rightarrow R) \vee (Q \rightarrow R)$
T	T	T	T	T	**T**	T	T	**T**
T	T	F	T	F	**F**	F	F	**F**
T	F	T	F	T	**T**	T	T	**T**
T	F	F	F	F	**T**	F	T	**T**
F	T	T	F	T	**T**	T	T	**T**
F	T	F	F	F	**T**	T	F	**T**
F	F	T	F	T	**T**	T	T	**T**
F	F	F	F	F	**T**	T	T	**T**

Exercises 1.3: Truth values of equivalent statements are in bold face. Relevant entries for other types of problems may also be in bold face. Column heading are sometimes abbreviated for space considerations, e.g., by the given operation as acting on the statements of the two previous columns.

1.

P	Q	$P \rightarrow Q$	P	$P \wedge Q$	$P \leftrightarrow (P \wedge Q)$
T	T	**T**	T	T	**T**
T	F	**F**	T	F	**F**
F	T	**T**	F	F	**T**
F	F	**T**	F	F	**T**

3.

P	Q	$P \wedge Q$	P	$(P \wedge Q) \rightarrow P$
T	T	T	T	**T**
T	F	F	T	**T**
F	T	F	F	**T**
F	F	F	F	**T**

P	Q	P	$P \vee Q$	$P \rightarrow (P \vee Q)$
T	T	T	T	**T**
T	F	T	T	**T**
F	T	F	T	**T**
F	F	F	F	**T**

5.

(a)
P	Q	$P \rightarrow Q$	P	$\wedge$	Q	$\rightarrow$
T	T	T	T	T	T	**T**
T	F	F	T	F	F	**T**
F	T	T	F	F	T	**T**
F	F	T	F	F	F	**T**

(b)
P	Q	$P \rightarrow Q$	$\sim Q$	$\wedge$	$\sim P$	$\rightarrow$
T	T	T	F	F	F	**T**
T	F	F	T	F	F	**T**
F	T	T	F	F	T	**T**
F	F	T	T	T	T	**T**

7. Valid.

P	Q	$\rightarrow$	$Q \rightarrow P$	$\wedge$	$P \leftrightarrow Q$	$\rightarrow$
T	T	T	T	T	T	**T**
T	F	F	T	F	F	**T**
F	T	T	F	F	F	**T**
F	F	T	T	T	T	**T**

9. Fallacy.

P	Q	$P \rightarrow Q$	Q	$\wedge$	P	$\rightarrow$
T	T	T	T	T	T	**T**
T	F	F	F	F	T	**T**
F	T	T	T	T	F	**F**
F	F	T	F	F	F	**T**

Answers to Odd-Numbered Problems

11. Valid.

P	Q	P → Q	~	P	∧	~Q	→
T	T	T	F	T	F	F	**T**
T	F	F	F	T	T	T	**T**
F	T	T	T	F	F	F	**T**
F	F	T	T	F	F	T	**T**

13. Valid. (All cases are vacuously true.)

15. Valid.

17. Valid.

P	Q	R	P↓Q	Q↓R	~R	∨	~P	↓
(truth table)								

19. Valid.

21. Valid.

23. Valid.

P	Q	R	P	Q∨R	↓	~R	∨	~P	↓
(truth table)									

25. Fallacy.   **29.** Valid.

27. Fallacy.   **31.** Valid.

33. Valid.

P	R	P → R	R → P	∧	P ↔ R	→
T	T	T	T	T	T	**T**
T	F	F	T	F	F	**T**
F	T	T	F	F	F	**T**
F	F	T	T	T	T	**T**

Exercises 1.4:

1. For every prison, there is a method of escape (that will get you out of that prison).

3. There exists a prison that every method of escape will get you out of.

5. There exists a method of escape and a prison for which the method will get you out of.

7. $(\exists p \in P)(\forall m \in M)[m$ will not get you out of $p]$. There exists a prison that no method of escape will get you out of.

9. $(\forall p \in P)(\exists m \in M)[m$ will not get you out of $p]$. For any prison there exists a method of escape that will not work with that prison.

11. $(\forall m \in M)(\forall p \in P)[m$ will not get you out of $p]$. No matter what the method of escape, or the prison, the method will not get you out of that prison.

13. $(\exists x \in R)[x \notin S]$.

15. $(\exists x, y \in R)(\forall r, t \in S)[rx + ty \neq 1]$.

17. (a) Negation: $(\forall x \in \mathbb{R})(x^2 \geq 0)$. Negation is true. (b) Negation: $(\exists x \in \mathbb{R})(|-x| = x)$. Negation is true (any $x \geq 0$ works). (c) Negation: $(\exists x \in \mathbb{R})(\forall y \in \mathbb{R})(y = 2x+1)$. Original statement is true.

19. (a) Negation: $(\exists x, y \in \mathbb{R})((x < y) \wedge (x^2 \geq y^2))$. (b) The negation is true, for example $x = -1$ and $y = 1$, or $x = -5$ and $y = 4$.

Exercises 1.5:

1.
$$x \in (A \cup (B \cup C))$$
$$\iff (x \in A) \vee (x \in B \cup C)$$
$$\iff (x \in A) \vee [(x \in B) \vee (x \in C)]$$
$$\iff [(x \in A) \vee (x \in B)] \vee (x \in C)$$
$$\iff [x \in (A \cup B)] \vee (x \in C)$$
$$\iff x \in (A \cup B) \cup C, \text{ q.e.d.}$$

3.
$$x \in A - (B \cap C)$$
$$\iff (x \in A) \wedge [\sim (x \in B \cap C)]$$
$$\iff (x \in A) \wedge [\sim ((x \in B) \wedge (x \in C))]$$
$$\iff (x \in A) \wedge [(\sim (x \in B)) \vee (\sim (x \in C))]$$
$$\iff [(x \in A) \wedge (\sim (x \in B))]$$
$$\vee [(x \in A) \wedge (\sim (x \in C))]$$
$$\iff (x \in A - B) \vee (x \in A - C)$$
$$\iff x \in (A - B) \cup (A - C), \text{ q.e.d.}$$

5.
$$x \in (B \cup C)'$$
$$\iff (x \in U) \wedge [\sim (x \in B \cup C)]$$
$$\iff (x \in U) \wedge [\sim ((x \in B) \vee (x \in C))]$$
$$\iff [(x \in U) \wedge [(\sim (x \in B)) \wedge (\sim (x \in C))]$$
$$\iff [(x \in U) \wedge (\sim (x \in B))] \wedge [(x \in U) \wedge (\sim (x \in C))]$$
$$\iff (x \in B') \wedge (x \in C')$$
$$\iff x \in B' \cap C', \text{ q.e.d.}$$

7. $A - (B - A) = A$.

A

$B - A$

$A - (B - A)$

9. $(A - B) \cap (B - A) = \varnothing$.

$A - B$

$B - A$

intersection

11. If $A \subset B$, then $B - A \neq \varnothing$.

13. $B' \subseteq A'$.

15. $A = \varnothing$ and $B = \varnothing$.

17. $\mathbb{I} = \mathbb{R} - \mathbb{Q}$.

19. Yes, since $A \triangle B = (A - B) \cup (B - A) = (B - A) \cup (A - B) = B \triangle A$.

21. $A \triangle A = \varnothing$, $A \triangle U = A'$, $A \triangle \varnothing = A$.

23. $A \triangle B = \{x \mid (x \in A - B) \vee (x \in B - A)\}$.

25. Conjecture: For a Venn Diagram with n sets, there are 2^n regions and 2^{2^n} different possible shadings.

27. $n_1 + n_2 + n_3 = (n_1 + n_2) + (n_2 + n_3) - (n_2)$. Here n_i represents the number of elements in the given region:

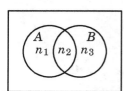

29. $\{1, 5, 7, 8\}$

30. $2^3 = 8$

33. $2^5 = 32$.

Answers to Odd-Numbered Problems

Exercises 2.1:

1. Let $\varepsilon > 0$. Set $\delta = \varepsilon/9$. Then $\delta > 0$ exists and $|x-2| < \delta \implies |f(x) - f(2)| = |9x - 11 - 7| = |9x - 18| = 9|x - 2| < 9\delta = 9 \cdot \frac{\varepsilon}{9} = \varepsilon$, q.e.d.

3. Let $\varepsilon > 0$. Set $\delta = \varepsilon/3$. Then $\delta > 0$ exists and $|x - 5| < \delta \implies |f(x) - f(5)| = |3x + 1 - 16| = |3x - 15| = 3|x - 5| < 3\delta = 3 \cdot \frac{\varepsilon}{3} = \varepsilon$, q.e.d.

5. Let $\varepsilon > 0$. Set $\delta = \varepsilon/|m|$. Then $\delta > 0$ exists and $|x - a| < \delta \implies |f(x) - f(a)| = |mx + b - (ma + b)| = |mx - ma| = |m| \cdot |x - a| < |m|\delta = |m| \cdot \frac{\varepsilon}{|m|} = \varepsilon$, q.e.d.

7. Let $\varepsilon > 0$. Set $\delta = \min\{1, \varepsilon/19\}$. Then $\delta > 0$ exists and $|x - 9| < \delta \implies |f(x) - f(9)| = |x^2 - 81| = |x + 9| \cdot |x - 9| \leq 19|x - 9| < 19\delta \leq 19 \cdot \frac{\varepsilon}{19} = \varepsilon$, q.e.d.

   (Note that $|x - 9| < \delta \implies |x - 9| < 1 \implies x \in (8, 10) \implies x + 9 \in (17, 19)$.)

9. Let $\varepsilon > 0$. Set $\delta = \min\{1, \varepsilon/25\}$. Then $\delta > 0$ exists and $|x - 2| < \delta \implies |f(x) - f(2)| = |5x^2 - 3 - 17| = |5x^2 - 20| = 5|x + 2| \cdot |x - 2| \leq 5 \cdot 5|x - 2| < 25\delta \leq 25 \cdot \frac{\varepsilon}{25} = \varepsilon$, q.e.d.

   (Note that $|x - 2| < \delta \implies |x - 2| < 1 \implies x \in (1, 3) \implies x + 2 \in (3, 5)$.)

11. Let $\varepsilon > 0$. Choose any $\delta > 0$. Then $|x - a| < \delta \implies |f(x) - f(a)| = |b - b| = 0 < \varepsilon$, q.e.d.

13. Let $\varepsilon > 0$. Set $\delta = \min\{1, \varepsilon/37\}$. Then $\delta > 0$ exists and $|x + 3| < \delta \implies |f(x) - f(-3)| = |x^3 - 27| = |x^2 - 3x + 9| \cdot |x + 3| \leq (|x^2| + |3x| + |9|)|x + 3| \leq (16 + 3(4) + 9)|x + 3| = 37|x + 3| < 37\delta \leq 37 \cdot \frac{\varepsilon}{37} = \varepsilon$, q.e.d.

   (Note we used $|x + 3| = |x - (-3)| < \delta \implies |x + 3| < 1 \implies x \in (2, 3)$, as well as the triangle inequality $|a + b + c| \leq |a| + |b| + |c|$.)

15. Let $\varepsilon > 0$. Set $\delta = \min\{a, \varepsilon\sqrt{a}\}$. Then $\delta > 0$ exists and $|x - a| < \delta \implies |f(x) - f(a)| = |\sqrt{x} - \sqrt{a}| = \left|\frac{(\sqrt{x} - \sqrt{a})(\sqrt{x} + \sqrt{a})}{\sqrt{x} + \sqrt{a}}\right| = \frac{|x - a|}{\sqrt{x} + \sqrt{a}} \leq \frac{|x - a|}{\sqrt{a}} < \frac{1}{\sqrt{a}}\delta \leq \frac{1}{\sqrt{a}} \cdot \varepsilon\sqrt{a} = \varepsilon$, q.e.d.

   (Note: we needed $\delta \leq a$ so that $|x - a| < \delta \implies x > 0$, so that $\sqrt{x}$ is defined.)

17. (a) Let $\varepsilon > 0$. Set $\delta = \varepsilon^3$. Then $\delta > 0$ exists and $|x - 0| < \delta \implies |f(x) - f(0)| = |\sqrt[3]{x} - 0| = \sqrt[3]{|x|} < \sqrt[3]{\delta} = \sqrt[3]{\varepsilon^3} = \varepsilon$, q.e.d.

   (b) Let $\varepsilon > 0$. Let δ be the minimum of 1 and $\varepsilon \cdot (7^{2/3} + 2 \cdot 7^{1/3} + 4)$. (Note that $|x - 8| < \delta \leq 1 \implies x \in (7, 9)$.) Then

$|x - 8| < \delta \implies |f(x) - f(8)| = |\sqrt[3]{x} - 2|$

$= |\sqrt[3]{x} - 2| \cdot \frac{|x^{2/3} + 2x^{1/3} + 4|}{|x^{2/3} + 2x^{1/3} + 4|}$

$= \frac{|x - 8|}{|x^{2/3} + 2x^{1/3} + 4|}$

$< \frac{\delta}{|7^{2/3} + 2 \cdot 7^{1/3} + 4|}$

$\leq \frac{\varepsilon \cdot (7^{2/3} + 2 \cdot 7^{1/3} + 4)}{(7^{2/3} + 2 \cdot 7^{1/3} + 4)}$

$= \varepsilon$, q.e.d.

19. No. Recall $P \longrightarrow \mathscr{F}$ is false anytime P is true, and $|f(x) - f(a)| < \varepsilon$ is false if $f(a)$ is undefined. Therefore $|x - a| < \delta \longrightarrow |f(x) - f(a)| < \varepsilon$ is false anyime $|x - a| < \delta$ is true. Therefore the continuity definition is not satisfied.

Exercises 2.2:

1. all $x \in \mathbb{R}$
3. $x > 7/3$
5. $-4 < x < 4$
7. all $x \in \mathbb{R}$
9. all $x \in \mathbb{R}$
11. all $x \in \mathbb{R}$
13. $x < 1$
15. all $x \in \mathbb{R}$

17. (a) no, (b) no, (c) no, (d) essential

19. (a) no, (b) no, (c) no, (d) removable

21. $A = 5$

23. Assume $C \neq 0$. Let $\varepsilon > 0$. Then $\varepsilon_1 = \varepsilon/|C| > 0$, so by the continuity of $f(x)$ at $x = a$ we have there exists $\delta_1 > 0$ so that $|x - a| < \delta_1 \implies |f(x) - f(a)| < \varepsilon_1$. Thus $|x - a| < \delta_1 \implies |Cf(x) - Cf(a)| = |C| \cdot |f(x) - f(a)| < |C|\varepsilon_1 = |C| \cdot \frac{\varepsilon}{|C|} = \varepsilon$, q.e.d. (with $\delta = \delta_1$).

25. (If) Suppose $f(x)$ is both left-continuous and right-continuous at $x = a$. Let $\varepsilon > 0$. Then there exists $\delta_1 > 0$ so that $x \in (a - \delta_1, a] \implies |f(x) - f(a)| < \varepsilon$. There also exists $\delta_2 > 0$ so that $x \in [a, a + \delta_2) \implies |f(x) - f(a)| < \varepsilon$. Choose $\delta = \min\{\delta_1, \delta_2\}$, so $\delta > 0$ exists, and $|x - a| < \delta \implies |x - a| < \delta_1 \wedge |x - a| < \delta_2 \implies x \in (a - \delta_1, a] \cup [a, a + \delta_2) \implies |f(x) - f(a)| < \varepsilon$, q.e.d.

(Only If) Suppose $f(x)$ is continuous at $x = a$. Then $x \in (a - \delta, a] \implies |x - a| < \delta \implies |f(x) - f(a)| < \varepsilon$. and so $f(x)$ is left-continuous. Furthermore, $x \in [a, a + \delta) \implies |x - a| < \delta \implies |f(x) - f(a)| < \varepsilon$, and so $f(x)$ is right-continuous, q.e.d.

Exercises 2.3:

1.

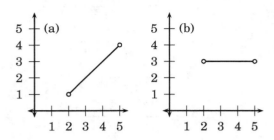

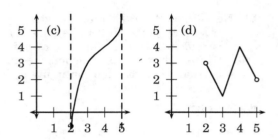

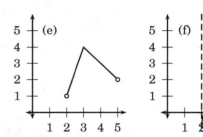

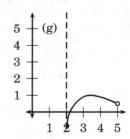

3.

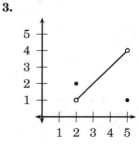

5. If $f(x) = x^5 - 8x^2 + 15x$, then $f(2) = 30$ while $f(3) = 216$. Since $f(x)$ is continuous on $[2,3]$, and 97 is between 30 and 216, there exists $x_0 \in (2,3)$ such that $f(x_0) = 97$.

7. $f(x) < 0$ on $(-3, 3)$.

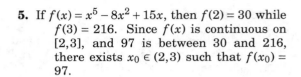

9. $f(x) = x^2 + 3x - 2 \geq 0$ on $\left(-\infty, \frac{1}{2}(-3-\sqrt{17})\right] \cup \left[\frac{1}{2}(-3+\sqrt{17}), \infty\right)$

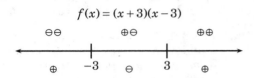

11. $x^2 + 5 > 0$ for all $x \in \mathbb{R}$.

13. $f(x) = \frac{x}{x+5} < 0$ on $(-5, 0)$.

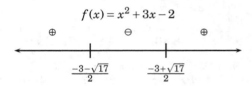

15. $f(x) = \frac{(x+4)(x-4)}{x^2+16} > 0$ on $(-\infty, -4) \cup (4, \infty)$.

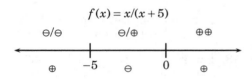

17. $f(x) = \frac{x(27-x)}{(x+6)(x+5)} < 0$ on $(-\infty, -6) \cup (-5, 0) \cup (27, \infty)$.

19. $f(x) = \frac{x^2 - 8x - 5}{(x+5)(x-7)} > 0$ on $(-\infty, -5) \cup (4-\sqrt{21}, 7) \cup (7, \infty)$.

Answers to Odd-Numbered Problems

$$f(x) = \frac{x^2-8x-5}{(x+5)(x-7)}$$

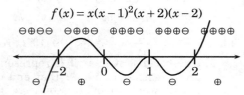

21. $f(x) = \frac{(x+1)(x-1)}{(x^2+2)(x-\sqrt{5})(x+\sqrt{5})} \leq 0$ on $(-\sqrt{5}, -1] \cup [1, \sqrt{5})$

$$f(x) = \frac{(x+1)(x-1)}{(x^2+2)(x-\sqrt{5})(x+\sqrt{5})}$$

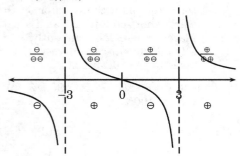

23. $f(x) = x(x-1)^2(x+2)(x-2)$

$$f(x) = x(x-1)^2(x+2)(x-2)$$

25. $f(x) = \frac{x}{(x+3)(x-3)}$

Exercises 2.4:

1. (a) -1, (b) -3, (c) DNE, (d) 1.

3. (a) discontinuous ($\lim_{x \to -2} f(x) = -1$, $f(-2) = 1$), (b) continuous (limit and function values are the same, -3), (c) discontinuous (limit DNE, cannot equal $f(3)$), (d) continuous (limit and function values are the same, 1).

5. 3.

7. DNE.

9. 3.

11. 0.

13. (a) 1/2, (b) 1/3, (c) 0, (d) DNE, (e) DNE, (f) 0. (g) Knowing the form is 0/0 was insufficient in knowing the limits value or existence; limits in (a)–(f) were all 0/0 form, but values and existence varied.

15. 10.

17. $-1/9$.

19. 4.

21. 1/32.

23. (a) 0, (b) -2, (c) 1/2, (d) DNE.

25. $-1/4$.

Exercises 2.5:

1. -3, DNE, DNE.

3. 3, 4, DNE.

5. 0, 0, 0.

7. No, no, yes.

9. Yes, yes, yes.

11. Yes, yes, yes.

13. $\lim_{x \to -1^+} f(x) = \lim_{x \to -1^+} [-2(x-1)] = 4$,

$\lim_{x \to -1^-} f(x) = \lim_{x \to -1^-} \left[\frac{3}{2}(x+3)\right] = 3$,

$\therefore \lim_{x \to -1} f(x)$ DNE.

No, no, yes. ($f(-1) = 4$).

15. $\lim_{x \to 1^+} f(x) = \lim_{x \to 1^+} (x-1) = 0$,

$\lim_{x \to 1^-} f(x) = \lim_{x \to 1^-} [-2(x-1)] = 0$,

$\therefore \lim_{x \to 1} f(x) = 0$.

Yes, yes, yes. ($f(1) = 0$).

17. (a) DNE, 0, DNE. (b) Right continuity only. (c) 0, DNE, DNE. (d) Left continuity only.

19. 2/7.

21. Does not exist. Left-side limit is 1/8; right-side limit is $-1/8$.

23. 0.

Exercises 2.6:

1. ∞

3. $-\infty$

5. ∞

7. DNE

9. 0

11. $-\infty$

13. DNE

15. ∞

17. $-1/6$

19. $-\infty$

21. DNE

23. (a)–(c)

$$f(x) = \frac{1}{(x^2+3)(x+\sqrt{3})(x-\sqrt{3})}$$

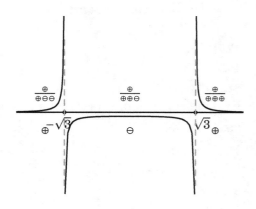

Exercises 2.7:

1. 0. (Form $0^+ \cdot B$.)

3. As $x \to 5^+$, since $x^2 - 25 > 0$ we have $\frac{x-1}{x^2-25} \le \frac{x+\cos x}{x^2-25} \le \frac{x+1}{x^2-25}$. $\lim_{x \to 5^+} \frac{x-1}{x^2-25} \xrightarrow{4/0^+} \infty$ and so $\lim_{x \to 5^+} \frac{x+\cos x}{x^2-25} = \infty$. (That the third function also has limit ∞ can also be mentioned.)

5. (a) $-\infty$, (b) ∞, (c) ∞, (d) $-\infty$.

7. Answers vary.

9. $\lim_{x \to 2} 4 = 4$, $\lim_{x \to 2}[(x-2)^2 + 4] \xrightarrow{\text{cont}} 4$, so $\lim_{x \to 2} f(x) = 4$ by the Sandwich Theorem.

11. DNE. (Too many places $x \sin \frac{1}{x}$ is negative.)

13. 0: $0 \le \sqrt{x^2 \sin^2 \frac{1}{x}} \le \sqrt{x^2}$, $\lim_{x \to 0} 0 = 0$, $\lim_{x \to 0} \sqrt{x^2} \xrightarrow{\text{cont}} 0$, therefore $\lim_{x \to 0} \sqrt{x^2 \sin^2 \frac{1}{x}} = 0$.

15. Form $0^+ \cdot \sin(\csc(\infty)) = 0^+ \sin(\text{DNE})$, so DNE. (Too many holes in the graph, where $\csc(1/x)$ is undefined, as $x \to 0^+$.)

17. 0: $-|\sqrt[3]{x}| \le \sqrt[3]{\sin\left(\frac{1}{x}\right)} \le |\sqrt[3]{x}|$. First and third functions have limit zero (by continuity), and therefore so does the second.

19. 1.   21. ∞ 23. DNE

25. ∞

Exercises 2.8:

1. ∞ 7. $-\infty$ 13. 3/2
3. ∞ 9. 0
5. ∞ 11. 0   15. $-\infty$

17. (a) 0, (b) 0, (c) ∞, (d) 1/3, (e) DNE.

19. ∞

21. Letter choices may vary. Here we assume $M, N \in \mathbb{R}$.

   (a) $(\forall M)(\exists N)[x > N \to f(x) > M]$,
   (b) $(\forall M)(\exists N)[x < N \to f(x) > M]$,
   (c) $(\forall M)(\exists N)[x > N \to f(x) < M]$,
   (d) $(\forall M)(\exists N)[x < N \to f(x) < M]$.

23. (a) $R_p = \frac{1}{\frac{1}{10} + \frac{1}{R}} = \frac{10R}{R+10}$, (b) 10Ω, reasonable because it is as if R represents infinite resistance, i.e., if that end of the circuit were "open" (disconnected), and no charges could pass through there, so all must pass through the 10Ω resistor. (c) 0, as if the resistor on the right were "short-circuiting," as if a wire were placed where R is currently, and virtually all current would cross that wire instead of the 10Ω resistor.

25. (a) 0, (b) 63.2%, 86.5%, 95.0%, (c) $q(t) \to CV$, or 100%.

27. (a) 0   (d) $e^0 = 1$
    (b) ∞ (e) $e^{-1} = \frac{1}{e}$.
    (c) 0

29. (a) ∞ (d) ∞
    (b) 0   (e) ∞
    (c) ∞ (f) $-\infty$

31. $\lim_{x \to \infty} \frac{p(x)}{q(x)} = \lim_{x \to \infty} \frac{a_n x^n \left(1 + \frac{a_{n-1}}{a_n x} + \cdots + \frac{a_0}{a_n x^n}\right)}{b_m x^m \left(1 + \frac{b_{m-1}}{b_m x} + \cdots + \frac{b_0}{b_m x^m}\right)} = \left(\lim_{x \to \infty} \frac{a_n x^n}{b_m x^m}\right) \cdot \left(\lim_{x \to \infty} \frac{1 + \frac{a_{n-1}}{a_n x} + \cdots + \frac{a_0}{a_n x^n}}{1 + \frac{b_{m-1}}{b_m x} + \cdots + \frac{b_0}{b_m x^m}}\right)$, since this second limit in the product exists and is finite. In fact this second limit is 1, which proves the theorem for the case $x \to \infty$. The case $x \to -\infty$ can be proved the same way.

Answers to Odd-Numbered Problems

Exercises 2.9:

1. 9
3. 9/7
5. ∞
7. 1

9. 1
11. 0
13. $8/\pi^2$
15. 1

17. 1
19. 1
21. 1
23. 0

25. DNE. Denominator is zero infinitely many times as $x \to 0$.

27. (a) $u = \csc x = \frac{1}{\sin x} \xrightarrow{1/0^+} \infty$ as $x \to 0^+$. Get
$$\lim_{u \to \infty} \frac{2u^2 + 3u + 11}{5u^2 + 4u + 7} = 2/5.$$
(b) Multiply by $(\sin^2 x)/(\sin^2 x)$.
$$\lim_{x \to 0^+} \frac{2 + 4\sin x + 11\sin^2 x}{5 + 4\sin x + 7\sin^2 x} = 2/5.$$

29. $\xequals{1/5 + \sin 0^+}$ $1/5$.
31. $\xequals{\infty + B}$ ∞.

Exercises 3.1:

1. -2
3. $4x$
5. $\frac{1}{2\sqrt{x}}$
7. $\frac{-5}{2\sqrt{9-5x}}$
9. $-x^{-3/2} = \frac{-1}{x\sqrt{x}}$
11. $\frac{1}{3}(x+1)^{-2/3}$
13. $\frac{1}{(x+1)^2}$
15. (a) $v(t) = -32t + 15$, (b) 15, -305, (c) $t = 15/32$, (d) $1505/64 \approx 23.52$
17. $y = 7 + 8(x-4)$
19. $y = 3 + \frac{1}{6}(x-9)$
21. (a) 2, (b) 2, (c) 0.8
23. $\frac{2}{3}x^{-1/3}$

Exercises 3.2:

1. $2x - 199$
3. $7t^6 - 19$
5. $-72y^7$
7. $5x^2 - 4x - \frac{8}{3}$
9. $5\cos x - 3\sin x$
11. $20y - 3 - 4\cos y$
13. $\frac{dx^{-1}}{dx} = (-1)x^{-1-1} = \frac{-1}{x^2}$
15. Omitted for space reasons are $f(-3) = -27$, $f'(-3) = 27$, $f(3) = 27$, $f'(3) = 27$, and tangent lines.

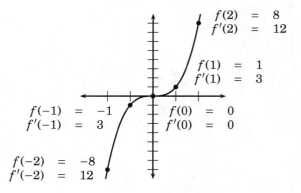

17. Assuming $f(x) = x^n$, $n = 1, 2, 3, \ldots$. Then

$$f'(x) = \lim_{\Delta x \to 0} \frac{(x + \Delta x)^n - x^n}{\Delta x}$$
$$= \lim_{\Delta x \to 0} \frac{((x + \Delta x) - (x))\left((x + \Delta x)^{n-1} + (x + \Delta x)^{n-2}x + \cdots + (x + \Delta x)x^{n-2} + x^{n-1}\right)}{\Delta x}$$
$$= \lim_{\Delta x \to 0} \frac{\Delta x \left((x + \Delta x)^{n-1} + (x + \Delta x)^{n-2}x + \cdots + (x + \Delta x)x^{n-2} + x^{n-1}\right)}{\Delta x}$$
$$= \lim_{\Delta x \to 0} \left((x + \Delta x)^{n-1} + (x + \Delta x)^{n-2}x + \cdots + (x + \Delta x)x^{n-2} + x^{n-1}\right)$$
$$= x^{n-1} + x^{n-1} + \cdots + x^{n-1} + x^{n-1}$$
$$= nx^{n-1}, \text{ q.e.d.}$$

19. $f'(x) = 4x^3 - 8x = 4x(x^2 - 2) = 0$ at $x = 0, \pm\sqrt{2}$. Local maximum at $(0,0)$. Local minima at $(\pm\sqrt{2}, -4)$.

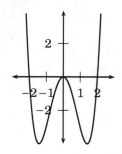

21. $f(x) = x^4 \Longrightarrow f'(x) = 4x^3 \Longrightarrow f'(2) = 32$.

23. $f(x) = \sin x \Longrightarrow f'(x) = \cos x \Longrightarrow f'(1) = \cos 1 \approx 0.5403$

25. $f(x) = \cos x \Longrightarrow f'(x) = -\sin x \Longrightarrow f'(0) = 0$.

27.
$$f'(x) = \lim_{\Delta x \to 0} \frac{\cos(x + \Delta x) - \cos x}{\Delta x}$$
$$= \lim_{\Delta x \to 0} \frac{\cos x \cos(\Delta x) - \sin x \sin(\Delta x) - \cos x}{\Delta x}$$
$$= \lim_{\Delta x \to 0} \left[\cos x \frac{\cos(\Delta x) - 1}{\Delta x} - \sin x \frac{\sin(\Delta x)}{\Delta x}\right]$$
$$= \cos x \cdot 0 - \sin x \cdot 1 = -\sin x, \text{ q.e.d.}$$

29. $f'(x) = 1 - 2\sin x = 0 \iff \sin x = \frac{1}{2}$. Local maxima at $\frac{\pi}{6} + 2n\pi$ where $n \in \{0, \pm 1, \pm 2, \ldots\}$, local minima at $\frac{5\pi}{6} + 2n\pi$, same n. Increasing from each minimum to closest adjacent maximum to its right, and decreasing from each maximum to closest adjacent minimum to its right. $\lim_{x \to \infty} f(x) \xrightarrow{B+\infty}$, $\lim_{x \to -\infty} f(x) \xrightarrow{B-\infty} -\infty$. x-axis is marked every $\pi/6$ units, while y-axis is marked every 1 unit.

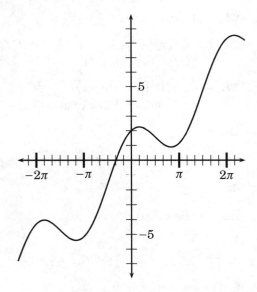

31. Continuous for all x (check two-sided limits at $x = 0, \pi$).
$$f'(x) = \begin{cases} 0 & \text{if } x > \pi \\ \cos x & \text{if } 0 \leq x < \pi \\ 1 & \text{if } x < 0 \end{cases}$$

$f'(x)$ DNE at $x = \pi$.

33.
$$\frac{du^2}{dx} = \frac{du^2}{du} \cdot \frac{du}{dx} = 2u \cdot \frac{du}{dx} = 2(2x+5)(2)$$
$$= 8x + 20.$$

35. $V = s^3 \Longrightarrow \frac{dV}{ds} = 3s^2 \Longrightarrow \frac{dV}{ds}\Big|_{s=10\,cm} = 300\,cm^2$, or more precisely, $300\,cm^3/cm$, in the units of $\Delta V / \Delta s$.

37. (B)

39. (A)

Exercises 3.3:

1. (a) feet/sec, (b) feet/sec^2, (c) feet/sec^3

3. (a) (All coordinates are $(\pm\sqrt{2}, \pm\sqrt{2})$, with $\sqrt{2} \approx 1.414$.

$t = 3\pi/4$		$t = \pi/4$
$dx/dt = -\sqrt{2}$		$dx/dt = -\sqrt{2}$
$dy/dt = -\sqrt{2}$		$dy/dt = \sqrt{2}$
$t = 5\pi/4$		$t = 7\pi/4$
$dx/dt = \sqrt{2}$		$dx/dt = \sqrt{2}$
$dy/dt = -\sqrt{2}$		$dy/dt = \sqrt{2}$

5. 300 (units are cm^3/cm)

7. (a) $4\pi r^2$, (b) $r = \sqrt{5}$ cm

9. (a) $I = 6t^2 + 3$, (b) 99

11. 16 watts

13. 44.27 m/sec (from comparing various potential and kinetic energies)

Exercises 3.4:

1. $-11x^{-12} = \frac{-11}{x^{12}}$

3. $\frac{4}{3}x^{1/3} = \frac{4}{3}\sqrt[3]{x}$

5. $-t^{-3} = \frac{-1}{t^3}$

7. $-99(1-9x)^{10}$

9. $\frac{5x^4}{\sqrt{2x^5-1}}$

11. $100(x+5)^{99}$

13. $100(3x^2 + 5x - 9)^{99}(6x + 5)$

15. $-3(x^4 - x + 1)^{-4}(4x^3 - 1) = \frac{-3(4x^3-1)}{(x^4-x+1)^4}$

17. $\frac{1}{2\sqrt{x+\sqrt{x+\sqrt{x}}}} \cdot \left[1 + \frac{1}{2\sqrt{x+\sqrt{x}}} \cdot \left(1 + \frac{1}{2\sqrt{x}}\right)\right]$

19. $-\sin\theta \cdot \frac{d\theta}{dt}$

21. $\cos(\cos x)$

23. $\frac{\cos\sqrt{x}}{2\sqrt{x}}$

25. $-3\cos^2 x \sin x$

27. 14

29. (a) $\frac{dy}{dx}\big|_{3/5} = \frac{-x}{\sqrt{1-x^2}}\big|_{3/5} = -3/4$. Tangent line:
$$y = \frac{4}{5} - \frac{3}{4}\left(x - \frac{3}{5}\right) = -\frac{3}{4}x + \frac{5}{4}.$$
(b) $\frac{dy}{dx}\big|_{(3/5, 4/5)} = \frac{-x}{y}\big|_{(3/5, 4/5)} = -\frac{3}{4}$. Same tangent line equation as in (a).

31. $\frac{dF}{dr} = -2kq_1q_2 r^{-3} = -2 \cdot \frac{kq_1q_2}{r^2} \cdot \frac{1}{r} = -2\frac{F}{r}$, q.e.d.

33.
$$\frac{dF}{dr}\bigg|_{r=0.5} = -2Gm_1m_2 r^{-3}\bigg|_{r=.5}$$
$$= -1.062 \times 10^{-6} m_1 m_2.$$

35. (a) $C'(x) = 0.009x^2 + 0.4x + 0.3$, (b) $C'(4) = 2.04400$, (c) 2.04403, (d) very close.

Exercises 3.5:

1.
$$\frac{d}{dx}\left[x^2 \cdot x^2\right] = x^2 \frac{dx^2}{dx} + x^2 \frac{dx^2}{dx}$$
$$= x^2 \cdot 2x + x^2 \cdot 2x = 4x^3,$$
$$\frac{d}{dx}[x^4] = 4x^3.$$

3.
$$\frac{d}{dx}[\cos x \sec x] = \cos x \cdot \frac{d\sec x}{dx} + \sec x \cdot \frac{d\cos x}{dx}$$
$$= \cos x \sec x \tan x + \sec x(-\sin x)$$
$$= \tan x - \tan x = 0,$$
$$\frac{d}{dx}[\cos x \sec x] = \frac{d\,1}{dx} = 0.$$

5.
$$\frac{d}{dx}\left[\frac{1}{\cos x}\right] = \frac{\cos x \frac{d\,1}{dx} - 1 \frac{d\cos x}{dx}}{(\cos x)^2} = \frac{0-(-\sin x)}{\cos^2 x}$$
$$= \frac{1}{\cos x} \cdot \frac{\sin x}{\cos x} = \sec x \tan x,$$
$$\frac{d}{dx}[\sec x] = \sec x \tan x.$$

7. $\frac{d}{dx}[\sin x \sec x] = \sin x \frac{d\sec x}{dx} + \sec x \frac{d\sin x}{dx}$
$= \sin x \sec x \tan x + \sec x \cos x$
$= \tan^2 x + 1 = \sec^2 x,$
$\frac{d\tan x}{dx} = \sec^2 x.$

9. $2\tan x \sec^2 x$

11. $\frac{\csc 2x \cot 2x}{\sqrt{1-\csc 2x}}$

13. $\frac{(x^2-3)(3x^2-7)-(x^3-7x+5)(2x)}{(x^2-3)^2} = \frac{x^4-2x^2-10x+21}{(x^2-3)^2}$

15.
$$\cos^3 x \cdot 2\sin x \cos x + \sin^2 x \cdot 3\cos^2 x(-\sin x)$$
$$= \sin x \cos^2 x(2\cos^2 x - 3\sin^2 x)$$
$$= \sin x \cos^2 x(2 - 5\sin^2 x)$$

17. $\frac{(x-1)(4x^3)-(x^4+3)(1)}{(x-1)^2} = \frac{3x^4-4x^3-3}{(x-1)^2}$

19. $\frac{(\cos x+2)(\cos x)-(\sin x+3)(-\sin x)}{(\cos x+2)^2} = \frac{1+2\cos x+3\sin x}{(\cos x+1)^2}$.

21. $\frac{\sqrt{x^2+1}\cdot 1 - x \cdot \frac{2x}{2\sqrt{x^2+1}}}{x^2+1} = \frac{1}{(x^2+1)^{3/2}}$

23. $-54\cot 9x \csc^2 9x + \frac{3\sin 6x}{\sqrt{\cos 6x+1}}$

25. $f_1' f_2 f_3 f_4 f_5 + f_1 f_2' f_3 f_4 f_5 + f_1 f_2 f_3' f_4 f_5 + f_1 f_2 f_3 f_4' f_5 + f_1 f_2 f_3 f_4 f_5'$

27. $\frac{d}{dx}[(f(x))^{-1}] = -1(f(x))^{-2}\frac{df(x)}{dx} = \frac{-f'(x)}{(f(x))^2}$. The Quotient Rule can also be used.

29. $\pi\sqrt{\frac{1}{3g}}$

31. (a) W/m^2, (b) W/m^3, (c) 1.74 m

33. $q = \frac{fp}{p-f} \implies \frac{dq}{dp} = \frac{-f^2}{(p-f)^2}$.

Exercises 3.6:

1. $-A/B$

3. $\frac{dy}{dx} = x/y$, $y = -4 - \frac{5}{4}(x-5) = -\frac{5}{4}x + \frac{9}{4}$

5. $(1 + \sec^2 x)/(3y^2 + \sec y \tan y)$

7. $(-y\cos x - \sin y)/(x\cos y + \sin x)$

9. y/x (after some multiplication and division of the equation one gets from differentiation)

11. (a) $\frac{4x(x^2-y^2)-6x(x^2+y^2)^2}{4y(x^2-y^2)+6y(x^2+y^2)^2} = \frac{x}{y} \cdot \frac{2(x^2-y^2)-3(x^2+y^2)^2}{2(x^2-y^2)+3(x^2+y^2)^2}$

(b) Slope is $\frac{7\sqrt{3}}{3}$. Line is $y = \frac{\sqrt{3}}{4} + \frac{7\sqrt{3}}{3}\left(x - \frac{1}{4}\right)$.

(c) $dy/dx \xrightarrow{0/2}_{0/1} 0$ the limit form depending on final expression for dy/dx.

(d) $\lim_{x \to 1^-} \frac{dy}{dx}$, $\lim_{x \to -1^+} \frac{dy}{dx}$ do not exist (graph has "vertical tangents" at those points).

(e) $\lim_{y \to 1^-} \frac{dy}{dx} = 0$.

13. (a)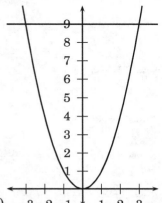

(b) $x = \pm 3$, (c) where graphs intersect (where heights of f and g are the same), (d) $x = 0$, (e) x-value where graphs have the same slope, (f) heights being equal is not the same as slopes being equal.

15. $xy^3 - x^2y = 8 \implies dy/dx = \frac{y(2x-y^2)}{3xy^2-x^2}$. (Find where $dy/dx = 0$ on the curve.) $y = 0$ is impossible from the original equation $xy^3 - x^2y = 8$. Therefore any such point will satisfy $y = \pm\sqrt{2x} \implies x > 0$ ($x \neq 0$ for the same reason $y \neq 0$, and because of the denominator in dy/dx.) From original equation (using $y = \sqrt{2x}$) we get $\sqrt{2}x^{5/2} = 8 = 2^{6/2}$ and so $x^{5/2} = 2^{5/2}$ and so $x = 2$, also implying $y = \sqrt{2(2)} = 2$, so only $(2,2)$ is a point of horizontal tangency.

17.
$g(x)y = f(x)$
$\implies g'(x)y + g(x)\frac{dy}{dx} = f'(x)$
$\implies \frac{dy}{dx} = f'(x)/g(x) - g'(x)y/g(x)$
$= f'(x)/g(x) - g'(x)[f(x)/g(x)]/g(x)$
$= \frac{f'(x)g(x) - g'(x)f(x)}{(g(x))^2}$, q.e.d.

Exercises 3.7:

1. $\frac{1}{2\sqrt{\sin^{-1} x} \cdot \sqrt{1-x^2}}$

3. $\frac{1}{2\sqrt{\tan^{-1} x} \cdot (x^2+1)}$

5. $\frac{1}{2\sqrt{\sec^{-1} x} \cdot |x|\sqrt{x^2-1}}$

7. $\frac{1}{\sqrt{1-x} \cdot 2\sqrt{x}}$

9. $\frac{1}{2\sqrt{x}(x+1)}$

11. $\frac{1}{2x\sqrt{x-1}}$ (Note: $x > 0$ necessarily.)

13. $\frac{-x}{|x|\sqrt{x^2-1}} + \csc^{-1} x$

15. $2x \tan^{-1} x$

17. $\frac{1}{x^2+9}$

19. $\frac{25}{|\frac{x}{5}|\sqrt{\frac{x^2}{25}-1}} \cdot \frac{1}{5} + \sqrt{25-x^2} + x \cdot \frac{1}{2\sqrt{25-x^2}} \cdot (-2x) = \frac{125}{|x|\sqrt{x^2-25}} + \sqrt{25-x^2} - \frac{x^2}{\sqrt{25-x^2}}$

21.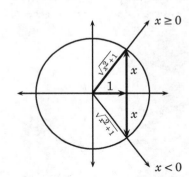

$\frac{d}{dx}\left[\sin^{-1}\left(\frac{x}{\sqrt{x^2+1}}\right)\right] = \frac{d \tan^{-1} x}{dx} = \frac{1}{x^2+1}$.

23. $\frac{d \sin^{-1} \frac{1}{x^2}}{dx} = \frac{d \csc^{-1}(x^2)}{dx} = \frac{-1}{x^2\sqrt{x^4-1}} \cdot 2x = \frac{-2}{x\sqrt{x^4-1}}$

25. $\frac{d}{dx} \cos^{-1}\left(\frac{x}{\sqrt{25+x^2}}\right) = \frac{d}{dx} \cot^{-1}\frac{x}{5} = \cdots = \frac{-5}{x^2+25}$

27.
- Case $x > 0$: Then $|x| = x$ and so $\frac{d}{dx}\sec^{-1}|x| = \frac{d}{dx}\sec^{-1} x = \frac{1}{|x|\sqrt{x^2-1}} = \frac{1}{x\sqrt{x^2-1}}$.
- Case $x < 0$: Then $|x| = -x$, and so $\frac{d}{dx}\sec^{-1}|x| = \frac{d}{dx}\sec^{-1}(-x) = \frac{1}{|-x|\sqrt{(-x)^2-1}} \cdot \frac{d(-x)}{dx} = \frac{-1}{|x|\sqrt{x^2-1}} = \frac{-1}{(-x)\sqrt{x^2-1}} = \frac{1}{x\sqrt{x^2-1}}$, q.e.d.

29. These say the same, though with the variable names changed. We can derive the other texts' version with implicit differentiation (starting with $x = f^{-1}(y)$) as we did when deriving derivatives of inverse trigonometric functions, or we can see it if we replace the dummy variable x with y:
$$g'(y) = \frac{1}{f'(g(y))}.$$
Since $x = g(y)$ the left-hand side is dx/dy. The right-hand side is $1/(dy/dx) = 1/f'(x)$ which is $1/f'(g(y))$.

31. $\frac{d}{dx}\left[x\sin^{-1}x + \sqrt{1-x^2}\right] = x \cdot \frac{1}{\sqrt{1-x^2}} + \sin^{-1}x + \frac{-2x}{2\sqrt{1-x^2}} = \frac{x}{\sqrt{1-x^2}} + \sin^{-1}x - \frac{x}{\sqrt{1-x^2}} = \sin^{-1}x$, q.e.d.

Exercises 3.8:

1. (a) $e^{\sin x}\cos x$; $e^x \cos e^x$, (b) $e^{\cos x}(-\sin x)$; $-e^x \sin e^x$, (c) $e^{\tan x}\sec^2 x$; $e^x \sec^2 e^x$, (d) $e^{\cot x}(-\csc^2 x)$; $-e^x \csc^2 e^x$, (e) $e^{\sec x}\sec x \tan x$; $e^x \sec e^x \tan e^x$, (f) $e^{\csc x}(-\csc x \cot x)$; $-e^x \csc e^x \cot e^x$.

3. $-e^{-x}$

5. $\frac{-2e^{1/x^2}}{x^3}$

7. $6x^2 \sin e^{x^3} \cos e^{x^3}$

9. $\frac{(x^2+1)\cdot 2e^{2x} - e^{2x}(2x)}{(x^2+1)^2} = \frac{e^{2x}(2x^2-2x+2)}{(x^2+1)^2}$

11. $\frac{e^x + e^{-x}}{2}$

13. xe^x

15. $(a > 1) \wedge (x_2 > x_1) \implies a^{x_2} - a^{x_1} = \underbrace{a^{x_1}}_{>0}(a^{x_2-x_1} - 1) > a^{x_1}(a^0 - 1) = 0$
$\implies a^{x_2} > a^{x_1}$.

17. $x > 0$

19. $a = 1$ is trivial; $a = 0$ is trivial for $x > 0$ and nonexistent for $x \leq 0$; $a < 0$ has infinitely many "holes," e.g., $x = 1/2, 3/4$, and any other power that represents an even root to an odd power, and also any irrational number x (see Exercise 16).

Exercises 3.9:

1. $\frac{4}{x}$

3. $\frac{5}{x}$

5. $\frac{1}{3x}$

7. $2x$

9. $\cot x$

11. $\sec x \csc x$

13. $\tan x$

15. $\frac{2\ln x}{x}$

17. $\frac{\cos(\ln x)}{x}$

19. $\frac{\sec(\ln|x|)\tan(\ln|x|)}{x}$

21. $\frac{1}{\ln(\ln x)} \cdot \frac{1}{\ln x} \cdot \frac{1}{x}$

23. $\frac{43x+69}{x^2+3x}$ (simplified)

25. $\frac{9}{9x-1} - \frac{2}{2x+4} = \frac{9}{9x-1} - \frac{1}{x+2} = \frac{19}{(9x-1)(x+2)}$

27. $\frac{2}{x} + 3\cot x + 8\tan 2x - \frac{2x}{3(x^2-9)}$

29. $\frac{d}{dx}\ln|\sec x + \tan x| = \frac{\sec x \tan x + \sec^2 x}{\sec x + \tan x} = \frac{\sec x(\tan x + \sec x)}{\sec x + \tan x} = \sec x$

31.
   (a) See 29 above.
   (b) $\frac{d}{dx}\ln|\csc x + \cot x| = \frac{1}{\csc x + \cot x} \cdot (-\csc x \cot x - \csc^2 x) = \frac{-\csc x(\cot x + \csc x)}{\csc x + \cot x} = -\csc x$.

33. $f(x) = (g(x))^n \implies f'(x) = n(g(x))^{n-1}g'(x) + (g(x))^n (0) \ln(g(x)) = n(g(x))^{n-1} \cdot g'(x)$.

35.
   (a) $y = \log_1 x \iff 1^y = x$ which requires $x = 1$ but then y can be any real number, and so the graph would be a vertical line (the set $\{(x,y) \mid (x=1) \wedge (y \in \mathbb{R})\}$), which does not represent a function.
   (b) The domain would be $\{1\}$.
   (c) (3.122) would imply $\log_1 x = (\ln x)/(\ln 1) = (\ln x)/0$, which is undefined.

37. $\frac{d}{dx}\left[x\sec^{-1}x - \ln\left|x+\sqrt{x^2-1}\right|\right] = x \cdot \frac{1}{x\sqrt{x^2-1}} + \sec^{-1}x - \frac{1}{x+\sqrt{x^2-1}} \cdot \left[1 + \frac{2x}{2\sqrt{x^2-1}}\right] = \frac{1}{\sqrt{x^2-1}} + \sec^{-1}x - \frac{\sqrt{x^2-1}+x}{\sqrt{x^2-1}\left(x+\sqrt{x^2-1}\right)} = \frac{1}{\sqrt{x^2-1}} + \sec^{-1}x - \frac{1}{\sqrt{x^2-1}} = \sec^{-1}x$.

Exercises 3.10:

1. $4^{\sin x}(\cos x)\ln 4$

3. $\frac{\sec x \csc x}{\ln 2}$

5. $\log_4 3 = \frac{\ln 3}{\ln 4}$

7. $\frac{2\cot x}{\ln 2} - \frac{4\tan x}{\ln 2}$

Answers to Odd-Numbered Problems

9. $y = a^x \implies \ln y = x \ln a \implies \frac{1}{y} \cdot \frac{dy}{dx} = \ln a \implies \frac{dy}{dx} = y \ln a \implies \frac{dy}{dx} = a^x \ln a$, q.e.d.

11. $y = \frac{u}{v} \implies \ln|y| = \ln|u| - \ln|v| \implies \frac{1}{y} \cdot \frac{dy}{dx} = \frac{1}{u}\frac{du}{dx} - \frac{1}{v} \cdot \frac{dv}{dx} \implies \frac{dy}{dx} = y\left[\frac{1}{u}\frac{du}{dx} - \frac{1}{v} \cdot \frac{dv}{dx}\right] = \frac{u}{v}\left[\frac{1}{u}\frac{du}{dx} - \frac{1}{v} \cdot \frac{dv}{dx}\right] = \frac{1}{v} \cdot \frac{du}{dx} - \frac{u}{v^2} \cdot \frac{dv}{dx} = \frac{v\frac{du}{dx} - u\frac{dv}{dx}}{v^2}$, q.e.d.

13. $y = \sec x = \frac{1}{\cos x} \implies \ln|y| = \ln 1 - \ln|\cos x| = -\ln|\cos x| \implies \frac{1}{y} \cdot \frac{dy}{dx} = -\frac{-\sin x}{\cos x} = \tan x \implies \frac{dy}{dx} = y\tan x = \sec x \tan x$, q.e.d.

15. $\sqrt{\frac{1+x}{1-x}}\left(\frac{1}{(1+x)(1-x)}\right)$

17. $(x^2+4)^3(x^3+6)^6\left[\frac{6x}{x^2+4} + \frac{18x^2}{x^3+5}\right]$

19. $e^{2x}\sin 5x \cos 9x (2 + 5\cot 5x - 9\tan 9x)$

21. $x^x(1+\ln x)$

23. $x^x \cdot x^{x^x-1} + x^{x^x}\ln x\left[x \cdot x^{x-1} + x^x \ln x\right]$, or $x^{x^x}\left[x^{x-1} + x^x(1+\ln x)\ln x\right]$

25. $\frac{d}{dx}\left[(f(x))^{g(x)}\right] = \frac{d}{dx}\left[e^{\ln(f(x)) \cdot g(x)}\right] = e^{\ln(f(x)) \cdot g(x)} \cdot \frac{d}{dx}[\ln(f(x)) \cdot g(x)] = (f(x))^{g(x)}\left[\ln f(x) \cdot g'(x) + g(x) \cdot \frac{f'(x)}{f(x)}\right] = (f(x))^{g(x)}\ln f(x) \cdot g'(x) + g(x)(f(x))^{g(x)-1}f'(x)$, q.e.d.

27. $\frac{d}{dx}10^x = x(10)^{x-1}\frac{d 10}{dx} + 10^x \ln 10 \cdot \frac{dx}{dx} = x(10)^{x-1}(0) + 10^x \ln 10 \cdot 1 = 10^x \ln 10$, q.e.d.

29. $\frac{d}{dx}(\sin x)^2 = 2(\sin x)^1 \cdot \frac{d \sin x}{dx} + (\sin x)^2 \ln(\sin x) \cdot \frac{d 2}{dx} = 2\sin x \cos x + \sin^2 x \ln(\sin x)(0) = 2\sin x \cos x$. (Here we naively assumed $\sin x > 0$.)

31. $y = \log_a x \implies a^y = x \implies \log_b a^y = \log_b x \implies y \log_b a = \log_b x \implies y = (\log_b x)/(\log_b a) \implies \log_a x = (\log_b x)/(\log_b a)$, q.e.d.

Exercises 3.11:

1.
   (a) $x^2 + y^2 = 9\sin^2 t + 9\cos^2 t = 9(\sin^2 t + \cos^2 t) = 9$. Thus the particle lies on the circle $x^2 + y^2 = 9$ (center $(0,0)$, radius 3.

   (b) Motion is clockwise around the circle.

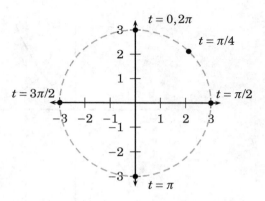

   (c) $v_x = 4\cos t$, $v_y = -3\sin t$, $a_x = -3\sin t$, $a_y = -3\cos t$.

3. (a) $12t^2 - 4t - 8$, (b) $24t - 4$, (c) 1 second, (d) 6 seconds.

5.
   (a) $F = \frac{d}{dt}(mv) = m\left(\frac{dv}{dt}\right) = ma$.

   (b) $F = \frac{d}{dt}(mv) = m \cdot \frac{dv}{dt} + v \cdot \frac{dm}{dt} = v\left(\frac{dm}{dt}\right) + ma$.

7.
   (a) $N_0\left(\frac{1}{2}\right)^{t/8.0197}$ (e) $-0.0864 N_0$
   (b) $0.919 N_0$
   (c) $0.421 N_0$ (f) $-0.794 N_0$
   (d) 34.66 days   (g) $-0.0364 N_0$

Exercises 3.12:

1. $\frac{1}{2\sqrt{x}}\cosh\sqrt{x}$, $\frac{\cosh x}{2\sqrt{\sinh x}}$.

3. $\frac{d}{dx}(\text{sech}\, x) = \frac{d}{dx}\frac{1}{\cosh x} = \frac{-\sinh x}{\cosh^2 x} = -\text{sech}\, x \tanh x$

5. $\frac{d}{dx}\coth x = \frac{d}{dx}\left[\frac{\cosh x}{\sinh x}\right] = \frac{\sinh x(\sinh x) - \cosh x(\cosh x)}{\sinh^2 x} = \frac{\sinh^2 x - \cosh^2 x}{\sinh^2 x} = \frac{-1}{\sinh^2 x} = -\text{csch}^2 x$

7.
   (a) 2   (d) 1/2
   (b) $-\sqrt{3}/2$
   (c) $-2\sqrt{3}/3$ (e) $-\sqrt{3}/3$

9.
   (a) -2.3124 (d) -0.2027
   (b) DNE   (e) DNE
   (c) DNE   (f) 0.1987

11. Use that $\text{csch}\, x = 1/\sinh x$:

688   *Answers to Odd-Numbered Problems*

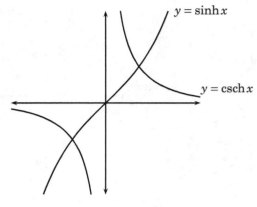

13. 1

15. Use, from the previous problem, that $\lim_{x\to\infty}\tanh x = 1$, $\lim_{x\to-\infty}\tanh x = -1$, and that $\frac{d}{dx}\tanh x = \text{sech}^2 x > 0$ on all of $\mathbb{R}$. Also $\tanh 0 = 0$.

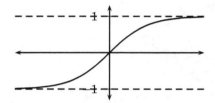

17. From the graph of $y = \tanh x$ above, we graph $y = \coth x = 1/\tanh x$:

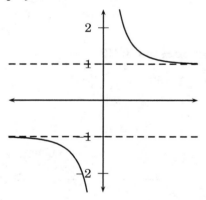

19. $\frac{d}{dx}\left[\cosh^{-1}x\right] = \frac{d}{dx}\ln\left(x+\sqrt{x^2-1}\right) =$
$\frac{1+\frac{2x}{2\sqrt{x^2-1}}}{x+\sqrt{x^2-1}} = \frac{1+\frac{x}{\sqrt{x^2-1}}}{x+\sqrt{x^2-1}} \cdot \frac{x-\sqrt{x^2-1}}{x-\sqrt{x^2-1}} =$
$\frac{x-\sqrt{x^2-1}+\frac{x^2}{\sqrt{x^2-1}}-x}{x^2-(x^2-1)} = -\sqrt{x^2-1} + \frac{x^2}{\sqrt{x^2-1}} =$
$\frac{(x^2-1)+x^2}{\sqrt{x^2-1}} = \frac{1}{\sqrt{x^2-1}}$, q.e.d.

21. $\text{csch}^{-1}x = \sinh^{-1}\left(\frac{1}{x}\right) = \ln\left(\frac{1}{x}+\sqrt{\left(\frac{1}{x}\right)^2+1}\right) =$
$\ln\left[\frac{x\left(\frac{1}{x}+\sqrt{\frac{1}{x^2}+1}\right)}{x}\right] = \ln\left[\frac{1+\sqrt{1+x^2}}{x}\right]$, q.e.d.

23. $\sinh A \cosh B + \cosh A \sinh B$
$= \left(\frac{e^A-e^{-A}}{2}\right)\left(\frac{e^B+e^{-B}}{2}\right) + \left(\frac{e^A+e^{-A}}{2}\right)\left(\frac{e^B-e^{-B}}{2}\right)$
$= \frac{e^{A+B}+e^{A-B}-e^{B-A}-e^{-A-B}}{4}$
$+ \frac{e^{A+B}-e^{A-B}+e^{B-A}-e^{-A-B}}{4}$
$= \frac{2e^{A+B}-2e^{-(A+B)}}{4} = \frac{e^{A+B}-e^{-(A+B)}}{2}$
$= \sinh(A+B)$, q.e.d.

25. $\cosh 2x = \sinh^2 x + \cosh^2 x = (\cosh^2 x - 1) + \cosh^2 x = 2\cosh^2 x - 1$.

$\cosh 2x = \sinh^2 x + \cosh^2 x = \sinh^2 x + (1 + \sinh^2 x) = 1 + 2\sinh^2 x$.

27.

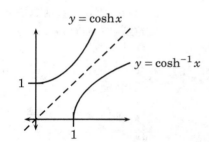

29. $f(x) = \sinh^{-1} x$ has derivative $f'(x) = \frac{1}{\sqrt{x^2+1}} > 0$ for all $x \in \mathbb{R}$, and so $\sinh^{-1} x$ is increasing on all of $\mathbb{R}$.

Answers to Odd-Numbered Problems   689

Exercises 4.1:

1. $(fg)'' = [(fg)']' = [f'g + fg']' = f''g + f'g' + f'g' + fg'' = f''g + 2f'g' + fg''$, q.e.d.

$(fg)''' = f'''g + 3f''g' + 3f'g'' + fg'''$.

3.

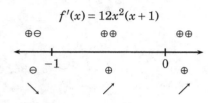

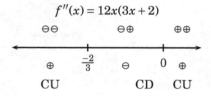

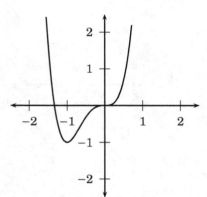

5.

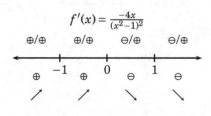

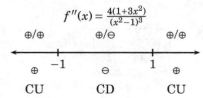

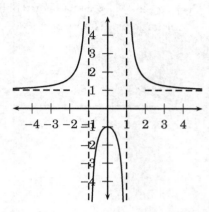

7. $f(x) = x + \frac{x}{x^2-1} \approx x$ (dashed oblique line) for $|x|$ large.

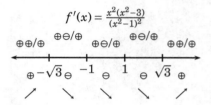

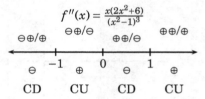

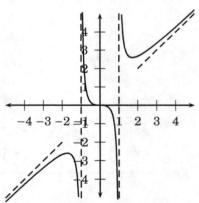

9. $f(x) = x+2+\frac{8}{x-3} \approx x+2$ (dashed oblique line) for large $|x|$. (Note $\sqrt{8} \approx 2.83$.)

$f'(x) = \frac{x^2-6x+1}{(x-3)^2}$

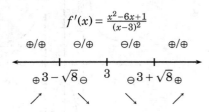

$f''(x) = \frac{16}{(x-3)^3}$

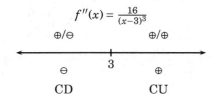

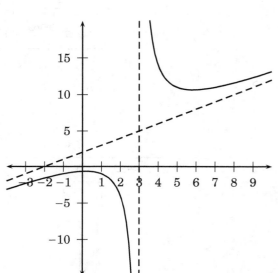

11.

$f'(x) = \frac{-2}{x^3}$

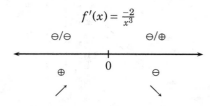

$f''(x) = \frac{6}{x^4}$

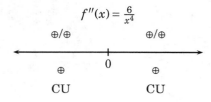

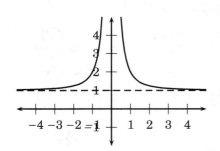

13. $f(x) = \frac{x^4-1}{x^2} = x^2 - \frac{1}{x^2} \approx x^2$ (dashed curve) for large $|x|$.

$f'(x) = \frac{2x^4+2}{x^3}$

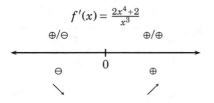

$f''(x) = \frac{2(x^2-3)}{x^4}$

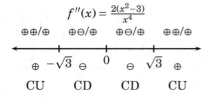

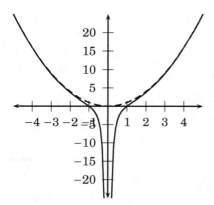

15. $f(x) = |(x-3)(x+1)|$. (Sign charts for f' and f'' will follow from the graph. $y = (x-3)(x+1) = x^2 - 2x - 3$ graphed with a dashed curve. Note $y' = 2x - 2 = 0$ at $x = 1$, the vertex of the dashed curve.)

Answers to Odd-Numbered Problems

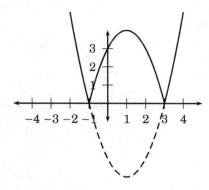

$f' > 0$ on $(-1,1) \cup (3,\infty)$
$f' < 0$ on $(-\infty,-1) \cup (1,3)$
$f'' > 0$ on $(-\infty,-1) \cup (3,\infty)$
$f'' < 0$ on $(-1,3)$

17.

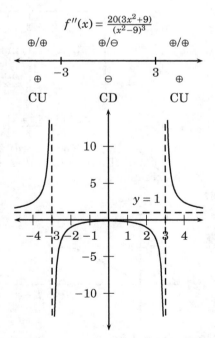

19. $f(x) = 0$ at $x = 1$, $x \to 0^+ \implies f(x) \to 0$ (assumed). Note $e^{-1} \approx 0.367879$.

$f'(x) = 1 + \ln x$

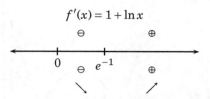

$f''(x) = 1/x > 0$ for all $x > 0$.

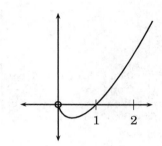

21. $(f(x) = x + \ln|x|$ has a VA at $x = 0)$

$f'(x) = 1 + \frac{1}{x} = \frac{x+1}{x}$

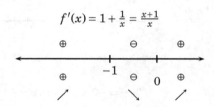

$f''(x) = \frac{-1}{x^2} < 0$ for all $x \neq 0$, so CD except at $x = 0$.

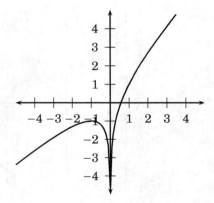

23. $f(x) = |(x+3)(x+1)|$. $y = (x+3)(x+1) = x^2 + 4x + 3$ graphed with a dashed curve (note $y' = 2x + 40$ at $x = -2$, the vertex of the dashed curve.) Sign charts for f', f'' follow from the graph.

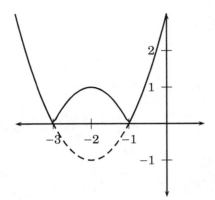

$f' > 0$ on $(-3,-2) \cup (-1,\infty)$
$f' < 0$ on $(-\infty,-3) \cup (-2,-1)$
$f'' > 0$ on $(-\infty,-3) \cup (-1,\infty)$
$f'' < 0$ on $(-3,-1)$

25.

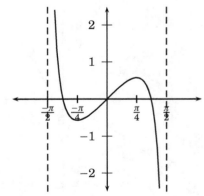

27. (a) $[4, 10]$, (b) $[3, 10]$

29. (a) none, (b) none, (c) $x = -3$, (d) none, (e) $x = -3$, (f) none, (g) $f(x) = (x+3)^n$ has an inflection point at $x = -3$ if n is odd and $n \geq 3$.

31.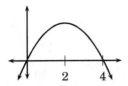

33.

(a)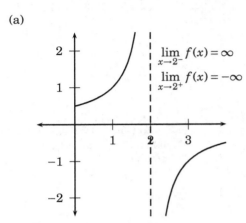

(b)

$\lim_{x \to 2^-} f(x) = \infty$
$\lim_{x \to 2^+} f(x) = \infty$

35. $f(x) = \dfrac{1}{\sigma\sqrt{2\pi}} e^{-\frac{1}{2}\left(\frac{x-\mu}{\sigma}\right)^2} \left[-\left(\dfrac{x-\mu}{\sigma}\right)\right]$. $f'(x) = 0 \iff x = \mu$, $f' > 0$ on $(-\infty,\mu)$ and $f' < 0$ on (μ,∞). Therefore $x = \mu$ is the global maximum.

37.

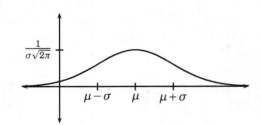

Exercises 4.2:

1. min $(-2, 1)$, max $\left(3, \frac{17}{2}\right)$

Answers to Odd-Numbered Problems

3. min $(0,0)$, max $\left(-1, \frac{1}{5}\right)$ and $\left(1, \frac{1}{5}\right)$

5. min $(0,0)$ and $(1,0)$, max $\left(\frac{1}{2}, 1\right)$

7. min $\left(\frac{-\pi}{2}, -2\right)$, max $\left(\tan^{-1}\frac{1}{4}, 2\sqrt{17}\right) \approx (.2450, 8.246)$

9. min $(-2, 1)$, max $(-1, 2)$

11. min $(-3, 0)$, $(0, 0)$ and $(3, 0)$, max $\left(\sqrt{6}, 6\sqrt{3}\right)$ and $\left(-\sqrt{6}, 6\sqrt{3}\right)$

13. min $(0,0)$ and $(16,0)$, max $\left(\frac{2}{3}\sqrt{6}, 4.46\right)$

15. $f'(x) = 2ax + b \implies x = \frac{-b}{2a}$ yields the critical point at $\left(\frac{-b}{2a}, f\left(\frac{-b}{2a}\right)\right)$.
$f''(x) = 2a > 0$ if $a > 0$, and crititcal point is a minimum.
$f''(x) = 2a < 0$ if $a < 0$, and crititcal point is a maximum.

17. $x = y = 14$

19.
 (a) $0, 30, 40, 60, 70, 80, 90, 100$
 (b) $I(x) = 10\sqrt{x} - x$ has its maximum at $x = 25$: $I(25) = 25$.

21. maximum area $200\,\text{ft}^2$. Two congruent given sides' lengths $10\,\text{ft}$, third side length $20\,\text{ft}$

23. $\frac{1}{2}\sqrt{7}$

25. $I = 3\,\text{Amp}$

27. around $\$57.01$

29. width $2/\sqrt{12} = \frac{2}{2\sqrt{3}} = 1/\sqrt{3} = \frac{1}{3}\sqrt{3}$, height $2/\sqrt{6} = (\sqrt{2}\sqrt{2}) \div (\sqrt{2}\sqrt{3}) = \sqrt{2/3} = \frac{1}{3}\sqrt{6}$ (so $h = w\sqrt{2}$)

31. brick side is $5\sqrt{10} \approx 15.8$ (in feet), and the perpendicular sides are $10\sqrt{10} \approx 31.6$, total cost $\approx \$1897$

33. $D = 2\pi\sqrt[3]{\dfrac{500}{\pi}}, \quad h = \dfrac{1000}{\pi\sqrt[3]{\left(\frac{500}{\pi}\right)^2}}$

Exercises 4.3:

1.

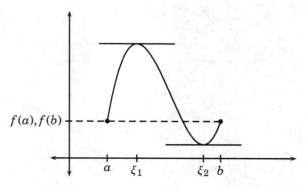

3. $\xi = 2$

5. $\xi = \pi/2$

7. $\xi = e - 1 \approx 1.718$

9. $\xi = (2.5)^2 = 6.25$

11. $\xi = \ln\left(\dfrac{e^2 - e^{-2}}{4}\right) \approx 0.595$

13. $f'(\pi/2)$ DNE, and so the condition that $f'(x)$ exists for all $x \in (0, \pi)$ is not met.

15. $60 < f(2) \le 84$

17.
 (a) $f(x) = 2x - \sin x$
 (b) $f(0) = 0$
 (c) $f'(x) = 2 - \cos x$ exists for all $x \in \mathbb{R}$
 (d) $f'(\xi) = 2 - \cos\xi = 0$ is impossible ($\cos\xi \in [-1, 1]$), and so the value of b does not exist.

19. $f'(\xi) = 2\xi = b + a = \dfrac{f(b) - f(a)}{b - a} \implies \xi = \dfrac{b+a}{2}$

Exercises 4.4:

1. $f(x) \approx 10 + \frac{1}{20}(x - 100)$

3. $f(x) \approx -(x + 1)$

5. $f(x) \approx \frac{\pi}{6} + \frac{2\sqrt{3}}{3}\left(x - \frac{1}{2}\right)$

7. $f(x) \approx x$

9.
 (a) $f(x) \approx x$
 (b) $f(x) \approx 1 + 2\left(x - \frac{\pi}{4}\right)$
 (c) $f(x) \approx -1 + 2\left(x + \frac{\pi}{4}\right)$

11. $f(x) \approx 4 + \frac{3}{4}(x+3) = \frac{3}{4}x + \frac{25}{4}$

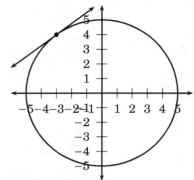

13. $dx^4 = 4x^3\,dx$

15. $d\sin\theta = \cos\theta\,d\theta$

17. $d\sec\xi = \sec\xi\tan\xi\,d\xi$

19. $d\sin 2x = 2\cos 2x\,dx$

21. $d\left(\dfrac{x}{1+\cos x}\right) = \dfrac{1+\cos x + x\sin x}{(1+\cos x)^2}\,dx$

23. $d(uv) = \dfrac{d(uv)}{dx}dx = \left(u\dfrac{dv}{dx}+v\dfrac{du}{dx}\right)dx = u\dfrac{dv}{dx}dx + v\dfrac{du}{dx}dx = u\,dv + v\,du$, q.e.d.

25. $df(u(x)) = \dfrac{df(u(x))}{dx}dx = \dfrac{df(u(x))}{d(u(x))}\cdot\dfrac{d(u(x))}{dx}\cdot dx = f'(u(x))u'(x)\,dx$, q.e.d.

27.
  (a) $0.271\left(\frac{4}{3}\pi\right) \approx 1.135$ (in m³)

  (b) $\Delta V = \frac{4}{3}\pi - 0.7\left(\frac{4}{3}\pi\right) \approx 1.257$ (in m³)

  (c) For a small change in radius, the change in volume would be approximately the surface area of the sphere times the small change in radius.

29. $dV = 12\pi(10)^2(0.2) = 240\pi \approx 753.98$ (in cm³); $\Delta V = 4\pi(10.2)^3 - 4\pi(10)^3 = 244.8\pi \approx 769.06$, which not too different.

31. with v constant, $v = \lambda f \Longrightarrow 0 = \lambda\,df + f\,d\lambda \Longrightarrow d\lambda = \dfrac{-\lambda}{f}df \approx -1.21\times 10^{-8}$ (in m), similar to $\Delta\lambda = -1.2\times 10^{-8}$.

Exercises 4.5:

1. $f(-2) = 024$, $f(-3) = 3$ and so a root exists in $[-3,-2]$. Letting $x_1 = -2.9$ gives

$$x_2 = -2.953544817$$
$$x_3 = -2.951376700$$
$$x_4 = -2.951373036$$
$$x_5 = -2.951373036$$

3. $f(4) = -8$, $f(5 = 1)$. Letting $x_1 = 4.8$ gives

$$x_2 = 4.9$$
$$x_3 = 4.898979592$$
$$x_4 = 4.898979485$$
$$x_5 = 4.898979486 \approx \sqrt{24}$$

5. (a)

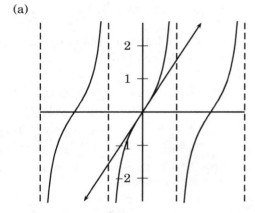

  (b) Newton's method only converges for $x_1 = 4.5$, and diverges for all others: $x_n \to 4.493613903$

  (c) Newton's method only converges for $x_1 = 7.7$, and diverges for all others: $x_n \to 7.725251837$

7. 3 years (2.8212 years)

9.

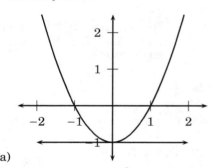

  (a)

  (b) $x_{n+1} = \frac{1}{2}x_n + \dfrac{1}{2x_n} = \dfrac{x_n^2+1}{2x_n}$

  (c)

$$x_1 = 2$$
$$x_2 = 1.25$$
$$x_3 = 1.025$$
$$x_4 = 1.000304878$$
$$x_5 = 1.000000046$$
$$x_6 = 1.000000000$$

  (d) If $x_1 = 1$, then $x_2 = x_3 = x_4 = \cdots = 1$.

Answers to Odd-Numbered Problems

(e) If $x_1 = 0$, then x_2 and so on are undefined. (The tangent line to the curve at x_1 is horizontal, so x_2 does not exist.)

11.

(a) $x_{n+1} = \frac{2}{3}x_n$

(b) $x_1 = 1$, $x_2 = 2/3$, $x_3 = \left(\frac{2}{3}\right)^2$, $x_4 = \left(\frac{2}{3}\right)^3$

(c) $x_n = \left(\frac{2}{3}\right)^{n-1}$

(d) $x_n = \left(\frac{2}{3}\right)^{n-1} \xrightarrow{(2/3)^\infty} 0$ as $n \to \infty$ (a zero of the function will occur at $x = 0$)

13. $f'(x) = 2x - 4\sin x$, with $x = 0$ as one solution in $[-\pi, \pi]$. Finding other zeroes: let $g(x) = 2x - 4\sin x$ and apply Newton's Method to g:

$x_1 = 3 \implies x_n \to 1.895494267$,
$x_1 = -3 \implies x_n \to -1.895494267$.

(a)

$f'(x) = 2x - 4\sin x$

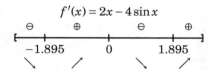

(b) $f''(x) = 2x - 4\cos x = 0$ at $x = \pi\pi/3$ (where $\cos x = \frac{1}{2}$)

$f''(x) = 2x - 4\cos x$

$\begin{array}{ccc} \oplus & \ominus & \oplus \\ \text{CU} & \frac{-\pi}{3} \quad \text{CD} \quad \frac{\pi}{3} & \text{CU} \end{array}$

(c)

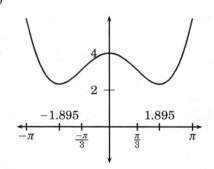

Exercises 4.6:

1. $3x^2 \cdot \frac{dx}{dt} + 3y^2 \cdot \frac{dy}{dt} = 0$

3. $\frac{dx}{dt} = \frac{1}{2\sqrt{A}} \cdot \frac{dA}{dt}$

5. $\frac{dA}{dt} = 2\pi r \cdot \frac{dr}{dt}$

7. $\frac{dS}{dt} = 8\pi r \cdot \frac{dr}{dt}$

9. $\frac{ds}{dt} = 12x \cdot \frac{dx}{dt}$

11. $\frac{dy}{dt} = \tan\theta \cdot \frac{dx}{dt} + x\sec^2\theta \cdot \frac{d\theta}{dt}$

13. $\frac{dP}{dt} = \frac{-k}{V^2} \cdot \frac{dV}{dt}$

15. $\frac{dP}{dt} = 2 \cdot \frac{dx}{dt} + 2 \cdot \frac{dy}{dt}$

17. $\frac{dV}{dt} = \frac{2}{3}\pi rh \cdot \frac{dr}{dt} + \frac{1}{3}\pi r^2 \cdot \frac{dh}{dt}$

19. $\frac{dm}{dt} = \frac{m_0 v}{c^2\sqrt{1-v^2/c^2}} \cdot \frac{dv}{dt}$

21. $0.56\,\text{cm}$

23. $\frac{d\theta}{dt} = \frac{3}{6500}\frac{\text{rad}}{\text{sec}} \approx 0.0005\,\text{rad/sec} \approx 0.028\,\text{degree/sec}$

25. $-30\sqrt{3}\,\text{cm}^2/\text{sec} \approx -52\,\text{cm}^2/\text{sec}$

27. $0.0076\,\text{in/min}$

29. $0.61\,\text{m/sec}$

31. $86.4\pi\,\text{in}^3/\text{sec} \approx 271.43\,\text{in}^3/\text{sec}$

33. $-20\,\text{cm}^2/\text{sec}$

35. $\frac{80\pi}{3}\,\text{cm}^3/\text{sec} \approx 83.78\,\text{cm}^3/\text{sec}$

37. $0.75\,\text{units/sec}$

39. $4.5\,\text{Volt/min}$

41. $-0.03\,\Omega/\text{sec}$

43. $\frac{dE}{dt} = \frac{1}{2}m \cdot 2v \cdot \frac{dv}{dt} = \frac{1}{2}(50\,\text{kg}) \cdot 2 \cdot \frac{30\,\text{m}}{\text{sec}} \cdot \frac{4\,\text{m}}{\text{sec}^2}$
$= 6000\,\text{watt}$

45. (a) $1.99 \times 10^{-7}\,\text{rad/sec}$
(b) $18.5\,\text{mile/sec}$

47. $0.08\,\text{mm/hr}$

49. (a) $P(0) = (KP_0(1))/(K = P_0(0)) + \frac{KP_0}{K} = P_0$

(b) $\lim_{t\to\infty} \frac{KP_0 e^{rt}}{K + P_0(e^{rt}-1)}$
$= \lim_{t\to\infty} \frac{KP_0}{Ke^{-rt} + P_0(1-e^{-rt})}$
$= \frac{KP_0}{0 + P_0(1-0)} = \frac{KP_0}{P_0} = K.$

(c) $\frac{dP}{dt} =$
$\dfrac{KP_0 r(K - P_0)}{\frac{K^2}{e^{rt}} + 2KP_0 - \frac{2KP_0}{e^{rt}} + P_0^2 e^{rt} - P_0^2 + \frac{P_0^2}{e^{rt}}}$

(d) Since $t \to \infty \implies e^{rt} \to \infty$, we have $\lim_{t\to\infty} \dfrac{dP}{dt} = 0$.

51. $0.6649°\text{F/hr}$

53. $3x - 3\ln x + 4y - 2\ln y = 7$

55. $\dfrac{dx}{dt} = .1775$ thousands/week. Most likely, as the number of prey x is increasing (cause), this allows the number of predators y to increase also (effect).

57. $\dfrac{dy}{dt} = -0.0891$ thousands/week. As the number of prey x is decreasing (cause), predators are more likely to weaken or starve and the number of predators y is thus decreasing (effect).

59. (a) $k = 33{,}788$ lb-in
(b) $6\,\text{in}^3/\text{sec}$

61. $1\,\text{mol}/(\text{L-min})$

63. $h = 0.5\tan s$

65. $0.342\,\text{m}$

67. $490(\tan s)\,\text{J}$

69. (a) $1.39\,\text{J}$, (b) $2.33\,\text{J}$, (c) $16.30\,\text{J}$, (d) approaching infinity

71. $\left(490\tan s + \dfrac{25}{18}\sec^4 s\right)\text{J}$

Exercises 5.1:

1. $\frac{d}{dt}\left[\frac{1}{2}e^{2t}\right] = \frac{1}{2}e^{2t}(2) = e^{2t}$, q.e.d.

3. $\frac{d}{dt}\left[\frac{1}{2}\sin 2t\right] = \frac{1}{2}\cos 2t \cdot 2 = \cos 2t$, q.e.d.

5. $\frac{d}{dx}\ln|\sec x + \tan x| = \frac{1}{\sec x + \tan x} \cdot (\sec x \tan x + \sec^2 x) = \frac{\sec x(\tan x + \sec x)}{\sec x + \tan x} = \sec x$, q.e.d.

7. $\frac{d}{dx}\left[\frac{1}{2}(x - \sin x \cos x)\right] = \frac{1}{2}[1 - (\sin x(-\sin x) + \cos x \cos x)] = \frac{1}{2}[1 + \sin^2 x - \cos^2 x] = \frac{1}{2}[1 - \cos^2 x + \sin^2 x] = \frac{1}{2}[\sin^2 x + \sin^2 x] = \frac{1}{2}(2\sin^2 x) = \sin^2 x$, q.e.d.

9. $\frac{d}{dx}[\ln|\sec x|] = \frac{1}{\sec x} \cdot \sec x \tan x = \tan x$, q.e.d.

11. $\frac{d}{dx}\tan^{-1}(\ln x) = \frac{1}{(\ln x)^2 + 1} \cdot \frac{1}{x} = \frac{1}{x(1+(\ln x)^2)}$, q.e.d.

13. $3x^{1/3} + C$

15. $\tan x + C$

17. $\frac{1}{7}x^7 + \frac{3}{5}x^5 + x^3 + x + C$

19. $\frac{2}{3}\sqrt{w} + C$

21. $5\cos x + C$

23. $-\cot t - t + C$

25. $2\cos t + 3t + 5$

27. $f(x) = x^3 + x^2 + 5x - 5$

29. $s(t) = \tan^{-1} t + \frac{\pi}{4}$

31. $f(x) = \sec x + 3$

33. $f(x) = 4x - 3$

35. $f(x) = -\cos x + \sin x + 3$

Exercises 5.2:

1. $1+2+3+4+5+6 = 21$

3. $1 + \frac{1}{2} + \frac{1}{4} + \frac{1}{8} + \frac{1}{16} + \frac{1}{32} = \frac{63}{32}$

5. $-1+1-1+1-1+1-1+1-1+1-1 = -1$

7. $2+2+2+2+2+2+2+2+2+3 = 23$

9. $\left(\frac{1}{3}-1\right) + \left(\frac{1}{4}-\frac{1}{2}\right) + \left(\frac{1}{5}-\frac{1}{3}\right) + \left(\frac{1}{6}-\frac{1}{4}\right) + \left(\frac{1}{7}-\frac{1}{5}\right) = -\frac{25}{21}$

11. $\sum_{i=1}^{200} i - \sum_{i=1}^{99} i = \frac{(200)(201)}{2} - \frac{(99)(100)}{2} = 15{,}105$

13. $2\sum_{i=1}^{20} + 20 = 2\frac{(20)(21)}{2} + 20 = 440$

15. $\sum_{i=1}^{20}[4i^3 - 12i^2 + 9i] = 4\frac{20^2(21)^2}{4} - 12\frac{(20)(21)(41)}{6} + 9\frac{(20)(21)}{2} = 143{,}850$

17. $\sum_{i=14}^{30}(i-1)^3 = 13^3 + 14^3 + \cdots + 29^3 = \sum_{i=13}^{29} i^3 = (1^3 + 2^3 + \cdots + 12^3 + 13^3 + \cdots + 29^3) - (1^3 + 2^3 + \cdots + 12^3) = \sum_{i=1}^{29} i^3 - \sum_{i=1}^{12} i^3 = \frac{29^2 30^2}{4} - \frac{12^2 13^2}{4} = 183{,}141$.

19. $x^2 = 312.5$, $s = 17.68$.

21.
  (i) $P_1 : \sum_{i=1}^{1} i^2 = 1^2 = 1 = \frac{1(2)(3)}{6}$ is true.
  (ii) Assume $P_n : \sum_{i=1}^{n} i^2 = \frac{n(n+1)(2n+1)}{6}$ is true.
  (iii) Consider $P_{n+1} : \sum_{i=1}^{n+1} i^2 = \sum_{i=1}^{n} i^2 + (n+1)^2 \xrightarrow{\text{by } P_n} \frac{n(n+1)(2n+1)}{6} + (n+1)^2 = (n+1)\left[\frac{n(2n+1)}{6} + (n+1)\right] = \frac{n+1}{6}[2n^2 + 7n + 6] = \frac{(n+1)(n+2)(2n+3)}{6} = \frac{(n+1)((n+1)+1)(2(n+1)+1)}{6}$, which shows P_{n+1} is true, so $P_n \Longrightarrow P_{n+1}$, q.e.d.

Exercises 5.3:

1. 1/2   7. 18   13. 0

3. 44/3   9. $\pi/2$

5. 49/3   11. 1   15. 29/2

17.
  (a)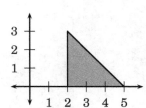
  (b) $A = L \cdot W = 5 \cdot 4 = 20$
  (c) $\int_1^6 4\,dx = 4(6) - 4(1) = 20$

19.
  (a)

  (b) $A = \frac{1}{2}bh = \frac{1}{2}(3)(3) = 9/2$
  (c) $\int_2^5 (5-x)\,dx = \left[5(5) - \frac{1}{2}(5)^2\right] - \left[5(2) - \frac{1}{2}(2)^2\right] = 9/2$

21.
  (a) (not to scale)

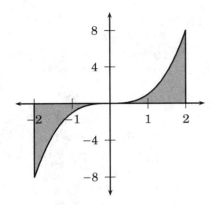

(b) the symmetry implies the net area is 0

(c) $\int_{-2}^{2} x^3 \, dx = \frac{1}{4}(2)^4 - \frac{1}{4}(-2)^4 = 0$

23.

(a) (not to scale)

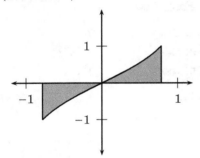

(b) the symmetry implies the net area is 0

(c) $\int_{-\pi/4}^{\pi/4} \tan x \, dx = \ln\sec|\pi/4| - \ln\sec|-\pi/4| = 0$

25. $\int_a^b f(x)\,dx + \int_b^c f(x)\,dx = F(b) - F(a) + F(c) - F(b) = F(c) - F(a) = \int_a^c f(x)\,dx$

27.

(a)

(b) $A = \int_0^L W \, dx$

(c) $A = \big|_0^L = WL - W(0) = WL$

(d) $A = LW$ from the geometry

29.

(a)

(b) $V = \pi r^2 h = \pi W^2 L$

(c) $V = \int_0^L \pi W^2 \, dx = \pi W^2 x \big|_0^L = \pi W^2 L - \pi W^2(0) = \pi W^2 L$.

(d) same

Exercises 5.4:

1.

(a) $\frac{d}{dx} \int_0^x t^2 \, dt = x^2$

(b) $\frac{d}{dx} \int_0^x t^2 \, dt = \frac{d}{dx}\left[\frac{1}{3}x^3 - \frac{1}{3}(0)^3\right] = x^2 - 0 = x^2$

3. e^{x^2}

5. $2x^5 \sin x^2$

7. $3t^{11} \sin^2 t^3$

9. $\cos x \left(\sin^3 x - 3\right)^5$

11. $\sin^2 x$

13. $-x^2$

15. $2x^2$

17. $-\frac{\sin x}{\cos^2 x + 1} - \frac{\cos x}{\sin^2 x + 1}$

19. 0

21. $\sin(x+2)^2 - \sin(x^2)$

23. $12x(4x^2 - 1)\ln x$

25. -6

27. -15

29. 1

31. 4

33. $f(-z) = \frac{1}{\sqrt{2\pi}} e^{-(-z)^2/2} = \frac{1}{\sqrt{2\pi}} e^{-z^2/2} = f(z)$, q.e.d.

35. 0.1359

37. 0.4772

39. 0.9544

41. 0.3413

43. 0

45. 1.1962

47. average is $2/\pi$, $\xi = \sin^{-1}(2/\pi) \approx 0.6901$

Answers to Odd-Numbered Problems

49. average is $\frac{100}{3}$, $\xi = \frac{10\sqrt{3}}{3}$

51. average is $\pi/4$, $\xi = \pm\sqrt{\frac{4}{\pi}-1} \approx \pm 0.5227$

53. average is $58/3$, $\xi = -1 + \frac{2\sqrt{39}}{3} \approx 3.16$

55.
    (a) $x^2 + x + 3$
    (b) $x^2 + x - 6$
    (c) (a) and (b) differ by a constant

57.
    (a) $f(h(x)) \cdot h'(x) - f(g(x)) \cdot g'(x)$
    (b) $f(h(x)) - f(g(x))$
    (c) These are the same if and only if $h'(x), g'(x) = 1$, i.e., $h(x) = x + C_1$ and $g(x) = x + C_2$.

59. No, areas can cancel: $\int_{-1}^{1} x\,dx = 0$ but $f(x)$ is not the zero function.

61. No, areas in the new integral cannot cancel: $\int_{-1}^{1} x\,dx = 0$ but $\int_{-1}^{1} |x|\,dx = \int_{-1}^{0}(-x)\,dx + \int_{0}^{1} x\,dx = \frac{1}{2} + \frac{1}{2} = 1$.

63. (a) 12, (b) −8

Exercises 5.5:

1. 2   **5.** 32   **9.** 4/3

3. 3/2   **7.** 32   **11.** 72

13. (a) 1, (b) $\frac{\pi}{2} - 1$

15. $\frac{4}{3}\pi r^3$

17.
    (a) vertical; See Example 5.5.8. page 583.
    (b) dy; same example as (a)
    (c) horizontal; See Example 5.5.7, page 582.
    (d) dx; same example as (c)
    (e) y; same example as (a)
    (f) x; same example as (c)

19. $V = \int_{-R}^{R} \pi\left(\sqrt{R^2 - x^2}\right)^2 dx = \pi \int_{-R}^{R} (R^2 - x^2)\,dx = \frac{4}{3}\pi R^3$

21.
    1. Disc Method
    2. $V = \pi \int_{-1}^{2} [(y+2)^2 - (y^2)^2]\,dy = \pi \int_{-1}^{2}(y^2 + 4y + 4 - y^4)\,dy$

23. $\frac{243\pi}{5} \approx 152.68$

25. $\frac{256\pi}{15} \approx 53.62$

27. $\frac{64\pi}{5} \approx 40.21$

29. $\int_{1}^{4} \sqrt{1 + \frac{16}{9}x^{2/3}}\,dx$

31.
    (a) $\int_{0}^{8} \sqrt{1 + 9/16}\,dx = \frac{5}{4}(8) - \frac{5}{4}(0) = 10$
    (b) $\sqrt{8^2 + 6^2} = \sqrt{100} = 10$

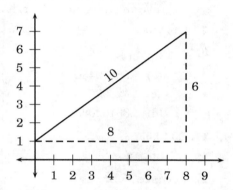

Exercises 5.6:

1. (a) $s(t) = -0.2t^3 - 0.02t^4 + C$, (b) $7.5\,\text{m}$

3.
    (a) $12\,\text{J}$
    (b) $12\,\text{J}$
    (c)

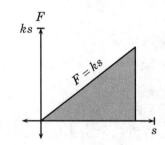

    (d) $A = \frac{1}{2}(s)(ks) = \frac{1}{2}ks^2$

5. (a) $U(x) = x^3 + x^2 + kx$, (b) $k = 14$.

7. (a) $\frac{100\,\text{lb}}{25\,\text{ft}} = 4\,\text{lb/ft}$,
    (b) $\int_{0}^{25} y \cdot 4\,\text{lb/ft}\,dy = 2y^2 \Big|_{0}^{25} = 1250\,\text{ft-lb}$

9.
    (a) Weight of an infinitesimal length of cable at angle θ will be density × length, or $10\,dl = 10 \cdot 5\,d\theta = 50\,d\theta$.
    (b) Height it is lifted is $5\sin\theta$, so $dW = (10 \cdot 5\,d\theta) \cdot 5\sin\theta = 250\sin\theta\,d\theta$.

(c) $\int_0^\pi 250\sin\theta\,d\theta = -250\cos\pi - (-250\cos 0) = 250 + 250 = 500$, in ft-lb.

Exercises 5.7:

1. (a) $e - 1 \approx 1.71828183$, (b) 1.72722191

3. 1.71831884

5. Both Simpson's Rule and the FTC give the exact answer of 1000/3 here, which is expected because Simpson's Rule uses areas under parabolas to approximate integrals, and the graph of $f(x)$ is a parabola.

7. 1.49436085

9. (a) $\ln 4$
   (b) 1.38629436
   (c) 1.42809066
   (d) 1.39162088

11. 1113.33

13. $x^2 - 4x + 3$

Exercises 5.8:

1. $\frac{1}{6}\sin^6 x + C$

3. $1/32$

5. $-\frac{1}{6}\cos^3 2x + C$

7. $-\frac{1}{42}(8-7x)^6 + C$

9. $-\frac{1}{2}\ln 3 \approx -0.54391$

11. 0

13. $2\sqrt{\tan x} + C$

15. $\frac{-1}{\ln x} + C$

17. $\frac{1}{3}\left(\tan^{-1}x\right)^3 + C$

19. $1/6$

21. $\frac{2}{11}\left(5 + \sqrt{x}\right)^{11} + C$

23. $\frac{1}{2}\ln\left(\frac{\pi}{3}\right) \approx 0.42365$

25. $\frac{1}{2}\sqrt{x^4+1} + C$

27. $\sqrt{x^2-9} + C$

29. $\ln|\sec x| + C = -\ln|\cos x| + C$

31. $\frac{2}{5}(1-x)^{5/2} - \frac{2}{3}(1-x)^{3/2} + C$
    $= \frac{-2}{15}(3x+2)(1-x)^{3/2} + C$

33. $\ln|x+1| + \frac{1}{x+1} + C$

35. $\frac{2}{27}\left(7\sqrt{10} - 2\right) \approx 1.4916$

37. $\frac{1}{6}(2x+1)^{3/2} - \frac{1}{2}(2x+1)^{1/2} + C$
    $= \frac{1}{3}(x-1)\sqrt{x+1} + C$

39. $\frac{4}{3}\pi \cdot 5^3 \approx 523.60$

41. $\ln|2 + \sqrt{3}| \approx 1.31696$

Exercises 5.9:

1. $\frac{d}{dx}[-\cos x] = -(-\sin x) = \sin x$

3. $\frac{d}{dx}[\ln|\sec x|] = \frac{1}{\sec x} \cdot \sec x \tan x = \tan x$

5. $\frac{d}{dx}[\ln|\sec x + \tan x|] = \frac{1}{\sec x + \tan x} \cdot (\sec x \tan x + \sec^2 x) = \frac{\sec x(\tan x + \sec x)}{\sec x + \tan x} = \sec x$

7. $\frac{1}{3}\sin x^3 + C$

9. $\frac{1}{5}\ln|\sec x^5| + C$

11. 0

13. $\frac{1}{3}\ln|\sec 3x + \tan 3x| + C$

15. $-2\ln\left|\csc\frac{x}{2} + \cot\frac{x}{2}\right| + C$

17. $\frac{-1}{\omega}\cos(\omega t) + C$

19. $\frac{1}{2}\ln\left|\sec(x^2+1) + \tan(x^2+1)\right| + C$

21. $1/2$

23. $\frac{1}{4}\ln\left(\frac{3}{2}\right)$

25. $-\frac{1}{120}\ln\left|\csc\left(6e^{5x^4}\right)\right| + C$

27. $-\frac{1}{6}\ln\left|\csc 3x^2 + \cot 3x^2\right| + C$

29. $x - 2\ln|\sec x + \tan x| + \tan x + C$

31. $\sin x + C$

33. $-2\ln|\csc x + \cot x| + 2\ln|\csc x| + \cos x + C$

35. $P = 2\pi/B \implies B = 2\pi/P$, and from Exercise 34 we have $\int_0^P \left|\sin\frac{2\pi x}{P}\right| dx = \int_0^{2\pi/B} |\sin Bx|\,dx = 4/B = \frac{4P}{2\pi} = \frac{2P}{\pi}$.

37. (a) $dx/dt = -2\pi fr \sin(2\pi ft)$

(b) $\int_{1/8}^{1/4} |v_x(t)|\,dt$
$= \int_{1/8}^{1/4} \underbrace{|-2\pi \cdot 1 \cdot 5\sin(2\pi \cdot 1 \cdot t)|}_{>0 \text{ on } [1/8, 1/4]} dt$
$= \int_{1/8}^{1/4} [10\pi \sin(2\pi t)]\,dt$
$= 5[-\cos(\pi/2) + \cos(\pi/4)] = \frac{5\sqrt{2}}{2} \approx 11.107$

(c) $\int_0^1 \underbrace{|-2\pi(.5)(6)\sin(2\pi(.5)t)|}_{\geq 0 \text{ on } [0,1]} dt$
$= \int_0^1 6\pi \sin(\pi t)\,dt$
$= -6\cos\pi + 6\cos 0 = 6 + 6 = 12$

(d) In that time interval, the angle varies from 0 to π (and the horizontal position x from 6 to -6 or $-r$ to r), and so one semicircle is traversed, and so the horizontal change is the length of one diameter (12 m).

39. $\frac{d}{dx}\ln|\sin x| = \frac{1}{\sin x} \cdot \cos x = \cot x$

41. $\frac{d}{dx}\ln|\csc x - \cot x| = \frac{-\csc x \cot x + \csc^2 x}{\csc x - \cot x} = \frac{\csc x(-\cot x + \csc x)}{\csc x - \cot x} = \csc x$

Exercises 5.10:

1. $\ln|\tan^{-1} x| + C$

3. $\ln|\sec(\ln x) + \tan(\ln x)| + C$

5. $\frac{1}{3}\ln 2 \approx 0.23105$

7. $\pi/12 \approx 0.26180$

9. $\sec^{-1}|\ln x| + C$

11. $e^{e^x} + C$

13. $-e^{1/x} + C$

15. $-2\sqrt{1 - \sec x} + C$

17. $\frac{-1}{21}\cot^7 3x + C$

19. $6/(\ln 2 \cdot \ln 3) \approx 7.8792$

21. $\frac{1}{\ln 5}\ln|\sec 5^x + \tan 5^x| + C$

23. $\frac{1}{2}\tan^{-1}(e^{2x}) + C$

25. $-\frac{2}{3}\sqrt{1 - e^{3x}} + C$

27. $\ln\left(\frac{\ln 3}{\ln 2}\right) \approx 0.46056$

29. $-\frac{1}{4}\sqrt{1 - 4x^2} + \frac{1}{2}\sin^{-1}(2x) + C$

31.
(a) new integral: $\int e^u \frac{1}{2} du = \frac{1}{2}e^u + C$
$= \frac{1}{2}e^{x^2} + C$
(b) new integral: $\int \frac{1}{2} du = \frac{1}{2}u + C$
$= \frac{1}{2}e^{x^2} + C$

33.
(a) $\frac{1}{2}\sin^2 x + C_1$
(b) $-\frac{1}{2}\cos^2 x + C_2$
(c) The functions $\frac{1}{2}\sin^2 x$, $-\frac{1}{2}\cos^2 x$ differ by a constant, and therefore have the same derivative $(\sin x \cos x)$: $\frac{1}{2}\sin^2 x = \frac{1}{2}(1 - \cos^2 x) = -\frac{1}{2}\cos^2 x + \frac{1}{2}$.

35.
(a) $\int_0^{20} P(t)\,dt = 300$ (megawatt-Hour)
(b) $\lim_{t\to\infty} P(t) = \lim_{t\to\infty} \frac{30}{e^{-0.6(t-100)} + 1}$
$\frac{30/(e^{-\infty}+1)}{} \quad \frac{30}{0+1} = 30$ (megawatt)

Exercises 5.11:

1. $\sin^{-1}\left(\frac{x}{\sqrt{5}}\right) + C$

3. $\frac{1}{2}\sec^{-1}|3x| + C$

5. $\frac{1}{\sqrt{3}}\sin^{-1}\left(\frac{x}{\sqrt{2}}\right) + C$

7. $\frac{1}{2\sqrt{5}}\tan^{-1}\left(\frac{e^{2x}}{\sqrt{5}}\right) + C$

9. $\sin^{-1}\left(\frac{1}{2}\tan x\right) + C$

11. $\frac{1}{\sqrt{5}}\tan^{-1}\left(\frac{x-2}{\sqrt{5}}\right) + C$

13. $\sec^{-1}|x + 2| + C$

15. $-\sin^{-1}\left(\frac{3-x}{3}\right) + C$

17. $\frac{d}{dx}\left[\frac{1}{a}\tan^{-1}\left(\frac{u}{a}\right)\right] = \frac{1}{a}\frac{1}{1+(u/a)^2}\cdot\frac{1}{a}$
$= \frac{1}{a^2(1+u^2/a^2)} = \frac{1}{a^2+u^2}$

19. $\sin^{-1}\left(\frac{x}{\sqrt{5}}\right) + C$

21. $\frac{1}{2}\sec^{-1}|3x| + C$

23. $\frac{1}{2\sqrt{5}}\tan^{-1}\left(\frac{e^{2x}}{\sqrt{5}}\right) + C$

25. $\frac{1}{\sqrt{3}}\sec^{-1}\left(\frac{e^x}{\sqrt{3}}\right) + C$

Index

absolute value function, 248
acceleration, 408
Ampére, André-Marie, 266
amperes, 266
and ($\wedge$), operation, 2, 4, 5
antiderivative, 507
antidifferentiation, 507
arc length, 588
arccosine, 343
Archimedes, 121
arcsecant, 347
arcsine, 340
arctangent, 345
area
    below a graph, 540, 541
    between curves, 577
    total (unsigned), 547
argument
    logical, 37
    of a logic operation, 4
average rate of change, 212, 407, 448–450, 452
average value of a function, 564
average velocity, 208, 565

bi-implication ($\leftrightarrow$), operation, 2, 4, 8
bisection method of approximation, 476
Boyle's Law, 505

catenary, 399
Cavalieri's Principle, 579
Cavalieri, Bonaventura, 579
Chain Rule, 273, 288
change of variable
    definite integrals, 639
    indefinite integrals, 629
    limits, 171
closed interval, 65

complement of a set, 68
concave down, 408, 410
concave up, 408, 410
concavity, 412
conjunction, 5
constant of integration, 509
continuity
    at a point, 77
    of compositions of functions, 164
    of polynomials, 92
    of rational functions, 94
    of trigonometric functions, 166
    on closed intervals, 107
    on open intervals, 105
    one-sided, 96
    versus limits, 123
contradiction, 12, 29
contrapositive (of an implication), 21
converse (of an implication), 39
coulomb, 266
Cramer's Rule, 499
Cramer, Gabriel, 499
critical point, 431
Cylindrical Shell Method, 584

de Coulomb, Charles-Augustin, 266
De Morgan's Laws (logic), 19
De Morgan's Laws (set theory), 71
decreasing function, 237, 238
definite integral, 509, 536, 539, 540
    as a limit of Riemann Sums, 536–538
    bounds on, 567
    elementary properties, 559–562
derivative, 212
    and multiplicative constants, 234
    as rate of change, 212
    higher-order, 408

implicit differentiation method, 319
of a constant, 234
of a polynomial, 235
of a power, 230
of a product, 297
of a quotient, 303
of a rational power, 280
of a sum, 232
of composite functions (Chain Rule), 273
of hyperbolic functions, 400, 401
of inverse hyperbolic functions, 401
of inverse trigonometric functions, 336
of trigonometric functions, 242–244, 283, 305

Descartes, René, xi
detachment, law of, 37
difference quotient, 212
differentiability, 212
 implies continuity, 229
differentials, 464
differentiation, see also derivatives
 implicit, 319
 logarithmic, 378
Disc/Washer Method, 580
discontinuity, 95
 essential, 102
 removable, 102
disjunction, 6
disjunctive syllogism, 41
displacement (net, as an integral), 549
distance traveled (as an integral), 549

e, 184, 360
Einstein, Albert, 55, 157, 537
empty set ($\varnothing$), 51
epsilon-delta (ε-δ)
 continuity definition, 77
 finite limit at a point, 123
 one-sided continuity, 96
equivalence, valid, 18
error types, 532
even function, 563
 integration, 564
excluded middle, 3
exclusive or (XOR), operation, 6

exponential decay, 358, 386, 395, 396
exponential functions, 183, 356
 algebraic properties, 367
 derivatives, 359–361
 derivatives for bases other than e, 384
 graphs, 183, 358
 integrals, 521
 limits, 183, 184, 359
 versus polynomials, 363
exponential growth, 358, 364, 386, 395, 503
Extreme Value Theorem (EVF), 107
Extreme Value Theorem (EVT), 107, 431

$\mathscr{F}$, 29
fallacy, 37
 of the converse, 39
 of the inverse, 42
Faraday, Michael, 267, 296
First Derivative Test, 410
fluid pressure, 605–609
force, 261
 due to pressure, 605
Franklin, Benjamin, 266
frustum of a cone, 592
Fundamental Theorem of Calculus
 First and Second FTC relationship, 556
 First FTC, 554
 Second FTC, 537

Gauss, Johann Carl Friedrich, 450
Greek alphabet, xix

Henry, Joseph, 296
Hooke's Law, 288, 602
Hooke, Robert, 288
hydrostatic pressure, 605–609
hyperbolic functions, 399
 derivatives, 400, 401
 derivatives of inverses, 401
 inverses and integration, 522

if and only if, 2, 9
if, then, 7
implication ($\rightarrow$), operation, 2, 4, 7
implication, valid, 33, 34
implicit differentiation, 319

implicit function, 318
inclusive or, 6
increasing function, 237, 238
indefinite integral, 508
indeterminate form (limit), 127
infinitesimals, 538
    and area, 574
    and geometry, 573
    versus Riemann Sums, 574
inflection point, 412
instantaneous rate of change, 212
instantaneous velocity, 210
integral
    definite, 509, 536, 539, 540
        bounds on, 567
            elementary properties, 559–562
    differentiation of, 554–558
    indefinite, 508, 509
    symbol $\int$, 508, 509
integration, 507
    by substitution, 629
Intermediate Value Theorem (IVT), 107, 108
intersection ($\cap$) of sets, 68
interval, 65
inverse (of an implication), 42
inverse cosine, 343
inverse functions, 334
inverse secant, 347
inverse sine, 340
inverse tangent, 345
inverse trigonometric functions, 335
    derivatives, 336
        integral forms yielding, 520, 663

joule, energy unit, 262
Joule, James, 262

kinetic energy, 262, 263, 600
Krantz, Steven, 9

Lagrange, Joseph-Louis, 212
law of detachment, 37
left-continuous, 96
Leibniz
    Gottfried Wilhelm, 231
    notation, 213, 231, 232, 273, 274

length of a curve, 588
lexicographical order, 3
limit, 121
    at infinity, 174
    by substitution, 198
    composition theorem, 164
    determinate forms, 127, 149, 153, 161,
        165, 174, 180, 181
    finite at a point, 123
    general theorems, 189
    indeterminate forms, 127, 137, 177
    infinite, 146, 147, 175
    one-sided finite at a point, 138
    Sandwich Theorem, 158
    versus continuity, 123
linear (secant line) interpolation, 448, 459
linear (tangent line) approximation, 459, 460
local maximum/minimum, 238–240, 410, 411
logarithmic differentation, 379
logarithmic differentiation, 378–382
logarithmic functions
    algebraic properties, 367, 373
    bases other than e, 383
    definition, 184
    derivative (base e), 370
    derivatives for bases other than e, 383
    graphs, 183
    limits, 183, 184
    natural (base e), 184, 366
logarithmic growth, 369
logic operations, 2, 4
logical equivalence, 17

marginal cost, 220–222
mathematical induction, 528
Mean Value Theorem (MVT)
    for derivatives, 448
    for integrals, 565
    uses in proofs, 451
modus ponens, 37
modus tollens, 40

natural logarithm, 184, 366
    derivative, 370
    graph, 369, 370

limits, 184, 369
necessary and sufficient for, 8
necessary for, 8
negation (~), operation, 2, 4
net displacement (as an integral), 549
Newton, 475
newton (unit), 262
Newton's Method, 475, 477, 478
    recursion formula, 477
Newton's Second Law of Motion, 263
Newton, Sir Isaac, xi, 231, 288
not, see negation
null set, see empty set

oblique asymptote, 421–423
odd function, 545
    integration, 563
Oersted, Hans Christian, 266
Ohm's Law, 255, 269, 270
Ohm, Georg Simon, 269
one-to-one functions, 334
only if, 2, 7, 9
open interval, 65
or ($\vee$), operation, 2, 4, 6

parallelepiped, 584
partition, 534, 535, 537, 538, 540, 543
percentage error, 532
piecewise-defined function
    continuity, 98, 101
    derivatives, 248–250, 252
    graphing, 101, 426, 427
    integrals, 545–548, 551
    limits, 132, 138, 144
Pinching Theorem, 158
potential energy, 263
power (physics), 265, 269, 272, 394, 395, 446, 502, 506
Power Rule
    derivatives, 230, 231, 280, 281, 283, 284
    integrals, 511, 512
price elasticity of demand, 285
prime (') notation, 212, 232, 234, 274
Product Rule, 297
proper subset, 66

quantifiers, 52
    negation, 56
Quotient Rule, 303

radioactive decay, 396, 398
rate of change
    average, 212
    instantaneous, 212
related rates, 485
relative maximum/minimum, see local maximum/minimum
Riemann Sum, 507, 532, 534, 535, 537
    approximation of an integral, 617
    right endpoint approximation, 618
Riemann, Bernhard, 537
right-continuous, 96
Rolle's Theorem, 449, 450, 457
Rolle, Michel, 450
Rubin, J.E., 37
Russell, Bertrand, 1

Sandwich Theorem, 158
secant line, 214
Second Derivative Test, 411
sets, 50
    complement, 68
    intersection ($\cap$), 68
    set difference, 68
    union ($\cup$), 68
Shell Method, see Cylindrical Shell Method
sigma (Σ) notation
    defined, 525
    properties, 526, 527
Simpson's Rule, 618, 621, 624, 626
    error bounds, 627
Simpson, Thomas, 475, 618
slant asymptote, see oblique asymptote
slope of a curve, 214
Special Relativity, 157
Squeeze Theorem, 158
statements, component, 3
statements, compound, 3
Suber, Peter, 37
subset, 61
    proper, 66

substitution
    definite integrals, 639
    indefinite integrals, 629
    limits, 171
sufficient for, 8
summation notation, see sigma notation
surface area of revolution, 592

$\mathscr{T}$, 29
tangent line, 214, 460
tautology, 12, 29, 34
torque, 609, 610
    and work, 611
Trapezoidal Rule, 621, 626
    error bounds, 627
trigonometric functions
    continuity, 166
    derivatives, 242, 305
    integrals, 518, 645
    inverses, 334, 520, 663
    special limits, 194, 197
truth tables, 3

union ($\cup$) of sets, 68

valid equivalence, 18
valid implication, 33
velocity
    average, 208
    instantaneous, 210
vertical asymptote, 102
volume
    by cross-sectional areas, 579
    Cylindrical Shell Method, 584
    Disc/Washer Method, 580

watt (unit), 265
Watt, James, 265
work, 261, 262, 605, 610, 611

x-intercept, 115, 150, 239
XOR, see exclusive or

y-intercept, 404

zero of a polynomial, 114
    multiplicity or degree, 115

Zu Gengzhi, 579
Zucker, Steven, 9

CPSIA information can be obtained
at www.ICGtesting.com
Printed in the USA
BVHW010040010719
552239BV00014B/228/P

9 781516 542277